Applied Environmental Science and Engineering for a Sustainable Future

Applied Environmental Science and Engineering for a Sustainable Future (AESE) series covers a variety of environmental issues and how they could be solved through innovations in science and engineering. Our societies thrive on the advancements in science and technology which pave the way for better standard of living. The adverse effect of such improvements is the deterioration of the environment. Thus, better catchment management in order to sustainably manage all types of resources (including water, minerals and others) is of paramount importance. Water and wastewater treatment and reuse, solid and hazardous waste management, industrial waste minimisation, soil restoration and agriculture as well as myriad of other topics needs better understanding and application. This book series aims at fulfilling such a task in coming years.

More information about this series at http://www.springer.com/series/13085

Jan Filip • Tomáš Cajthaml • Petra Najmanová •
Miroslav Černík • Radek Zbořil
Editors

Advanced Nano-Bio Technologies for Water and Soil Treatment

Springer

Editors
Jan Filip
Regional Centre of Advanced
Technologies and Materials
Palacký University Olomouc
Olomouc, Czech Republic

Tomáš Cajthaml
Institute of Microbiology of the Czech Academy
of Sciences
Prague, Czech Republic

Petra Najmanová
Department of Biotechnology
University of Chemistry and Technology
Prague, Czech Republic

Miroslav Černík
Institute for Nanomaterials, Advanced
Technologies and Innovation
Technical University of Liberec
Liberec, Czech Republic

Radek Zbořil
Regional Centre of Advanced
Technologies and Materials
Palacký University Olomouc
Olomouc, Czech Republic

ISSN 2570-2165 ISSN 2570-2173 (electronic)
Applied Environmental Science and Engineering for a Sustainable Future
ISBN 978-3-030-29839-5 ISBN 978-3-030-29840-1 (eBook)
https://doi.org/10.1007/978-3-030-29840-1

This Springer imprint is published by the registered company Springer Nature Switzerland AG.
The registered company address is: Gewerbestrasse 11, 6330 Cham, Switzerland

Foreword

It is most fitting that this foreword is being written literally on the eve of the very first field-scale demonstration, two decades ago, of the efficacy of nanoscale zero-valent iron (nZVI) as a remediation technology for the treatment of contaminated groundwater at a manufacturing site in Trenton, New Jersey, USA. Memorialized in an *Environ. Sci. Technol.* journal article co-authored with Professor Wei-xian Zhang, the "Father" of the nZVI technology in 2001, this initial work was, in the clarity of retrospection, modest in scope and observations. We found that a kilo of nZVI slurry could be gravity-fed into the surficial sand-dominated aquifer impacted by trichloroethene, that nZVI aggregated very quickly, that the potential for subsurface travel seemed rather limited, and that multiple injections might be required to clean up a site, among others. What we could not have anticipated at that time was the spark this trailblazing work would eventually represent. Since those humble beginnings at Lehigh University, scores of research groups all over the world and some remediation practitioners are working with improved versions of the nZVI technology and a new generation of novel nanomaterials with a common goal—to identify, develop, and apply nanotechnology-based remediation agents to enhance environmental quality, especially that of soils and groundwater which can profoundly impact our potable water supplies. Among the most prolific contributors to this burgeoning community of nanoremediation researchers and practitioners are my colleagues from the Czech Republic who have edited this comprehensive and soon-to-be impactful tome.

The Editors, Principal Investigators, and Subject Matter Experts who contributed to *Advanced Nano-Bio Technologies for Water and Soil Treatment* are among the burgeoning field of applied environmental nanotechnology's most impactful contributors. Several of these scientists and engineers played key roles in the pivotal EU-led NanoRem consortium of 28 universities, national laboratories, and industry that, from 2013 to 2017, spearheaded the basic research, development, and application of a spectrum of promising nanomaterials for environmental remediation. Over the past 5 years, continued academic interest and, to some degree, the commercial development of key "nanoremediation" technologies are most encouraging. As is demonstrated throughout the book, nanotechnology has the potential to enhance the performance and effectiveness of traditional remediation remedies by significantly accelerating the rate of contaminant transformation owing to smaller particle sizes. It can expand the spectrum of contaminant classes that can be treated as evidenced in the ability of catalyst-doped nZVI to degrade chlorinated benzenes, whereas iron powders and turnings are largely ineffective. Moreover, the diminution of particle size can enable improved and targeted delivery of remedial agents to subsurface contaminated areas that were previously difficult to reach or inaccessible. Nevertheless, against this largely optimistic backdrop, considerable work remains to fully characterize and appropriately vet the efficacy of these novel nanomaterials, assess the implications of their usage with respect to potential receptors, and conduct robust cost-benefit analyses as many of these technologies lack track records of performance in the field.

Naturally, the book begins with a part (Part I) on reductive technologies, showcasing the standard-bearer nZVI, which now has been showcased in more than 100 field-scale demonstrations around the world. It contrasts the many variations on nZVI with other reducing strategies (e.g., utilization of dithionate) and describes significant new enhancements associated with the application of DC electric fields to help drive nanoremediation agents through low-permeability subsurface formations such as clays. Part II introduces new nano-oxidation technologies, including high valence ferrates, which may provide exciting new water treatment applications. In Part III, the Editors focus on the integration of nanotechnology into the biotreatment of waters and groundwaters. The ex situ treatment of soils impacted by persistent organic pollutants such as polycyclic aromatic hydrocarbons using ligninolytic fungi and enzymes is addressed in Part IV. The Editors shift gears with Part V and focus on the implications of using nanoremediation—that is, they address the ecotoxicological impacts on receptors associated with the exposure to nanomaterials in the field. Part VI ties together the overarching observations and conclusions of the spectrum of nano- and nanobiotechnologies covered in the book and forecasts the future prospects of these technologies. Included is discussion on applications for emerging contaminants, new regulatory developments, and how these technologies might fit into new water security and quality strategies. I applaud the Editors, Chapter Authors, and Subject Matter Experts on their contributions and earnestly believe that this book will prove to be an invaluable reference for environmental remediation researchers and practitioners alike.

Senior Consultant, Geosyntec Consultants, Inc., Princeton, NJ, USA
10 May 2019

Daniel W. Elliott, Ph.D., BCEEM

Preface

One of the major issues that are currently dealt with all around the world is the depletion of clean/drinking water resources along with losing fertile soil, which would satisfy the burgeoning demand for food supply due to a growing population. Therefore, contamination from industry, environmental accidents, or improper wastewater treatment requires a fast, efficient, and cost-effective action to take. Advanced nanotechnologies, biotechnologies, or their combinations could represent a highly promising ecological and economical alternative to traditional remediation techniques. Due to the diverse character of the target pollutants, the key processes typically involve oxidation, reduction, sorption, and/or biological degradation. In this book, we aim to bridge theory and practice by sharing our experience with eliminating a wide range of pollutants from various resources utilizing innovative nanotechnologies, biotechnologies, and their possible combinations. What has not been omitted is evaluating the toxicity of both emerging pollutants and industrial nanoparticles. All the above-mentioned topics represent the core of an 8-year-long project aimed at applied research entitled "Environmentally friendly nanotechnologies and biotechnologies in water and soil treatment" (NanoBioWat) supported by the Technology Agency of the Czech Republic (project no. TE01020218). The following academic and industrial partners actively participated in the project as well as in the production of this book (all partners are based in the Czech Republic): Palacký University Olomouc, Regional Centre of Advanced Technologies and Materials; Technical University of Liberec, The Institute for Nanomaterials, Advanced Technology and Innovation; Institute of Microbiology of the Czech Academy of Sciences; AECOM CZ s.r.o.; AQUATEST a.s.; DEKONTA, a.s.; GEOtest, a.s.; LAC, s.r.o.; and MEGA, a.s. Leading researchers and experts from the particular fields, being either members of the above-mentioned consortium or based at other institutes, were asked to make their contributions to this book.

This book is organized into five topical parts and covers the most recent findings in the particular fields: (i) *Reductive technologies* for water treatment: this part deals with reductive remedial technologies applicable mainly to an in situ treatment of inorganic and organic contaminants. Nanoscale zero-valent iron is a major reagent under study, yielding numerous results from various sites under various conditions. Other chemical reductants, such as dithionate, are discussed as well. The enhancement of either natural or chemical processes by DC electric field as a very promising method to accelerate and increase the efficiency of the remedial process along with reducing the cost is tackled as well. (ii) *Oxidative technologies* for water treatment: this part includes a basic overview of various innovative oxidation technologies applicable to water treatment with a strong focus on technologies based on iron compounds in high-valent states (co-called ferrates IV, V, and VI), including the properties of ferrates, their synthesis and applicability. Similarly, radical reactions and photooxidations are covered and discussed regarding their applicability to remediation techniques. (iii) *Biotechnologies for water treatment*: this part provides the overview of modern and advanced methods based on the application of microorganisms and their compartments, especially the combination of microbes or enzymes with nanotechnology applications. A special attention is also paid to recent findings concerning bioelectrical processes participating in the remediation processes. The presented results of nano-bio and bio-nano approaches demonstrate the feasibility and high efficiency of the combined methods. (iv) *Biotechnologies for soil treatment*: this part includes the overview of ex situ bioremediation treatment of contaminated soil. New details about mycoremediation technology using ligninolytic fungi for biodegradation of soil and groundwater contaminated with persistent organic pollutants (POPs) are discussed. The use of a composting technology for polycyclic aromatic hydrocarbons (PAHs) removal from contaminated soil is outlined with respect to its practical application. The last chapter of this part is dedicated to the techniques of bioremediation, including enzymes, biosurfactants, or genetically modified organisms use in real applications. (v) *Ecotoxicology* of both environmental pollutants and nanomaterials used for remediation: this part comprises theoretical support regarding novel findings on ecotoxicity of pollutants and nanomaterials. The importance of this part is underpinned by the fact that there is still lack of a suitable, comprehensive, and standardized set of tests for ecotoxicological evaluation of the novel nanomaterials; further research in this direction is needed.

Each part (i–iv) is organized as follows: it contains chapters focused on general description of the particular technologies followed by several field studies, 10 altogether, demonstrating the applicability of the particular technology. Moreover, the book has a concluding chapter dealing with **future prospects** for techniques treating contaminants of emerging concern in water and soils/sediments. Conclusions and suggestions made not only within this chapter but also throughout the whole book could be of interest to scientists and, primarily, practitioners who deal with water quality. Rising population is a phenomenon that entails different issues ranging from sustainable sources of clean water to cultivating soil for agricultural activities and feeding animals. The last part of the book contains a collection of five **technical chapters** (appendices) providing technical details on actions taken in relation with a pilot/full-scale application of key nano-/biotechnologies. Each chapter focuses on one specific aspect of the implementation of the selected technology/material such as nanoscale zero-valent iron injection into groundwater, field-scale contaminant monitoring, and nanoparticle migration and transformation. Here we also cover protocols on (eco)toxicological assessment of nanoparticles and evaluation of changes in the microbial communities prior to and after nanoremediation.

Although other previously published papers and books (or book chapters) tackle certain aspects of advanced nano-/biotechnologies, this is the first time a complete and comprehensive treatise on the latest progress in innovative technologies has been published along with clear demonstration of the applicability of the particular methods on the basis of the results yielded in the pilot tests. Therefore, this multidisciplinary book will be suitable for broad readership including environmental scientists, practitioners, policymakers, and toxicologists and, of course, students of diverse fields involving material science, chemistry, biology, geology, hydrogeology, engineering, etc.

Olomouc, Czech Republic Jan Filip
Prague, Czech Republic Tomáš Cajthaml

Acknowledgements

Special thanks go to Monika Klimparová and Zdenka Červenková (both from the Regional Centre of Advanced Technologies and Materials, Palacký University, Olomouc, The Czech Republic) for proofreading and technical improvements, respectively, which significantly improved the overall quality of the book. This book was produced with the assistance of the Technology Agency of the Czech Republic since it was one of the outcomes of the project entitled NanoBioWat ("Environmentally friendly nanotechnologies and biotechnologies in water and soil treatment"; project No. TE01020218), solved within the programme of the Competence Centres.

Contents

Part VII Technical Chapters

About the Editors

Dr. Jan Filip received his Master's and PhD degrees in Mineralogy from Masaryk University in Brno, the Czech Republic (2002 and 2008, respectively). His dissertation focused on the crystal chemistry of borosilicate minerals and OH defects in nominally anhydrous minerals. In 2005, he joined the research group of Prof. Radek Zbořil's and currently leads the group "Environmental nanotechnologies" at the Regional Centre of Advanced Technologies and Materials, Palacký University Olomouc, the Czech Republic. He has been on fellowships at different universities and institutes abroad, e.g., the University of Vienna, Austria, or the Swedish Museum of Natural History, Stockholm, Sweden, and actively participates in various national and international collaborative projects. His research interests cover development, optimization, and application of zero-valent iron nanoparticles, ferrates, and iron oxide nanoparticles in technologies for water treatment, identification of nanoparticles in the environment, application of X-ray-based techniques (X-ray diffraction, X-ray fluorescence, X-ray photoelectron spectroscopy, and related techniques), and Mössbauer spectroscopy in materials science. He is the author and co-author of four book chapters, more than 105 papers in refereed journals that have received about 2000 citations, and several other applied results including verified technologies, utility models, and prototypes. He is a named inventor of five patents. His H-index has reached 23.

Regional Centre of Advanced Technologies and Materials, Palacký University Olomouc, Olomouc, Czech Republic

Prof. Tomáš Cajthaml completed his PhD study in 2002. His thesis focused on fungal degradative mechanisms of persistent organic pollutants. He has served at various foreign institutes, e.g., the Helmholtz Centre for Environmental Research, Leipzig, Germany (8 months) or the Plant Macromolecule Research Centre (CERMAV), Grenoble, France (4 months). He has been the Head of the Laboratory of Environmental Technology, Institute of Microbiology, the Czech Academy of Sciences, since 2006. Since 2015 he has been a professor at Charles University and the Director of the Institute for Environmental Studies, Faculty of Science, Charles University. He is a member of several scientific organizations and a delegate from the Czech Republic at European Federation for Biotechnology. He is also on the Grant Review Committees of the Technology Agency of the Czech Republic and the Czech Science Foundation. He has been the author and principal investigator of more than 20 national and international grant projects. His research interests include persistent organic pollutants, endocrine disrupters, microbial bioremediation, analysis of organopollutants, bioavailability and degradation of relevant pollutants, identification of biodegradation products of pollutants, and supercritical and subcritical extraction techniques and their applications. Professor Cajthaml is the author and co-author of more than 180 papers. His publications have received about 4100 citations. His H-index has reached 37. He is also an author and co-author of five national patents and several other applied results including verified technologies, utility models, and designs.

Institute of Microbiology, Czech Academy of Sciences, Prague, Czech Republic
Institute for Environmental Studies, Faculty of Science, Charles University, Prague 2, Czech Republic

Ing. Petra Najmanová received her Master's degree at the University of Chemistry and Technology, Prague, the Czech Republic, in 2002. In 2003, she joined DEKONTA, an international company dealing with environmental services, where she works until today. She has 15 years of professional experience gained in the Czech Republic as well as abroad in the fields of biodegradation, biological treatment, soil cleaning, remediation, and bioremediation. She participates in a number of research projects focused on, among others, innovations of biological remediation technologies. Petra Najmanová also engages in Project Development Assistance (Serbia, Moldova, Philippines) and full-scale remediation projects around the world (Israel, Kazakhstan, Bosnia, Romania, etc.). She has led a commercially accredited laboratory for 10 years. She has been the Head of the R&D Department at DEKONTA since 2018. Currently, she is doing her PhD at the University of Chemistry and Technology, Biotechnology Department, Prague, the Czech Republic. Her dissertation focuses on thermally enhanced bioremediation of chlorinated contaminants in groundwater. She is a named inventor of five patents and the author of many other applied results including verified technologies and utility models. She is the author or co-author of several papers in refereed journals.

DEKONTA, a.s., Czech Republic
University of Chemistry and Technology, Prague, Czech Republic

Miroslav Cernik completed his engineering study and CSc dissertation at the Czech Technical University in Prague, Faculty of Nuclear Sciences and Physical Engineering, Department of Nuclear Chemical Engineering, the Czech Republic in

1987 and 1991, respectively. In 1994, he received his PhD from the Institute of Terrestrial Ecology, Department of Soil Geochemistry of Swiss Federal Institute of Technology, Switzerland. He worked for the company AQUATEST, the Czech Republic as an expert in mathematical modeling; later, he became Director of Research and Development of AQUATEST company. In 2004, he joined Technical University of Liberec, Faculty of Mechatronics, Informatics and Interdisciplinary Studies, the Czech Republic. In 2007, he habilitated with a thesis "The use of elemental Fe nanoparticles for reduction of contaminants in-situ." Since 2012 he has worked at the Institute for Nanomaterials, Advanced Technology and Innovation (CxI), Technical University of Liberec, the Czech Republic as the Director of Research and Head of Department of Nanomaterials in the Natural Sciences. In 2014, he became Professor of applied sciences in engineering. He leads a team of more than 40 scientists and participates in international and national projects, mostly aimed at applications of new remedial methods. In the project NANOREM (large collaborative project of 7. FP EU), he was responsible for a work package of zero-valent iron research and production. His research interests include the description and application of nanotechnologies and nanomaterials in remediation technologies, risks of nanomaterials, and other in situ remediation technologies. He is the author of over 130 scientific papers in international scientific journals and proceedings recognized by WOS. He participated in more than 50 international conferences including several invited lectures. He is the author of five books or book chapters, five patents, and several other applied results including verified technologies, utility models, and prototypes. His WOS papers have received a total number of 1900 citations and his H-index has reached 21.

Institute for Nanomaterials, Advanced Technology and Innovation, Technical University of Liberec, Liberec, Czech Republic

Prof. Radek Zbořil received his PhD from Palacký University Olomouc, the Czech Republic in 2000. He was on different fellowships at, e.g., the University of Delaware, Newark, Delaware, USA, or the University of Tokyo, Japan. Since 2010 he has been a professor at Palacký University Olomouc and the General Director of the Regional Centre of Advanced Technologies and Materials. In 2011, he received an award from the Ministry of Education, Youth and Sports of the Czech Republic for an outstanding performance in research, experimental development, and innovations. Professor Zbořil is the principal investigator of more than 30 national and international projects with the total funds of over EUR 64 million allocated to Palacký University Olomouc. Professor Zbořil is the author and co-author of more than 520 papers. His publications have received over 23,000 citations and his H-index has reached 66. Professor Zbořil features on the Highly Cited Researchers 2018 list. His research interests include nanomaterials—iron, iron oxide, noble metals, carbon-based (graphene derivatives, carbon dots) nanostructures, magnetic nanoparticles, and hybrid systems and their advanced nanoarchitecture and applications in water treatment; energy materials; antimicrobial technologies; catalysis; biosensing; medicine (MRI, drug delivery); and biotechnologies. Professor Zbořil is a named inventor of several European and US patents, e.g., the Development of large-scale technology for the production of zero-valent iron nanoparticles (European patent No. 2164656) or Process of whey protein separation from milk medium and apparatus for its implementation (European patent No. 2873329).

Regional Centre of Advanced Technologies and Materials, Palacký University Olomouc, Olomouc, Czech Republic

Contributors

Bhavya Arora Department of Chemistry, Green Chemistry Network Centre, University of Delhi, Delhi, India

Luděk Bláha RECETOX, Faculty of Science, Masaryk University in Brno, Brno, Czech Republic

Enric Brillas Laboratori d'Electroquímica dels Materials i del Medi Ambient, Departament de Química Física, Facultat de Química, Universitat de Barcelona, Barcelona, Spain

Tomáš Cajthaml Institute of Microbiology of the Czech Academy of Sciences, Prague, Czech Republic
Institute for Environmental Studies, Faculty of Science, Charles University, Prague, Czech Republic

Miroslav Černík Institute for Nanomaterials, Advanced Technologies and Innovation, Technical University of Liberec, Liberec, Czech Republic
AQUATEST a.s., Prague, Czech Republic

Philippe F.-X. Corvini Institute for Ecopreneurship, School of Life Sciences, University of Applied Sciences and Arts Northwestern Switzerland, Muttenz, Switzerland

Lucie Coudert Research Institute on Mines and Environment (RIME), Université du Québec en Abitibi-Témiscamingue (UQAT), Rouyn-Noranda, QC, Canada

Claire Coutris Division of Environment and Natural Resources, Norwegian Institute of Bioeconomy Research, Ås, Norway

Stefano Covino Institute for Environmental Studies, Faculty of Science, Charles University, Prague, Czech Republic
Department of Chemistry, Biology and Biotechnology, University of Perugia, Perugia, Italy

Monika Čvančarová Institute for Ecopreneurship, School of Life Sciences, University of Applied Sciences and Arts Northwestern Switzerland, Muttenz, Switzerland

Matteo Daghio Department of Earth and Environmental Sciences, University of Milano-Bicocca, Milan, Italy
Department of Agriculture, Food, Environment and Forestry, University of Florence, Florence, Italy

Iva Dolinová Institute for Nanomaterials, Advanced Technologies and Innovation, Technical University of Liberec, Liberec, Czech Republic

Sriparna Dutta Department of Chemistry, Green Chemistry Network Centre, University of Delhi, Delhi, India

Jan Filip Regional Centre of Advanced Technologies and Materials, Palacký University Olomouc, Olomouc, Czech Republic

Andrea Franzetti Department of Earth and Environmental Sciences, University of Milano-Bicocca, Milan, Italy

Manoj B. Gawande Regional Centre of Advanced Technologies and Materials, Palacký University Olomouc, Olomouc, Czech Republic

Martin Halecký Department of Biotechnology, University of Chemistry and Technology, Prague, Czech Republic

Michal Hegedüs GEOtest, a.s., Brno, Czech Republic

Monika Heřmánková DEKONTA a.s., Stehelčeves, Czech Republic
AECOM CZ, s.r.o., Praha, Czech Republic

Tereza Hnátková DEKONTA a.s., Stehelčeves, Czech Republic
Department of Applied Ecology, Faculty of Environmental Sciences, Czech University of Life Sciences Prague, Prague, Czech Republic

Jakub Hofman RECETOX, Faculty of Science, Masaryk University in Brno, Brno, Czech Republic

Pavel Hrabák Institute for Nanomaterials, Advanced Technologies and Innovation, Technical University of Liberec, Liberec, Czech Republic

Jaroslav Hrabal MEGA a.s., Stráž pod Ralskem, Czech Republic

Petra Innemanová DEKONTA a.s., Stehelčeves, Czech Republic
Institute for Environmental Studies, Faculty of Science, Charles University, Prague, Czech Republic

Jia-Qian Jiang Department of Civil Engineering & Environmental Management, SCEBE, Glasgow Caledonian University, Glasgow, Scotland, UK

Erik J. Joner Division of Environment and Natural Resources, Norwegian Institute of Bioeconomy Research, Ås, Norway

Vladislav Knytl DEKONTA a.s., Stehelčeves, Czech Republic

Jan Kolařík Regional Centre of Advanced Technologies and Materials, Palacký University Olomouc, Olomouc, Czech Republic

Evguenii Kozliak Department of Chemistry, University of North Dakota, Grand Forks, ND, USA

Jan Kukačka DEKONTA a.s., Stehelčeves, Czech Republic

Petr Kvapil Photon Water Technology s.r.o., Liberec, Czech Republic

Petr Lacina GEOtest, a.s., Brno, Czech Republic

Ondřej Lhotský DEKONTA a.s., Prague, Czech Republic
Institute for Environmental Studies, Faculty of Science, Charles University, Prague, Czech Republic

Marta I. Litter Instituto de Investigación e Ingeniería Ambiental, Universidad Nacional de San Martín, CONICET, Buenos Aires, Argentina

Libor Machala Regional Centre of Advanced Technologies and Materials, Palacký University Olomouc, Olomouc, Czech Republic

Tomáš Mackuľak Institute of Chemical and Environmental Engineering, Faculty of Chemical and Food Technology, Slovak University of Technology, Bratislava, Slovakia

Petr Maršík Laboratory of Plant Biotechnologies, Institute of Experimental Botany, Academy of Sciences of the Czech Republic, Prague, Czech Republic

Pierre Matz Solvay SA, Brussels, Belgium

Catherine N. Mulligan Building, Civil, and Environmental Engineering, Concordia University, Montreal, QC, Canada

Petra Najmanová DEKONTA a.s., Stehelčeves, Czech Republic
Department of Biotechnology, University of Chemistry and Technology, Prague, Czech Republic

Carmen Mihaela Neculita Research Institute on Mines and Environment (RIME), Université du Québec en Abitibi-Témiscamingue (UQAT), Rouyn-Noranda, QC, Canada

Jan Němeček Institute for Nanomaterials, Advanced Technologies and Innovation, Technical University of Liberec, Liberec, Czech Republic

Jaroslav Nosek Institute for Nanomaterials, Advanced Technologies and Innovation, Technical University of Liberec, Liberec, Czech Republic
AQUATEST a.s., Prague, Czech Republic

Petr Parma Institute for Nanomaterials, Advanced Technologies and Innovation, Technical University of Liberec, Liberec, Czech Republic

Eleni Petala Regional Centre of Advanced Technologies and Materials, Palacký University Olomouc, Olomouc, Czech Republic

Tanapon Phenrat Department of Civil Engineering, Environmental Engineering Program, Naresuan University, Phitsanulok, Thailand

Martin Pivokonský Institute of Hydrodynamics of the Czech Academy of Sciences, Prague, Czech Republic

Robert Raschman DEKONTA a.s., Stehelčeves, Czech Republic

Eric Rosa Research Institute on Mines and Environment (RIME), Université du Québec en Abitibi-Témiscamingue (UQAT), Rouyn-Noranda, QC, Canada
Groupe de Recherche sur l'Eau Souterraine (GRES—Groundwater Research Group), UQAT, Amos, QC, Canada

Dietmar Schlosser Department of Environmental Microbiology, Helmholtz Centre for Environmental Research – UFZ, Leipzig, Germany

Jaroslav Semerád Institute of Microbiology of the Czech Academy of Sciences, Prague, Czech Republic
Institute for Environmental Studies, Faculty of Science, Charles University, Prague, Czech Republic

Michal Šereš DEKONTA a.s., Stehelčeves, Czech Republic
Institute for Environmental Studies, Faculty of Science, Charles University, Prague, Czech Republic

Alena Ševců Institute for Nanomaterials, Advanced Technologies and Innovation, Technical University of Liberec, Liberec, Czech Republic

Patrick Shahgaldian Institute for Chemistry and Bioanalytics, School of Life Sciences, University of Applied Sciences and Arts Northwestern Switzerland, Muttenz, Switzerland

Rakesh Kumar Sharma Department of Chemistry, Green Chemistry Network Centre, University of Delhi, Delhi, India

Ignasi Sirés Laboratori d'Electroquímica dels Materials i del Medi Ambient, Departament de Química Física, Facultat de Química, Universitat de Barcelona, Barcelona, Spain

Petra Skácelová Regional Centre of Advanced Technologies and Materials, Palacký University Olomouc, Olomouc, Czech Republic
NANO IRON, s.r.o., Židlochovice, Czech Republic

Jan Slunský NANO IRON, s.r.o., Židlochovice, Czech Republic

Petr Soudek Laboratory of Plant Biotechnologies, Institute of Experimental Botany, Academy of Sciences of the Czech Republic, Prague, Czech Republic

Roman Špánek Institute for Nanomaterials, Advanced Technologies and Innovation, Technical University of Liberec, Liberec, Czech Republic

Jana Steinová Institute for Nanomaterials, Advanced Technologies and Innovation, Technical University of Liberec, Liberec, Czech Republic

Vojtěch Stejskal Institute for Nanomaterials, Advanced Technologies and Innovation, Technical University of Liberec, Liberec, Czech Republic
Photon Water Technology s.r.o., Liberec, Czech Republic
AQUATEST a.s., Prague, Czech Republic

Tatiana Stella Department of Earth and Environmental Sciences (DISAT), University of Milano-Bicocca, Milan, Italy

Nikola Vacková Faculty of Science, Charles University, Prague, Czech Republic
AQUATEST a.s., Prague, Czech Republic

Tomáš Vaněk Laboratory of Plant Biotechnologies, Institute of Experimental Botany, Academy of Sciences of the Czech Republic, Prague, Czech Republic

Adriana Velosa Regional Centre of Advanced Technologies and Materials, Palacký University Olomouc, Olomouc, Czech Republic

Roman Vokáč AECOM CZ, s.r.o., Praha, Czech Republic

Jan Vymazal Department of Applied Ecology, Faculty of Environmental Sciences, Czech University of Life Sciences Prague, Prague, Czech Republic

Stanisław Wacławek Institute for Nanomaterials, Advanced Technologies and Innovation, Technical University of Liberec, Liberec, Czech Republic

Petr Zajíček Regional Centre of Advanced Technologies and Materials, Palacký University Olomouc, Olomouc, Czech Republic

Josef Zeman Department Geological Sciences, Faculty of Science, Masaryk University in Brno, Brno, Czech Republic

Part I
Reductive Technologies

Chapter 1
Geochemical Principles of Reductive Remediation Processes

Miroslav Černík and Josef Zeman

Abstract Water is a solvent that is absolutely essential to sustain life. It is one of the world's most precious resources. The uniqueness consists in its polarity, high boiling point, and some other important properties. Several of the water parameters, for example, alkalinity or redox potential, were discussed in this chapter. The environmental significance of these basic parameters is enormous because they are affecting the stability and mobility of substances in the ecosystem, and other parameters are either directly determined or strongly influenced by them. In addition, further important geochemical aspects, presented, for example, in the graphical form, i.e., the pε-pH diagram (Pourbaix diagram) or redox ladder, were thoroughly explained in this chapter. Subsequently, the in situ chemical reduction (ISCR) techniques were discussed with the special emphasis on the nano zero-valent iron (nZVI). The nZVI reactions and thermodynamics behind them were also reviewed here.

Keywords Geochemistry · Redox potential · Nano zero-valent iron (nZVI) · In situ chemical reduction (ISCR)

1.1 Introduction

Water is the carrier of the entire geochemical system of natural exogenous processes. Water has a number of unique properties that determine the behavior of the rock matrix and biogenic components. Its basic properties are a high dielectric constant (Malmberg and Maryott 1956), high surface tension, and high heat capacity. Natural water contains dissolved substances, some of which have reduction or oxidation

M. Černík (✉)
Institute for Nanomaterials, Advanced Technologies and Innovation, Technical University of Liberec, Liberec, Czech Republic

AQUATEST a.s., Prague, Czech Republic
e-mail: miroslav.cernik@tul.cz

J. Zeman
Department Geological Sciences, Faculty of Science, Masaryk University in Brno, Brno, Czech Republic

J. Filip et al. (eds.), *Advanced Nano-Bio Technologies for Water and Soil Treatment*, Applied Environmental Science and Engineering for a Sustainable Future, https://doi.org/10.1007/978-3-030-29840-1_1

properties. This creates a geochemical balance between the solvent (water), the dissolved substance, and the rock matrix. Altering these conditions also leads to changes in the water chemistry. A significant change in the environment is alkalinity, which in natural waters is mainly a function of the sum of carbonate species and OH^-. Therefore, the geochemical system has a certain inertia and, to some degree, can buffer natural oscillations or cause anthropogenically induced changes.

The basic chemical processes used in remediation are mainly changes in acid-alkaline conditions, redox conditions, or the concentration of species causing precipitation. Since the two basic components of chemical reactions are protons and electrons, their exchanges in redox processes and acid-alkaline reactions are fundamental in nature. These processes, which are observed as a reaction of mineral precipitation, weathering, dissolution, or organic matter decomposition, etc., are fundamental reactions occurring on the surface of the earth's crust because of the contact between rock, water, and the atmosphere, and determine the parameters of the surrounding environment. While the significance of pH for the toxicity and mobility of hazardous substances (heavy metals, radionuclides, organic substances, etc.) has been known for a long time, the importance of the redox potential has been significantly underestimated or even neglected except for the oxidation zone of ore deposits. It is commonly assumed that pH is a critical parameter for the state of the natural environment so the other parameters (redox potential, content of dissolved solids in the aqueous environment, concentration of gases) play just a minor role. However, the results of systematic studies on the natural environment and natural processes show that the decisive factors in most cases are redox processes and pH, and the other parameters are either directly determined or strongly influenced by these processes.

1.2 Stability of Redox Conditions

The stable redox environment in the geosphere is caused by biota that enables photosynthetic processes. This biota synthesizes complex organic substances from carbon dioxide using solar energy according to the following simplified equation of photosynthesis (Eq. 1.1)

$$CO_2 + H_2O \rightarrow CH_2O + O_2 \tag{1.1}$$

where CH_2O is a general formula for organic matter (like glucose $C_6H_{12}O_6$). During the process, 472 kJ/mol of solar energy is consumed (Stumm and Morgan 1995) and the energy is "stored" and subsequently used to secure the life processes of the biota. When looking at the oxidation state (formal charge, valence) of the substance or compound, reduction of carbon occurs through photosynthesis from oxidation state (+IV) to (0) and an oxidation of oxygen from oxidation state –II to 0:

$$C^{+IV}O_2^{-II} + H_2O \rightarrow C^0H_2O + O_0^2 \tag{1.2}$$

Photosynthesis ensures a constant renewal of elemental oxygen, the main oxidant in the geosphere, and produces organic compounds, which enter the soil and water after the biota dies. These organic substances then act as reducing agents through respiration or decay (natural decomposition of organic substances). This process is actually the above-mentioned photosynthetic reaction (Eq. 1.2) in reverse; oxygen is consumed releasing carbon dioxide, water, and energy.

Another very important component of the natural geochemical system is iron. Iron is the most widespread transient metal element and the second most widespread metal on Earth. The content of iron in the Earth's crust is 62 g/kg. In nature, it is mainly found in valence states of Fe^{2+} and Fe^{3+}. Under oxidation conditions, stable iron is in the valence state of Fe^{3+}; under reducing conditions, it is in the valence state of Fe^{2+}, and minerals containing Fe^{3+} are also common. Besides oxidation of organic matter, a change in the valence state of Fe is the main process that generates electron transfer in nature. Electrons are generated during oxidation and consumed during reduction. Fe^{4+} compounds are very unstable and have no practical significance, Fe^{5+} and Fe^{6+} compounds are used as very strong oxidizing agents but are not found in nature. Pure iron rarely occurs in nature (in volcanites). Iron in low valence states (Fe^{0}) can be used as a reducing agent.

In the case of iron (ZVI nanoparticles), it is known that water and dissolved oxygen cause iron oxidation, which leads to corrosion (rusting) by the following Eq. (1.3):

$$Fe^{0} + \tfrac{1}{2}O_2 + H_2O \rightarrow Fe^{2+} + 2OH^{-} \tag{1.3}$$

The very name of the redox process—oxidation-reduction reaction—implies that there are two processes occurring—oxidation and reduction, which are inseparable, simultaneous, and dependent. This is a contrast to acid-base reactions, where protons (H^{+} ions) are released by, e.g., acid dissolution in water, solvated and stable in a water environment.

Oxygen participates in the above process in the reduction part in which electrons are consumed and the oxidation state of oxygen reduces according to the formal reaction (Eq. 1.4).

$$\tfrac{1}{2}O_0^2 + H^{+} + 2e^{-} \rightarrow O^{-II}H^{-} \tag{1.4}$$

Reversely, iron participates in the oxidation process in which electrons are released and the iron oxidation state increases.

$$Fe^{0} \rightarrow Fe^{2+} + 2e^{-} \tag{1.5}$$

By combining these two equations—reduction (Eq. 1.4) and oxidation (Eq. 1.5) respectively—we obtain the final equation for iron rusting, where two electrons are exchanged between the iron and oxygen. For such a heterogeneous process, the surface area of the solid reagent is the key parameter for the reaction rate. This has a practical significance only in the case of systems described by a single process because comparison with a more complex process is ambiguous (Tiehm et al. 1997).

1.3 Quantitative Expression of Redox Potential

Redox potential is an intensive parameter of the environment and expresses the overall potential of the system for oxidation-reduction processes, just as pH expresses the acidity of acid-base reactions. Usually, the redox potential for a general oxidation-reduction reaction (Stumm and Morgan 1995).

$$\mathrm{Ox} + ne^{-} \leftrightarrow \mathrm{Red} \tag{1.6}$$

is derived from the well-known Nernst equation (Eq. 1.7).

$$\mathrm{Eh} = \mathrm{Eh}^{\circ} + \frac{RT}{nF} \ln \frac{a_{\mathrm{Ox}}}{a_{\mathrm{Red}}} \tag{1.7}$$

where E_{h} is the potential of the redox reaction (Eq. 1.6). E^0 is the standard potential, i.e., potential in unit activities of oxidized (Ox) and reduced (Red) forms of the substance, R is the universal gas constant, T is thermodynamic temperature, F is Faraday's constant, n is the number of electrons that are exchanged during the redox reaction between Ox and Red species of the redox pair, a_{Ox} and a_{Red} are activities of oxidized and reduced species, respectively.

Analogically to pH, which expresses the activity of protons in a solution in the form of a negative decadic logarithm, it is possible to express Eh as the pε, negative decadic logarithm of the activity a_{e-} of electrons in the environment:

$$\mathrm{p}\varepsilon = -\log a_{\mathrm{e}^-} \tag{1.8}$$

with the relation

$$\mathrm{Eh} = \frac{2.303RT}{F} \mathrm{p}\varepsilon = 0.059\mathrm{p}\varepsilon \tag{1.9}$$

where a_{e-} is the activity of electrons.

To measure the potential of electrochemical reactions, it is necessary to use two electrodes between which the potential is measured. One electrode measures the observed response, the other serves as a reference. Platinum is a good material for the electrodes (Schuettler 2007) because it is resistant in most environments and is not subject to its own redox reactions. The reference standard hydrogen electrode (SHE) is a plate coated with spongy black platinum, which is saturated by gaseous hydrogen H_2 at a pressure of 101.325 kPa and is immersed in a solution with a unit activity of H^+.

The electrochemical reaction.

$$\frac{1}{2}H_2(g) \rightarrow H^+(aq) + e^- \tag{1.10}$$

on the reference electrode under standard conditions is arbitrarily given a potential (Eh) of 0 V. For practical reasons, redox potential is measured using other reference electrodes with a stable potential instead of SHE. The most commonly used reference electrodes are Ag/AgCl (silver chloride) (Ives and Janz 1961) and Hg/Hg_2Cl_2 (calomel) electrodes. Silver chloride electrodes consist of a silver wire coated with AgCl precipitate, which is immersed in a KCl solution with a specific concentration. Based on the KCl concentration in the solution, the potential of the electrode ranges from 197 mV (saturated KCl) to 288.1 mV (0.1 M KCl) and is temperature-dependent. The saturated calomel electrode whose potential is about 247 mV (at 20 °C) works in a similar way. The measured potential must then be recalculated to SHE by adding the above-mentioned values depending on the reference electrode used and the temperature. The values not corrected to SHE are sometimes reported as ORP (oxidation-reduction potential) compared to E_h or ORP_H related to SHE.

1.4 Stability of Water and Eh-pH Diagrams

The acidity and redox properties of an aqueous environment are the basic parameters affecting the stability and mobility of substances in this environment (Violante et al. 2010). These properties and their influence on the state of a substance can be summarized in the form of Eh-pH (or pε-pH) diagrams. The redox conditions (like pH conditions) of an aqueous environment cannot acquire unlimited values as they are limited by the reaction of water with other substances. Since water in most cases is in contact with the atmosphere (either with the normal atmosphere containing oxygen or with the soil atmosphere containing carbon dioxide or hydrogen), the processes limiting the Eh values are the oxidation or reduction reactions of water producing these gases. Both types of reactions are very similar in principle (Table 1.1). Oxidation can be expressed as a reaction in which water loses the

Table 1.1 The basic oxidation-reduction reactions of water

Water oxidation	Water reduction
$H_2O - 2e^- \leftrightarrow \frac{1}{2}O_2(g) + 2H^+$	$H_2O + e^- \leftrightarrow \frac{1}{2}H_2(g) + OH^-$
Equilibrium constant (log form):	Equilibrium constant (log form):
$-\log K = \log a\,H_2O - \frac{1}{2}\log p\,O_2 + 2p\varepsilon + 2pH$	$\log K = \log K_W + \frac{1}{2}\log p\,H_2 + p\varepsilon + pH$
pε: Water oxidation	pε: Water reduction
$p\varepsilon = 20.78 - pH$	$p\varepsilon = -pH$

Note, $p\,O_2$ and $p\,H_2$ are the partial pressures of the gases present and K_W is autoprotolysis constant of water electrons and oxygen gas is formed; and reduction as a reaction where water acquires electrons and hydrogen gas is formed.

The pε values were calculated by the following conditions: In both equations of equilibrium constants, there is a partial pressure of the gas involved. The partial pressure can be equal to 1 and its log = 0. The equilibrium constant for the oxidation of water at 25 °C and pressure 0.1 MPa has a value of log $K = -41.56$ and the water has a unit activity; for the reduction process the log $K = 0$ according to (Eq. 1.10). These curves are the limits of the pε-pH graph of water stability because higher and lower values for water oxidation and reduction, respectively, are not thermodynamically possible.

On the pε-pH diagram (Pourbaix diagram) these are the parallel lines with a slope − 1, which intersect the vertical axis at pH = 0 at a value of 20.78 (upper limit) and 0 (lower limit). In Fig. 1.1 these curves are shown on the Eh-pH and pε-pH diagrams and the areas of conditions characteristic of various natural environments are indicated. It should be noted that the value of the redox potential without expressing the pH cannot determine whether the conditions are oxidizing or reducing. For example, a redox potential of +400 mV in acidic water represents reducing conditions (peat bogs), but in alkaline water, it represents oxidizing conditions (saline lakes, saline soils).

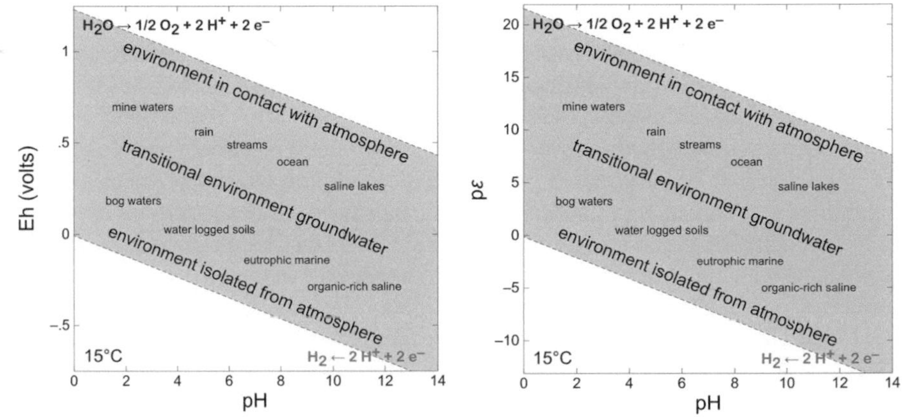

Fig. 1.1 pH values and redox potential in Eh and pε scales for various natural environments. Under conditions that lie outside the defined boundaries, the water is unstable, and oxidation occurs through the release of gaseous oxygen (upper limit) or reduction through the release of hydrogen gas (lower limit)

Table 1.2 Dissolved compounds, which govern thc redox potential in natural waters

Compound	Change of charge	Number of exchanged e^-
$O_2 \leftrightarrow H_2O$	$O^0 \leftrightarrow O^{-2}$	2 e^-
$NO_3^- \leftrightarrow NH_4^+$	$N^{+V} \leftrightarrow N^{-III}$	8 e^-
$MnO_2 \leftrightarrow Mn^{2+}$	$Mn^{+IV} \leftrightarrow Mn^{+II}$	2 e^-
$Fe^{3+} \leftrightarrow Fe^{2+}$	$Fe^{+III} \leftrightarrow Fe^{+II}$	1 e^-
$H_2O \leftrightarrow H_2$	$H^{+I} \leftrightarrow H^0$	1 e^-
$CO_2 \leftrightarrow CH_4$	$C^{+IV} \leftrightarrow C^{-IV}$	8 e^-
$SO_4^{2-} \leftrightarrow H_2S$	$S^{+VI} \leftrightarrow S^{-II}$	8 e^-

1.5 Problems of Measurement and Interpretation of Redox Potentials

The redox potential of an environment is not an absolutely specified property. It is determined by the substances present in the environment. Substances that can release (reduction agent) or accept electrons (oxidizing agent) change the electron activity in the environment, thus changing the Eh. The most important substances in the environment determining the redox potential are presented in Table 1.2. Taking into consideration the fact that many of these substances are multi-electron transition components or elements, the overall reaction occurs as a multistage cascade of partial reactions with equilibriums between forms of substances (species) in various stages of oxidation state. Calculation of the redox potential based on the analytical concentration of individual redox pairs is very difficult for many reasons (Pitter 2009):

- the presence of several redox pairs simultaneously,
- slow reaching thermodynamic equilibrium,
- dynamic stationary state of the system with a low proportion of oxidized or reduced components,
- multi-electron transition of electrons in redox pairs.

In many cases, the redox potential in natural water is controlled by a redox pair of divalent and trivalent iron

$$Fe^{2+} \leftrightarrow Fe^{3+} + e^- \tag{1.11}$$

with the following relationship applied for the activity of the electrons

$$p\varepsilon = -\log K + \log \frac{a_{Fe^{3+}}}{a_{Fe^{2+}}} \tag{1.12}$$

where $\log K = -13.01$ (under standard conditions). The corresponding value of Eh can be expresses by Eq. (1.9). The analytically determined concentrations of Fe^{+II} and Fe^{+III} are usually taken for calculating Eh. However, under real conditions, both oxidation states of iron occur not only as free ions Fe^{2+} and Fe^{3+} but also as many hydroxo- and other complexes (sulfate, chloride, etc.). The activities of these ions are then significantly different from the analytical total concentrations (often almost negligible) of the valence forms and therefore the Eh calculation is incorrect.

This behavior can be documented using a simple example: in a solution at pH = 4.88 and Eh = 481.8 mV, there is 20 μmol/L of dissolved iron (1.12 mg/L). Analytically, the concentrations of dissolved iron in a divalent form of 17.4 μmol/L (87.2% of the total concentration of dissolved iron) and in a trivalent form of 2.4 μmol/L were determined. The redox potential calculated from these total concentrations using the relations (Eq. 1.12) and (Eq. 1.9) gives Eh = 720 mV, a significantly different value from the input value of 481.1 mV. This also shows an oxidizing environment, while the actual redox potential corresponds to anoxic or mildly reducing conditions (Fig. 1.1). Under the given conditions, both cations undergo a complexation reaction with the hydroxyl anion and form hydroxocomplexes. While free Fe^{2+} ions are prevalent, constituting 99.63% of its total Fe^{+II} content ($Fe(OH)^{+}$ being in the minority), for a trivalent iron hydroxocomplex, $Fe(OH)^{2+}$ constitutes 92.96% of its overall content and free Fe^{3+} ions contribute only negligibly to the total content (less than 0.01%). If we use the free concentrations of the participating substances calculated by the speciation model, we obtain the true value of 481.8 mV for the redox potential.

1.6 Geochemical Processes in Water

The main geochemical parameters and the stability of the environment are determined by pH and the oxidative reduction potential of Eh. The stability of pH in natural waters is determined by the carbonate system and exchange reactions of clay minerals. The acidity of the natural environment is determined by the CO_2 cycle and alkalinity by dissolving and weathering limestone and silicate rocks. In the case of Eh, the oxidation capacity reservoir is not only the oxygen itself (in the atmosphere and dissolved in the water), but also oxidized substances (e.g., nitrates). Reduction capacity is determined primarily by dead organic matter and also by reduced substances (e.g., sulfides). Putting aside the crucial redox reactions of biogenic processes (photosynthesis vs. respiration and decomposition), it is possible to describe the most important redox actions in exogenous processes by the following reactions:

Oxygen reduction/decomposition of water.

$$\tfrac{1}{2}O_2 + 2e^- + 2H^+ \leftrightarrow H_2O \tag{1.13}$$

Weathering/crystallization of pyrite.

$$FeS_2 + 8H_2O \leftrightarrow Fe^{2+} + 2SO_4^{2-} + 14e^- + 16H^+ \tag{1.14}$$

Oxidation/reduction of nitrogen.

$$NH_4^+ + 2H_2O \leftrightarrow NO_2^- + 6e^- + 8H^+ \quad (1.15)$$

$$NO_2^- + H_2O \leftrightarrow NO_3^- + 2e^- + 2H^+ \quad (1.16)$$

Oxidation/reduction of Fe^{2+} according to (Eq. 1.11) and subsequent hydrolysis

$$Fe^{3+} + 3H_2O \leftrightarrow Fe(OH)_3 + 3H^+ \quad (1.17)$$

The dissolution of pyrite as an important reservoir of reduction capacity is illustrated in Fig. 1.2.

Changes in the pH of an environment may be caused by redox reactions but not vice versa. Nevertheless, changes in the pH are usually buffered by the carbonate system in the natural processes. The buffer capacity is the ability of the solution to withstand the addition of an acid or alkali and maintain a near constant pH. It is highest in situations where concentrations of acids and conjugated (linked) bases are comparable. A solution is least buffered in a situation where the concentration of the acid and the conjugated base differs most (one form is mostly present in the solution). Therefore, acidity and alkalinity are important parameters of the natural environment and can fundamentally influence the remedial action conducted by chemically supported technologies.

The Eh value determines the activity of electrons in the natural environment. The greatest electron activity (pε) is in a reducing environment and decreases in an increasing oxidation environment. In natural systems, the general sequence of reduction processes—the so-called redox ladder—applies when a tiered profile of the pε develops, which is valid for a certain time and position until the relevant oxidant is consumed.

The general classification of a redox environment is usually determined by the source of oxygen. In an oxic environment, the source of the oxidation capacity is O_2

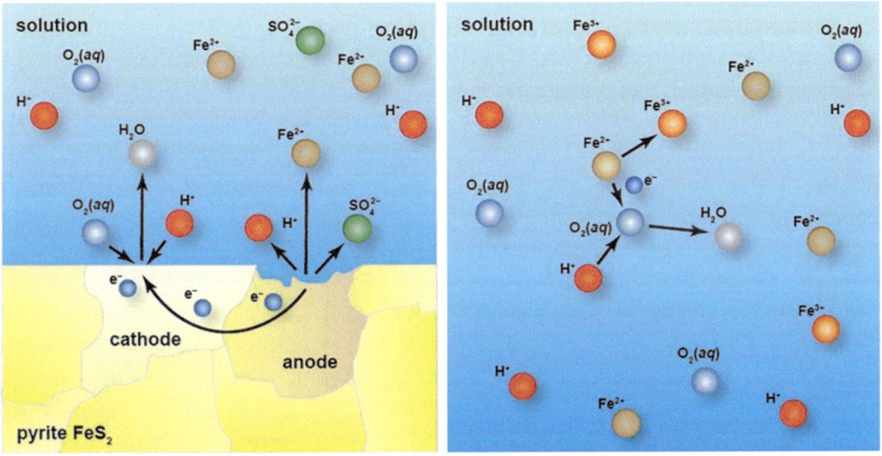

Fig. 1.2 Dissolution of pyrite

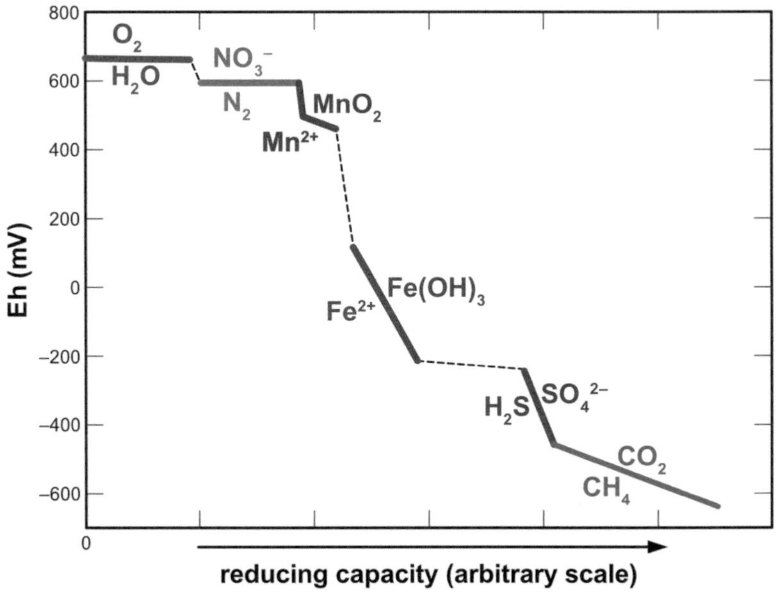

Fig. 1.3 Redox ladder of sequential oxidation of species in natural waters

or NO_3^-. In a suboxic environment, it is MnO_2 or $Fe(OH)_3$, and in a reducing environment it is SO_4^{2-} or CO_2 (Fig. 1.3).

1.7 The Principle of Remedial Reduction Technologies

For reduction methods, reagents that easily release electrons are used as a reduction agent, thus adjusting the conditions in the groundwater on the lower limit of water stability or even below it (Fig. 1.1). If there are substances that are capable of accepting electrons in the system, oxidation-reduction processes are then in progress. The most commonly used reagents are sodium dithionites ($Na_2S_2O_4$) (Chung 1981), calcium polysulfide (CaS_x) (Wazne et al. 2007), sodium metabisulfite ($Na_2S_2O_5$) (Chang 2003), sodium hydrosulfide (NaHS), sodium sulfite (Na_2SO_3) (Bianco Prevot et al. 2018), ferrous sulfate (Fe_2SO_4) (Mončeková et al. 2016), and currently metallic elements (zero-valent), namely Fe^0 (Tosco et al. 2014) iron, in the form of chips or nanoparticles. In addition, organic waste from food production, e.g., whey, which contains the anion of lactic acid ($CH_3CH(OH)COO^-$), can be biologically degraded to CO_2 with the release of electrons. The reaction is similar to the general degradation of organic compounds caused by oxygen, which is the opposite of photosynthesis expressed by Eq. (1.1). The oxidation reaction for the above-mentioned reducing agents can be expressed by the following equations

$$S_2O_4^{2-} + 4H_2O \rightarrow 6e^- + 8H^+ + 2SO_4^{2-} \quad (1.18)$$

$$S_2O_5^{2-} + 3H_2O \rightarrow 4e^- + 6H^+ + 2SO_4^{2-} \quad (1.19)$$

$$HS^- + 4H_2O \rightarrow 8e^- + 9H^+ + SO_4^{2-} \quad (1.20)$$

$$SO_3^{2-} + H_2O \rightarrow 2e^- + SO_4^{2-} + 2H^+ \quad (1.21)$$

$$Fe^{2+} \rightarrow e^- + Fe^{3+} \quad (1.22)$$

$$Fe^0 \rightarrow 2e^- + Fe^{2+} \quad (1.23)$$

$$CH_3CH(OH)COO^- + 3H_2O \rightarrow 12e^- + 11H^+ + 3CO_2 \quad (1.24)$$

Electrons are released in all of these reactions, causing a reduction of the target contaminant. An example can be a reduction of chlorinated hydrocarbon of general formula RX

$$R\text{–}X + nH^+ + me^- \rightarrow R\text{–}H + X^- \quad (1.25)$$

e.g., of TCE

$$C_2HCl_3 + 5H^+ + 8e^- \rightarrow C_2H_6 + 3Cl^- \quad (1.26)$$

As opposed to oxidation, the product is not carbon dioxide but non-chlorinated ethylene.

1.8 Principle of nZVI Application

Iron in a metallic form is not stable in contact with atmosphere and water (Torrey et al. 2015) and undergoes chemical oxidation (rusting). Figure 1.4 shows the E_h-pH stability diagram for dissolved iron of activity 10^{-6}, which corresponds to a concentration of about 0.06 mg/L. As the figure shows, under these conditions, the solution only retains divalent iron because during its oxidization to trivalent iron, the Fe^{3+} is precipitated in the form of goethite, oxyhydroxide (FeOOH), or magnetite (ferrous-ferric oxide, Fe_3O_4). Figure 1.4b shows that the area of the elementary iron Fe^0 stability lies below the lower limit of the water stability and the iron will always oxidize in contact with oxygenated water according to Eq. (1.3). The area of the Fe^{2+} stability and solid iron oxyhydroxide (goethite) overlaps and the redox equilibrium in the solution will be set by the reaction.

$$Fe^{2+} + 2H_2O \leftrightarrow e^- + FeOOH + 3H^+ \quad (1.27)$$

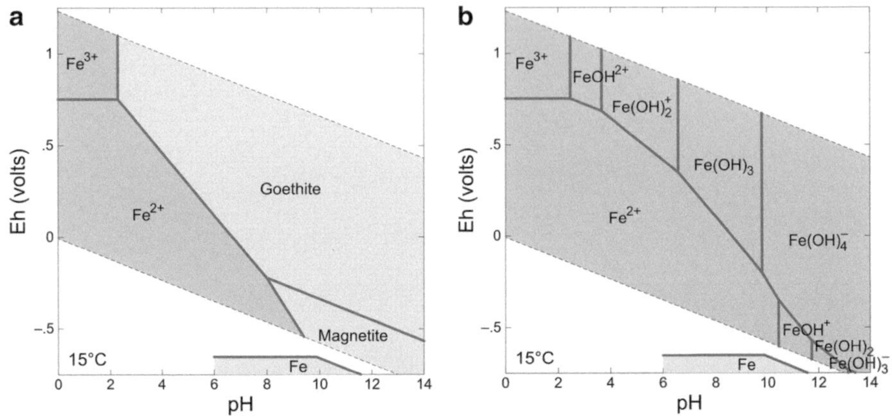

Fig. 1.4 (**a**) Eh-pH stability diagram for dissolved iron (iron activity is set to 10^{-6}, which corresponds to 0.06 mg/L), (**b**) Eh-pH diagram of species of dissolved iron indicating the stability of Fe^0

It is clear that a detailed thermodynamic analysis of a specific environment together with the analytical data on the composition of water, substrate, and contaminants can significantly contribute to determining the optimum remedial strategy.

The reaction of ZVI (and especially nZVI) in water has several steps due to the change in oxidation-reduction conditions of water. Initially, dissolved oxygen causes iron oxidation by Eq. (1.3). In the case of nZVI, which has an extremely large specific surface, the reaction (Eq. 1.3) is rapid. Iron precipitates in the form of ferric oxyhydroxide (FeOOH) or directly as ferric hydroxide $Fe(OH)_3$. The solubility of oxygen depends on the concentration of dissolved substances and decreases as the dissolved substances increase (Pitter and Chudoba 1990). At 10 °C and atmospheric pressure (101.3 kPa) the solubility of oxygen in water is 11.3 mg/L. After consumption of all the oxygen present and a decrease in the redox potential, groundwater reacts with the iron under anoxic conditions through the process of corrosion with hydrogen production

$$Fe^0 + 2H_2O \rightarrow Fe^{2+} + 2OH^- + H_2 \tag{1.28}$$

Since the reaction (Eq. 1.28) is slower, the loss of elemental iron is not as intense as it is during aerobic corrosion (Eq. 1.3). On the other hand, hydrogen generated during the anaerobic corrosion can promote the growth of anaerobic microorganisms that can also dehalogenate chlorinated hydrocarbons. The growth of microorganisms on the surface of the iron is influenced by the porosity of the material, hence the size of the reactive surface. In strongly reducing environments consisting of cinder pig, there is often an increase in concentrations of methane and light gaseous hydrocarbons (e.g., propane, butane) (Pitter and Chudoba 1990). This may be caused by either a reduction of the CO_2 present in the groundwater or by the hydrogenation of carbon, which is present in the iron as an admixture produced by hydrogen during anaerobic corrosion.

Because of iron corrosion under both aerobic and anaerobic conditions, there is an increase in the OH^- concentration resulting in an increase in the pH value of the system. This often rises to values of around 10 depending on the buffer capacity of the groundwater (carbonate content in particular). Dissolved carbon dioxide and bicarbonate also inhibit the increase in pH by the following reactions

$$H_2CO_3 + 2OH^- \rightarrow CO_3^{2-} + 2H_2O \quad (1.29)$$

$$HCO_3^- + OH^- \rightarrow CO_3^{2-} + H_2O \quad (1.30)$$

and help to buffer the system, which is important for both chemical and biological reduction processes.

In the real environment of natural water, other complexes (sulfate, carbonate, chloride, etc.) are present and must be included in the balance calculation, and therefore the speciation model is essential for an estimation of redox conditions.

Zero-valent iron can reduce chlorinated compounds to harmless components by a combination of Eqs. (1.25) and (1.28). The oxidation-reduction reaction is:

$$2Fe^0 + R\text{–}Cl + 3H_2O \rightarrow 2Fe^{2+} + R\text{–}H + 3OH^- + H_2 + Cl^- \quad (1.31)$$

This process can be accomplished in several ways by nZVI. The major processes are as follows (Fig. 1.5):

1. R-Cl reduction at Fe surface

$$Fe^0 + R\text{–}Cl + H^+ \rightarrow Fe^{2+} + R\text{–}H + Cl^- \quad (1.32)$$

2. R-Cl reduction by Fe^{2+} oxidation to Fe^{3+}

$$2Fe^{2+} + R\text{–}Cl + H^+ \rightarrow 2Fe^{3+} + R\text{–}H + Cl^- \quad (1.33)$$

3. Fe^0 reaction with water and hydrogen production

$$Fe^0 + 2H_2O \rightarrow Fe^{2+} + H_2 + 2OH^- \quad (1.34)$$

4. R-Cl reduction by hydrogen

$$H_2 + R\text{–}Cl \rightarrow R\text{–}H + H^+ + Cl^- \quad (1.35)$$

All these processes play a role in R-Cl reduction.

From a thermodynamic point of view, a transfer of electrons according to Eq. (1.31) in one step is not very probable and is possible only if the R-Cl molecule

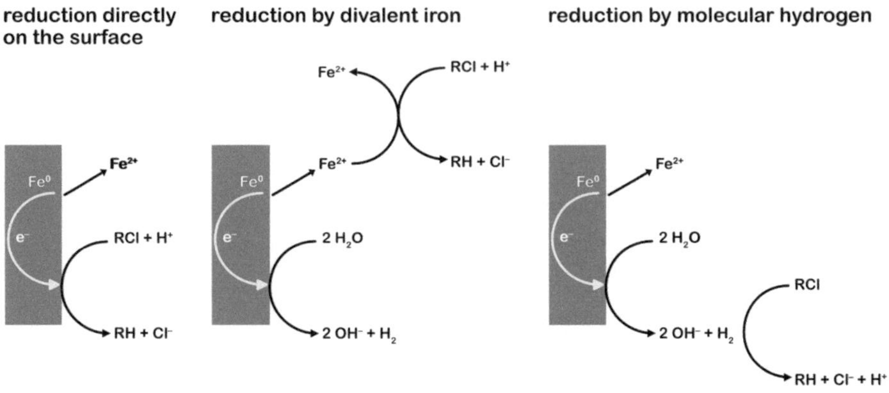

Fig. 1.5 Principles of reduction of R-Cl by Fe compounds

remains on the iron surface for a sufficiently long period of time. There are two reaction paths considered—sequential hydrogenolysis and β-elimination (Janda et al. 2004; Liu et al. 2005).

1.8.1 *Sequential Hydrogenolysis*

During sequential hydrogenolysis, C-Cl bonds in halogenated hydrocarbons are gradually replaced by the C-H bonds under sufficiently strong reducing conditions to form Cl^- and less chlorinated hydrocarbon according to Eq. (1.25). The feasibility of hydrogenolysis depends on the type of hydrocarbon and decreases from aliphatic to aromatic hydrocarbons. Hydrogenolysis is also influenced by the presence and location of substituents and heteroatoms.

1.8.2 *Reductive β-Elimination*

During reductive β-elimination of halogenated organic substances, multiple bonds (e.g., triple) between carbon atoms are created, and halogens are simultaneously removed from molecules in the form of halides. For the multiple bond formation, the halogen atoms must be bound to two adjoining carbon atoms. The reaction of halogenated ethene (e.g., 1,2-DCE) can be generally described by the equation

$$CHX = CHX + 2e^- \rightarrow CH \equiv CH + 2X^- \tag{1.36}$$

where acetylene is the final product.

References

Bianco Prevot A, Ginepro M, Peracaciolo E, Zelano V, De Luca DA (2018) Chemical vs bio-mediated reduction of hexavalent chromium. An in-vitro study for soil and deep waters remediation. Geoderma 312:17–23. https://doi.org/10.1016/j.geoderma.2017.09.032

Chang L-Y (2003) Alternative chromium reduction and heavy metal precipitation methods for industrial wastewater. Environ Prog Sustain Energy 22(3):174–182. https://doi.org/10.1002/ep.670220315

Chung SK (1981) Mechanism of sodium dithionite reduction of aldehydes and ketones. J Org Chem 46(26):5457–5458. https://doi.org/10.1021/jo00339a057

Ives DJG, Janz GJ (eds) (1961) Reference electrodes: theory and practice. Academic, New York

Janda V, Vasek P, Bizova J, Belohlav Z (2004) Kinetic models for volatile chlorinated hydrocarbons removal by zero-valent iron. Chemosphere 54(7):917–925. https://doi.org/10.1016/j.chemosphere.2003.08.033

Liu Y, Majetich SA, Tilton RD, Sholl DS, Lowry GV (2005) TCE dechlorination rates, pathways, and efficiency of nanoscale iron particles with different properties. Environ Sci Technol 39 (5):1338–1345. https://doi.org/10.1021/es049195r

Malmberg CG, Maryott AA (1956) Dielectric constant of water from 0° to 100° C. J Res Natl Bur Stand 56(1):2641

Mončeková M, Novotný R, Koplík J, Kalina L, Bílek V, Šoukal F (2016) Hexavalent chromium reduction by ferrous sulphate heptahydrate addition into the Portland clinker. Proced Eng 151:73–79. https://doi.org/10.1016/j.proeng.2016.07.382

Pitter P (2009) Hydrochemie, 4th edn. VŠCHT Praha, Prague

Pitter P, Chudoba J (1990) Biodegradability of organic substances in the aquatic environment. CRC Press, Boca Raton

Schuettler M (2007) Electrochemical properties of platinum electrodes in vitro: comparison of six different surface qualities. In: 2007 29th Annual International Conference of the IEEE Engineering in Medicine and Biology Society, 22–26 Aug. 2007, pp 186–189. https://doi.org/10.1109/IEMBS.2007.4352254

Stumm W, Morgan JJ (1995) Aquatic chemistry: chemical equilibria and rates in natural waters, 3rd edn. Wiley, New York

Tiehm A, Stieber M, Werner P, Frimmel FH (1997) Surfactant-enhanced mobilization and biodegradation of polycyclic aromatic hydrocarbons in manufactured gas plant soil. Environ Sci Technol 31(9):2570–2576. https://doi.org/10.1021/es9609967

Torrey JD, Killgore JP, Bedford NM, Greenlee LF (2015) Oxidation behavior of zero-valent iron nanoparticles in mixed matrix water purification membranes. Environ Sci Water Res Technol 1 (2):146–152. https://doi.org/10.1039/C4EW00068D

Tosco T, Petrangeli Papini M, Cruz Viggi C, Sethi R (2014) Nanoscale zerovalent iron particles for groundwater remediation: a review. J Clean Prod 77:10–21. https://doi.org/10.1016/j.jclepro.2013.12.026

Violante A, Cozzolino V, Perelomov L, Caporale AG, Pigna M (2010) Mobility and bioavailability of heavy metals and metalloids in soil environments. J Soil Sci Plant Nutr 10(3):268–292. https://doi.org/10.4067/S0718-95162010000100005

Wazne M, Jagupilla SC, Moon DH, Jagupilla SC, Christodoulatos C, Kim MG (2007) Assessment of calcium polysulfide for the remediation of hexavalent chromium in chromite ore processing residue (COPR). J Hazard Mater 143(3):620–628. https://doi.org/10.1016/j.jhazmat.2007.01.012

Chapter 2
Nanoscale Zero-Valent Iron Particles for Water Treatment: From Basic Principles to Field-Scale Applications

Tanapon Phenrat, Petra Skácelová, Eleni Petala, Adriana Velosa, and Jan Filip

Abstract Reductive technologies of groundwater and soil treatment, based on nanoscale zero-valent iron (nZVI) particles, have been recognized and generally accepted as modern remediation tools for elimination of broad range of both organic and inorganic environmental contaminants, mainly at sites where fast and efficient removal of persistent and emerging organic and inorganic pollutants is needed. In this chapter, we summarize the basic principles and chemical pathways of the nZVI interaction with water and contaminants, recent approaches to nZVI modifications enhancing their reactivity and longevity (including electrostatic and steric stabilization of nZVI, synthesis and applicability of bimetallic particles and nZVI-based nanocomposites, emulsification of nZVI particles, and combination of nZVI with electrokinetics), and present an overview on field-scale applications of nZVI for remediation purposes all over the world. The main aim is to demonstrate the diverse properties of nZVI particles and their possible limitations for water treatment.

Keywords Nanoparticles · Iron · nZVI · Groundwater · Surface and chemical modification · Nanocomposites · Reaction mechanisms · Field-scale applications

---

T. Phenrat
Department of Civil Engineering, Environmental Engineering Program, Naresuan University, Phitsanulok, Thailand

P. Skácelová
Regional Centre of Advanced Technologies and Materials, Palacký University Olomouc, Olomouc, Czech Republic

NANO IRON, s.r.o, Židlochovice, Czech Republic

E. Petala · A. Velosa · J. Filip (✉)
Regional Centre of Advanced Technologies and Materials, Palacký University Olomouc, Olomouc, Czech Republic
e-mail: jan.filip@upol.cz

J. Filip et al. (eds.), *Advanced Nano-Bio Technologies for Water and Soil Treatment*, Applied Environmental Science and Engineering for a Sustainable Future,
https://doi.org/10.1007/978-3-030-29840-1_2

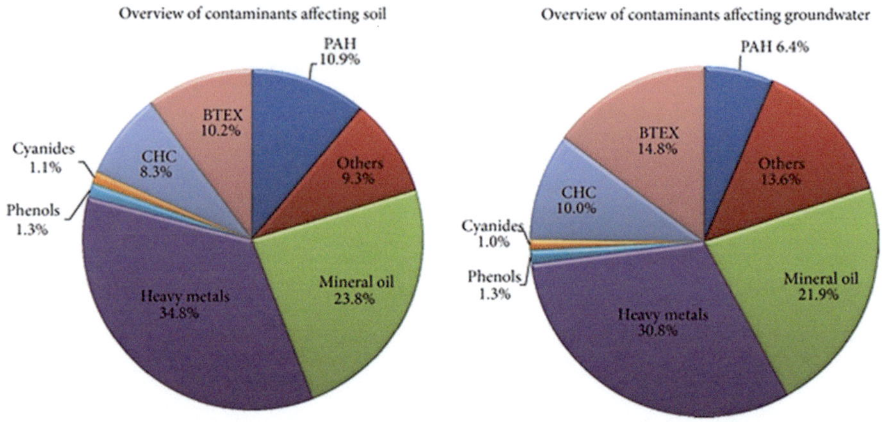

Fig. 2.1 Prevalence of soil and groundwater contaminants in Europe (adapted from Panagos et al. 2013 with permission)

2.1 Groundwater and Soil Contamination as a Worldwide Problem and Opportunity for Nanotechnologies

Groundwater and soil contamination caused by various toxic substances has posed a serious threat to human and ecological health while it has been of great concern to the general public worldwide for more than 40 years. In 2011, approximately 1.17 million potentially contaminated sites and 0.127 million contaminated sites were identified in the European Economic Area (European Environment Agency 2014). Figure 2.1 illustrates the statistics on the prevalence of soil and groundwater contaminants in Europe (Panagos et al. 2013). Over the last three decades, several European countries have introduced national policies and practices for the management of contaminated sites (Rodrigues et al. 2009). Although, in 2002, the EU Thematic Strategy for Soil Protection was proposed by the European Commission with the objective to protect soils across the EU, no specific regulations for soil protection at the EU level have been introduced so far. 'While the Commission in May 2014 decided to withdraw the proposal for a Soil Framework Directive, the Seventh Environment Action Programme, which entered into force on 17 January 2014, recognizes that soil degradation is a serious challenge (European Commission 2019). However, remediation activities in European countries are still going on. By 2018, 58,340 sites had been remediated (European Environment Agency 2014). A substantial budget is used for contaminated site management. For example, annual national expenditures for managing contaminated sites are, on average, EUR 10.7 per capita (an average of 0.041% of the national gross domestic product—GDP). Out of this, 42% comes from public budgets and around 81% is used for remediation measures, while 15% is used for sites.

Similarly, in the US, soil and groundwater contamination is a serious issue. Infamous contamination cases, such as Times Beach, Love Canal, Woburn, and the Valley of Drums, as well as their health effects (US EPA 2000, 2018a; Swartjes

Table 2.1 Prevalent contaminants of concern (COC) at National Priorities List (NPL) sites (adapted from Siegrist et al. 2011 with permission)

Contaminant of concern (COC)	Priority as a COC at US hazardous waste sites (ATSDR[a] ranking)	Prevalence at US NPL sites		US drinking water standard (μg/L)
		Number of NPL sites with COC present	Sites with COC present as a % of total NPL sites	
Arsenic	1	1149	68	10
Lead	2	1272	76	15
Mercury	3	714	49	2
Vinyl chloride	4	616	37	2
Polychlorinated biphenyls	5	500	31	0.5
Benzene	6	1000	59	5
Cadmium	7	1014	61	5
Polycyclic aromatic hydrocarbons	8	600	42	–
Benzo(*a*)pyrene	9	–	–	0.2
Chloroform	11	717	50	100
Trichloroethene	16	852	60	5
Chromium	18	1127	68	100
Tetrachloroethene	33	771	54	5
Pentachlorophenol	45	313	20	1
Carbon tetrachloride	47	425	26	5
Xylene (total)	58	840	51	10,000
Toluene	71	959	60	1000
Methylene chloride	80	882	56	5
1,1,1-trichloroethane	97	823	50	200
Ethylbenzene	99	829	49	700

[a]*ATSDR* Agency for Toxic Substances and Disease Registry

2011) generated environmental cleanup regulations, including the Comprehensive Environmental Response, Compensation, and Liability Act (CERCLA) for abandoned sites and corrective action of the Resource Conservation and Recovery Act (RCRA) for active sites and generated remediation engineering (Applegate and Laitos 2006; US EPA 2018a). Progressively, hundreds of thousands of sites with different degrees of contamination have been identified in the US. Table 2.1 summarizes prevalent contaminants of concern (COCs) at National Priorities List (NPL) sites. In 2018, the NPL featured 1566 contaminated sites. 381 of them have been sufficiently restored, posing a minimal risk, and proposed for site deletion (US EPA 2018b). Nevertheless, over the next decades, cleanup research and business will still be active in restoring the remainder as well as newly spilled sites. The US EPA

(2011) estimated that between the years 2004 and 2033 the cost for remediation may be as high as US$209 billion, while the number of cleanup sites could be as high as 294,000.

For Asian countries, remediation is at an infant stage, although soil and groundwater contamination has long been threatening the quality of life. For example, in China, according to the Ministry of Land Resources, in 2007, over 10% of cultivated land was substantially contaminated with heavy metals from mining and metallic smelting (approx. two million ha), oil extraction and refining (approx. five million ha), solid-waste stockpiles (i.e., open dumps; approx. 50,000 h.), improper handling of industry gaseous emission, wastewater discharge, and processing residue (approx. ten million ha), and sewage irrigation (approx. 3.3 million ha) (Li et al. 2015). In 2014, the Chinese government published a national soil survey report showing that 16.1% of all soil and 19.4% of cultivated land was contaminated with both organic and inorganic chemical contaminants. The total area of contaminated soil was roughly 25 million ha, while 3.5 million ha of farmland was so contaminated that no agriculture should be allowed on it (The Economist 2017). Cadmium and arsenic were found in 40% of the affected land. China undertook 28, 40, and 32 remediation projects for the years 2013, 2014, and 2015, respectively. The government estimated that, with 3.33 million ha of contaminated farmland already identified, the total budget remediation could reach as much as US$ 157,000 million (based on 2018 USD) (Gu and Stanway 2017).

Currently, more than 59 remediation technologies based on at least one of 14 different types of treatment processes are available in various stages of development and application (Van Deuren et al. 2002). Nevertheless, there is always enough room for nanotechnology to step in as an innovative in situ remediation technology. This gives birth to nanoscale zerovalent iron particle (nZVI), one of the nanotechnologies that is most frequently applied for groundwater and soil remediation. The primary utilization of ZVI is a passive remediation technology called permeable reactive barrier (Fig. 2.2). For this technique, bulk zerovalent iron (ZVI) in the form

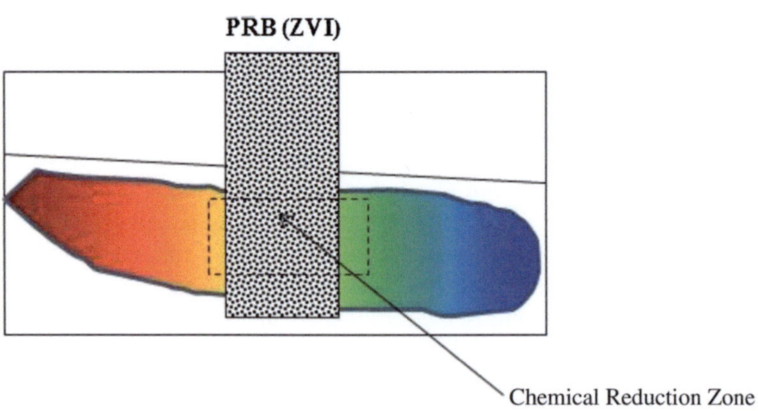

Fig. 2.2 Permeable reactive barrier using bulk ZVI

Table 2.2 Contaminants of concern amendable by ZVI and nZVI as well as their standard redox potentials (E^0) in aqueous solution at 25 °C (Bard et al. 1985; O'Carroll et al. 2013)

Aqueous solution	Half reactions	E^0 (V)
Chromium (Cr)	$CrO_4^{2-} + 8H^+ + 3e^- \leftrightarrow Cr^{3+} + 4H_2O$	1.51
Chromium (Cr)	$CrO_7^{2-} + 14H^+ + 6e^- \leftrightarrow 2Cr^{3+} + 7H_2O$	1.36
Platinum (Pt)	$Pt^{2+} + 2e^- \leftrightarrow Pt$	1.19
Palladium (P)	$Pd^{2+} + 2e^- \leftrightarrow Pd$	0.92
Mercury (Hg)	$Hg^{2+} + 2e^- \leftrightarrow Hg$	0.86
Silver (Ag)	$Ag^+ + e^- \leftrightarrow Ag$	0.80
Arsenic (As^V)	$H_3AsO_4 + 2H^+ + 2e^- \leftrightarrow HAsO_2 + 4H_2O$	0.56
Copper (Cu)	$Cu^{2+} + 2e^- \leftrightarrow Cu$	0.34
Uranium (U)	$UO_2^{2+} + 4H^+ + 2e^- \leftrightarrow U^{4+} + 2H_2O$	0.27
Arsenic (As^{III})	$H_3AsO_3 + 3H^+ + 3e^- \leftrightarrow As + 3H_2O$	0.24
Copper (Cu)	$Cu^{2+} + e^- \leftrightarrow Cu^+$	0.16
Lead (Pb)	$Pb^{2+} + 2e^- \leftrightarrow Pb$	−0.13
Nickel (Ni)	$Ni^{2+} + 2e^- \leftrightarrow Ni$	−0.25
Cadmium (Cd)	$Cd^{2+} + 2e^- \leftrightarrow Cd$	−0.40
Iron (Fe)	$Fe^{2+} + 2e^- \leftrightarrow Fe$	−0.44
Zinc (Zn)	$Zn^{2+} + 2e^- \leftrightarrow Zn$	−0.76
Barium (Ba)	$Ba^{2+} + 2e^- \leftrightarrow Ba$	−2.92
1,2-Dichloroethane	$ClH_2C - CH_2Cl + 2e^- \leftrightarrow H_2C = CH_2 + 2Cl^-$	0.74
Carbon tetrachloride (CT)	$CCl_4 + H^+ + 2e^- \leftrightarrow CHCl_3 + Cl^-$	0.67
Tetrachloroethylene (PCE)	$Cl_2C = CHCl + H^+ + 2e^- \leftrightarrow Cl_2C = CH_2 + Cl^-$	0.57
Trichloroethylene (TCE)	$Cl_2C = CHCl + H^+ + 2e^- \leftrightarrow Cl_2C = CH_2 + Cl^-$	0.53
Vinyl chloride (VC)	$ClHC = CH_2 + H^+ + 2e^- \leftrightarrow H_2C = CH_2 + Cl^-$	0.45
1,1-Dichloroethane (1,1-DCE)	$Cl_2C = CH_2 + H^+ + 2e^- \Leftrightarrow ClHC = CH_2 + Cl^-$	0.42

of iron filings has been used as a reducing agent to build permeable reactive barriers (PRB) since 1994 (Reynolds et al. 1990; Gillham and O'Hannesin 1994).

The fundamental chemistry of ZVI in an aqueous environment from a reaction viewpoint is summarized in Eqs. (2.1–2.3). For environmental remediation purposes, Fe^0 can be oxidized by contaminants of concerns (COCs) (such as trichloroethene—TCE) in Eq. (2.1)) (as electron acceptors) as long as COCs have an $E_H{}^0$ greater than −0.447 V (see Table 2.2 for more examples of COCs treatable by ZVI and nZVI). As a result of the electron transfer, in most cases, ZVI transforms such COCs to more environmentally benign by-products or immobilized state. This transformation involves reductive dechlorination of chlorinated organics (trichloroethylene [TCE], tetrachloroethylene [PCE], and vinyl chloride [VC]) and immobilization of metals. In the meantime, Fe^0 can also react with water (or H^+) to produce H_2 gas (Eq. 2.3), which is a competing reaction to the reductive treatment and is strongly controlled by the availability of H^+ (i.e., pH).

$$Fe^{2+} + 2e^- \rightarrow Fe^0, \quad E_H^0 = -0.447\,V \tag{2.1}$$

$$TCE + ne^- + (n - 3)H^+ \rightarrow product + 3Cl^- \tag{2.2}$$

$$H^+ + e^- \rightarrow H* \rightarrow ½H_2 \tag{2.3}$$

Although the oxidation of Fe^0 to Fe^{2+} (Eq. 2.1) is usually assumed, in environmentally relevant applications (i.e., groundwater at a natural pH), the transformation of Fe^0 core to the iron oxide shell, such as magnetite (Fe_3O_4) (Eq. 2.4) and maghemite (Fe_2O_3), is often observed (Liu et al. 2005; Reinsch et al. 2010).

$$3Fe^0 + 4H_2O \rightarrow Fe_3O_4 + 8H^+ + 8e^- \tag{2.4}$$

PRB is an established and effective technique to intercept contaminant plume; however, it is a passive technique meaning that the site owner has to maintain the PRB as long as the contaminant source exists and keeps generating toxic plume, which can last decades or even a century. This situation called for an active technique which can actively manage the source zone and expedite site closure. As shown in Fig. 2.3, the nanoscale conceptually allows ZVI to become an active technique that can be intentionally delivered to attack the source zone, thus speeding up the remediation.

The small size of nZVI not only offers the potential for injection into the subsurface for in situ remediation (Elliott and Zhang 2001; Schrick et al. 2004) but also results in an increasing fraction of atoms at the surface, excess surface energy, and high surface area (Wang and Zhang 1997). These properties lead to higher contaminant degradation/immobilization rates per mass of the remediation agents compared to bulk materials. Moreover, polymeric surface modification and supporting materials can be used to modify nZVI in order to enhance dispersion stability as well as to selectively target some specific COCs such as dense non-aqueous phase liquid (chlorinated organics). NZVI-focused research has progressed, over the past two decades, from laboratory development to field scale applications. This historical perspective has recently been reviewed by Phenrat and Lowry (2019). They divide the nZVI research and development into eight major topics, including (1) nZVI synthesis and reactivity, (2) aggregation/agglomeration, (3) transport/delivery/deposition, (4) polymer modification, (5) CMC modification, (6) toxicity, (7) sulfidation, and (8) weak magnetic/electromagnetic field, all of which are essential vehicles for effective in situ subsurface remediation. Figure 2.4 shows statistical results of peer-reviewed journal papers and the total citations of each major field. Noticeably, the number of peer-reviewed papers and citations in this field has increased quadratically so far. In 2017 alone, there were 262 peer-reviewed journal papers and 8094 citations in the nZVI field of study. Obviously, nanotechnology for subsurface remediation has been an active field of study for two decades and will continue to be active as long as contamination exists.

This chapter aims to summarize the recent development of nZVI technology including chemical pathways of pollutants treated by nZVI, modification and enhancement of nZVI, and pilot applications of nZVI as published in the scientific literature before 2018. For more details on nZVI research and development see the recently published book dealing with this topic Phenrat and Lowry (2019).

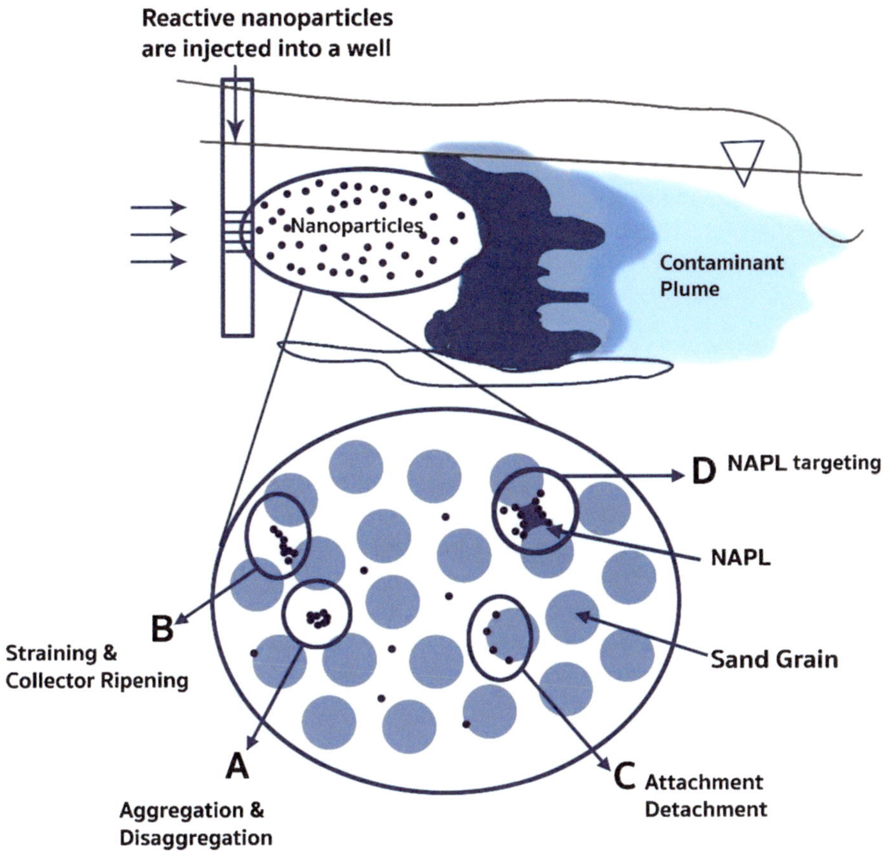

Fig. 2.3 Idealized conceptual mode of delivering nZVI for in situ non-aqueous phase liquid (DAPL) source-zone remediation. Particle mobility and contaminant targetability are needed for effective remediation (adapted from Phenrat and Lowry 2019 with permission)

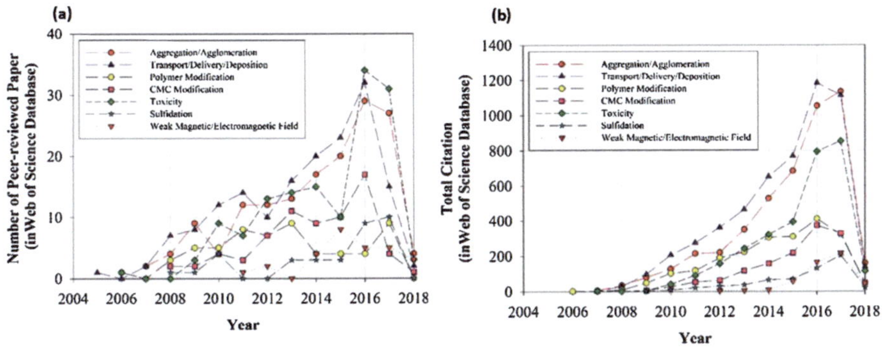

Fig. 2.4 (**a**) Total number of peer-reviewed papers and (**b**) total citations per year for each major field of nZVI research based on the Web of Science Database from 2001 to 2018 (February)

2.2 Chemical Pathways of Pollutants Removal by Zerovalent Iron

Under anaerobic conditions, Fe^0 can be oxidized by H_2O or H^+ yielding Fe^{2+} and H_2, both of which are also potential reducing agents for contaminants. There are two main dehalogenation reactions by which the organic compounds can be reduced by ZVI: hydrogenolysis (replacement of a halogen atom by a hydrogen) and reductive elimination, in which two halide ions are released. In both reactions, there is a net transfer of two electrons and they can be mediated either by atomic hydrogen transfer or by direct electron transfer.

The mechanism and reactivity of chlorinated compounds reduction by ZVI is somewhat controversial and still not very well understood. While some papers show an increased reactivity with an increasing number of halogen atoms in the organic compounds, others show the opposite trend. It seems that these conflicting data result from differences in the materials—ZVI produced by borohydride reduction versus ZVI produced by hydrogen reduction from iron minerals—as well as from differences in the experimental conditions (Elsner and Hofstetter 2011).

Wang and Farrell (2003), for example, observed that TCE reduction occurred almost exclusively by atomic hydrogen transfer at low pH values and by atomic hydrogen transfer and direct electron transfer at neutral pH values, while PCE reacted mainly via direct electron transfer at both low and neutral pH values. However, in acid conditions and micromolar concentrations, TCE reaction rates were faster than those of PCE due to faster reduction of TCE by atomic hydrogen transfer, while in neutral environment and millimolar concentrations, PCE reaction rates were faster than those of TCE. This variation of relative reaction rates was explained by a lower contribution of the atomic hydrogen reaction mechanism with increasing pH values and pollutant concentrations.

In the case of chlorinated methanes, the degradation pathways by reaction with zerovalent iron may differ from the chlorinated ethenes. Song and Carraway (2006), for example, observed that CCl_4 (CT), $CHCl_3$, and CH_2Cl_2 degradation rates were not affected by changing the hydrogen concentration in water or reaction atmosphere. So, in contrast to TCE degradation by nano ZVI synthesized by borohydride (Liu et al. 2005), no catalytic hydrogenation was found to be the degradation route for any of these compounds. Actually, CH_2Cl_2 is the main degradation product during CCl_4 reduction by ZVI and it is considered a as the final product since no degradation was observed by ZVI reaction. However, CH_4 is also produced but in very low amounts and is claimed to be generated directly through CCl_4 via a concerted elimination steps mechanism mediated by carbon radicals and carbanions.

Li and Farrell (2001) published an electrochemical investigation of the rate limiting mechanisms for TCE and CCl_4 reduction and concluded that rates of CT reduction were limited by the rate of outer-sphere electron transfer, while rates of TCE reduction were not limited by rates of electron transfer. Reduction via an outer-sphere mechanism requires only physical adsorption of CT on or near the ZVI surface. Then the production of chlorinated byproducts from chloroalkanes would be explained by a stepwise dechlorination process in which a brief interaction of this compound with the ZVI surface would promote an electron transfer one at time. On the other hand, the TCE reduction mechanism starts with a chemisorption step that controls the overall rate of reaction before the electron transfer.

The degradation of chlorinated ethanes has been observed to be dependent on the number of chlorine atoms as well as on their position in the molecule. Song and Carraway (2005) studied the degradation of a series of chlorinated ethanes and observed that the reactivity increased by increasing chlorination. They also reported that among tri and tetrasubstituted compounds, the reactivity was higher for compounds with chlorine atoms more localized in only one carbon, e.g., 1,1,1-TCA > 1,1,2-TCA. The proposed explanation for this difference in the reactivity is the shifting of the mechanism, in which the 1,1,1-TCA reacts by a concerted pathway involving α-elimination and hydrogenolysis, while 1,1,2-TCA reduction would proceed by β-elimination.

With the aim of increasing the reactivity; mobility; and transport in subsurface, along with the inhibition of ZVI passivation, many research groups have been studying the effect of adding metal catalysts (Schrick et al. 2004); coating nanoparticles with polymers/surfactants (Wang et al. 2010); supporting nanoparticles on different substrates (Jia et al. 2011); and applying ZVI in water-in-oil emulsions (Berge and Ramsburg 2009).

Schrick et al. (2004), using Fe^0/Ni bimetallic nanoparticles, observed a TCE degradation rate constant that was 50 times faster than pure ZVI, indicating that the bimetallic nanoparticles were more efficient in degrading TCE when compared to the monometallic iron nanoparticles.

The introduction of a second catalytic metal could also prevent toxic byproduct formation by dehalogenating TCE via hydrogen reduction rather than via electron transfer and even enables the reduction of persistent compounds like monochlorophenols, which cannot be reduced by ZVI itself (Morales et al. 2002). On the other hand, the presence of different metals increases water promoted corrosion of iron, which can result in lower reactivity and lifetime. Yan et al. (2010) observed that Pd-doped ZVI nanoparticles immersed in water for 24 h did not exhibit any metallic iron component in the XPS spectrum, while the Fe^0 peak, although with decreased intensity, remained observable for pure ZVI nanoparticles kept on the same conditions. Besides, in a reactivity TCE reduction study with Pd-ZVI nanoparticles, the apparent reaction rate constant decreased from 5.7 1/h for the fresh particles to 0.96 1/h upon 24 h aging.

In bimetallic systems, the atomic hydrogen adsorbed on the reductant surface (H_{ads}) is postulated to be responsible for bimetal reactivity and the generation of H_{ads} species has been proposed by different ways. It could be produced by the dissociative chemisorption of H_2, itself generated by water reduction, or as an intermediate to H_2 generation. It is also suggested that absorbed atomic hydrogen (H_{abs}) within the metal additive lattice, instead of surface-adsorbed atomic hydrogen (H_{ads}), would represent the reactive entity in iron-based bimetallic systems (Cook 2009).

ZVI can also be used for oxidative degradation of organic compounds, either providing Fe^{2+} to Fenton process (Eq. 2.5) or by reacting with oxygen itself through a superoxide radical ($O_2^{-\bullet}$) intermediate leading to the production of hydrogen peroxide in situ (Eq. 2.6).

$$Fe^{2+} + H_2O_2 \rightarrow Fe^{3+} + OH^- + {}^{\bullet}OH \quad (2.5)$$

$$Fe^0 + O_2 + 2H^+ \rightarrow Fe^{2+} + H_2O_2 \quad (2.6)$$

It is expected that during the oxidative process the compounds are not only dehalogenated but mineralized as well. The degradation process at this time is claimed to be promoted by hydroxyl radical (${}^{\bullet}OH$), which is a non-specific and powerful oxidant ($E^0 = 2.73$ V).

Keenan and Sedlak (2008) evaluated the effect of different ligands (oxalate, nitrilotriacetic acid (NTA), or ethylenediaminetetracetic acid (EDTA) on nZVI/H_2O/O_2 system and observed that all of them promoted an increase in the oxidant yield by limiting iron precipitation and accelerating the rates of key reactions like ferrous iron oxidation by oxygen and hydrogen peroxide.

On the other hand, Correia de Velosa and Pupo Nogueira (2013) also evaluated the effect of some ligands (EDTA, glycine, citrate, oxalate and DTPA) on 2,4-Dichlorophenoxyacetic acid (2,4-D) degradation by nZVI/H_2O/O_2 and reported that the only effective ligands on the catalysis of 2,4-D oxidation were EDTA and DTPA. They also postulate that the catalysis process is run by the ligand-Fe(II) species at pH values lower than 5 and by Fe(II) species at pH higher than this.

Presenting a standard potential of −440 mV, ZVI is considered a potential reductant agent for any other metal holding reduction potentials more positive than it. This property makes ZVI an interesting material for removal of heavy metals like Ni(II), Cu(II), Cr(VI), Pb(II), e.g., from groundwater or wastewater matrices.

The main mechanisms by which heavy metals are removed from solution in the ZVI/H_2O system are reduction, adsorption, and precipitation/co-precipitation, according to the metal. For example, the reduction of Cu(II) (Eqs. 2.7 and 2.8) or As(V) (Eq. 2.9) by ZVI is more thermodynamically favorable than precipitation and sorption, as well as it is less affected by pH change and the presence of ligands (Li et al. 2017).

$$Cu^{2+} + Fe \rightarrow Fe^{2+} + Cu \quad (2.7)$$

$$2Cu^{2+} + Fe + H_2O \rightarrow Fe^{2+} + Cu_2O + 2H^+ \quad (2.8)$$

$$H_2AsO_4^- + Fe + 3H^+ \rightarrow Fe^{2+} + H_3AsO_3 + H_2O \quad (2.9)$$

On the other hand, if the metal/metalloid is present like an oxyanion, such as AsO_4^{3-} or SeO_4^{2-}, then the precipitation by Fe^{2+} (Eq. 2.10) becomes important. This can be even the main cause of As removal, for example.

$$3Fe^{2+} + 2HAsO_4^{2-} \rightarrow Fe_3(AsO_4)_2(s) + 2H^+ \quad (2.10)$$

It is postulated that the Cr(VI) removal by ZVI occurs via immediate adsorption on the surface of the materials followed by electron transfer (reduction) through the oxidation of Fe^0 to Fe^{3+} (Eq. 2.11). The just generated ions (Fe^{3+} and Cr^{3+}) are then removed from the solution by precipitation of mixed hydroxides (Eq. 2.12) (Fu et al. 2014).

$$Cr_2O_7^{2-} + 2Fe + 14H^+ \rightarrow 2Cr^{3+} + 2Fe^{3+} + 7H_2O \tag{2.11}$$

$$Cr^{3+} + Fe^{3+} + 6OH^- \rightarrow Cr(OH)_3(s) + Fe(OH)_3(s) \tag{2.12}$$

Ling et al. (2017) followed the removal of several metals (Ag(I), Ni(II), Cr(VI), As(V), Cs(I) and Zn(II)) by nZVI via high-sensitivity X-ray energy-dispersive spectroscopy-scanning transmission electron microscopy (XEDS-STEM). Since the studied metals presented very different electrochemical and coordination properties, they were used like probes, in order to understand the reactive pathways. Some of their conclusions are the following: strong oxidizing agents like Cr (VI) react by diffusion and encapsulation in the core of nZVI, while metal cations with a reduction potential close to or more negative than that of ZVI, such as Cs (I) and Zn(II), are removed by sorption or surface-complex formation.

From the application point of view, it is important to evaluate the effect of common ions on metal/metalloid removal by ZVI. Smedley and Kinniburgh (2002), for example, reported that high concentrations of phosphate in groundwater can inhibit As(V) removal by sorption since phosphate competes for sites on hydrous ferric oxides. The presence of Ca^{2+} or humic acid alone did not affect the Cr (VI) removal by ZVI in batch studies; however, the presence of bicarbonate ions increased it (Liu et al. 2009b).

The presence of nitrate, an oxidant usually reduced by ZVI, on Pb^{2+} solutions influences drastically its removal. At a low concentration of nitrate, the removal of Pb^{2+} by precipitation is increased by the pH increase promoted by nitrate reduction. However, in excess of nitrate (in relation to ZVI), the ferrite particles responsible for adsorption of Pb^{2+} are dissolved by the nitrate driven oxidation of Fe^{2+} and Pb^{2+} cations are remobilized to solution (Su et al. 2014).

Actually, nitrate can impact the performance of ZVI-driven reductions by two main ways: since it can also be reduced by ZVI, it competes with the target compounds for the reactive sites on iron particles and it works like a passivating agent, which leads to the generation of an Fe(III) (oxyhydr)oxide shell that inhibits the reactivity and decreases the lifetime of ZVI. Liu et al. (2007), for example, reported that an increase in the nitrate concentration led to the inhibition of TCE reduction up to seven-fold when the nitrate concentration reached 5 mM.

Other anions, like silicate or bicarbonate, also inhibit the degradation processes. Although these anions cannot be reduced by ZVI, they can complex to iron surface generating compounds like FeH_3SiO_4, $FeH_2SiO_4^-$, and $\equiv FeHSiO_4^{2-}$ or $FeCO_3$, decreasing the access of target compound to iron surface by forming a film or protective layer. On the other hand, chlorine and sulfate ions play corrosive roles in attacking and breaking the iron oxide layers and exposing the bare metal to the target compounds. Then, these anions usually increase the reactivity of ZVI system (Sun et al. 2016).

2.3 Modification of nZVI Particles and Enhancement of Their Reactivity

In spite of their high surface activity and remarkable environmental uptake capacity, most synthetic ZVI nanoparticles display some significant disadvantages from the technological/chemical engineering and cost/benefit perspectives. Therefore, drawbacks such as the strong tendency towards aggregation, fast oxidation, and rapid sedimentation can eliminate their long-term reactivity and decrease their maximum capacity.

Generally, it has been reported that the limited mobility of nZVI particles in saturated porous media is attributed to two reasons; firstly, nZVI can be filtered from the solution by attaching to aquifer materials and secondly, agglomeration and aggregation remarkably immobilize the particles (Phenrat et al. 2007). Agglomeration and aggregation phenomena decrease the specific surface area, and in turn, affects the reactivity and mobility in the subsurface and porous media, such as sand and soil (Ponder et al. 2000; Sun et al. 2007). The strong tendency towards nZVI aggregation is to a large extent attributed to their unstable colloidal nature and long-ranged attractive magnetic interactions between the particles (Phenrat et al. 2009).

Gravitational sedimentation of nZVI particles is a result of the aggregation effect and can be a good indicator of the colloidal stability of the particles. When nanoparticles are dispersed in a media, they can remain stable for very long time under some conditions. These conditions demand that the diffusion flux of nanoparticles overcomes the sedimentation flux. The diffusion flux of nanoparticles, that opposes gravity, is inversely proportional to the particle size, while the sedimentation flux is proportional to the square of the particle radius (Phenrat et al. 2008). When nanoparticles start to aggregate to bigger clusters in the range of micrometers, they settle as a result of the fact that the sedimentation flux turns to be bigger than the diffusion flux. Thus, their delivery and mobility is limited.

The effects of the fast oxidation of nZVI particles include a gradual loss of reactivity and permeability (Keum and Li 2004). While metallic iron exists in an aqueous environment, dissolved oxygen (DO) and water initialize its oxidation-dissolution. The whole reaction involves the formation of soluble ionic products or insoluble oxides/hydroxides. In other words, in aqueous environment, the nZVI oxidation creates species such as soluble $Fe^{2+}_{(aq)}$, H_2 and various precipitates, e.g., $Fe(OH)_2$, $Fe(OH)_3$, Fe_3O_4, Fe_2O_3, FeOOH, $Fe_5HO_8{\cdot}4H_2O$, and green rusts (Crane and Scott 2012). In this manner, when the precipitates are formed as a layer on the surface of nZVI and reach a critical thickness, the reactivity of nZVI is inhibited. In this point, the reactivity is eliminated while the surface of metallic iron is blocked to interact with any media or with the pollutants. These factors concerning the oxidation of the particles can be critical for the overall efficacy of this material since nZVI can react, be oxidized, consumed, or blocked before it reaches the target and remediates a desired site.

For all the above-mentioned reasons, the modification and/or stabilization of the nZVI nanoparticles, which can lead to steady or even enhanced remediation ability and improved mobility, is deemed necessary. In the past few years, different approaches and a large number of research works have led to a new field of modification of nZVI particles. Therefore, new various methods have been developed in order to synthesize more active (Wu and Ritchie 2006), stable (Sun et al. 2007; Siskova et al. 2012), and mobile nZVI particles (Tiraferri et al. 2008; Kim et al. 2009) to simplify the synthetic procedure; reduce the cost; and finally enhance the efficacy (Zhang et al. 2012; Alessi and Li 2001), availability, and applicability to large-scale (Chen et al. 2012) and delivery capabilities (Kanel and Choi 2007).

Surface modification by electrostatic and steric stabilization, such as coating with polyelectrolytes or nonionic surfactants (Alessi and Li 2001), dispersing the particles in oil–water emulsions (Quinn et al. 2005), use of a support material for their synthesis (Wu et al. 2012; Ponder et al. 2001) and bimetallic particles with iron and a second less reactive metal, are strategies that have been investigated thoroughly. These methods have proved that it is possible to overcome the magnetic attraction between the iron nanoparticles, change the surface or interfacial properties, increase mobility and stability, minimize aggregation, and significantly increase the reactivity.

2.3.1 Electrostatic and Steric Stabilization

Surface modification by electrostatic and steric stabilization can provide conditions that overcome the magnetic attraction between the iron nanoparticles and change the surface or interfacial properties (Fig. 2.5), thus enhancing the colloidal stability and mobility. By electrostatic stabilization the surface charge is changed and repulsive forces overcome the affinity of aggregation. On the other hand, steric stabilization is typically attained by the adsorption of long-chain hydrophilic polymers whose long loops and tails extend out into the solution (e.g., surfactants) (Tiraferri et al. 2008). Combined electro-steric stabilization is also promising in terms of the use of ionic polymeric molecules that provide good dispersion and high efficacy (Sun et al. 2007). Such ionic polyelectrolytes have been applied successfully in the past (see Fig. 2.6) including polyaspartate (PAP), carboxymethyl cellulose (CMC), and polystyrene sulfonate (PSS). The modified nZVI particles in these cases have exhibited mobility in porous media due to slow desorption of polymeric surface modifiers. Every time, according to the contaminant that is targeted, the election and usage of

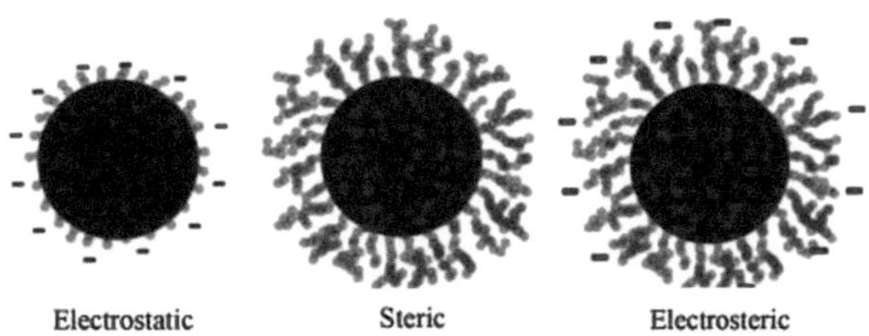

Fig. 2.5 Schematic representation of electrostatic, steric and electrosteric stabilization of nanoparticles (adapted from Tang and Lo 2013 with permission)

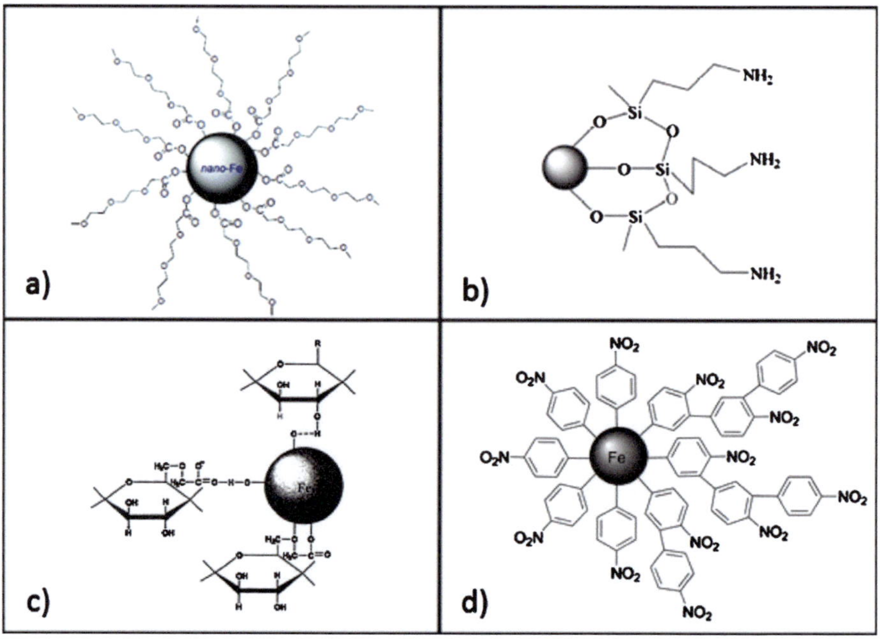

Fig. 2.6 Examples of nZVI surface modified by different surfactants (**a**) methoxyethoxyethoxyacetic acid (MEEA) (adapted from Kharisov et al. 2012 with permission), (**b**) 3-aminopropyltriethoxysilane (APS) (adapted from Liu et al. 2009a with permission), (**c**) carboxymethyl cellulose (CMC) (adapted from Zhao and He 2011 with permission), (**d**) aromatic diazonium salts (ADSs) (adapted from Guselnikova et al. 2015 with permission)

the appropriate surfactant can dramatically enhance the capacity and the colloidal stability. The higher efficiency is attributed not only to the lower extent of agglomeration of the iron particles, but also to the increased local concentration and sorption of the contaminant on the surface of iron (Alessi and Li 2001; Zhang et al. 2002). It has been reported that even the hydrodynamic diameter of the modified particles was increased after modification, their colloidal stability was enhanced, and the activation energy in the transformation of toxic compounds was decreased (Li et al. 2006; Saleh et al. 2005; Lien and Zhang 1999).

2.3.2 *Bimetallic Particles*

Bimetallic particles of iron with a second less reactive, i.e., usually noble, metal, e.g., Pd/Fe, Ni/Fe, Pt/Fe, Ag/Fe, Cu/Fe, have exhibited a significantly high efficacy for the degradation of many compounds (Chen et al. 2008). The incorporated metal, even in a small amount, can substantially enhance the overall nZVI reaction rate by acting as a catalyst for electron transfer and hydrogenation (Li et al. 2006). The additive metal lowers the activation energy of the reaction and increases the reaction rate. Thus, bimetallic particles have been involved in applications where bare nZVI usually shows a slow reaction rate, e.g., aromatics and polychlorinated biphenyls (PCBs) (O'Carroll et al. 2013). The doped metals that are chosen in these systems have a much higher redox potential (E^0) than iron. In that way a galvanic cell is formed, i.e., iron acts as an anode, and electron release reactions are accelerated, i.e., the reduction of pollutants is promoted in a higher rate (Elliott and Zhang 2001). Moreover, the additive metal can play an important role in prevention of the surface precipitation of iron oxide products, thus hindering the inhibition of the reactivity of iron. Thereby, iron particles can be more stable on air while those that have lost their surface activity by the degradation of toxic contaminants can be reactivated.

Bimetallic particles can be formed through different methods and different structures can be obtained as it is depicted in the Fig. 2.7. Among the most popular ways of synthesis are counted coating and growing of metal ions around nZVI particles by co-reduction, electrochemical synthesis and inverse micelles, independent nucleation and growth of two kinds of metal species, and the deposition of the noble metal onto the surface of nZVI particles (Liu et al. 2014; Ferrando et al. 2008).

However, a critical point in their application is their potential structural change over the time and the potential environmental risk concerning the addition of one more metal in the aquifer horizon, especially in the case of Ni/Fe nanoparticles.

2.3.3 *Emulsification*

When there is a need of nZVI application to treatment of dense nonaqueous phase liquid (DNALP) source zones, the best way to be delivered is in emulsified oil–water suspensions (Fig. 2.8). Emulsified nano zerovalent iron (EnZVI) is a biodegradable emulsion, composed of a surfactant, biodegradable vegetable oil, water, and

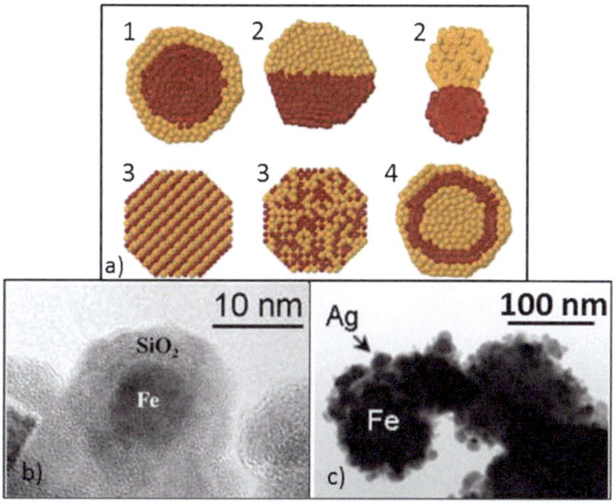

Fig. 2.7 (**a**) Schematic representation of bimetallic NPs with severe possible structures (*1*) core–shell segregated structures (*2*) heterostructure (*3*) intermetallic or alloyed structures and (*4*) multishell structures (adapted from Liu et al. 2014; Ferrando et al. 2008 with permission) and examples of dimetallic particles based on nZVI (**b**) Si/Fe (adapted from Fernández-Pacheco et al. 2006 with permission), (**c**) Ag/Fe (adapted from Marková et al. 2013 with permission)

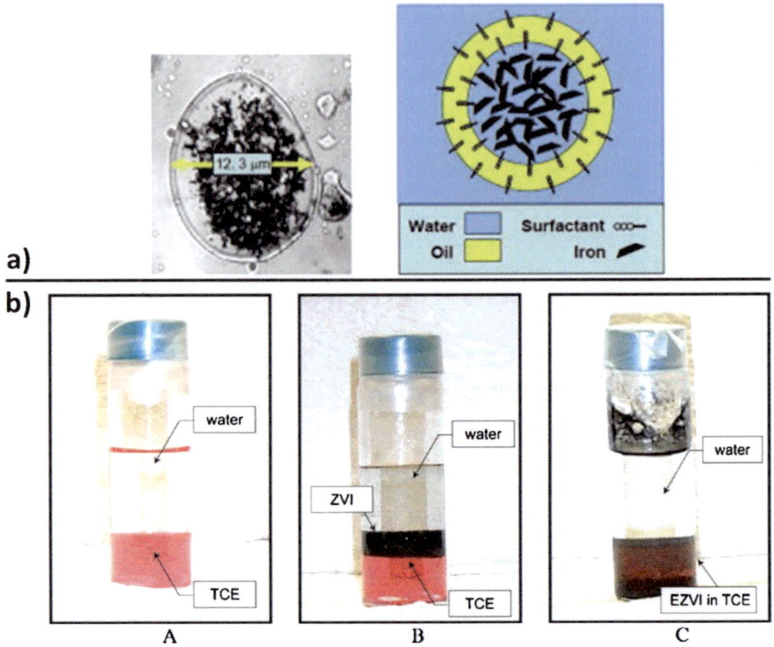

Fig. 2.8 (**a**) Magnified image and schematic illustration of EnZVI (Su et al. 2012) (**b**) Photos showing the dispersion of ZVI and EnZVI in a mixture of water and trichloroethene, TCE, solution dyed with Sudan IV, (*A*) are shown the two distinct phases of water and TCE, (*B*) distinct phases of water and TCE while the ZVI that was added remains separated from the TCE phase, (*C*) distinct phases of water and TCE but EZVI is dispersed in the TCE phase (adapted from Quinn et al. 2005 with permission)

nanoscale or micro-scale zerovalent iron particles in an emulsion (Su et al. 2012; Yang and Chang 2011). In that case, nZVI dispersed in water is surrounded by a liquid oil membrane forming droplets; the droplets then shield nZVI from direct contact with the media into which it is applied to. EnZVI possesses hydrophobic properties, making it miscible with DNALP contaminants. In that context, when the oil emulsion droplet comes in contact with DNALP, DNALP can easily diffuse into the interior place of the droplet where it can interact with nZVI and be degraded. The degradation by-products subsequently diffuse from the droplet out to the aqueous phase (O'Hara et al. 2006). In these systems, vegetable oil and surfactants sequesters some of the organic contaminants and can further serve as electron donors facilitating the total degradation process (Su et al. 2012; Singh and Misra 2015). Application of EnZVI on a large scale has been reported in several real cleanup scenarios mainly focused on organochlorine contaminants. For example, volatile organic compounds, e.g., trichloroethylene (TCE) and tetrachroroethylene (PCE) (Lee et al. 2007), and pesticides, e.g., atrazine and cyanazine (Waria et al. 2009), have been successfully and efficiently removed from subsurface aqueous systems. Namely, in USA almost 10% of the applied nZVI processes is related to the EnZVI, while, in Europe, no such a field application has been reported so far (Mueller et al. 2012). Su et al. (2012)

described the pneumatic and direct injection as two different delivery methods of EnZVI, providing promising results relating to the dechlorination and reductive dechlorination of chlorinated ethenes. Extensive tests by O'Hara et al. (2006) revealed the synergetic removal effect of EnZVI, while the combination of sequestration and abiotic degradation mechanisms driven by the oil emulsion and nZVI, respectively, was found. Moreover, there have been promising results relating to the stability and targetability of nZVI (Sheu et al. 2015). Dong et al. (2015) showed that nZVI in emulsion appears to form fewer aggregates in comparison with bare nZVI, while the oil concentration can be optimized, which leads to an efficient and long-term reactivity toward toxic compounds removal. Furthermore, in a pilot-scale study where a nZVI-emulsified colloidal substrate was used for PCE degradation, Sheu et al. (2016) not only observed the efficient dechlorination after 130 days of operation, but also noticed that the use of the nZVI-emulsified colloidal substrate increased the popolutation of *Dehalococcoides* spp. (DHC) and *Desulfitobacterium* spp. (DSB). The increase of DHC and DSB population promotes the PCE dechlorination process (Grostern and Edwards 2006). It has to be noted that there are some challenges that have to be dealt with in relation to the injection of such particles to a subsurface, e.g., due to the viscosity of these emulsions (Bhattacharjee and Ghoshal 2016). The type of surfactant and the concentrations of all the components, e.g., oil, could be tuned in order to maintain the highest remediation performance. For instance, it was found that ionic surfactants are more preferable than nonionic surfactants since they can lead, to some extent, to the enhancement of the degradation ability of nZVI (Cook 2009).

2.3.4 Using Electrokinetics

Electrokinetic (EK) remediation technology has been applied successfully to many pilot-scale sites for soil and underwater treatment of various organic and inorganic contaminants (Virkutyte et al. 2002; Gomes et al. 2012). An electrokinetic system includes a pair of electrodes and direct-current power imbedded in the soil. A low intensity electric current passes between a cathode and an anode. In that way, different ions, charged particles, and water are moving towards the opposite charged electrode. Mechanisms such as electro-migration, electro-osmosis, electrophoresis, and electrolysis take place (Mulligan et al. 2001; Weng et al. 2006). The contaminants and their byproducts can be collected by electroplating or precipitation/co-precipitation at the electrodesby the generated H_2 and OH^-. In Fig. 2.9a are depicted the electrokinetic phenomena that can occur involving the movement of electricity, charged particles and fluids.

The integration of both techniques, electrokinetics and nZVI, can couple the advantages and increase the benefits regarding water remediation. Among the overall reactions that occur during the application of the electrokinetics, electrolysis reaction of water causes the creation of H^+ and OH^- ions, with the first one to be more mobile. The protons (H^+) are very beneficial to the reaction with ferrous ions

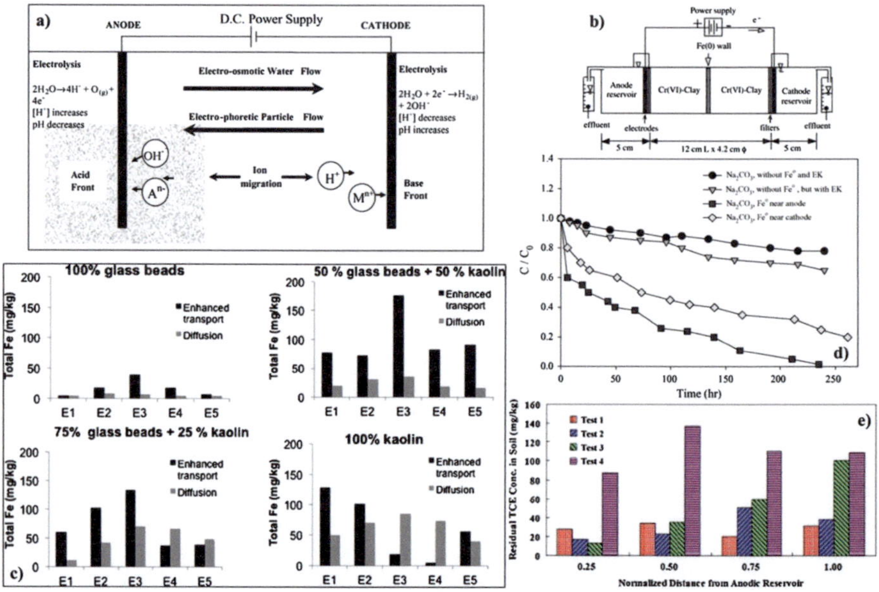

Fig. 2.9 (**a**) Basics of electrokinetic phenomena (adapted from Glendinning et al. (2007) with permission), (**b**) Model of the combination of nZVI and electrokinetics process setup (adapted from Weng et al. (2007) with permission), (**c**) Total iron distribution on the electrophoretic cell (adapted from Gomes et al. (2013) with permission) (**d**) The effect of electrokinetics on water chemistry (adapted from Chang and Cheng (2006) with permission) (**e**) The effect of electrokinetics on TCE removal (adapted from Yang and Chang (2011) with permission)

and in that way they eliminate formation of the iron oxide-passivation layer on nZVI surface, thus extending the operational life of nZVI (Chang and Cheng 2006). Moreover, electro-migration phenomena provide inhibition of sedimentation and aggregation of nZVI, i.e., their mobility and advection is increased (Černík et al. 2019). Subsequently, long-term and high reactivity of nZVI is guaranteed. In Fig. 2.9b is depicted a model example of the merge of the two techniques for hexavalent chromium (Cr(VI)) removal, where nZVI is positioned as a reactive wall and constant electric potential gradient was applied in order to move the electrolyte solution within the soil cell. Weng et al. (2007) showed the successful incorporation of nZVI and EK by presenting the synergetic effect of this combination in hexavalent chromium reduction. Chowdhury et al. (2012) investigated the impact of an external electric field on nZVI mobility in two different porous media. Due to the surface properties of nZVI, the nZVI injection into the media was preferably applied near the cathode. The obtained results revealed the potential of the minimization of nZVI oxidation and the enhancement of nZVI migration when electrokinetics was applied.

Moreover, electrokinetics offers the flexibilty to involve any nZVI based material in such processes. For instance, Reddy et al. (2011) studied the transport and reactivity of aluminum lactate modified nZVI in dinitrotoluene contaminated soils under applied electric potential, showing the altered properties both in mobility and reactivity terms. Yang and Chang (2011) analyzed the removal efficiency of emulsified nZVI combined with electrokinetic remediation technique, providing promising insights into the application of such integrated methods. Gomes et al. (2013) tested the transport properties of polymer coated nZVI when it is combined with EK, indicating that nZVI mobility in various porous media can be enhanced with the use of direct current.

2.3.5 nZVI Supported on Various Materials

Immobilization and stabilization of highly active nZVI can be achieved by synthetically entrapping them onto a matrix. Aggregation phenomena are reported to be considerably reduced when a matrix or the support material is used for the synthesis of nZVI. In this case, nZVI possesses a higher specific area, colloidal stability, homogeneous dispersion, and narrower size distribution (Ponder et al. 2000). Therefore, supported nZVI display higher activity compared with non-supported systems (Ponder et al. 2001). Additionally, a support material can serve as a “host” for the byproducts that are formed during the nZVI reaction with the pollutants, e.g., degraded pollutant/Fe(III) hydroxides precipitates. Thus, the surface of nZVI remains longer and highly active.

The synthesis and investigation of the stability and mobility of nZVI has been achieved with great results in the presence of various support materials such as clays (Zhang et al. 2012; Üzüm et al. 2009; Wu et al. 2012; Olson et al. 2012), polymer resins (Ponder et al. 2000, 2001), amorphous silica and silica sand (Oh et al. 2007; Dorathi and Kandasamy 2012), exfoliated graphite (Zhang et al. 2006), water-soluble starch (He and Zhao 2005), cationic exchange membranes (Kim et al. 2008), cellulose acetate membranes (Meyer et al. 2004), alginate bead (Kim et al. 2010), activated carbon (Mackenzie et al. 2012), mesoporous silica (OMS) (Li et al. 2011), mesoporous silica microspheres (Qiu et al. 2011), nanostructured silica SBA-15 (Saad et al. 2010), zeolites (Zhang et al. 2002), chitosan (Geng et al. 2009; Liu et al. 2012), graphene, and ordered mesoporous carbon (Ling et al. 2012). In Table 2.3 are listed some of the support materials and their functions that have been used according the literature; in Fig. 2.10 are shown some results that compare the efficacy of bare nZVI with supported nZVI.

Besides the benefits that arise from limiting the aggregation of nZVI, synergetic effects can boost the total degradation ability of these materials even more (Table 2.4). When an appropriate support material is used, absorption; reduction; or photocatalytic properties can be added and multiply the effects. Thus, the use of a support material is a matter of high technological importance because it can extend the use and sustainability of nZVI. Moreover, the range of the targeted pollutants can be widened, involving nZVI in a larger variety of environmental scenarios.

Table 2.3 Various nZVI support materials and their function (Zhou et al. 2016)

Modified material	Function
Silica	Protect the particles from intersystem dipolar interactions; provide numerous hydroxyl groups on the surface as active sites
Quercetin	Form complexes with some metal ions
Silver zeolite	High reactivity (e.g., antimicrobial activity) and good thermal stability; cation exchange and adsorption
Bifunctional polymer (outer PEO and inner PPO)	Outer hydrophilic region for colloidal stability and an inner hydrophobic region for solubilization of organic compounds
AMT-TMSPT	To form a stable complex with soft transition metal ions (e.g., Ag and Cd).
CTAB and CPC	Retain analytes by strong hydrophobic and electrostatic interactions
PV3A	Reduce particle size, ζ-potential, isoelectric point
Graphene	Have excellent mechanical, electrical, thermal, and optical properties and very high specific surface area; to form a strong π-π stacking interaction with the benzene ring
PDA	Offer biocompatibility, dispersibility in water, multifunctional groups (amino and catechol groups), and provide π-π stacking interaction to targets
PDMS	Full of functional groups such as hydrocarbyl, high biocompatibility, hydrophobic
PNIPAM	Thermosensitive, change the morphology by controlling temperature to accomplish the release of target compounds
MOFs	Possess extremely large surface area, low density, microporosity, easily designed or modified to have different pore sizes, and regarded as promising candidates for storage, separation, and catalysis
Chitosan	Hydrophilic, biocompatible, and biodegradable; full of amino groups, can form a chelate complexes with heavy metal ions
Clay	Large surface area, hydrophilic, widely available, inexpensive and safe

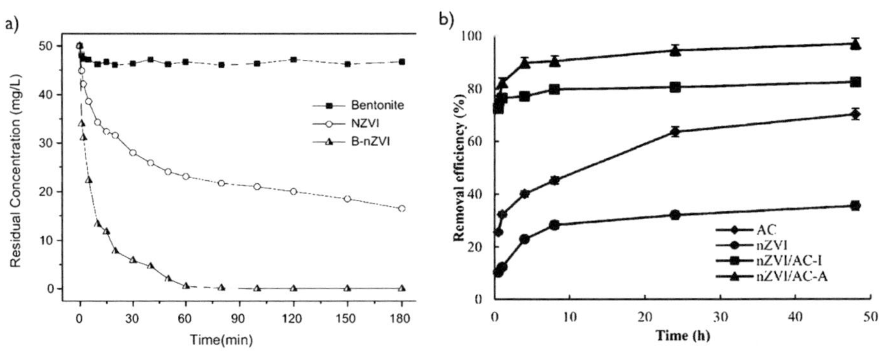

Fig. 2.10 Enhanced removal ability of nZVI (**a**) towards Cr(VI), B-nZVI: nZVI supported on bentonite (adapted from Shi et al. 2011 with permission), (**b**) towards hexachlorobenzene (HCB), nZVI/AC: nZVI supported on activated carbon (I and A refer to different processes followed for loading of iron on activated carbon) (adapted from Chen et al. 2014 with permission)

Table 2.4 Summary of examples of the nZVI modifications and their properties

Modified nZVI	Beneficial features/added properties
nZVI supported on mesoporous silica (Petala et al. 2013)	Elimination of agglomeration, high surface area and reactivity
nZVI supported on biochar (Dong et al. 2017)	Alleviation of passivation, biochar acts as a scavenger for byproducts such as Cr(III)/Fe (III) hydroxides, enhanced remediation ability
nZVI modified with sodium dodecyl sulfate (SDS, an anionic surfactant) (Huang et al. 2015)	High stabilization (less aggregation and sedimentation), higher remediation ability
nZVI/chitosan (Jin et al. 2016)	Avoidance of agglomeration and air-oxidation
nZVI/titanium oxide (Petala et al. 2016)	Synergetic photocatalytic and reductive properties
nZVI/activated carbon (Tseng et al. 2011)	Combination of physical adsorption capacity and dechlorination destructive capacities
nZVI/amphiphilic polysiloxane graft copolymers (Krajangpan et al. 2012)	Higher remediation ability, higher dispersibility, colloidal stability

2.4 Remediation Using nZVI

While nZVI has been widely used for site remediation in USA since the beginning of the new millennium, the number of full-scale remediations employing nZVI is lower in Europe (Mueller et al. 2012). Nevertheless, there is an increasing tendency to use nZVI as a novel in situ reduction technology and this material has become a well-known reagent for many environmental consultants. Although RNIP particles (Toda Kogyo Corp., Japan) have been applied to a few pilot-scale applications over the past 20 years, NANOFER nZVI (NANO IRON s.r.o., Czech Republic) is the only commercially produced nanoscale ZVI used for groundwater remediation in Europe nowadays. It has been used to treat over twenty sites, mainly in Belgium, France, Switzerland, Spain, Portugal, Italy, Denmark and the Czech Republic (Bardos et al. 2018).

Despite the great number of laboratory studies focused on pollutant removal by nZVI published in the form of scientific papers, the number of the articles documenting the full-scale or at least pilot-scale applications of nZVI is much smaller. Their summary is provided in Table 2.5. What can account for such a lack of literary sources could be the (hydro)geological as well as geochemical complexity of groundwater and soil on site (i.e., leading to much higher degree of uncertainty than in case of laboratory tests under the well-defined conditions) complicating interpretation of the observed results (Litter et al. 2018).

The first documented pilot application of nZVI (in form of Fe/Pd bimetallic nanoparticles) was performed by the pioneers of this technology, D. W. Elliott and W. Zhang, in 2001 (Elliott and Zhang 2001). In the area of an active industrial zone in Trenton, New Jersey, 1.7 kg of nZVI was fed by gravity through infiltration wells within 2 days. The concentration of PCE (target contaminant; initial concentration up to 800 μg/L) was reduced by 96% in 1 month after the application. Since that time, bimetallic or bare nZVI has been used in plenty of pilot tests in USA but only some of them have been summarized with the results published in scientific journals (see Table 2.5). In most cases, nanoparticles were prepared directly on site using reduction of ferric salt by borohydride and only small amount of nZVI (less than 20 kg) was applied. The researchers focused not only on the remediation efficiency of nanoparticles (Glazier et al. 2003; Bennett et al. 2010; Chowdhury et al. 2015) but also on their migration ability and influence on microbial communities (Bennett et al. 2010; Kocur et al. 2014, 2015).

A pilot-scale application of nZVI in Marine Corps Depot former dry cleaners on Parris Island, South Carolina, was unique by its scale among other published pilot tests. Total amount of 275 kg of RNIP nZVI was emulsified with corn oil and surfactant in order to enhance its mobility in the subsurface. The emulsion was applied deploying two methods—pneumatic injection and direct injection. The authors described the effects of nZVI on CHCs levels in the first paper (Su et al. 2012), and transport and transformation of the nanoparticles in the second one (Su et al. 2013).

The use of nZVI for Cr(VI) reduction in a pilot in situ remediation was recently reported (Němeček et al. 2015). The polluted site is located in the northern part of the Czech Republic where potassium dichromate spread from the leather processing plant. Prior to the treatment of the site, the Cr(VI) concentration in the soil, accounting for up to 46 mg/kg, prevailed over those in the groundwater (3 mg/L). The contaminated aquifer is situated in the quaternary sands and gravels with clayey admixtures and the water table was found 4.5–5.5 m bgl (aquifer thickness about 5 m). The groundwater flows with a velocity 0.2–2 m per day and discharges into a river 500 m away. In August 2012, 120 kg of nZVI (NANOFER 25 from NANO IRON s.r.o., Czech Republic) was applied to the saturated contaminated zone through three injection wells situated perpendicularly to the groundwater flow with spacing of 2.8 m (i.e., configuration of the injected geochemical barrier). Another four monitoring wells (1 of them up-gradient and 3 down-gradient) has been drilled at the site and the evolution of selected physico-chemical parameters and changes in Cr(VI) contamination level were monitored before and after the nZVI. The nZVI application resulted in an immediate decrease in the redox potential down to −484 mV (within 1 day), followed by a rapid decrease in the concentration of both Cr(VI) and total Cr. The rapid reductive effect of nZVI was observed at a distance of 7 m down-gradient.

A similar approach, which seems to be highly promising due to the partial regeneration of the applied nZVI through successive abiotic and biotic reduction steps, was applied by the same authors at another site in the Czech Republic (Němeček et al. 2016). In this case, the aquifer is located in quaternary sandy gravel with silty admixtures and overlaid by clay and clayey loam. It is approx. 4 m thick and drained by the groundwater flow equal to 1.5 m per day into a river at the distance of 430 m. The groundwater was contaminated with Cr(VI) (4–57 mg/L) and TCE/cis-DCE (400–6000 μg/L of sum of CHCs) as a consequence of the historical chromium coating industrial activities. Two types of nZVI particles (20 kg of NANOFER STAR followed by another 20 kg of NANOFER 25S, both from NANO IRON s.r.o., Czech Republic) were applied by direct push technology during

Table 2.5 Summary of the pilot applications of nZVI as published in the scientific literature before 2018

Year of publication	Name of the site	State, province	Type of nZVI (additional substrate)	Amount of ZVI	Application method	Target contaminant	Aquifer description	Reference
2001	Trenton	New Jersey, USA	Borohydride-nZVI with palladium	1.7 kg	Fed by gravity through infiltration wells	TCE		Elliott and Zhang (2001)
2003	Research Triangle Park	North Carolina, USA	Borohydride-nZVI	11.2 kg	Pressure injection through permanent well	CHCs	Distinct, stacked, and repeating sedimentary packages consisting of sandstone grading and fining upward into siltstone	Glazier et al. (2003) and Zhang (2003)
2010	San Francisco Bay	California, USA	Borohydride-nZVI and borohydride-Fe/Pd (0.1 wt%) (CMC)	0.248 kg	Single well push-pull test	TCE	Holocene and Pleistocene alluvial sediments, predominantly silts and clays, and coarse-grained sediments	Bennett et al. (2010)
2010	Kaohsiung	Taiwan	Borohydride-Fe/Pd (1 wt%), Fe/Pd (1 wt%) from Lehigh nanotech, LLC	60 kg	Fed by gravity through infiltration wells	VC	Unconfined aquifer, composed of medium to coarse sand and few silt	Wei et al. (2010)
2012, 2013	Parris Island	South Carolina, USA	RNIP (corn oil)	275 kg	Pneumatic injection and direct-push	CHCs	Shallow, unconfined aquifer generally consisting of permeable, fine to medium, Pleistocene age sand	Su et al. (2012, 2013)
2014, 2015	Sarnia	Ontario, Canada	Borohydride-nZVI (CMC)	0.75 kg	Fed by gravity through infiltration wells	CHCs	Brown weathered clay, underlain by gray clay	Kocur et al. (2014, 2015)

(continued)

Table 2.5 (continued)

Year of publication	Name of the site	State, province	Type of nZVI (additional substrate)	Amount of ZVI	Application method	Target contaminant	Aquifer description	Reference
2015	London	Ontario, Canada	Borohydride-nZVI (CMC)	0.142 kg	Fed by gravity through infiltration wells	TCE	Sand and gravel fill or brown reworked silty sand containing sand and silt	Chowdhury et al. (2015)
2014, 2015	Hrádek nad Nisou	Czech Republic	NANOFER 25 (whey)	120 kg	Pressure injection through permanent wells	Cr(VI)	Quaternary sands and gravels with clayey admixtures	Němeček et al. (2014, 2015)
2016	—	Czech Republic	NANOFER 25S, NANOFER STAR (whey)	40 kg	Direct-push	Cr(VI), CHCs	Quaternary sandy gravels with silty admixture	Němeček et al. (2016)
2016	Wonju	South Korea	NANOFER 25S	30 kg	Injection through a packer	TCE	Unconsolidated soil layer, weathered rock, and an intact Jurassic biotite granite bottom zone	Ahn et al. (2016)
2017	Hruševo	Slovenia	NANOFER 25, NANOFER STAR, borohydride-nZVI	From 0.5 to 5 g/L in a dynamic operation mode	Treatment of effluent from an existing waste water treatment plant	Heavy metals, nitrogen in various forms, bacteria	—	Oprčkal et al. (2017)
2017	Jiangxi	China	Borohydride-nZVI	Approx. 0.462 g/L in a continuous operation mode (running for 120 days)	nZVI reactors connected to the existing waste water treatment plant	Heavy metals	—	Li et al. (2017)

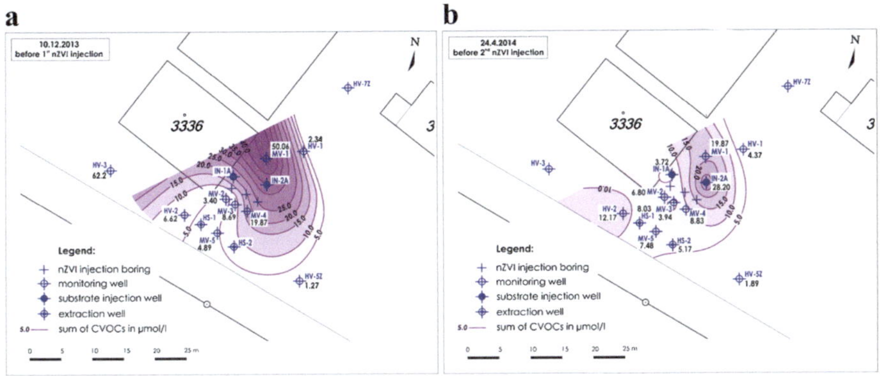

Fig. 2.11 Evolution of chlorinated VOCs (CHCs) concentration in the groundwater (a) before and (b) after the first nZVI application (adapted from Němeček et al. 2016 with permission)

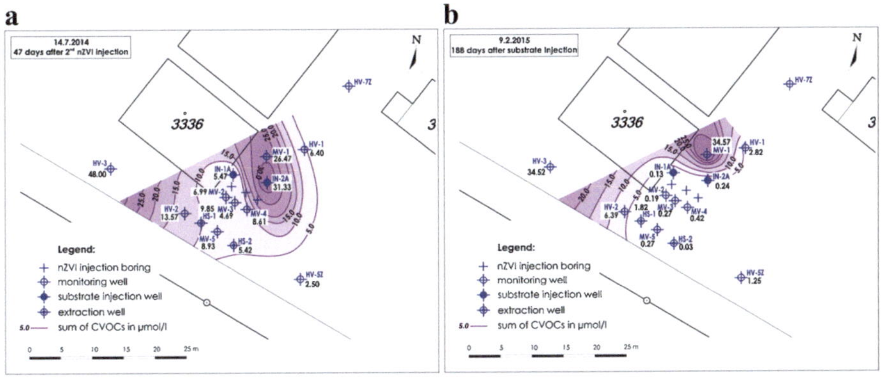

Fig. 2.12 Evolution of CVOCs (CHCs) concentration in the groundwater (a) before and (b) after the second nZVI application (adapted from Němeček et al. 2016 with permission)

the period of 4 months into the three boreholes situated perpendicularly to the groundwater flow. In the second step, the whey was applied a few months later using a circulation system resulting in 60 mg/L of TOC in the groundwater. The application of nZVI caused a decrease in the redox to below −400 mV, subsequently stabilized at 50–100 mV after the injection of whey. The evolution of chlorinated volatile organic compounds concentration during the pilot test is presented in Figs. 2.11 and 2.12.

Another pilot application of NANOFER particles was performed in South Korea (Ahn et al. 2016). nZVI was injected into the groundwater in order to treat TCE source in the aquifer of the Road Maintenance Office. Except the evaluation of the nZVI effectivity in the groundwater, the authors focused on the effect of nitrate and DO on TCE reduction and studied also reactive lifetime of injected nZVI particles. The application resulted in removal of >95% of the total TCE in 60 days. The undesirable reactions of nZVI with dissolved oxygen and nitrates were identified as contributing to Fe(0) consumption to a large extent. The reactive lifetime of the nanoparticles at the site was found to be at least 103 days but some residual Fe (0) was proved to be present on site even after 165 days.

Not only were pilot trials on the in situ groundwater remediation reported, but also the usage of nZVI for waste water treatment was documented. A Slovenian research team investigated the effects of different nZVI particles (NANOFER 25, NANOFER STAR and borohydride-nZVI) on metal, bacteria and content of nitrogen in a real effluent water from a small biological wastewater treatment plant (Oprčkal et al. 2017). This work showed that there can be a risk of releasing trace elements into the water if these are bound to nZVI particles. Therefore, careful optimization of the iron load, and of the mixing and settling times, is necessary for the efficient process optimization. Despite in-house nZVI being the most efficient at inactivating pathogenic bacteria, these nanoparticles are not appropriate for remediation since it was found that the high levels of B and Na, originating from the $NaBH_4$ used for their synthesis, contaminated the remediated water. The NANOFER 25 slurry most effectively removed potentially toxic elements and at optimal mixing and settling times (400 and 180 min, respectively) effectively disinfected effluent water at a low (0.5 g/L) iron load.

In another study, a full-scale application of nZVI for heavy metal removal from waste water of the Jiangxi Copper Company, China, has been reported (Li et al. 2017). The waste water was contaminated predominantly with Cu, Ni, Zn, Pb, As, Se and Sb. Borohydride-reduced nZVI was applied in 5 m $\times$ 5 m $\times$ 3 m separated modules connected to the existing waste water treatment plant. The performance of the technology over 120 days proved that nZVI acted as a highly efficient reagent (>99.5% removal of key metals with capacities of 245 mg and 226 mg of As and Cu per 1 g of nZVI, respectively) providing a low redox potential and subsequent separation of metals. The advantage of such a technology lies in the fact that nZVI keeps a low redox condition in the closed reactors, lowering the required nZVI demand over time (>12 months) with partial recycling of nZVI.

References

Ahn J-Y, Kim C, Kim H-S, Hwang K-Y, Hwang I (2016) Effects of oxidants on *in situ* treatment of a DNAPL source by nanoscale zero-valent iron: a field study. Water Res 107:57–65. https://doi.org/10.1016/j.watres.2016.10.037

Alessi DS, Li Z (2001) Synergistic effect of cationic surfactants on perchloroethylene degradation by zero-valent iron. Environ Sci Technol 35(18):3713–3717. https://doi.org/10.1021/es010564i

Applegate JS, Laitos JG (2006) Environmental law: RCRA, CERCLA, and the management of hazardous waste. Foundation Press, New York

Bard AJ, Parsons R, Jordan J (eds) (1985) Standard potentials in aqueous solution, 1st edn. CRC Press, New York

Bardos P, Merly C, Kvapil P, Koschitzky H-P (2018) Status of nanoremediation and its potential for future deployment: risk-benefit and benchmarking appraisals. Remediat J 28(3):43–56. https://doi.org/10.1002/rem.21559

Bennett P, He F, Zhao D, Aiken B, Feldman L (2010) In situ testing of metallic iron nanoparticle mobility and reactivity in a shallow granular aquifer. J Contam Hydrol 116(1–4):35–46. https://doi.org/10.1016/j.jconhyd.2010.05.006

Berge ND, Ramsburg CA (2009) Oil-in-water emulsions for encapsulated delivery of reactive iron particles. Environ Sci Technol 43(13):5060–5066. https://doi.org/10.1021/es900358p

Bhattacharjee S, Ghoshal S (2016) Phase transfer of palladized nanoscale zerovalent iron for environmental remediation of trichloroethene. Environ Sci Technol 50(16):8631–8639. https://doi.org/10.1021/acs.est.6b01646

Černík M, Nosek J, Filip J, Hrabal J, Elliott DW, Zbořil R (2019) Electric-field enhanced reactivity and migration of iron nanoparticles with implications for groundwater treatment technologies: proof of concept. Water Res 154:361–369. https://doi.org/10.1016/j.watres.2019.01.058

Chang J-H, Cheng S-F (2006) The remediation performance of a specific electrokinetics integrated with zero-valent metals for perchloroethylene contaminated soils. J Hazard Mater 131 (1–3):153–162. https://doi.org/10.1016/j.jhazmat.2005.09.026

Chen L-H, Huang C-C, Lien H-L (2008) Bimetallic iron–aluminum particles for dechlorination of carbon tetrachloride. Chemosphere 73(5):692–697. https://doi.org/10.1016/j.chemosphere.2008.07.005

Chen L, Jin S, Fallgren PH, Swoboda-Colberg NG, Liu F, Colberg PJS (2012) Electrochemical depassivation of zero-valent iron for trichloroethene reduction. J Hazard Mater 239–240:265–269. https://doi.org/10.1016/j.jhazmat.2012.08.074

Chen W-F, Pan L, Chen L-F, Wang Q, Yan C-C (2014) Dechlorination of hexachlorobenzene by nano zero-valent iron/activated carbon composite: iron loading, kinetics and pathway. RSC Adv 4(87):46689–46696. https://doi.org/10.1039/C4RA06760F

Chowdhury AIA, O'Carroll DM, Xu Y, Sleep BE (2012) Electrophoresis enhanced transport of nano-scale zero valent iron. Adv Water Resour 40:71–82. https://doi.org/10.1016/j.advwatres.2012.01.014

Chowdhury AIA, Krol MM, Kocur CM, Boparai HK, Weber KP, Sleep BE, O'Carroll DM (2015) nZVI injection into variably saturated soils: field and modeling study. J Contam Hydrol 183:16–28. https://doi.org/10.1016/j.jconhyd.2015.10.003

Cook SM (2009) Assessing the use and application of zero-valent iron nanoparticle technology for remediation at contaminated sites. Report for the U.S. Environmental Protection Agency. Available via https://clu-in.org/download/studentpapers/Zero-Valent-Iron-Cook.pdf. Accessed 13 Jul 2019

Correia de Velosa A, Pupo Nogueira RF (2013) 2,4-Dichlorophenoxyacetic acid (2,4-D) degradation promoted by nanoparticulate zerovalent iron (nZVI) in aerobic suspensions. J Environ Manag 121:72–79. https://doi.org/10.1016/j.jenvman.2013.02.031

Crane RA, Scott TB (2012) Nanoscale zero-valent iron: future prospects for an emerging water treatment technology. J Hazard Mater 211–212:112–125. https://doi.org/10.1016/j.jhazmat.2011.11.073

Dong J, Wen C, Liu D, Zhang W, Li J, Jiang H, Qin C, Hong M (2015) Study on degradation of nitrobenzene in groundwater using emulsified nano-zero-valent iron. J Nanopart Res 17:31. https://doi.org/10.1007/s11051-014-2829-9

Dong H, Deng J, Xie Y, Zhang C, Jiang Z, Cheng Y, Hou K, Zeng G (2017) Stabilization of nanoscale zero-valent iron (nZVI) with modified biochar for Cr(VI) removal from aqueous solution. J Hazard Mater 332:79–86. https://doi.org/10.1016/j.jhazmat.2017.03.002

Dorathi PJ, Kandasamy P (2012) Dechlorination of chlorophenols by zero valent iron impregnated silica. J Environ Sci 24(4):765–773. https://doi.org/10.1016/S1001-0742(11)60817-6

Elliott DW, Zhang W-X (2001) Field assessment of nanoscale bimetallic particles for groundwater treatment. Environ Sci Technol 35(24):4922–4926. https://doi.org/10.1021/es0108584

Elsner M, Hofstetter TB (2011) Current perspectives on the mechanisms of chlorohydrocarbon degradation in subsurface environments: insight from kinetics, product formation, probe molecules, and isotope fractionation. In: Tratnyek PG, Grundl TJ, Haderlein SB (eds) Aquatic redox

chemistry, ACS symposium series, vol 1071. American Chemical Society, Washington, DC, pp 407–439. https://doi.org/10.1021/bk-2011-1071.ch019
European Commission (2019) Soil – Environment – European Commission. http://ec.europa.eu/environment/soil/index_en.htm. Accessed 13 July 2019
European Environment Agency (2014) Progress in management of contaminated sites. European Environment Agency, Copenhagen
Fernández-Pacheco R, Arruebo M, Marquina C, Ibarra R, Arbiol J, Santamaría J (2006) Highly magnetic silica-coated iron nanoparticles prepared by the arc-discharge method. Nanotechnology 17:1188–1192. https://doi.org/10.1088/0957-4484/17/5/004
Ferrando R, Jellinek J, Johnston RL (2008) Nanoalloys: from theory to applications of alloy clusters and nanoparticles. Chem Rev 108(3):845–910. https://doi.org/10.1021/cr040090g
Fu F, Dionysiou D, Liu H (2014) The use of zero-valent iron for groundwater remediation and wastewater treatment: a review. J Hazard Mater 267:194–205. https://doi.org/10.1016/j.jhazmat.2013.12.062
Geng B, Jin Z, Li T, Qi X (2009) Kinetics of hexavalent chromium removal from water by chitosan-Fe^0 nanoparticles. Chemosphere 75(6):825–830. https://doi.org/10.1016/j.chemosphere.2009.01.009
Gillham RW, O'Hannesin SF (1994) Enhanced degradation of halogenated aliphatics by zero-valent iron. Groundwater 32(6):958–967. https://doi.org/10.1111/j.1745-6584.1994.tb00935.x
Glazier R, Venkatakrishnan R, Gheorghiu F, Walata L, Nash R, Zhang W (2003) Nanotechnology takes root. Civ Eng 73(5):64–69. https://cedb.asce.org/CEDBsearch/record.jsp?dockey=0136224
Glendinning S, Lamont-Black J, Jones CJFP (2007) Treatment of sewage sludge using electrokinetic geosynthetics. J Hazard Mater 139(3):491–499. https://doi.org/10.1016/j.jhazmat.2006.02.046
Gomes HI, Dias-Ferreira C, Ribeiro AB (2012) Electrokinetic remediation of organochlorines in soil: enhancement techniques and integration with other remediation technologies. Chemosphere 87(10):1077–1090. https://doi.org/10.1016/j.chemosphere.2012.02.037
Gomes HI, Dias-Ferreira C, Ribeiro AB, Pamukcu S (2013) Enhanced transport and transformation of zerovalent nanoiron in clay using direct electric current. Water Air Soil Pollut 224:1710. https://doi.org/10.1007/s11270-013-1710-2
Grostern A, Edwards EA (2006) Growth of *Dehalobacter* and *Dehalococcoides* spp. during degradation of chlorinated ethanes. Appl Environ Microbiol 72(1):428–436. https://doi.org/10.1128/AEM.72.1.428-436.2006
Gu H, Stanway D (2017) China needs patience to fight costly war against soil pollution: government. Reuters
Guselnikova OA, Galanov AI, Gutakovskii AK, Postnikov PS (2015) The convenient preparation of stable aryl-coated zerovalent iron nanoparticles. Beilstein J Nanotechnol 6:1192–1198. https://doi.org/10.3762/bjnano.6.121
He F, Zhao D (2005) Preparation and characterization of a new class of starch-stabilized bimetallic nanoparticles for degradation of chlorinated hydrocarbons in water. Environ Sci Technol 39(9):3314–3320. https://doi.org/10.1021/es048743y
Huang D-L, Chen G-M, Zeng G-M, Xu P, Yan M, Lai C, Zhang C, Li N-J, Cheng M, He X-X, He Y (2015) Synthesis and application of modified zero-valent iron nanoparticles for removal of hexavalent chromium from wastewater. Water Air Soil Pollut 226:375. https://doi.org/10.1007/s11270-015-2583-3
Jia H, Gu C, Boyd SA, Teppen BJ, Johnston CT, Song C, Li H (2011) Comparison of reactivity of nanoscaled zero-valent iron formed on clay surfaces. Soil Sci Soc Am J 75(2):357–364. https://doi.org/10.2136/sssaj2010.0080nps
Jin X, Zhuang Z, Yu B, Chen Z, Chen Z (2016) Functional chitosan-stabilized nanoscale zero-valent iron used to remove acid fuchsine with the assistance of ultrasound. Carbohydr Polym 136:1085–1090. https://doi.org/10.1016/j.carbpol.2015.10.002
Kanel SR, Choi H (2007) Transport characteristics of surface-modified nanoscale zero-valent iron in porous media. Water Sci Technol 55(1–2):157–162. https://doi.org/10.2166/wst.2007.002

Keenan CR, Sedlak DL (2008) Ligand-enhanced reactive oxidant generation by nanoparticulate zero-valent iron and oxygen. Environ Sci Technol 42(18):6936–6941. https://doi.org/10.1021/es801438f

Keum Y-S, Li QX (2004) Reduction of nitroaromatic pesticides with zero-valent iron. Chemosphere 54(3):255–263. https://doi.org/10.1016/j.chemosphere.2003.08.003

Kharisov BI, Rasika Dias HV, Kharissova OV, Jiménez-Pérez VM, Olvera Pérez B, Muñoz Flores B (2012) Iron-containing nanomaterials: synthesis, properties, and environmental applications. RSC Adv 2(25):9325–9358. https://doi.org/10.1039/C2RA20812A

Kim H, Hong H-J, Lee Y-J, Shin H-J, Yang J-W (2008) Degradation of trichloroethylene by zero-valent iron immobilized in cationic exchange membrane. Desalination 223(1–3):212–220. https://doi.org/10.1016/j.desal.2007.03.015

Kim H-J, Phenrat T, Tilton RD, Lowry GV (2009) Fe^0 nanoparticles remain mobile in porous media after aging due to slow desorption of polymeric surface modifiers. Environ Sci Technol 43 (10):3824–3830. https://doi.org/10.1021/es802978s

Kim H, Hong H-J, Jung J, Kim S-H, Yang J-W (2010) Degradation of trichloroethylene (TCE) by nanoscale zero-valent iron (nZVI) immobilized in alginate bead. J Hazard Mater 176 (1–3):1038–1043. https://doi.org/10.1016/j.jhazmat.2009.11.145

Kocur CM, Chowdhury AI, Sakulchaicharoen N, Boparai HK, Weber KP, Sharma P, Krol MM, Austrins L, Peace C, Sleep BE, O'Carroll DM (2014) Characterization of nZVI mobility in a field scale test. Environ Sci Technol 48(5):2862–2869. https://doi.org/10.1021/es4044209

Kocur CMD, Lomheim L, Boparai HK, Chowdhury AIA, Weber KP, Austrins LM, Edwards EA, Sleep BE, O'Carroll DM (2015) Contributions of abiotic and biotic dechlorination following carboxymethyl cellulose stabilized nanoscale zero valent iron injection. Environ Sci Technol 49 (14):8648–8656. https://doi.org/10.1021/acs.est.5b00719

Krajangpan S, Kalita H, Chisholm BJ, Bezbaruah AN (2012) Iron nanoparticles coated with amphiphilic polysiloxane graft copolymers: dispersibility and contaminant treatability. Environ Sci Technol 46(18):10130–10136. https://doi.org/10.1021/es3000239

Lee Y-C, Kwon T-S, Yang J-S, Yang J-W (2007) Remediation of groundwater contaminated with DNAPLs by biodegradable oil emulsion. J Hazard Mater 140(1–2):340–345. https://doi.org/10.1016/j.jhazmat.2006.09.036

Li T, Farrell J (2001) Electrochemical investigation of the rate-limiting mechanisms for trichloroethylene and carbon tetrachloride reduction at iron surfaces. Environ Sci Technol 35 (17):3560–3565. https://doi.org/10.1021/es0019878

Li X-Q, Elliott DW, Zhang W-X (2006) Zero-valent iron nanoparticles for abatement of environmental pollutants: materials and engineering aspects. Crit Rev Solid State Mater Sci 31 (4):111–122. https://doi.org/10.1080/10408430601057611

Li J, Li H, Zhu Y, Hao Y, Sun X, Wang L (2011) Dual roles of amphiphilic triblock copolymer P123 in synthesis of α-Fe nanoparticle/ordered mesoporous silica composites. Appl Surf Sci 258 (2):657–661. https://doi.org/10.1016/j.apsusc.2011.07.037

Li XN, Jiao WT, Xiao RB, Chen WP, Chang AC (2015) Soil pollution and site remediation policies in China: a review. Environ Rev 23(3):263–274. https://doi.org/10.1139/er-2014-0073

Li S, Wang W, Liang F, W-x Z (2017) Heavy metal removal using nanoscale zero-valent iron (nZVI): theory and application. J Hazard Mater 322:163–171. https://doi.org/10.1016/j.jhazmat.2016.01.032

Lien H-L, Zhang W-X (1999) Transformation of chlorinated methanes by nanoscale iron particles. J Environ Eng 125(11):1042–1047. https://doi.org/10.1061/(ASCE)0733-9372(1999)125:11(1042)

Ling X, Li J, Zhu W, Zhu Y, Sun X, Shen J, Han W, Wang L (2012) Synthesis of nanoscale zero-valent iron/ordered mesoporous carbon for adsorption and synergistic reduction of nitrobenzene. Chemosphere 87(6):655–660. https://doi.org/10.1016/j.chemosphere.2012.02.002

Ling L, Huang X, Li M, Zhang W (2017) Mapping the reactions in a single zero-valent Iron nanoparticle. Environ Sci Technol 51(24):14293–14300. https://doi.org/10.1021/acs.est.7b02233

Litter MI, Quici N, Meichtry M (eds) (2018) Iron nanomaterials for water and soil treatment, 1st edn. Pan Stanford Publishing, Singapore

Liu Y, Majetich SA, Tilton RD, Sholl DS, Lowry GV (2005) TCE dechlorination rates, pathways, and efficiency of nanoscale iron particles with different properties. Environ Sci Technol 39 (5):1338–1345. https://doi.org/10.1021/es049195r

Liu Y, Phenrat T, Lowry GV (2007) Effect of TCE concentration and dissolved groundwater solutes on NZVI-promoted TCE dechlorination and H2 evolution. Environ Sci Technol 41:7881–7887. https://doi.org/10.1021/es0711967

Liu Q, Bei Y, Zhou F (2009a) Removal of lead(II) from aqueous solution with amino-functionalized nanoscale zero-valent iron. Cent Eur J Chem 7(1):79–82. https://doi.org/10.2478/s11532-008-0097-1

Liu T, Rao P, Lo IMC (2009b) Influences of humic acid, bicarbonate and calcium on Cr (VI) reductive removal by zero-valent iron. Sci Total Environ 407(10):3407–3414. https://doi.org/10.1016/j.scitotenv.2009.01.043

Liu T, Wang Z-L, Zhao L, Yang X (2012) Enhanced chitosan/Fe^0-nanoparticles beads for hexavalent chromium removal from wastewater. Chem Eng J 189–190:196–202. https://doi.org/10.1016/j.cej.2012.02.056

Liu W-J, Qian T-T, Jiang H (2014) Bimetallic Fe nanoparticles: recent advances in synthesis and application in catalytic elimination of environmental pollutants. Chem Eng J 236:448–463. https://doi.org/10.1016/j.cej.2013.10.062

Mackenzie K, Bleyl S, Georgi A, Kopinke F-D (2012) Carbo-iron – an Fe/AC composite – as alternative to nano-iron for groundwater treatment. Water Res 46(12):3817–3826. https://doi.org/10.1016/j.watres.2012.04.013

Marková Z, Šišková KM, Filip J, Čuda J, Kolář M, Šafářová K, Medřík I, Zbořil R (2013) Air stable magnetic bimetallic Fe–Ag nanoparticles for advanced antimicrobial treatment and phosphorus removal. Environ Sci Technol 47(10):5285–5293. https://doi.org/10.1021/es304693g

Meyer DE, Wood K, Bachas LG, Bhattacharyya D (2004) Degradation of chlorinated organics by membrane-immobilized nanosized metals. Environ Prog 23(3):232–242. https://doi.org/10.1002/ep.10031

Morales J, Hutcheson R, Cheng IF (2002) Dechlorination of chlorinated phenols by catalyzed and uncatalyzed Fe(0) and Mg(0) particles. J Hazard Mater 90(1):97–108. https://doi.org/10.1016/S0304-3894(01)00336-3

Mueller NC, Braun J, Bruns J, Černík M, Rissing P, Rickerby D, Nowack B (2012) Application of nanoscale zero valent iron (NZVI) for groundwater remediation in Europe. Environ Sci Pollut Res 19(2):550–558. https://doi.org/10.1007/s11356-011-0576-3

Mulligan CN, Yong RN, Gibbs BF (2001) Remediation technologies for metal-contaminated soils and groundwater: an evaluation. Eng Geol 60(1–4):193–207. https://doi.org/10.1016/S0013-7952(00)00101-0

Němeček J, Lhotský O, Cajthaml T (2014) Nanoscale zero-valent iron application for in situ reduction of hexavalent chromium and its effects on indigenous microorganism populations. Sci Total Environ 485–486:739–747. https://doi.org/10.1016/j.scitotenv.2013.11.105

Němeček J, Pokorný P, Lacinová L, Černík M, Masopustová Z, Lhotský O, Filipová A, Cajthaml T (2015) Combined abiotic and biotic in-situ reduction of hexavalent chromium in groundwater using nZVI and whey: a remedial pilot test. J Hazard Mater 300:670–679. https://doi.org/10.1016/j.jhazmat.2015.07.056

Němeček J, Pokorný P, Lhotský O, Knytl V, Najmanová P, Steinová J, Černík M, Filipová A, Filip J, Cajthaml T (2016) Combined nano-biotechnology for in-situ remediation of mixed contamination of groundwater by hexavalent chromium and chlorinated solvents. Sci Total Environ 563–564:822–834. https://doi.org/10.1016/j.scitotenv.2016.01.019

O'Carroll D, Sleep B, Krol M, Boparai H, Kocur C (2013) Nanoscale zero valent iron and bimetallic particles for contaminated site remediation. Adv Water Resour 51:104–122. https://doi.org/10.1016/j.advwatres.2012.02.005

Oh YJ, Song H, Shin WS, Choi SJ, Kim Y-H (2007) Effect of amorphous silica and silica sand on removal of chromium(VI) by zero-valent iron. Chemosphere 66(5):858–865. https://doi.org/10.1016/j.chemosphere.2006.06.034

O'Hara S, Krug T, Quinn J, Clausen C, Geiger C (2006) Field and laboratory evaluation of the treatment of DNAPL source zones using emulsified zero-valent iron. Remediat J 16(2):35–56. https://doi.org/10.1002/rem.20080

Olson MR, Sale TC, Shackelford CD, Bozzini C, Skeean J (2012) Chlorinated solvent source-zone remediation via ZVI-clay soil mixing: 1-year results. Groundwater Monit Remediat 32 (3):63–74. https://doi.org/10.1111/j.1745-6592.2011.01391.x

Oprčkal P, Mladenovič A, Vidmar J, Mauko Pranjić A, Milačič R, Ščančar J (2017) Critical evaluation of the use of different nanoscale zero-valent iron particles for the treatment of effluent water from a small biological wastewater treatment plant. Chem Eng J 321:20–30. https://doi.org/10.1016/j.cej.2017.03.104

Panagos P, Van Liedekerke M, Yigini Y, Montanarella L (2013) Contaminated sites in Europe: review of the current situation based on data collected through a European network. J Environ Public Health 2013:158764. https://doi.org/10.1155/2013/158764

Petala E, Dimos K, Douvalis A, Bakas T, Tucek J, Zbořil R, Karakassides MA (2013) Nanoscale zero-valent iron supported on mesoporous silica: characterization and reactivity for Cr (VI) removal from aqueous solution. J Hazard Mater 261:295–306. https://doi.org/10.1016/j.jhazmat.2013.07.046

Petala E, Baikousi M, Karakassides MA, Zoppellaro G, Filip J, Tuček J, Vasilopoulos KC, Pechoušek J, Zbořil R (2016) Synthesis, physical properties and application of the zero-valent iron/titanium dioxide heterocomposite having high activity for the sustainable photocatalytic removal of hexavalent chromium in water. Phys Chem Chem Phys 18(15):10637–10646. https://doi.org/10.1039/C6CP01013J

Phenrat T, Lowry GV (eds) (2019) Nanoscale zerovalent iron particles for environmental restoration: from fundamental science to field scale engineering applications, 1st edn. Springer, Cham. https://doi.org/10.1007/978-3-319-95340-3

Phenrat T, Saleh N, Sirk K, Tilton RD, Lowry GV (2007) Aggregation and sedimentation of aqueous nanoscale zerovalent iron dispersions. Environ Sci Technol 41(1):284–290. https://doi.org/10.1021/es061349a

Phenrat T, Saleh N, Sirk K, Kim H-J, Tilton RD, Lowry GV (2008) Stabilization of aqueous nanoscale zerovalent iron dispersions by anionic polyelectrolytes: adsorbed anionic polyelectrolyte layer properties and their effect on aggregation and sedimentation. J Nanopart Res 10 (5):795–814. https://doi.org/10.1007/s11051-007-9315-6

Phenrat T, Kim H-J, Fagerlund F, Illangasekare T, Tilton RD, Lowry GV (2009) Particle size distribution, concentration, and magnetic attraction affect transport of polymer-modified Fe^0 nanoparticles in sand columns. Environ Sci Technol 43(13):5079–5085. https://doi.org/10.1021/es900171v

Ponder SM, Darab JG, Mallouk TE (2000) Remediation of Cr(VI) and Pb(II) aqueous solutions using supported, nanoscale zero-valent iron. Environ Sci Technol 34(12):2564–2569. https://doi.org/10.1021/es9911420

Ponder SM, Darab JG, Bucher J, Caulder D, Craig I, Davis L, Edelstein N, Lukens W, Nitsche H, Rao L, Shuh DK, Mallouk TE (2001) Surface chemistry and electrochemistry of supported zerovalent iron nanoparticles in the remediation of aqueous metal contaminants. Chem Mater 13 (2):479–486. https://doi.org/10.1021/cm000288r

Qiu X, Fang Z, Liang B, Gu F, Xu Z (2011) Degradation of decabromodiphenyl ether by nano zero-valent iron immobilized in mesoporous silica microspheres. J Hazard Mater 193:70–81. https://doi.org/10.1016/j.jhazmat.2011.07.024

Quinn J, Geiger C, Clausen C, Brooks K, Coon C, O'Hara S, Krug T, Major D, Yoon W-S, Gavaskar A, Holdsworth T (2005) Field demonstration of DNAPL dehalogenation using emulsified zero-valent iron. Environ Sci Technol 39(5):1309–1318. https://doi.org/10.1021/es0490018

Reddy KR, Darko-Kagya K, Cameselle C (2011) Electrokinetic-enhanced transport of lactate-modified nanoscale iron particles for degradation of dinitrotoluene in clayey soils. Sep Purif Technol 79(2):230–237. https://doi.org/10.1016/j.seppur.2011.01.033

Reinsch BC, Forsberg B, Penn RL, Kim CS, Lowry GV (2010) Chemical transformations during aging of zerovalent iron nanoparticles in the presence of common groundwater dissolved constituents. Environ Sci Technol 44(9):3455–3461. https://doi.org/10.1021/es902924h

Reynolds GW, Hoff JT, Gillham RW (1990) Sampling bias caused by materials used to monitor halocarbons in groundwater. Environ Sci Technol 24(1):135–142. https://doi.org/10.1021/es00071a017

Rodrigues SM, Pereira ME, Ferreira da Silva E, Hursthouse AS, Duarte AC (2009) A review of regulatory decisions for environmental protection: part I — challenges in the implementation of national soil policies. Environ Int 35(1):202–213. https://doi.org/10.1016/j.envint.2008.08.007

Saad R, Thiboutot S, Ampleman G, Dashan W, Hawari J (2010) Degradation of trinitroglycerin (TNG) using zero-valent iron nanoparticles/nanosilica SBA-15 composite (ZVINs/SBA-15). Chemosphere 81(7):853–858. https://doi.org/10.1016/j.chemosphere.2010.08.012

Saleh N, Phenrat T, Sirk K, Dufour B, Ok J, Sarbu T, Matyjaszewski K, Tilton RD, Lowry GV (2005) Adsorbed triblock copolymers deliver reactive iron nanoparticles to the oil/water interface. Nano Lett 5(12):2489–2494. https://doi.org/10.1021/nl0518268

Schrick B, Hydutsky BW, Blough JL, Mallouk TE (2004) Delivery vehicles for zerovalent metal nanoparticles in soil and groundwater. Chem Mater 16(11):2187–2193. https://doi.org/10.1021/cm0218108

Sheu YT, Chen SC, Chien CC, Chen CC, Kao CM (2015) Application of a long-lasting colloidal substrate with pH and hydrogen sulfide control capabilities to remediate TCE-contaminated groundwater. J Hazard Mater 284:222–232. https://doi.org/10.1016/j.jhazmat.2014.11.023

Sheu YT, Lien PJ, Chen KF, Ou JH, Kao CM (2016) Application of NZVI-contained emulsified substrate to bioremediate PCE-contaminated groundwater – a pilot-scale study. Chem Eng J 304:714–727. https://doi.org/10.1016/j.cej.2016.06.126

Shi L-N, Lin Y-M, Zhang X, Chen Z-L (2011) Synthesis, characterization and kinetics of bentonite supported nZVI for the removal of Cr(VI) from aqueous solution. Chem Eng J 171(2):612–617. https://doi.org/10.1016/j.cej.2011.04.038

Siegrist RL, Crimi M, Simpkin TJ (eds) (2011) In situ chemical oxidation for groundwater remediation, SERDP ESTCP environmental remediation technology, vol 3, 1st edn. Springer, New York. https://doi.org/10.1007/978-1-4419-7826-4

Singh R, Misra V (2015) Stabilization of zero-valent iron nanoparticles: role of polymers and surfactants. In: Aliofkhazraei M (ed) Handbook of nanoparticles. Springer, pp 1–19. https://doi.org/10.1007/978-3-319-13188-7_44-1

Siskova K, Tucek J, Machala L, Otyepkova E, Filip J, Safarova K, Pechousek J, Zboril R (2012) Air-stable nZVI formation mediated by glutamic acid: solid-state storable material exhibiting 2D chain morphology and high reactivity in aqueous environment. J Nanopart Res 14:805. https://doi.org/10.1007/s11051-012-0805-9

Smedley PL, Kinniburgh DG (2002) A review of the source, behaviour and distribution of arsenic in natural waters. Appl Geochem 17(5):517–568. https://doi.org/10.1016/S0883-2927(02)00018-5

Song H, Carraway ER (2005) Reduction of chlorinated ethanes by nanosized zero-valent iron: kinetics, pathways, and effects of reaction conditions. Environ Sci Technol 39(16):6237–6245. https://doi.org/10.1021/es048262e

Song H, Carraway ER (2006) Reduction of chlorinated methanes by nano-sized zero-valent iron. Kinetics, pathways, and effect of reaction conditions. Environ Eng Sci 23(2):272–284. https://doi.org/10.1089/ees.2006.23.272

Su C, Puls RW, Krug TA, Watling MT, O'Hara SK, Quinn JW, Ruiz NE (2012) A two and half-year-performance evaluation of a field test on treatment of source zone tetrachloroethene and its chlorinated daughter products using emulsified zero valent iron nanoparticles. Water Res 46(16):5071–5084. https://doi.org/10.1016/j.watres.2012.06.051

Su C, Puls RW, Krug TA, Watling MT, O'Hara SK, Quinn JW, Ruiz NE (2013) Travel distance and transformation of injected emulsified zerovalent iron nanoparticles in the subsurface during two and half years. Water Res 47(12):4095–4106. https://doi.org/10.1016/j.watres.2012.12.042

Su Y, Adeleye AS, Zhou X, Dai C, Zhang W, Keller AA, Zhang Y (2014) Effects of nitrate on the treatment of lead contaminated groundwater by nanoscale zerovalent iron. J Hazard Mater 280:504–513. https://doi.org/10.1016/j.jhazmat.2014.08.040

Sun Y-P, Li X-Q, Zhang W-X, Wang HP (2007) A method for the preparation of stable dispersion of zero-valent iron nanoparticles. Colloids Surf A Physicochem Eng Asp 308(1–3):60–66. https://doi.org/10.1016/j.colsurfa.2007.05.029

Sun Y, Li J, Huang T, Guan X (2016) The influences of iron characteristics, operating conditions and solution chemistry on contaminants removal by zero-valent iron: a review. Water Res 100:277–295. https://doi.org/10.1016/j.watres.2016.05.031

Swartjes FA (ed) (2011) Dealing with contaminated sites: from theory towards practical application, 1st edn. Springer, Dordrecht. https://doi.org/10.1007/978-90-481-9757-6

Tang SCN, Lo IMC (2013) Magnetic nanoparticles: essential factors for sustainable environmental applications. Water Res 47(8):2613–2632. https://doi.org/10.1016/j.watres.2013.02.039

The Economist (2017) The most neglected threat to public health in China is toxic soil. The Economist Newspaper Limited

Tiraferri A, Chen KL, Sethi R, Elimelech M (2008) Reduced aggregation and sedimentation of zero-valent iron nanoparticles in the presence of guar gum. J Colloid Interface Sci 324 (1–2):71–79. https://doi.org/10.1016/j.jcis.2008.04.064

Tseng H-H, Su J-G, Liang C (2011) Synthesis of granular activated carbon/zero valent iron composites for simultaneous adsorption/dechlorination of trichloroethylene. J Hazard Mater 192(2):500–506. https://doi.org/10.1016/j.jhazmat.2011.05.047

US EPA (2000) Superfund: 20 years of protecting human health and the environment. Office of Solid Waste and Emergency Response, Washington, DC

US EPA (2011) Beneficial effects of the superfund program. Office of Superfund Remediation and Technology Innovation, Washington, DC

US EPA (2018a) Superfund: CERCLA Overview. https://www.epa.gov/superfund/superfund-cercla-overview. Accessed 10 Dec 2018

US EPA (2018b) Superfund: National Priorities List (NPL). https://www.epa.gov/superfund/superfund-national-priorities-list-npl. Accessed 10 Dec 2018

Üzüm Ç, Shahwan T, Eroğlu AE, Hallam KR, Scott TB, Lieberwirth I (2009) Synthesis and characterization of kaolinite-supported zero-valent iron nanoparticles and their application for the removal of aqueous Cu^{2+} and Co^{2+} ions. Appl Clay Sci 43(2):172–181. https://doi.org/10.1016/j.clay.2008.07.030

Van Deuren J, Lloyd T, Chhetry S, Liou R, Peck J (2002) Remediation technologies screening matrix and reference guide, version 4.0. FRTR. https://frtr.gov/matrix2/top_page.html. Accessed 10 Dec 2018

Virkutyte J, Sillanpää M, Latostenmaa P (2002) Electrokinetic soil remediation — critical overview. Sci Total Environ 289(1–3):97–121. https://doi.org/10.1016/S0048-9697(01)01027-0

Wang J, Farrell J (2003) Investigating the role of atomic hydrogen on chloroethene reactions with iron using Tafel analysis and electrochemical impedance spectroscopy. Environ Sci Technol 37 (17):3891–3896. https://doi.org/10.1021/es0264605

Wang C-B, Zhang W-X (1997) Synthesizing nanoscale iron particles for rapid and complete dechlorination of TCE and PCBs. Environ Sci Technol 31(7):2154–2156. https://doi.org/10.1021/es970039c

Wang W, Zhou M, Jin Z, Li T (2010) Reactivity characteristics of poly(methyl methacrylate) coated nanoscale iron particles for trichloroethylene remediation. J Hazard Mater 173(1–3):724–730. https://doi.org/10.1016/j.jhazmat.2009.08.145

Waria M, Comfort SD, Onanong S, Satapanajaru T, Boparai H, Harris C, Snow D, Cassada DA (2009) Field-scale cleanup of atrazine and cyanazine contaminated soil with a combined chemical-biological approach. J Environ Qual 38(5):1803–1811. https://doi.org/10.2134/jeq2008.0361

Wei Y-T, Wu S-C, Chou C-M, Che C-H, Tsai S-M, Lien H-L (2010) Influence of nanoscale zero-valent iron on geochemical properties of groundwater and vinyl chloride degradation: a field case study. Water Res 44(1):131–140. https://doi.org/10.1016/j.watres.2009.09.012

Weng C-H, Lin TY, Chu S-H, Yuan C (2006) Laboratory-scale evaluation of Cr(VI) removal from clay by electrokinetics incorporated with Fe(O) barrier. Pract Period Hazard Toxic Radioact Waste Manage 10(3):171–178. https://doi.org/10.1061/(ASCE)1090-025X(2006)10:3(171)

Weng C-H, Lin Y-T, Lin TY, Kao CM (2007) Enhancement of electrokinetic remediation of hyper-Cr(VI) contaminated clay by zero-valent iron. J Hazard Mater 149(2):292–302. https://doi.org/10.1016/j.jhazmat.2007.03.076

Wu L, Ritchie SMC (2006) Removal of trichloroethylene from water by cellulose acetate supported bimetallic Ni/Fe nanoparticles. Chemosphere 63(2):285–292. https://doi.org/10.1016/j.chemosphere.2005.07.021

Wu P, Li S, Ju L, Zhu N, Wu J, Li P, Dang Z (2012) Mechanism of the reduction of hexavalent chromium by organo-montmorillonite supported iron nanoparticles. J Hazard Mater 219–220:283–288. https://doi.org/10.1016/j.jhazmat.2012.04.008

Yan W, Herzing AA, Li X-Q, Kiely CJ, Zhang W-X (2010) Structural evolution of Pd-doped nanoscale zero-valent iron (nZVI) in aqueous media and implications for particle aging and reactivity. Environ Sci Technol 44(11):4288–4294. https://doi.org/10.1021/es100051q

Yang GCC, Chang Y-I (2011) Integration of emulsified nanoiron injection with the electrokinetic process for remediation of trichloroethylene in saturated soil. Sep Purif Technol 79(2):278–284. https://doi.org/10.1016/j.seppur.2011.03.004

Zhang W-X (2003) Nanoscale iron particles for environmental remediation: an overview. J Nanopart Res 5(3–4):323–332. https://doi.org/10.1023/A:1025520116015

Zhang P, Tao X, Li Z, Bowman RS (2002) Enhanced perchloroethylene reduction in column systems using surfactant-modified zeolite/zero-valent iron pellets. Environ Sci Technol 36 (16):3597–3603. https://doi.org/10.1021/es015816u

Zhang H, Jin Z-H, Han L, Qin C-H (2006) Synthesis of nanoscale zero-valent iron supported on exfoliated graphite for removal of nitrate. Trans Nonferrous Metals Soc China 16:s345–s349. https://doi.org/10.1016/S1003-6326(06)60207-0

Zhang Y, Li Y, Li J, Sheng G, Zhang Y, Zheng X (2012) Enhanced Cr(VI) removal by using the mixture of pillared bentonite and zero-valent iron. Chem Eng J 185–186:243–249. https://doi.org/10.1016/j.cej.2012.01.095

Zhao D, He F (2011) Preparation and application of stabilized iron nanoparticles for dechlorination of chlorinated hydrocarbons in soils, sediments, and ground water. United States Patent

Zhou Q, Li J, Wang M, Zhao D (2016) Iron-based magnetic nanomaterials and their environmental applications. Crit Rev Environ Sci Technol 46(8):783–826. https://doi.org/10.1080/10643389.2016.1160815

Chapter 3
Other Chemical Reductive Methods

Jan Něměček, Stanisław Wacławek, and Miroslav Černík

Abstract Chromium is used ubiquitously in various industrial applications as it has several advantages. It can provide, e.g., a robust and rust-resistant coating for various materials. However, the most dangerous form—chromium(VI)—is soluble in water and therefore can trigger numerous health problems. One of the most common ways to remediate Cr(VI) is to reduce it to a less toxic and mobile form, i.e., chromium(III). This can be achieved in different ways; nevertheless, this chapter will mainly cover the in situ reduction of Cr(VI) with sulfur compounds. Two sites contaminated with Cr(VI) will be presented, and actions related to their decontamination will be described. The design of the planned techniques will be assessed, in the bench-scale studies first. Afterwards remediation in the field scale will be shown as a model example of how to manage and operate remedial interventions for sites contaminated with Cr(VI) based on the in situ reduction with the sulfur compounds.

Keywords In situ methods · Sulfur compounds · Chromium(VI) · In situ chemical reduction (ISCR)

3.1 Introduction

In general, in situ chemical reduction (ISCR) is a method based on chemical reduction of selected contaminants by a reagent, used as a source of electrons. Besides zero-valent iron, discussed elsewhere, the most commonly used reagents are dissolved sulfur compounds in different oxidation states: sulfides S(–II) are represented by sodium hydrosulfide (NaHS), polysulfides S_x(–II) by calcium

---

J. Něměček · S. Wacławek
Institute for Nanomaterials, Advanced Technologies and Innovation, Technical University of Liberec, Liberec, Czech Republic

M. Černík (✉)
Institute for Nanomaterials, Advanced Technologies and Innovation, Technical University of Liberec, Liberec, Czech Republic

AQUATEST a.s., Prague, Czech Republic
e-mail: miroslav.cernik@tul.cz

J. Filip et al. (eds.), *Advanced Nano-Bio Technologies for Water and Soil Treatment*, Applied Environmental Science and Engineering for a Sustainable Future,
https://doi.org/10.1007/978-3-030-29840-1_3

polysulfide (CaS_x), dithionites S(+III) by sodium dithionite ($Na_2S_2O_4$), sulfites S (+IV) by sodium metabisulfite ($Na_2S_2O_5$) and sodium sulfite (Na_2SO_3). These compounds act only on inorganic contaminants, which are clearly dominated by hexavalent chromium. Another type of components is iron in a low oxidation state Fe(+II) in, e.g., ferrous sulfate ($FeSO_4$). The advantage of these compounds is, unlike ZVI, their solubility in water and therefore better transport in groundwater.

Besides dithionite discussed in a case study below, the most used compounds are calcium polysulfide and sodium metabisulfite. Calcium polysulfide reacts with Cr(+VI) according to Eq. (3.1) (Graham et al. 2006).

$$2CrO_4^{2-} + CaS_x + 10H^+ \rightarrow 2Cr(OH)_3(s) + 3S(s) + 3Ca^{2+} + 2H_2O \quad (3.1)$$

The advantage of calcium polysulfides lies in their stability, which is important in low permeable environments where more time is needed for the reducing solution to act, or in a heterogeneous environment, which requires longer reactivity due to slow diffusion of Cr(+VI) from less permeable zones. Polysulfides react at a neutral pH and therefore compared to dithionite, it is not necessary to infiltrate a buffer. Moreover, it is not recommended to apply calcium polysulfide in acidic environments ($pH < 5$) due to the possible formation of hydrogen sulfide. Moreover, the cost of calcium polysulfides is significantly higher than that of dithionite solution.

Chromium reduction by sodium metabisulfite is favorable in an acidic environment. At a pH below 2.5, it takes from 15 to 20 min, independently of the process conditions. As the pH increases, the speed significantly decreases, which is a disadvantage. While sodium metabisulfite needs acidic conditions to reduce chromium, subsequent precipitation of $Cr(OH)_3$ runs at an alkaline pH. The mechanism of reduction of hexavalent chromium can be described by the following equation (Schindewolf 1976):

$$3S_2O_5^{2-} + 4CrO_4^{2-} + 14H^+ \rightarrow 2Cr_2(SO_4)_3 + 7H_2O \quad (3.2)$$

3.2 Reduction of Hexavalent Chromium by Sodium Dithionite

Sodium dithionite ($Na_2S_2O_4$) contains $(S_2O_4)^{2-}$ anion, in which sulfur occurs as S(+III). Sodium dithionite is unstable in aqueous environments; it spontaneously decomposes (disproportionate into sulfites $(SO_3)^{2-}$ and thiosulfate $(S_2O_3)^{2-}$ (Amonette et al. 1994):

$$2Na_2S_2O_4 + H_2O \rightarrow Na_2S_2O_3 + 2NaHSO_3 \quad (3.3)$$

In alkaline solution, it has strong reducing properties and is disproportionate to sulfites $(SO_3)^{2-}$ and sulfides S(–II) by the reaction:

$$3Na_2S_2O_4 + 6NaOH \rightarrow 5Na_2SO_3 + Na_2S + 3H_2O \quad (3.4)$$

Both of the produced species are reducing agents and tend to release electrons. Therefore, in order to increase the reducing properties, the pH is, during the application, maintained to be >11 with a buffer solution of potassium carbonate and bicarbonate (Fruchter et al. 2000).

Sodium dithionite can react with hexavalent chromium either directly or indirectly. The direct reaction relies on the direct oxidation of S(+III) to S(+VI) and simultaneous reduction of Cr(+VI) to Cr(+III):

$$2CrO_4^{2-} + S_2O_4^{2-} + 2H_2O + 2H^+ \rightarrow 2Cr(OH)_3(s) + 2SO_4^{2-} \tag{3.5}$$

Indirect reaction relies on the reduction of Cr(+VI) by Fe(+II), which is generated from naturally occurred Fe(+III) by the dithionite reduction (US EPA 2000):

$$S_2O_4^{2-} + 2Fe^{3+}(s) + 2H_2O \rightarrow 2Fe^{2+}(s) + 2SO_3^{2-} + 4H^+ \tag{3.6}$$

$$HCrO_4^- + 7H^+ + 3Fe^{2+}(s) \rightarrow Cr^{3+}(s) + 4H_2O + 3Fe^{3+}(s) \tag{3.7}$$

Chromium hydroxide $Cr(OH)_3$ is formed during the direct reduction; however, a permeable reactive zone with Fe(+II) is formed during the indirect reduction, where residual Cr(VI) content in groundwater flowing into it is reduced and $Cr(OH)_3$ or the mixed mineral $Cr_{(1\text{-}x)}Fe_x(OH)_3$ are formed (Palmer and Puls 1994; Henderson 1994).

Sulfites, sulfates, and thiosulfates created by the decomposition of dithionite in the environment may further react to form a substance similar to gypsum: $CaSO_3{\cdot}½H_2O$, $CaSO_4{\cdot}2H_2O$, and $CaS_2O_3{\cdot}6H_2O$ (Amonette et al. 1994). This leads to the stabilization of the produced sulfates; therefore, they do not need to be pumped out in cases where high groundwater salinity is not acceptable because of increased sensitivity of the aquifer. On the other hand, insoluble precipitates can decrease the permeability around the application wells, which represents a disadvantage of the precipitation of these substances.

3.2.1 Site Description

The method of in situ reduction with sodium dithionite was applied to remediating land contaminated by an accidental release of wastewater from a chromium plating factory. The bedrock formed by the granite massif is in its upper part, weathered and overlapped by sandy eluvium, clayey loams, and backfill. The contaminated area is in a saturated zone of sandy eluvium and weathered granite and in a fracture system of bedrock. The groundwater is partially drained by a surface watercourse, which flows approximately 150 m away from the source of the contamination, and rises seasonally at the base of the slope above the watercourse, and part of the groundwater is drained by an unsealed sewerage collector. The maximum concentration of Cr(VI) in groundwater (2700 mg/L) was found in well HS-1, which is located 5 m

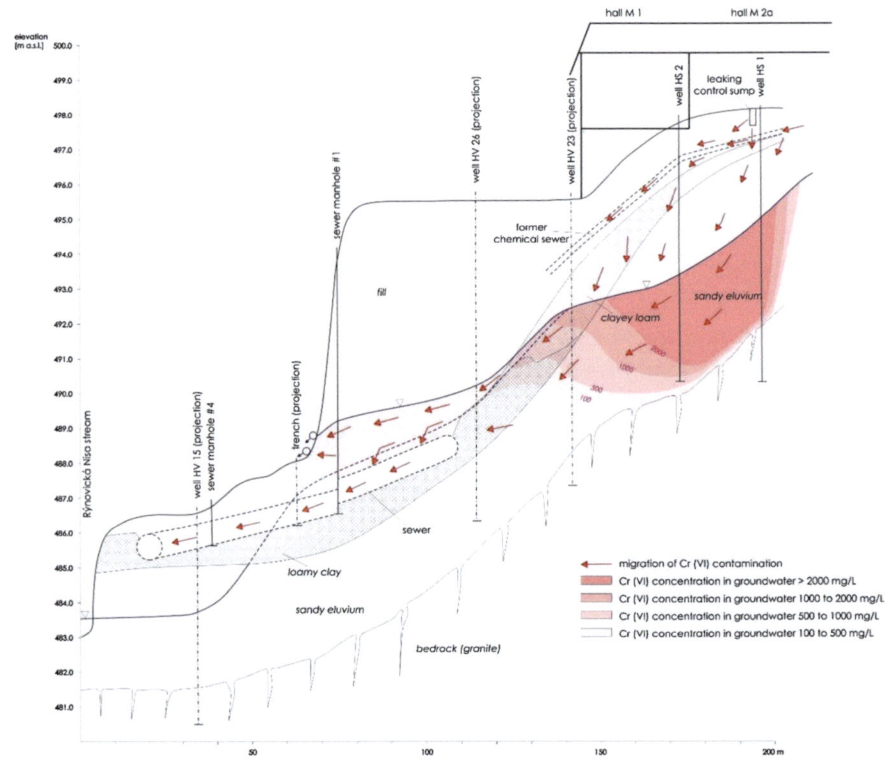

Fig. 3.1 Schematic cross-section of the contaminated area with source of contamination and transport pathways

away from the source of the contamination (Fig. 3.1). Migration of the contamination in the unsaturated zone is determined, at least in some sections, along preferential pathways, probably along the route of the former chemical sewer or other utilities.

The observed Cr(VI) plumes narrow considerably in the direction of the flow and continue towards the area of the old backfill (buried valley side). The initial content of Cr(VI) in the seepage at the base of the slope of the backfill reached 500–600 mg/L. The groundwater flowing through the natural rock environment to the watercourse had a significantly lower initial content of Cr(VI) (16 mg/L) than the aquifer in the backfill. On the basis of the results of the survey, the total amount of Cr(VI) in the rock environment was estimated to be approximately 6.3 tons, approximately 5.2 tons of which was in the source area (2200 m^2).

After the source of the contamination was found and eliminated in an unsealed sewerage collector, emergency pumping was initiated to prevent the contamination from further spreading, and the most contaminated groundwater was extracted. The volume of Cr(VI) recovered after 20 months of remedial pumping (until the in situ chemical reduction was launched) was approximately 1.1 tons (~20%). The remedial pumping decreased the levels of Cr(VI) to 29% of the initial levels of the contamination, this representing an average value for the wells located in the source area. The concentrations, however, decreased very slowly or stagnated at the level of hundreds of mg/L, which is significantly higher than the preliminarily set remediation target level for this area (20 mg/L). It was therefore necessary to proceed with in situ chemical reduction of Cr(VI).

3.2.2 Laboratory Tests

In order to select the optimal reductive agent, laboratory column tests of chemical reduction were conducted with two potentially suitable reducing agents—sodium metabisulfite ($Na_2S_2O_5$) and sodium dithionite ($Na_2S_2O_4$). Four parallel columns were filled with soil contaminated by Cr(VI) and were washed by:

- sodium metabisulfite solution (1 g/L),
- sodium metabisulfite solution (1 g/L) and whitewash (1.2 g/L),
- sodium dithionite solution (1 g/L) and
- distilled water (to simulate the effectiveness of remedial pumping).

The concentrations of Cr(VI) in the water at the outlet of the columns are shown in Fig. 3.2. The volumes of Cr_{total} and Cr(VI) in the soil columns after they were washed are given in Table 3.1.

The results of the column tests show that the fastest decrease in Cr_{total} and Cr(VI) in the water at the column outlet was achieved by washing the columns with sodium dithionite solution followed by sodium metabisulfite solution, distilled

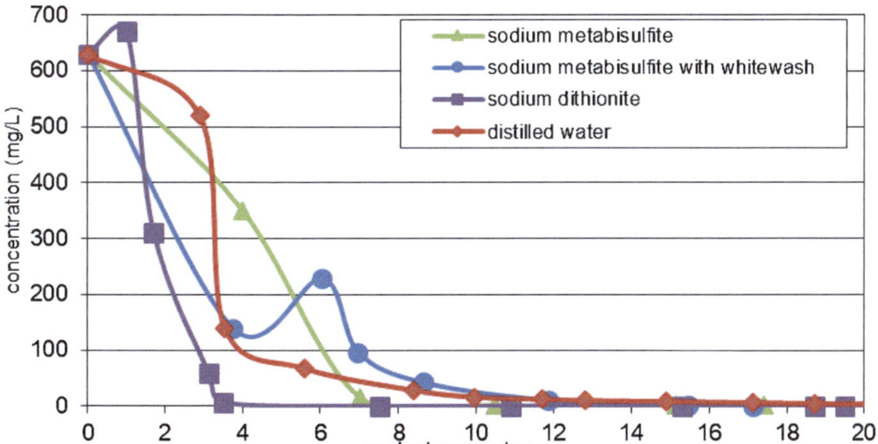

Fig. 3.2 Concentrations of Cr(VI) at the column outlet after they had been washed with sodium metabisulfite solution (pH 3.32), sodium metabisulfite with whitewash (pH 8.03), sodium dithionite solution (pH 4.2), and distilled water (pH 7.44)

Table 3.1 Concentrations of Cr_{total} and Cr(VI) in the soil after washing

Reducing agent	Cr_{total} (mg/kg)	Decrease in Cr_{total} (%)	Cr (VI) (mg/kg)	Decrease in Cr(VI) (%)
Initial concentration	950	–	24	–
After washing with sodium metabisulfite solution	670	29	0.11	>99
After washing with sodium metabisulfite solution and milk of lime	740	22	0.87	96
After washing with sodium dithionite	690	27	<0.1	>99
After washing with distilled water	510	46	7	71

water, and sodium metabisulfite solution and whitewash. The number of pore volumes needed to achieve the remedial limit for Cr(VI) of 20 mg/L are as follows:

- sodium dithionite: 3.4 pore volumes;
- sodium metabisulfite: 7 pore volumes;
- distilled water: 9.4 pore volumes;
- sodium metabisulfite solution and whitewash: 11.8 pore volumes.

Chemical analysis of the soil confirmed that washing with the tested reducing agents led to reduction of Cr(VI) by up to over 99%. The reduction of Cr_{total} was not so significant (mostly around 25%). This suggests that the resulting Cr(III) was immobilized to a greater extent in the rock environment before it was extracted. In contrast, the largest decrease in Cr_{total} (46%) was achieved by washing with distilled water, which is due to the fact that chromium remains mobile and migrates out of the column. Resulting from the column tests, the most suitable reducing agent for the in situ reduction of Cr(VI) at the site is sodium dithionite.

3.2.3 Operational Application

After verifying the methods in the laboratory, operational applications in the source area of contamination were performed. In the main phase, the reducing solution was infiltrated into 30 wells; into four of them even twice. On average, 25.5 m^3 of the reducing solution was infiltrated into the wells followed by 12.6 m^3 of clean water. However, the amount of solution applied to the individual wells differed depending on the well absorption capacity. The infiltration solution contained 10–14 g/L of sodium dithionite and the buffer solution composed of 40–57 g/L of carbonate and 2.0–2.3 g/L of bicarbonate. Groundwater extraction was performed to prevent the contaminated groundwater spreading during the infiltration of the reduction solution, and to provide a better distribution of the solution in the remedial area.

The post-remedial monitoring of the groundwater was performed 2 weeks, 2 months, and 4 months after the infiltration was completed. The final monitoring showed that the concentration of Cr(VI) in most of the wells was below 0.1 mg/L and

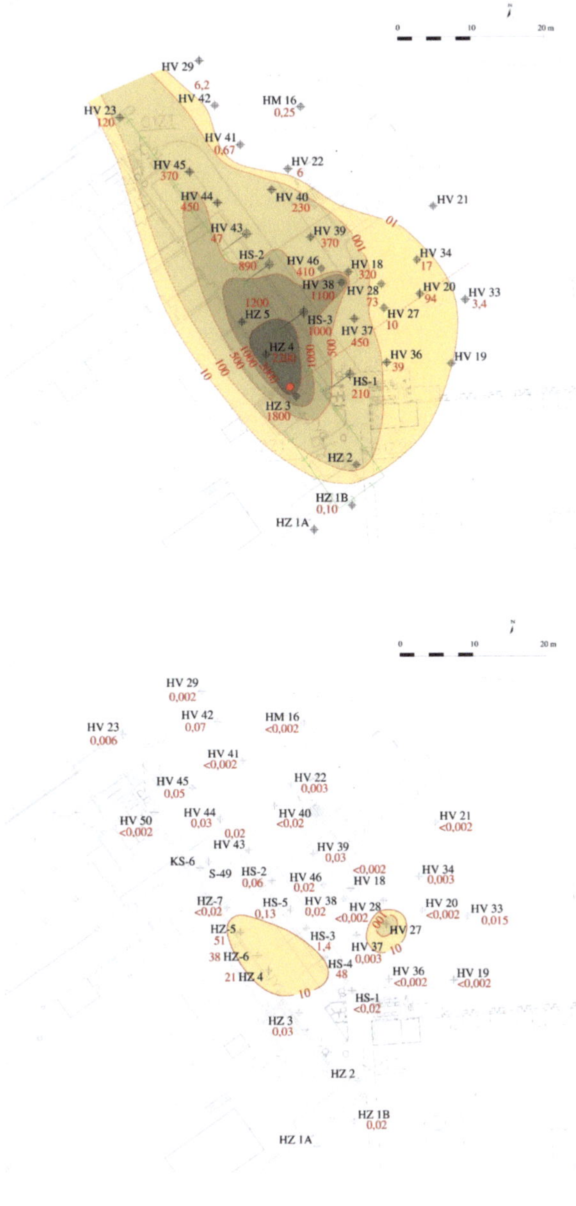

Fig. 3.3 Concentration of Cr(VI) in groundwater before infiltration (above) and 4 months after completion of the main infiltrating phase (below)

the contamination removal efficiency was over 99.9% (Fig. 3.3). There was a reduction in the contaminated groundwater in the remedial area by 92%. At six wells, however, the monitoring detected a rebound effect, i.e., an increase in the Cr(VI) concentration. The maximum increase was found 4 months after the infiltration and was in wells HZ-5 (51 mg/L), HV-27 (170 mg/L), and HS-4 (48 mg/L). The rebound effect was probably caused by diffusion and/or desorption of Cr(VI) from the very low-permeable environment of weathered granite to more permeable layers.

The dithionite infiltrated into the saturated zone gradually decomposed into sulfate, whose concentration in the groundwater increased to approximately 1000–5000 mg/L. After the infiltration was completed, there was a downward trend in the concentrations of sulfates in most of the wells. A more significant decrease was revealed for the concentrations of potassium and sodium as a result of sorption processes; in the case of bicarbonates, it was a result of the neutralization of hydrogen ions and weathering of carbonates such as calcite and rhodochrosite.

Due to Cr(VI) rebounding in the above-mentioned wells, there was an additional phase of infiltration of the reducing solution. The reducing agent was the same as for the main phase, i.e., sodium dithionite (14 g/L) with K_2CO_3 (56 g/L) and $KHCO_3$ (2.8 g/L) buffers. In total, 270 m^3 of the solution and 94 m^3 of pure water were infiltrated into 12 wells. The concentration of Cr(VI) in the groundwater in the source area was significantly below 0.1 mg/L two and four months after the completion of this infiltration phase. Post-remedial monitoring was conducted at monthly intervals over the following 2 years in ten selected wells in the source area. During the reporting period, the target limit was only sporadically exceeded in a maximum of two wells.

3.3 Reduction of Hexavalent Chromium by Metabisulfite

The Permon Křivoklát site with proven massive contamination of groundwater and soil by Cr(VI) was selected to test the efficiency of the sodium metabisulfite method. This contamination comes from an inadequately sealed sewerage system in the chrome production plant. The total amount of the released contaminant was not recorded and the investigation work prior to the remediation found Cr(VI) at the concentration of 60 mg/L in the groundwater (well HV-22, Fig. 3.4).

Geological and hydrogeological conditions of the site are complicated. The area has two different hydrogeological rock types—upper fluvial sand-gravel sediments (maximum thickness of 8 m—porous permeability) and lower silty shale (fissure permeability) with local spilites and tuffs. Rock of a Proterozoic age is relatively impermeable or semipermeable underlying rock with fluvial sediments. The transmissivity coefficient of both rock types ranges in the order of 10^{-5} m^2/s and the hydraulic conductivity is in the order of 10^{-4}–10^{-6} m/s, which corresponds to rather weakly permeable to lightly permeable rock.

The groundwater level in these rocks is unconfined to slightly confined and has been documented (mainly in periods of intense precipitation) at about 7 m below ground level (base of the sandy gravel). The data from the monitoring equipment/system showed significant levels of oscillation, which probably caused the high concentration of dissolved oxygen (on the saturation level).

Before the application of metabisulfite, laboratory tests were conducted using three different reducing agents:

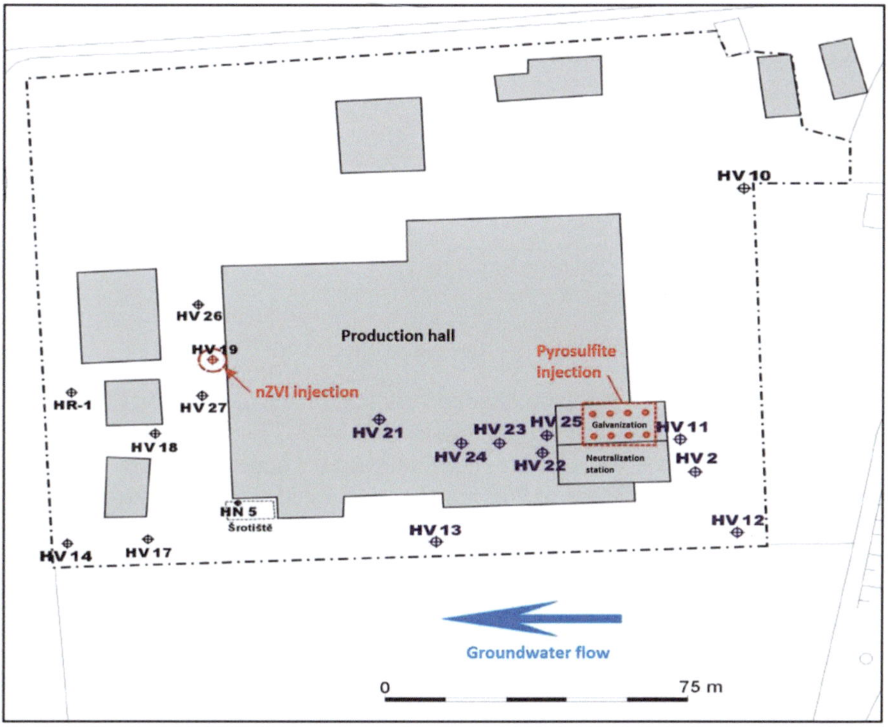

Fig. 3.4 Schematic location of Permon site

- $FeSO_4 \cdot 7H_2O$—ferrous sulfate heptahydrate,
- $Na_2S_2O_5$—sodium metabisulfite,
- Na_2SO_3—sodium sulfite.

The reagents were added in three concentrations (0.5, 2, and 10 g/L). The initial concentration of Cr(VI) was approximately 40 mg/L. The monitored parameters were concentrations of Cr(VI), Cr_{total}, pH, SO_3^{2-}, and SO_4^{2-}. The main emphasis during the evaluation of the test was placed on the reaction time and effectiveness of the Cr(VI) reduction. A significant effect on the groundwater chemistry arose from the behavior of the pH. The results of the Cr(VI) reduction depending on the amount of reagents added are shown for each reagent in Table 3.2.

The results show that $FeSO_4 \cdot 7H_2O$ has very promising reducing properties at the lowest concentrations, and the pH was also optimized. In this case, it is necessary to allow for a greater formation of precipitates that could cause decrease in the hydraulic conductivity in the well and fissures in the rock environment, thereby reducing the effectiveness of the method. Sodium sulfite Na_2SO_3 showed good reduction properties only at a concentration over 2 g/L. There are limitations on the use of sulfite due to its higher pH, which causes higher Cr(VI) migration as Cr(VI) tends to sorb at a lower pH.

Table 3.2 Results of laboratory experiments of the Cr(VI) removal. The initial concentration Cr(VI) was 40 mg/L, detection limit 0.02 mg/L

Substance	Concentration (g/L)	Cr(VI) (mg/L)	pH
$FeSO_4 \cdot 7H_2O$	0.5	0.27	5.18
	2	<0.02	5.22
	10	<0.02	3.62
$Na_2S_2O_5$	0.5	<0.02	4.38
	2	<0.02	3.50
	10	<0.02	3.77
Na_2SO_3	0.5	36.0	7.25
	2	1.2	8.10
	10	0.2	8.69

Sodium metabisulfite $Na_2S_2O_5$ showed very good reducing effects. The concentration of Cr(VI) dropped below the detection limit of 0.02 mg/L. One of the limiting factors, however, is the reaction rate. For this reason, a concentration of 1 g/L was selected for the column tests as the reaction was rapid and there were no large formations of precipitates. In addition, the pH values that range in the acidic region may promote sorption of Cr(VI) and thus increase the efficiency of stabilization of chromium contamination. It should be noted that sodium metabisulfite is nontoxic and harmless. It is commonly used in industrial chemistry and finishing. Given the above-described properties, it was therefore chosen as the reducing agent for the pilot test.

3.3.1 *Pilot Test at the Site*

The pilot test at the Permon Křivoklát site (Fig. 3.5) consisted of the gradual infiltration of pure water (for 7 days), sodium metabisulfite solution (30 days), and clean water once more (7 days). The infiltration was made into a system of eight wells in the area of a former galvanization plant and wells HV-22 to HV-25 located in the groundwater flow direction were monitored. In the initial period, a total of 370 m^3 of pure water (53 m^3 per 24 h) was infiltrated. As shown in Fig. 3.5, the concentration of Cr(VI) in groundwater in this phase decreased only in wells HV-22 and HV-24. In the other monitored wells (HV-23 and HV-25), the concentrations even gradually increased. This is most likely caused by washing out the unsaturated zone containing Cr(VI). In the case of well HV-22, probably located at a fracture zone, there was a rapid washing of contamination from the preferential pathway and decrease in the Cr(VI) concentration to 0.55 mg/L. Due to a rebound effect, values increased to approximately 15 mg/L 14 days after the start of the test.

In the infiltration phase a total of 750 kg of dissolved sodium metabisulfite (concentration 0.45 g/L) was infiltrated. The results show that the concentrations of Cr(VI) decreased in wells HV-22 and HV-25 (located closest to the infiltration wells in the groundwater flow direction) after about 11 days from the beginning of the infiltration. The initial increase in the Cr(VI) concentration is likely to be connected with the washing of unsaturated zone. Due to lack of knowledge about

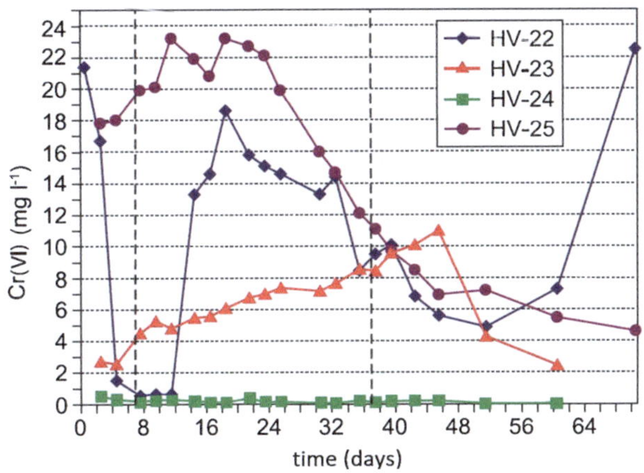

Fig. 3.5 Development of the Cr(VI) concentrations (vertical black dashed lines indicate the beginning and the end of dosing, respectively)

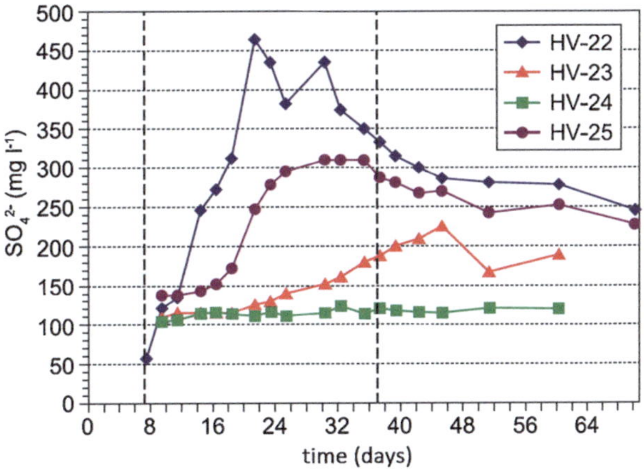

Fig. 3.6 Development of concentrations of sulfates (vertical black dashed lines indicate the beginning and the end of dosing, respectively)

the volumes and distribution of contamination in the source zone, a more accurate interpretation is not possible. This hypothesis also supports the development of concentrations of sulfates, which is shown in Fig. 3.6. There is an apparent gradual increase in the concentrations of sulfates as the unsaturated zone became saturated.

In the case of well HV-23, which also has an elevated concentration of Cr(VI) but is situated further away from the infiltration wells, the decrease in the concentrations occurred as late as approximately 38 days after the beginning of the infiltration. This was most likely caused by the retention of water in the rock environment. This has been confirmed by the concentrations of sulfates.

In general, the effect of sodium metabisulfite infiltration can be evaluated as positive, because its steady distribution in the rock environment brought a gradual reduction of the Cr(VI) concentration.

After the pilot infiltration of sodium metabisulfite, the water infiltration and dilution was performed for a period of approximately 7 days. The aim was to verify whether there was no rapid increase in the Cr(VI) concentration after the dosing was concluded. The Cr(VI) concentrations in the groundwater decreased approximately 24 days after the sodium metabisulfite infiltration was concluded. Thereafter, there was an increase in the Cr(VI) concentration in well HV-22 (rebound effect); this was probably due to the consumption of sodium metabisulfite and release of Cr(VI) from the source zones that still remained contaminated (amount of infiltrated agent was limited).

References

Amonette JE, Szecsody JE, Schaef HT, Templeton JC, Gorby YA, Fruchter JS (1994) Abiotic reduction of aquifer materials by dithionite: a promising in-situ remediation technology. Pacific Northwest Laboratory, Richland

Fruchter JS, Cole CR, Williams MD, Vermeul VR, Amonette JE, Szecsody JE, Istok JD, Humphrey MD (2000) Creation of a subsurface permeable treatment zone for aqueous chromate contamination using in situ redox manipulation. Ground Water Monit Rem 20(2):66–77. https://doi.org/10.1111/j.1745-6592.2000.tb00267.x

Graham MC, Farmer JG, Anderson P, Paterson E, Hillier S, Lumsdon DG, Bewley RJF (2006) Calcium polysulfide remediation of hexavalent chromium contamination from chromite ore processing residue. Sci Total Environ 364(1–3):32–44. https://doi.org/10.1016/j.scitotenv.2005.11.007

Henderson T (1994) Geochemical reduction of hexavalent chromium in the Trinity Sand aquifer. Groundwater 32(3):477–486. https://doi.org/10.1111/j.1745-6584.1994.tb00665.x

Palmer CD, Puls RW (1994) Natural attenuation of hexavalent chromium in groundwater and soils. EPA/540/5-94/505

Schindewolf U (1976) Taschenbuch der Abwasserbehandlung fur die metallverarbeitende Industrie, Band 1: Chemie. Von L. Hartinger. Carl Hanser Verlag, München-Wien 1976. 1. Aufl., 404 S., 187 Abb. u. 72 Tab., Ln., DM 96. Chem Ing Tech 48(12):1221–1221. https://doi.org/10.1002/cite.330481228

US EPA (2000) In situ treatment of soil and groundwater contaminated with chromium. Technical resource guide. Cincinnati, Ohio

Chapter 4
Combination of Electrokinetics and nZVI Remediation

Miroslav Černík, Jaroslav Hrabal, and Jaroslav Nosek

Abstract Electrogeochemical processes as a whole remedial technology have a considerable application potential. They may be applied at sites with complex geological conditions where hydraulic intervention is inefficient or unreliable. In the case of ClE remediation, electrochemical processes can substantially accelerate not only the time necessary for the process of reductive dechlorination, but they also significantly reduce the necessary dose of principal reagents-especially costly nanoscale zero-valent iron (nZVI) particles. It is of operational importance to accelerate the transport of nZVI and its oxidation products in the electric field and to ensure the homogeneous distribution of nZVI in the required space by suppressing the aggregation of the particles. By arranging the electrodes, it is possible to accelerate the migration or, in contrast, stabilize the particles at a predetermined location. For example, this effect can be used to operate geochemical reactive barriers that can reliably replace hydraulic barriers at certain sites. At the model site, a decrease in the measured concentrations of ClE in relation to the connection of a direct current is demonstrated. These concentrations were measured in the monitoring boreholes that were not directly used to infiltrate the reactants and are located between the electrodes. This demonstrates a spatial reduction in the concentrations throughout the treated area.

Keywords Zero-valent iron · Nanoparticles · DC field · Chlorinated hydrocarbons · Dechlorination · Groundwater · Site applications

---

M. Černík (✉) · J. Nosek
Institute for Nanomaterials, Advanced Technologies and Innovation, Technical University of Liberec, Liberec, Czech Republic

AQUATEST a.s., Prague, Czech Republic
e-mail: miroslav.cernik@tul.cz

J. Hrabal
MEGA a.s., Stráž pod Ralskem, Czech Republic

J. Filip et al. (eds.), *Advanced Nano-Bio Technologies for Water and Soil Treatment*, Applied Environmental Science and Engineering for a Sustainable Future,
https://doi.org/10.1007/978-3-030-29840-1_4

4.1 Introduction

Electrokinetic remediation, also known as electromigration, electrokinetic soil processing, electroreclamation, or electrochemical decontamination, uses a direct electric current (DC) to remediate organic compounds, heavy metals, radionuclides, or mixed organic and inorganic waste from soils, groundwater, etc. (Lima et al. 2017) When the DC is applied, several phenomena occur, such as electroosmosis, electrophoresis, and electromigration, which affect the migration of substances in both ionic and colloidal forms (Ho et al. 1999a; Gomes et al. 2015). Moreover, due to the existing protonic and hydroxyl groups on the surface of the minerals forming the sedimentary rocks, the sediment surface has an electric charge, which depends on the pH and ionic strength of the surrounding solution (electrolyte) (Saleh et al. 2006). The application of an external electric field leads to the release of ions from the electric double layer and their migration towards the opposite charged electrode (electromigration). With increasing pH (around the cathode), the surface charge of soil particles changes from positive, through neutral, to negative. An acidic environment is produced in the area of the anode, which triggers desorption of the contaminants from the soil surface and pores. In an environment with low hydraulic conductivity or high heterogeneity, the released contaminants are transported in the electric field by electrokinetics (Ho et al. 1999b).

Electrogeochemical processes can be applied to chemicals that can undergo oxidation-reduction changes or changes that are initialized by these reactions. Typical example of it is reduction or oxidation of metals and by changing their valence they precipitate from groundwater or substances that can be decomposed or chemically modified by these processes (a typical example is reductive dehalogenation of chlorinated ethenes-ClE). While in the case of metal coprecipitation, the rock environment may become gradually clogged with newly formed minerals; in the case of hydrogenation of ClE, this effect will be minimal as the chemically treated contaminant does not form a mineral phase. However, this may occur by the interaction of groundwater with the rock matrix.

Remediation technologies are very often based on a combination of various processes and the use of different reactive materials. Zero-valent iron is a promising reactant for the removal of various contaminants from groundwater because of a high reductive capacity and eco-friendliness, i.e., production of nontoxic iron oxides after the removal of the pollution (Černík et al. 2019; Johnson et al. 2009; Kanel and Choi 2007; Macé et al. 2006). Under the appropriate conditions, the use of nZVI as a remediation agent can be highly effective, but the effect lasts for a relatively short time and acts over a relatively small radius from the application point (Bennett et al. 2010). There is abundant evidence that the mobility of nZVI in the porous media is limited under almost all conditions (Bennett et al. 2010). Nanoscale ZVI also has a tendency to aggregate quickly and settle into pores mainly because of magnetic attractive forces (Lowry and Johnson 2004). Therefore, nanoscale iron particles are often modified by several methods, e.g., they are coated with various polymers (Park et al. 2009; Schrick et al. 2004; Wang et al. 2010; Zhang et al. 2009), a protective layer made of biodegradable oil (Quinn et al. 2005), natural gums (Padil Vinod et al. 2017), etc. Another method for improving the effectiveness of nZVI is to use it in combination with another remedial method, e.g., biodegradation (Klimkova et al. 2011; Němeček et al. 2016).

A new way of increasing the effectiveness of nZVI is to support the remediation with a DC electric field (Černík et al. 2019) mentioned above. The use of a direct current can help to overcome the poor mobility of nZVI (Gomes et al. 2015). With an increasing pH (around the cathode), the surface charge of the nZVI particles changes from positive to negative. The larger the magnitudes of the surface charge, the larger the repulsive forces and smaller the tendency for aggregation. Moreover, the negative surface charge forestalls nanoparticle adsorption on the negatively charged minerals (Morais et al. 1976; Sposito 1998). Similarly, the field induces migration of contaminants from their source to the treatment zone.

The influence of ZVI and DC can be interpreted by a conceptual geochemical model. Unaffected groundwater, generally, has moderate E_h and slightly positive and neutral pH. Sometimes, the water shows a more reductive E_h, indicating the ongoing biological reductive dechlorination of contaminants or other biologically enhanced reductive processes. The application of nZVI leads to a sharp decrease in the E_h, sometimes to the limit of water stability, and an increase in the pH depending on the buffer capacity (mainly carbonate content) of the aquifer. After depletion of the reduction capacity of nZVI (by its oxidation), the conditions return to a moderate oxidation state (the kinetics of the process depend on the balance of the substances with an oxidation potential brought into the reaction zone, and not on the level of contamination). As it was mentioned above, the application of the DC field changes the conditions in the reaction zone, depending on the distance of the electrodes. Anoxic conditions are not usually reached on the cathode because of the increase in the pH, even when there is a significant decrease in the E_h. Conversely, in the vicinity of the anode, despite the increase in the E_h, anoxic conditions are often reached as a result of a significant decrease in the pH (Fig. 4.1).

4.2 Electrokinetic Remediation

4.2.1 Basic Principles of the Action of an Electric Field on Water Parameters

The content of dissolved substances in groundwater determines its electrical conductivity, which determines the current density, i.e., the number of electrons that can be transmitted through the rock environment at a certain voltage. However, if the practical decomposition voltage on an electrode is exceeded, electrolysis of water occurs according to the following equations (Fig. 4.2):

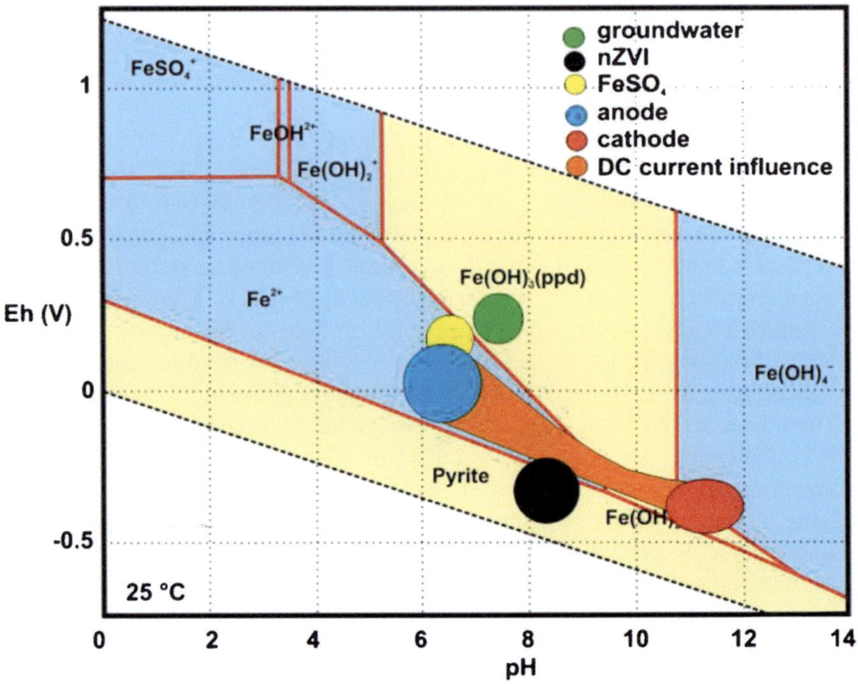

Fig. 4.1 Pourbaix diagram of conditions for different remedial actions

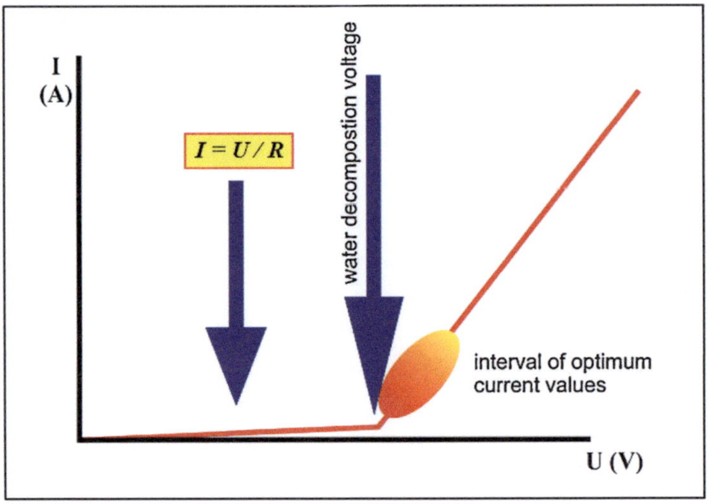

Fig. 4.2 Dependence of the current on the voltage when the practical decomposition voltage is exceeded

$$\text{Cathode}: \quad 2H_2O + 2e^- \rightarrow H_2 + 2OH^- \tag{4.1}$$

$$\text{Anode}: \quad 2H_2O \rightarrow O_2 + 4H^+ + 4e^- \tag{4.2}$$

Water on the cathode reduces the hydroxide ions, which increases the pH. Oxidation of the anion of oxygen on the anode produces protons that lower the pH. This process increases the conductivity of the water and decreases the electrical resistance of the environment, which increases the current density in the electric field.

The function of the electrode material is mainly discussed in the electrosynthesis of organic compounds. Direct transfer of electrons from the cathode to the organic compound preferentially occurs with the use of *sp* metals (Pb, Hg, Sn, Ga, Ti, Zn, Cd, Bi, Al, In, etc.) that require the greatest overvoltage for hydrogen liberation. Absorption of hydrogen on these *sp* metals has generally not been observed (or minimally), and cathode electrosynthesis can be controlled by the formation of oxonium ions, from which hydrogen is excreted on the cathode. The use of *sp* metals also favors the production of anionic radicals and their secondary chemical reactions leading to hydrodimerization, cathode pairing, and the formation of organometallic compounds. On the other hand, the lowest hydrogen overvoltage for hydrogen excretion is observed when *d* metals, which have the highest occupied outer *dd* orbitals, are used (Pt, Ru, Ni, Pd, Rh, Fe, Co, etc.). Conversely, hydrogen absorption on these metals is significant and hydrogen on the surface of a cathode made from d metal acts as a hydrogenation agent (Torii 2006).

The use of certain metals is excluded for environmental toxicity (Hg, Pb, Cd, etc.) and others for economic reasons (Pt, Pd). Other materials, like stainless steel, are also not applicable or applicable with limits because of the oxidation-reduction processes. Stainless Steel releases alloying additives (Cr, Mo, etc.), which are very toxic for the environment. Compared to noble metals and stainless steel, pure Fe has a significantly lower electrode potential and in galvanic series belongs to the least noble metals. Anodic corrosion, which causes oxidation and consequently degradation of the anode, must then also be considered. However, the produced iron oxides are natural compounds, and also cost of Fe is affordable.

4.2.2 Aquarium Test

As can be seen from Eqs. (4.1) and (4.2), the electrolysis of water leads to changes in the pH and E_h around the electrodes. To estimate the profile of these parameters in 3D, an experiment was created in a non-draining 3D reactor-an aquarium. Fine siliceous sand of a 0.5–2 mm fraction simulating aquifer was placed in a $30 \times 30 \times 150$ cm glass aquarium filled with tap water. According to Fig. 4.3, 24 stainless steel tubes, which were perforated at the bottom, were placed into the container for the sampling of water samples. Finally, 25×25 cm titanium plate electrodes in a distance of 1 m were placed and connected through a 24 V source.

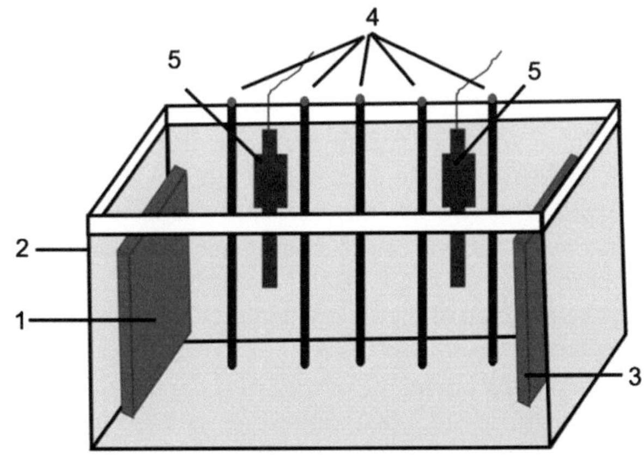

Fig. 4.3 Aquarium tests: block diagram (*1*-anode, *2*-glass aquarium, *3*-cathode, *4*-sampling points, *5*-E_h/pH electrodes)

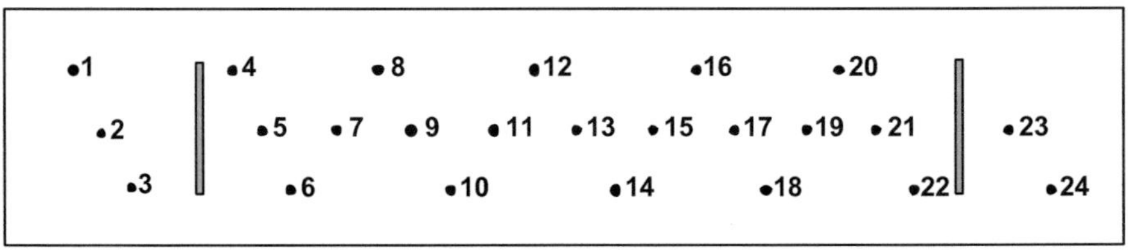

Fig. 4.4 Aquarium tests: diagram of distribution of sampling points (vertical view)

The entire system was closed with a lid to prevent access to air for better simulation of the anoxic underground environment.

In order to measure the physical and chemical parameters, a WTW Multi 3430 Multimeter with a Sentix®940, Sentix®900, and TetraCon® 925 sensor was used to measure the pH, E_h, and conductivity, respectively. The experiment was performed for 34 days to stabilize most of the parameters. Sampling points 4, 5, 6, 7, 9, 11, 13, 15, 17, 19, 21, and 22 (see Fig. 4.4) were used.

The results determined during 34 days showed kinetics of water electrolysis and distribution of its products (Fig. 4.5). The pH measurements displayed a decrease in the pH on the anode from the original value of approximately 7 to 1.2, and, on the other hand, an increase in the pH on the cathode up to 13.5. A pronounced borderline between the acidic and alkaline areas was located in the central part of the reactor, where the products migrating to the oppositely charged electrode were neutralized.

In the case of E_h, there were also significant changes in the reactor profile. In the vicinity of the anode, where the pH decreased, the E_h significantly increased. By contrast, in the vicinity of the cathode, where the pH increased, the E_h decreased significantly to −600 mV. Moreover, the pronounced pH borderline in the centre of the reactor is not created for the E_h. When we plotted the pH and E_h profiles measured horizontally in the centre of the reactor at the end of the experiment (Fig. 4.6a), we saw the interesting profiles discussed above; while the pH created a sharp interface, the E_h changed more or less linearly.

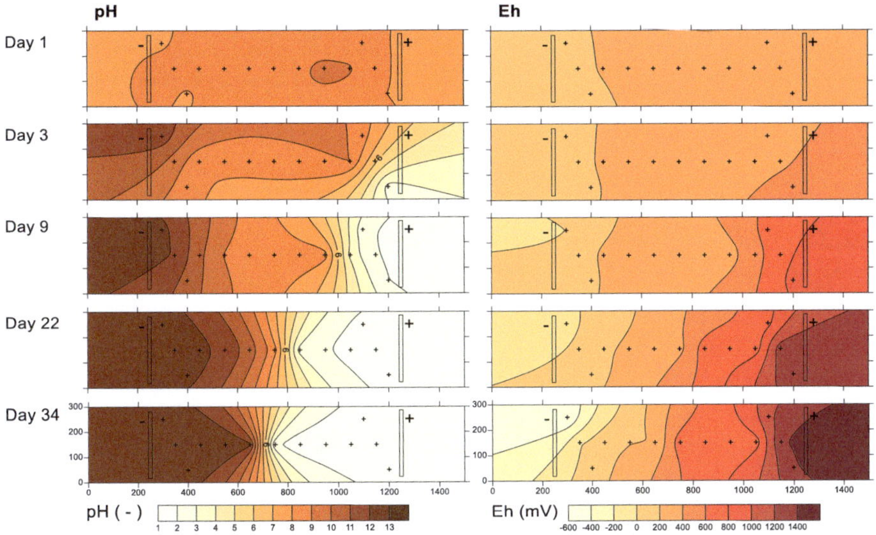

Fig. 4.5 Time development of pH and E_h in the 3D reactor

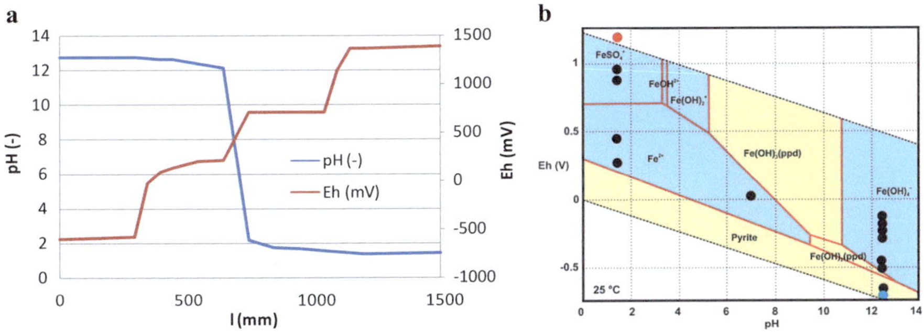

Fig. 4.6 pH and E_h profiles in the centre of 3D reactor (a) and stability diagram (b)

When the individual measurements were plotted on a general iron stabilization chart (this procedure is not entirely legitimate because there is no Fe present in the system), there were several changes in the oxidation and reducing environment from the point of view of oxidation reduction by forms of Fe in the space between the anode and the cathode. The cathode (Fig. 4.6b) and the anode were located on the borderline of the stability of water (partial pressure of oxygen and hydrogen are equal to atmospheric pressure, respectively). At increasing distances from the cathode, the E_h increased with a constant pH. The environment gradually changed from a reduction zone to an oxidation zone. Dramatic change in the pH in the middle of the aquarium caused again change of the oxidation form of Fe back to Fe(II), followed by a gradual change to Fe(III) at a fixed low pH. Therefore, in real applications, it cannot be unequivocally stated that a reduction environment will be developed for the cathode and oxidation for the anode, and the groundwater condition cannot be determined simply by the size (or sign) of E_h. Real oxidation-reduction conditions in the space between the electrodes will depend on a number of geochemical processes.

4.2.3 Site Application of DC in Hořice

As in the case of the previously described reactor, it is possible to demonstrate the effect of an electric current directly on groundwater in an in situ arrangement. A pilot polygon was located at the site of the Karbox plant (Hořice in Podkrkonoší, Czech Republic). The site, located in a town, was contaminated in the 1970s and 1980s by chlorinated ethenes, which were released from an industrial activity. The total area of contamination is approximately 120 m × 60 m. Geologically, the site is part of the Bohemian Cretaceous Basin divided into Cenomanian and lower Turonian layers. The Quaternary cover is formed by a 6-meter-thick layer of loess and secondary loess separated from Turonian sediments by about a meter-thick layer of silty clays. The groups of strata are tectonically afflicted and vertically shifted. The uppermost and contaminated aquifer is in the 10-m-thick Quaternary sediments mainly represented by re-deposited sandy silty clays of a very low permeability (transmissivity of 10^{-7} m^2 s^{-1}). The contamination was located in five separate hotspots, which were considerably elongated in the direction of the groundwater flow. A 6 × 6-meter test polygon was defined around the well IS10.

The site consists of five contaminated spots elongated in the direction of the groundwater flow. The chlorinated ethenes concentration in groundwater reached 60 mg L^{-1}. The experiment run around the well IS-10, which is a hotspot for total concentration of ClE and contains the highest concentrations of all daughter products of PCE sequential dechlorination (Fig. 4.7). Only the PCE highest concentration is

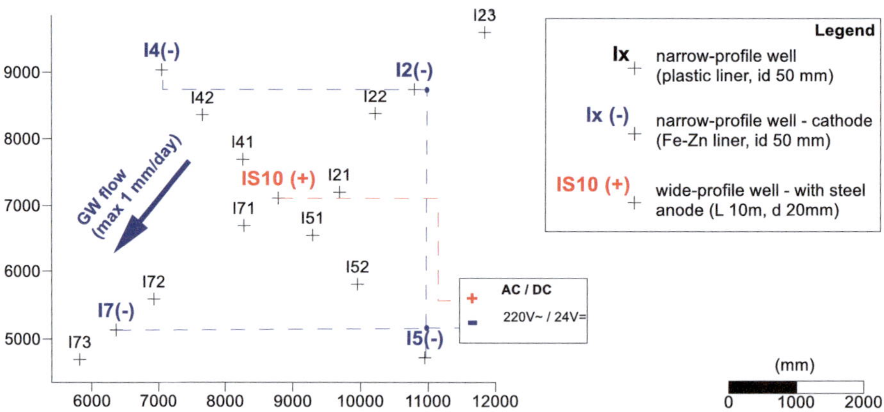

Fig. 4.7 Location of the electrodes and monitoring boreholes in the test polygon Hořice

Table 4.1 Concentration of ClE (μg/L) in the well IS10 prior the experiment ($t = 8$ days), at the end of DC run ($t = 87$ days) and 85 days later ($t = 172$ days)

Time	8 d	87 d	172 d
Sum of ClE	27,709	14,183	23,063
PCE	5471	3402	5215
TCE	3731	830	1213
DCE	17,430	9205	15,974
VC	1077	747	661

PCE tetrachloroethene, *TCE* trichloroethene, *DCE* dichloroethene, *VC* vinyl chloride

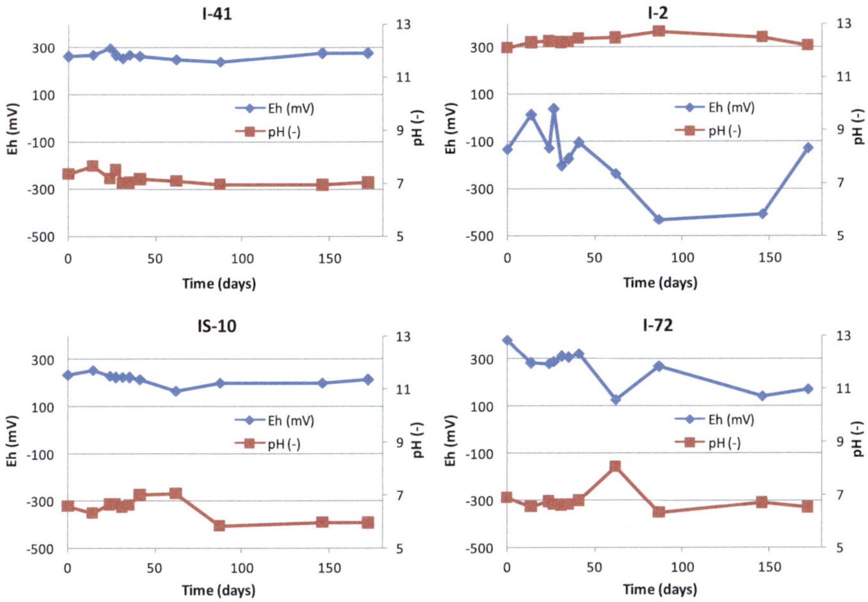

Fig. 4.8 Changes in E_h and pH values in selected boreholes (IS-10 represents the anode, I-2 cathode, I-41 and I-72 boreholes in the wider area of the cathode)

shifted to the surroundings of I22, as a result of an inflow of the PCE contaminated water from NE direction (I23 well). The original ($t = 8$ days) concentration of ClE was 27.7 mg/L, with a majority of DCE (Table 4.1).

Steel electrodes (four cathodes and one anode) were inserted into the aquifer to the depth of the impermeable substrate (length 10 m) and the field was generated by a 24 V AC/DC converter with a maximum output of 300 W. Eleven monitoring boreholes were drilled to monitor the changes in the parameters (Fig. 4.7). This arrangement makes it possible to study the pH and E_h profiles between the cathodes and anodes in different directions of groundwater flow, which will also have a significant impact on these profiles.

The results of the long-term monitoring are presented in Figs. 4.8 and 4.9. In the area of the cathodes (I-2), the E_h values changed from the initial values of around +200 mV almost immediately after the electrical current is connected. An

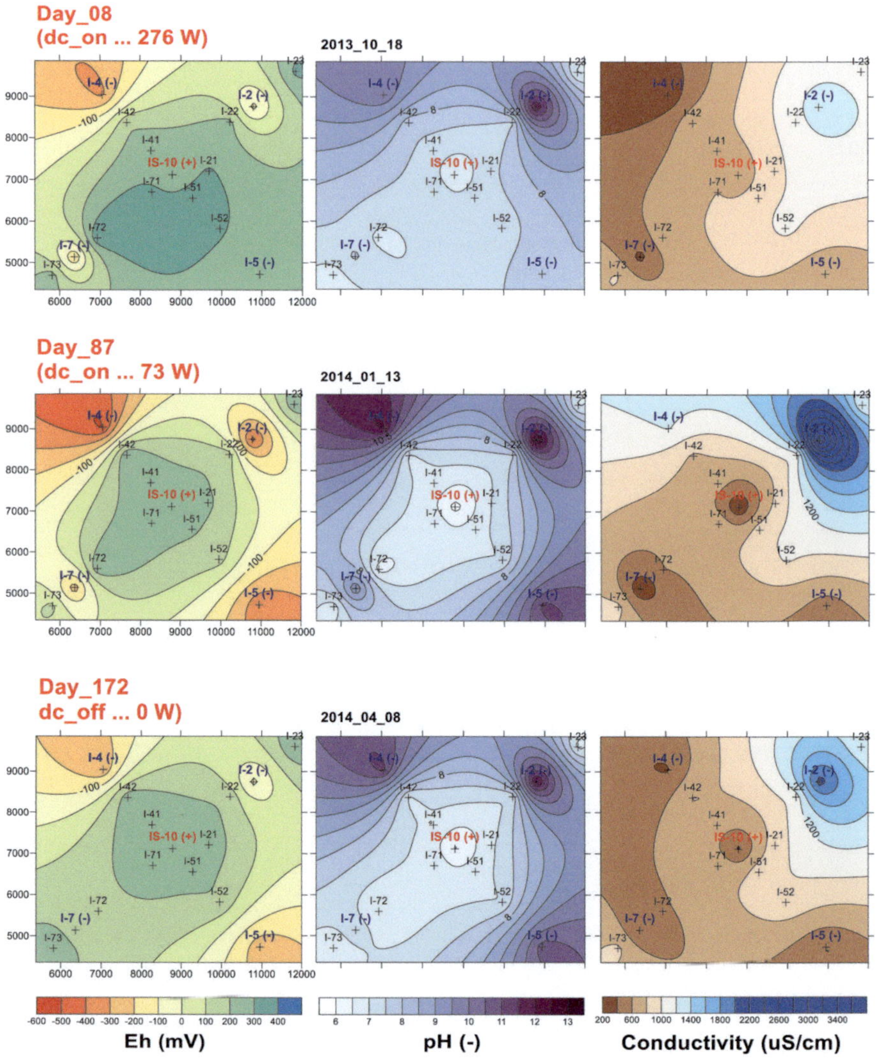

Fig. 4.9 Course of areal distribution of measured physico-chemical parameters

instantaneous decrease in the E_h, which gradually continued to –400 mV, but later (from day 150) an increase due to electrode passivation was observed. Oppositely, on the anode (IS-10), the change in the E_h is very small and the E_h is maintained around +200 mV for the whole experiment. The E_h on the electrode near the anode (I-41) behaves similarly to the anode and well I72 near the cathode similarly to the cathode, but the decrease in the E_h is not so significant. The pH in all of the boreholes, with the exception of the cathode (I-2), varied between 6 and 8; in the area of the cathode, the pH reached extreme values above 12.

Changes in the pH occurred only after exceeding the decomposition voltage, where the electrolytic decomposition of groundwater occurred, and the water with the changed pH subsequently spread through the aquifer because of dispersion, electromigration, and groundwater flow.

The long-term conditions of E_h and pH throughout the polygon stabilized approximately on the boundary of the oxidation and reduction conditions, i.e., outside the stability field of $Fe(OH)_3$ (Fig. 4.6b). The stability field of $Fe(OH)_2$ or $Fe(OH)_4^-$ at the cathode is achieved at a pH greater than 11. The groundwater around anode does not reach a very low pH and the conditions correspond to the stability field of dissolved Fe^{2+} ions, i.e., the reduction conditions are created. This state can be interpreted not only by the anode dissolution and subsequent release of Fe^{2+} into the groundwater, but also by the gradual dissolving of iron from the rock matrix. In the treated area, the content of the dissolved iron, which can enter into reactions with the contaminant, gradually increased. Moreover, changes in the E_h are independent of the groundwater flow.

The spread of the pH changes in the aquifer is then slower because of these processes (Fig. 4.8). After disconnecting the DC current (at day 88), the system had relatively large inertia due to rapid consumption of the electrodes.

Figure 4.9 shows areal representation of E_h, pH, and conductivity at the pilot test polygon in three time periods (8, 87, and 172 days). In all three cases, the situation is stable in time with well distinguished areas around anode and cathodes.

Due to the application of DC field, overall concentration of ClE decreased significantly (Fig. 4.10, Table 4.1). Around anode (IS10), the ClE total concentration decreased to 50% of the original value, comparable with the decrease in all individual ClEs. However, the drop in ClE is not permanent and very strong rebound effect was observed ($t = 172$ d). This rebound is probably due to the release of ClE from contaminated sediments rather than by the inflow of the contaminated water from the surroundings. The sum of ClE in the well IS10 returned to 83% of the original value with a slightly increased ration of DCE (from 63 to 69%).

4.3 Synergic Action of nZVI and a DC Field

4.3.1 Laboratory Reactor Test

In order to study the oxidation-reduction processes of the contaminant removal, laboratory reactor tests are the most reliable experiments. There are different setups of the reactors (Fig. 4.11). Glass reactors are equipped with well-sealed inputs or outputs (ports). Mechanical mixers, sampling points, or electrodes are permanently installed inside the reactor, so that the sampling process itself is realized without access to air in the reactor. The experiment can be run in a simple reactor, where the samples of contaminated water can be taken in regular intervals or the reactor is directly connected to gas chromatography. In case of electrokinetic experiments, the reactor can be equipped with the DC electrodes. The most exact measurement, where

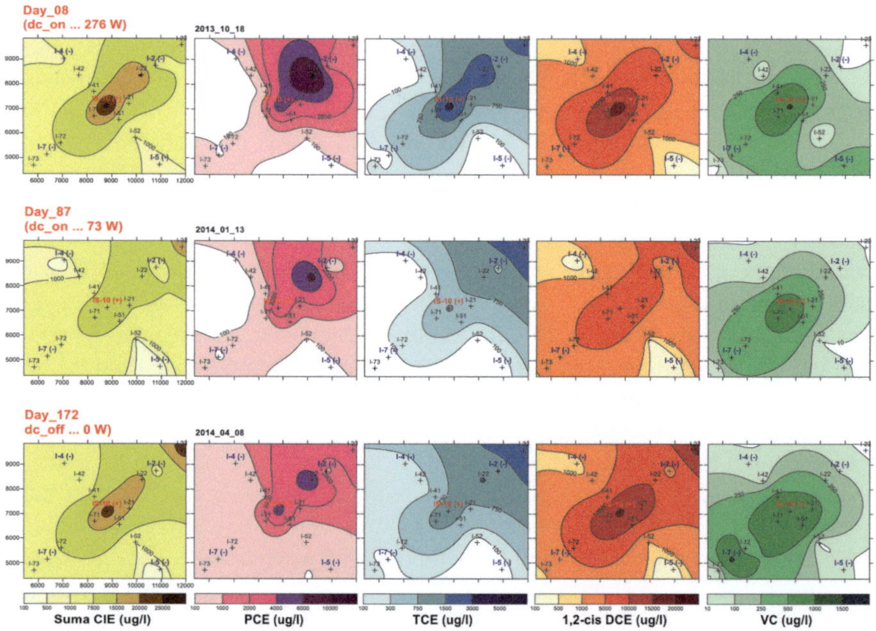

Fig. 4.10 Course of areal distribution of measured ClE

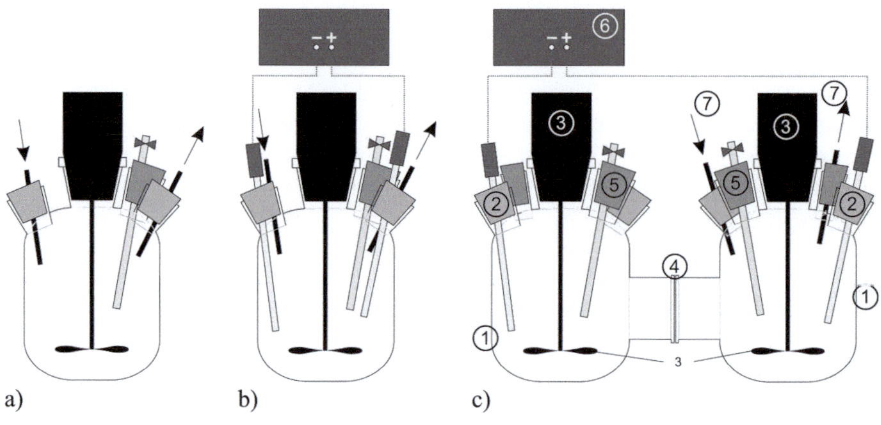

Fig. 4.11 Possible setups of reactors with direct connection to gas chromatograph: (**a**) simple reactor; (**b**) reactor with electrodes; (**c**) double reactor separated by ionic membrane. *1*-reactor, *2*-electrode, *3*-mechanical mixer, *4*-ionic membrane, *5*-sampling port, *6*-DC unit, *7*-input/output for the gas chromatograph

single effects of oxidation and reduction are studied, is provided in a double reactor (a cathode and anode compartment) separated by a semipermeable electrically conductive membrane. Reactors are designed as a closed mixed system with contaminated water and a remediation agent. One electrode is placed in each compartment and the compartments are connected directly to a gas chromatograph to determine the products of decay. This setup enables to confirm the above-mentioned processes during DC and nZVI application in aquifer.

The laboratory test focused on the demonstration and quantification of the increased reactivity caused by the addition of the DC electric field for dehalogenation of water contaminated by chlorinated hydrocarbons. For testing the groundwater from sites contaminated by chlorinated ethenes, namely PCE (concentration of 1586 mg/L); TCE (866 mg/L); and 1,2-cis-DCE (583 mg/L), two types of nZVI were selected—commercial products NANOFER STAR (NANO IRON, Czech Republic) and STAR_DC (NANO IRON, Czech Republic). STAR_DC is a product developed especially for applications of nZVI enhanced by the DC electric field. It mainly consists of nZVI with about 3–5% of magnetite, the specific surface is higher than 10 m^2/g and it is used without any surface modification.

The laboratory experiments were run in simple reactors under various conditions of the nZVI and DC combinations and there was also a reactor without any nZVI and electricity in order to observe tightness of the reactors and natural degradation rate of each contaminant (control). The experiments were run for 24 days with sampling after 1, 3, 8, 14, and 24 days. The DC unit (type Velleman PS1503SB) generated 5 V voltage field in the reactor. In frame of this laboratory reactor tests, there were six separate reactors in the following configuration:

- *Control*: observation of experimental error and possible uncertainties (volatility of volatile organic compounds—VOCs, manipulation with samples, natural degradation, etc.);
- *Control + DC*: observation of DC effect alone
- *STAR 0.3 g/L*: nZVI NANOFER STAR, concentration 0.3 g/L;
- *STAR 0.3 g/L + DC*: nZVI NANOFER STAR, concentration 0.3 g/L enhanced by DC electric field;
- *STAR_DC 0.8 g/L*: composite material STAR_DC, concentration 0.8 g/L;
- *STAR_DC 0.8 g/L + DC*: composite material STAR_DC, concentration 0.8 g/L enhanced by DC electric field.

During the laboratory tests with Aargau groundwater, a strong reductive effect of direct electric current was observed. Added value of the DC electric field to the ClE reduction is shown in each of the three pairs of reactors—with or without DC application. The results from the reactor with a NANOFER STAR concentration of 0.3 g/L and without DC showed very similar courses to the results from the Control. The reductive effect of NANOFER_DC in a concentration of 0.8 g/L is also rather weak. Twentyfour days after the nZVI application, there was only 50% of ClE in the total reduction (Fig. 4.12).

Efficiency of the ClE reduction in reactors with DC enhancement was far better than without DC. Actually, the reductive potential when only DC was used without nZVI was better than in the experiments with nZVI but without DC. The highest degradation potential was documented in the reactor with 0.8 g/L STAR_DC enhanced by DC, where, after 3 days from the application, no TCE and DCE and

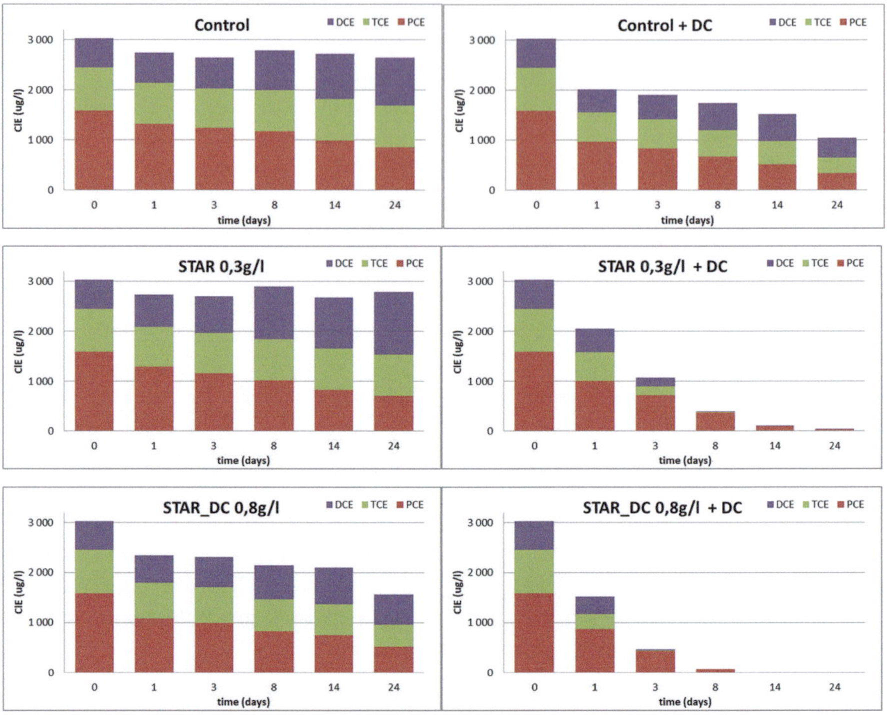

Fig. 4.12 Kinetics of chlorinated ethenes (ClE) decrease in reactors with different conditions, Aargau groundwater, ZVI (STAR or STAR-DC) at concentration 0.3 and 0.8 g/L, DC on/off

only 30% of PCE were observed. Eight days after the nZVI application, there was only 67 μg/L of ClE in total, compared to 3035 μg/L prior to the application, which counts for 98% reduction. The results from the reactor with 0.3 g/L NANOFER STAR and the DC field were also very promising, although 3 days after the nZVI application DCE, TCE, and PCE were still present; but after 8 days, 87% of the ClE reduction was determined.

The results of the experiments in a double reactor separated by ionic membrane show that the most significant process for ClE dechloration is not a direct contact of a ClE molecule with nZVI but reduction by electrons released as a result of Fe^{2+} oxidation to Fe^{3+}.

In the contaminated water with the nZVI addition, no decomposition products of ClE were observed on the cathode, only on the anode during the intensive dissolution of the surface of the electrode (Fig. 4.13). The nZVI particles in the cathode space were protected from oxidation by an excess of electrons (a strongly reducing environment). The contact with the nZVI did not lead to decomposition products. From this, it can be concluded that the contact of chlorinated ethene molecules with the nZVI in the presence of hydrogen did not cause their hydrogenation. For the successful hydrogenation of ClE in DC-based applications, it is essential to establish

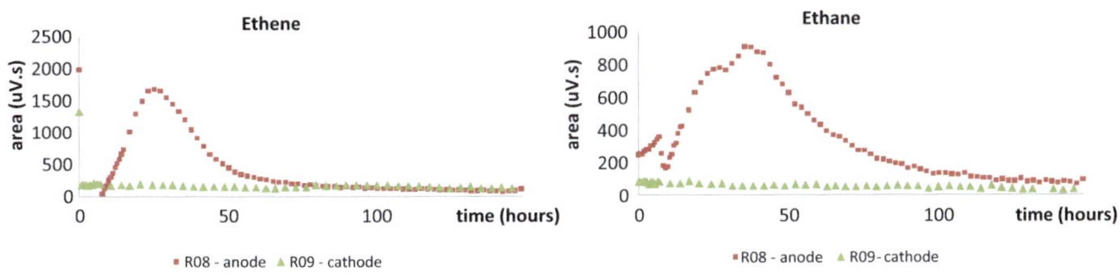

Fig. 4.13 Evolution of dechlorination products of ClE in reactor

a mildly reducing environment approximately on the boundary between the stability of Fe^{2+} and Fe^{3+} ions.

4.3.2 *Field Application of the Combined Method*

The field application of the combined method was performed at the Hořice site. From a hydrogeological point of view, the Cenomanian aquifer is the most important at the site. Its permeability is fissure-porous, with transmissivity of up to 10^{-3} m^2/s, and the aquifer is artesian in part of its profile. The confining layer above the aquifer is located about 20 m below the ground level. Another aquifer is Turonian, and has fissure permeability with transmissivity of the order of 10^{-5} m^2/s. The aquifer is recharged mainly from the overflow between the Turonian and the Cenomanian aquifers. The upper aquifer is a Quaternary aquifer, which is bound to flooded sandy-gravel loams. The permeability is porous and very low, the transmissivity is in the order of 10^{-7} m^2/s. ClE contamination is present in all three of the aquifers, the highest being in the Quaternary.

The first remedial work took place in 2008 (month 0, M0) using nZVI (RNIP, TODA, Japan) only. In total, 27 kg were applied to 11 wells in test polygon IS4. Due to the very low hydraulic conductivity of the Quaternary sediments, uniform saturation was very unlikely and very heterogeneous nZVI concentration was observed. The second application of nZVI was made in M12, again spatially, but this time using the product NANOFER 25S (NANOIRON, Czech Republic).

The original contamination in well IS4 (M0) was composed predominantly of DCE in the amount of 7,300 µg/L with a minor content of other chlorinated ethenes, i.e., VC (800 µg/L), PCE (470 µg/L), and TCE (310 µg/L), so that the total concentration of ClE was 8880 µg/L. During the two subsequent applications of nZVI, there was a decrease in the concentrations of chlorinated hydrocarbons; but, before the installation of the DC field (and before the third infiltration of nZVI, M24), the total concentration of chlorinated hydrocarbons returned practically back to the original value, due to the well-known rebound effect. The concentration of the principal contaminant DCE was 7,040 µg/L, VC 2,330 µg/L, and there was a negligible content of PCE (10 µg/L) and TCE (10 µg/L).

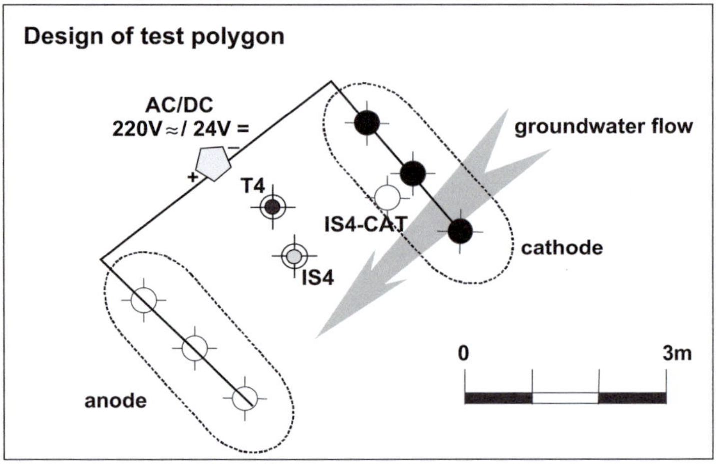

Fig. 4.14 Diagram of electrode connections in one of the remediation polygons

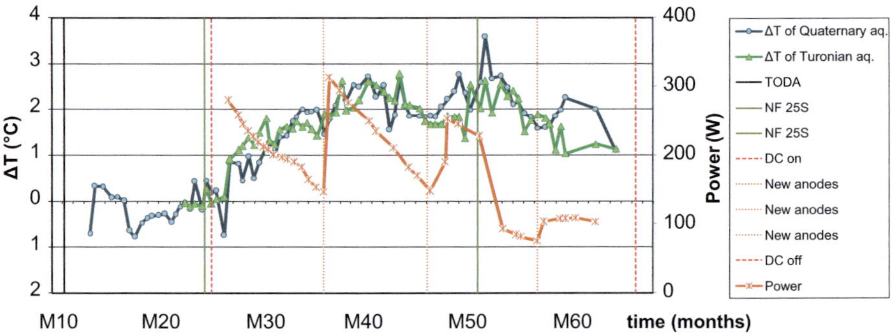

Fig. 4.15 Groundwater temperature affected by the DC current

Two cathodes and three downstream anodes in the perpendicular lines of the groundwater flow were located at the 2 m downstream of the groundwater drainage monitoring wells (pairs-where T4 borehole is cased to Quaternary aquifer and IS4 to Turonian without communication with Quaternary) (Fig. 4.14). Ten-meter-long steel bars with a diameter of 20 mm were used as the electrodes. In order to collect groundwater samples for measuring the concentrations of ClE and chemistry changes near the area of the cathode, motoring boreholes (28 mm diameter tubes perforated up to 3 m from the bottom) were placed 20 cm downstream of the middle cathode IS4-CAT. The power source was set to a constant voltage of 24 V.

The DC current is fluctuating in time because of anode oxidation and their replacement between 300 and 650 W. The electric power caused heating of the groundwater, which is demonstrated in Fig. 4.15 as a course of temperature increase in the Turonian and Quaternary aquifers (temperature differences of T-4 and IS-19 wells and IS-4 and Q-19 wells, respectively). Before the experiment started,

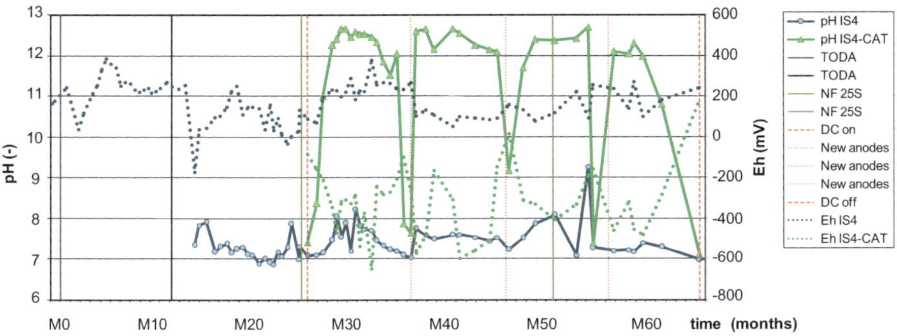

Fig. 4.16 Development of pH and E_h in the test polygon

temperature fluctuations were inside ±0.5 °C interval. However, after the DC was switched on, an increase in the temperature of up to 2.5 °C was recorded in both aquifers. Moreover, the temperature increase is proportional to the electric power. During the first year, the temperature increased by about 2 °C compared with the surrounding aquifers, but then the first stagnation was observed, as a result of a decrease in the electric power of the oxidized anodes. After the anode was replaced, the current increased and consequently the temperature rose up to 2.5 °C. To keep such an elevated temperature in a flowing aquifer is difficult; with another anode oxidation the temperature difference decreased. Again, the anode replacement increased the differences to about 2.5 °C, with a local maximum of 3.5 °C in a Quaternary aquifer in a winter period. The anode destruction caused a temperature decrease again and it remained until the end of the experiment.

4.3.2.1 Evaluation of the pH Course

Well IS4 monitors the situation in the middle of the electrode lines. During the whole monitoring period, the pH value ranged between 7 and 8 independently of the installed technologies (Fig. 4.16). In the area of the cathode (IS4-CAT), the pH was strongly dependent on the DC field causing a reduction in the water and thus the formation of OH^- ions. During the initial period of approximately 6 weeks, there was a gradual increase in the pH up to a value greater than 12.5, where the pH stabilized and was not affected by the gradual degradation of the anodes. In M35, i.e., 10 months after the DC field was connected, there was a sudden decrease in the pH due to anode dysfunction. After the anodes were replaced, there was again a rapid increase in the pH back to a constant value of approximately 12.5. This process was repeated during the whole pilot test.

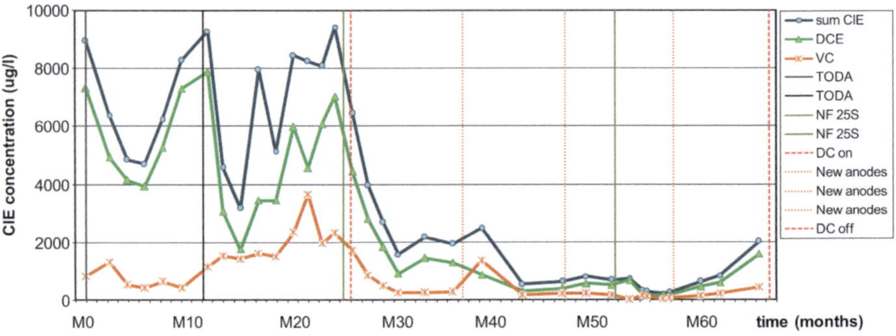

Fig. 4.17 Course of the concentrations of ClE in monitoring well IS4

4.3.2.2 Evaluation of the Course of E_h

The nZVI application prior to electrokinetics caused a temporal decrease in the E_h values. In well IS4, the first and second applications of nZVI led to a decrease from the original value of +440 mV to values around 0 mV and then a gradual return to positive values (Fig. 4.16). After the DC field was connected, there was a significant gradual decrease in the area of the cathode (well IS4-CAT). The decrease after 2.5 months was to −650 mV and the value remained low for about 8 months before it increased owing to the anode degradation. This process was repeated after every replacement of the anodes. At the end of the experiment (M63) the E_h increased to +160 mV, signalizing the end of the electrode's life and simultaneously the nZVI reductive power. The effects of the changes in the E_h were also observed while monitoring well IS4, although there was only a slight decrease.

4.3.2.3 Evaluation of the Course of Total ClE

The test polygon is located within an area with a more or less constant concentration of ClE and constant inflow of contaminated water. Therefore, after each application of nZVI the ClE concentrations returned to their original values sooner or later, when the reductive power of the nZVI was exhausted (Fig. 4.17). An absolute reduction in the DCE concentration, and thus in the total ClE, was similar after both nZVI applications, i.e., a decrease to approximately a half the initial concentration. Also, the duration of the reduction was similar, i.e., approximately 1 year. The only significant difference was the representation of the individual ClEs, and in particular VC. The concentration of VC mimicked the decrease in the DCE during the first application of nZVI, but persisted at a low level during the periods when the DCE concentrations returned to their original values. VC began to increase prior to the second application of nZVI, but no decrease was observed and, conversely, an increase lasted until the third application.

During the first 6 month after connecting the DC field, there was a significant decrease in DCE, VC, and total ClE concentrations. Since well IS4 was not directly used for infiltration of nZVI and it was located in the centre of the DC field, it demonstrated the effect of spatial reduction of contaminants throughout the treated area. Decreases were relatively rapid in this low permeable environment and therefore cannot be completely interpreted by the groundwater flow from the cathode, where hydrogenation of ClE may occur directly on the electrode. Unlike the application of nZVI only, the ClE concentrations remained low and did not rebound during the whole period when the DC field was used. Replacement of the anodes and additional infiltration of nZVI had no effect on the ClE concentrations. The entire system was shut down in M63, when the electrodes were no longer active. The ClE concentrations subsequently increased because of the flow of contaminated water from the surroundings.

4.3.2.4 Changes in Geochemical Conditions

The geochemical conditions in the rock environment change by a direct current. An essential fact is that the E_h-pH conditions in the rock environment shift beyond the field of stability of goethite, the most common mineral phase of Fe^{3+}. This phase is insoluble under normal conditions. The electric field maintains nZVI and its decomposition products in the field of stability of Fe^{2+}, which allows the prolongation of the nZVI activity in the structure. The geochemical conditions in the monitoring boreholes at the site most frequently correspond to the field of stability of Fe $(HCO_3)^-$ and near the cathode $Fe(OH)_4^-$. Part of the Fe^{2+} is captured in the pyrite structure, including the sulfides entering the treated area in the groundwater. In the given environment, the water is in contact with marlstone and the products of its weathering, so it is significantly enriched with the carbonate ion. From the point of view of the final target mineral phase, the iron is transformed in the rock environment in addition to pyrite, magnetite, siderite, and the occurrence of iron hydroxide is not excluded, as demonstrated by the results of the laboratory experiments. Magnetite and siderite are partially soluble in a decreasing pH to normal values so they can re-dose the environment with Fe^{2+} ions, which plays an important role in the process of the reductive dechlorination of ClE.

References

Bennett P, He F, Zhao D, Aiken B, Feldman L (2010) In situ testing of metallic iron nanoparticle mobility and reactivity in a shallow granular aquifer. J Contam Hydrol 116:35–46. https://doi.org/10.1016/j.jconhyd.2010.05.006

Černík M, Nosek J, Filip J, Hrabal J, Elliott DW, Zbořil R (2019) Electric-field enhanced reactivity and migration of iron nanoparticles with implications for groundwater treatment technologies: proof of concept. Water Res 154:361–369. https://doi.org/10.1016/j.watres.2019.01.058

Gomes HI, Rodríguez-Maroto JM, Ribeiro AB, Pamukcu S, Dias-Ferreira C (2015) Numerical prediction of diffusion and electric field-induced iron nanoparticle transport. Electrochim Acta 181:5–12. https://doi.org/10.1016/j.electacta.2014.11.157

Ho SV, Athmer C, Sheridan PW, Hughes BM, Orth R, McKenzie D, Brodsky PH, Shapiro A, Thornton R, Salvo J, Schultz D, Landis R, Griffith R, Shoemaker S (1999a) The Lasagna technology for in situ soil remediation. 1. Small field test. Environ Sci Technol 33 (7):1086–1091. https://doi.org/10.1021/es980332s

Ho SV, Athmer C, Sheridan PW, Hughes BM, Orth R, McKenzie D, Brodsky PH, Shapiro AM, Sivavec TM, Salvo J, Schultz D, Landis R, Griffith R, Shoemaker S (1999b) The Lasagna technology for in situ soil remediation. 2. Large field test. Environ Sci Technol 33 (7):1092–1099. https://doi.org/10.1021/es980414g

Johnson RL, O'Brien Johnson G, Nurmi JT, Tratnyek PG (2009) Natural organic matter enhanced mobility of nano zerovalent iron. Environ Sci Technol 43(14):5455–5460. https://doi.org/10.1021/es900474f

Kanel SR, Choi H (2007) Transport characteristics of surface-modified nanoscale zero-valent iron in porous media. Water Sci Technol 55(1–2):157–162. https://doi.org/10.2166/wst.2007.002

Klimkova S, Cernik M, Lacinova L, Filip J, Jancik D, Zboril R (2011) Zero-valent iron nanoparticles in treatment of acid mine water from in situ uranium leaching. Chemosphere 82 (8):1178–1184. https://doi.org/10.1016/j.chemosphere.2010.11.075

Lima AT, Hofmann A, Reynolds D, Ptacek CJ, Van Cappellen P, Ottosen LM, Pamukcu S, Alshawabekh A, O'Carroll DM, Riis C, Cox E, Gent DB, Landis R, Wang J, Chowdhury AIA, Secord EL, Sanchez-Hachair A (2017) Environmental electrokinetics for a sustainable subsurface. Chemosphere 181:122–133. https://doi.org/10.1016/j.chemosphere.2017.03.143

Lowry GV, Johnson KM (2004) Congener-specific dechlorination of dissolved PCBs by microscale and nanoscale zerovalent iron in a water/methanol solution. Environ Sci Technol 38 (19):5208–5216. https://doi.org/10.1021/es049835q

Macé C, Desrocher S, Gheorghiu F, Kane A, Pupeza M, Cernik M, Kvapil P, Venkatakrishnan R, Zhang W-X (2006) Nanotechnology and groundwater remediation: a step forward in technology understanding. Remediat J 16(2):23–33. https://doi.org/10.1002/rem.20079

Morais FI, Page AL, Lund LJ (1976) The effect of pH, salt concentration, and nature of electrolytes on the charge characteristics of Brazilian tropical soils. Soil Sci Soc Am J 40(4):521–527. https://doi.org/10.2136/sssaj1976.03615995004000040022x

Němeček J, Pokorný P, Lhotský O, Knytl V, Najmanová P, Steinová J, Černík M, Filipová A, Filip J, Cajthaml T (2016) Combined nano-biotechnology for in-situ remediation of mixed contamination of groundwater by hexavalent chromium and chlorinated solvents. Sci Total Environ 563–564:822–834. https://doi.org/10.1016/j.scitotenv.2016.01.019

Padil Vinod VT, Wacławek S, Senan C, Kupčík J, Pešková K, Černík M, Somashekarappa HM (2017) Gum karaya (*Sterculia urens*) stabilized zero-valent iron nanoparticles: characterization and applications for the removal of chromium and volatile organic pollutants from water. RSC Adv 7(23):13997–14009. https://doi.org/10.1039/C7RA00464H

Park H, Kanel SR, Choi H (2009) Arsenic removal by nano-scale zero valent iron and how it is affected by natural organic matter. In: Environmental applications of nanoscale and microscale reactive metal particles, ACS Symposium Series, vol 1027. American Chemical Society, Washington, DC, pp 135–161. https://doi.org/10.1021/bk-2009-1027.ch008

Quinn J, Geiger C, Clausen C, Brooks K, Coon C, O'Hara S, Krug T, Major D, Yoon W-S, Gavaskar A, Holdsworth T (2005) Field demonstration of DNAPL dehalogenation using emulsified zero-valent iron. Environ Sci Technol 39(5):1309–1318. https://doi.org/10.1021/es0490018

Saleh N, Sirk K, Liu Y, Phenrat T, Dufour B, Matyjaszewski K, Tilton RD, Lowry GV (2006) Surface modifications enhance nanoiron transport and NAPL targeting in saturated porous media. Environ Eng Sci 24(1):45–57. https://doi.org/10.1089/ees.2007.24.45

Schrick B, Hydutsky BW, Blough JL, Mallouk TE (2004) Delivery vehicles for zerovalent metal nanoparticles in soil and groundwater. Chem Mater 16(11):2187–2193. https://doi.org/10.1021/cm0218108

Sposito G (1998) On points of zero charge. Environ Sci Technol 32(19):2815–2819. https://doi.org/10.1021/es9802347

Torii S (2006) Electroorganic reduction synthesis, vol 1. Wiley-VCH, Weinheim

Wang W, Zhou M, Jin Z, Li T (2010) Reactivity characteristics of poly(methyl methacrylate) coated nanoscale iron particles for trichloroethylene remediation. J Hazard Mater 173(1):724–730. https://doi.org/10.1016/j.jhazmat.2009.08.145

Zhang X, Lin Y-M, Chen Z-L (2009) 2,4,6-trinitrotoluene reduction kinetics in aqueous solution using nanoscale zero-valent iron. J Hazard Mater 165(1):923–927. https://doi.org/10.1016/j.jhazmat.2008.10.075

Chapter 5
Field Study I: In Situ Chemical Reduction Using Nanoscale Zero-Valent Iron Materials to Degrade Chlorinated Hydrocarbons

Vojtěch Stejskal and Nikola Vacková

Abstract The presented study provides comparison of three separate nZVI field tests using NANOFER STAR, NZVI-C3, and NANOFER STAR modified by carboxymethyl cellulose (CMC). These field tests were conducted on sites contaminated with chlorinated ethenes. All three applications were performed via a direct-push technique on the same site one after another. The particular groundwater was monitored for chlorinated ethenes and their nontoxic degradation products. The test took about 90 days in each case. The most successful material in the study was NANOFER STAR, degrading 84% of present trichloroethene or 60% of vinyl chloride. Other materials used in the study achieved far lower efficiency. NZVI-C3, a carbon-iron material fabricated in laboratory volumes by Palacký University Olomouc, was characteristic for the combination of reduction and sorption. Modification of nanoscale zero-valent iron particles by CMC resulted in not only higher mobility of the material but also a significant decrease in its reactivity.

Keywords Nanoremediation · nZVI · In situ chemical reduction · Chlorinated hydrocarbons

---

V. Stejskal (✉)
Institute for Nanomaterials, Advanced Technologies and Innovation, Technical University of Liberec, Liberec, Czech Republic

Photon Water Technology s.r.o., Liberec, Czech Republic

AQUATEST a.s., Prague, Czech Republic
e-mail: vojtech.stejskal@photonwater.com

N. Vacková
Faculty of Science, Charles University, Prague, Czech Republic

AQUATEST a.s., Prague, Czech Republic

J. Filip et al. (eds.), *Advanced Nano-Bio Technologies for Water and Soil Treatment*, Applied Environmental Science and Engineering for a Sustainable Future,
https://doi.org/10.1007/978-3-030-29840-1_5

5.1 Introduction

Chlorinated ethenes are common contaminants, which have been used in a range of industrial application (Paul and Smolders 2014). They represent one of the most persistent organic pollutants due to their strong carbon-chlorine bond (Adetutu et al. 2015).

A basic treatment technology applied commonly on many sites is a pump-and-treat remediation system, which consists of pumping of polluted groundwater and its ex situ treatment. Due to relatively high cost of ex situ treatment, there has been a rapid development of in situ processes that do not create any residual waste unlike in situ chemical reduction (ISCR) and oxidation (ISCO) or biodegradation (Stroo and Ward 2010). The application of ISCR using nanoscale zero-valent iron particles for contaminated groundwater remediation is nowadays one of the standard methods for in situ treatment or stabilization of chlorinated hydrocarbons, selected inorganic anions, heavy metals, and others. This method has been successfully applied to many sites in the Czech Republic and worldwide.

The effectiveness of ISCR by nZVI depends on the geological and hydrogeological conditions, which affect both the reactivity of nanoparticles with contaminants and their spreading in the subsurface. Practically, it is very important to choose a good material. This work is focused on comparison of three different materials by their efficiency:

- NANOFER STAR, a commercially available product
- NZVI-C3, nanoparticles fabricated by Palacký University Olomouc
- NANOFER STAR modified by carboxymethyl cellulose to increase mobility of nZVI

5.2 Materials and Methods

5.2.1 Site Description

The studied area is located in one of the largest chemical plants in the Czech Republic with history of 140 years of chemical production. A long-term chemical and metallurgical production and incautious manipulation with dangerous materials led to chemical contamination of the site.

Field tests with nZVI were performed in one of the most contaminated areas of the site. The aquifer is formed by Quaternary terrace that consists of fluvial gravels and sands with various amounts of clayey material within the whole profile. The terrace lies 4–14 m under the surface. The lower part of the terrace consists of gravel and gravelly sand with basalt pebbles of the average size of 15 cm or quartz pebbles of 10 cm (Kvapil et al. 2014).

Hydraulic conductivity of the aquifer varies between 6×10^{-5} and 2×10^{-4} m/s, which corresponds to the intermediate and higher values of hydraulic conductivity.

Areas with lower permeability are filled with Quaternary fluvial clay and loam and show a hydraulic conductivity around 10^{-6} m/s (Kvapil et al. 2011).

The studied area is contaminated especially with trichloroethene (TCE) reaching maximum molar concentrations of 400 μmol/L, i.e., 78 mg/L. Due to the TCE decomposition to less chlorinated hydrocarbons, the studied area is subsequently also contaminated with dichloroethene—DCE (500 μmol/L, i.e., 48 mg/L) and vinyl chloride (VC) 200 μmol/L, i.e., 12.5 mg/L).

5.2.2 *Methodology of Groundwater Monitoring and Analytical Methods*

All water samples were taken under dynamic conditions after stabilization of oxidation–reduction potential (ORP) and pH. The groundwater was pumped out of the wells by peristaltic pump M610 (Solinst, Canada). Reading of the ORP and pH values was conducted at the end of every sampling using a WTW 3430 Multimeter (WTW, Germany) and recalculated against a standard hydrogen electrode.

Concentration of chlorinated hydrocarbons was determined according to EPA Method 8260B and concentrations of ethene and ethane as nontoxic products of chlorinated hydrocarbons' degradation were analyzed in accordance with the methodology described by Lewin et al. (1990).

The degree of dechlorination was calculated using molar concentrations (mmol/L) of all present parameters according to the equation (Agathos and Reineke 2003; Dolinová et al. 2016)

$$degree\ of\ dechlorination = \frac{[TCE] + 2[DCE] + 3[VC] + 4[ethene] + 4[ethane]}{4[PCE] + [TCE] + [DCE] + [VC] + [ethene] + [ethane]} \cdot 100\ [\%] \tag{5.1}$$

5.2.2.1 NANOFER Star

NANOFER STAR (NANO IRON, s.r.o., Czech Republic) is an nZVI powder developed in collaboration with the Regional Centre of Advanced Technologies and Materials. It consists of an Fe(0) surface stabilized nanoparticles. The average particle size is <50 nm, the specific surface area of the particles is >25 m^2/g.

Its main features are surface stabilization, transportability, air-stability, and reactivity. Owing to these properties, the material is safe to store, transport, handle, and process. Its storage time is practically unlimited if stored in closed packaging in a cool and dry environment. It was developed with the aim to create an nZVI material that could be prepared on site and directly applied to groundwater.

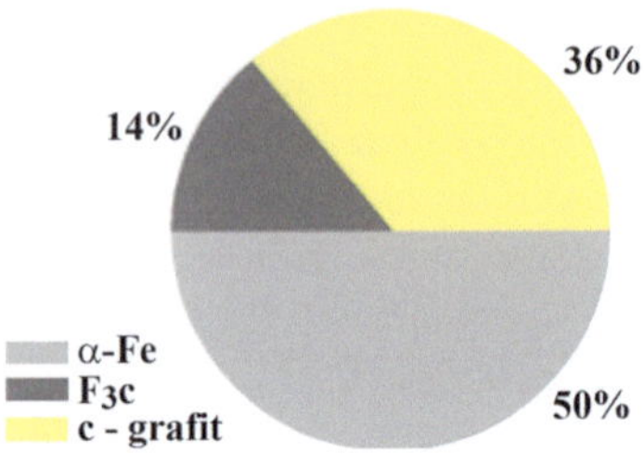

Fig. 5.1 Proportional representation of components in NZVI-C3. (Reproduced from Vacková 2018)

5.2.2.2 NZVI-C3

NZVI-C3 is a type of an nZVI powder produced usually on a laboratory scale by Palacký University Olomouc, Czech Republic. Particles of NZVI-C3 consist of Fe (0), Fe_3C, and C (Fig. 5.1). The specific surface area is approximately 28.5 m^2/s. Carbon should theoretically adsorb molecules of contaminants, whilst nanoscale zero-valent iron particles chemically reduce the contamination.

5.2.3 Methodology of Zero-Valent Iron Nanoparticles Applications

Each of the three applications of nanoscale zero-valent iron particles were performed using the same technology—direct-push injection via temporarily cased injection probes. The application pressure during such injections usually reaches 0.3–2 MPa depending on the properties of the particular layers. The five injection points during all the three applications were positioned in the same way—DP-1 to DP-5, see Fig. 5.2. The DP probes were cemented after each injection to avoid possible daylighting of the applied agent during the next injection. The application depth of all probes during all the injection campaigns was settled to 5–7 m below ground level—the most contaminated zone according to the performed soil air sampling. The injection flow rate during each injection was about 2–3 m^3/h.

The first injected nZVI material was NANOFER STAR. In total, 40 kg of an nZVI powder was injected in the form of suspension at a concentration of 2 g/L, making it 20 m^3 altogether. The injection of NANOFER STAR was performed at the beginning of April 2015. The second injection campaign was held in August 2015—the injected material was NZVI-C3 carbon-coated nZVI, only 20 kg of powder was injected in a volume of 10 m^3 (a concentration of 2 g/L). The last injection (NANOFER STAR nZVI with microscale zero-valent iron and carboxymethyl cellulose—CMC) was realized in September 2016. A NANOCOMPOSITE material made of 60 kg of ZVI and 20 kg of CMC was applied in the form of suspension. 20 m^3 of suspension of 3 g/L of ZVI and 1 g/L of CMC was injected altogether.

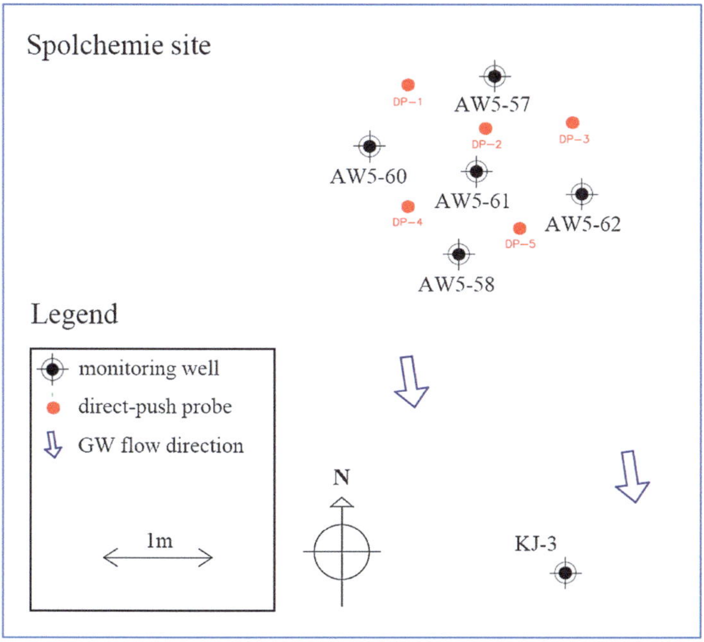

Fig. 5.2 Situation of injection probes and monitoring wells. (Adapted from Vacková 2018)

5.3 Results

5.3.1 NANOFER Star

Figure 5.3 illustrates the effect of the first application. The reduction potential dropped in boreholes AW5–57, AW5–58, AW5–60, AW5–61, and AW5–62 immediately after the injection. The strongest decrease was recorded in AW5–58 (E_H decreased from 110 mV to −370 mV). The values of E_H started to increase and they rose for 30 days of the injection. Approximately after 30 days, the redox conditions in the groundwater remained constant (Vacková 2018).

The major contaminants in the pilot test area were TCE, DCE, and VC. There was an imperceptible concentration of tetrachloroethene registered; therefore, it was not plotted on the graphs below. To determine the reduction of chlorinated ethenes (ClE), products of degradation (ethane and ethene) were investigated. Contaminants as well as products of degradation are plotted together on Figs. 5.4–5.6. For better understanding and comparison, the graphs were created in the same scale (maximum value 1400 μmol/L).

The NANOFER STAR application caused a sudden decrease in the contaminant concentrations. Figure 5.4 (Vacková 2018) represents the sharpest decrease of 98%, which was detected in the borehole AW5–57 7 days after the nZVI application. The total concentration of ClE went down from 1294 μmol/L to 20.5 μmol/L. This

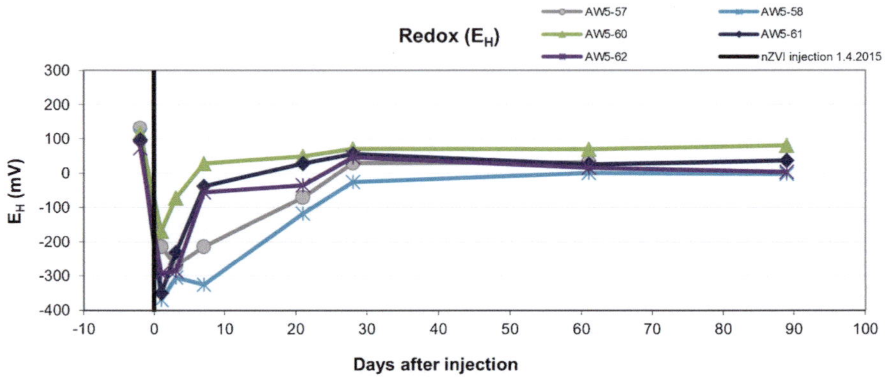

Fig. 5.3 Effect of the application of NANOFER STAR application on E_H values. (Adapted from Vacková 2018)

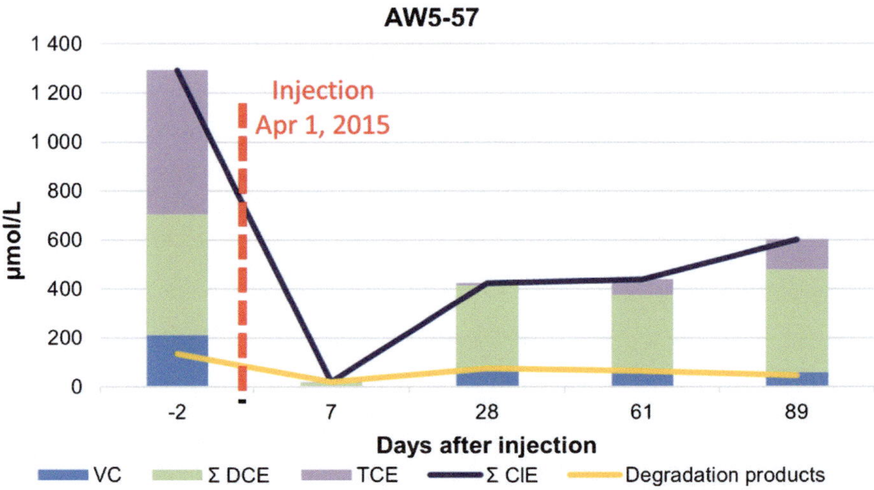

Fig. 5.4 Reduction of ClE in AW5–57 after the application of NANOFER STAR. (Adapted from Vacková 2018)

reduction was most likely caused by the combined effect of dilution and reduction of ClE contamination present in the groundwater. One month after the injection, the process of dilution did not play an important role in the decrease in the ClE concentrations. Therefore, the following reduction of contamination probably resulted from the dechlorination process.

For the determination of the reductive potential of the applied nZVI reduction ratio, periods of 2 and 3 months after the injection were important, which represented 66% and 54%, respectively, in the case of the total ClE concentrations. The nZVI application showed very high reductive efficiency in TCE, which rebounded to only 21% 3 months after the nZVI injection, compared with its pre-application concentration—79% reduction of TCE was monitored in AW5–57. Its moderate increase at

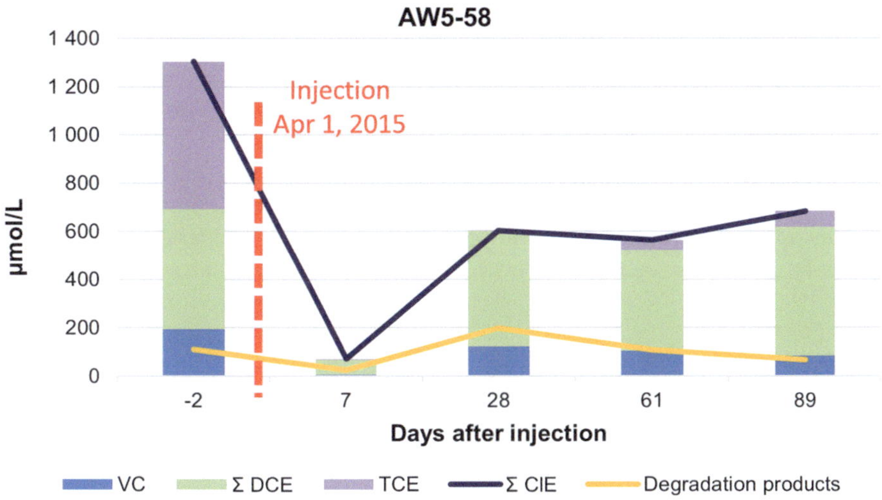

Fig. 5.5 Reduction of ClE in AW5–58 after the application of NANOFER STAR. (Adapted from Vacková 2018)

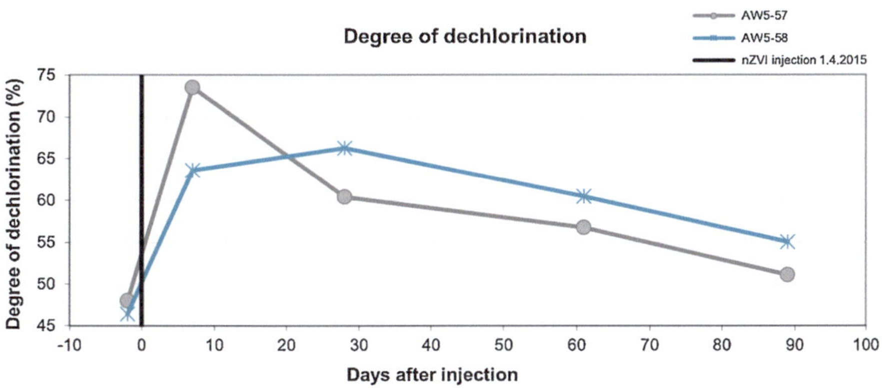

Fig. 5.6 Effect of the application of NANOFER STAR on the degree of dechlorination. (Adapted from Vacková 2018)

the end of the monitoring period could have signified an inflow of new water contaminated mainly with TCE. The samples of groundwater from the time of 3 months after the nZVI injection showed 72% reduction of VC. Considering the highest toxicity and high mobility of VC among ClE, this is a very positive result for the NANOFER STAR application.

The second sharpest decrease of total ClE 94.5% 1 week after the injection was measured in the borehole AW5–58. Figure 5.5 (Vacková 2018) demonstrates that the total concentration of ClE contamination fell from 1305 μmol/L to 72 μmol/L. Similarly to AW5–57, the sudden reduction in AW5–58 was caused by the combination of dilution and reduction of ClE contamination present in the groundwater.

The reduction ratio 2 and 3 months after the injection, which was 57% and 48%, respectively, in the case of total ClE concentrations, showed a high reductive potential. The nZVI application showed very high reductive influence on TCE, which accounted only for 11% 3 months after the nZVI injection, compared with its pre-application concentration—89% reduction of TCE was monitored in AW5–58. A similar course was observed in VC concentrations—the samples of groundwater from the time of 3 months after the nZVI injection showed 57% reduction of VC.

Figure 5.6 shows an ongoing process of dechlorination in the most contaminated boreholes AW5–57 and AW5–58. In borehole AW5–57, the degree of dechlorination was increasing only during the first week after the application. It went up from 48% to 74%. After that, the process attenuated and the final value of the degree accounted for 51% (Vacková 2018).

In the borehole AW5–58, the rising trend lasted up to 28 days during which dechlorination went from 46% to 66%. After that, the process attenuated similarly to AW5–57 and the final value of the degree accounted for 55%.

In the borehole AW5–60, immediately after the injection, there was a drop in the degree of dechlorination by 8%, which could have resulted from the dilution of groundwater by the injected suspension of nZVI. The rising trend was delayed and started 7 days after the injection when it went up from the low of 50%. Two months after the injection time, it reached the high of 67%. After that, a decreasing trend was observed.

Overall, the post-application values remained higher than the pre-application, which reflects that the reduction process had a long duration of 89 days. The course of the degree of dechlorination indicates an ongoing process of dechlorination.

5.3.2 *NZVI-C3*

Figure 5.7 illustrates the effects of the NZVI-C3 nZVI application on the redox potential. It decreased significantly and immediately only in the borehole AW5–60 where the values of E_H decreased from 90 mV to −100 mV. In the other boreholes, the E_H values mainly oscillated around the original positive values (Vacková 2018).

Compared with the previous application, the main contaminant in the subsurface was not TCE but DCE. Furthermore, the total concentration of ClE decreased by 27% after the first application. The results of the second injection of nZVI were, for easier comparison, plotted on the graphs in the same scale (max value 1000 μmol/L).

Figures 5.8 and 5.9 demonstrate the effect of the NZVI-C3 nZVI application on the concentration of ClE in the subsurface. The injection caused a short-term decrease in ClE (mainly of DCE) in every single borehole. Compared with the first injection in April 2015, the contaminant reduction was more moderate. The

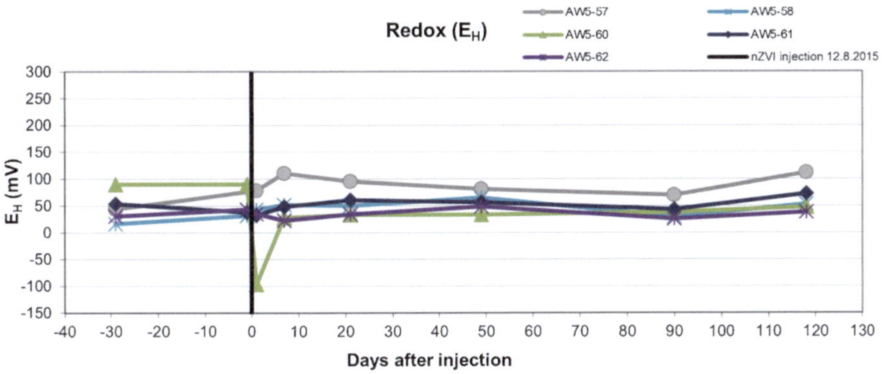

Fig. 5.7 Effect of the application of NZVI-C3 nZVI on E_H values. (Adapted from Vacková 2018)

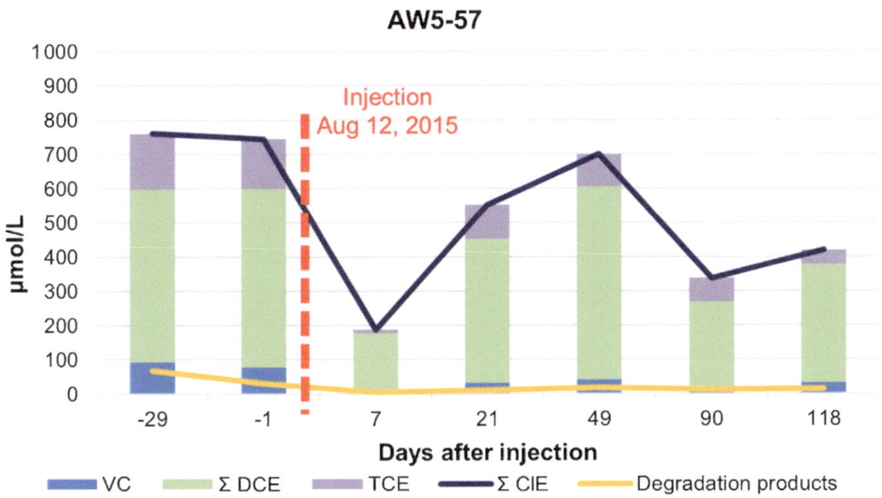

Fig. 5.8 Reduction of ClE in AW5–57 after the NZVI-C3 nZVI application. (Adapted from Vacková 2018)

strongest fall in DCE was observed in the most contaminated boreholes AW5–57 and AW5–58.

In AW5–57, the total concentration of ClE dropped by 75% during the first 7 days after the injection (Fig. 5.8, Vacková 2018). The ClE concentration decreased from 760 µmol/L (before the injection) to 188 µmol/L and then rose back gradually up to 699 µmol/L 49 days after the injection. The sudden drop in AW5–57 was most likely caused by the combined effect of dilution and reduction of ClE contamination present in the groundwater. Subsequently, an increase in the ClE concentration was observed 21 days after the injection. The effect of dilution probably was not present anymore. It seems that a new inflow of contaminated groundwater increased the concentration of ClE.

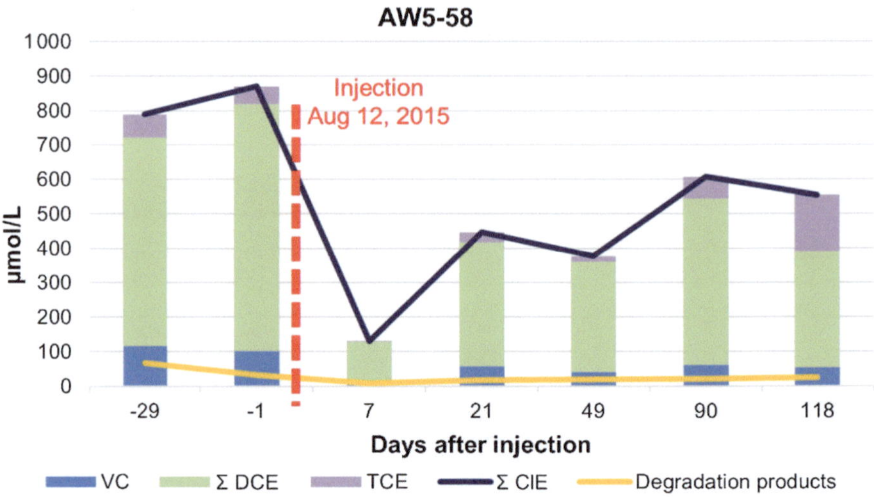

Fig. 5.9 Reduction of ClE in AW5–58 after the NZVI-C3 nZVI application. (Adapted from Vacková 2018)

To determine the reductive potential of NZVI-C3 particles, the reduction ratio 3 and 4 months after the injection was conclusive. The total concentration of ClE decreased by 54% and 44%, respectively. The application showed very high and long-term reductive influence on TCE, the concentration of which accounted only for 29% TCE reduction 4 months after the nZVI injection, compared with its pre-application concentration. In total, the TCE reduction of 71% was monitored in AW5–57. Long-term reduction of DCE and VC was also observed. The samples of groundwater from the time of 4 months after the NZVI-C3 nZVI injection showed 34% reduction of DCE and 60% reduction of VC.

The strongest concentration reduction caused by the NZVI-C3 application was observed in the most contaminated borehole AW5–58 (Fig. 5.9, Vacková 2018). The concentrations of ClE dropped after 7 days by 85%. The concentration of the contamination decreased from 871 µmol/L to 131 µmol/L and then rose back up to 607 µmol/L 90 days after the injection. The decrease in the total concentration of ClE was most likely caused by both the dilution of contaminant and by the reduction of ClE. Within the time of 21 days, an increase in the ClE concentration was observed. It seems that the dilution process terminated and a new inflow of contaminated groundwater with a high concentration of TCE was observed. NZVI-C3 nZVI was present in the borehole; however, its concentration was decreasing. Therefore, it is probable that the reductive reactions or the sorption process grew weaker and led to the increase in the ClE concentration. The final reduction in AW5–58 accounted for 36% of the ClE contamination at the end of the monitoring.

In contrast to the borehole AW5–57, the level of the TCE contamination tripled within the monitoring period. Its concentrations went up from 52 µmol/L before the injection to 164 µmol/L within 4 months of the injection. On the other hand, the

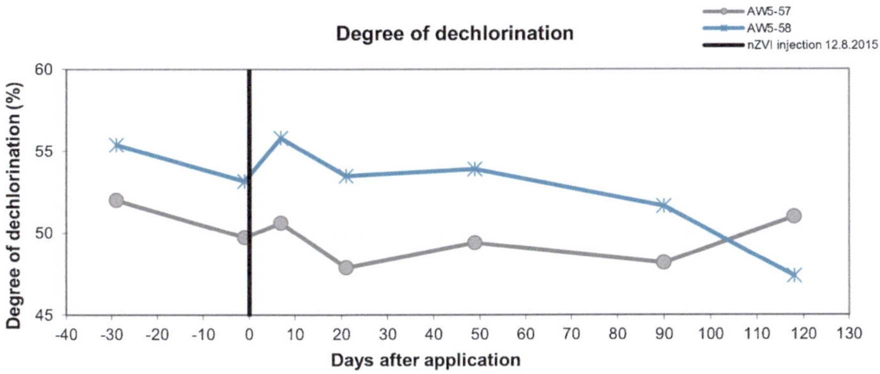

Fig. 5.10 Effect of the application of NZVI-C3 nZVI on degree of dechlorination. (Adapted from Vacková 2018)

application was successful in terms of DCE and VC reduction. At the end of the monitoring period, the reduction of DCE accounted for 53% and VC for 47%.

The parameter degree of dechlorination only confirmed the assumption that the process of dechlorination was not set up after this application of NZVI-C3 nZVI. Insignificant short increases were probably caused by the effect of dilution when 10 m^3 of the application suspension was injected into the subsurface and replaced the original groundwater (Fig. 5.10, Vacková 2018).

The results of this application show that the reduction of ClE was not accompanied by an increase in the degradation products; however, in general, the application was effective because of a significant decrease in the contamination concentration. This decrease was most likely caused predominantly by the process of sorption of ClE on the molecules of carbon-nZVI in the subsurface.

5.3.3 *Nanocomposite with Carboxymethyl Cellulose*

Figure 5.11 illustrates the effect of the NANOCOMPOSITE with the CMC application. The reduction potential decreased in boreholes AW5–57, AW5–58, AW5–60, AW5–61, and AW5–62 immediately after the injection. The largest decrease was recorded in AW5–60 (E_H decreased from 110 mV to −265 mV). Nevertheless, the reduction was not permanent. E_H values started to increase after 13 days. They reached the pre-application value approximately 20–27 days after the injection. In the borehole AW5–57, the values of E_H were oscillating after the injection. The reduction process in KJ-3 was delayed because of the distance; however, the influence of the nZVI particles was noticeable for 7 days.

In AW5–57, the total concentration of ClE dropped during the first 20 days after the injection by 22%. The combination of dilution and reduction of ClE probably led to a decrease in the total contamination from 414 μmol/L (before the injection) to

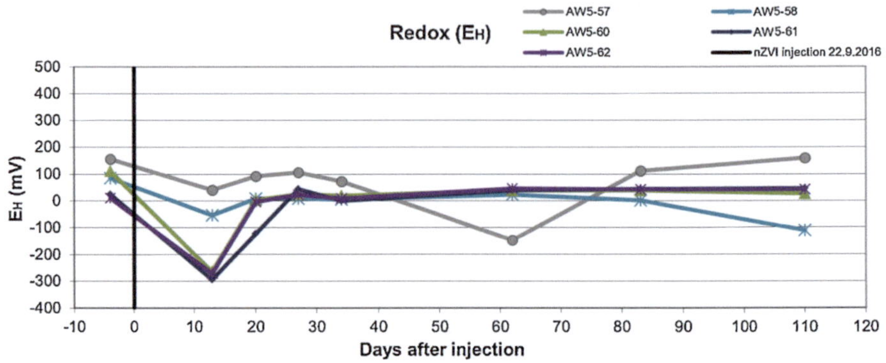

Fig. 5.11 Effect of the application of NANOFER STAR nZVI with CMC on E_H values. (Adapted from Vacková 2018)

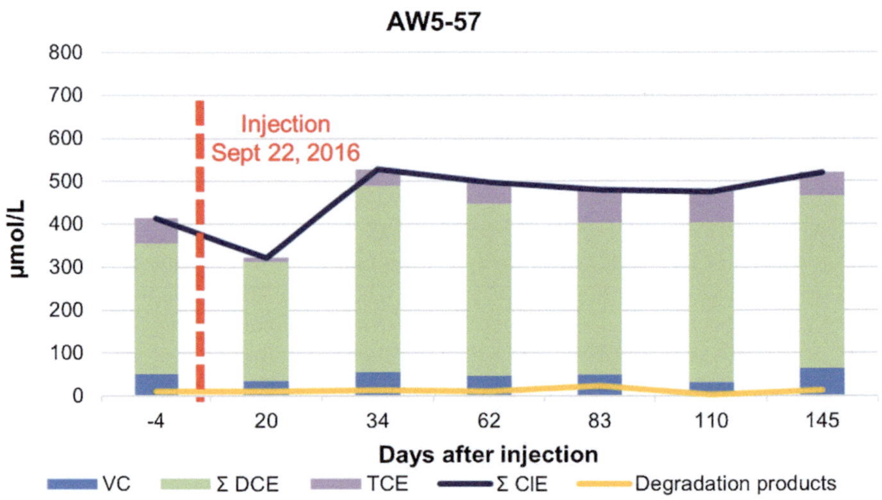

Fig. 5.12 Reduction of ClE in AW5–57 after the NANOCOMPOSITE with CMC application. (Adapted from Vacková 2018)

322 µmol/L. Then the concentration rose to 528 µmol/L 34 days after the injection and stayed constant. The reductive conditions in the subsurface were not strong enough to establish a long-term reduction of main contaminants (Fig. 5.12, Vacková 2018). Therefore, the final reduction of ClE in AW5–57 was negative and increased by 25% at the end of the monitoring.

The concentration of TCE was reduced by 7% compared to the pre-application values within 5 months of the application. The concentrations of DCE and VC rose by 14% and 26%, respectively.

In the borehole AW5–58, the reduction of ClE was even flatter (Fig. 5.13, Vacková 2018). After the injection, a slight decrease in the total ClE (14%) was registered. Considering the detection of a maximum concentration of iron on site,

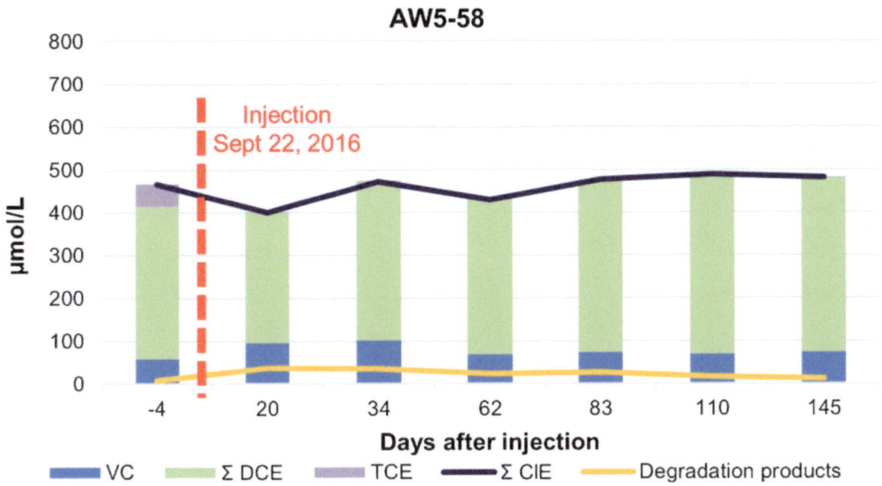

Fig. 5.13 Reduction of ClE in AW5–58 after the NANOCOMPOSITE with CMC application. (Adapted from Vacková 2018)

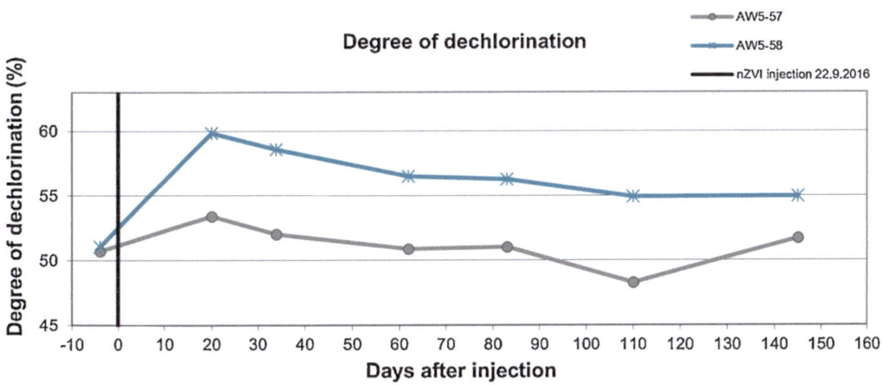

Fig. 5.14 Effect of the application of NANOCOMPOSITE with CMC on degree of dechlorination. (Adapted from Vacková 2018)

more significant decrease in the ClE concentration was expected. One month later, the concentration rose back to the pre-application values and remained almost without a change until the end of the monitoring process. Approximately 5 months after the injection, the total concentrations of ClE were little higher than before the application (by 3%). Even though the total reduction of ClE in AW5–58 was not effective, the nZVI application had a reductive impact on the concentrations of TCE, which went down by 97% in total.

The parameter degree of dechlorination, showed in Fig. 5.14 (Vacková 2018), demonstrates a reduction effect of the NANOCOMPOSITE with the CMC injection. The reduction process was confirmed in all boreholes. Even though the concentration of ClE was reduced mainly only over the first 20 days after the injection, the values of the degree of dechlorination greatly increased (above pre-application values for whole 146 days of the field test). The sharpest increase after the injection was observed in the borehole AW5–58, where the degree of dechlorination rose from 51% to 60% during the first 20 days. After that, the degree decreased gradually and leveled off at around 55%.

5.4 Discussion

5.4.1 Effects on E_H

On the basis of the decrease in E_H after the injection, the application of NANOFER STAR seems to be the most effective. The reductive process was observed in four out of six boreholes and had the strongest course. The effect of this injection lasted up to 30 days, after that E_H remained constant. The NZVI-C3 application had almost no influence on the reduction potential. A significant decrease was registered only in the borehole AW5–60, which is located to the west of the injection centre. The NANOCOMPOSITE with the CMC application caused a dip of reduction potential; however, the reduction was not as strong as during the NANOFER STAR application. In this case, the reductive conditions lasted for 13 days.

5.4.2 Evaluation of the Contaminant Reduction

The main contaminants in the pilot test area were TCE and DCE. After each application, concentrations of ClE plummeted immediately; however, the decrease in the ClE concentration was usually not long-term. It seems that the decrease in TCE caused production of DCE molecules. Therefore, in most of the boreholes, the concentration of DCE rose gradually after the injection.

After the NANOFER STAR application, the sharpest decrease in ClE was monitored in AW5–57 and AW5–58 boreholes and lasted for 7 days. The NZVI-C3 application caused a sharp, short-term drop in ClE in AW5–58, AW5–60, and AW5–62 boreholes. After 7 days, the values of ClE recovered quickly. The NANOCOMPOSITE with CMC application had the smallest impact on the ClE concentrations. There was a slight fall in TCE after the injection, accompanied with a drop in DCE 20 days after the injection.

Table 5.1 shows the efficiency of each nZVI application in the ClE reduction (counted for concentrations of all boreholes). The concentration of 4926 μmol/L of ClE, which is the sum of ClE on the pilot test area measured before the injections started, was considered as 100%.

We can see that the NANOFER STAR application was the most efficient because of the 39% decrease in the total ClE concentrations. The nZVI application showed

Table 5.1 Efficiency of nZVI injections. (Reproduced from Vacková 2018)

nZVI application	Mean of sum ClE (μmol/L)	Relative reduction (%)
NANOFER STAR – Before injection	4926	100
NANOFER STAR – 7 days after injection	830	83
NANOFER STAR – 89 days after injection	2987	**39**
NZVI-C3 – Before injection	3605	100
NZVI-C3–7 days after injection	1205	66
NZVI-C3–90 days after injection	2681	**26**
NANOFER STAR with CMC – Before injection	2410	100
NANOFER STAR with CMC – 20 days after injection	1765	27
NANOFER STAR with CMC – 83 days after injection	2207	**8**

very high reductive influence on the TCE (decrease of 84%) and on VC (decrease of 60%) on the whole pilot test area. The decrease in the TCE concentration lasted up to 1 month and even after its concentration remained low.

Compared with the other injections, the NZVI-C3 injection caused 26% decrease of total contaminant. The reduction of TCE turned out well significantly in the borehole AW5–57, where it reached 71%. The concentration of DCE was oscillating during the monitoring period and its reduction varied between 25% to 53% in the pilot test area (with the exception of the borehole AW5–60) 4 months after the injection. Overall, the reduction of ClE was not long-term. One week after the injection, the contamination started rising again. Moreover, no significant increase in the VC was observed, as it would be expected because of the degradation of DCE to VC. It was probably caused by the process of sorption. Thanks to the carbon-iron structure of nanoparticles, the contaminants were likely inhibited by sorption on the carbon coat. It is important to point out that the process of sorption does not serve as the final solution because of a possible process of desorption.

The NANOCOMPOSITE with CMC caused only 8% decrease in the contaminants. The reduction of TCE and DCE was observed in several boreholes. The application had a long-term reductive impact on the TCE concentrations in AW5–58 (reduction of 97%). The concentration of VC was not affected by the application and its values were fluctuating during the monitored period.

5.4.3 Degree of Dechlorination Evaluation

The degree of dechlorination is an important parameter to be considered in this evaluation. It is a statistical parameter that expresses changes of contaminants as well as products of dechlorination. The increasing trend of the degree of dechlorination indicates an ongoing process of dechlorination.

The most significant increase of the degree was observed after the NANOFER STAR application. There was a short increase of the degree of dechlorination in boreholes AW5–57 (25%), AW5–58 (17%), during the first 7 days. After 1 week, the degree of dechlorination decreased slowly. In the borehole AW5–58, the rising trend lasted up to 28 days when the degree increased to 66% from previous 46% before the injection. Overall, the values of the degree remained above the pre-application values for 2 months, which reflects the duration of the process of dechlorination.

In contrast, there were no signs of the ongoing process of dechlorination after the NZVI-C3 application. It seems that it was caused by the inhibition when the contaminant was adsorbed on carbon coat of the particles. Therefore, no products of dechlorination emerged.

Not very strong increase in the degree of dechlorination lasting up to 20 days was shown after the NANOCOMPOSITE with CMC. The addition of CMC into the suspension of nZVI, which prevents iron from oxidation, probably decreased the reactivity of the particles. Longer but less intensive course of the degree of dechlorination was observed. Therefore, the process of dechlorination was probably present in the subsurface but was not as striking as it was after the previous injection of bare NANOFER STAR nZVI.

5.5 Conclusions

Three separate nZVI applications determined remedial potential of particular materials. The NANOFER STAR nZVI application, which caused the temporary reduction of 39%, was considered the most efficient. It proved the applicability of NANOFER STAR nZVI for remediation of ClE. NANOFER STAR application showed a very high reductive potential on TCE (decrease of 84%) and VC (decrease of 60%) throughout the whole pilot test.

The NZVI-C3 material showed its ability for ClE dechlorination, especially by sorption on carbon. The process of sorption increases the applicability of this method in sites contaminated by pollutants that need more complex method of remediation such as heavy metals. The addition of CMC into the suspension of nZVI leads to the improvement of mobility of nZVI particles but also deterioration of reductive potential of nZVI due to occupation of its surface by CMC. Therefore, CMC cannot be recommended as a convenient and economical stabilizer especially in case of relatively expensive nZVI.

References

Adetutu EM, Gundry TD, Patil SS, Golneshin A, Adigun J, Bhaskarla V, Aleer S, Shahsavari E, Ross E, Ball AS (2015) Exploiting the intrinsic microbial degradative potential for field-based *in situ* dechlorination of trichloroethene contaminated groundwater. J Hazard Mater 300:48–57. https://doi.org/10.1016/j.jhazmat.2015.06.055

Agathos S, Reineke W (eds) (2003) Biotechnology for the environment: soil remediation. Focus on biotechnology, vol 3B, 1st edn. Springer, Dordrecht

Dolinová I, Czinnerová M, Dvořák L, Stejskal V, Ševců A, Černík M (2016) Dynamics of organohalide-respiring bacteria and their genes following *in-situ* chemical oxidation of chlorinated ethenes and biostimulation. Chemosphere 157:276–285. https://doi.org/10.1016/j.chemosphere.2016.05.030

Kvapil P, Gaňa P, Růžička P (2011) Souhrnná zpráva provozu sanace podzemních vod mraku 5 za rok 2010: SPOLCHEMIE Ústí nad Labem. Aquatest a.s., Praha

Kvapil P, Valerián R, Růžička P, Gaňa P (2014) Souhrnná zpráva za rok 2013: Opatření vedoucí k nápravě starých ekologických zátěží v areálu společnosti SPOLCHEMIE Ústí nad Labem. Aquatest a.s., Praha

Lewin K, Blakey NC, Cooke DA (1990) The validation of methodology in the determination of methane in water – final report. Water Research Centre, Marlow

Paul L, Smolders E (2014) Inhibition of iron (III) minerals and acidification on the reductive dechlorination of trichloroethylene. Chemosphere 111:471–477. https://doi.org/10.1016/j.chemosphere.2014.04.057

Stroo HF, Ward CH (eds) (2010) In situ remediation of chlorinated solvent plumes, SERDP ESTCP environmental remediation technology, 1st edn. Springer, New York. https://doi.org/10.1007/978-1-4419-1401-9

Vacková N (2018) Comparison of effectiveness of three application of zero-valent iron nanoparticles for remediation of groundwater polluted by chlorinated ethenes. Master's thesis. Charles University, Faculty of Science, Prague, Czech Republic

Chapter 6
Field Study II: Pilot Application of nZVI/DC-Combined Methods at Aargau Site

Vojtěch Stejskal, Jaroslav Nosek, Miroslav Černík, Petr Kvapil, and Pierre Matz

Abstract Combination of nZVI and electrical field was successfully tested in laboratory and afterwards on several sites in the Czech Republic. With successful references we went abroad to Aargau site in Switzerland (November 2017). Aargau site is characteristic by deep aquifer with bottom in 16–17 meters of depth and 3–4 m of thickness. It is contaminated by chlorinated ethenes in concentrations from 12 to 70 mg/L in total, (especially tetrachloroethene and trichloroethene). In Switzerland we injected NANOFER STAR DC, a product designed for application with direct current (DC). In total, 25 m^3 of 10 g/L nZVI was injected via packer system into 5 wells creating a permeable reactive barrier. A system of electrodes was installed directly into the monitoring wells due to very deep groundwater level. Nine-month-long monitoring was held on Aargau site including a tracer test confirming direction of groundwater passing through the barrier. DC application has a positive impact on efficiency and longevity of used nZVI particles. Permeable reactive barrier in Switzerland was able to degrade inflowing chlorinated hydrocarbons concentration in range of 20–69 mg/L with efficiency 92–98% for first 7 months since nZVI application and up to 9 months with efficiency 74%. These results showed a significant improvement of efficiency of this method compared to the conventional use of nZVI.

V. Stejskal (✉)
Institute for Nanomaterials, Advanced Technologies and Innovation, Technical University of Liberec, Liberec, Czech Republic

Photon Water Technology s.r.o., Liberec, Czech Republic

AQUATEST a.s., Prague, Czech Republic
e-mail: vojtech.stejskal@photonwater.com

J. Nosek · M. Černík
Institute for Nanomaterials, Advanced Technologies and Innovation, Technical University of Liberec, Liberec, Czech Republic

AQUATEST a.s., Prague, Czech Republic

P. Kvapil
Photon Water Technology s.r.o., Liberec, Czech Republic

P. Matz
Solvay SA, Brussels, Belgium

J. Filip et al. (eds.), *Advanced Nano-Bio Technologies for Water and Soil Treatment*, Applied Environmental Science and Engineering for a Sustainable Future,
https://doi.org/10.1007/978-3-030-29840-1_6

Keywords Chlorinated hydrocarbons · Electrokinetics · Chemical reduction · Nanoremediation · nZVI · Tracer test

6.1 Introduction

The use of nZVI for contaminated groundwater remediation is nowadays one of standard methods for in situ treatment or stabilization of chlorinated hydrocarbons, selected inorganic anions, heavy metals and others. The effectiveness of this method depends on the geological and hydrogeological conditions, which affect both the reactivity of nanoparticles with contaminants and their spreading in the subsurface. In the case of low permeability environment, the migration of nanoparticles is very limited and the efficiency of the method might be therefore reduced. The application of nanoparticles supported by direct current (DC) electric field of certain field intensity leads to a significant increase of lifetime and total efficiency of this combined remediation technique.

The site selected for the pilot test of nZVI enhanced by electrokinetics is located in canton of Aargau, Switzerland. The site is situated a few hundred meters from the river Rhine. Until 2004 a mercury cell chlor-alkali electrolysis plant operated at the site, and chlorinated solvents were produced in a dedicated production unit from 1940s to 1970s. The soil and groundwater around this production and storage units have been contaminated by accidentally spilled chlorinated products or by contaminated wastewater leaching from the sewer. The main contaminants found in the area are tetrachloroethene (PCE), trichloroethene (TCE), hexachloroethane (HCA), and hexachlorobutadiene (HCBu).

Within the site, the groundwater flows to the west towards the river. The pilot test area consists of unconfined alluvial sediments of sand and gravel, but stones and boulders are also abundant. The marlstone bedrock (Opalinus Clay) is found at 16.5 m bgl and the average water table at 13 m bgl. The hydraulic conductivity is high causing groundwater flow velocity in order of meters per day (as determined based on the tracer test).

6.2 Materials and Methods

The monitoring system on the Aargau site consisted of nine monitoring/application wells and three wells equipped by micropump system (Fig. 6.1). Each micropump well had two sampling horizons. The monitoring/application wells were approximately 16–17 m deep and had an open screen only at the bottom 70–100 cm. Micropump wells collected water from different depth horizons between 15 and 16.5 m below the ground level. The wells and their use in the monitoring campaigns are listed in Table 6.1.

As the diameter of the monitoring wells is only 50 mm DIN, it was necessary to install permanent sampling pumps into the wells together with electrodes. The

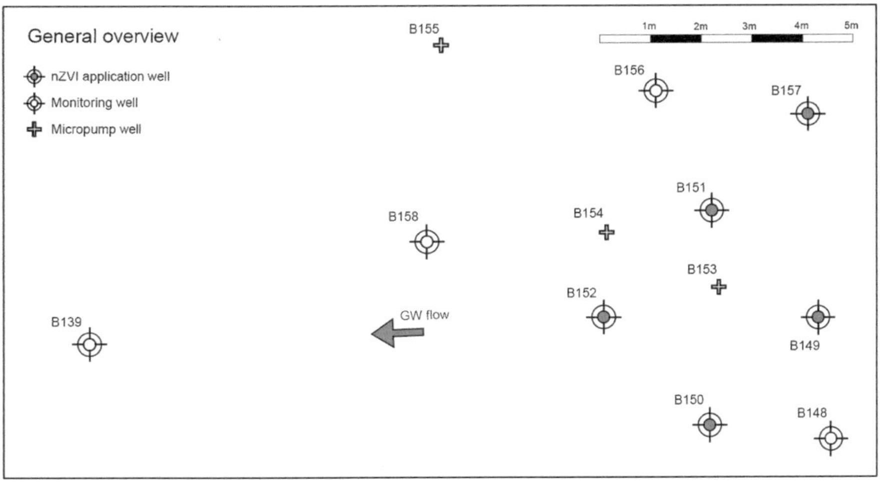

Fig. 6.1 General overview of the site

Table 6.1 Wells and their use during the field application

Use of well	List of wells
Monitoring and application wells (cased wells)	B139, B148, B149, B150, B151, B152, B156, B157, B158
Micropump wells	B153, B154, B155

groundwater was monitored by conducting groundwater level measurement; the measurement of ORP, pH, conductivity, temperature, and dissolved oxygen; and analysis of chlorinated hydrocarbons (CHC), total iron, bromides, and nontoxic degradation products of in situ chemical reduction. The samples were always taken the moment the above-mentioned parameters were stabilized.

6.3 Tracer Test

A tracer test was performed with the aim to trace the migration pathway of the contaminants, estimate travel velocity, and dispersion parameters in the highly contaminated groundwater of B149, where CHC exceeded 60 mg/L in total. Three kilograms of KBr diluted in 1000 L of technological water was applied. The concentration of bromide cation in the solution was 2 g/L. The application of the tracer into B149 was performed on a stable water level. The natural groundwater level was increased by 50 cm and the continuous application took 23 h.

The isolines of tracer at different times (day 1, day 2, day 6, and day 13; Figs. 6.2, 6.3, 6.4 and 6.5) show relatively constant progress of the tracer movement. Figure 6.6 illustrates the tracer-breakthrough curves for Br^- tracer. The trends of the tracer-

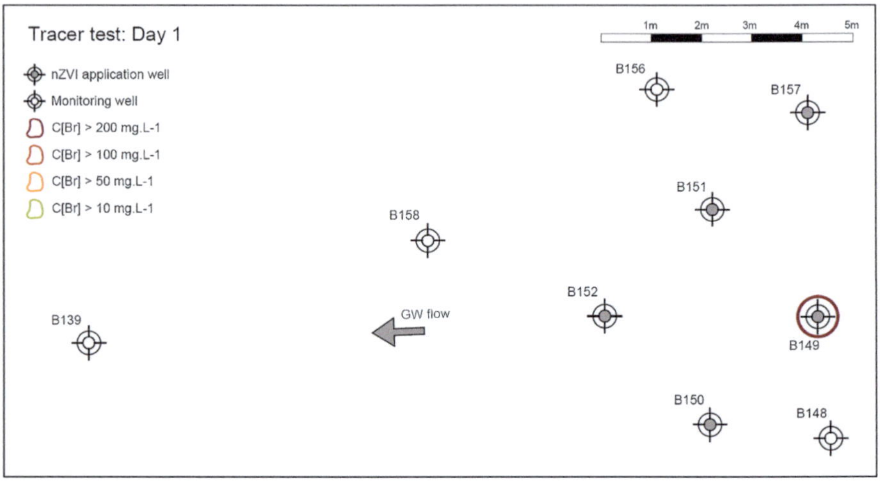

Fig. 6.2 Isolines of tracer concentrations in the start of the application (day 1)

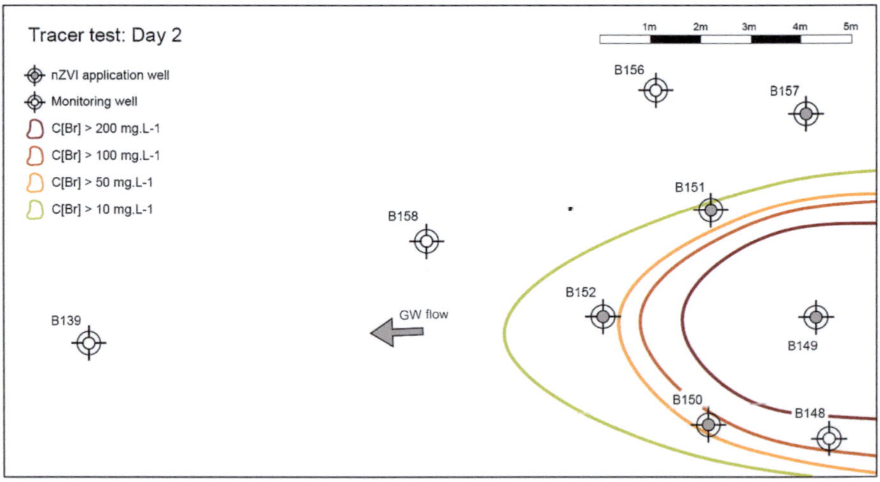

Fig. 6.3 Isolines of tracer concentrations at day 2

breakthrough curves are different for each well. In the case of the B152 well, a gradual decline can be seen after reaching the peak tracer concentration of 245 mg/L 9 days after the start of the tracer test. For well B139, the peak tracer concentration is significantly lower (64 mg/L on 13th day) and the whole peak is much broader, which is due to a longer distance and bigger dispersion.

The parameters determined by site monitoring during the tracer test were evaluated by QTracer2 software and are enlisted in the following output parameters (Table 6.2).

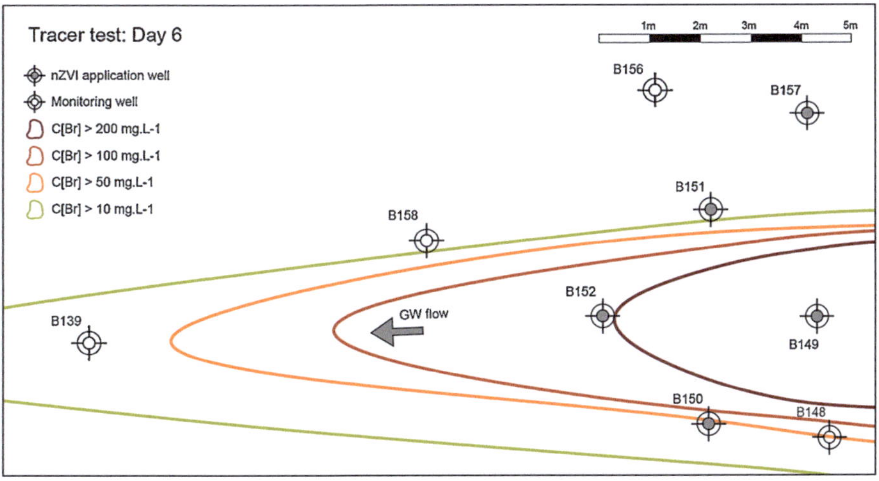

Fig. 6.4 Isolines of tracer concentrations at day 6

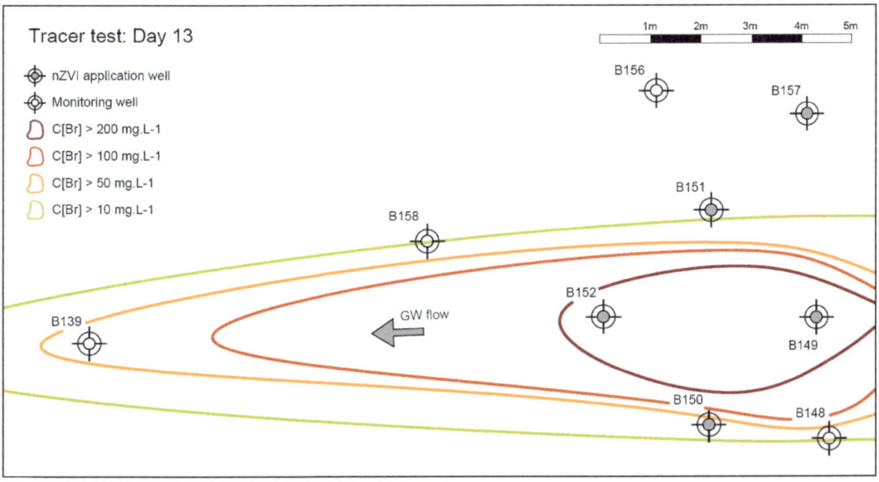

Fig. 6.5 Isolines of tracer concentrations at day 13

The overall recovery of the injected tracer is 4.31%, from which approximately 57% of the tracer was recovered in the B152 well and 43% in the B139 well. In general, the tracer progressively moved from east to west following the direction of the groundwater flow within the area of interest. The closer proximity of the B152 well to the application well (B149) is demonstrated by several factors: the first arrival of the tracer (reaching the B152 well 1 day after the start of the test) and by both the peak tracer concentration and time to peak tracer concentration (245 mg/L on 9th day). On the other side, the tracer achieved a peak concentration of 64 mg/L on the 13th day after the start of the tracer test, first arrival of the tracer into B139 was

Fig. 6.6 Time-dependent tracer concentration at two selected wells

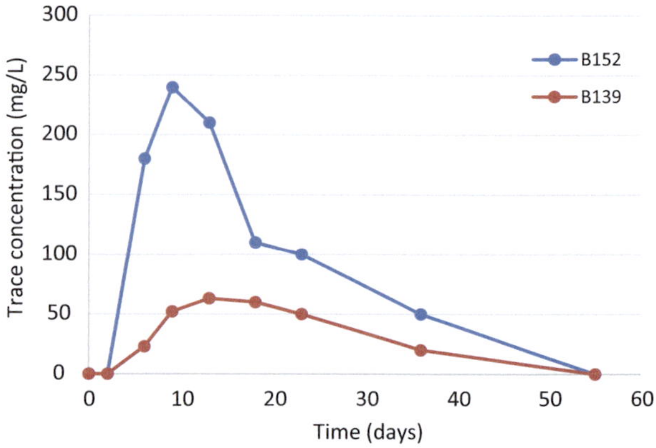

Table 6.2 Outputs from the Qtracer2 software for the B139 and B152 wells

Parameter	B152	B139
Recovery of injected tracer	2.46%	1.85%
Distance from input to outflow point	4.23 m	14.42 m
First arrival	1 day	Not defined
Time to peak tracer concentration	9 days	13 days
Peak tracer concentration	245 mg/L	64 mg/L
Mean tracer transit time	18.3 days	21.4 days
Mean tracer velocity	0.23 m/day	0.70 m/day
Péclet number	4.44	6.84
Longitudial dispersion coefficient	0.26×10^{-5} m^2/s	1.7×10^{-5} m^2/s

5 days after its application, but it was only due to lack of monitoring between day 0 and day 5. The mean tracer transit time, which marked the time frame of the tracer occurrence during its underground movement, was slightly greater in the case of the B139 well (at 21.4 days), while its value was recorded at 18.3 days for the B152 well. A significantly larger difference was observed in the values of the mean tracer velocity, which was three times greater for the B139 well (at 0.70 m/day) than in the case of the B152 well (0.23 m/day). The travel time to well B139, which is about three times as far as well B152, was only slightly longer than to B152. This effect can be caused by a preferential zone (trough filled by gravel). The migration does not take place directly from the source well to B139, and, therefore, the tracer is diluted (concentration about four times smaller than the B152 well). The dispersion coefficient, calculated on the basis of the travel velocity and dispersivity, is an order of magnitude greater in the case of the B139 well (0.17×10^{-4} vs 0.26×10^{-5} m^2/s) compared the B152 well, for the same reason. Also the Péclet number value of 6.84 for B139 represents the domination of advection compared with mixed diffusion and advection for the B152 well, which also indicates preferential advective flow to B139.

6.4 nZVI Application and System Installation

Pilot injection of nZVI was performed in November 2017. The packer injection was put into wells B149, B150, B151, B152, and B157. Apart from these wells, nZVI was visually observed also in well B148 during the injection, this confirmed a good distribution of nanoparticles in the subsurface (and confirmed existence of preferential migration zones). 250 kg of NANOFER DC in total (NANO IRON, s.r.o., Czech Republic) was injected into the subsurface in concentration of 10 g/L with a total injected volume of 25 m^3. Into each injection point, 5000 L of nZVI suspension was injected. A reactive barrier (RB) of electrokinetically enhanced nanoscale zero-valent iron particles was created.

There were three separate polygons for DC application. Each polygon had its own power supply.

- The first polygon consisted of one anode (B149) and one cathode (B148);
- The second consisted of two anodes (B150, B158) and one cathode (B152);
- And the third polygon consisted also of two anodes (B151, B157) and one cathode (B156).

All electrodes were permanently installed in the above-mentioned wells, connected by copper wiring to the DC units. The wire-connected wells with the DC units on the ground were protected by plastic cover. The DC units had their own kWh counters with automatized logging and saved the data throughout the DC application.

6.5 Results

6.5.1 Physicochemical Parameters of Groundwater

As there were another two nZVI injections performed in the past, the groundwater of the Aargau site was in reductive conditions even before the current in situ injection. However, the nZVI application intensified the reductive conditions in the groundwater—the redox potential decreased in cased wells from −200 mV (average in October 2017) to −400 mV (average in October 2018). Dissolved oxygen decreased significantly in all monitoring points until May or July 2018 (Fig. 6.7).

6.5.2 Contamination

The courses of contamination concentrations in the inflow area of RB are documented in Fig. 6.8. This zone is the place where contaminated water enters the pilot test area. The increases in the concentration of particular contaminants are

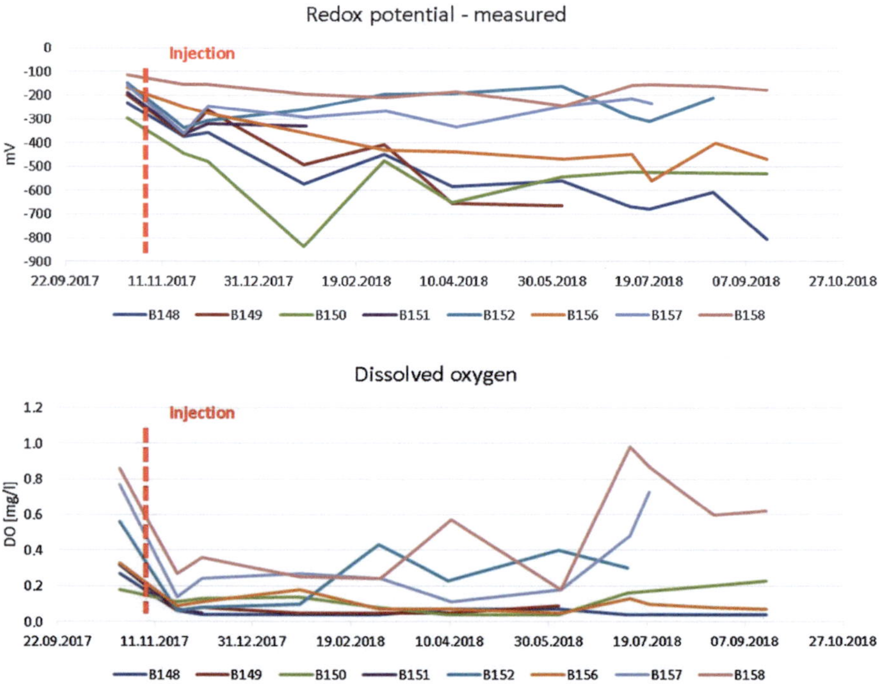

Fig. 6.7 Results of redox potential and dissolved oxygen measurements

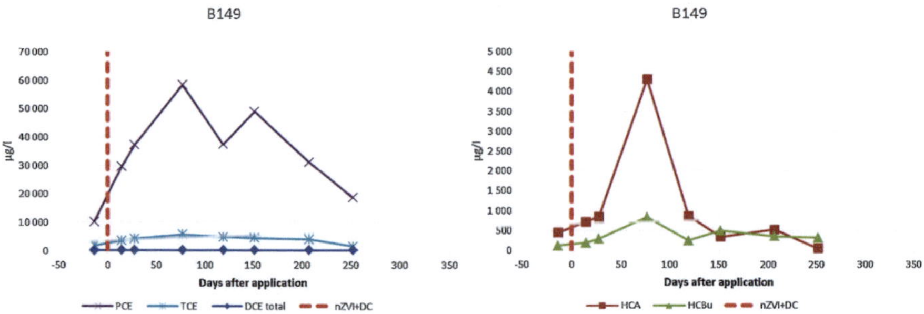

Fig. 6.8 CHC concentrations in B149 – inflow into RB

partially caused by the application itself, which has a significant impact on the actual dynamics of the groundwater; it could also be caused by the release of contaminants temporarily caught in the unsaturated zone. The continual increase in the CHC contamination in well B149 for 75 days after the nZVI application can be caused by the combination of the above-mentioned factors or it could also be the result of an inflow of new contamination from an area further upstream of RB. Due to the lack of monitoring wells further upstream of RB, this claim is difficult either to prove or refute.

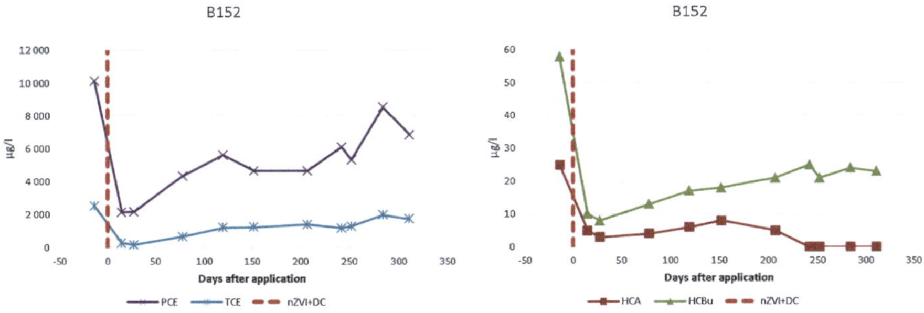

Fig. 6.9 CHC concentrations in B152 – central part of RB

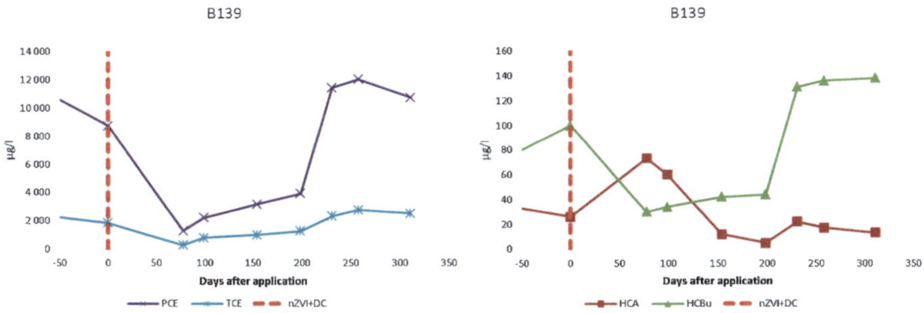

Fig. 6.10 CHC concentrations in B139 – outflow of RB

The CHC concentrations increased from 12 mg/L to 69 mg/L whilst 90% of the CHC amount was represented by PCE. The decreasing level of the groundwater did not allow sampling of B149 after the day 250 until the end of the pilot test.

Figure 6.9 shows decreases in the central area of RB (represented by B152). In this area, the present contaminants are in contact with nanoscale zero-valent iron and direct current already enough time to be reduced. The concentration of the chlorinated ethenes decreased significantly after the system installation. However, due to the flow of the contaminated water from the inflow, the CHC concentrations increased. For this reason, the concentrations of the chlorinated ethenes, HCA, and HCBu decreased significantly in this area. This is a very positive result of the application. HCA (originally in concentration of 25 µg/L) was reduced completely to values below the lower detection limit of laboratory analysis. HCBu shows reduction (311 days after injection) of about 65% at the end of the pilot test.

The outflow area of the reactive barrier is characterized by well B139, which is situated approximately 14 meters downgradient of B149. The groundwater in this area is not under a direct influence of the electrokinetically enhanced nZVI particles; B139 represents the treated water leaving the application area of RB.

This decrease continues significantly—the CHC concentrations decreased from 11 mg/L to 1.8 mg/L in total (Fig. 6.10). A similar effect of increasing the CHC concentration as in the case of B152 was observed here, but the duration of the groundwater treatment was longer.

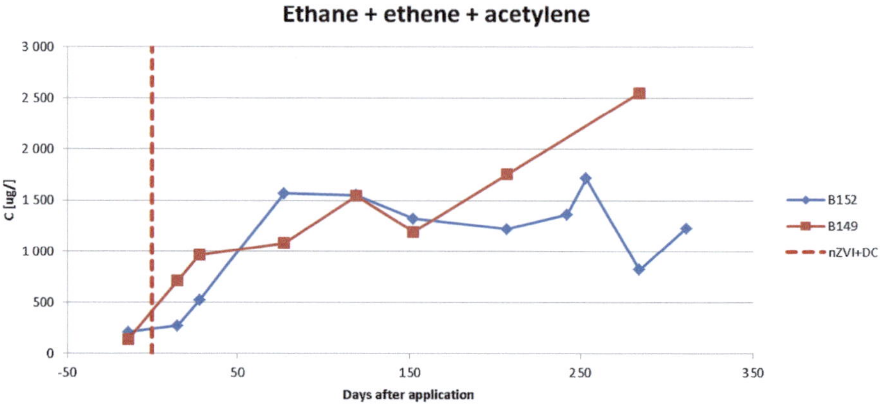

Fig. 6.11 Ethane, ethene, and acetylene summary concentrations in B149 and B152

Figure 6.11 documents courses of nontoxic degradation products (sum of ethane, ethane, and acetylene) of dechlorination in wells B149 and B152—wells situated in the inflow and central area of RB; thus, this is the place where the greatest production of ethane, ethene, and acetylene is expected.

The concentrations of nontoxic degradation products increased equally on both wells during the first 155 days after application—B149 increases from 135 μg/L to 1200 μg/L and B152 from 205 μg/L to 1500 μg/L. B152 afterwards stagnates in time until the end of the pilot test 311 days after the nZVI application, in contrast to B149, where the concentration of the degradation products continuously increased in time until the end of the pilot test. At the end of the pilot test, the concentrations of the degradation products were observed in the inflow area 19 times higher than before the injection of nZVI, respectively 6–7 times higher in the central area.

Documented increases of ethane, ethene, and acetylene confirmed degradation of chlorinated hydrocarbons to nontoxic products of its decomposition.

6.5.3 Mass Balance

Prior to the technology installation, we carried out multiple sampling of CHC (May 2016, August 2016, February 2017, June 2017, and October 2017—just before the application), which showed that the concentrations in all the monitoring points were throughout the 18-month period of sampling relatively stable in the whole area. The total CHC concentration was around 10 mg/L. Differences of CHC concentrations between B139, B149, and B152 in the particular periods were in order of analytical error (max. 10–15%).

Figure 6.12 shows the data collected after the application of nZVI and creation of reactive barrier. The concentration scale and monitoring points are the same as in the previous graph, only the sampling period is shorter. There are results of laboratory

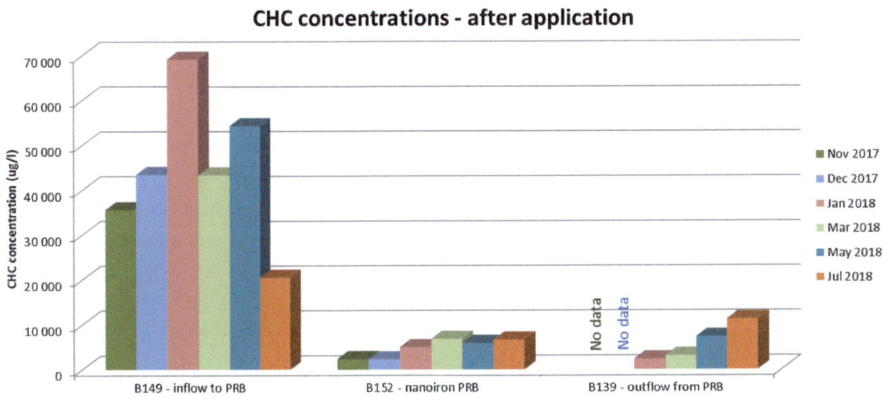

Fig. 6.12 Overview of CHC concentrations after nZVI application

analyses performed on the samples collected in November 2017, December 2017, January 2018, March 2018, May 2018, and July 2018. Unfortunately, no samples of B139 were collected in November and December 2017.

We have observed a significant increase of CHC concentrations in the inflow area of RB. Such an important and continuous increase of contaminants is difficult to explain without a possibility of monitoring the groundwater in the wider area from where the contaminated water is flowing. It is possible to suggest several different processes that could play a role in the observed CHC increase:

- Washing of the unsaturated zone and release of CHC temporarily fixed in this zone
- Dilution of CHC from free phase present on the aquifer bottom
- Increase of CHC concentration due to drought and decreasing groundwater level
- Inflow of contaminated groundwater

There are probably more reasons for such an increase. The fact that even such elevated concentrations of CHC were almost completely treated on RB is a very positive result of the application:

- Comparison of the inflow and outflow concentrations of CHC for the first 6 months after the injection proved its reduction between 87% and 97% (sum of CHC)
- After 8 months, the reduction ratio of the summary CHC concentration between the inflow and outflow decreased to 46%, even though the water flowing into RB contained PCE (90%) and TCE (7%) and on the RB's outflow, the water was contaminated with a mixture of PCE (35%), TCE (11%), and DCE (53%)

Unfortunately, it was not possible to collect relevant groundwater samples from this well after July 2018 because of a low inflow of groundwater into well B149.

For estimating the amounts of CHC treated, it is also necessary to estimate the thickness of the aquifer that depends on the groundwater level. Because of permanently installed electrodes and sampling pumps in all the monitoring wells except

B139, this well was used for calculating the aquifer thickness in time. The groundwater level in the well is decreasing in time, mean value for the interval of measurement is 3.5 m.

The inflow groundwater velocity was defined by a tracer test performed in the summer 2018—0.7 m/day between B149 and B139. Effective porosity of sand/gravel Quaternary aquifers was about 30% (0.3), this value was used for mass balance calculation. The total inflow in m^3/day into each segment was calculated by multiplication of width, thickness, inflow velocity, and effective porosity.

Calculated mass balance gives the following results. In the B149 segment, there was injected 100 kg (two application wells) of nZVI degraded 18.9 kg of CHC. In B148 segment, there was only one application well, and therefore only 50 kg of nZVI applied, which degraded approximately 8.5 kg of CHC—about a half of what was degraded in the B149 segment. In the northern part of the pilot site, there was injected 100 kg of nZVI and if we assume a similar efficiency of electrokinetically enhanced nZVI, total degraded amount is about 45.5 kg of CHC on the whole site.

6.6 Conclusions

The combined technology of nanocomposite and electrokinetics was successfully applied on Aargau site in northern Switzerland in November 2017.

After creation of a diffused reactive barrier, there was a significant increase of CHC concentration in the inflow area observed. A continual increase lasted 75 days, and the CHC concentration increased from 12 mg/L to 69 mg/L.

With the aim to track the water flow of such contaminated water, we conducted a tracer test. The applied tracer was found in B152 and further in B139 confirming a contaminated groundwater flow from B149 through the RB to B139 with the estimated velocity of 0.7 m/day.

The reduction process of CHC on the nZVI reactive barrier lasted for the first 6 months after the RB installation and the values of CHC reduction were in the range of 87–97%. In July 2018, (8 months after the application) reduction of about 46% was still observed. The CHC composition shows that while the inflowing groundwater was contaminated mainly with PCE (90%), the outflowing groundwater was contaminated with DCE (53%). The concentrations of the nontoxic degradation products of the in situ chemical reduction increased up to 2500 μg/L in the RB area, that means 19 times higher compared with the state prior to the test.

On the basis of the results of the performed tracer test and the measured reduction of CHC contamination, the mass balance of degraded CHC was estimated. As the tracer test described well only the southern part of the pilot test area, the calculation of the contamination mass balance was performed only in this part, where 150 kg of nZVI was applied. The principal result from the southern part is that each 50 kg of the injected nZVI degraded 9.1 kg of CHC, thus degrading approximately 45 kg of CHC in total.

Part II
Oxidative Technologies

Chapter 7
Introduction to Oxidative Technologies for Water Treatment

Marta I. Litter

Abstract Chemical oxidation and advanced oxidation technologies are important processes for the abatement of refractory and/or toxic pollutants in wastewaters. The species participating in advanced oxidation technologies are powerful radicals, such as the hydroxyl radical, $HO^{\bullet}$, which can attack virtually all organic compounds at very high reaction rates. In this chapter, oxidative technologies, such as permanganate, ozonation, Fenton and related processes, electrochemical oxidation, γ-radiolysis and electron-beam treatment, nonthermal plasma, persulfate, wet air oxidation, supercritical water oxidation, ultrasound, zero-valent iron, ferrate, water photolysis under vacuum ultraviolet, periodate, chlorine and heterogeneous photocatalysis, some of them combined with UV light, are briefly described, together with examples of their applications. The most important and recent papers are cited together with review papers. These oxidative technologies can process wastewaters resistant to conventional treatments and are complementary to them.

Keywords Chemical oxidation · Advanced oxidation processes · Fenton · Ozonation

7.1 Introduction

Very strict government environmental regulations have led, over the last few decades, to the development of new and more effective water purification technologies. In most cases, anthropogenically polluted water can be efficiently treated by biological methods, activated carbon adsorption or other adsorbents, or by conventional physical and chemical treatments (flocculation, filtration, thermal oxidation, etc.). Nevertheless, in some cases, these procedures are not adequate for reaching the

M. I. Litter (✉)
Instituto de Investigación e Ingeniería Ambiental, Universidad Nacional de San Martín, CONICET, Buenos Aires, Argentina
e-mail: marta.litter@gmail.com; litter@cnea.gov.ar

J. Filip et al. (eds.), *Advanced Nano-Bio Technologies for Water and Soil Treatment*, Applied Environmental Science and Engineering for a Sustainable Future,
https://doi.org/10.1007/978-3-030-29840-1_7

degree of purity required by law or by the final use of the water (e.g., drinking water, irrigation, recreational activities). In addition, some of the contaminants containing, e.g., aromatic, sulfur, nitrogen, oxygen, double-bond groups, or high molecular compounds are difficult to be transformed by these technologies. Groundwater and soils contaminated with organic compounds such as solvents are often difficult to remediate. The standard technology for groundwater treatment, known as pump-and-treat remediation, is expensive and takes years to complete. Moreover, the contaminants held in low permeability zones are often difficult to access by standard pump-and-treat systems, thus prolonging the time required for remediation and increasing the cost of the remediation. Conventional technologies such as separation by gravity, centrifugation, coagulation, flotation, adsorption, biological methods, filtration methods, thermal oxidation, etc., generate byproducts like sludge, syrups, or salts that need further treatment or that can be expensive and dangerous to handle. Furthermore, some types of effluents cannot be treated directly by biological technologies because of the presence of high levels of recalcitrant compounds, like aromatic hydrocarbons.

In those cases, chemical oxidation technologies are efficient processes to treat systems containing these recalcitrant pollutants. Chemical oxidation has been largely applied to wastewater treatment by means of oxidants such as hydrogen peroxide, potassium permanganate, ozone, and combinations of ozone with other oxidants. These systems have the potential to offer rapid (weeks to months) removal of the contaminants. Other more powerful processes efficient at treating water polluted with very resistant contaminants are the so-called advanced oxidation technologies or processes (AOTs, AOPs). They have yielded very good results in industrialized countries and are beginning to be employed in developing regions (Legrini et al. 1993; Huang et al. 1993; US EPA 1998; Calgon Carbon Corporation 1996; Bolton and Cater 1994; Domènech et al. 2004; Litter 2005; Oppenländer 2003; Wang and Xu 2012; Boczkaj and Fernandes 2017; Tijani et al. 2014; Stefan 2018a).

AOTs concept was initially established by Glaze (1987) and Glaze et al. (1987, 1992), who defined AOTs as processes involving generation and use of powerful transitory species, principally the hydroxyl radical ($HO^•$), able to change in the structure of chemical species. These species can be generated photochemically (including the use of solar light) or by other forms of energy, and have enough potential to oxidize organic matter (OM). The most important oxidant is $HO^•$, which is the most energetic species after fluorine (Table 7.1) and can attack virtually all organic compounds. $HO^•$ reacts 10^6–10^{12} times more rapidly than alternative oxidants. Some AOTs (heterogeneous photocatalysis, radiolysis, etc.), can also produce reducing species, allowing the transformation of pollutants that are difficult to be oxidized, such as some metal ions or halogenated compounds. However, reduction technologies will not be dealt with in this chapter.

$HO^•$ are able to react by hydrogen abstraction from aliphatic carbon atoms (Eq. 7.1), electrophilic addition to double bonds or aromatic rings (Eq. 7.2), and electron transfer reactions (Eq. 7.3). $HO^•$ can also oxidize metals or metalloids such as As(III). Another active oxygen species like the superoxide radical, $O_2^{•-}$ and its conjugate acid form, the hydroperoxyl (also known as perhydroxyl) radical, $HO_2^•$, are also produced in many AOTs, but they are by far less active than $HO^•$.

Table 7.1 Redox potentials of some oxidants (Legrini et al. 1993)

Species	E^0 (V, 25 °C)[a]
Fluorine	3.03
Hydroxyl radical	2.80
Ferrate	2.20
Ozone	2.08
Persulfate	2.01
Hydrogen peroxide	1.78
Periodate	1.60
Chlorine dioxide	1.57
Permanganate	1.51
Hypochlorous acid	1.49
Perchlorate	1.39
Chlorine	1.36
Dissolved oxygen	1.23
Bromine	1.09
Iodine	0.54

[a]Redox potentials referred to standard hydrogen electrode (SHE)

$$RH + HO^{\bullet} \rightarrow R^{\bullet} + H_2O \tag{7.1}$$

$$R - CH = CH_2 + HO^{\bullet} \rightarrow R - C^{\bullet}H - CH_2OH \tag{7.2}$$

$$RX + HO^{\bullet} \rightarrow RX^{\bullet +} + HO^{-} \tag{7.3}$$

Advanced oxidation technologies are listed in Table 7.2 and are usually divided into non-photochemical and photochemical processes. Each method has advantages and disadvantages that can be exploited given the specific setting of a site or wastewater impacted with a contaminant.

The majority of AOTs can be applied to the remediation and detoxification of low or medium water volumes. Ground-, surface, and wastewaters can be treated. Toxic pollutants such as pesticides, heavy metals, and others can be treated in small-scale mobile treatment units, easy to install in industrial plants. The methods can be used alone or combined with other AOTs or with conventional methods. The use of modular units allows selecting the best technology or combination of technologies to treat specific wastewater. AOTs can also be applied to pollutants in air and soils; they may even allow disinfection or sterilization of bacteria, viruses, and other microorganisms.

AOTs offer several advantages over conventional methods of treatment. One of the most important ones is that the pollutants are not only transferred from one phase to another (as in air stripping or activated carbon treatment), but they are also chemically transformed, which leads, in many cases, to complete mineralization (destruction) of the pollutant. As a consequence, AOTs are very useful to treat refractory pollutants resistant to other treatments, for example, biological technologies. AOTs can treat contaminants at very low levels (ppb), and no reaction byproducts are generally formed. The technologies are also useful to improve the

Table 7.2 Advanced oxidation technologies and other related processes

Non-photochemical processes	Photochemical processes
Oxidation with permanganate	Water photolysis under vacuum ultraviolet (VUV) irradiation
Ozonation in alkaline conditions	UV/hydrogen peroxide
Ozonation with hydrogen peroxide (O_3/H_2O_2)	UV/O_3
Fenton and related processes (Fe^{2+}/H_2O_2)	Photo-Fenton and related processes
Electrochemical oxidation	UV/periodate
γ-Radiolysis and electron-beam treatment	UV/persulfate
Non-thermal plasma	UV/chlorine
Persulfate	Heterogeneous photocatalysis
Wet air oxidation (WAO)	
Supercritical water oxidation (SCWO) (Destaillats et al. 2000a, b)	
Electrohydraulic discharge–ultrasound (US)	
Zero-valent iron (ZVI)	
Ferrate (K_2FeO_4, Fe(VI))	

Adapted from Litter (2005)

organoleptic properties of water or can be just used to discolor dark industrial wastes. In most cases, they consume much less energy than some conventional methods, for example, incineration. Usually, these technologies do not generate sludge, which would require additional treatment or disposal. Nevertheless, it must be taken into account that wastes with relatively high chemical oxygen demand (COD) contents (>5.0 g/L) or with very recalcitrant pollutants cannot be suitably treated by AOTs because they would require large amounts of expensive reagents or electrical energy for irradiation (Andreozzi et al. 1999).

As total destruction of the pollutant is not always required, AOTs are especially useful in two cases: (a) as a pretreatment to transform recalcitrant pollutants into more biodegradable compounds, or (b) as a posttreatment to polish waters before their discharge to the receptor bodies (Scott and Ollis 1995). The main idea of this combination is to use the more expensive technology only during the first or the final step of the treatment in order to reduce the cost.

AOTs can provide effective technological solutions for water treatment; the applicability can cover areas such as: (1) industrial effluents (distillery, agrochemical, kraft-bleaching, pulp and paper, textile dyes, oilfield and metal-plating wastes); (2) hazardous effluents (hospital and slaughterhouse wastes); (3) removal of pathogens and persistent (endocrine-disrupting) pharmaceutical residues from municipal wastewater treatment plant (WWTP) effluents; (4) removal of organic micropollutants such as pesticides, heavy metals (lead, etc.), and arsenic; (5) conditioning and stabilization of biological sludge from WWTPs (Comninellis et al. 2008).

AOTs can be improved through coupling various processes, for example, UV/H_2O_2, UV/O_3, O_3/H_2O_2, $UV/O_3/H_2O_2$, $UV/TiO_2/H_2O_2$, US/Fenton's reagent, UV/Fenton's reagent, WAO/H_2O_2, and electrolysis/Fenton's reagent, because the rate of the formation of reactive species increases compared with the respective individual processes, sometimes with synergistic changes. Sequential application of various AOPs is another way to enhance selectivity. A separation treatment prior to the application of an AOP can be designed—transferring pollutants from the liquid to another phase so that they can be treated more easily. Effluents containing volatile organics or solids may be first subjected respectively to stripping or coagulation, flocculation, sedimentation, and filtration in order to remove these compounds from the effluent prior to the AOP treatment. AOPs can be applied as a pretreatment stage to enhance the biodegradability and reduce the toxicity, followed by a biological posttreatment (Comninellis et al. 2008).

AOTs have actually a variable development and commercialization degree, in constant change as technological advances take place. Currently, UV/H_2O_2, UV/O_3, UV/H_2O_2/O_3, UV/Fenton, and UV/TiO_2 are totally or partially commercialized.

In what follows, permanganate, Fenton, ozonation, ferrate, radical reactions, and photooxidation will be briefly described. Since there is a large amount of references to the subject, only the most important and recent papers have been included, together with review papers to be consulted.

7.2 Non-photochemical Oxidative Methods

7.2.1 Permanganate

Permanganate is one of the most well-recognized chemical oxidants within the environmental industry and has been widely used for treatment of pollutants in drinking water and wastewater applications for over 50 years. The use of permanganate in groundwater treatment applications (in situ chemical oxidation—ISCO) is also a proven, well-documented technology.

Permanganate is traditionally available as sodium or potassium salt. The initial oxidation reactions are independent of pH, but sometimes the pH influences the type of intermediate products (Yan and Schwartz 2000), and the conversion of intermediate oxidation products to CO_2 occurs more rapidly under acidic pH conditions. The main advantage is that final benign reaction products such as carbon dioxide, water, and inorganic salts (e.g., chlorides) are produced via direct electron exchange processes.

Potassium permanganate has an affinity for organic compounds containing carbon–carbon double bonds, aldehyde groups, and hydroxyl groups. The permanganate ion takes electron density from the π bonds in chlorinated alkenes, creating a bridged oxygen compound known as the cyclic hypomanganate ester (Yan and Schwartz 2000). This intermediate ester is unstable and further reacts by a number of mechanisms including hydroxylation, hydrolysis, or cleavage. Under normal pH and temperature conditions, the primary oxidation reactions involve spontaneous cleavage of the carbon–carbon bond. Once the double bond is broken, the highly unstable carbonyl groups are immediately converted to CO_2 through either hydrolysis or further oxidation by the permanganate ion. Manganese dioxide (MnO_2) is the end product of the reduction of the permanganate oxyanion under neutral to basic pH conditions by the following reaction:

$$MnO_4^- + 2H_2O + 3e^- \rightarrow MnO_2 + 4OH^- \quad (7.4)$$

At low pH, reduction of MnO_4^- proceeds to Mn(II):

$$MnO_4^- + 8H^+ + 5e^- \rightarrow Mn^{2+} + 4H_2O \quad (7.5)$$

The Mn(II) cation is soluble in water at concentrations above the regulatory limit (50 μg/L) when chloride or sulfate counteranions are present.

Permanganate is easily handled, readily available, and it is a strong and versatile oxidizing agent with a relatively high oxidizing power (Table 7.1), reacting with a wide range of organic compounds over a wide range of pH. Permanganate oxidation involves primarily electron transfer in contrast with the Fenton's reaction and the persulfate reactions that also involve free radical reaction. Advantages of using permanganate over other chemical oxidants include: (1) unlike Fenton reactions, the reactions are not exothermic, (2) pH control is not an issue, (3) catalysts are not required, (4) free radical scavengers such as carbonates are not required, and (5) being a mild oxidant, it can be used in association with bioremediation. Permanganates have been successfully used in the oxidation of alkanes, alkenes, aromatic hydrocarbons, ketones, aldehydes, PAHs, etc. (Achugasim et al. 2014 and references therein).

7.2.2 *Ozonation*

Ozone has been used for treating drinking water since the beginning of the twentieth century (Gerrity et al. 2018), because of its high power to oxidize contaminants and the innocuous decay products generated in the process (oxygen and water). This makes the method less toxic compared with other treatments that use, e.g., Cl_2 or chromic acid. The compound is a powerful oxidant (Table 7.1) and an efficient bactericide.

Ozone can react by a direct reaction (slow and selective) with an organic compound P:

$$O_{3(g)} \rightarrow O_{3(l)} + P \rightarrow P_{ox} \quad k \approx 1x10^3 \ 1/(M\,s) \quad (7.6)$$

or by a radical reaction (rapid and non-selective), accelerated in alkaline media (Huang et al. 1993; Domènech et al. 2004; Litter 2005; Oppenländer 2003; Glaze et al. 1987, 1992; Hoigné and Bader 1976, 1983; Beltrán et al. 1994a, b; Legube and Karpel Vel Leitner 1999):

$$2O_3 + 2H_2O \rightarrow 2HO^{\bullet} + O_2 + 2HO_2^{\bullet} \quad k \approx 10^8 x 10^{10}\ 1/(M\ s) \tag{7.7}$$

In the direct reaction, ozone reacts with alkenes through the electrophilic attack of the double bond carbon in a fashion similar to permanganate. Ozone bridges the carbon–carbon double bond to form unstable ozonide intermediates, which decompose into smaller oxidation species up to CO_2 and H_2O or stable refractory compounds, such as acetic acid or oxalic acid. Ozone does not react well with chlorinated alkanes, particularly chlorinated methanes. This route leads to a very limited mineralization of the organic compounds, and its use for removal of pollutants must be reinforced by modifying the method.

Ozone can generate $HO^{\bullet}$ via catalytic decomposition of water, becoming an AOT (Huang et al. 1993; Boczkaj and Fernandes 2017; Glaze et al. 1987; Hoigné and Bader 1976; Beltrán et al. 1994a, b; Andreozzi et al. 1999; Roche et al. 1994):

$$O_3 + HO^- \rightarrow O_2 + HO_2^- \tag{7.8}$$

$$HO_2^- + O_3 \rightarrow O_3^{\bullet -} + HO_2^{\bullet} \tag{7.9}$$

$$HO_2^{\bullet} \leftrightarrows O_2^{\bullet -} + H^+ \tag{7.10}$$

$$O_2^{\bullet -} + O_3 \rightarrow O_3^{\bullet -} + O_2 \tag{7.11}$$

$$O_3^{\bullet -} + H^+ \rightarrow HO_3^{\bullet} \tag{7.12}$$

$$HO_3^{\bullet} \rightarrow HO^{\bullet} + O_2 \tag{7.13}$$

$$O_3 + HO^{\bullet} \leftrightarrows O_2 + HO_2^{\bullet} \tag{7.14}$$

$$O_3 + HO_2^{\bullet} \leftrightarrows 2\,O_2 + HO^{\bullet} \tag{7.15}$$

As the species formed in the process, especially $HO^{\bullet}$, have a higher oxidant ability than O_3, the indirect pathway is less selective and can be initiated by HO_2^-, $HCOO^-$, Fe^{2+}, humic substances, or mainly by HO^-. For this reason, ozonation is sensibly more effective in alkaline media, presenting an optimal efficiency around pH 9.

Fe(II), Fe(III), Mn(II), Ni(II), Co(II), Ag(I) salts or solid oxides (goethite, Fe(III)/Al_2O_3, ZnO, MnO_2, TiO_2, Cu/Al_2O_3, or Cu/TiO_2) can be added to enhance the technology (Catazone process) (Boczkaj and Fernandes 2017; Legube and Karpel Vel Leitner 1999).

The combination of both direct and indirect routes enhances substantially OM degradation. The process depends on the composition of the mixture to be treated, the pH of the solution, and the ozone dose. As mineralization takes place, bicarbonate and carbonate ions are produced; these species are $HO^{\bullet}$ scavengers and, for this reason, the pH should be carefully controlled. Intermediate oxidation products, like acetic and oxalic acids, often resist mineralization.

Ozone is industrially applied to water treatment either alone or in combination with hydrogen peroxide and/or activated carbon. Improvements of the ozone technology, including combinations with catalysts and AOTs have been described (Kasprzyk-Hordern et al. 2003; von Gunten 2003a, b). O_3 is characterized by a short lifetime and low solubility in aqueous media.

Ozonation is a very well-known commercialized technology for water treatment. It has been successfully used in the discoloration of kaolin and cellulose pulp and, in general, in the treatment of extremely polluted aqueous effluents. Ozonation is a good pretreatment before a biological treatment because complex organics are transformed into aldehydes, ketones, or carboxylic acids, all easily biodegradable. Ozonation is versatile, and it allows combining with other conventional methods or AOTs. Ozone can be simply produced in situ by electric discharge in current of oxygen or air, leaving neither odors nor residual tastes. In contrast, from the operational point of view, the use of ozone is complicated because there are mass transfer limitations due to the difficult access of the gaseous molecule to the aqueous phase (Roche et al. 1994). To overcome the low solubility, the process requires efficient stirring; some techniques like fixed beds of porous glass or metals, solid catalysts, stirring, line mixers, contact towers and increase of retention time by large bubble columns or diffusers have also been explored to enhance the process (Boczkaj and Fernandes 2017). Another way to improve the process is to increase the retention time in the reactor by large bubble columns or to increase the solubility of ozone by increasing the pressure to several atmospheres. However, any additional modifications raise the investment costs. Furthermore, a rather high O_3/pollutant molar ratio (more than 5:1) is generally needed for the complete destruction of the compound, which makes the treatment even more expensive. Further to this, in some cases, the method does not lead to the complete mineralization. The temperature must be controlled because of the risk of volatilization of the initial or the intermediate compounds. Final degassing devices in the circuit are necessary to deplete the ozone completely, which will be deleterious in a potential biological posttreatment; this also increases the costs. In conclusion, the use of ozone implies high capital costs and additional equipment for destroying the remaining ozone, together with safety and health problems and mass transfer limitation due to the low solubility of O_3 in water and danger of escape of volatile organic compounds (VOCs) to the atmosphere caused by the bubbling of the reagent.

Recent advances for ozone applications in treatment of recalcitrant contaminants have been driven in endocrine disrupting compounds, pharmaceutical, pesticides, and personal care products (Gerrity et al. 2018).

7.2.3 O_3/H_2O_2

Combination of hydrogen peroxide and O_3 produces an enhancement of ozonation (Domènech et al. 2004; Litter 2005; Hoigné and Bader 1976) in a process called

Perozone® (or peroxone), where both direct and indirect ozone oxidation of organic compounds take place (Fischbacher et al. 2018).

As known, H_2O_2 is a weak acid, a powerful oxidant (Table 7.1), and an unstable compound, which disproportionates with a maximal rate at the pH of its pK_a:

$$H_2O_2 \leftrightarrows HO_2^- + H^+ \quad pK_a = 11.6 \tag{7.16}$$

$$H_2O_2 + 2e^- + 2H^+ \rightarrow 2H_2O \tag{7.17}$$

$$H_2O_2 \rightarrow H_2O + ½\,O_2 \tag{7.18}$$

$$H_2O_2 + HO_2^- \rightarrow H_2O + O_2 + HO^- \tag{7.19}$$

The redox potential increases with the decrease of pH (Neyens and Baeyens 2003). In alkaline medium, H_2O_2 will react with HO^- to produce perhydroxyl ions:

$$H_2O_2 + HO^- \rightarrow HO_2^- + H_2O \tag{7.20}$$

Therefore, an acidic medium is more favorable to using H_2O_2 processes for water treatment. H_2O_2 has been widely used for removing low levels of pollutants from wastewaters (chlorine, nitrites, sulfites, hypochlorites, etc.) and it is also used as a disinfectant (Neyens and Baeyens 2003). However, when low H_2O_2 concentrations are used (to avoid increasing costs), reaction rates are low, and this makes its use ineffective in treating high levels of refractory pollutants, such as highly chlorinated aromatic compounds and some inorganic compounds (e.g., cyanides).

O_3 decomposition is initiated by H_2O_2 through an electron-transfer reaction producing $HO^•$ (Eq. 7.21) (Huang et al. 1993); alternatively, this reaction can be interpreted as the activation of H_2O_2 by ozone (Gurol and Akata 1996):

$$O_3 + H_2O_2 \rightarrow HO^• + O_2 + HO_2^• \tag{7.21}$$

The major effects of combining O_3 and H_2O_2 result from the increase of the oxidation efficiency by conversion of O_3 to $HO^•$ and the improvement of O_3 transfer from the gaseous to the liquid phase (Boczkaj and Fernandes 2017; Roche and Prados 1995). In basic medium, H_2O_2 will generate HO_2^-, which can promote the decomposition of O_3 to $HO^•$ with higher effectiveness.

The process can treat organic pollutants at very low concentrations (μg/L), at an optimal O_3/H_2O_2 molar ratio $\cong$ 2:1.

Although the process is expensive, it is fast and effective in decomposing organochlorinated compounds (trichloroethylene (TCE), tetrachloroethylene, etc.). It is very convenient in the posttreatment of water submitted to disinfection treatments using chlorine or chlorine dioxide because it can decompose potential byproducts such as THM or related compounds. This approach can be applied to degradation of pesticides (Beltrán et al. 1994a, b).

7.2.4 Fenton and Related Reactions

The Fenton process is the production of $HO^{\bullet}$ by reaction of H_2O_2 and Fe(II) in solution. It is based on the very well-known Fenton's experiments from the end of the nineteenth century, which demonstrated that this system was able to oxidize organic compounds (Fenton 1894). The mechanism was later proposed by Haber and Weiss according to reaction 7.22 and following reactions (Huang et al. 1993; Domènech et al. 2004; Litter 2005; Walling 1975; Pignatello et al. 2006; Babuponnusami and Muthukumar 2014; Litter and Slodowicz 2017; Miller et al. 2018):

$$Fe^{2+}{}_{aq} + H_2O_2 \rightarrow Fe^{3+}{}_{aq} + HO^- + HO^{\bullet} \quad k = 76\ \mathrm{L/(mol\ s)} \tag{7.22}$$

$$Fe^{3+} + H_2O_2 \rightarrow Fe(OOH)^{2+} + H^+ \rightarrow Fe^{2+} + HO_2{}^{\bullet} + H^+ \tag{7.23}$$

$$Fe^{2+} + HO^{\bullet} \rightarrow Fe^{3+} + HO^- \tag{7.24}$$

$$Fe^{3+} + HO_2{}^{\bullet} \rightarrow Fe^{2+} + H^+ + O_2 \tag{7.25}$$

$$Fe^{2+} + HO_2{}^{\bullet} \rightarrow Fe^{3+} + HO_2{}^- \tag{7.26}$$

$$H_2O_2 + HO^{\bullet} \rightarrow HO_2{}^{\bullet} + H_2O \tag{7.27}$$

$$2HO^{\bullet} \rightarrow H_2O_2 \tag{7.28}$$

$$2HO_2{}^{\bullet} \rightarrow H_2O_2 + O_2 \tag{7.29}$$

$$HO^{\bullet} + HO_2{}^{\bullet} \rightarrow H_2O + O_2 \tag{7.30}$$

Generally, the cycle of the continuous Fe(II)/Fe(III) generation takes place until either of the reagents (Fe(II) or H_2O_2) has been consumed. $HO^{\bullet}$ can oxidize Fe^{2+}, Eq. 7.24, but this reaction is unproductive regarding its utility for transformation of pollutants. Otherwise, $HO^{\bullet}$ are able to react with organic compounds by Eqs. 7.1, 7.2 and 7.3, as described in Sect. 7.1.

Besides Fe(II), other transition metal ions such as Fe(III), Cu(I), or Mn(II) can promote similar processes, which are then called Fenton-like or Fenton type.

Fe^{3+} and Fe^{2+} can oxidize or reduce organic radicals, or the radicals can recombine (Neyens and Baeyens 2003; Tang and Tassos 1997):

$$R^{\bullet} + Fe^{3+} \rightarrow R^+ + Fe(II) \tag{7.31}$$

$$R^{\bullet} + Fe^{2+} \rightarrow R^- + Fe^{3+} \tag{7.32}$$

$$2R^{\bullet} \rightarrow R - R \tag{7.33}$$

The maximum catalytic activity of the Fe(II)/Fe(III)-H_2O_2 system is at pH about 2.8–3.0. At low pH, the complexation of Fe(III) with H_2O_2 (Eq. 7.23) is inhibited

(Safarzadeh-Amiri et al. 1996b); at pH > 1 and in the presence of excess H_2O_2, the decomposition of H_2O_2 occurs:

$$2H_2O_2 \rightarrow O_2 + 2H_2O \tag{7.34}$$

At pH > 5, the formation of relatively inactive iron oxohydroxides, precipitation of ferric hydroxide, and the acceleration of the H_2O_2 self-decomposition are favored (Kremer 1999), decreasing the oxidative power of the process. Due to the existence of a lower amount of free iron ions, less $HO^{\bullet}$ are generated; besides, the oxidation potential of this radical diminishes when the pH is increased (Pignatello et al. 2006; Parsons 2004).

After the work of Walling (1975), the radical mechanism described by Eqs. 7.22, 7.23, 7.24, 7.25, 7.26, 7.27, 7.28, 7.29 and 7.30 has been broadly accepted for reactions in acidic media. However, a second mechanism was suggested (Ikehata and Gamal El-Din 2006; Pignatello et al. 1999), where the ferryl ion (FeO^{2+}, involving Fe(IV)) (Jacobsen et al. 1998; Kremer 1999, 2003) is the transitory species instead of $HO^{\bullet}$ (Eq. 7.35). FeO^{2+} is able to oxidize organic compounds.

$$Fe^{2+} + H_2O_2 \rightarrow FeO^{2+} + H_2O \tag{7.35}$$

The participation of other intermediates and mechanistic pathways has been also proposed (Pignatello et al. 2006).

The common Fenton process occurs at room temperature and atmospheric pressure. The reagents are easy to acquire, store and handle, and they are environmentally friendly. In the laboratory, the metal is traditionally added as pure ferrous salts, but at a higher scale the use of these salts becomes prohibitively expensive, and normally $Fe_2(NH_4)_2SO_4$, which contains 20% of active iron, is used. There are not mass transfer limitations in the process because it is a homogeneous system with all the reagents in the solution. The design of the reactors for technological application is rather simple (Domènech et al. 2004). Nevertheless, the formation of solid sludge due to the precipitation of iron oxides, which needs additional separation and causes disposal problems, also the wastage of H_2O_2, and the need of high Fe^{2+} concentrations (40–80 ppm) with the continuous or intermittent addition of both reagents represent important disadvantages of the treatment (Pignatello et al. 2006; Mukherjee et al. 2016; Iurascu et al. 2009). Excess of Fe^{2+}, H_2O_2, perhydroxyl radical, or halogens (if present) can act as $HO^{\bullet}$ scavengers. The degradation rate increases with Fe^{2+} concentration, but no effect is observed above a certain value; however, a large amount of iron ions should be avoided because it contributes to increasing the content of total dissolved salts in the effluent stream (Gogate and Pandit 2004a). Generally, the reaction rate is very high until complete H_2O_2 depletion. Theoretically, the H_2O_2/substrate molar ratio needed for destruction of soluble compounds oscillates between 2 and 10. However, in practice, this ratio may be sometimes as high as 1000 because of the presence in environmental samples of other $HO^{\bullet}$ competing species. Obviously, H_2O_2 must be completely eliminated before introducing the effluent into a biological treatment (Gogate and Pandit 2004a). At the end of the process, it is common to alkalinize the waters, with simultaneous addition of a flocculant to eliminate the remaining iron (Pignatello 1992).

The application of the Fenton process for destroying toxic organics began in 1960 (Huang et al. 1993).The method is effective in treating waters containing pharmaceuticals, dyes, insecticides, water from petroleum refineries and fuel terminals, nitroaromatics and explosives such as trinitrotoluene (TNT), chlorinated aliphatic and aromatic compounds (chlorobenzene, pentachlorophenol, phenols, chlorinated phenols, octachloro-*p*-dioxin, polychlorinated biphenyls (PCBs)), formaldehyde, etc. Only some compounds cannot be attacked by this reagent: acetone, acetic acid, oxalic acid, or paraffins. The technology has been successfully applied to the COD reduction of municipal water and groundwater, in treating leachates, and as a pretreatment of non-biodegradable compounds. It has been effectively applied as a good oxidant of herbicides, hexadecane, or dieldrin in the area of soil treatment (e.g., Safarzadeh-Amiri et al. 1996b; Bigda 1995; Lin and Lo 1997; Watts et al. 2002; Quan et al. 2003).

Generally, a complete mineralization cannot be attained in Fenton processes because resistant intermediates such as carboxylic acids, which react very slowly with $HO^{\bullet}$, are formed, and the unproductive Eq. 7.24 predominates. Occasionally, products that are more toxic than the initial ones are formed; this obliges to careful monitoring of the process until the complete depletion of these noxious byproducts is achieved (Sedlak and Andren 1991).

As said, the use of Fe(II)/Fe(III) in the solution for Fenton processes suffers from considerable disadvantages, such as the formation of a high amount of iron sludge that has to be removed, the requirement of large amounts of Fe^{2+} (ca. 50–80 ppm), and a high H_2O_2/Fe^{2+} molar ratio. In addition, the acidification of the effluents before the reaction and the neutralization of the treated solutions before disposal should be part of the process. Due to these problems, in the mid-1990s, heterogeneous Fenton catalysts, i.e., solid iron-containing compounds or solid materials rich in iron, started to be developed (Pignatello et al. 2006; Mukherjee et al. 2016; Tang and Chen 1996; Fajerwerg and Debellefontaine 1996; Pera-Titus et al. 2004; Navalon et al. 2010; Li et al. 2015; Dhakshinamoorthy et al. 2012; Zhao et al. 2016). Several iron-containing materials such as clays, activated carbon, silicas, zeolites, hydrotalcite-like compounds, fly ashes, iron-oxide minerals, iron oxide nanocatalysts, metal-exchange resins and layered materials, Nafion films or Nafion resins containing iron, iron-coated pumice particles, and iron immobilized aluminates have been tested; additionally, iron-containing polyethylene copolymers, alginate gel beads, structured silica fabrics, or brick grain were also essayed (Domènech et al. 2004; Pera-Titus et al. 2004; Dhakshinamoorthy et al. 2012 and references therein). Zero-valent state metals have also been proposed, especially zero-valent iron (ZVI), which will be treated in Sect. 7.2.11. Pure and metal-modified iron oxides such as magnetite, goethite, hematite, etc., can be used as heterogeneous Fenton catalysts because they are relatively less expensive, allow magnetic separation, and present both high stability and improved adsorption capacity (Teel et al. 2001; Pouran et al. 2014).

Generally, solid materials allow the use of higher pH conditions, so their separation from the waste stream is possible.

Fe(II) catalyzes O_3 degradation giving the ferryl intermediate, which can directly oxidize the organic pollutant or evolve to $HO^{\bullet}$ (Litter 2005 and references therein):

$$Fe^{2+} + O_3 \rightarrow FeO^{2+} + O_2 \tag{7.36}$$

$$FeO^{2+} + H_2O \rightarrow Fe^{3+} + HO^{\bullet} + HO^{-} \tag{7.37}$$

7.2.5 Electrochemical Oxidation

The application of electrical current (2–20 A) between two suitable electrodes (Domènech et al. 2004; Brillas et al. 1998; Zhou et al. 2018) immersed in water produces primary chemical reactions, with $HO^{\bullet}$ generation, which can be used to oxidize pollutants:

$$H_2O \rightarrow HO^{\bullet} + H^{+} + e^{-} \tag{7.38}$$

$$O_2 + 2H^{+} + 2e^{-} \rightarrow H_2O_2 \tag{7.39}$$

The efficiency of the system can be improved by adding Fe(II), and this process is known then as electro-Fenton (EF) (Nidheesh et al. 2018). If an Fe sacrificial anode is used, it provides stoichiometric amounts of Fe for the Fenton reaction, and the process is known as electrocoagulation (EC) (or peroxicoagulation) (Zhou et al. 2018; Fernandes et al. 2015 and references therein).

EF processes can be homogeneous or heterogeneous. Homogeneous EF processes use soluble forms of iron; the most common ones are salts such as ferrous sulfate, ferric chloride, etc., which generate ferrous or ferric ions in water and undergo Fenton reaction with the in situ generated H_2O_2. The heterogeneous EF process uses solid iron catalysts, generally very slightly soluble or insoluble in water; the most common ones are iron oxides. Generated H_2O_2 reacts with ferrous ions, resulting in the generation of $HO^{\bullet}$ as in Eq. 7.22. The generated ferric ions undergo cathodic reduction (Eq. 7.40) and regenerate Fe^{2+}; conventional Fenton chain reactions also take place (Nidheesh et al. 2018).

$$Fe^{3+} + e^{-} \rightarrow Fe^{2+} \tag{7.40}$$

In the EF process, the increase in solution pH is counterbalanced by the generation of protons coming from the water oxidation at the anode (Eq. 7.41), and from the generation of carboxylic acids coming from the degradation of the pollutants.

$$2H_2O \rightarrow O_2 + 4H^+ + 4e^- \tag{7.41}$$

The process is accepted worldwide because of its high efficiency, in situ generation of H_2O_2, negligible or total absence of sludge production, higher Fe^{2+} ion regeneration rate, etc. EF processes can be homogeneous and heterogeneous depending on the solubility of the iron forms having been used. The most commonly used homogeneous catalysts are ferrous sulfate, ferric chloride, etc. Heterogeneous EF processes use insoluble solid catalysts, such as iron oxides.

EC uses consumable electrodes to supply ions to the solution, allowing the contaminants to form agglomerates. The coagulating ions are produced in situ through three stages: (i) formation of the coagulants by electrolytic oxidation of the sacrificial electrode, (ii) destabilization of the contaminants and particulate suspension, and breaking of emulsions, and (iii) aggregation of the destabilized phases to form flocs. The contaminants present in the solution are treated either through chemical reactions and precipitation, or by physical and chemical attachment to colloidal materials generated by the electrode corrosion. The coagulated particles can be separated by sedimentation or electroflotation. Iron and aluminum are the most widely used EC electrode materials since they are cheap, readily available, and effective. With iron as the anode, the oxidation produces Fe^{2+}, which is dissolved into the solution, leading to $Fe(OH)_n$ ($n = 2$ or 3) flocs. The flocs have strong affinity for colloids, dispersed particles, and ionic species, and they cause flocculation, generating bigger particles. In the case of Al, the electrolytic dissolution of the anode produces Al^{3+} and $Al(OH)_2^+$, which are transformed initially into $Al(OH)_3$ and finally polymerized to $Al_n(OH)_{3n}$ (Fernandes et al. 2015).

7.2.6 Gamma Radiolysis and Processes with Electron Beams

These processes are based on the generation of highly reactive electrons, radical ions, and neutral radicals by exposing the target water to the beams of mass particles or high energy electromagnetic waves (Domènech et al. 2004; Makogon et al. 1998). Gamma rays (Gammacell, ^{60}Co) (Chaychian et al. 1999), X-rays, or electron beam accelerators (Van-de-Graaff or linear (LINAC)) (Oppenländer 2003; US EPA 1997) can be used. When the electron beam enters the solution, the electrons lose their energy by non-elastic collisions with the water molecules, and reactive species are generated:

$$H_2O \rightarrow e_{aq}^- + H^\bullet + HO^\bullet + H_2 + H_2O_2 + H^+ \tag{7.42}$$

The first three species are the primary products of water radiolysis. The aqueous electron, e_{aq}^-, and the hydrogen atom, $H^\bullet$, are strong reductants, which attack the organic matter by different mechanisms; while e_{aq}^- produces halogen atom abstraction ($E^0 = 2.77$ V), $H^\bullet$ produces hydrogen addition or abstraction. $HO^\bullet$ acts as an oxidant, as in the other AOTs.

The method is ideal for the treatment of VOCs and semivolatile organic compounds (SVOCs) in groundwaters, wastewaters, drinking water, and leachates. The compounds that have been widely studied are the halogenated species such as PCBs, easily oxidizable and attacked by $HO^•$ (Chaychian et al. 1999).These compounds can be mineralized or degraded to products of lower molecular weight. The method does not generate sludge or other wastes that would need ulterior treatment, or toxic compounds such as dioxins. In contrast, if the radiation doses are low, aldehydes, organic acids, and resistant SVOCs can be formed. The process requires high electrical consumption; it is not economically effective for high concentrations of pollutants. Industrial scale installations are currently being developed. The first full-scale application in the treatment of emulsifiers was reported at the Voronezh rubber plant in Russia (Pikaev 2001) and, since then, some other plants have been installed (Wojnárovits et al. 2018). The combination with ozone increases the efficiency because of the fast formation of additional reactive species.

Radiation technology has been proved useful for the discoloration and degradation of dyes (Chen et al. 2008; Ma et al. 2007), sewage sludge processing (Park et al. 2009), oxidization of organic pollutants (Al-Sheikhly et al. 2006; Hu and Wang 2007), removal of pesticides (Basfar et al. 2007), and for the decomposition of pharmaceutical compounds (Sanchez-Polo et al. 2009). The application of radiation technology to sewage sludge processing and degradation of dyes has been reviewed in detail (Rauf and Ashraf 2009; Wang and Wang 2007; Wojnárovits and Takács 2008).

7.2.7 Non-thermal Plasma

Plasma is considered the fourth state of the matter and it contains ions and free electrons (electrical gas). Plasma can be generated in non-thermal form from an electrical discharge or bombardment of gas with an electron beam of high energy; the energy of the electrons in the plasma is around 10 eV, which is equivalent to high temperatures. Such plasma is a good source of highly reductive and oxidative reactive species such as $O(^3P)$, $HO^•$, N, H, NH, CH, O_3, $O_2(^1\Delta)$, the own plasma electrons, etc. The presence of these species allows using the method in multiple applications: SO_x and NO_x removal from gas exhausts, decomposition of aliphatic and halogenated aliphatic hydrocarbons, gases of industrial exhausts and incinerators, treatment of VOCs in soils and groundwaters (with previous transfer to the vapor phase by pumping), treatment of polluted soils with VOCs (with previous application of heat and fluidization in an inert gas), etc. (Domènech et al. 2004; Rosocha and Korzekwa 1999).

Two types of treatment processes can be employed to apply plasma methods to degradation of pollutants: (a) indirect plasma, represented by ozone generators whereby the plasma is used to generate the oxidant, which is then delivered to a separate reactor for water treatment; (b) direct plasma, which contacts directly the liquid phase containing the contaminants (Rosocha and Korzekwa 1999; Mededovic Thagard and Locke 2018).

The technology does not generate toxic byproducts, such as dioxins or furans, it operates at pressure and temperatures close to the ambient, it does not require fuel (it minimizes secondary residues), and it can simultaneously remove dangerous organics and emissions like SO_x/NO_x. It does not require catalysts. However, this technology has not reached yet sufficient level of development to be used in practice, but some work on the design of the reactors has been recently attempted (Stratton et al. 2015).

7.2.8 *Oxidation in Sub- and Supercritical Water*

These technologies allow the oxidation of the pollutants in a mixture of water with oxygen, or air at high pressures and temperatures (Domènech et al. 2004; Kronholm and Riekkola 1999; Zhang and Chuang 1999; Martino and Savage 1999; Tungler et al. 2015).The operating process under subcritical conditions is called wet air oxidation (WAO); it works at pressures between 1000–22,000 kPa and temperatures between 150–370 °C. The mechanism involves the primary carbonization of the organic compounds and the further reaction with $HO^{\bullet}$ produced in the catalytic transformation of dissolved O_2 (DO) on the surface of the carbon center, Eqs. 7.43 and 7.44. Nitrogen, halogens, and sulfur moieties are also mineralized, Eqs. 7.45, 7.46 and 7.47. Equation 7.48 shows the origin of the main byproducts.

$$O_2 \rightarrow H_2O_2 \rightarrow O_2 \text{ and } HO^{\bullet} \tag{7.43}$$

$$HO^{\bullet} + C \rightarrow CO_2 + H_2O \tag{7.44}$$

$$N \rightarrow NH_3, NO_3^{-} \text{ or elemental N} \tag{7.45}$$

$$X \rightarrow X^{-} \tag{7.46}$$

$$S \rightarrow SO_4^{2-} \tag{7.47}$$

$$C + HO^{\bullet} \rightarrow \text{low molecular weight organic acids} \tag{7.48}$$

Figure 7.1 shows a typical flow diagram of a WAO process (Mishra et al. 1995).

Since the oxidation reactions are exothermic, sufficient energy may be released in the reactor to allow the WAO system to operate without any additional heat input at or above a chemical oxygen demand (COD > ~10,000 mg/L), and the process can be self-supported; in comparison, incineration requires COD between 300,000 and 400,000 mg/L. The addition of an oxidant such as O_2, H_2O_2, or potassium persulfate (Kronholm and Riekkola 1999) improves the efficiency. Any kind of wastes can be treated this way, including sludge, water wastes of high COD, biorecalcitrant wastes, municipal sewage sludge, distillery wastes, paper pulp, black liquors of textile waters, wastewaters with CN^{-} and nitriles, wastes adsorbed on the carbon used in treatment of effluents (for regeneration), etc. Polyphenols, perchlorophenols,

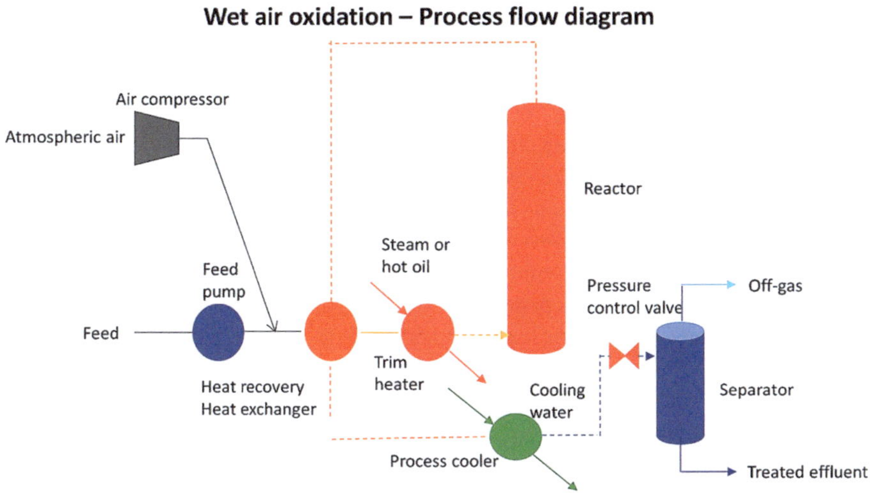

Fig. 7.1 Typical flow diagram of a WAO process. (Adapted from Mishra et al. 1995

aliphatics, polychlorinated aliphatic compounds and aromatic hydrocarbons (toluene, pyrene, and substituted aromatics) can be destroyed. The method is excellent for destruction of carbohydrates. In contrast with the incineration, there is little interaction with the environment and the process can be easily coupled with a biological treatment.

It should be mentioned that mineralization is generally incomplete because of the formation of carboxylic acids, alcohols, aldehydes, or ketones of low molecular weight, which represents a disadvantage. Acetic and propionic acids are very resistant and require catalysts for their destruction, and the same occurs with halogenated aromatic compounds (1,2-dihalobenzenes, PCBs, perchlorophenols). The low solubility of O_2 in water generates mass transfer problems, which can limit the efficiency. The main disadvantage is the necessity of employing very expensive construction materials.

The process known as catalyzed wet air oxidation (CWAO) (Zhang and Chuang 1999) uses homogeneous and heterogeneous catalysts to improve WAO efficiency. As catalysts, homogeneously dissolved Cu(II) salts, Cu/activated carbon, Cu supported in γ-alumina, Mn(V) and activated Cr compounds, Co/Bi, Mn/Cu, MnO_2, CuO, Fe_2O_3, and Zn oxides have been used. The best results have been obtained with Pd-Pt-Ce/alumina catalysts. The main problem of this technology is the final separation of the catalyst. The method can be improved by Fe^{2+}/H_2O_2 injection (Zhang and Chuang 1999; Mantzavinos et al. 1996; Martino and Savage 1999). CWAO has been applied to many different model effluents, but relatively few studies have been devoted to real and complex industrial wastes (Tungler et al. 2015).

The oxidation with supercritical water (SCWO) (Domènech et al. 2004; Martino and Savage 1999; Vadillo et al. 2013) uses water in supercritical conditions of pressure and temperature, higher than those of the critical point of water,[1] between 450 and 700 °C and pressures close to 25 MPa. In the supercritical state, the water behaves as a fluid of relatively low viscosity, density, and dielectric constant. Under supercritical conditions, the number of hydrogen bridges decreases, the solubility of organic compounds and gases increases, and the solubility of the electrolytes decreases. These reasons make SCWO an exceptional treatment system, with an oxidation efficiency higher than 99.99% in very short contact times (5–60 s), not requiring any additional treatment of the gaseous products. Due to the high solubility of O_2 in supercritical water, there are no mass transfer problems. In addition, as the surface tension is null, O_2 can penetrate into the smallest pores and can oxidize any organic substance. On the other hand, it is possible to remove inorganic compounds by precipitation. The process can be energetically self-supported if organics in concentrations higher than 5% are treated, and the combustion heat can be recovered as process heat at high temperature or power. However, the process requires severe operative conditions; moreover, it is not attractive for diluted waters, and dibenzofurans and dioxins can be produced. Similarly to WAO, SCWO needs expensive especial construction materials for high temperatures and pressures.

SCWO can be applied to the treatment of aqueous conventional organic wastes, sludges, chlorinated and nitrogenated solvents, chlorinated, phosphorous and nitrogen pesticides and herbicides, PCBs, dyes, plasticizers, plastics, byproducts of the chemical and pharmaceutical industry, military products such as explosives, propellants or war gases, etc. It is highly promising for treating hazardous wastes as well. However, SCWO treatments at the pilot plant scale of real wastewaters are scarce, and the application of this technology to industrial wastewaters has two main drawbacks: corrosion and salt deposition; some other problems should be also solved related to management of biphasic wastes, presence of suspended solids, high costs, etc. Therefore, currently, the industrial scale-up and commercialization of the process is still a critical issue. Improvements, especially in the design of the proper reactors, are imperative. The design of SCWO plants has to include an energy recovery system in order to achieve economic feasibility of the process. Although some commercial plants were built in the past, only two of them are still in operation. Technical solutions to decreasing the capital and the operating costs in order to achieve full commercial development of this technology are vital (Vadillo et al. 2013).

7.2.9 *Electrohydraulic Discharge – Ultrasound*

This technology uses high power ultrasound (from 15 kHz up to 1 MHz), taking advantage of the electrohydraulic cavitation, i.e., the growing and cyclic collapse of gas bubbles under these conditions. The gas implodes and very high local

[1]Water critical points: $T_c = 374$ °C and $P_c = 22.1$ MPa.

temperatures and pressures (4000–10,000 K and 100,000–1,000,000 kPa in the center of the collapsed bubbles) are reached (Huang et al. 1993; Domènech et al. 2004; Destaillats et al. 2000a, b, 2003; Gogate and Pandit 2004a; Henglein 1987; Hoffmann et al. 1996; Adewuyi 2001; Sadjadi 2014; Pflieger et al. 2015; Pankaj 2010; Larpparisudthi et al. 2018). The degradation of the organic matter by sonolysis takes place through three processes: reactions of supercritical H_2O (see Sect. 7.2.8), direct pyrolysis, and reactions with the radicals generated by thermal reaction, or by other reactions in the presence of oxygen or nitrogen atmosphere (Eqs. 7.49, 7.50, 7.51, 7.52, 7.53, 7.54, 7.55, 7.56, 7.57, 7.58, 7.59, 7.60, 7.61, 7.62, 7.63, 7.64, 7.65 and 7.66). Equation 7.51 shows H_2O_2 formation, another useful oxidant together with $HO^•$.

$$H_2O +)))) \rightarrow H^• + HO^• \tag{7.49}$$

$$H^• + H^• \rightarrow H_2 \tag{7.50}$$

$$2HO^• \rightarrow H_2O_2 \tag{7.51}$$

$$O_2 +)))) \rightarrow 2O^{•(a)} \tag{7.52}$$

$$O^• + H_2O \rightarrow 2\,HO^• \tag{7.53}$$

$$O_2 + H^• \rightarrow HO_2{}^• \tag{7.54}$$

$$HO_2 + H^• \rightarrow H_2O_2 \tag{7.55}$$

$$2HO_2{}^• \rightarrow H_2O_2 + O_2 \tag{7.56}$$

$$O_2 + 2HO_2{}^• \rightarrow 2\,H_2O_2 \tag{7.57}$$

$$O^• + H_2O \rightarrow H_2O_2 \tag{7.58}$$

$$H^• + H_2O \rightarrow H_2 + HO^• \tag{7.59}$$

$$H^• + O_2 \rightarrow HO_2{}^• \tag{7.60}$$

$$N_2 + O^• \rightarrow NO + N^• \tag{7.61}$$

$$N_2 +)))) \rightarrow 2\,N^{•(a)} \tag{7.62}$$

$$N^• + O_2 \rightarrow NO + O^• \tag{7.63}$$

$$N^• + O^• \rightarrow NO \tag{7.64}$$

$$NO + O^• \rightarrow NO_2 \tag{7.65}$$

$$N^• + HO^• \rightarrow\rightarrow HNO_2 \rightarrow HNO_3 \tag{7.66}$$

Figure 7.2 shows the formation and collapse of a cavitation bubble, and three reaction zones (i.e., the cavity interior, the gas–liquid interface, and the bulk solution) in the cavitation process (Wang and Xu 2012).

The technology is simple and economically competitive, and the degradation is complete in short time (minutes to hours). The method is very good for treating

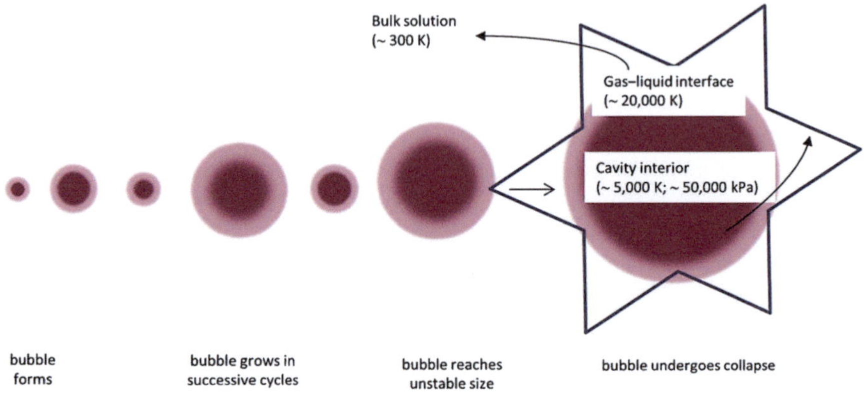

Fig. 7.2 Formation and collapse of a cavitation bubble with the three reaction zones. (Adapted from Wang and Xu 2012)

volatile compounds because the combustion takes place directly inside the collapsing hot bubbles. It has been applied to abstract chlorine from chlorobenzene and *o*-dichlorobenzene, for decomposition of aliphatics and chlorinated aliphatics (CCl_4, $CHCl_3$), for *p*-nitrophenol and TNT degradation, for decomposition and discoloration of azo dyes, for degradation of Triton X-100 and related surfactants, for total decomposition of H_2S, parathion, MTBE, etc. It can be improved by H_2O_2, O_3, or Fe (II) addition and by elimination of O_2 and operation in inert atmosphere (Ar) to promote reductions (Meichtry et al. 2018).

The scale-up of ultrasound systems has been studied a great deal and there is sufficient evidence that the application of the method is possible in the short term. It is a very economical technology in comparison with other oxidizing technologies (Pankaj 2010; Gogate and Pandit 2004b). Practical application of the ultrasonic process for wastewater treatment is often limited by its low mineralization efficiency, formation of intermediate products, setup costs, and noise in the operation. However, the combination of ultrasound and Fenton reagent, i.e., sono-Fenton oxidation, has a great potential for rapid destruction of refractory organics in a short span of time through the mechanisms of thermal destruction and removal of free $HO^•$ (Ma 2012). The sono-Fenton process utilizes the advantages of these two methods to generate more $HO^•$. As cavitation operates also in terms of pyrolysis, this can help in removing some compounds refractory to $HO^•$. Moreover, the mass transfer resistances associated with the Fenton-based processes can be eliminated by the turbulent conditions present in the reactor. The produced Fe^{3+} can react with H_2O_2 through reaction 7.23, and the cycle continues but enhanced by ultrasound (Bagal and Gogate 2014). However, from an engineering point of view, further investigation is necessary for commercializing the sono-Fenton system (Ma 2012).

The synergistic effects of sonolysis combined with ozonation (sonozone process) have been described for azobenzene and methyl orange oxidation (Destaillats et al. 2000a). This process notably increases the transformation rate. The increase in the mass transfer coefficient of O_3 due to mechanical effects (better mixing and breaking of the gas bubbles) favors O_3 dissolution in water; the thermal decomposition of ozone in the collapsing bubbles seems to be the main mechanism for destruction of pollutants through $HO^•$ generation (Olson and Barbier 1994).

$$O_3 \rightarrow O_2 + O(^3P) \tag{7.67}$$

$$O(^3P) + H_2O \rightarrow 2HO^• \tag{7.68}$$

7.2.10 *Persulfate and Sulfate Radical Related Advanced Oxidation Processes*

Due to the nonselectivity of AOTs based on $HO^•$ generation, these technologies are limited when $HO^•$ have to react with coexisting water constituents with rate constants similar to those of the target pollutant. In recent years, sulfate radical related AOPs (S-AOPs) have been increasingly studied and proposed (Liu et al. 2013). These processes generate and use sulfate radicals ($SO_4^{•-}$) of strong oxidation potential (2.6 V) (Deng and Ezyske 2011; Oh et al. 2016; Karpel Vel Leitner 2018; Yang et al. 2015). $SO_4^{•-}$ have a higher standard reduction potential than $HO^•$ at neutral pH, indicating superiority in degrading several organic compounds. At acidic pH, both $HO^•$ and $SO_4^{•-}$ demonstrate almost similar reduction potentials but, in general, $SO_4^{•-}$ are more selective and oxidative than $HO^•$.

$SO_4^{•-}$ have a longer half-life than $HO^•$ because they undergo preferentially electron transfer reactions while $HO^•$ can react by hydrogen abstraction, which is less dominant. To explain the difference in oxidizing power for these two radicals at neutral pH and their similarity at acidic pH, it has been suggested that the difference lies in the abilities of their redox partners as leaving groups, i.e., the bisulfate and sulfate ions for $SO_4^{•-}$ and the water molecule for $HO^•$ (Anipsitakis and Dionysiou 2004a). S-AOPs are efficient for destruction of pesticides, perfluorocarboxylic acids, and cyanotoxins, and it is reported that both aliphatic and aromatic acids undergo more efficient mineralization by $SO_4^{•-}$ than by $HO^•$, being less influenced by competing constituents, such as alkalinity and natural organic matter (NOM) in real waters (Yang et al. 2015).

Two types of processes generating $SO_4^{•-}$ are described, persulfate or peroxidisulfate (PDS) and peroxymonosulfate (PMS).

PDS ($S_2O_8^{2-}$) is a strong oxidant comparable to O_3 (Table 7.1). However, when PDS is used alone, the efficiency of the treatment is low or null, and the combination with other catalyst or energy sources to activate $SO_4^{•-}$ generation is needed, such as heat, chemicals, transition metal catalysts, elevated pH, UV irradiation and ultrasound (Deng and Ezyske 2011; Oh et al. 2016). Subsequently, $SO_4^{•-}$ may initiate production of other intermediate highly reactive oxygen species (ROS) such as $HO^•$ (Eq. 7.72) (Kolthoff and Miller 1951).

$$S_2O_8^{2-} + \text{heat} \rightarrow 2SO_4^{\bullet -} \quad E_a = 33.4\ \text{kcal/mol} \tag{7.69}$$

$$S_2O_8^{2-} + H_2O \rightarrow 2HSO_4^{-} + HO^{\bullet} \tag{7.70}$$

$$S_2O_8^{2-} + 2OH^{-} \rightarrow 2SO_4^{2-} + 2HO^{\bullet} \tag{7.71}$$

$$SO_4^{\bullet -} + H_2O \rightarrow HO^{\bullet} + HSO_4^{-} \tag{7.72}$$

These ROS can initiate a series of radical propagation and termination chain reactions where the organics are partially or fully decomposed. Sulfate radicals rapidly degrade a broad spectrum of organic pollutants, such as TCE, 1,1,1-trichloroethane, trichloroethane (TCA), MTBE, and diphenylamine. It is the newest oxidant used for in situ chemical oxidation (ISCO) for groundwater and soil clean-up (Deng and Ezyske 2011).

PDS is highly soluble in water and it is not expensive. It has high solubility and long lifetime, and can be used for a wide range of organic contaminants. In the oxidation process, sulfate ions will be generated as the end-product, which leads to an increase in the salt content in the effluent, but sulfate is practically inert and it is not considered to be a pollutant; PDS can be regenerated from sulfate electrolytically for reuse in water (Lau et al. 2007).

PMS (HSO_5^-) has a higher oxidizing potential (1.82 V) than H_2O_2 (1.76 V), and it is more efficient than PDS for degradation processes (Cai et al. 2015). OXONE is the commercial name of the triple salt $2KHSO_5 \cdot KHSO_4 \cdot K_2SO_4$, which is the source of the strong peroxymonosulfate oxidant ($KHSO_5$). PMS can be activated to generate $SO_4^{\bullet -}$ by UV, heterogeneous catalysts, and transition metal ions, such as Mn^{2+}, Fe^{2+}, and especially Co^{2+}.

$$HSO_5^{-} + M^{n+} \rightarrow SO_4^{\bullet -} + M^{(n+1)+} + OH^{-} \tag{7.73}$$

$$SO_4^{\bullet -} + M^{n+} \rightarrow M^{(n+1)+} + SO_4^{2-} \tag{7.74}$$

$$M^{n+} + HO^{\bullet} \rightarrow M^{(n+1)+} + OH^{-} \tag{7.75}$$

However, due to the adverse effects of Co(II) on human health, PMS activated by iron or UV irradiation was considered a more environmentally friendly technology (Guan et al. 2011).

An enhancement of the PMS system by ozonation has been observed, operating by simultaneous generation of $HO^{\bullet}$ and $SO_4^{\bullet -}$ (Yang et al. 2015).

In parallel, US has been used to activate of PDS and PMS, enhancing the degradation efficiency of contaminants (Cai et al. 2015).

7.2.11 Zero-Valent Iron

The use of iron in its elemental state (Fe^0, ZVI) as a reducing agent was used many years ago to treat recalcitrant compounds (e.g., halogenated olefins such as TCE), or metal ions (for example, Cr(VI) to Cr(III)) (Powell et al. 1995; Warren et al. 1995; Deng et al. 1999, 2000; Su and Puls 2001; Litter et al. 2018a and references therein). ZVI has been recognized as a potential tool for the removal of contaminants owing to its exceptional properties: it is abundant (iron is the fourth most abundant element in the earth's crust), non-toxic, cheap, easy to be produced, and the process requires little maintenance. Although ZVI is an effective reductant, it offers also good properties to degrade or oxidize pollutants, when combined with H_2O_2 in Fenton-type processes. ZVI can treat metals, halogenated organics, nitroaromatics, metalloids such as As, nitrate, dyes, phenolic compounds, etc. The use of the technology has enormously increased in the last years. The first study on the environmental use of ZVI, describing the kinetics of reduction of Cr(VI) by metallic iron was published in 1982 (Gould 1982), and the first field application of ZVI in the permeable reactive barrier (PRB) technology for in situ remediation of contaminated groundwater was reported in 1991 (O'Hannesin and Gillham 1992). Metallic iron in the form of iron microparticles (granular or powdered, >0.1 mm in diameter) has been used in packed bed reactors and PRBs intersecting the contamination plume (Blowes et al. 2000; Litter et al. 2010). Since then, the use of iron as a reactive material in PRB has been extensively reported (Khudenko 1985; Senzaki 1991; Gillham and O'Hannesin 1994) (see, e.g., Litter et al. 2014, 2018a; Balko and Tratnyek 1998; Cantrell et al. 1995; Domga et al. 2015; Gillham 1993; Orth and Gillham 1995; Matheson and Tratnyek 1994; Senzaki and Kumagai 1988). Over the last two decades, the number of publications has increased exponentially (Fu et al. 2014), including some reviews (Fu et al. 2014; Noubactep 2008, 2016a, b; Cundy et al. 2008; Guan et al. 2015).

In recent years, the use of nanoparticulate zero-valent iron (nZVI) has been reported. nZVI possesses inherent characteristics, different from those of the macroscopic or bulk iron forms (Litter et al. 2018a; Kharisov et al. 2016; Crane and Scott 2012; Noubactep and Caré 2010), especially because the reduction of the particle size of Fe(0) materials from mm to 10–100 nm increases the surface area and thus the chemical reactivity, leading to improved catalytic properties (Zhang 2003). nZVI particles exhibit a typical core–shell structure, where the core consists primarily of a nucleus of zero-valent iron and an external oxide shell, composed of Fe(II) and Fe (III), formed as a result of the corrosion and oxidation of the metallic iron (Kharisov et al. 2016). The outer layer acts not only as protection of the nucleus but also as an adsorption surface for pollutants and participates on the charge transfer processes between Fe(0) and the compounds in solution (Scherer et al. 1999).

The use of nanoparticles is convenient for in situ applications. In addition, other metals such as palladium or nickel are usually added to increase the reduction rate, forming a bimetallic nanometal.

The use of ZVI in a Fenton-type process together with H_2O_2 for the degradation of pollutants (known as advanced Fenton process (AFP)) can avoid or reduce the problems arising in the homogeneous Fenton processes. Oxidation occurs through iron ions released into solution or via reactions taking place between solutes and surface-bound species. The working pH can be higher, but the precipitation of iron hydroxides/oxides can be avoided because few iron ions are present in the aqueous phase; an easy separation of the catalyst after the application can be done. Although in AFP the oxidation reactions are much slower than the performed reactions in homogeneous systems, the heterogeneous reactions consume generally less H_2O_2.

Regarding the removal mechanisms, electrochemical reactions dealing with iron corrosion have to be considered in AFP (Ghauch 2015). A detailed mechanism involves oxidation of Fe(0) by H^+ in the absence of O_2 to yield Fe^{2+}, which is continuously formed and available for reaction with H_2O_2 (Eq. 7.22), producing $HO^{\bullet}$ before being involved in precipitation reactions (Fu et al. 2014).

$$Fe(0) + 2H^+ \rightarrow Fe^{2+} + H_2 \tag{7.76}$$

$$Fe(0) + 2Fe^{3+} \rightarrow 3Fe^{2+} \tag{7.77}$$

$$Fe^{2+} + H_2O_2 \rightarrow Fe^{3+} + HO^{\bullet} \tag{7.22}$$

$$Fe(0) + H_2O_2 + 2H^+ \rightarrow Fe^{2+} + 2H_2O \tag{7.78}$$

$$R^{\bullet} + Fe^{3+} \rightarrow R^+ + Fe(II) \tag{7.31}$$

$$Fe^{2+} + 2H_2O \rightarrow Fe(OH)_2 \downarrow +2H^+ \tag{7.79}$$

$$Fe^{2+} + H_2O_2 \rightarrow FeO^{2+} + H_2O \tag{7.80}$$

In the presence of oxygen, the following additional equations take place, with faster recycling of Fe^{3+} at the iron surface, Eq. 7.87 (Bremner et al. 2006):

$$Fe(0) + O_2 + 2H^+ \rightarrow Fe^{2+} + H_2O_2 \tag{7.81}$$

$$2Fe(0) + O_2 + 4H^+ \rightarrow 2Fe^{2+} + 2H_2O \tag{7.82}$$

$$2Fe(0) + O_2 + 2H_2O \rightarrow 2Fe^{2+} + 4HO^- \tag{7.83}$$

$$6Fe^{2+} + O_2 + 6H_2O \rightarrow 2Fe_3O_4 \downarrow + 12H^+ \tag{7.84}$$

$$4Fe^{2+} + O_2 + 10H_2O \rightarrow 4Fe(OH)_3 \downarrow + 8H^+ \tag{7.85}$$

$$4Fe^{2+} + O_2 + 6H_2O \rightarrow 4FeOOH \downarrow + 8H^+ \tag{7.86}$$

$$2Fe^{3+} + Fe \rightarrow 3Fe^{2+} \tag{7.87}$$

Then, the course of reactions would follow the Fenton typical mechanisms to yield oxidized products or the total mineralization of the organic compound. At neutral pH values, Fe^{2+} oxidation by O_2 most likely produces the ferryl ion (Fe(IV)) (Eq. 7.35) (Litter and Slodowicz 2017). Figure 7.3 shows a simplified scheme of heterogeneous Fenton reactions with ZVI.

For examples regarding the application of ZVI or nZVI in AFP reactions, the readers can consult Litter and Slodowicz (2017). Among the oxidative processes with AFP, As(III) oxidation to As(V) is interesting because the iron corrosion products constitute an excellent adsorbent for As(V) (Morgada et al. 2009).

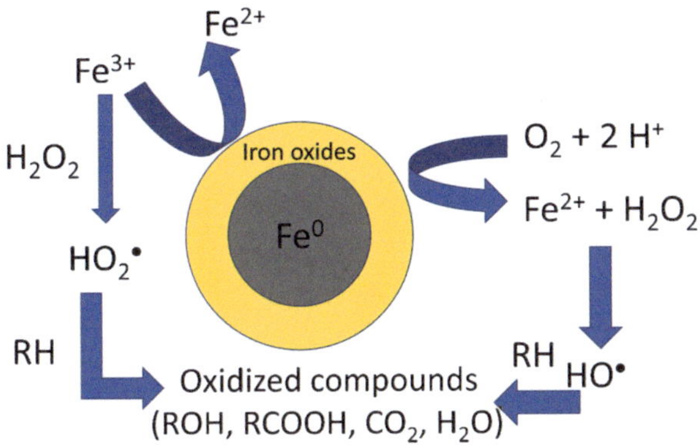

Fig. 7.3 Simplified scheme of heterogeneous Fenton reactions with ZVI

The advanced Fenton process can also be improved by US. Solid particles provide additional nuclei for the cavitation phenomena, increasing the number of cavitation events in the reactor and hence the overall degradation extent (Bagal and Gogate 2014).

7.2.12 *Ferrate*

Ferrates are oxo-compounds of high-valent iron species ($Fe^{VI}O_4^{2-}$, $Fe(V)O_4^{3-}$, Fe (V), and $Fe(IV)O_4^{4-}$). FeO_4^{2-} is a powerful oxidizing agent both in acidic and basic solutions.

$$FeO_4^{2-} + 8H^+ + 3e^- \rightarrow Fe^{3+} + 4H_2O \quad E^0 = +2.20\ V \tag{7.88}$$

$$FeO_4^{2-} + 4H_2O + 3e^- \rightarrow Fe(OH)_3 + 5OH^- \quad E^0 = +0.72\ V \tag{7.89}$$

The redox potential of ferrate(VI) ions in acidic conditions is higher than that of other oxidants/disinfectants (e.g., higher than that of ozone, Table 7.1). In aqueous solutions, ferrate(VI) exhibits a distinctive red-violet color.

$Fe^{VI}O_4^{2-}$ has aroused considerable interest as an environmentally friendly oxidant and disinfectant in remediation processes (Sharma et al. 1999, 2015; Sharma 2002; Jiang and Lloyd 2002; Jiang 2014; Rai et al. 2018). Interestingly, Fe (VI) species exhibit coagulant properties with simultaneous oxidizing power, and these dual properties (coagulation and oxidation/disinfection) are superior for treating pollutants when compared with conventional coagulants (i.e., ferric sulfate and aluminum sulfate), producing also small volumes of sludge. It has been significantly used as a green oxidant in synthetic organic transformations, water oxidation catalysis, waste remediation, high-capacity battery cathodes, destruction of microorganisms, organic and inorganic contaminants, elimination of suspended/colloidal materials and reduction of phosphate concentrations in sewage treatment. Reactions are very fast (seconds to minutes). Ferrate can be used to treat emerging micropollutants, radionuclides, pharmaceuticals, endocrine disruptors, dyes and persistent organic pollutants (POPs). It can also remove metals (arsenite, arsenate, and metals in M−cyanide complexes, with M = Cd, Cu, and Zn), and nonmetals (Sharma et al. 2015; Jiang 2014; Rai et al. 2018; Jiang et al. 2018). Ferrate (VI) exhibits several additional desirable properties including the formation of few byproducts, the absence of mutagenic or carcinogenic byproducts, a high oxidizing power over the entire pH range, and the ability to produce stable salts. There is also a tremendous interest in exploiting potassium ferrate(VI) as a coagulant to deactivate harmful microorganisms (bacteria and viruses).

One of the drawbacks of ferrate is its preparation. Various strategies, including wet chemical, electrochemical, and thermal approaches have been proposed and optimized. The most common way is dry oxidation by heating/melting of various iron oxide-containing minerals under strong alkaline and oxygen flow conditions. Electrochemical methods that employ anodic oxidation using iron or an alloy as the anode and NaOH or KOH as the electrolyte, and wet oxidation of an Fe(III) salt under strong alkaline conditions using hypochlorite or chlorine as the oxidant are other alternatives. However, each method has its advantages and disadvantages, but an effort to improve the ways of the synthesis has been made recently, which can be consulted in Rai et al. (2018).

Numerous reactions occur in the ferrate reaction systems: (i) the generation of Fe (V) and Fe(IV) through one-electron and two-electron transfer processes, (ii) production of radical species that can also generate Fe(V) and Fe(IV) species, (iii) further reactions of Fe(V) and Fe(IV) with contaminants, (iv) self-decompositions of Fe(VI), Fe(V), and Fe(IV) species, (v) reactions of ferrates with ROS, $O_2^{\bullet -}$ and H_2O_2, produced from self-decompositions. Among the inorganic contaminants, cyanides and sulfide react via a one-electron transfer step, while oxy-compounds of sulfur, selenium, arsenic, and nitrogen probably go through an initial two-electron transfer step. Regarding oxidation of organic compounds by Fe (VI), a range of electron equivalents per Fe(VI) (i.e., oxidation capacity) is reported. More aspects and examples can be found in Sharma et al. (2015).

In conclusion, Fe(VI) provides different oxidation capacities to oxidize contaminants while Fe(VI) is reduced to Fe(III) or Fe(II). In some cases, as POPs of fluorocompounds, Fe(VI) cannot degrade the contaminant, but Fe(V) and Fe(IV), using solid compounds such as K_3FeO_4 or Na_4FeO_4 present higher reactivity to degrade these molecules. As a consequence, ferrates are highly promising and environmentally friendly agents exhibiting multimodal activity due to their high oxidation capacity simultaneously combined with disinfection and coagulation. However, some aspects such as the high pH of the water being treated with an alkaline ferrate solution remains to be a great concern. Some problems such as an easier ferrate synthesis, stability control, mechanisms of action, large-scale production of Fe(VI), and large-scale applications have yet to be solved (Jiang et al. 2018).

7.3 Photochemical Technologies

7.3.1 Use of Light Irradiation for Water Purification

Usually, light appreciably increases the reaction rate of AOTs in comparison with the same technology in the absence of illumination. As source of light, high-pressure mercury or xenon arc lamps, with good emission in the near-UV range, can be used. Some applications require short-UV irradiation, and in this case, cheap and easily available germicide lamps can be used. Operating costs are reduced because of a lower power consumption to generate $HO^{\bullet}$ compared with other rather more expensive AOPs. If solar light can be used, a consequent saving of electrical power will be produced, with safer industrial installations. As light will be totally directed to the system, the photochemical industrial equipment can be more compact, and smaller tanks will be employed. The use of light also increases the flexibility of the system, allowing the use of a variety of oxidants and operability conditions. Another advantage of the photochemical technologies is that no drastic changes are needed in the effluents as, for instance, in alkaline ozonation. It is worthwhile to point out, however, that the light-mediated AOPs, especially homogeneous processes, are not adequate for treating mixtures of substances of high absorbance, or containing high amounts of solids in suspension, because the quantum efficiency decreases by loss of light, by dispersion, and/or by competitive light absorption.

Although the use of direct irradiation with UVC light can lead to the transformation of the molecules, direct photolysis is generally not useful for the treatment of pollutants. Direct photolysis is important only for compounds that react very slowly with $HO^{\bullet}$ or do not react at all, as nitrophenols, NO_2^-, halogenated compounds (Burrows et al. 2002), trihalomethanes (THM), chloromethanes, chloroethanes, chlorinated aromatics, and chlorinated phenols. The use of 254-nm irradiation is well documented in the literature (Legrini et al. 1993; Calgon Carbon Corporation 1996), and it is effective to discolor textile dyes at low concentrations; when direct photolysis is compared with other processes such as 254-nm UV/TiO_2 and combined TiO_2 photocatalysis/activated carbon, it was demonstrated that, at low dye concentrations (5–10 mg/L), the photolytic treatment is 2–3 times faster than the other processes for color removal (Gomes da Silva and Faria 2003).

Irradiation with KrCl excimer lamps (222 nm) is used for chlorinated aliphatics as CCl_4 or 1,1,1-trichloroethane because the rupture of the C–Cl bond takes place at 210–230 nm. Generally, the technology is combined with other conventional methods. Limitations of the process are: (i) low efficiency, (ii) application only to compounds absorbing at 200–300 nm, (iii) only one target compound can be treated with reasonably good results. The mechanism and products of UV radiation decomposition have been described for important pollutants such as DDT, lindane, PCP, TNT, and atrazine (Golimowski and Golimowska 1996 and references therein).

In many cases, direct photolysis may be favored in the presence of oxygen and substances that can act as photosensitizers. Sensitizers (Sens) are compounds that absorb visible light and are excited to a higher energy state from which an energy transfer occurs, the excess of energy being transferred to other molecules present in the system (Faust et al. 1999). In this sense, some dyes like Rose Bengal (RB), phthalocyanines, or methylene blue promote singlet oxygen ($^1O_2$) formation in excellent quantum yields (Wilkinson et al. 1993); singlet oxygen is a powerful oxidant able to attack OM and microorganisms (US EPA 1998; García 1994; Meng et al. 2017):

$$\text{Sens} + h\nu \rightarrow {}^1\text{Sens}* \rightarrow {}^3\text{Sens} \tag{7.90}$$

$${}^3\text{Sens} + {}^3O_2 \rightarrow \text{Sens} + {}^1O_2 \tag{7.91}$$

$${}^1O_2 + A \rightarrow AO_2 \tag{7.92}$$

For water purification, the efficiency is strongly dependent on the production rate of $^1O_2$ in the aqueous solution. This process has not been commercialized yet; one of the main restrictions is the requirement of removing the dye from the water after the treatment. For this reason, attempts of immobilization to different supports have been made, but this process leads to a decrease in the efficiency of $^1O_2$ production (Schaap et al. 1975). More research is needed to be done in order to improve this technology on account of the potential the system has for effective disinfection of drinking water.

7.3.2 Vacuum-Ultraviolet Photolysis of Water

In this process, light of wavelengths lower than the UVC (<190 nm) is used, corresponding to the vacuum-ultraviolet range (VUV) (Legrini et al. 1993; Litter 2005; Oppenländer 2003). Xe excimer lamps ($\lambda_{exc} = 172$ nm) are the most used light sources. Excitation with these energies produces the homolytic breakage of chemical bonds, degrading OM in condensed and gaseous phases. Fluorinated and chlorinated hydrocarbons can be converted to dehalogenated compounds (Legrini et al. 1993; US EPA 1998). However, the application of direct VUV is limited, and VUV water photolysis (Eq. 7.93) is the most important use for degradation of pollutants.

$$H_2O + h\nu \rightarrow HO^\bullet + H^\bullet \tag{7.93}$$

This way, $HO^\bullet$ and $H^\bullet$ are generated in situ, without the addition of external agents. As water has high-absorption cross-section, the total incident VUV radiation is absorbed within a very narrow layer around the lamp axle (Lopez et al. 2000; Alapi et al. 2018; Zoschke et al. 2014). Quantum yield of reaction 7.93 varies between 0.33 at 185 nm and 0.72 at 147 nm (Heit et al. 1998). The process produces also aqueous electrons, which are strong reductants, with a lower quantum yield (0.05), almost independent of the wavelength at 160–190 nm (Lopez et al. 2000).

$$H_2O + h\nu \rightarrow HO^{\bullet} + H^+ + e^-{}_{aq} \quad (7.94)$$

In the presence of oxygen, $HO_2^{\bullet}$ and $O_2^{\bullet-}$ are quickly generated:

$$O_2 + H^{\bullet} \rightarrow HO_2^{\bullet} \quad k_{HO2}{}^{\bullet} = 1 \times 10^{10}\ 1/(M\ s) \quad (7.95)$$

$$O_2 + e^-{}_{aq} \rightarrow O_2^{\bullet-} \quad k_{O2}{}^{\bullet-} = 2 \times 10^{10}\ 1/(M\ s) \quad (7.96)$$

The produced oxidants ($HO^{\bullet}$, $HO_2^{\bullet}$, $O_2^{\bullet-}$) and reductants ($H^{\bullet}$, $e^-{}_{aq}$, $HO_2^{\bullet}$, $O_2^{\bullet-}$) allow simultaneous reductions and oxidations. Degradation of pollutants in water and in currents of air (with high humidity) is possible with this technology; it can be used for treating compounds difficult to be oxidized (e.g., chlorinated and fluorinated hydrocarbons such as $ClCH_3$), and in the production of ultrapure water. As VUV lamps present high radiant power, and because of the high cross-section absorption of water at these wavelengths, the process presents high efficiency. No addition of chemical agents is necessary and the technology is simple and competitive. Nevertheless, O_2 and high power supply are required together with quartz reactors. The technology has not been commercialized yet, and it is presently at the development stage.

The development of VUV light sources has opened new possibilities for the in situ generation of $HO^{\bullet}$; the VUV photolysis of water is being actively researched into compared to other AOPs. Various studies have shown that the irradiation of water by 185 nm VUV light results in a fast degradation of organic micropollutants and a sequential decomposition of larger NOM molecules. Additionally, the UV emission of low-pressure mercury vapor lamps at 254 nm leads to a simultaneous disinfection of the water. The main inorganic byproduct in drinking water—nitrite—is cause for concern; it is formed from the nitrate present in the raw water. Especially, the combination of VUV irradiation of water and the generation of O_3 by VUV in the gas phase offers many opportunities for the future water treatment. The combination of VUV and ozone generated by the same lamp enhances the oxidation and disinfection efficiency of the system and minimizes the formation of nitrite (Zoschke et al. 2014).

However, at present, only a few practical applications of VUV irradiation at 185 nm exist. The low penetration of the VUV light is the main technical limitation of VUV irradiation. Additionally, no standards for the VUV irradiation exist and the complex reaction mechanisms involving $HO^{\bullet}$ makes the studies conducted in different reactor systems or different water matrices difficult to be compared. In most practical cases, VUV irradiation is not a real alternative to the conventional processes in water treatment, such as adsorption on activated carbon, or to other advanced oxidation processes because of the outlined limitations of the VUV (Zoschke et al. 2014). However, VUV irradiation offers new possibilities for special applications like the preparation of ultrapure water or as a main treatment process for decentralized systems. As for other AOPs, the most appropriate application in large-scale water treatment is their use as a pretreatment to enhance biodegradability. When UV and VUV photolysis were compared for the pretreatment and posttreatment of coking wastewater, it was concluded that UV irradiation was better in improving the biodegradability of the effluent, while VUV was more effective in photolyzing the residual organic compounds and inorganic N-species in the bio-treated effluent (Xing et al. 2015).

7.3.3 UV/H_2O_2

The oxidizing power of H_2O_2 can be sensibly improved by $HO^•$ generation through cleavage of the O–O union with photons of enough energy (>213 kJ/mol, the energy bond, corresponding to $\lambda < 280$ nm). The reaction has a low quantum yield ($\phi_{H2O2} = 0.5$) due to strong recombination of the radicals in solution (Litter 2005; Boczkaj and Fernandes 2017; Lopez et al. 2000; Bircher et al. 1997), and produces almost quantitatively one $HO^•$ per quantum of radiation absorbed in the 200–300 nm range, i.e., $\phi_{HO•} = 0.5$:

$$H_2O_2 + h\nu \rightarrow 2HO^• \tag{7.97}$$

For a complete description of the mechanistic pathways, Stefan (2018b) can be consulted. H_2O_2 photolysis is usually performed with low- or medium-pressure mercury vapor lamps. Almost 50% of the energetic consumption is lost in the form of heat or emissions lesser than 185 nm, which are absorbed by the quartz jacket. Generally, cheap germicidal lamps (254 nm) are used; however, as H_2O_2 absorption is maximal at 220 nm, it is more convenient to use Xe/Hg lamps that, although more expensive, emit in the 210–240 nm range.

Together with H_2O_2 ($\varepsilon = 18.6$ 1/(M cm) at 254 nm), other species absorb photons at these wavelengths, and they can act as light filters. However, the direct photolysis (if possible) of the contaminants may enhance the oxidative destruction. A turbulent flow is needed to continuously renew the solution in the surroundings of the lamp because the UV radiation decays exponentially towards the solution bulk (Legrini et al. 1993; Litter 2005).

In neutral or slightly alkaline pH media, the concentration of the conjugate anion of hydrogen peroxide (HO_2^-) increases (reaction 7.16). As this species has a higher absorption coefficient ($\varepsilon_{254} = 240$ 1/(M cm)) than H_2O_2, light absorption and $HO^•$ production is increased under these conditions, making the photochemical process more efficient than at acidic pH (Legrini et al. 1993; Beltrán et al. 1996). However, higher pH values promote the formation of bicarbonate and carbonate ions (coming from the mineralization of OM or already present in the waters), which inhibits the oxidative action of $HO^•$:

$$HO^• + HCO_3^- \rightarrow CO_3^{•-} + H_2O \tag{7.98}$$

$$HO^• + CO_3^{2-} \rightarrow CO_3^{•-} + HO^- \tag{7.99}$$

However, this competence takes place in all AOTs involving $HO^•$ in carbonated solutions. Changes due to mineralization processes affect reaction rates and attention should be paid to this in almost all AOTs (Gogate and Pandit 2004a).

In UV/H_2O_2 processes, the rate of degradation of compounds depends on the concentration of H_2O_2. Generally, the rate increases with H_2O_2 concentration up to an optimum value; at higher concentrations, an inhibitory effect has been found (Lopez et al. 2000; López Cisneros et al. 2002). When $HO^•$ concentrations are high, recombination of these radicals is a competitive reaction, which regenerates H_2O_2 (reverse of reaction 7.97); other competitive reactions occur according to reactions in Eqs. 7.100, 7.101, 7.102 and 7.103, from which reactions 7.100 and 7.103 consume $HO^•$ and inhibit oxidation, while $HO_2^•$ radicals are much less reactive than $HO^•$ (Baxendale and Wilson 1957).

$$HO^{\bullet} + H_2O_2 \rightarrow HO_2^{\bullet} + H_2O \tag{7.100}$$

$$HO_2^{\bullet} + H_2O_2 \rightarrow HO^{\bullet} + H_2O + O_2 \tag{7.101}$$

$$2HO_2^{\bullet} \rightarrow H_2O_2 + O_2 \tag{7.102}$$

$$HO_2^{\bullet} + HO^{\bullet} \rightarrow H_2O + O_2 \tag{7.103}$$

Figure 7.4 shows a scheme for the reactions occurring in UV/H_2O_2 processes (Legrini et al. 1993).

In order to avoid this drawback, the optimal H_2O_2 concentration must be determined through treatability tests, as an excess can retard the degradation; this depends on the concentration and chemical nature of the pollutants.

When low-pressure mercury vapor lamps are used, a high H_2O_2 concentration has to be added to produce enough $HO^•$, and the process is less effective. This limitation can be overcome using high-intensity UV lamps.

The advantages of the UV/H_2O_2 process are the following: H_2O_2 is a commercially accessible oxidant, it is thermally stable, and can be easily stored on site, It presents an infinite solubility in water, and produces 2 $HO^•$ per each H_2O_2 molecule, i.e., it is an effective source of $HO^•$. In contrast with processes using ozone, there are no mass transfer problems because the UV/H_2O_2 system is homogeneous, the capital investment is low and the operation is simple. However, as the cross-section absorption at 254 nm of H_2O_2 is low, high concentrations of the reagent should be used, with a continuous monitoring of the concentration during the process in order to avoid its depletion. The method is not efficient for waters exhibiting high absorbance at $\lambda < 300$ nm or containing compounds that can compete with the generation of $HO^•$, requiring a high amount of H_2O_2 in those cases.

UV/H_2O_2 is one of the oldest AOPs. It is useful for the treatment of industrial effluents, especially from dye industry (López Cisneros et al. 2002; Ince et al. 1997). Organochlorinated aliphatics, aromatics, phenols and chlorinated phenols, pesticides, and pharmaceuticals are among the compounds that can be degraded (Legrini et al. 1993; Gogate and Pandit 2004a; Lopez et al. 2003).

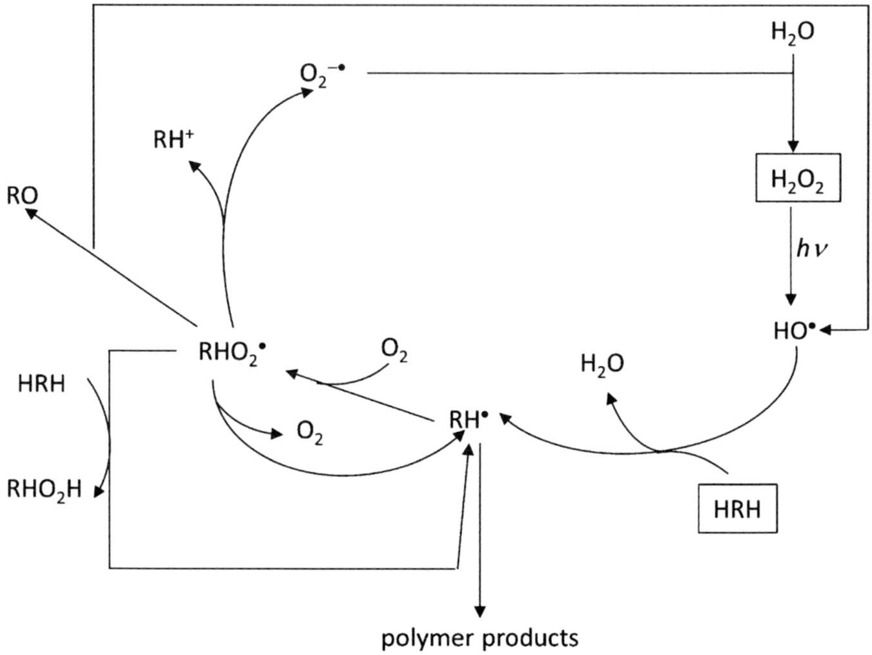

Fig. 7.4 Scheme of the reactions occurring in UV/H_2O_2 systems. (Adapted from Legrini et al. 1993)

At present, UV/H_2O_2 is a commercial technology and it has already been installed in large-scale drinking water applications (Kruithof et al. 2007; Sichel et al. 2011). The main drawback of the technology is the relatively high consumption of electrical energy leading to considerable treatment costs.

The combination with US or a pretreatment with O_3 highly improve the UV/H_2O_2 (Fung et al. 1999; Arslan and Balcioglu 2001). The combination UV/H_2O_2/O_3 is another alternative and will be described later.

Full-scale UV/H_2O_2 AOP installations have been described using Trojan technology (Stefan 2018b).

Recently, the concept of Electrical Energy per Order (E_{EO}), a figure of merit for evaluating the energy requirements of UV AOPs in the degradation of contaminants has been evaluated for the proper application at bench-, pilot-, and full-scale, using sucralose as a standard substance for reactor comparison (Keen et al. 2018). E_{EO} was proposed as a new figure of merit (a numeric descriptor of process efficiency and a valuable design parameter) for UV reactors to replace or complement more ambiguous performance parameters, such as cost per unit volume. E_{EO} was introduced in 1996 to evaluate AOPs (Bolton et al. 1996), later published as a report by the IUPAC Photochemistry Commission (Bolton et al. 2001) and in a recently published book (Collins and Bolton 2016). The E_{EO} is the electrical energy necessary to reduce the concentration of a contaminant by one order of magnitude (90% reduction) in a unit volume of water. The E_{EO} involves only the electrical energy input to the process in order to achieve the degradation target and it can be applied to any other AOP that uses UV irradiation, but it does not give explanations about the mechanism or details of the process. It is a more fundamental concept than the previous parameters and, if properly reported, it could be used to compare different systems based on reports and papers from different years and laboratories, regardless of the energy costs in a given year or the different experimental setups used.

7.3.4 Photoinduced Ozonation (UV/O_3)

UV irradiation of O_3 in water produces H_2O_2 quantitatively:

$$O_3 + h\nu + H_2O \rightarrow H_2O_2 + O_2 \tag{7.104}$$

The generated H_2O_2 is photolyzed (Eq. 7.97), generating $HO^{\bullet}$ radicals, and also reacts with the excess of O_3, according to Eq. 7.21. This method might be considered in principle only as an expensive form of generating H_2O_2 and then $HO^{\bullet}$. Indeed, the process is a combination of UV/H_2O_2 and O_3/H_2O_2, but the advantage is that O_3 has a higher absorption coefficient than H_2O_2 ($\varepsilon_{254} = 3{,}300$ 1/(M cm)), and such method can be used to treat water with high UV absorption background. The efficiency is higher than that of O_3 or direct UV, and the reactor does not need to be in quartz because UVB light (280–315 nm) can be used. If $\lambda < 310$ nm is used, photolysis of O_3 takes place, generating additional $HO^{\bullet}$ and other oxidants, with the subsequent increase of the efficiency (Peyton and Glaze 1988):

$$O_3 + h\nu \rightarrow O_2\left({}^1\Delta g\right) + O\left({}^1D\right) \tag{7.105}$$

$$O\left({}^1D\right) + H_2O \rightarrow 2HO^{\bullet} \tag{7.106}$$

Generally, an increase of the O_3 concentration increases the degradation rate of the pollutant, as demonstrated for the case of atrazine. Although direct ozonation can contribute, 87% of the oxidation process proceeds, in this case, through the radical pathway (Beltrán et al. 1994a).

In contrast with the results in the absence of light, alkaline pH reduces the reaction rate, as it has been observed in the case of 2,6-DNT degradation. The decrease on the rate is due to the dissociation of $HO^{\bullet}$ in the less active oxygen anion radical (Eq. 7.107) and to the lower solubility of O_3 at high pH (Beltrán et al. 1998).

$$HO^{\bullet} \rightarrow O^{\bullet -} + H^+ \tag{7.107}$$

Although ozonation is improved under UV light, it was found that the use of high initial concentrations of O_3 (1000 mg/L) (without irradiation) was more effective than the UV/O_3 combination to treat formulated pesticides like atrazine, alachlor, carbofuran, etc., because of the presence of large amounts of $HO^{\bullet}$ scavengers in the formulations (Chiron et al. 2000).

It was recently demonstrated that solar light is also because it enhances ozonation, as proved in the degradation of two model organic compounds, phenol and malic acid. This process has been called Heliozon. Rates of OM removal were also higher, and faster and complete mineralization was achieved even at large initial TOC values (as large as 49,000 ppm). This represents a possible way to increase O_3 reactivity at a low cost. The simultaneous presence of sunlight and Fe(II) in solution produced also a beneficial effect in the mineralization, less effective with other metal ions like Cu (II), Ni(II), Mn(II), and Co(II) (Sánchez et al. 2003).

The method has been applied to potable water, to highly contaminated wastewater, in disinfection, discoloration of waters of the paper industry, degradation of chlorinated aliphatic hydrocarbons, etc. In Gurol and Akata (1996), the first applications of the technology are mentioned. Ozonation is greatly improved when UV irradiation is combined with TiO_2 (see Sect. 7.3.12).

The use of O_3, as indicated in Sect. 7.2.2, requires high capital costs and poses problems to safety and health.

7.3.5 $UV/O_3/H_2O_2$

Addition of light to the H_2O_2/O_3 process produces a net increase on the efficiency. The thermal process is accelerated, especially the very slow reaction in Eq. 7.21. In the combination of UV with H_2O_2/O_3, $HO^\bullet$ generation is enhanced:

$$2O_3 + H_2O_2 + h\nu \rightarrow 3O_2 + 2HO^\bullet \tag{7.108}$$

The three separate processes, UV/H_2O_2, UV/O_3, and $UV/H_2O_2/O_3$, have shown to be very effective for decontamination of groundwater and for soil remediation (Huang et al. 1993; US EPA 1998; Domènech et al. 2004; Litter 2005). In contrast to UV/O_3 and UV/H_2O_2 technologies, commercially available (US EPA 1998), $UV/H_2O_2/O_3$ application studies are at present only at the pilot-plant scale.

O_3 is fairly stable in dry air and has a half-life of several hours in low concentration. In water, O_3 half-life is several minutes, and because it is very reactive in an aqueous environment, it can promote oxidations 10 to 1000 times faster (Hoigné and Bader 1983) than most oxidants used in water treatment. Due to its short half-life, O_3 cannot be compressed and stored; instead, it must be generated on site and used immediately. Electrical generation is the only practical and safe method for large-scale applications, and it can be generated by corona discharge generators.

7.3.6 Photo-Fenton Process

Fenton processes can be highly improved by UV/visible irradiation ($\lambda < 600$ nm). As opposed to dark Fenton processes, where Fe^{3+} ions are accumulated in the system and the reaction cannot proceed once Fe^{2+} ions are totally consumed, under light, Fe^{2+} ions are regenerated from Fe^{3+} by photoreduction. The enhancement was explained by four main reasons (Litter and Slodowicz 2017; Faust and Hoigné 1990):

1. photolysis of iron(III) hydroxocomplexes (λ < 580 nm) yields extra $HO^{\bullet}$ and regenerates Fe(II) (Eqs. 7.109 and 7.110) by an MLCT reaction (Bauer and Fallmann 1997; Bauer et al. 1999; Pliego et al. 2015):

$$\left[\mathrm{Fe(OH)}^{2+}\right] + h\nu \rightarrow \mathrm{Fe}^{2+} + \mathrm{HO}^{\bullet} \tag{7.109}$$

$$[\mathrm{Fe(H_2O)}]^{3+} + h\nu \rightarrow \mathrm{Fe}^{2+} + \mathrm{HO}^{\bullet} + \mathrm{H}^{+} \tag{7.110}$$

2. photogenerated Fe(II) participates in the Fenton reaction (Eq. 7.23) to produce additional $HO^{\bullet}$, increasing the oxidation rate in comparison with the dark Fenton process (Gogate and Pandit 2004a; Pignatello 1992);
3. if $\lambda < 310$ nm is used, photolysis of H_2O_2 takes place as an additional $HO^{\bullet}$ source (Bigda 1995):

$$\mathrm{H_2O_2} + h\nu \rightarrow 2\mathrm{HO}^{\bullet} \tag{7.111}$$

4. photolysis of Fe(III) chelates (Fe_3L) formed between Fe^{3+} and the organic substrate, its degradation intermediates, or other possible ligands present in the reaction medium makes the use of the photons up to the visible spectrum efficient (Legrini et al. 1993; Lin and Lo 1997):

$$[\mathrm{Fe_3L}] + h\nu \rightarrow \left[\mathrm{Fe^{3+}L}\right]* \rightarrow \mathrm{Fe}^{2+} + \mathrm{L}^{\bullet} \tag{7.112}$$

Depending on the ligand, the Fe(III) complexes have different light absorption properties, and reactions in Eqs. 7.109, 7.110, and 7.112 take place with different quantum yields and at different wavelengths. Fe(III)–carboxylate complexes have much higher quantum yields than Fe(III)–water complexes and promote the reaction intensively. However, the presence of iron complexes in the media can lead to a lesser photodegradation of organic contaminants due to their stronger capacity to absorb radiation (Safarzadeh-Amiri et al. 1997; de Oliveira et al. 2007). In addition, the total amount of iron needed and the sludge generation processes are considerably reduced in the photo-Fenton system (Hermosilla et al. 2009).

The best performance of the photo-Fenton process is at pH around 3, for the same reasons indicated for the dark Fenton processes. At pH 2.8, the dominant iron species in solution is $[Fe(OH)]^{2+}$, which is also the most photoactive Fe(III)–water complex. $Fe(OH)^{2+}$ and $[Fe(H_2O)_5(HO)]^{2+}$ are also more soluble, making iron precipitation less possible, and they are more photoactive to radiation in the 280–405 nm range, allowing the use of sunlight (Bauer et al. 1999; Kim et al. 1997).

Acidic conditions around pH 3 also favor the conversion of the $HO^{\bullet}$ scavengers carbonate and bicarbonate into carbonic acid, which has a low reactivity against $HO^{\bullet}$ (Legrini et al. 1993). At circumneutral pH, it has been suggested that the oxidant species are both $HO^{\bullet}$ and ferryl (Pouran et al. 2015). The participation of Fe(V)=O species in conjunction with visible absorbing iron species has been suggested (Pignatello et al. 1999).

Although the method is efficient, H_2O_2 has to be added continuously and acidic conditions have to be maintained. The most frequent uses have been the treatment of industrial water, soils, and leachates, nitroaromatics, polychlorinated phenols, herbicides (2,4-D and 2,4,5-T), and pesticides. The method is useful to treat high-strength organic wastewaters (Pouran et al. 2015).

The photoelectro-Fenton method (Boye et al. 2003) complements photo-Fenton and EF reaction, but the photoelectrochemical process is more powerful than the EF process (Brillas et al. 2003a).

As said for the dark Fenton processes, ZVI or nZVI can be used also as solid iron materials in photo-Fenton reactions, and the first reports were published in 1998 for pesticides, finding that there was no significant difference in the degradation rate between the UV/Fe(0)/H_2O_2 and the UV/Fe^{2+}/H_2O_2 systems (Doong and Chang 1998a, b). Later, reports on treatment of dyes (Kusic et al. 2006; Devi et al. 2009), and total petroleum hydrocarbons from a diesel fuel (Dehghani et al. 2014) were published. Fenton and photo-Fenton oxidation of petroleum aromatic hydrocarbons using nZVI was recently reported (Fard et al. 2013).

The performance of nZVI in the heterogeneous Fenton process was compared with EF and PEF processes for phenol degradation, finding that PEF was the most efficient process (Babuponnusami and Muthukumar 2012).

7.3.7 Photoferrioxalate and Other Fe(III) Complexes

Oxalic acid forms complexes with Fe(III) that absorb strongly from 254 to 450 nm, being active up to 510 nm (Hatchard et al. 1956; Pozdnyakov et al. 2008). The absorption corresponds to a LMCT band, with ε_{max} values around 10^3–10^4 1/(M cm). Photolysis of trisoxalatoferrate(III) (ferrioxalate) constitutes the most used chemical actinometer; the quantum yield of Fe^{2+} formation is high ($\phi = 1.0$–1.2) and almost independent of the wavelength (Hatchard et al. 1956).

If H_2O_2 is added, the photochemical reduction of the Fe(III) complex will be coupled to a Fenton reaction (Eq. 7.22) (Domènech et al. 2004; Litter 2005; Safarzadeh-Amiri et al. 1997; Zuo and Hoigné 1992). Thus, the use of illuminated mixtures of H_2O_2 and ferrioxalate is very efficient for the photodegradation of organic contaminants; the energy required to treat the same volume of the selected wastewater is ca. 20% of the energy required by the common photo-Fenton system (Safarzadeh-Amiri et al. 1997; Safarzadeh-Amiri et al. 1996a, b; Nogueira and Jardim 1999).

The main reactions in the photoferrioxalate/H_2O_2 system are described by the following reactions (Lee et al. 2003; Quici et al. 2005):

$$Fe(C_2O_4)_3{}^{3-} + h\nu \rightarrow Fe^{2+} + 2C_2O_4{}^{2-} + C_2O_4{}^{\bullet -} \quad (7.113)$$

$$C_2O_4{}^{\bullet -} \rightarrow CO_2{}^{\bullet -} + CO_2 \quad (7.114)$$

$$CO_2{}^{\bullet -} + O_2 \rightarrow O_2{}^{\bullet -} + CO_2 \quad (7.115)$$

$$CO_2{}^{\bullet -} + Fe(C_2O_4)_3{}^{3-} \rightarrow Fe^{2+} + 3C_2O_4{}^{2-} + CO_2 \quad (7.116)$$

$$HO_2{}^{\bullet}\ (\text{or } O_2{}^{\bullet -}) + Fe^{2+} + H^+\ (2H^+) \rightarrow Fe^{3+} + H_2O_2 \quad (7.117)$$

$$HO_2{}^{\bullet}\ (\text{or } O_2{}^{\bullet -}) + Fe(C_2O_4)_3{}^{3-} \rightarrow Fe^{2+} + 3C_2O_4{}^{2-} + O_2 + H^+ \quad (7.118)$$

$$HO_2{}^{\bullet}\ (\text{or } O_2{}^{\bullet -}) + Fe(OH)^{2+} \rightarrow Fe^{2+} + O_2 \quad (7.119)$$

$$HO_2{}^{\bullet} + HO_2{}^{\bullet}\ (\text{or } O_2{}^{\bullet -} + H^+) \rightarrow H_2O_2 + O_2 \quad (7.120)$$

Reaction in Eq. 7.118 is the predominant one at high [H_2O_2] (in the mM range) and acidic pH, while at low [H_2O_2] (μM range), reaction in Eq. 7.117 is the preferred way. Then, the Fenton reaction (Eq. 7.22) takes place.

The method is useful to treat waters with high absorbance at $\lambda < 300$ nm because the high ferrioxalate absorption cross-section in the 200 to 400 nm range and solar light can be used. The reagents are totally water soluble, and there are no mass transfer limitations. The process is cheap and the oxidant is accessible.

Ferrioxalate technology has been used for the treatment of aromatic and chloroaromatic hydrocarbons, chlorinated ethylenes, ethers, alcohols, ketones, and other compounds. Nevertheless, total mineralization is seldom attained. Other iron carboxylates were tested (EDTA, NTA, citrate, etc.) (Safarzadeh-Amiri et al. 1996a, b; Nogueira et al. 1999).

7.3.8 *Photo-Fenton and Ozonation*

The combination of photo-Fenton and O_3 yields a high destruction efficiency of organic compounds like phenol (Canton et al. 2003), 2,4-D (Brillas et al. 2003a, b), aniline, or 2,4-chlorophenol (Kasprzyk-Hordern et al. 2003 and references therein). As said in Sect. 7.2.2, metal ions catalyze O_3 decomposition. The combination of ozone with UV light and iron as catalyst improves the oxidative capability of the system owing to the regeneration of Fe(III). In the presence of visible light, Fe(III) ions can be reduced to Fe(II) by a photo-Fenton process, closing a loop mechanism where Fe species act as catalysts while generating additional $HO^{\bullet}$ and ferryl radicals. Irradiation with UV light also causes $HO^{\bullet}$ generation by the direct UV/O_3 pathway and photo-Fenton reactions. The interaction of Fe(III) and ligand species in solution, which ends in photochemically active complexes, can also take place in these complex systems (Litter 2005).

However, the information is still rare regarding the effects of UV light on the ZVI system. In the case of nitrate reduction by Fe(0), a detrimental effect of 254-nm irradiation on Fe(II) dissolution and nitrate removal was reported. It seems that the role of UV light is strongly dependent on the solution composition (Liao et al. 2003). These processes deserve profound further research.

7.3.9 UV/Chlorine

Chlorine is commonly used in disinfection processes, e.g., for inactivating pathogens. Hypochlorous acid is a weak acid ($pK_a = 7.5$) and, at pH < 5, the dominant species is HOCl, whereas at pH > 10, OCl^- is the main species. Chlorine is added during the primary stages of drinking water treatment to help to control taste and odor, color, and bacterial growth in filtration beds. However, the detection of known or suspected carcinogenic disinfection byproducts in distribution systems, from treatment of raw waters containing even moderate levels of NOM, has led to a closer inspection and careful monitoring of chlorine dosages and residuals (Watts and Linden 2007; Chan et al. 2012).

Free chlorine absorbs UVC and/or UVB photons with quantum yields greater than 1.0. Nowell and Hoigné (1992) confirmed that the predominant active species from the UV/chlorine process are $HO^•$, enabling the process to become a potential AOP. On the other hand, OCl^- absorption is relatively strong in the UV region and the maximum of the photolysis is at 292 nm, but the high-wavelength tail significantly overlaps the UV end of the solar spectrum. The following reactions occur in the UV/chlorine process (Jin et al. 2011):

$$Cl_2 + H_2O \rightarrow HOCl + HCl \quad (7.121)$$

$$HOCl \leftrightarrows H^+ + OCl^- \quad (7.122)$$

$$HOCl + h\nu \rightarrow HO^• + Cl^• \quad (7.123)$$

$$OCl^- + h\nu \rightarrow O^{•-} + Cl^• \quad (7.124)$$

$$O^{•-} + H_2O \rightarrow HO^• + OH^- \quad (7.125)$$

In alkaline solution, the products of OCl^- photolysis include singlet and triplet oxygen. The potential of UV/HOCl under mildly acidic conditions has been qualified as an alternative to the UV/H_2O_2 AOP (Watts and Linden 2007). However, the higher maximum $HO^•$ production-yield factors in the UV/H_2O_2 process make it more efficient than in the UV/chlorine one for the removal of organic compounds, and more investigation should be done (Jin et al. 2011).

A study compares UV/chlorine with UV/H_2O_2 focusing on the economical and energy saving potential of the process. The design of the process considerably reduces costs, energy consumption, and byproduct generation from UV/HOCl AOPs. Energy reductions of 30–75%, depending on the specific compound, were observed compared with the UV/H_2O_2 process. As energy consumption generates the big part of the process costs up to 30–50%, cost savings can be expected (Sichel et al. 2011).

7.3.10 UV/Periodate

Periodic acid, H_5IO_6, and periodate, IO_4^-, are strong oxidants (see Table 7.1). Irradiation of periodate solutions under short-UV light generates several radicals ($IO_3^•$, $HO^•$, $IO_4^•$) and other oxidative species (IO_3^-, HOI, I_2, H_2O_2, O_3). The oxidation of a system containing this reagent under UV light is less selective but more efficient than other AOTs. The proposed mechanism can be very complex (Weavers et al. 1997).

With this technology, a wide variety of compounds at low concentrations can be destroyed. It can be used for discoloration of dye-containing waters and for treatment of other wastewaters. For a better effectiveness, waters should have a low absorbance. So far, there are no legislated discharge requirements for iodine compounds, from which I_2 and I^- are the more toxic (but still of low toxicity). Iodine can be recovered by ionic exchange, and periodate can be electrochemically regenerated. The technology is faster than other photochemical AOTs and seems very promising. Recently *p*-nitrophenol was treated with UV-activated potassium periodate leading to 79.5% degradation after 60 min at the optimum conditions: 30 mg/L of the substrate, potassium periodate concentration 386.3 mg/L, pH 6.2 (Saien and Bazkiaei 2018).

7.3.11 UV/Persulfate

The absorption spectrum of PDS is very similar to that of H_2O_2, characterized by a gradual increase in absorption without a noticeable maximum in the accessible wavelength region. At 254 nm, $\varepsilon = 20$ L/(mol cm), and a dissociative transition is reached, leading to the scission of the –O–O– linkage. $\phi_{SO4•-}$ is lower than 2, corresponding to a quantum yield of PDS consumption of below unity (0.58–0.79) (Mark et al. 1990). As said before (Sect. 7.2.10), since the reactions of PDS are generally slow at normal temperature, the photochemical activation of PDS to $SO_4^{•-}$, working at ambient temperature, has been proposed (Eq. 7.126), followed by reactions in Eqs. 7.70, 7.71 and 7.72 (Lau et al. 2007; Anipsitakis and Dionysiou 2004b):

$$S_2O_8{}^{2-} + h\nu \rightarrow 2SO_4{}^{\bullet -} \tag{7.126}$$

Due to the adverse effects of some metal catalysts such as Co(II) on human health, PMS activated by UV irradiation was considered as a better, environmentally friendly and applicable technology (Guan et al. 2011). PMS presents absorption maxima at $\lambda = 248$ nm and at $\lambda = 254$ nm, and after irradiation at $\lambda = 254$ nm, $HO^{\bullet}$ and $SO_4^{\bullet-}$ are generated through the homolytic cleavage of the peroxo bond (Karpel Vel Leitner 2018; Brienza and Katsoyiannis 2017). Quantum yield of $SO_4^{\bullet-}$ formation is much larger than that of $HO^{\bullet}$ formation from H_2O_2 (Karpel Vel Leitner 2018). $SO_4^{\bullet-}$ is transformed into $HO^{\bullet}$ in strong alkaline media. PMS is activated to produce both $SO_4^{\bullet-}$ and $HO^{\bullet}$ radicals, when the peroxide bond (–O–O–) is homolytically cleaved, as reported in the following equations (Brienza and Katsoyiannis 2017):

$$HSO_5^{-} + h\nu \rightarrow SO_4^{\bullet -} + HO^{\bullet} \quad (7.127)$$

$$HSO_5^{-} + M^{n+} \rightarrow SO_4^{\bullet -} + M^{(n+1)+} + OH^{-} \quad (7.73)$$

$$HSO_5^{-} + M^{n+} \rightarrow M^{(n+1)+} + HO^{\bullet} \quad (7.128)$$

The following order of efficiency has been reported: UV/PDS > UV/PMS > UV/H_2O_2, supported by the absorbance of the oxidants at 254 nm and the higher photosensitivity. For catalyzed UV/M/Ox technologies, tested for several metals, the order was: UV/Fe(III)/H_2O_2 > UV/Fe(II)/H_2O_2 > UV/Co(II)/PMS > UV/Ag(I)/PDS (Anipsitakis and Dionysiou 2004b).

7.3.12 *Heterogeneous Photocatalysis*

Heterogeneous photocatalysis is a process based either on direct or indirect absorption of visible or UV radiant energy by a solid, normally a wide bandgap semiconductor. In the interfacial region between the excited solid and the solution, destruction or removal of contaminants takes place, with no chemical change of the catalyst. Several reviews can be found about this technology (e.g., Legrini et al. 1993; US EPA 1998; Calgon Carbon Corporation 1996; Litter 2005; Fujishima and Zhang 2006; Lee and Park 2013; Ohtani 2010).

When a semiconductor particle is excited by light of energy higher than that of the bandgap, electron–hole pairs are created, and electrons and holes migrate to the surface where they react with adsorbed species, acceptors (A) or donors (D) (Fig. 7.5) (Mills and Le Hunte 1997). Electron–hole pairs that cannot separate and/or react with surface species, recombine with energy dissipation. Various materials are candidates to act as photocatalysts as, for example, TiO_2, ZnO, CdS, iron oxides, WO_3, ZnS, etc. These materials are economically available, and many of them participate in chemical processes in the nature. Besides, most of these materials can be excited with light of wavelength in the range of the solar spectrum ($\lambda > 310$ nm); this increases the interest of a possible use of sunlight. So far, the most investigated photocatalysts are metallic oxides, particularly TiO_2; this

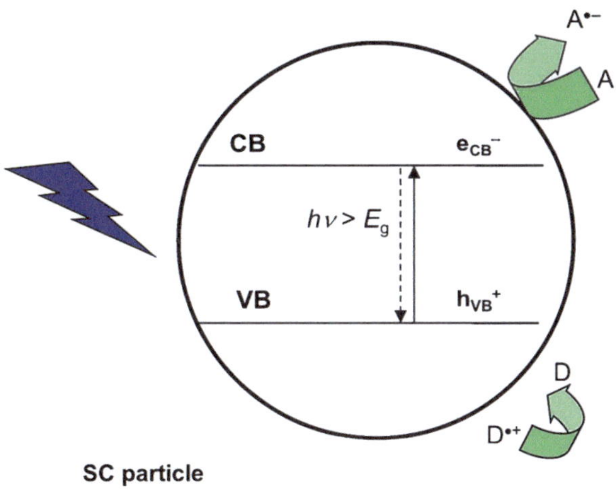

Fig. 7.5 Scheme of a heterogeneous photocatalytic process

semiconductor presents a high chemical stability and can be used in a wide pH range, being able to produce electronic transitions by light absorption in the near ultraviolet range (UVA). The general mechanism for TiO_2 is:

$$TiO_2 + h\nu \rightarrow e_{CB}{}^{-} + h_{VB}{}^{+} \tag{7.129}$$

$$A + e_{CB}{}^{-} \rightarrow A^{\bullet -} \tag{7.130}$$

$$D + h_{VB}{}^{+} \rightarrow D^{\bullet +} \tag{7.131}$$

$$H_2O_{ads}/OH_{surf}{}^{-} + h_{VB}{}^{+} \rightarrow HO_{surf}{}^{\bullet} \tag{7.132}$$

$$2\,HO_{surf}{}^{\bullet} \rightarrow H_2O_2 \tag{7.133}$$

$$H_2O_2 + h_{VB}{}^{+} \rightarrow HO_2{}^{\bullet} + H^{+} \tag{7.134}$$

In this way, $HO^{\bullet}$ and other ROS are generated, and they participate in oxidation processes of pollutants. The driving force for the electron transfer process in the interface is the difference of the energy between the levels of the semiconductor and the redox potential of the species close to the particle surface. The photogenerated holes give rise to the oxidation of a donor D to $D^{\bullet +}$ while the electrons of the conduction band lead to the reduction of an acceptor A to $A^{\bullet -}$. The most usual semiconductors present oxidative valence bands (redox potentials from +1 to +3.5 V) and moderately reductive conduction bands (+0.5 to −1.5 V) (Morrison 1980). Thus, in the presence of redox species close or adsorbed to the semiconductor particle and under illumination, simultaneous oxidation and reduction reactions can take place in the semiconductor–solution interface. The holes react with adsorbed substances, in particular with adsorbed water or OH^{-} ions, generating $HO^{\bullet}$ radicals and/or other radicals, as in other AOTs. Normally, in environmental applications, the photocatalytic processes take place in aerobic environments, and the adsorbed oxygen is the principal electron acceptor species:

$$O_2 + e_{CB}^{-} \rightarrow O_2^{\bullet -} \quad (7.135)$$

The efficiency of the photocatalytic reaction depends on different factors. One of the most critical aspects is the high probability of electron–hole recombination, which competes with the separation of the photogenerated charges. On the other hand, as there is no physical separation between the anodic reaction site (oxidation by holes) and the cathodic one (reduction by electrons), back reactions can be of importance. The low efficiency is one of the most severe limitations of heterogeneous photocatalysis. In addition, the use of UV light restricts the technology when TiO_2 is used, but efforts to extend the photocatalytic activity to the visible range have already been made (Banerjee et al. 2014; Litter et al. 2018b; Pelaez et al. 2012; Dong et al. 2015). Solar applications have been reviewed (Blanco-Galvez et al. 2007).

Heterogeneous photocatalysis over TiO_2 can be also combined with other AOTs. For example, addition of Fe(III) and H_2O_2 combines UV/TiO_2 with photo-Fenton; in this way, the destruction of some resistant pollutants can be improved.

Combination of UV/TiO_2 and O_3 (photocatalytic ozonation) is also possible (Mehrjouei et al. 2015). O_3 acts there as a powerful oxidant in place of O_2, which has a slow electron transfer from TiO_2 (reaction in Eq. 7.135) (Kasprzyk-Hordern et al. 2003; Gilbert 2002). In the presence of TiO_2, O_3 generates $HO^{\bullet}$ through the formation of an ozonide radical ($O_3^{\bullet -}$) in the adsorption layer:

$$O_3 + \text{photocatalyst} \rightarrow O^{\bullet} + O_2 \quad (7.136)$$

$$O_3 + h\nu \rightarrow O^{\bullet} + O_2 \quad (7.137)$$

$$H_2O_2 + h\nu \rightarrow 2HO^{\bullet} \quad (7.138)$$

$$O_3 + e_{CB}^{-} \rightarrow O_3^{\bullet -} \quad (7.139)$$

$$O_2 + e_{CB}^{-} \rightarrow O_2^{\bullet -} \quad (7.140)$$

$$H_2O_2 + e_{CB}^{-} \rightarrow HO^{\bullet} + OH^{-} \quad (7.141)$$

$$O_3^{\bullet -} + H^{+} \rightarrow HO_3^{\bullet} \quad (7.12)$$

$$HO_3^{\bullet} \rightarrow HO^{\bullet} + O_2 \quad (7.13)$$

$HO^{\bullet}$ generation from O_3 is pH-dependent and it increases as the pH decreases. This fact avoids the use of high alkaline pH to induce $HO^{\bullet}$ formation from O_3. In many cases, the extents of mineralization achieved this way are higher than when using single AOTs.

Photocatalytic ozonation is among the most expensive treatment technologies, and, therefore, its use for the removal of biodegradable pollutants from water is not justifiable. The particular importance of this oxidation method lies in the ability to destroy poorly biodegradable organic compounds such as pharmaceutics, surfactants, detergents, dyes, herbicides and pesticides, aromatic and aliphatic organohalogens, saturated aliphatic carboxylic acids, nitroaromatics, or in the ability to improve the biological degradability of wastewater samples containing these compounds.

Another interesting combination is heterogeneous photocatalysis with US, because this process hinders the inactivation of the catalyst by reaction intermediates, which usually block the catalyst. US also reduces mass transfer limitations occurring in the case of immobilized catalysts (see Gogate and Pandit 2004a for a detailed description of this combined process).

Photocatalysis can be combined also with ferrate (see, e.g., Sharma et al. 2015 and Sharma et al. 2010).

7.4 Conclusions

Chemical oxidation processes, and particularly AOPs, represent a powerful means for the abatement of refractory and/or toxic pollutants in wastewaters. Different AOP techniques have been developed and their use either alone or combined allows choosing the most appropriate procedure for the treatment of specific systems. These technologies can process wastewaters resistant to conventional treatments and are complementary to them.

A generalization on the application of an AOT can never be made. Each effluent must be previously characterized, and treatability tests at laboratory scale must be performed to choose the most appropriate method. Knowledge of kinetics, with establishment of the limiting step and limiting reagent(s), and comparison with other conventional treatments should be available before applying the technology. Obviously, a complex chemical composition always presents a greater difficulty than simple mixtures, and $HO^{\bullet}$ scavengers are usually the main source of efficiency reduction.

In general, AOPs are more adequate for treating small flows (or volumes) and not too high concentrations of pollutants. Small COD contents, not higher than 5 g/L, can be suitably treated. Higher concentrations would require high concentrations of expensive reagents and/or high electrical power consumption.

Each AOT has an optimum working pH value and the pH variation during the reaction must be continuously controlled. At the end of the process, another pH adjustment will be needed in many cases before the biological treatment or to comply with local regulations before discharging the effluent to the receiving bodies.

The use of toxicological tests (Microtox, Amphitox, etc.) to control the formation of noxious byproducts along the process is mandatory. The purpose is to use the technology until toxicity is reduced to a certain level, beyond which a conventional, less expensive method can bring about the mineralization process with the obvious reduction of costs.

Although a lot of research has been done to understand mechanistic and kinetic aspects of AOTs, which can be improved in the future by new investigations, some requirements are still needed for the wide commercialization, referred to reactor optimization and modeling.

In conclusion, considerable effort should be made to improve and adapt AOPs for water and soil treatment, combining the studies with fundamental work for the thorough understanding of the mechanisms and kinetic aspects related to the systems.

Acknowledgement This work was partially supported by Agencia Nacional de Promoción Científica y Tecnológica of Argentina, PICT-2015-208 and BioCriticalMetals – ERAMIN 2015 grants.

References

Achugasim O, Ojinnaka CM, Osuji LC (2014) Potassium permanganate as an oxidant in the remediation of soils polluted by Bonny light crude oil. Sky J Soil Sci Environ Manage 3(2):4–19

Adewuyi YG (2001) Sonochemistry: environmental science and engineering applications. Ind Eng Chem Res 40(22):4681–4715. https://doi.org/10.1021/ie010096l

Alapi T, Schrantz K, Arany E, Kozmér Z (2018) Vacuum UV radiation-driven processes. In: Stefan MI (ed) Advanced oxidation processes for water treatment: fundamentals and applications. IWA Publishing, London, pp 195–240

Al-Sheikhly M, Poster DL, An J-C, Neta P, Silverman J, Huie RE (2006) Ionizing radiation-induced destruction of benzene and dienes in aqueous media. Environ Sci Technol 40(9):3082–3088. https://doi.org/10.1021/es052533j

Andreozzi R, Caprio V, Insola A, Marotta R (1999) Advanced oxidation processes (AOP) for water purification and recovery. Catal Today 53(1):51–59. https://doi.org/10.1016/S0920-5861(99)00102-9

Anipsitakis GP, Dionysiou DD (2004a) Radical generation by the interaction of transition metals with common oxidants. Environ Sci Technol 38(13):3705–3712. https://doi.org/10.1021/es035121o

Anipsitakis GP, Dionysiou DD (2004b) Transition metal/UV-based advanced oxidation technologies for water decontamination. Appl Catal B 54(3):155–163. https://doi.org/10.1016/j.apcatb.2004.05.025

Arslan I, Balcioglu IA (2001) Advanced oxidation of raw and biotreated textile industry wastewater with O_3, H_2O_2/UV-C and their sequential application. J Chem Technol Biotechnol 76(1):53–60. https://doi.org/10.1002/1097-4660(200101)76:1<53::AID-JCTB346>3.0.CO;2-T

Babuponnusami A, Muthukumar K (2012) Removal of phenol by heterogenous photo electro Fenton-like process using nano-zero valent iron. Sep Purif Technol 98:130–135. https://doi.org/10.1016/j.seppur.2012.04.034

Babuponnusami A, Muthukumar K (2014) A review on Fenton and improvements to the Fenton process for wastewater treatment. J Environ Chem Eng 2(1):557–572. https://doi.org/10.1016/j.jece.2013.10.011

Bagal MV, Gogate PR (2014) Wastewater treatment using hybrid treatment schemes based on cavitation and Fenton chemistry: a review. Ultrason Sonochem 21(1):1–14. https://doi.org/10.1016/j.ultsonch.2013.07.009

Balko BA, Tratnyek PG (1998) Photoeffects on the reduction of carbon tetrachloride by zero-valent iron. J Phys Chem B 102(8):1459–1465. https://doi.org/10.1021/jp973113m

Banerjee S, Pillai SC, Falaras P, O'Shea KE, Byrne JA, Dionysiou DD (2014) New insights into the mechanism of visible light photocatalysis. J Phys Chem Lett 5(15):2543–2554. https://doi.org/10.1021/jz501030x

Basfar AA, Mohamed KA, Al-Abduly AJ, Al-Kuraiji TS, Al-Shahrani AA (2007) Degradation of diazinon contaminated waters by ionizing radiation. Radiat Phys Chem 76(8–9):1474–1479. https://doi.org/10.1016/j.radphyschem.2007.02.055

Bauer R, Fallmann H (1997) The photo-Fenton oxidation – a cheap and efficient wastewater treatment method. Res Chem Intermed 23(4):341–354. https://doi.org/10.1163/156856797X00565

Bauer R, Waldner G, Fallmann H, Hager S, Klare M, Krutzler T, Malato S, Maletzky P (1999) The photo-fenton reaction and the TiO_2/UV process for waste water treatment – novel developments. Catal Today 53(1):131–144. https://doi.org/10.1016/S0920-5861(99)00108-X

Baxendale JH, Wilson JA (1957) The photolysis of hydrogen peroxide at high light intensities. Trans Faraday Soc 53:344–356. https://doi.org/10.1039/TF9575300344

Beltrán FJ, García-Araya JF, Acedo B (1994a) Advanced oxidation of atrazine in water – II. Ozonation combined with ultraviolet radiation. Water Res 28(10):2165–2174. https://doi.org/10.1016/0043-1354(94)90028-0

Beltrán FJ, González M, Rivas J, Marín M (1994b) Oxidation of mecoprop in water with ozone and ozone combined with hydrogen peroxide. Ind Eng Chem Res 33(1):125–136. https://doi.org/10.1021/ie00025a017

Beltrán FJ, Ovejero G, Rivas J (1996) Oxidation of polynuclear aromatic hydrocarbons in water. 3. UV radiation combined with hydrogen peroxide. Ind Eng Chem Res 35(3):883–890. https://doi.org/10.1021/ie950363l

Beltrán FJ, Encinar JM, Alonso MA (1998) Nitroaromatic hydrocarbon ozonation in water. 2. Combined ozonation with hydrogen peroxide or UV radiation. Ind Eng Chem Res 37(1):32–40. https://doi.org/10.1021/ie970426v

Bigda RJ (1995) Consider Fenton's chemistry for wastewater treatment. Chem Eng Prog 91(12):62–66

Bircher KG, Lem W, Simms KM, Dussert BW (1997) Combination of UV oxidation with other treatment technologies for the remediation of contaminated water. J Adv Oxid Technol 2(3):435–441. https://doi.org/10.1515/jaots-1997-0311

Blanco-Galvez J, Fernández-Ibáñez P, Malato-Rodríguez S (2007) Solar photocatalytic detoxification and disinfection of water: recent overview. J Sol Energy Eng 129(1):4–15. https://doi.org/10.1115/1.2390948

Blowes DW, Ptacek CJ, Benner SG, McRae CWT, Bennett TA, Puls RW (2000) Treatment of inorganic contaminants using permeable reactive barriers. J Contam Hydrol 45(1–2):123–137. https://doi.org/10.1016/S0169-7722(00)00122-4

Boczkaj G, Fernandes A (2017) Wastewater treatment by means of advanced oxidation processes at basic pH conditions: a review. Chem Eng J 320:608–633. https://doi.org/10.1016/j.cej.2017.03.084

Bolton JR, Cater SR (1994) Homogeneous photodegradation of pollutants in contaminated water: an introduction. In: Helz GR, Zepp RG, Crosby DG (eds) Aquatic and surface photochemistry. CRC Press, Boca Raton, p 467

Bolton JR, Bircher KG, Tumas W, Tolman CA (1996) Figures-of-merit for the technical development and application of advanced oxidation processes. J Adv Oxid Technol 1(1):13–17. https://doi.org/10.1515/jaots-1996-0104

Bolton JR, Bircher KG, Tumas W, Tolman CA (2001) Figures-of-merit for the technical development and application of advanced oxidation technologies for both electric- and solar-driven systems (IUPAC Technical Report). Pure Appl Chem 73(4):627–637. https://doi.org/10.1351/pac200173040627

Boye B, Dieng MM, Brillas E (2003) Anodic oxidation, electro-Fenton and photoelectro-Fenton treatments of 2,4,5-trichlorophenoxyacetic acid. J Electroanal Chem 557:135–146. https://doi.org/10.1016/S0022-0728(03)00366-8

Bremner DH, Burgess AE, Houllemare D, Namkung K-C (2006) Phenol degradation using hydroxyl radicals generated from zero-valent iron and hydrogen peroxide. Appl Catal B 63 (1–2):15–19. https://doi.org/10.1016/j.apcatb.2005.09.005
Brienza M, Katsoyiannis IA (2017) Sulfate radical technologies as tertiary treatment for the removal of emerging contaminants from wastewater. Sustainability 9(9):1604. https://doi.org/10.3390/su9091604
Brillas E, Mur E, Sauleda R, Sánchez L, Peral J, Domènech X, Casado J (1998) Aniline mineralization by AOP's: anodic oxidation, photocatalysis, electro-Fenton and photoelectro-Fenton processes. Appl Catal B 16(1):31–42. https://doi.org/10.1016/S0926-3373(97)00059-3
Brillas E, Baños MÁ, Garrido JA (2003a) Mineralization of herbicide 3,6-dichloro-2-methoxybenzoic acid in aqueous medium by anodic oxidation, electro-Fenton and photoelectro-Fenton. Electrochim Acta 48(12):1697–1705. https://doi.org/10.1016/S0013-4686(03)00142-7
Brillas E, Calpe JC, Cabot P-L (2003b) Degradation of the herbicide 2,4-dichlorophenoxyacetic acid by ozonation catalyzed with Fe^{2+} and UVA light. Appl Catal B 46(2):381–391. https://doi.org/10.1016/S0926-3373(03)00266-2
Burrows HD, Canle LM, Santaballa JA, Steenken S (2002) Reaction pathways and mechanisms of photodegradation of pesticides. J Photochem Photobiol B 67(2):71–108. https://doi.org/10.1016/S1011-1344(02)00277-4
Cai C, Zhang H, Zhong X, Hou L (2015) Ultrasound enhanced heterogeneous activation of peroxymonosulfate by a bimetallic Fe–Co/SBA-15 catalyst for the degradation of Orange II in water. J Hazard Mater 283:70–79. https://doi.org/10.1016/j.jhazmat.2014.08.053
Calgon Carbon Corporation (1996) The AOT handbook: advanced oxidation technologies. Calgon Carbon Corporation, Ontario
Canton C, Esplugas S, Casado J (2003) Mineralization of phenol in aqueous solution by ozonation using iron or copper salts and light. Appl Catal B 43(2):139–149. https://doi.org/10.1016/S0926-3373(02)00276-X
Cantrell KJ, Kaplan DI, Wietsma TW (1995) Zero-valent iron for the in situ remediation of selected metals in groundwater. J Hazard Mater 42(2):201–212. https://doi.org/10.1016/0304-3894(95)00016-N
Chan PY, El-Din MG, Bolton JR (2012) A solar-driven UV/chlorine advanced oxidation process. Water Res 46(17):5672–5682. https://doi.org/10.1016/j.watres.2012.07.047
Chaychian M, Silverman J, Al-Sheikhly M (1999) Ionizing radiation induced degradation of tetrachlorobiphenyl in transformer oil. Environ Sci Technol 33(14):2461–2464. https://doi.org/10.1021/es9900914
Chen Y-P, Liu S-Y, Yu H-Q, Yin H, Li Q-R (2008) Radiation-induced degradation of methyl orange in aqueous solutions. Chemosphere 72(4):532–536. https://doi.org/10.1016/j.chemosphere.2008.03.054
Chiron S, Fernandez-Alba A, Rodriguez A, Garcia-Calvo E (2000) Pesticide chemical oxidation: state-of-the-art. Water Res 34(2):366–377. https://doi.org/10.1016/S0043-1354(99)00173-6
Collins J, Bolton JR (2016) Advanced oxidation handbook. American Water Works Association, Denver
Comninellis C, Kapalka A, Malato S, Parsons SA, Poulios I, Mantzavinos D (2008) Advanced oxidation processes for water treatment: advances and trends for R&D. J Chem Technol Biotechnol 83(6):769–776. https://doi.org/10.1002/jctb.1873
Crane RA, Scott TB (2012) Nanoscale zero-valent iron: future prospects for an emerging water treatment technology. J Hazard Mater 211–212:112–125. https://doi.org/10.1016/j.jhazmat.2011.11.073
Cundy AB, Hopkinson L, Whitby RLD (2008) Use of iron-based technologies in contaminated land and groundwater remediation: a review. Sci Total Environ 400(1–3):42–51. https://doi.org/10.1016/j.scitotenv.2008.07.002
de Oliveira IS, Viana L, Verona C, Vargas Fallavena VL, Nunes Azevedo CM, Pires M (2007) J Hazard Mater 146(3):564–568. https://doi.org/10.1016/j.jhazmat.2007.04.057

Dehghani M, Shahsavani E, Farzadkia M, Reza Samaei M (2014) Optimizing photo-Fenton like process for the removal of diesel fuel from the aqueous phase. J Environ Health Sci Eng 12:87. https://doi.org/10.1186/2052-336X-12-87

Deng Y, Ezyske CM (2011) Sulfate radical-advanced oxidation process (SR-AOP) for simultaneous removal of refractory organic contaminants and ammonia in landfill leachate. Water Res 45(18):6189–6194. https://doi.org/10.1016/j.watres.2011.09.015

Deng B, Burris DR, Campbell TJ (1999) Reduction of vinyl chloride in metallic iron−water systems. Environ Sci Technol 33(15):2651–2656. https://doi.org/10.1021/es980554q

Deng N, Luo F, Wu F, Xiao M, Wu X (2000) Discoloration of aqueous reactive dye solutions in the UV/Fe^0 system. Water Res 34(8):2408–2411. https://doi.org/10.1016/S0043-1354(00)00099-3

Destaillats H, Colussi AJ, Joseph JM, Hoffmann MR (2000a) Synergistic effects of sonolysis combined with ozonolysis for the oxidation of azobenzene and methyl orange. J Phys Chem A 104(39):8930–8935. https://doi.org/10.1021/jp001415+

Destaillats H, Hung H-M, Hoffmann MR (2000b) Degradation of alkylphenol ethoxylate surfactants in water with ultrasonic irradiation. Environ Sci Technol 34(2):311–317. https://doi.org/10.1021/es990384x

Destaillats H, Hoffmann MR, Wallace HC (2003) Sonochemical degradation of pollutants. In: Tarr MA (ed) Chemical degradation methods for wastes and pollutants: environmental and industrial applications. CRC Press, Boca Raton, pp 5-1–5-33

Devi LG, Rajashekhar KE, Raju KSA, Kumar SG (2009) Kinetic modeling based on the non-linear regression analysis for the degradation of Alizarin Red S by advanced photo Fenton process using zero valent metallic iron as the catalyst. J Mol Catal A Chem 314(1–2):88–94. https://doi.org/10.1016/j.molcata.2009.08.021

Dhakshinamoorthy A, Navalon S, Alvaro M, Garcia H (2012) Metal nanoparticles as heterogeneous Fenton catalysts. ChemSusChem 5(1):46–64. https://doi.org/10.1002/cssc.201100517

Domènech X, Jardim W, Litter M (2004) Tecnologías avanzadas de oxidación para la eliminación de contaminantes. In: Blesa MA, Sánchez Cabrero B (eds) Eliminación de contaminantes por fotocatálisis heterogénea. CIEMAT, Madrid, pp 7–34

Domga R, Togue-Kamga F, Noubactep C, Tchatchueng J-B (2015) Discussing porosity loss of Fe^0 packed water filters at ground level. Chem Eng J 263:127–134. https://doi.org/10.1016/j.cej.2014.10.105

Dong S, Feng J, Fan M, Pi Y, Hu L, Han X, Liu M, Sun J, Sun J (2015) Recent developments in heterogeneous photocatalytic water treatment using visible light-responsive photocatalysts: a review. RSC Adv 5(19):14610–14630. https://doi.org/10.1039/C4RA13734E

Doong R-a, Chang W-h (1998a) Photoassisted iron compound catalytic degradation of organophosphorous pesticides with hydrogen peroxide. Chemosphere 37(13):2563–2572. https://doi.org/10.1016/S0045-6535(98)00038-1

Doong R-a, Chang W-h (1998b) Photodegradation of parathion in aqueous titanium dioxide and zero valent iron solutions in the presence of hydrogen peroxide. J Photochem Photobiol A 116(3):221–228. https://doi.org/10.1016/S1010-6030(98)00292-5

Fajerwerg K, Debellefontaine H (1996) Wet oxidation of phenol by hydrogen peroxide using heterogeneous catalysis Fe-ZSM-5: a promising catalyst. Appl Catal B 10(4):L229–L235. https://doi.org/10.1016/S0926-3373(96)00041-0

Fard MA, Torabian A, Bidhendi GRN, Aminzadeh B (2013) Fenton and photo-Fenton oxidation of petroleum aromatic hydrocarbons using nanoscale zero-valent iron. J Environ Eng 139(7):966–974. https://doi.org/10.1061/(ASCE)EE.1943-7870.0000705

Faust BC, Hoigné J (1990) Photolysis of Fe (III)-hydroxy complexes as sources of OH radicals in clouds, fog and rain. Atmos Environ Part A 24(1):79–89. https://doi.org/10.1016/0960-1686(90)90443-Q

Faust D, Funken K-H, Horneck G, Milow B, Ortner J, Sattlegger M, Schäfer M, Schmitz C (1999) Immobilized photosensitizers for solar photochemical applications. Sol Energy 65(1):71–74. https://doi.org/10.1016/S0038-092X(98)00099-1

Fenton HJJ (1894) LXXIII. – Oxidation of tartaric acid in presence of iron. J Chem Soc Trans 65:899–910. https://doi.org/10.1039/CT8946500899
Fernandes A, Pacheco MJ, Ciríaco L, Lopes A (2015) Review on the electrochemical processes for the treatment of sanitary landfill leachates: Present and future. Appl Catal B 176–177:183–200. https://doi.org/10.1016/j.apcatb.2015.03.052
Fischbacher A, Lutze HV, Schmidt TC (2018) Ozone/H_2O_2 and ozone/UV processes. In: Stefan MI (ed) Advanced oxidation processes for water treatment: fundamentals and applications. IWA Publishing, London, pp 163–194
Fu FL, Dionysiou DD, Liu H (2014) The use of zero-valent iron for groundwater remediation and wastewater treatment: a review. J Hazard Mater 267:194–205. https://doi.org/10.1016/j.jhazmat.2013.12.062
Fujishima A, Zhang X (2006) Titanium dioxide photocatalysis: present situation and future approaches. C R Chim 9(5–6):750–760. https://doi.org/10.1016/j.crci.2005.02.055
Fung PC, Huang Q, Tsui SM, Poon CS (1999) Treatability study of organic and colour removal in desizing/dyeing wastewater by UV/US system combined with hydrogen peroxide. Water Sci Technol 40(1):153–160. https://doi.org/10.2166/wst.1999.0034
García NA (1994) New trends in photobiology: Singlet-molecular-oxygen-mediated photodegradation of aquatic phenolic pollutants. A kinetic and mechanistic overview. J Photochem Photobiol B 22(3):185–196. https://doi.org/10.1016/1011-1344(93)06932-S
Gerrity D, Rosario-Ortiz FL, Wert EC (2018) Application of ozone in water and wastewater treatment. In: Stefan MI (ed) Advanced oxidation processes for water treatment: fundamentals and applications. IWA Publishing, London, pp 123–162
Ghauch A (2015) Iron-based metallic systems: an excellent choice for sustainable water treatment. Freiberg Online Geosci 38:1–80
Gilbert E (2002) Influence of ozone on the photocatalytic oxidation of organic compounds. Ozone Sci Eng 24(2):75–82. https://doi.org/10.1080/01919510208901598
Gillham RW (1993) Cleaning halogenated contaminants from groundwater. US Patent 5,266,213, 30 November 1993
Gillham RW, O'Hannesin SF (1994) Enhanced degradation of halogenated aliphatics by zero-valent iron. Groundwater 32(6):958–967. https://doi.org/10.1111/j.1745-6584.1994.tb00935.x
Glaze WH (1987) Drinking-water treatment with ozone. Environ Sci Technol 21(3):224–230. https://doi.org/10.1021/es00157a001
Glaze WH, Kang J-W, Chapin DH (1987) The chemistry of water treatment processes involving ozone, hydrogen peroxide and ultraviolet radiation. Ozone Sci Eng 9(4):335–352. https://doi.org/10.1080/01919518708552148
Glaze WH, Beltran F, Tuhkanen T, Kang J-W (1992) Chemical models of advanced oxidation processes. Water Pollut Res J Can 27(1):23–42
Gogate PR, Pandit AB (2004a) A review of imperative technologies for wastewater treatment I: oxidation technologies at ambient conditions. Adv Environ Res 8(3–4):501–551. https://doi.org/10.1016/S1093-0191(03)00032-7
Gogate PR, Pandit AB (2004b) A review of imperative technologies for wastewater treatment II: hybrid methods. Adv Environ Res 8(3–4):553–597. https://doi.org/10.1016/S1093-0191(03)00031-5
Golimowski J, Golimowska K (1996) UV-photooxidation as pretreatment step in inorganic analysis of environmental samples. Anal Chim Acta 325(3):111–133. https://doi.org/10.1016/0003-2670(96)00034-7
Gomes da Silva C, Faria JL (2003) Photochemical and photocatalytic degradation of an azo dye in aqueous solution by UV irradiation. J Photochem Photobiol A 155(1–3):133–143. https://doi.org/10.1016/S1010-6030(02)00374-X
Gould JP (1982) The kinetics of hexavalent chromium reduction by metallic iron. Water Res 16(6):871–877. https://doi.org/10.1016/0043-1354(82)90016-1

Guan Y-H, Ma J, Li X-C, Fang J-Y, Chen L-W (2011) Influence of pH on the formation of sulfate and hydroxyl radicals in the UV/peroxymonosulfate system. Environ Sci Technol 45 (21):9308–9314. https://doi.org/10.1021/es2017363

Guan X, Sun Y, Qin H, Li J, Lo IMC, He D, Dong H (2015) The limitations of applying zero-valent iron technology in contaminants sequestration and the corresponding countermeasures: the development in zero-valent iron technology in the last two decades (1994–2014). Water Res 75:224–248. https://doi.org/10.1016/j.watres.2015.02.034

Gurol MD, Akata A (1996) Kinetics of ozone photolysis in aqueous solution. AIChE J 42 (11):3283–3292. https://doi.org/10.1002/aic.690421128

Hatchard CG, Parker CA, Bowen EJ (1956) A new sensitive chemical actinometer – II. Potassium ferrioxalate as a standard chemical actinometer. Proc R Soc A 235(1203):518–536. https://doi.org/10.1098/rspa.1956.0102

Heit G, Neuner A, Saugy P-Y, Braun AM (1998) Vacuum-UV (172 nm) actinometry. The quantum yield of the photolysis of water. J Phys Chem A 102(28):5551–5561. https://doi.org/10.1021/jp980130i

Henglein A (1987) Sonochemistry: historical developments and modern aspects. Ultrasonics 25 (1):6–16. https://doi.org/10.1016/0041-624X(87)90003-5

Hermosilla D, Cortijo M, Huang CP (2009) Optimizing the treatment of landfill leachate by conventional Fenton and photo-Fenton processes. Sci Total Environ 407(11):3473–3481. https://doi.org/10.1016/j.scitotenv.2009.02.009

Hoffmann MR, Hua I, Höchemer R (1996) Application of ultrasonic irradiation for the degradation of chemical contaminants in water. Ultrason Sonochem 3(3):S163–S172. https://doi.org/10.1016/S1350-4177(96)00022-3

Hoigné J, Bader H (1976) The role of hydroxyl radical reactions in ozonation processes in aqueous solutions. Water Res 10(5):377–386. https://doi.org/10.1016/0043-1354(76)90055-5

Hoigné J, Bader H (1983) Rate constants of reactions of ozone with organic and inorganic compounds in water – I: non-dissociating organic compounds. Water Res 17(2):173–183. https://doi.org/10.1016/0043-1354(83)90098-2

Hu J, Wang J (2007) Degradation of chlorophenols in aqueous solution by γ-radiation. Radiat Phys Chem 76(8–9):1489–1492. https://doi.org/10.1016/j.radphyschem.2007.02.058

Huang CP, Dong C, Tang Z (1993) Advanced chemical oxidation: its present role and potential future in hazardous waste treatment. Waste Manage 13(5–7):361–377. https://doi.org/10.1016/0956-053X(93)90070-D

Ikehata K, Gamal El-Din M (2006) Aqueous pesticide degradation by hydrogen peroxide/ultraviolet irradiation and Fenton-type advanced oxidation processes: a review. J Environ Eng Sci 5 (2):81–135. https://doi.org/10.1139/s05-046

Ince NH, Stefan MI, Bolton JR (1997) UV/H_2O_2 degradation and toxicity reduction of textile azo dyes: Remazol Black-B, a case study. J Adv Oxid Technol 2(3):442–448. https://doi.org/10.1515/jaots-1997-0312

Iurascu B, Siminiceanu I, Vione D, Vicente MA, Gil A (2009) Phenol degradation in water through a heterogeneous photo-Fenton process catalyzed by Fe-treated laponite. Water Res 43 (5):1313–1322. https://doi.org/10.1016/j.watres.2008.12.032

Jacobsen F, Holcman J, Sehested K (1998) Reactions of the ferryl ion with some compounds found in cloud water. Int J Chem Kinet 30(3):215–221. https://doi.org/10.1002/(SICI)1097-4601(1998)30:3<215::AID-KIN7>3.0.CO;2-V

Jiang J-Q (2014) Advances in the development and application of ferrate(VI) for water and wastewater treatment. J Chem Technol Biotechnol 89(2):165–177. https://doi.org/10.1002/jctb.4214

Jiang J-Q, Lloyd B (2002) Progress in the development and use of ferrate(VI) salt as an oxidant and coagulant for water and wastewater treatment. Water Res 36(6):1397–1408. https://doi.org/10.1016/S0043-1354(01)00358-X

Jiang J-Q, Stanford C, Petri M (2018) Practical application of ferrate(VI) for water and wastewater treatment – site study's approach. Water Energy Nexus 1(1):42–46. https://doi.org/10.1016/j.wen.2018.05.001

Jin J, El-Din MG, Bolton JR (2011) Assessment of the UV/chlorine process as an advanced oxidation process. Water Res 45(4):1890–1896. https://doi.org/10.1016/j.watres.2010.12.008

Karpel Vel Leitner N (2018) Sulfate radical ion – based AOPs. In: Stefan MI (ed) Advanced oxidation processes for water treatment: fundamentals and applications. IWA Publishing, London, pp 429–460

Kasprzyk-Hordern B, Ziółek M, Nawrocki J (2003) Catalytic ozonation and methods of enhancing molecular ozone reactions in water treatment. Appl Catal B 46(4):639–669. https://doi.org/10.1016/S0926-3373(03)00326-6

Keen O, Bolton J, Litter M, Bircher K, Oppenländer T (2018) Standard reporting of Electrical Energy per Order (EEO) for UV/H_2O_2 reactors (IUPAC Technical Report). Pure Appl Chem 90 (9):1487–1499. https://doi.org/10.1515/pac-2017-0603

Kharisov BI, Kharissova OV, Rasika Dias HV, Ortiz Méndez U, Gómez de la Fuente I, Peña Y, Vázquez Dimas A (2016) Iron-based nanomaterials in the catalysis. In: Norena LE, Wang J-A (eds) Advanced catalytic materials. IntechOpen, Rijeka, pp 35–68. https://doi.org/10.5772/61862

Khudenko BM (1985) Mechanism and kinetics of cementation processes. Water Sci Technol 17 (4–5):719–731. https://doi.org/10.2166/wst.1985.0174

Kim S-M, Geissen S-U, Vogelpohl A (1997) Landfill leachate treatment by a photoassisted fenton reaction. Water Sci Technol 35(4):239–248. https://doi.org/10.2166/wst.1997.0128

Kolthoff IM, Miller IK (1951) The chemistry of persulfate: I. The kinetics and mechanism of the decomposition of the persulfate ion in aqueous medium. J Am Chem Soc 73(7):3055–3059. https://doi.org/10.1021/ja01151a024

Kremer ML (1999) Mechanism of the Fenton reaction. Evidence for a new intermediate. Phys Chem Chem Phys 1(15):3595–3605. https://doi.org/10.1039/A903915E

Kremer ML (2003) The Fenton reaction. Dependence of the rate on pH. J Phys Chem A 107 (11):1734–1741. https://doi.org/10.1021/jp020654p

Kronholm J, Riekkola M-L (1999) Potassium persulfate as oxidant in pressurized hot water. Environ Sci Technol 33(12):2095–2099. https://doi.org/10.1021/es9805977

Kruithof JC, Kamp PC, Martijn BJ (2007) UV/H_2O_2 treatment: a practical solution for organic contaminant control and primary disinfection. Ozone Sci Eng 29(4):273–280. https://doi.org/10.1080/01919510701459311

Kusic H, Koprivanac N, Srsan L (2006) Azo dye degradation using Fenton type processes assisted by UV irradiation: a kinetic study. J Photochem Photobiol A 181(2–3):195–202. https://doi.org/10.1016/j.jphotochem.2005.11.024

Larpparisudthi OA, Mason TJ, Paniwnyk L (2018) Ultrasound wave-based AOPs. In: Stefan MI (ed) Advanced oxidation processes for water treatment: Fundamentals and applications. IWA Publishing, London, pp 461–492

Lau TK, Chu W, Graham NJD (2007) The aqueous degradation of butylated hydroxyanisole by UV/$S_2O_8^{2-}$: study of reaction mechanisms via dimerization and mineralization. Environ Sci Technol 41(2):613–619. https://doi.org/10.1021/es061395a

Lee S-Y, Park S-J (2013) TiO_2 photocatalyst for water treatment applications. J Ind Eng Chem 19 (6):1761–1769. https://doi.org/10.1016/j.jiec.2013.07.012

Lee Y, Jeong J, Lee C, Kim S, Yoon J (2003) Influence of various reaction parameters on 2,4-D removal in photo/ferrioxalate/H_2O_2 process. Chemosphere 51(9):901–912. https://doi.org/10.1016/S0045-6535(03)00044-4

Legrini O, Oliveros E, Braun AM (1993) Photochemical processes for water treatment. Chem Rev 93(2):671–698. https://doi.org/10.1021/cr00018a003

Legube B, Karpel Vel Leitner N (1999) Catalytic ozonation: a promising advanced oxidation technology for water treatment. Catal Today 53(1):61–72. https://doi.org/10.1016/S0920-5861(99)00103-0

Li R, Zhang L, Wang P (2015) Rational design of nanomaterials for water treatment. Nanoscale 7 (41):17167–17194. https://doi.org/10.1039/C5NR04870B

Liao C-H, Kang S-F, Hsu Y-W (2003) Zero-valent iron reduction of nitrate in the presence of ultraviolet light, organic matter and hydrogen peroxide. Water Res 37(17):4109–4118. https://doi.org/10.1016/S0043-1354(03)00248-3

Lin SH, Lo CC (1997) Fenton process for treatment of desizing wastewater. Water Res 31 (8):2050–2056. https://doi.org/10.1016/S0043-1354(97)00024-9

Litter MI (2005) Introduction to photochemical advanced oxidation processes for water treatment. In: Boule P, Bahnemann DW, Robertson PKJ (eds) Environmental photochemistry Part II, The handbook of environmental chemistry, vol 2M. Springer, Berlin/Heidelberg, pp 325–366. https://doi.org/10.1007/b138188

Litter MI, Slodowicz M (2017) An overview on heterogeneous Fenton and photoFenton reactions using zerovalent iron materials. J Adv Oxid Technol 20(1):20160164. https://doi.org/10.1515/jaots-2016-0164

Litter MI, Morgada ME, Bundschuh J (2010) Possible treatments for arsenic removal in Latin American waters for human consumption. Environ Pollut 158(5):1105–1118. https://doi.org/10.1016/j.envpol.2010.01.028

Litter MI, Cortina JL, Fiúza AMA, Futuro A, Tsakiroglou C (2014) In-situ technologies for groundwater treatment: the case of arsenic. In: Bundschuh J, Holländer HM, Ma LQ (eds) In-situ remediation of arsenic-contaminated sites. CRC Press, London, pp 1–34

Litter MI, Quici N, Meichtry M (eds) (2018a) Iron nanomaterials for water and soil treatment. Pan Stanford Publishing Pte Ltd., Singapore

Litter MI, San Román E, Grela MA, Meichtry JM, Rodríguez HB (2018b) Sensitization of TiO_2 by dyes: a way to extend the range of photocatalytic activity of TiO_2 to the visible region. In: Ghosh S (ed) Visible light-active photocatalysis: nanostructured catalyst design, mechanisms, and applications. Weinheim, Wiley-VCH Verlag. https://doi.org/10.1002/9783527808175.ch10

Liu X, Zhang T, Zhou Y, Fang L, Shao Y (2013) Degradation of atenolol by UV/peroxymonosulfate: kinetics, effect of operational parameters and mechanism. Chemosphere 93(11):2717–2724. https://doi.org/10.1016/j.chemosphere.2013.08.090

López Cisneros R, Gutarra Espinoza A, Litter MI (2002) Photodegradation of an azo dye of the textile industry. Chemosphere 48(4):393–399. https://doi.org/10.1016/S0045-6535(02)00117-0

Lopez JL, García Einschlag FS, González MC, Capparelli AL, Oliveros E, Hashem TM, Braun AM (2000) Hydroxyl radical initiated photodegradation of 4-chloro-3,5-dinitrobenzoic acid in aqueous solution. J Photochem Photobiol A 137(2–3):177–184. https://doi.org/10.1016/S1010-6030(00)00357-9

Lopez A, Bozzi A, Mascolo G, Kiwi J (2003) Kinetic investigation on UV and UV/H_2O_2 degradations of pharmaceutical intermediates in aqueous solution. J Photochem Photobiol A 156(1–3):121–126. https://doi.org/10.1016/S1010-6030(02)00435-5

Ma Y-S (2012) Short review: current trends and future challenges in the application of sono-Fenton oxidation for wastewater treatment. Sustain Environ Res 22(5):271–278

Ma H, Wang M, Yang R, Wang W, Zhao J, Shen Z, Yao S (2007) Radiation degradation of Congo Red in aqueous solution. Chemosphere 68(6):1098–1104. https://doi.org/10.1016/j.chemosphere.2007.01.067

Makogon O, Fliount R, Asmus K-D (1998) Formation and degradation of halogenated Organic acids. Radiation versus photocatalytically induced processes. J Adv Oxid Technol 3(1):11–21. https://doi.org/10.1515/jaots-1998-0104

Mantzavinos D, Hellenbrand R, Livingston AG, Metcalfe IS (1996) Reaction mechanisms and kinetics of chemical pretreatment of bioresistant organic molecules by wet air oxidation. Paper presented at the 1st international conference on Oxidation Technologies for Water and Wastewater Treatment, Goslar, Germany, 12–15 May 1996

Mark G, Schuchmann MN, Schuchmann H-P, von Sonntag C (1990) The photolysis of potassium peroxodisulphate in aqueous solution in the presence of *tert*-butanol: a simple actinometer for 254 nm radiation. J Photochem Photobiol A 55(2):157–168. https://doi.org/10.1016/1010-6030(90)80028-V

Martino CJ, Savage PE (1999) Total organic carbon disappearance kinetics for the supercritical water oxidation of monosubstituted phenols. Environ Sci Technol 33(11):1911–1915. https://doi.org/10.1021/es981201u

Matheson LJ, Tratnyek PG (1994) Reductive dehalogenation of chlorinated methanes by iron metal. Environ Sci Technol 28(12):2045–2053. https://doi.org/10.1021/es00061a012

Mededovic Thagard S, Locke BR (2018) Electrical discharge plasma for water treatment. In: Stefan MI (ed) Advanced oxidation processes for water treatment: fundamentals and applications. IWA Publishing, London, pp 493–534

Mehrjouei M, Müller S, Möller D (2015) A review on photocatalytic ozonation used for the treatment of water and wastewater. Chem Eng J 263:209–219. https://doi.org/10.1016/j.cej.2014.10.112

Meichtry JM, Slodowicz M, Cancelada L, Destaillats H, Litter MI (2018) Sonochemical reduction of Cr(VI) in air in the presence of organic additives: what are the involved mechanistic pathways? Ultrason Sonochem 48:110–117. https://doi.org/10.1016/j.ultsonch.2018.05.014

Meng A-N, Chaihu L-X, Chen H-H, Gu Z-Y (2017) Ultrahigh adsorption and singlet-oxygen mediated degradation for efficient synergetic removal of bisphenol A by a stable zirconium-porphyrin metal-organic framework. Sci Rep 7:6297. https://doi.org/10.1038/s41598-017-06194-z

Miller CJ, Wadley S, Waite TD (2018) In: Stefan MI (ed) Advanced oxidation processes for water treatment: fundamentals and applications. IWA Publishing, London, pp 297–332

Mills A, Le Hunte S (1997) An overview of semiconductor photocatalysis. J Photochem Photobiol A 108(1):1–35. https://doi.org/10.1016/S1010-6030(97)00118-4

Mishra VS, Mahajani VV, Joshi JB (1995) Wet air oxidation. Ind Eng Chem Res 34(1):2–48. https://doi.org/10.1021/ie00040a001

Morgada ME, Levy IK, Salomone V, Farías SS, López G, Litter MI (2009) Arsenic (V) removal with nanoparticulate zerovalent iron: effect of UV light and humic acids. Catal Today 143 (3–4):261–268. https://doi.org/10.1016/j.cattod.2008.09.038

Morrison SR (1980) Electrochemistry at semiconductor and oxidized metal electrodes. Plenum Press, New York

Mukherjee R, Kumar R, Sinha A, Lama Y, Saha AK (2016) A review on synthesis, characterization, and applications of nano zero valent iron (nZVI) for environmental remediation. Crit Rev Environ Sci Technol 46(5):443–466. https://doi.org/10.1080/10643389.2015.1103832

Navalon S, Alvaro M, Garcia H (2010) Heterogeneous Fenton catalysts based on clays, silicas and zeolites. Appl Catal B 99(1–2):1–26. https://doi.org/10.1016/j.apcatb.2010.07.006

Neyens E, Baeyens J (2003) A review of classic Fenton's peroxidation as an advanced oxidation technique. J Hazard Mater 98(1–3):33–50. https://doi.org/10.1016/S0304-3894(02)00282-0

Nidheesh PV, Olvera-Vargas H, Oturan N, Oturan MA (2018) Heterogeneous electro-Fenton process: principles and applications. In: Zhou M, Oturan MA, Sirés I (eds) Electro-Fenton process: new trends and scale-up. The handbook of environmental chemistry. Springer, Singapore, pp 85–110. https://doi.org/10.1007/698_2017_72

Nogueira RFP, Jardim WF (1999) Solar photodegradation of water contaminants using potassium ferrioxalate. J Adv Oxid Technol 4(2):223–226

Nogueira RFP, Alberici RM, Mendes MA, Jardim WF, Eberlin MN (1999) Photocatalytic degradation of phenol and trichloroethylene: on-line and real-time monitoring via membrane introduction mass spectrometry. Ind Eng Chem Res 38(5):1754–1758. https://doi.org/10.1021/ie980497+

Noubactep C (2008) A critical review on the process of contaminant removal in Fe^0–H_2O systems. Environ Technol 29(8):909–920. https://doi.org/10.1080/09593330802131602

Noubactep C (2016a) Designing metallic iron packed-beds for water treatment: a critical review. Clean Soil Air Water 44(4):411–421. https://doi.org/10.1002/clen.201400304

Noubactep C (2016b) No scientific debate in the zero-valent iron literature. Clean Soil Air Water 44 (4):330–332. https://doi.org/10.1002/clen.201400780

Noubactep C, Caré S (2010) On nanoscale metallic iron for groundwater remediation. J Hazard Mater 182(1–3):923–927. https://doi.org/10.1016/j.jhazmat.2010.06.009
Nowell LH, Hoigné J (1992) Photolysis of aqueous chlorine at sunlight and ultraviolet wavelengths – I. Degradation rates. Water Res 26(5):593–598. https://doi.org/10.1016/0043-1354(92)90232-S
Oh W-D, Dong Z, Lim T-T (2016) Generation of sulfate radical through heterogeneous catalysis for organic contaminants removal: current development, challenges and prospects. Appl Catal B 194:169–201. https://doi.org/10.1016/j.apcatb.2016.04.003
O'Hannesin SF, Gillham RW (1992) A permeable reaction wall for in situ degradation of halogenated organic compounds. Paper presented at the 45th Canadian Geotechnical Society conference, Toronto, Ontario, 26–28 October 1992
Ohtani B (2010) Photocatalysis A to Z – what we know and what we do not know in a scientific sense. J Photochem Photobiol C 11(4):157–178. https://doi.org/10.1016/j.jphotochemrev.2011.02.001
Olson TM, Barbier PF (1994) Oxidation kinetics of natural organic matter by sonolysis and ozone. Water Res 28(6):1383–1391. https://doi.org/10.1016/0043-1354(94)90305-0
Oppenländer T (2003) Photochemical purification of water and air: advanced oxidation processes (AOPs): principles, reaction mechanisms, reactor concepts. Wiley-VCH, Weinheim
Orth WS, Gillham RW (1995) Dechlorination of trichloroethene in aqueous solution using Fe^0. Environ Sci Technol 30(1):66–71. https://doi.org/10.1021/es950053u
Pankaj (2010) Aqueous inorganic sonochemistry. In: Pankaj, Ashokkumar M (eds) Theoretical and experimental sonochemistry involving inorganic systems. Springer, Dordrecht, pp 213–271. https://doi.org/10.1007/978-90-481-3887-6_9
Park W, Hwang M-H, Kim T-H, Lee M-J, Kim IS (2009) Enhancement in characteristics of sewage sludge and anaerobic treatability by electron beam pre-treatment. Radiat Phys Chem 78 (2):124–129. https://doi.org/10.1016/j.radphyschem.2008.09.010
Parsons S (ed) (2004) Advanced oxidation processes for water and wastewater treatment. IWA Publishing, London
Pelaez M, Nolan NT, Pillai SC, Seery SK, Falaras P, Kontos AG, Dunlop PSM, Hamilton JWJ, Byrne JA, O'Shea K, Entezari MH, Dionysiou DD (2012) A review on the visible light active titanium dioxide photocatalysts for environmental applications. Appl Catal B 125:331–349. https://doi.org/10.1016/j.apcatb.2012.05.036
Pera-Titus M, García-Molina V, Baños MA, Giménez J, Esplugas S (2004) Degradation of chlorophenols by means of advanced oxidation processes: a general review. Appl Catal B 47 (4):219–256. https://doi.org/10.1016/j.apcatb.2003.09.010
Peyton GR, Glaze WH (1988) Destruction of pollutants in water with ozone in combination with ultraviolet radiation. 3. Photolysis of aqueous ozone. Environ Sci Technol 22(7):761–767. https://doi.org/10.1021/es00172a003
Pflieger R, Chave T, Vite G, Jouve L, Nikitenko SI (2015) Effect of operational conditions on sonoluminescence and kinetics of H_2O_2 formation during the sonolysis of water in the presence of Ar/O_2 gas mixture. Ultrason Sonochem 26:169–175. https://doi.org/10.1016/j.ultsonch.2015.02.005
Pignatello JJ (1992) Dark and photoassisted Fe^{3+}-catalyzed degradation of chlorophenoxy herbicides by hydrogen peroxide. Environ Sci Technol 26(5):944–951. https://doi.org/10.1021/es00029a012
Pignatello JJ, Liu D, Huston P (1999) Evidence for an additional oxidant in the photoassisted Fenton reaction. Environ Sci Technol 33(11):1832–1839. https://doi.org/10.1021/es980969b
Pignatello JJ, Oliveros E, MacKay A (2006) Advanced oxidation processes for organic contaminant destruction based on the Fenton reaction and related chemistry. Crit Rev Environ Sci Technol 36 (1):1–84. https://doi.org/10.1080/10643380500326564
Pikaev AK (2001) Application of pulse radiolysis and computer simulation for the study of the mechanism of radiation purification of polluted water. Res Chem Intermed 27(7–8):775–786. https://doi.org/10.1163/15685670152622059

Pliego G, Zazo JA, Garcia-Muñoz P, Munoz M, Casas JA, Rodriguez JJ (2015) Trends in the intensification of the Fenton process for wastewater treatment: an overview. Crit Rev Environ Sci Technol 45(24):2611–2692. https://doi.org/10.1080/10643389.2015.1025646
Pouran SR, Abdul Raman AA, Wan Daud WMA (2014) Review on the application of modified iron oxides as heterogeneous catalysts in Fenton reactions. J Cleaner Prod 64:24–35. https://doi.org/10.1016/j.jclepro.2013.09.013
Pouran SR, Aziz ARA, Daud WMAW (2015) Review on the main advances in photo-Fenton oxidation system for recalcitrant wastewaters. J Ind Eng Chem 21:53–69. https://doi.org/10.1016/j.jiec.2014.05.005
Powell RM, Puls RW, Hightower SK, Sabatini DA (1995) Coupled iron corrosion and chromate reduction: mechanisms for subsurface remediation. Environ Sci Technol 29(8):1913–1922. https://doi.org/10.1021/es00008a008
Pozdnyakov IP, Kel OV, Plyusnin VF, Grivin VP, Bazhin NM (2008) New insight into photochemistry of ferrioxalate. J Phys Chem A 112(36):8316–8322. https://doi.org/10.1021/jp8040583
Quan HN, Teel AL, Watts RJ (2003) Effect of contaminant hydrophobicity on hydrogen peroxide dosage requirements in the Fenton-like treatment of soils. J Hazard Mater 102(2–3):277–289. https://doi.org/10.1016/S0304-3894(03)00214-0
Quici N, Morgada ME, Piperata G, Babay P, Gettar RT, Litter MI (2005) Oxalic acid destruction at high concentrations by combined heterogeneous photocatalysis and photo-Fenton processes. Catal Today 101(3–4):253–260. https://doi.org/10.1016/j.cattod.2005.03.002
Rai PK, Lee J, Kailasa SK, Kwon EE, Tsang YF, Ok YS, Kim K-H (2018) A critical review of ferrate(VI)-based remediation of soil and groundwater. Environ Res 160:420–448. https://doi.org/10.1016/j.envres.2017.10.016
Rauf MA, Ashraf SS (2009) Radiation induced degradation of dyes – an overview. J Hazard Mater 166(1):6–16. https://doi.org/10.1016/j.jhazmat.2008.11.043
Roche P, Prados M (1995) Removal of pesticides by use of ozone or hydrogen peroxide/ozone. Ozone Sci Eng 17(6):657–672. https://doi.org/10.1080/01919512.1995.10555777
Roche P, Volk C, Carbonnier F, Paillard H (1994) Water oxidation by ozone or ozone/hydrogen peroxide using the "Ozotest" or "Peroxotest" methods. Ozone Sci Eng 16(2):135–155. https://doi.org/10.1080/01919519408552418
Rosocha LA, Korzekwa RA (1999) Advanced oxidation and reduction processes in the gas phase using non-thermal plasmas. J Adv Oxid Technol 4(3):247–264
Sadjadi S (2014) Nanocatalytic wastewater treatment system for the removal of toxic organic compounds. In: Kharisov BI, Kharissova OV, Dias HVR (eds) Nanomaterials for environmental protection. Wiley, Hoboken, pp 403–427. https://doi.org/10.1002/9781118845530.ch24
Safarzadeh-Amiri A, Bolton JR, Cater SR (1996a) Ferrioxalate-mediated solar degradation of organic contaminants in water. Sol Energy 56(5):439–443. https://doi.org/10.1016/0038-092X(96)00002-3
Safarzadeh-Amiri A, Bolton JR, Cater SR (1996b) The use of iron in advanced oxidation processes. J Adv Oxid Technol 1(1):18–26. https://doi.org/10.1515/jaots-1996-0105
Safarzadeh-Amiri A, Bolton JR, Cater SR (1997) Ferrioxalate-mediated photodegradation of organic pollutants in contaminated water. Water Res 31(4):787–798. https://doi.org/10.1016/S0043-1354(96)00373-9
Saien J, Bazkiaei MFV (2018) Homogenous UV/periodate process in treatment of *p*-nitrophenol aqueous solutions under mild operating conditions. Environ Technol 39(14):1823–1832. https://doi.org/10.1080/09593330.2017.1340348
Sánchez L, Domènech X, Casado J, Peral J (2003) Solar activated ozonation of phenol and malic acid. Chemosphere 50(8):1085–1093. https://doi.org/10.1016/S0045-6535(02)00699-9
Sanchez-Polo M, López-Peñalver J, Prados-Joya G, Ferro-García MA, Rivera-Utrilla J (2009) Gamma irradiation of pharmaceutical compounds, nitroimidazoles, as a new alternative for water treatment. Water Res 43(16):4028–4036. https://doi.org/10.1016/j.watres.2009.05.033

Schaap AP, Thayer AL, Blossey EC, Neckers DC (1975) Polymer-based sensitizers for photooxidations. II. J Am Chem Soc 97(13):3741–3745. https://doi.org/10.1021/ja00846a030
Scherer MM, Balko BA, Tratnyek PG (1999) The role of oxides in reduction reactions at the metal-water interface. In: Mineral-water interfacial reactions, ACS symposium series, vol 715. American Chemical Society, Washington, DC, pp 301–322. https://doi.org/10.1021/bk-1998-0715.ch015
Scott JP, Ollis DF (1995) Integration of chemical and biological oxidation processes for water treatment: review and recommendations. Environ Prog Sustain Energy 14(2):88–103. https://doi.org/10.1002/ep.670140212
Sedlak DL, Andren AW (1991) Oxidation of chlorobenzene with Fenton's reagent. Environ Sci Technol 25(4):777–782. https://doi.org/10.1021/es00016a024
Senzaki T (1991) Removal of chlorinated organic compounds from wastewater by reduction process: III treatment of trichloroethylene with iron powder. Kogyo Yosui 391:29–35
Senzaki T, Kumagai Y (1988) Removal of chlorinated organic compounds from wastewater by reduction process: treatment of 1, 1, 2, 2-tetrachloroethane with iron powder. Kogyo Yosui 357:2–7
Sharma VK (2002) Potassium ferrate (VI): an environmentally friendly oxidant. Adv Environ Res 6 (2):143–156. https://doi.org/10.1016/S1093-0191(01)00119-8
Sharma VK, Rivera W, Joshi VN, Millero FJ, O'Connor D (1999) Ferrate(VI) oxidation of thiourea. Environ Sci Technol 33(15):2645–2650. https://doi.org/10.1021/es981083a
Sharma VK, Graham NJD, Li X-Z, Yuan B-L (2010) Ferrate(VI) enhanced photocatalytic oxidation of pollutants in aqueous TiO_2 suspensions. Environ Sci Pollut Res 17(2):453–461. https://doi.org/10.1007/s11356-009-0170-0
Sharma VK, Zboril R, Varma RS (2015) Ferrates: Greener oxidants with multimodal action in water treatment technologies. Acc Chem Res 48(2):182–191. https://doi.org/10.1021/ar5004219
Sichel C, Garcia C, Andre K (2011) Feasibility studies: UV/chlorine advanced oxidation treatment for the removal of emerging contaminants. Water Res 45(19):6371–6380. https://doi.org/10.1016/j.watres.2011.09.025
Stefan MI (ed) (2018a) Advanced oxidation processes for water treatment: fundamentals and applications. IWA Publishing, London
Stefan MI (2018b) UV/Hydrogen peroxide process. In: Stefan MI (ed) Advanced oxidation processes for water treatment: fundamentals and applications. IWA Publishing, London, pp 7–122
Stratton GR, Bellona CL, Dai F, Holsen TM, Mededovic Thagard S (2015) Plasma-based water treatment: conception and application of a new general principle for reactor design. Chem Eng J 273:543–550. https://doi.org/10.1016/j.cej.2015.03.059
Su C, Puls RW (2001) Arsenate and arsenite removal by zerovalent iron: kinetics, redox transformation, and implications for in situ groundwater remediation. Environ Sci Technol 35 (7):1487–1492. https://doi.org/10.1021/es001607i
Tang WZ, Chen RZ (1996) Decolorization kinetics and mechanisms of commercial dyes by H_2O_2/iron powder system. Chemosphere 32(5):947–958. https://doi.org/10.1016/0045-6535(95)00358-4
Tang WZ, Tassos S (1997) Oxidation kinetics and mechanisms of trihalomethanes by Fenton's reagent. Water Res 31(5):1117–1125. https://doi.org/10.1016/S0043-1354(96)00348-X
Teel AL, Warberg CR, Atkinson DA, Watts RJ (2001) Comparison of mineral and soluble iron Fenton's catalysts for the treatment of trichloroethylene. Water Res 35(4):977–984. https://doi.org/10.1016/S0043-1354(00)00332-8
Tijani JO, Fatoba OO, Madzivire G, Petrik LF (2014) A review of combined advanced oxidation technologies for the removal of organic pollutants from water. Water Air Soil Pollut 225:2102. https://doi.org/10.1007/s11270-014-2102-y
Tungler A, Szabados E, Hosseini AM (2015) Wet air oxidation of aqueous wastes. In: Samer M (ed) Wastewater treatment engineering. IntechOpen, Rijeka, pp 153–178. https://doi.org/10.5772/60935

US EPA (1997) Innovative technology evaluation report: high voltage environmental applications, Inc., Electron beam technology. EPA/540/R-96/504. National Risk Management Research Laboratory, Office of Research and Development, Cincinnati

US EPA (1998) Handbook on advanced photochemical oxidation processes. EPA/625/R-98/004

Vadillo V, Sánchez-Oneto J, Portela JR, Martínez de la Ossa EJ (2013) Problems in supercritical water oxidation process and proposed solutions. Ind Eng Chem Res 52(23):7617–7629. https://doi.org/10.1021/ie400156c

von Gunten U (2003a) Ozonation of drinking water: Part I. Oxidation kinetics and product formation. Water Res 37(7):1443–1467. https://doi.org/10.1016/S0043-1354(02)00457-8

von Gunten U (2003b) Ozonation of drinking water: Part II. Disinfection and by-product formation in presence of bromide, iodide or chlorine. Water Res 37(7):1469–1487. https://doi.org/10.1016/S0043-1354(02)00458-X

Walling C (1975) Fenton's reagent revisited. Acc Chem Res 8(4):125–131. https://doi.org/10.1021/ar50088a003

Wang JL, Wang JZ (2007) Application of radiation technology to sewage sludge processing: a review. J Hazard Mater 143(1–2):2–7. https://doi.org/10.1016/j.jhazmat.2007.01.027

Wang JL, Xu LJ (2012) Advanced oxidation processes for wastewater treatment: formation of hydroxyl radical and application. Crit Rev Environ Sci Technol 42(3):251–325. https://doi.org/10.1080/10643389.2010.507698

Warren KD, Arnold RG, Bishop TL, Lindholm LC, Betterton EA (1995) Kinetics and mechanism of reductive dehalogenation of carbon tetrachloride using zero-valence metals. J Hazard Mater 41(2–3):217–227. https://doi.org/10.1016/0304-3894(94)00117-Y

Watts MJ, Linden KG (2007) Chlorine photolysis and subsequent OH radical production during UV treatment of chlorinated water. Water Res 41(13):2871–2878. https://doi.org/10.1016/j.watres.2007.03.032

Watts RJ, Stanton PC, Howsawkeng J, Teel AL (2002) Mineralization of a sorbed polycyclic aromatic hydrocarbon in two soils using catalyzed hydrogen peroxide. Water Res 36(17):4283–4292. https://doi.org/10.1016/S0043-1354(02)00142-2

Weavers LK, Hua I, Hoffmann MR (1997) Degradation of triethanolamine and chemical oxygen demand reduction in wastewater by photoactivated periodate. Water Environ Res 69(6):1112–1119. https://doi.org/10.2175/106143097X125849

Wilkinson F, Helman WP, Ross AB (1993) Quantum yields for the photosensitized formation of the lowest electronically excited singlet state of molecular oxygen in solution. J Phys Chem Ref Data 22(1):113–262. https://doi.org/10.1063/1.555934

Wojnárovits L, Takács E (2008) Irradiation treatment of azo dye containing wastewater: an overview. Radiat Phys Chem 77(3):225–244. https://doi.org/10.1016/j.radphyschem.2007.05.003

Wojnárovits L, Takács E, Szabó L (2018) Gamma-ray and electron beam-based AOPs. In: Stefan MI (ed) Advanced oxidation processes for water treatment: fundamentals and applications. IWA Publishing, London, pp 241–296

Xing R, Zheng Z, Wen D (2015) Comparison between UV and VUV photolysis for the pre- and post-treatment of coking wastewater. J Environ Sci 29:45–50. https://doi.org/10.1016/j.jes.2014.10.003

Yan YE, Schwartz FW (2000) Kinetics and mechanisms for TCE oxidation by permanganate. Environ Sci Technol 2000(34):2535–2541. https://doi.org/10.1021/es991279q

Yang Y, Jiang J, Lu X, Ma J, Liu Y (2015) Production of sulfate radical and hydroxyl radical by reaction of ozone with peroxymonosulfate: a novel advanced oxidation process. Environ Sci Technol 49(12):7330–7339. https://doi.org/10.1021/es506362e

Zhang W-x (2003) Nanoscale iron particles for environmental remediation: an overview. J Nanopart Res 5(3–4):323–332. https://doi.org/10.1023/A:1025520116015

Zhang Q, Chuang KT (1999) Treatment of combined bleach plant effluents via wet oxidation over a Pd–Pt–Ce/alumina catalyst. Environ Sci Technol 33(20):3641–3644. https://doi.org/10.1021/es990063l

Zhao X, Liu W, Cai Z, Han B, Qian T, Zhao D (2016) An overview of preparation and applications of stabilized zero-valent iron nanoparticles for soil and groundwater remediation. Water Res 100:245–266. https://doi.org/10.1016/j.watres.2016.05.019

Zhou M, Oturan MA, Sirés I (eds) (2018) Electro-Fenton process. New trends and scale-up. The handbook of environmental chemistry. Springer, Singapore. https://doi.org/10.1007/978-981-10-6406-7

Zoschke K, Börnick H, Worch E (2014) Vacuum-UV radiation at 185 nm in water treatment – a review. Water Res 52:131–145. https://doi.org/10.1016/j.watres.2013.12.034

Zuo Y, Hoigné J (1992) Formation of hydrogen peroxide and depletion of oxalic acid in atmospheric water by photolysis of iron(III)-oxalato complexes. Environ Sci Technol 26 (5):1014–1022. https://doi.org/10.1021/es00029a022

Chapter 8
Ferrates as Powerful Oxidants in Water Treatment Technologies

Libor Machala, Petr Zajíček, Jan Kolařík, Tomáš Mackuľak, and Jan Filip

Abstract Over the past decades, soluble compounds with iron in high oxidation state(s) (so-called ferrates) have turned out to be a highly promising and "green" oxidants deployed in water treatment technologies. Ferrates(IV, V, VI) have good oxidation properties, as indicated by the magnitude of the oxidation-reduction potential (an acidic reaction with an oxidation-reduction potential of 2.20 V, in the basic environment it is 0.72 V), making them interesting for oxidation of inorganic compounds including metals, the decomposition of broad range of organic pollutants including various types of micropollutants found in wastewater (pesticides, pharmaceuticals, and drugs), and for disinfection. The important advantage of ferrates is also their minimal environmental load as the reactions result in the formation of benign iron hydroxides. The only limitation of using ferrates in practice is their demanding production as well as their instability due to the high electron affinity of hexavalent iron.

Keywords Iron · High-valent states · Ferrates · Synthesis · Organic pollutants · Micropollutants · Pharmaceutical compounds · Heavy metals · Arsenic · Coagulation · Disinfection

8.1 Introduction

The occurrences of diverse emerging contaminants (heavy metals, volatile organic carbons, pesticides, personal care products, pharmaceuticals, organics, illicit drugs, and pathogenic microbes), particularly in soil and groundwater, have a negative impact on the environment. Therefore, the contamination of ground and surface

L. Machala · P. Zajíček · J. Kolařík · J. Filip (✉)
Regional Centre of Advanced Technologies and Materials, Palacký University Olomouc, Olomouc, Czech Republic
e-mail: jan.filip@upol.cz

T. Mackuľak
Institute of Chemical and Environmental Engineering, Faculty of Chemical and Food Technology, Slovak University of Technology, Bratislava, Slovakia

J. Filip et al. (eds.), *Advanced Nano-Bio Technologies for Water and Soil Treatment*, Applied Environmental Science and Engineering for a Sustainable Future,
https://doi.org/10.1007/978-3-030-29840-1_8

water with various pollutants has been recognized as a global problem that needs cost-effective and environmentally friendly solution(s). One of the ways to overcome the above-mentioned problem is to use ferrate(VI) that offers desirable chemical properties, such as extremely high redox potential (+2.2 V in acidic and + 0.7 V in alkaline solution, respectively) and benign reaction product(s). Thanks to these unique properties, ferrate(VI) is able to eliminate microorganisms, viruses, inorganic, and organic (both natural and synthetic) compounds (Jiang and Lloyd 2002; Kubiňáková et al. 2017; Mackul'ak et al. 2016; Rai et al. 2018).

The most common oxidation states in which iron can be usually found are Fe(0), Fe(II), and Fe(III). Other, unusually high oxidation states, e.g., +IV, +V, and +VI, have also been observed $[Fe^{VI}O_4]^{2-}$. The first mention of ferrates(VI) dates back to 1702. Later on, in 1715, K_2FeO_4 was prepared by oxidation of iron in molten KNO_3. During the smelting of iron ore with potassium carbonate and after the dissolution of residues in alkaline solution, the violet coloration was observed. Some researchers assumed that the violet color of this product was caused by the presence of $Fe^{VI}O_3^{-II}$ compound (Mácová et al. 2009). In an aqueous environment, ferrates(VI) possess dark violet color and the ion $[Fe^{VI}O_4]^{2-}$ has tetrahedral structure (Jiang and Lloyd 2002).

A lot of new information has emerged in the literature over the past years relating to the preparation of ferrate(VI), its application to the removal of various organic, inorganic compounds, radionuclides, and the mechanism of ferrate action (Feng et al. 2016; Híveš et al. 2016; Jiang and Lloyd 2002; Jiang 2014; Kralchevska et al. 2016a; Kubiňáková et al. 2015a, b; Mackul'ak et al. 2016; Rai et al. 2018; Sharma et al. 2016a, b). In this chapter, we summarize the current knowledge of a ferrate synthesis using various reaction routes, ferrate stability, and applicability to the elimination of broad range of inorganic and organic compounds. The disinfection properties of ferrates are also mentioned.

8.2 Synthesis of Ferrates(IV, V, VI)

There are several approaches to synthetizing potassium ferrate(VI) (K_2FeO_4), potassium ferrate(V) (K_3FeO_4), and sodium ferrate(IV) (Na_4FeO_4) (Mura et al. 2017; Sharma et al. 2015). Particularly wet chemical methods, electrochemical methods, and thermal processes (Fig. 8.1) are used. When selecting a suitable method of synthesis, it is necessary to consider its experimental difficulty and compromise between purity and the amount of the as-prepared ferrate. Generally, wet chemical synthesis and electrochemical synthesis result in a high-purity ferrate sample, but the amounts of the synthesized material are relatively low. Therefore, such an approach is not suitable for large-scale production, i.e., application of ferrates to water treatment. On the other hand, thermal syntheses enable to prepare much higher amounts of ferrates (up to kilograms per one synthetic cycle), but their purity is relatively low (approx. 30–50 wt.% of K_2FeO_4 in sample) mainly due to self-decomposition of ferrates at high temperatures.

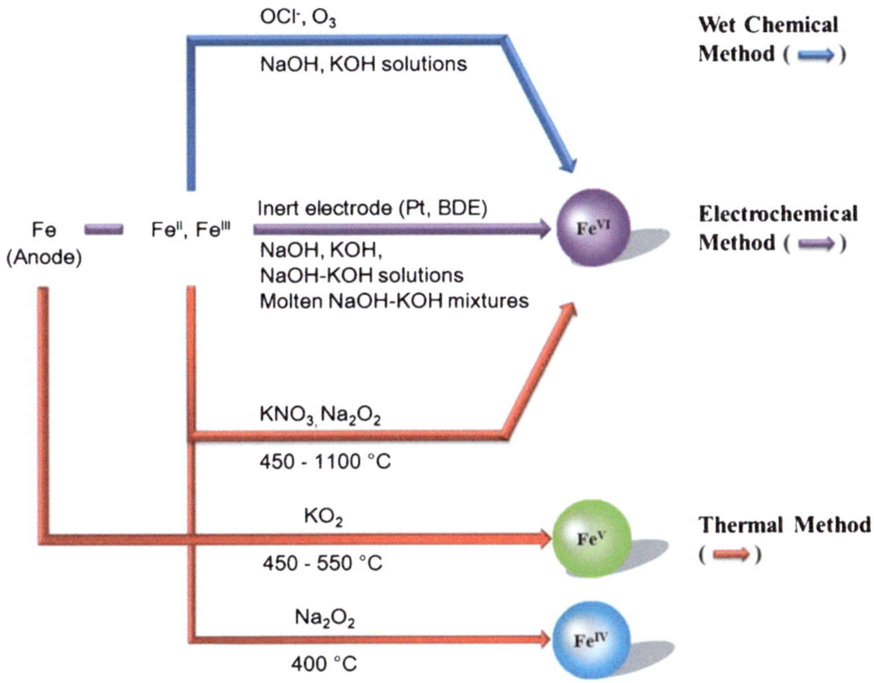

Fig. 8.1 Possible approaches to the synthesis of ferrates. (Reprinted with permission from Sharma et al. (2015). Copyright (2015) American Chemical Society)

Potassium ferrate(VI) (K_2FeO_4) as the most often applied ferrate or sodium ferrate(VI) (Na_2FeO_4) can be prepared by a wet chemical synthesis. Iron(III) oxides or their salts (e.g., ferric chloride or potassium nitrate) are oxidized by hyperchlorite ion (OCl^-) in highly alkaline solution, prepared by NaOH resulting in sodium ferrate (VI) according to the following reaction:

$$2Fe(OH)_3 + 3NaOCl + 4NaOH \rightarrow 2Na_2Fe^{VI}O_4 + 3NaCl + 5H_2O \quad (8.1)$$

Potassium ferrate(VI) can then be obtained through precipitation from the solution of Na_2FeO_4 by adding KOH according to the reaction:

$$Na_2Fe^{VI}O_4 + 2KOH \rightarrow K_2Fe^{VI}O_4 + 2NaOH \quad (8.2)$$

The purity of ferrates prepared by this method is very high (approx. 98%). However, it is worth mentioning that this wet chemical method of synthesis is not environmentally friendly as it results in high concentrations of NaOH and residues of NaOCl in reaction solutions.

Concerning a wet chemical approach, the electrochemical anodic oxidation of iron or cast iron in concentrated hydroxide solutions or melts provides much more environmentally friendly process employing an iron electrode (Fe(0)), salts of Fe (II) or Fe(III), and/or iron oxides. In the case of a melt, a suitable temperature for ferrate(VI) synthesis is up to 200 °C. Generally, in aqueous solutions, temperatures range from 20 to 70 °C (Mácová et al. 2009). Using electrochemical methods, high-purity ferrates are easily prepared. The drawback of the method lies in the overlap of potentials of the oxidation of Fe(III) to Fe(VI) and oxygen evolution. The yield of the synthesis depends on temperature, composition of the iron precursors, and the pH of the alkaline solution. As an example, the electrochemical synthesis of ferrate (VI) was studied in a eutectic NaOH-KOH molten salt and in a highly alkaline mixed NaOH-KOH aqueous solution (Híveš et al. 2008). Moreover, electrochemical *DC* (direct current) and *AC* (alternating current) techniques were found to be suitable for interfacial and metal dissolution studies in both aqueous and molten systems (Híveš et al. 2016). Transpassive iron dissolution in alkaline solutions is discussed less frequently. It has been found that in an alkaline environment, Fe is dissolved as ferrous(II), ferric(III) and, finally, ferrate(VI) species depending on the experimental conditions (Híveš et al. 2008, 2014, 2016; Hrnčiariková et al. 2013; Kerekeš et al. 2014; Kubiňáková et al. 2015a, b; Mácová et al. 2009; Rai et al. 2018; Sun et al. 2016; Yu and Licht 2008).

Thermal processes were applied to synthesize sodium ferrate(IV) (Na_4FeO_4), potassium ferrate(V) (K_3FeO_4), or potassium ferrate(VI) (K_2FeO_4). Sodium ferrate (IV) can be prepared by a solid-state reaction of Fe_2O_3 and Na_2O_2 powders mixed in the theoretical molar ratio of Fe:Na = 1:2 (nevertheless an excess of Na_2O_2 in the mixture is recommended to achieve complete oxidation of Fe_2O_3 within the sample) according to the reaction:

$$Fe_2O_3 + 3Na_2O_2 \rightarrow 2Na_2FeO_4 + Na_2O \tag{8.3}$$

The homogeneous mixture is heated at 400 °C for 1 h. When using a stream of oxygen during the synthesis, the temperature of 370 °C is sufficient. After annealing, the crucible containing the sample must be rapidly cooled down. The purity of Fe (IV) is more than 90%. It is highly recommended to carry out the synthesis in a glove box since both the sodium peroxide precursor and the sodium ferrate(IV) product are highly hygroscopic. Even Mössbauer spectroscopy analysis, which is typically used for verification of the purity of the as-prepared samples, is recommended to be conducted directly in the glove box (Fig. 8.2).

Potassium ferrate(V) (K_3FeO_4) can be synthesized by a two-step solid state reaction starting from homogenous mixture of Fe(III) hydroxide and KNO_3 (Machalová Šišková et al. 2016) with almost equal weights (e.g., several grams). In the first reaction step, the mixture was put into a corundum annealing crucible and the crucible was covered with a lid. The closed crucible was then put into the furnace and annealed at 950 °C for 30 min. After it was annealed, the closed crucible was removed from the furnace and put into a desiccator where it was left to cool down with $KFeO_2$ as a resulting product. Then the as-synthesized $KFeO_2$ was added to KNO_3 powder. Again, the mixture was homogenized in the grinding mortar. The homogenized mixture was then put into a corundum annealing crucible and was

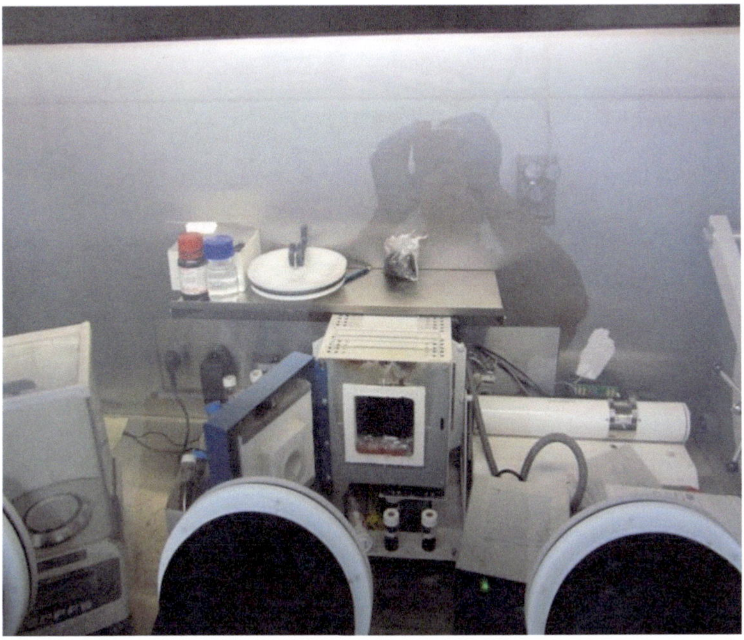

Fig. 8.2 A glove box with a furnace for the synthesis of sodium ferrate(IV). A room temperature Mössbauer spectrometer is installed directly inside the glove box (white tube on the right side)

covered with a lid. The crucible was then put into the furnace and annealed at 1000 °C for 30 min at an N_2 flow of 150 L/h. Before the termination of annealing, another metallic annealing crucible was cooled by liquid nitrogen; afterwards, liquid nitrogen was poured to the crucible. After removing the crucible from the furnace, the melt was poured into the crucible filled with liquid nitrogen. The as-prepared K_3FeO_4 is stable below the level of liquid nitrogen. After it was frozen, K_3FeO_4 was transferred to a porcelain dish with nitrogen and was put into a glove box. Again, more than 90% of Fe(V) (atomic percent of Fe) can be achieved. When the mixture of Fe(III) oxide and KNO_3 is heated at 1100 °C, potassium ferrate(VI) (K_2FeO_4) can be prepared; however, it will yield only 30% purity.

8.3 Experimental Methods for the Characterization of As-Prepared Ferrates and Determination of Their Purity

The relative amount of a ferrate in liquid and solid phases can be determined by using a variety of different analytical techniques (Mura et al. 2017; Sharma et al. 2015), such as volumetric, electrochemical (cyclic voltammetry and potentiometry), and spectroscopic techniques (Fourier-transform infrared spectroscopy—FTIR, ^{57}Fe Mössbauer spectroscopy). The concentration of ferrates in the solution can be

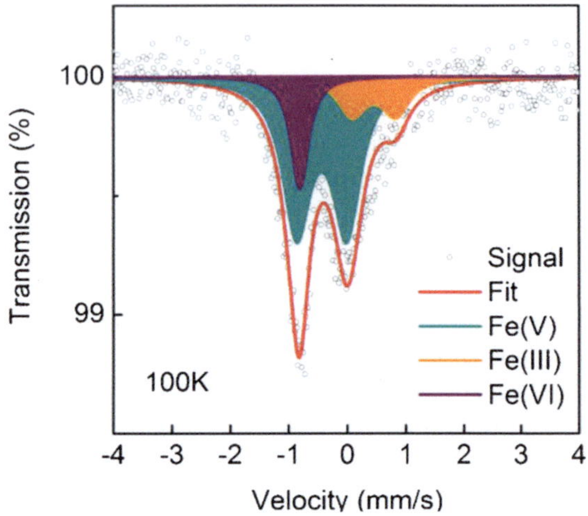

Fig. 8.3 A typical Mössbauer spectrum of Fe(VI), Fe(V), and Fe(III) species formed together during a transformation of ferrate(V) in ethanol. The sample was treated by rapid-freeze technique and spectrum measured at 100 K

monitored by UV-VIS spectrometry. The key experimental tool for characterization of solid samples of ferrates is Mössbauer spectroscopy on isotope ^{57}Fe. The method is selective to the Fe element so it is a unique tool for identification/characterization of Fe-containing phases. Importantly, Mössbauer spectroscopy is sensitive to valence state of iron through isomer shift value as one of the hyperfine parameter. Particularly, Fe(V) and Fe(VI) atoms show negative isomer shift values, which are very different from common Fe(II) or Fe(III) states (see Fig. 8.3). Ferrate samples often contain Fe(III) admixtures, which are amorphous and/or nanocrystalline. Low temperature and external magnetic field Mössbauer spectroscopy provide very detailed characterization of these Fe(III) phases. Even though Mössbauer spectroscopy is designed for characterization of solid samples, it is possible to analyze frozen solutions as well. Relative areas of the spectral components correspond to relative amounts of corresponding Fe-containing phases. It is worth mentioning that these relative amounts are expressed in atomic percent of Fe atoms. Therefore, the values must be recalculated when we need mass ratios of the phases (moreover, chemical formulas of the phases are needed).

8.4 Stability of Solid Ferrates at High Temperatures and in a Humid Air

The issue of thermal stability of ferrates is crucial for the thermal synthesis as well as for possible applications at elevated temperatures. Generally, the instability of ferrates at high temperatures significantly decreases yield and purity of ferrates prepared the thermal processes. The focus will now be shifted to the mostly used ferrate(VI), K_2FeO_4. A thorough study conducted by Machala et al. (2007) showed that potassium ferrate(VI) starts to decompose above 190 °C towards potassium ferrite ($KFeO_2$) and potassium oxides. After it is cooled down to room temperature, $KFeO_2$ becomes unstable and reacts with air H_2O and CO_2 with the formation of Fe_2O_3 nanoparticles and $KHCO_3$ carbonate (Machala et al. 2015a). Conventional experimental techniques confirmed the presence of neither Fe(V) nor Fe (IV) intermediates during the thermal decomposition of ferrates. A later study (Machala et al. 2015b) employing nuclear forward scattering of the synchrotron radiation proved the presence of several percent of Fe(IV) and Fe(V) during the decomposition of potassium ferrate(VI).

From the point of view of storage and handling of ferrates, the issue of stability of potassium ferrate(VI) in air at room temperature is crucial as well. In the study of Machala et al. (2009), the kinetics of solid-state transformation (aging) of potassium ferrate(VI) under various air-humidity conditions (55–95% relative humidity) at room temperature were studied by in situ ^{57}Fe Mössbauer spectroscopy. The kinetic data showed a significant increase in the decomposition rate with increasing air humidity (Fig. 8.4). Crystals of $KHCO_3$ and amorphous $Fe(OH)_3$ nanoparticles were identified as the transformation products. The decomposition kinetics was very unusual with two almost linear decay steps in case of lower humidity levels (55–70%). The first slow decay was probably observed due to the formation of the narrow compact layer of nanoparticulate $Fe(OH)_3$ reaction product. This layer limits the access of both H_2O and CO_2 participating in the reaction as the gaseous reactants. The second decay with a much faster rate showed a nearly positive linear relationship with the humidity.

8.5 Stability of Ferrates in Aqueous Solution

For applications of ferrates to water treatment, an aqueous solution of dissolved ferrate at a certain concentration should be usually prepared in advance. Therefore, it is necessary to know the kinetics and mechanism of ferrates' self-decay in water under different conditions. Stability of ferrates in water depends on many factors such as ferrate concentrations, pH, temperature, or presence of inorganic ions. Generally, ferrates tend to reduce to Fe(III) and/or charge disproportionation takes place. Concerning ferrate(VI), after it is dissolved in water, 5–10% of the ferrate are decomposed to Fe(III) immediately (during 1–2 s). From the literature (Lee et al. 2014), it is known that ferrate(VI) undergoes a dimerization process with Fe(IV) as an intermediate state; however, this reaction is probably too fast to identify Fe(IV) by conventional experimental techniques. The reduction of ferrate(VI) is accompanied by evolution of oxygen and increasing of pH up to 13. During the several-minute-long ferrate(VI) transformation in water, the pH value decreases slowly to approx. 10 and the content of ferrate(VI) slightly decreases by approx. 3%. Just before its application to water treatment, it is necessary to decrease pH in order to have almost neutral conditions along with enhancing the reactivity of the ferrate. The transformation of potassium ferrate(VI) in water can be described by chemical equation

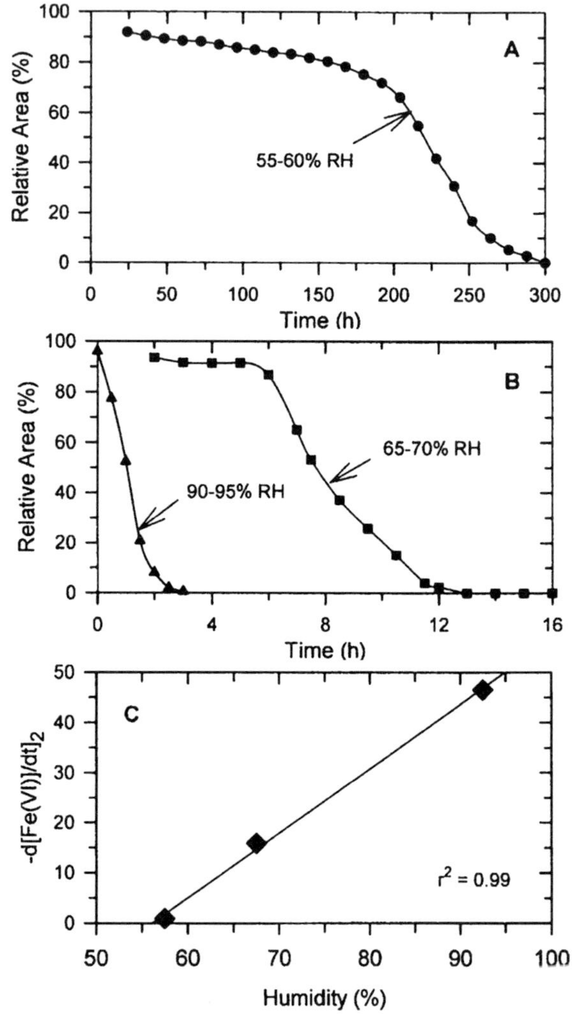

Fig. 8.4 Kinetics of transformation of solid potassium ferrate(VI) at room temperature in air expressed as dependence of Mössbauer spectral area of the ferrate subspectrum on the relative humidity ((**a**) lowest *RH*; (**b**) higher *RH* levels; (**c**) dependence of the second step decay rate on the relative humidity). (Reprinted with permission from Machala et al. (2009). Copyright (2009) Wiley-VCH Verlag GmbH & Co. KGaA, Weinheim)

$$2K_2FeO_4 + 5H_2O \rightarrow 2Fe(OH)_3 + 4KOH + 3/2O_2 \tag{8.4}$$

Dissolving ferrate(V) in water leads to fast disproportionation of Fe(V) into Fe (III) and Fe(VI). Formation of Fe(IV) during the process was not confirmed. Charge disproportionation and reduction of the ferrate compete in the process; therefore, the Fe(III)/Fe(VI) ratio is not typically 1:2 but quite close to 1:1. Thus, the transformation can be described by chemical equation

$$2K_3FeO_4 + 3/2H_2O \rightarrow KFeO_2 + K_2FeO_4 + 3KOH + 1/4O_2 \tag{8.5}$$

The formation of Fe(VI) during the transformation of ferrate(V) in water has an important application aspect. As it has been already mentioned, potassium ferrate (V) can be synthesized by the solid-state method on a large scale (contrary to ferrate (VI)). But after ferrate(V) is dissolved in water, the final oxidizing agent is Fe(VI).

Concerning sodium ferrate(IV), after it has been dissolved in water, immediate transformation to Fe(III) takes place. The process is accompanied by rapid increase of pH up to 13 and evolution of oxygen. The corresponding chemical equation describing the transformation can be written as

$$Na_4FeO_4 + 3/2H_2O \rightarrow NaFeO_2 + 3NaOH + 1/4O_2 \quad (8.6)$$

It is worth mentioning that due to this rapid decay, the applicability of sodium ferrate(IV) to water treatment is very limited. This was confirmed, e.g., in the study by Machalová Šišková et al. (2016), where very low efficiency in the degradation of estrogenic hormones was observed in case of sodium ferrate(IV), in comparison with potassium ferrates(V) or (VI).

8.6 Effect of Buffering Inorganic Ions on Stability of Ferrates

The study by Kolář et al. (2016) provided a systematic kinetics investigation of the decay of ferrate(VI) in the presence of inorganic buffering ions (borate, phosphate, and carbonate) at a pH range from 6.0 to 9.0. Detailed kinetic analysis revealed that carbonate anion enhanced the Fe(VI) transformation rate, compared to phosphate and borate ions. The order of the Fe(VI) decay rate in neutral solution condition was carbonate > phosphate $\geq$ borate. In alkaline solution, the decay rates of Fe(VI) were similar for the studied buffering ions. The results indicated that carbonate ions induced the formation of ultrasmall iron(III) oxyhydroxide nanoparticles (<5 nm), which contributed to the increased decay of Fe(VI) because of their larger specific surface area, where the catalytic reactions took place. On the other hand, the observed low reactivity of borate with Fe(VI) demonstrates that borate is the least reactive buffer for studies of Fe(VI) reactivity in neutral solutions.

8.7 Degradation of Organic Pollutants by Ferrate

One of the fresh challenges in water treatment is to develop new methods for elimination of organic pollutants, especially for persistent compounds. Various conventional methods such as adsorption, filtration, sedimentation, oxidation processes, and photocatalytical methods require relatively high operational costs and yield inadequate efficiency. Moreover, currently the emphasis is put on green technologies employing environmentally friendly agents and chemicals used during the processes. Monitoring intermediates and residuals of pollutants as well as testing toxicity of the final products of reactions are more and more required. In this way, ferrates represent effective and promising agents for water treatment of various organic pollutants.

The majority of studies on organic pollutants degradation by ferrates focuses on the use of ferrate(VI). This is a sensible approach because the stability of high-valent iron species in a water solution is dramatically different. While ferrate(VI) is relatively stable, from minutes to days, depending on the solution composition and physicochemical conditions, ferrate(IV) and ferrate(V) decay in seconds resulting in an already mentioned stable ferrate(VI) and trivalent iron products. Hence, ferrate (VI) reacts with organic pollutants although the primary oxidative agents used at the beginning were ferrate(IV) or ferrate(V). This also plays its role in focusing on iron (VI) within ferrate research. On the other hand, it is a well-known fact that ferrate (V) is more reactive than ferrate(VI).

From the kinetic perspective, the reactivity of ferrate(VI) with organic pollutants is described as second-order with the second-order rate constants from 3.0×10^{-2} to 1.7×10^{7} 1/(M s) (see Table 8.1). Only in the case of an excess of ferrate or pollutants, the system behaves as the first-order. In this special case, the apparent second-order rate constant is often examined as a function of pH (Sharma 2013). The reactivity with the pollutants (X) can be described by several pathways. The first possibility is reduction of Fe(VI) to Fe(V) by electron transfer with formation of radical (Eq. 8.7). Another pathway is that the two-electron transfer creates radicals (Eq. 8.8) or dimer (Eq. 8.9). The formed radical species react with ferrates (Eq. 8.10). Finally, oxygen atom transfer can be observed (Eq. 8.11). This pathway was confirmed for instance for aniline forming Fe(IV) and phenylhydroxylamine. Consequently, reactions involving the described products (Eqs. 8.7, 8.8, 8.9, 8.10 and 8.11) are initiated and influenced by the inherent structure of each pollutant. In addition, self-decomposition of ferrate can take place, leading to the formation of Fe (IV), Fe(V), and reactive oxygen species as $O_2^{\cdot-}$ and H_2O_2. The reactive oxygen species can participate in the reaction with pollutants.

$$\mathrm{Fe(VI)} + \mathrm{X} \rightarrow \mathrm{Fe(V)} + \mathrm{X} \tag{8.7}$$

$$\mathrm{Fe(VI)} + \mathrm{X} \rightarrow \mathrm{Fe(IV)} + 2\mathrm{X} \tag{8.8}$$

$$\mathrm{Fe(VI)} + \mathrm{X} \rightarrow \mathrm{Fe(IV)} + \mathrm{X}_2 \tag{8.9}$$

$$\mathrm{Fe(VI)} + \mathrm{X} \rightarrow \mathrm{Fe(IV)}/\mathrm{Fe(V)} + \text{product X} \tag{8.10}$$

$$\mathrm{Fe(VI)} + \mathrm{X} \rightarrow \mathrm{Fe(IV)} + \mathrm{XO} \tag{8.11}$$

The reactions can be influenced by various parameters. One of the most important one is pH. A large number of organic pollutants react rapidly with ferrates under neutral or slightly acidic pH, for example, substituted anilines, tetrabromobisphenol A, atrazine and more (Han et al. 2018; Sun et al. 2018; Zajíček et al. 2015). This is caused by the presence of $HFeO_4^{2-}$, which is a preferable species

Table 8.1 Selected organic pollutants and rate constants for reaction with ferrate (Adapted from Rai et al. 2018)

Organic pollutant	pH	Rate constant *k* (1/M s)
Hydroquinone	9.0	2.0×10^5
Cysteine	12.4	7.6×10^2
2-mercaptoethanesulfonic acid	9.0	3.0×10^4
2-mercaptobenzoic acid	10.0	2.5×10^4
3-mercaptopropionic acid	9.0	1.3×10^4
Methyl hydrazine	9.0	9.8×10^3
Aniline	9.0	6.2×10^3
p-toluidine	9.0	1.3×10^3
p-aminobenzoic acid	9.0	4.3×10^1
p-nitroaniline	9.0	3.0×10^1
Thiourea	9.0	3.4×10^3
Benzenesulfinate	9.0	1.4×10^2
Methionine	9.0	1.3×10^2
Cystine	12.4	1.2×10^2
Glycine	8.0	1.0×10^2
Trioxane	9.0	5.8×10^1
Glyoxylic acid	8.0	7×10^2
Dimethylamine	8.0	2.0×10^2
Iminodiacetic acid	8.0	1.0×10^2
Diethylsulfide	8.0	1.0×10^2
Thiodiethanol	8.0	1.0×10^2
Phenol	9.0	8.0×10^1
Nitriloacetic acid	8.0	2.0×10^0
N-methyliminodiacetic acid	8.0	2.0×10^0
Dimethylsulfoxide	8.0	1.0×10^0
Diethylamine	8.0	7.0×10^{-1}
Formaldehyde	8.0	5.0×10^{-1}
Acetaldehyde	8.0	4.0×10^{-1}
Glycolic acid	8.0	4.0×10^{-1}
Oxalic acid	8.0	1.0×10^{-1}
Ethyl alcohol	8.0	8.0×10^{-2}
Isopropyl alcohol	8.0	6.0×10^{-2}
Methyl alcohol	8.0	3.0×10^{-2}
4-methylphenol	7.0	6.9×10^2
Estradiol	7.0	7.7×10^2
Bisphenol A	7.0	6.4×10^2
Phenol	7.0	7.7×10^1
Octylphenol	9.1	1.8×10^3
Buten-3-ol	7.0	1.2×10^1
Ibuprofen	9.0	4.0×10^{-1}
Carbamazepine	8.0	1.6×10^1
Benzophenone-3	8.0	3.4×10^2

(continued)

Table 8.1 (continued)

Organic pollutant	pH	Rate constant *k* (1/M s)
Triclosan	7.0	4.1×10^2
Ciprofloxacin	7.0	4.7×10^2
Ciprofloxacin	8.0	1.1×10^2
Sulfamethizole	7.0	1.3×10^3
Sulfamethoxazole	7.0	1.3×10^3
Tetracycline	7.0	3.0×10^2

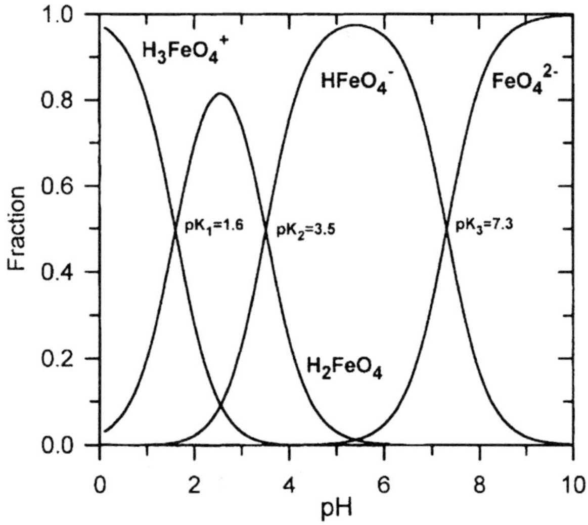

Fig. 8.5 Ferrate species under different pH (Adapted from Sharma 2013)

of ferrate under pH ranging from 4 to 7 (Fig. 8.5). This protonated species exhibits higher oxidation strength because it has larger spin density on oxo-ligands as an unprotonated counterpart. This was investigated by density functional theory (DFT) calculations (Kamachi et al. 2005). On the other hand, the rate constant of another organic pollutants can be limited by acidic pH. For example, this was observed for glycine and serine (Noorhasan et al. 2010). Another example can be phenol with optimal pH 9 for reaction with ferrate. A pH value in combination with reaction ratios can also influence the formation of reaction products. It was observed that oxidation of hydrogen sulfate by ferrate(VI) at pH 7 provided only thiosulfate as the final product; however, at higher pH and higher stoichiometric ratios, sulfite and sulfate were formed as well (Sharma 2002).

Numerous reactions of ferrate(VI) with various aliphatic and aromatic organic compounds were published (Table 8.1) (Rai et al. 2018). Ferrate easily reacts with electron-rich organic moieties such as olefins, amines, phenols, anilines, etc. In addition, adsorption on precipitating reaction products of ferrate can occur. The reactivity of ferrate was investigated by not only simple hydrocarbons, phenol, or chlorinated ethylenes but also complex organic pollutants with a more complicated structure. Contamination of drinking water sources with micropollutants, contamination of surface water with runoffs, and contamination of wastewaters with effluents from industry or less efficient wastewater treatment plants represent rather varied occurrence of pollutants, which can be potentially treated by ferrates.

One of the most discussed groups of pollutants occurring in surface- and wastewater is estrogens. The reactivity of ferrates with estrogens was tested and, except for ferrate(IV), an efficient removal/degradation of estrogens was observed (Machalová Šišková et al. 2016). The intermediates confirmed oxidation steps in the degradation. The hydroxyl groups of the parent molecule were attacked by the ferrate, which led to quinone-like structures, and subsequently the opening of an aromatic ring was observed. The product of degradation was malonic acid, finally mineralized to water and carbon dioxide (Fig. 8.6).

Another abundantly occurring pollutants are pesticides and herbicides. Most of them constitute a threat to the environment due to endocrine disrupting properties and bioaccumulation. As an example of this pollutant family, total atrazine and iodosulfuron degradation can be mentioned. Although the kinetics of the reaction was fast, some of the intermediates were determined (Zajíček et al. 2015). The oxidation of alkyl chain by ferrate(VI) generated a carbonyl group in structure of the intermediate phase. Ferrate predominantly attacks the ethyl group rather than the isopropyl group in the pollutant structure (Fig. 8.7). Generally, the formation of molecules with higher molar mass (oxygen insertion into the structure) is relatively common. This was, for instance, observed also during carbamazepine oxidation (Hu et al. 2009). For iodosulfuron degradation, products formed from the fission of the parent molecule (the both S-N and C-N bond cleavage) were confirmed.

In addition to carbamazepine or estrogens, lots of different pharmaceutical compounds constitute a threat to the purity of water and for the environment. Long-term exposition can cause toxic effects; and in the case of antibiotics, related increasing bacterial resistance was observed. The commonly discussed point is that these compounds are not fully removed from water in wastewater treatment plants. Ferrate(VI) exhibited effective oxidation properties to pharmaceuticals such as ciprofloxacin, enrofloxacin, sulfamethoxazole, ibuprofen, trimethoprim, propranolol, flurbiprofen. For most of them, optimal pH range was at neutral or slightly basic, which is acceptable for water treatment and does not increase operational costs. As another example of pharmaceuticals degradation, breaking of isoxazole ring was observed for sulfamethoxazole and conversion of amino group to nitro group or nitroso group. The determined final products were thus less toxic than the parent compound (Sharma et al. 2006).

Fe(VI) was also identified as an efficient oxidant to remediate several antibiotics such as β-lactam (cloxacillin, ampicillin, amoxicillin, penicillin G, cephalosporin) with the attack on the thioether moiety, tetracycline, and fluoroquinolone (Feng et al. 2016, 2017; Sharma et al. 2013, 2016a). Penicillin G and cephalosporin were converted initially to sulfoxide derivatives and finally to sulfones. Moreover, in the case of cephalosporin, also C-N bond can be attacked by a ferrate(VI) ion to form ammonia (Sharma 2010; Sharma et al. 2013). During the degradation of aliphatic amines by Fe(VI), dealkylation products or respective hydroxylamines are formed,

Fig. 8.6 Degradation of estrone by ferrate (VI) (Adapted from Machalová Šišková et al. 2016)

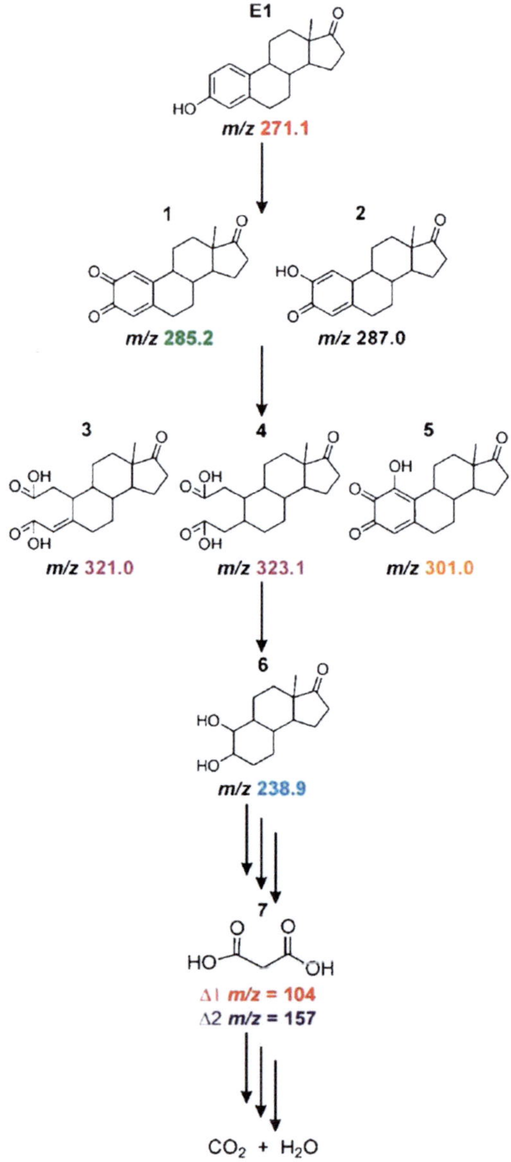

e.g., in the case of propranolol, ciprofloxacin, metoprolol, tramadol, atenolol (Rai et al. 2018). In the case of β-blockers, due to their different structure, different reaction pathways are observed. Generally, initial cleavage of an aromatic ring was observed. Carbonyl products were formed with subsequent transformation to a hydroxylamine derivative after the attack of ferrate on the amine moiety (Sharma et al. 2016a). Some analgesics react with ferrate(VI) very slowly. In case of tramadol, estimated degradation half-time is several minutes. Initially, ferrate reacts

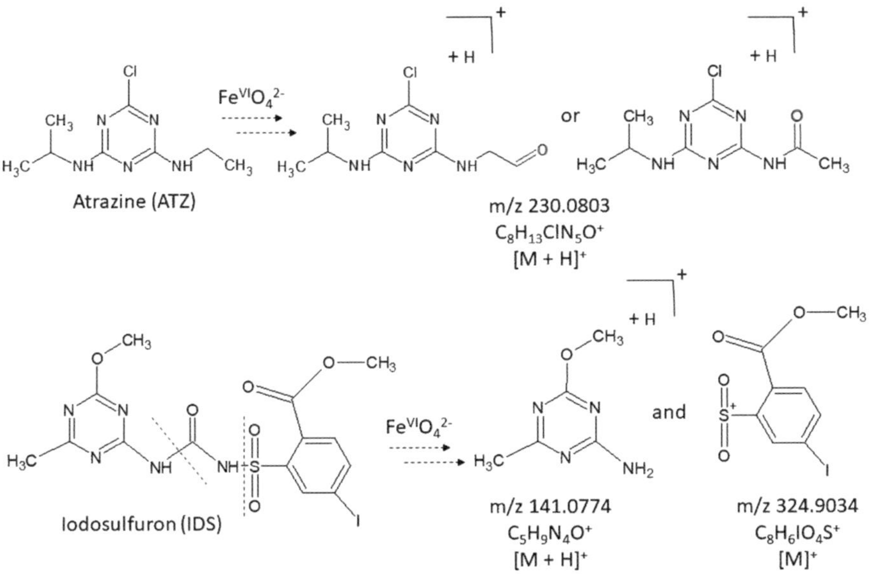

Fig. 8.7 Mechanism of atrazine and iodosulfuron degradation by ferrate(VI) (Adapted from Zajíček et al. 2015)

with amine group of tramadol with subsequent dealkylation yielding aldehyde and/or carboxylic groups (Sharma et al. 2016a).

Reactions of ferrate(VI) with sulfur- and nitrogen-containing pharmaceuticals are completed in a couple of seconds or minutes and the detected products are less toxic than the parent compounds (Sharma 2010). However, the related mechanism for the degradation is ambiguous and can hardly be elucidated empirically due to the rapid oxidation process and unstable intermediates for experimental trapping (Yu et al. 2017). In the case of sulfamethoxazole, DFT calculations were included in order to elucidate how ferrate(VI) initiated oxidative degradation process and to understand the reaction mechanisms (Yu et al. 2017). According to DFT study in the case of sulfamethoxazole, ferrate provides, at the initial step, oxygen atoms for the oxidation of aniline and isoxazole moieties. These theoretical findings are partially, not fully, supported by the experimental results (Casbeer et al. 2013; Hightower et al. 2012). The DFT calculations also show that a nucleophilic attack is a favorable pathway for ferrate(VI) to initiate oxidation of sulfamethoxazole, which has not been proposed previously. From the previous comments, it is clear that the degradation mechanism of micropollutants is very complicated and several oxidation products are formed during oxidation by ferrate.

Probably, the main problem is that the oxidation of drugs by ferrate(VI) can occur by several possible pathways (Sharma 2013). It is well known that several oxidation steps probably occur during the oxidation of a micropollutant by Fe(VI), which usually includes one- or two-electron transfer of Fe(VI) to form Fe(V) and Fe(IV), respectively, with Fe(II) and Fe(III) species as the final products. There is also a high probability of consecutive reactions of Fe(V) and Fe(IV), respectively, with other (micro)pollutants in the system; self-decompositions of all Fe(VI), Fe(V), and Fe (IV) species; reactions between all iron species from iron(VI) to iron(II) and/or oxygen species (i.e., H_2O_2) formed from either self-decompositions or during oxidation reaction of Fe(VI) with micropollutants (Sharma et al. 2015). Moreover, radicals of oxidized compounds formed during the degradation reaction can further react with mother compounds, with Fe(VI), Fe(V), Fe(IV), or with other radicals from different substances etc.

Dyes, hardly biodegradable (in)organic substances, can be also present in wastewaters (Lubello and Gori 2004). They can have toxic, carcinogenic, teratogenic, and mutagenic properties, some of them inhibiting the photosynthesis and biochemical pathways of aquatic animals and plants (Rai et al. 2018). Ferrate(VI) has been found to have great potential for their effective degradation. During the oxidation process, less toxic products are formed (Barışçı et al. 2016).

8.8 Removal of Heavy Metals and Metalloids by Ferrates

Heavy metals associated with exposure of highly toxic compounds of lead, cadmium, chromium, mercury, arsenic and others represent a severe threat to human health. Though toxicity and negative health effects of heavy metals have long been recognized, metals are widely used in a range of products including household appliances, paints, agriculture, motor vehicles, and electrical components. Contamination of all water types represents a growing danger in many countries all over the world because of the severe impact of such compounds on the population's health. For example, Bangladesh, some parts of India, Nepal, and Vietnam are countries where arsenic and heavy metals contamination of surface and drinking water is particularly large, showing an upsurge of poisoning cases and, overall, an increased life-risk for a large fraction of human and animal populations. Apart from contamination of toxic metals, phosphates often enter the environment. The presence of high levels of phosphorus input into water causes eutrophication that adversely affects aquatic ecosystems and water quality. The specific effects of excess phosphorus include blooms of harmful algae that can release toxins, create dead zones (or hypoxia in water), cause fish death, and foul taste and odor of drinking water (Faridmarandi and Naja 2014; Järup 2003).

For efficient removal of metal ions, the remediation process should combine the properties of an oxidant, high adsorption capacity (with the low sorbent/metal ion ratio), and it should also enable at least partial incorporation of metal ion into the structure of the reaction products, thus preventing the metal(loid) leaching back to water. The ferrates can meet all the mentioned criteria and can act as a highly efficient constituent in various water treatment technologies.

Several studies have explored the use of ferrate(VI) for removal of inorganic pollutants, in particular toxic heavy metals from various water samples (Table 8.2). The removal efficiency can reach, in the case of some heavy metals (Zn, Cu, Mn), almost 100% (Rai et al. 2018). Fe(VI) has the greatest potential to remove also other

Table 8.2 Removal of inorganic pollutants by ferrate(VI)

Pollutant	Effective ratio	Solution	Reference
PO_4^{3-}	~3:1 (mass)	Micropollutants	Lee et al. (2009)
Radionuclides	1000:1 (molar)	Hard fresh water/ seawater	Petrov et al. (2016)
I^-	1:1 (molar)	DI water	Kralchevska et al. (2016b)
Sb^{3+}	~7.7:1 (mass)	DI water	Lan et al. (2016)
$Cd(CN)_4^{2-}$/Ni $(CN)_4^{2-}$	4:1 (mass)	DI water	Yngard et al. (2008)

DI deionized

metal ions, such as As(III), Pb(II), Cr(III), or Hg(II) via co-precipitation and/or oxidation (Jiang et al. 2006). Moreover, ferrate(VI) has proven its ability to remove some natural or artificial radionuclides (^{238}U, ^{226}Ra, ^{232}Th, ^{137}Cs, ^{90}Sr, ^{239}Pu, etc.) from various environmental samples (mines, waters) by their sorption in aqueous environment with concomitant co-precipitation with iron(III) oxy-hydroxy species followed by sedimentation. Interestingly, potassium ferrate effectively acted as a scavenger to remove a mixture of alpha- and beta-emitting radionuclides ($^{137}Cs(I)$, $^{90}Sr(II)$, $^{152}Eu(III)$, $^{243}Am(III)$, $^{239}Pu(IV)$, $^{237}Np(V)$, $^{239}Np(V)$, and $^{238+233}U(VI)$) from water samples under laboratory conditions (Petrov et al. 2016; Rai et al. 2018).

Considering the high toxicity of arsenic compounds, many studies focused on their removal by ferrates (Jain et al. 2009; Lee et al. 2003). For instance Prucek et al. (2013) explained kinetics and mechanisms of arsenite and arsenate removing by ferrate(VI) (Fig. 8.8). A well-known fact is that sorption of metal ions is dependent on the pH value. This effect is connected with speciation of ions and surface charge. In the case of arsenic, decreasing the pH value sharply improved efficiency of arsenic removal. In these experiments, the sorption kinetics of arsenic removal was very fast and sorption equilibrium was reached approximately within 2 min. Ferrate (VI) removed almost all arsenic at an Fe/As weight ratio of 2/1 (at pH 6.6). The very important conclusion is that ferrate(VI) has the same efficiency in arsenite and arsenate removals. This is due to strong oxidizing properties of ferrate(VI) that oxidized arsenite to arsenate. The oxidation of arsenite was confirmed using X-ray photoelectron spectroscopy—XPS (Fig. 8.8d). On the basis of XPS results (i.e., As 3d high resolution spectra), only As(V) was detected. Therefore, arsenite can be effectively removed from water by ferrate(VI) even though this form of arsenic is difficult to remove from water by other processes. The combination of XPS and Mössbauer spectroscopy enables to reveal the mechanisms of As removal, which include two crucial aspects. Arsenic is partially incorporated into the crystal structure of solid precipitate of iron(III), which leads to an increase in the removal efficiency. Thus, arsenic is strongly bound to sorbent and is partially protected against leaching back into the environment. The remaining arsenic is adsorbed onto the surface of iron(III) oxide nanoparticles. Additionally, most of the formed phases are magnetically active; they can be separated easily from the medium by the application of an external magnetic field.

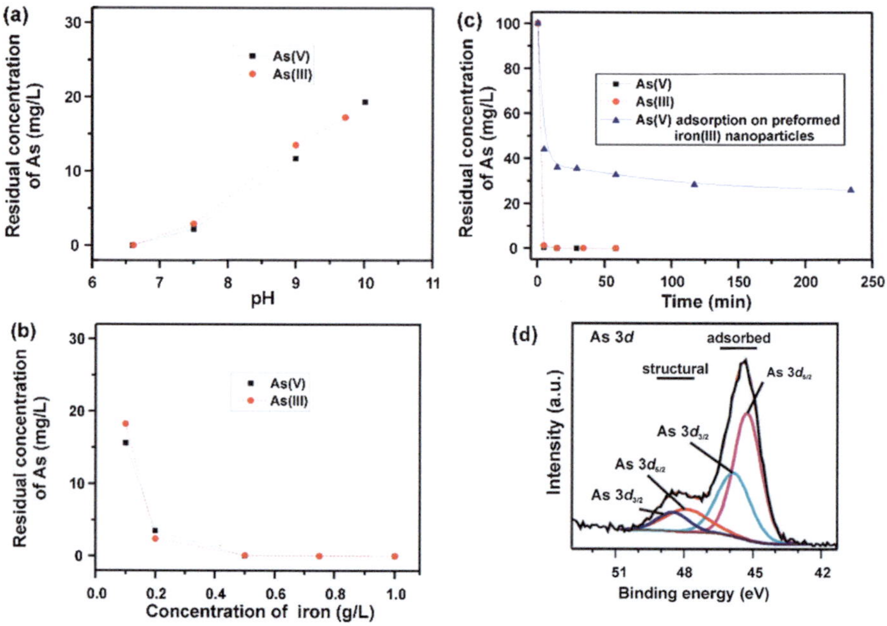

Fig. 8.8 Removal of arsenic by ferrate(VI). (**a**) Dependences of residual arsenic concentrations on the pH value of the reaction mixture (initial conditions: As = 0.1 g/L; Fe = 0.5 g/L); (**b**) dependences of residual arsenic concentrations on the concentration of iron (initial conditions: As = 0.1 g/L; the used pH = 6.6); (**c**) kinetics of arsenites and arsenates removal (initial conditions: As = 0.1 g/L; Fe = 0.5 g/L; and pH = 6.6); and (**d**) high-resolution As 3d core level photoelectron spectra of arsenate removal by ferrate(VI) sample. (Adapted from Prucek et al. 2013)

Further, removal of a wide range of heavy metals by ferrate(VI) has been studied. A thorough study by Prucek et al. (2015) showed that metal ions, such as Cd(II), Ni (II), Co(II), Cu(II), and Al(III), can be removed from water very effectively (Fig. 8.9). At an Fe/metal ion weight ratio of 1/0 or less, complete removal of Co (II), Cu(II), and Al(III) ions was observed. The ratio of 2/0 was required to remove Ni(II) ions completely from water. In the case of Cd(II) ions, the removal was ~70% even at an Fe/Cd(II) weight ratio of 15/1. These results were obtained at neutral or slightly acidic pH values. An increase in the pH enhanced the removal efficiency of heavy metal ions, whereas the removal efficiency of Al(III) decreased with increasing pH, which can be explained in terms of a more pronounced transformation of Al (III) ions to soluble $Al(OH)_4$. The mechanisms of metal ions removal were confirmed by the combination of XPS and Mössbauer spectroscopy techniques. The removal mechanism includes different processes for each particular metal. In the case of copper, cobalt, and nickel, ions are readily removed predominantly through the simultaneous formation of particular metal ferrite and γ-Fe_2O_3/γ-FeOOH core/shell nanoparticles embedding a part of heavy metals in their crystal lattice (i.e., octahedral sites); only a minor part of these elements is adsorbed on the surfaces of γ-Fe_2O_3/γ-FeOOH core/shell nanoparticles. Aluminum is partially incorporated in

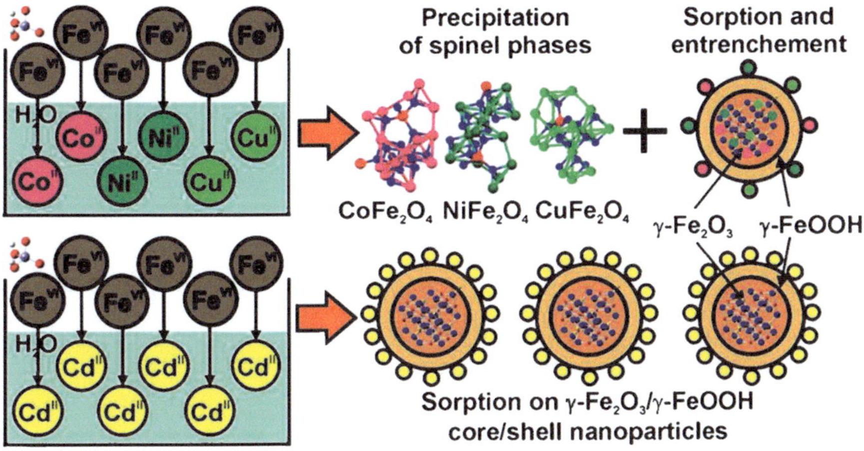

Fig. 8.9 The schematic illustration mechanisms of heavy metals removal by ferrate(VI). (Prucek et al. 2015, Copyright (2015) American Chemical Society)

the tetrahedral positions of the γ-Fe_2O_3 lattice and some portion of Al is adsorbed. Finally, cadmium would be treated by ferrate(VI) just at high Fe dosages with the predominant sorption. Therefore, there are three crucial aspects of the application of ferrate(VI) to the removal of heavy metals including: (i) the formation of metal ferrites ($CuFe_2O_4$, $CoFe_2O_4$, and $NiFe_2O_4$); (ii) structural incorporation of heavy metal ions into the crystal lattice of iron(III) oxide nanoparticles; (iii) the fast kinetic of metal removal even at very low ferrate(VI) dosages.

Similar results were achieved when ferrate(VI) was applied to metal cyanides, namely $K_2[Zn(CN)_4]$, $K_2[Cd(CN)_4]$, $K_2[Ni(CN)_4]$, and $K_3[Cu(CN)_4]$. In this case, toxic cyanides are degraded by oxidation and simultaneously the metals ions are adsorbed from water on the surface of precipitated iron(III). (Filip et al. 2011; Yngard et al. 2008)

Several studies described the critical influence of phosphates ions on the sorption of metals ions from the aquatic environment. The mechanism of interaction of phosphates, when over-concentrated in drinking water, with ferrate(VI) and the efficiency in phosphate removal was studied by Kralchevska et al. (2016a). From the kinetic data of phosphate elimination (Fig. 8.10) a fast reaction of ferrate (VI) (within 2 min) depending on pH values is evident. The effect of pH is similar to the case of arsenic removal. When the pH decreases, the phosphate removal efficiency increases. This study showed that removal of phosphates by ferrate (VI) from water can be achieved efficiently at a low Fe/P mass ratio of 5/1 at pH ~7.0. Moreover, it is evident from the experimental data (mainly low-temperature/in-field Mössbauer spectra, Fig. 8.10d–f) that phosphates, treated by ferrate(VI), solely adsorb onto the surface of iron(III) oxide/oxyhydroxide nanoparticles formed from ferrate(VI). The observed extent of leaching (or desorption) of phosphate from the solid phase(s) is low. This mechanism is explained on the basis of the comparison of

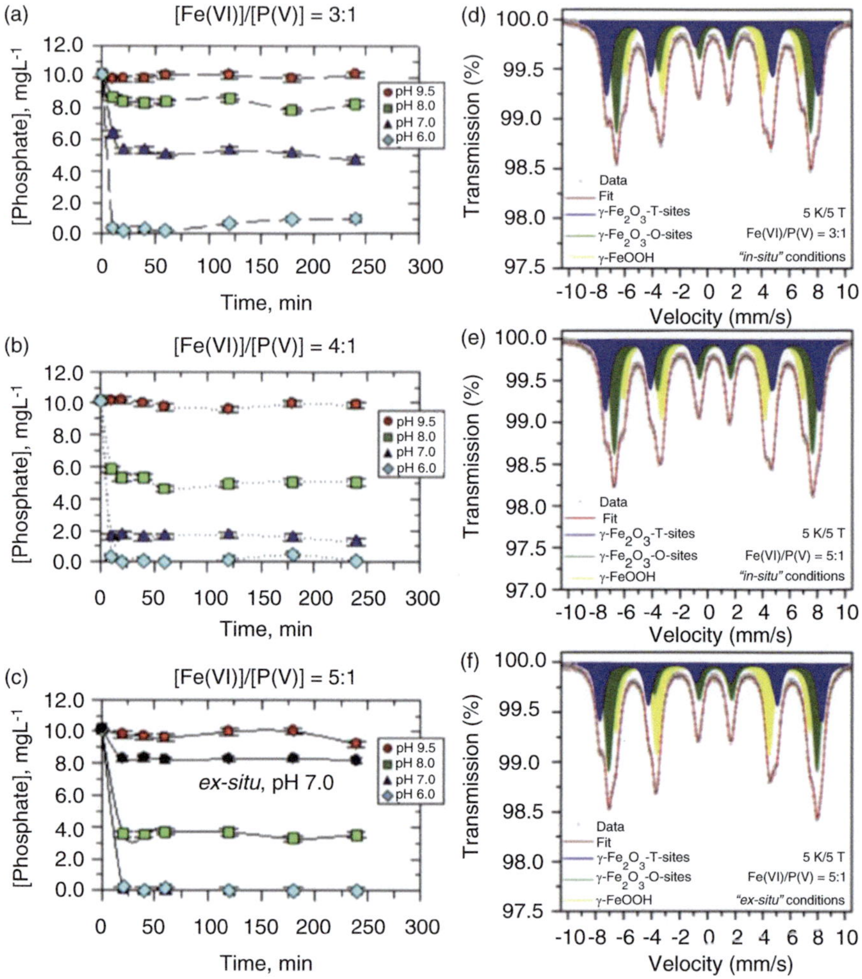

Fig. 8.10 Residual phosphate concentrations as a function of the reaction time and pH of the reaction mixture at Fe/P mass ratio of (**a**) 3/1, (**b**) 4/1, and (**c**) 5/1. (**d–f**) In-field (5 T) low-temperature (5 K) ^{57}Fe Mössbauer spectra of variously treated phosphate (Fe/P = 3/1, 5/1 and at pH ~7.0 for sorption on preformed iron oxides). (Adapted from Kralchevska et al. 2016a)

ferrate-treated phosphate with phosphate sorption on preformed iron(III) oxide/oxyhydroxide nanoparticles.

The efficiency and mechanism of heavy metals and arsenic removal by ferrate (VI) is dependent on the chemistry of the real water. Therefore, Kolařík et al. (2018) focused on the description of the influence of commonly present inorganic ions and organic carbon in waters on metal removal by ferrate(VI). Ions such as chlorides, nitrates, carbonates, and sulfates as well as humic acid had marginal or no effect on the efficiency in removing arsenates by ferrate(VI) (Fig. 8.11). By contrast, the

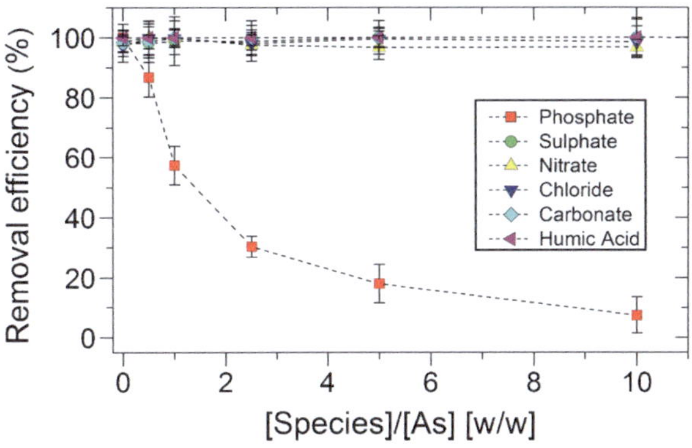

Fig. 8.11 Effect of concentrations of particular ions and humic acid on removal of arsenates by ferrate(VI) (initial conditions: As = 10 mg/L, Fe = 30 mg/L, pH = 6.6, and time = 24 h). (Adapted from Kolařík et al. 2018)

presence of phosphates ions had a negative influence on the efficiency in arsenates removal by ferrate(VI). With the increased concentration of phosphates, the efficiency in arsenates removal was found to decrease. The three crucial aspects of phosphates concerning the arsenate removal mechanism have been identified: (i) At a low P/As weight ratio (up to 1/1), the incorporation of arsenates ions into the crystal structure of the γ-Fe_2O_3 nanoparticles was suppressed. Arsenates were mainly adsorbed on the surface of γ-Fe_2O_3/γ-FeOOH nanoparticles. (ii) Increasing the P/As weight ratio (more than 1/1) led to the competition between arsenates and phosphates sorption on the surface of γ-Fe_2O_3/γ-FeOOH nanoparticles. With increased concentration of phosphates ions, the arsenates amount on the surface of γ-Fe_2O_3/γ-FeOOH nanoparticles was reduced. (iii) Further increase in the phosphates concentration, the complexation of iron(III) ions with phosphates ions occurred, which led to the reduction of arsenates removal efficiency due to a lower content of precipitated γ-Fe_2O_3/γ-FeOOH nanoparticles. In real waters, the arsenic contamination is usually present with phosphates, it is therefore necessary to know the content of phosphates in arsenic contaminated waters for its effective elimination.

8.9 Disinfection Properties of Ferrates

The ability of biological processes such as nitrification and denitrification to eliminate the resistant types of bacteria is typically limited or minimal (Kümmerer 2009). Chemical disinfection using chlorine is one of the easiest and the most frequent means for water treatment. However it can, on one hand, remove microorganisms and micropollutants from water but, on the other hand, harmful or hazardous by-products are formed during this process (Rai et al. 2018). To overcome this problem, other treatment processes and agents, e.g., ferrate(VI), are used/tested to eliminate microorganisms in water. During the water treatment by Fe(VI), no mutagenic and carcinogenic by-products are produced as demonstrated by Ames test (Jiang 2014; Jiang and Lloyd 2002; Rai et al. 2018). Ferrate(VI) can achieve very high (even 100%) efficiency in removing coliform bacteria from wastewaters due to its high oxidation potential at relatively low dosage (Jiang and Lloyd 2002; Kubiňáková et al. 2017; Rai et al. 2018).

Disinfection properties of ferrate were tested on various germs (*Bacillus cereus*, *Streptococcus bovis*, *Staphylococcus aureus*, *Shigella flexneri*, *Enterococcus faecalis*, and *Salmonella typhimurium*) and virus coliphage f2 (*Picornaviridae*) (Rai et al. 2018; Kubiňáková et al. 2017; Birošová et al. 2014; Sharma et al. 2015). This virus is usually present in sewage water and is highly resistant to chlorination (Kazama 1995). The algae caused a growing problem not only in cooling water but also in the production of drinking water. Algae removal from water is difficult because of their small size and the low specific gravity. Alga *Cladophora aegagropila* has been effectively eliminated by electrochemically synthetized potassium ferrate. The mechanism of action of Fe(VI) on the cells of algae can be evaluated by microscopic examination. Any apparent damage to the cell wall proved that ferrate(VI) penetrated into the cell and subsequently caused damage to internal organelles (Kubiňáková et al. 2017).

References

Barışçı S, Särkkä H, Sillanpää M, Dimoglo A (2016) The treatment of greywater from a restaurant by electrosynthesized ferrate (VI) ion. Desalin Water Treat 57(24):11375–11385. https://doi.org/10.1080/19443994.2015.1038740

Birošová L, Mackul'ak T, Bodík I, Ryba J, Škubák J, Grabic R (2014) Pilot study of seasonal occurrence and distribution of antibiotics and drug resistant bacteria in wastewater treatment plants in Slovakia. Sci Total Environ 490:440–444. https://doi.org/10.1016/j.scitotenv.2014.05.030

Casbeer EM, Sharma VK, Zajickova Z, Dionysiou DD (2013) Kinetics and mechanism of oxidation of tryptophan by ferrate(VI). Environ Sci Technol 47(9):4572–4580. https://doi.org/10.1021/es305283k

Faridmarandi S, Naja GM (2014) Phosphorus and water budgets in an agricultural basin. Environ Sci Technol 48(15):8481–8490. https://doi.org/10.1021/es500738v

Feng MB, Wang XH, Chen J, Qu RJ, Sui YX, Cizmas L, Wang ZY, Sharma VK (2016) Degradation of fluoroquinolone antibiotics by ferrate(VI): effects of water constituents and oxidized products. Water Res 103:48–57. https://doi.org/10.1016/j.watres.2016.07.014

Feng M, Wang Z, Sharma VK (2017) Synergetic effect of the oxidation of fluoroquionolone antibiotics by a combined use of ferrate(VI) and peroxymonosulfate. Paper presented at the 253rd American Chemical Society National Meeting & Exposition, San Francisco, April 2–6, 2017

Filip J, Yngard RA, Siskova K, Marusak Z, Ettler V, Sajdl P, Sharma VK, Zboril R (2011) Mechanisms and efficiency of the simultaneous removal of metals and cyanides by using ferrate(VI): crucial roles of nanocrystalline iron(III) oxyhydroxides and metal carbonates. Chem Eur J 17(36):10097–10105. https://doi.org/10.1002/chem.201100711

Han Q, Dong W, Wang H, Liu T, Tian Y, Song X (2018) Degradation of tetrabromobisphenol A by ferrate(VI) oxidation: performance, inorganic and organic products, pathway and toxicity control. Chemosphere 198:92–102. https://doi.org/10.1016/j.chemosphere.2018.01.117

Hightower SM, Lorenz BB, Bernard JG, Johnson MD (2012) Oxidation of phosphorus centers by ferrate(VI): spectral observation of an intermediate. Inorg Chem 51(12):6626–6632. https://doi.org/10.1021/ic3001812

Híveš J, Benová M, Bouzek K, Sitek J, Sharma VK (2008) The cyclic voltammetric study of ferrate (VI) formation in a molten Na/K hydroxide mixture. Electrochim Acta 54(2):203–208. https://doi.org/10.1016/j.electacta.2008.08.009

Híveš J, Gál M, Kerekeš K (2014) High oxidation state of iron in molten hydroxides. Chem Eng Trans 41:49–54. https://doi.org/10.3303/CET1441009

Híveš J, Gál M, Kerekeš K, Kubiňáková E, Mackuľak T (2016) Electrochemical ferrates (VI) preparation and wastewater treatment. In: Ferrites and ferrates: chemistry and applications in sustainable energy and environmental remediation, ACS Symposium Series, vol 1238. American Chemical Society, Washington, DC, pp 221–240. https://doi.org/10.1021/bk-2016-1238.ch008

Hrnčiariková L, Kerekeš K, Híveš J, Gál M (2013) The influence of anode composition on the electrochemical ferrate (VI) production in molten KOH. Int J Electrochem Sci 8(6):7768–7778

Hu L, Martin HM, Arce-Bulted O, Sugihara MN, Keating KA, Strathmann TJ (2009) Oxidation of carbamazepine by Mn(VII) and Fe(VI): reaction kinetics and mechanism. Environ Sci Technol 43(2):509–515. https://doi.org/10.1021/es8023513

Jain A, Sharma VK, Mbuya OS (2009) Removal of arsenite by Fe(VI), Fe(VI)/Fe(III), and Fe(VI)/Al(III) salts: effect of pH and anions. J Hazard Mater 169(1):339–344. https://doi.org/10.1016/j.jhazmat.2009.03.101

Järup L (2003) Hazards of heavy metal contamination. Br Med Bull 68(1):167–182. https://doi.org/10.1093/bmb/ldg032

Jiang J-Q (2014) Advances in the development and application of ferrate(VI) for water and wastewater treatment. J Chem Technol Biotechnol 89(2):165–177. https://doi.org/10.1002/jctb.4214

Jiang J-Q, Lloyd B (2002) Progress in the development and use of ferrate(VI) salt as an oxidant and coagulant for water and wastewater treatment. Water Res 36(6):1397–1408. https://doi.org/10.1016/S0043-1354(01)00358-X

Jiang J-Q, Panagoulopoulos A, Bauer M, Pearce P (2006) The application of potassium ferrate for sewage treatment. J Environ Manag 79(2):215–220. https://doi.org/10.1016/j.jenvman.2005.06.009

Kamachi T, Kouno T, Yoshizawa K (2005) Participation of multioxidants in the pH dependence of the reactivity of ferrate(VI). J Org Chem 70(11):4380–4388. https://doi.org/10.1021/jo050091o

Kazama F (1995) Viral inactivation by potassium ferrate. Water Sci Technol 31(5–6):165–168. https://doi.org/10.1016/0273-1223(95)00259-P

Kerekeš K, Hrnčiariková L, Híveš J, Gál M (2014) On the mechanism of electrochemical transpassive dissolution of Fe-based anodes in binary hydroxide media. J Electrochem Soc 161(1):C62–C68. https://doi.org/10.1149/2.075401jes

Kolář M, Novák P, Šišková KM, Machala L, Malina O, Tuček J, Sharma VK, Zbořil R (2016) Impact of inorganic buffering ions on the stability of Fe(vi) in aqueous solution: role of the carbonate ion. Phys Chem Chem Phys 18(6):4415–4422. https://doi.org/10.1039/C5CP07543B

Kolařík J, Prucek R, Tuček J, Filip J, Sharma VK, Zbořil R (2018) Impact of inorganic ions and natural organic matter on arsenates removal by ferrate(VI): understanding a complex effect of phosphates ions. Water Res 141:357–365. https://doi.org/10.1016/j.watres.2018.05.024

Kralchevska RP, Prucek R, Kolařík J, Tuček J, Machala L, Filip J, Sharma VK, Zbořil R (2016a) Remarkable efficiency of phosphate removal: ferrate(VI)-induced *in situ* sorption on core-shell nanoparticles. Water Res 103:83–91. https://doi.org/10.1016/j.watres.2016.07.021

Kralchevska RP, Sharma VK, Machala L, Zboril R (2016b) Ferrates(FeVI, FeV, and FeIV) oxidation of iodide: formation of triiodide. Chemosphere 144:1156–1161. https://doi.org/10.1016/j.chemosphere.2015.09.091

Kubiňáková E, Gál M, Kerekeš K, Híveš J (2015a) Electrochemical preparation of ferrates in NaOH melts. Chem List 109(9):714–717

Kubiňáková E, Kerekeš K, Gál M, Híveš J (2015b) Electrolytic ferrate preparation in various hydroxide molten media. J Appl Electrochem 45(9):1035–1042. https://doi.org/10.1007/s10800-015-0841-0

Kubiňáková E, Híveš J, Gál M, Fašková A (2017) Effect of ferrate on green algae removal. Environ Sci Pollut Res 24(27):21894–21901. https://doi.org/10.1007/s11356-017-9846-z

Kümmerer K (2009) Antibiotics in the aquatic environment – a review – part II. Chemosphere 75 (4):435–441. https://doi.org/10.1016/j.chemosphere.2008.12.006

Lan B, Wang Y, Wang X, Zhou X, Kang Y, Li L (2016) Aqueous arsenic (As) and antimony (Sb) removal by potassium ferrate. Chem Eng J 292:389–397. https://doi.org/10.1016/j.cej.2016.02.019

Lee Y, Um I-h, Yoon J (2003) Arsenic(III) oxidation by iron(VI) (ferrate) and subsequent removal of arsenic(V) by iron(III) coagulation. Environ Sci Technol 37(24):5750–5756. https://doi.org/10.1021/es034203+

Lee Y, Zimmermann SG, Kieu AT, von Gunten U (2009) Ferrate (Fe(VI)) application for municipal wastewater treatment: a novel process for simultaneous micropollutant oxidation and phosphate removal. Environ Sci Technol 43(10):3831–3838. https://doi.org/10.1021/es803588k

Lee Y, Kissner R, von Gunten U (2014) Reaction of ferrate(VI) with ABTS and self-decay of ferrate (VI): kinetics and mechanisms. Environ Sci Technol 48(9):5154–5162. https://doi.org/10.1021/es500804g

Lubello C, Gori R (2004) Membrane bio-reactor for advanced textile wastewater treatment and reuse. Water Sci Technol 50(2):113–119. https://doi.org/10.2166/wst.2004.0102

Machala L, Zboril R, Sharma VK, Filip J, Schneeweiss O, Homonnay Z (2007) Mössbauer characterization and in situ monitoring of thermal decomposition of potassium ferrate(VI), K_2FeO_4 in static air conditions. J Phys Chem B 111(16):4280–4286. https://doi.org/10.1021/jp068272x

Machala L, Zboril R, Sharma VK, Filip J, Jancik D, Homonnay Z (2009) Transformation of solid potassium ferrate(VI) (K_2FeO_4): mechanism and kinetic effect of air humidity. Eur J Inorg Chem 2009(8):1060–1067. https://doi.org/10.1002/ejic.200801068

Machala L, Filip J, Prucek R, Tucek J, Frydrych J, Sharma VK, Zboril R (2015a) Potassium ferrite ($KFeO_2$): synthesis, decomposition, and application for removal of metals. Sci Adv Mater 7 (3):579–587. https://doi.org/10.1166/sam.2015.2143

Machala L, Procházka V, Miglierini M, Sharma VK, Marušák Z, Wille H-C, Zbořil R (2015b) Direct evidence of Fe(v) and Fe(iv) intermediates during reduction of Fe(vi) to Fe(iii): a nuclear forward scattering of synchrotron radiation approach. Phys Chem Chem Phys 17 (34):21787–21790. https://doi.org/10.1039/C5CP03784K

Machalová Šišková K, Jančula D, Drahoš B, Machala L, Babica P, Alonso PG, Trávníček Z, Tuček J, Maršálek B, Sharma VK, Zbořil R (2016) High-valent iron (FeVI, FeV, and FeIV) species in water: characterization and oxidative transformation of estrogenic hormones. Phys Chem Chem Phys 18(28):18802–18810. https://doi.org/10.1039/C6CP02216B

Mackul'ak T, Birošová L, Bodík I, Grabic R, Takáčová A, Smolinská M, Hanusová A, Híveš J, Gál M (2016) Zerovalent iron and iron(VI): effective means for the removal of psychoactive pharmaceuticals and illicit drugs from wastewaters. Sci Total Environ 539:420–426. https://doi.org/10.1016/j.scitotenv.2015.08.138

Mácová Z, Bouzek K, Híveš J, Sharma VK, Terryn RJ, Baum JC (2009) Research progress in the electrochemical synthesis of ferrate(VI). Electrochim Acta 54(10):2673–2683. https://doi.org/10.1016/j.electacta.2008.11.034

Mura S, Malfatti L, Greppi G, Innocenzi P (2017) Ferrates for water remediation. Rev Environ Sci Biotechnol 16(1):15–35. https://doi.org/10.1007/s11157-016-9416-8

Noorhasan N, Patel B, Sharma VK (2010) Ferrate(VI) oxidation of glycine and glycylglycine: kinetics and products. Water Res 44(3):927–935. https://doi.org/10.1016/j.watres.2009.10.003

Petrov VG, Perfiliev YD, Dedushenko SK, Kuchinskaya TS, Kalmykov SN (2016) Radionuclide removal from aqueous solutions using potassium ferrate(VI). J Radioanal Nucl Chem 310 (1):347–352. https://doi.org/10.1007/s10967-016-4867-5

Prucek R, Tuček J, Kolařík J, Filip J, Marušák Z, Sharma VK, Zbořil R (2013) Ferrate(VI)-induced arsenite and arsenate removal by in situ structural incorporation into magnetic iron(III) oxide nanoparticles. Environ Sci Technol 47(7):3283–3292. https://doi.org/10.1021/es3042719

Prucek R, Tuček J, Kolařík J, Hušková I, Filip J, Varma RS, Sharma VK, Zbořil R (2015) Ferrate (VI)-prompted removal of metals in aqueous media: mechanistic delineation of enhanced efficiency via metal entrenchment in magnetic oxides. Environ Sci Technol 49(4):2319–2327. https://doi.org/10.1021/es5048683

Rai PK, Lee J, Kailasa SK, Kwon EE, Tsang YF, Ok YS, Kim K-H (2018) A critical review of ferrate(VI)-based remediation of soil and groundwater. Environ Res 160:420–448. https://doi.org/10.1016/j.envres.2017.10.016

Sharma VK (2002) Potassium ferrate(VI): an environmentally friendly oxidant. Adv Environ Res 6 (2):143–156. https://doi.org/10.1016/S1093-0191(01)00119-8

Sharma VK (2010) Oxidation of nitrogen-containing pollutants by novel ferrate(VI) technology: a review. J Environ Sci Health, Part A: Tox Hazard Subst Environ Eng 45(6):645–667. https://doi.org/10.1080/10934521003648784

Sharma VK (2013) Ferrate(VI) and ferrate(V) oxidation of organic compounds: kinetics and mechanism. Coord Chem Rev 257(2):495–510. https://doi.org/10.1016/j.ccr.2012.04.014

Sharma VK, Mishra SK, Nesnas N (2006) Oxidation of sulfonamide antimicrobials by ferrate (VI) [$Fe^{VI}O_4^{2-}$]. Environ Sci Technol 40(23):7222–7227. https://doi.org/10.1021/es060351z

Sharma VK, Liu F, Tolan S, Sohn M, Kim H, Oturan MA (2013) Oxidation of β-lactam antibiotics by ferrate(VI). Chem Eng J 221:446–451. https://doi.org/10.1016/j.cej.2013.02.024

Sharma VK, Zboril R, Varma RS (2015) Ferrates: greener oxidants with multimodal action in water treatment technologies. Acc Chem Res 48(2):182–191. https://doi.org/10.1021/ar5004219

Sharma VK, Chen L, Zboril R (2016a) Review on high valent Fe^{VI} (ferrate): a sustainable green oxidant in organic chemistry and transformation of pharmaceuticals. ACS Sustain Chem Eng 4 (1):18–34. https://doi.org/10.1021/acssuschemeng.5b01202

Sharma VK, Tolan S, Bumbálek V, Macova Z, Bouzek K (2016b) Stability of ferrate(VI) in 14 M NaOH-KOH mixtures at different temperatures. In: Ferrites and ferrates: chemistry and applications in sustainable energy and environmental remediation, ACS Symposium Series, vol 1238. American Chemical Society, Washington, DC, pp 241–253. https://doi.org/10.1021/bk-2016-1238.ch009

Sun XH, Zhang Q, Liang H, Ying L, Meng XX, Sharma VK (2016) Ferrate(VI) as a greener oxidant: electrochemical generation and treatment of phenol. J Hazard Mater 319:130–136. https://doi.org/10.1016/j.jhazmat.2015.12.020

Sun S, Liu Y, Ma J, Pang S, Huang Z, Gu J, Gao Y, Xue M, Yuan Y, Jiang J (2018) Transformation of substituted anilines by ferrate(VI): kinetics, pathways, and effect of dissolved organic matter. Chem Eng J 332:245–252. https://doi.org/10.1016/j.cej.2017.08.116

Yngard RA, Sharma VK, Filip J, Zboril R (2008) Ferrate(VI) oxidation of weak-acid dissociable cyanides. Environ Sci Technol 42(8):3005–3010. https://doi.org/10.1021/es0720816

Yu XW, Licht S (2008) Advances in electrochemical Fe(VI) synthesis and analysis. J Appl Electrochem 38(6):731–742. https://doi.org/10.1007/s10800-008-9536-0

Yu H, Chen JW, Xie HB, Ge P, Kong QW, Luo Y (2017) Ferrate(VI) initiated oxidative degradation mechanisms clarified by DFT calculations: a case for sulfamethoxazole. Environ Sci Process Impacts 19(3):370–378. https://doi.org/10.1039/c6em00521g

Zajíček P, Kolář M, Prucek R, Ranc V, Bednář P, Varma RS, Sharma VK, Zbořil R (2015) Oxidative degradation of triazine- and sulfonylurea-based herbicides using Fe(VI): the case study of atrazine and iodosulfuron with kinetics and degradation products. Sep Purif Technol 156:1041–1046. https://doi.org/10.1016/j.seppur.2015.08.024

Chapter 9
Radical Reactions and Their Application for Water Treatment

Pavel Hrabák and Stanisław Wacławek

Abstract Basic chemistry and water treatment of broad range of oxidants and related radical species are covered in this chapter. A general introduction to oxidants and radicals is followed by detailed sections on chlorine species, advanced oxidation processes, persulfates, and non-consensual radical mechanisms. Further, detailed information on oxidant applicability and activation, oxidant-specific recalcitrant pollutants and commonly formed by-products is provided. To assess the suitability of the specific oxidants for real water conditions, matrix components interferences are discussed. Considerable attention is paid to chemistry of innovative oxidants (persulfates) and to the controversial aspects of superoxide radical anion reactivity with carbon tetrachloride.

Keywords AOP · Hydroxyl radical · Sulfate radical · Superoxide radical · Ozone · Reactive chlorine · Persulfates · Carbon tetrachloride

9.1 Introduction

Free radicals are, from the chemistry point of view definition, atoms or molecules with one or more unpaired electrons in bonding orbitals whereby paramagnetic metal ions are not normally regarded as radicals (IUPAC 1997). This electron arrangement leads, apart from a few important exceptions (see below), to high reactivity of free radicals. Oxidants generally possess the ability to attract the whole bonding electron pair from their reaction counterparts. Free radicals, in contrast, often mediate one-electron transfer reactions. Basic instrumentation for researching radical-driven reactions is electron paramagnetic resonance (EPR) and stopped-flow spectroscopic techniques (Hayyan et al. 2016; Qian and Buettner 1999).

Water-stable free radicals—molecular oxygen (O_2) and chlorine dioxide (ClO_2)—are examples of such important exceptions. Both O_2 and ClO_2 are well

P. Hrabák (✉) · S. Wacławek
Institute for Nanomaterials, Advanced Technologies and Innovation, Technical University of Liberec, Liberec, Czech Republic
e-mail: pavel.hrabak@tul.cz

J. Filip et al. (eds.), *Advanced Nano-Bio Technologies for Water and Soil Treatment*, Applied Environmental Science and Engineering for a Sustainable Future, https://doi.org/10.1007/978-3-030-29840-1_9

water-soluble gases. Oxygen, which is actually a bi-radical, is often employed as a cheap electron-pair acceptor for oxidation of undesirable inorganic water components like Fe^{2+} or Mn^{2+} ions. However, molecular oxygen oxidative capabilities towards organic pollutants are very limited. This feature is called spin restriction of oxygen. For the reaction with molecular oxygen, both electrons need to be of the same spin, which is the condition biomolecules or organic pollutants rarely satisfy (Halliwell and Gutteridge 1984). Dissolved ClO_2 has been researched for a growing number of water treatment applications (Aieta and Berg 1986; Huber et al. 2005), which will be further discussed in Sect. 9.2.

Some oxidants such as H_2O_2, O_3, or $S_2O_8{}^{2-}$ are precursors of a number of other free radicals. The umbrella term reactive oxygen species (ROS) also encompasses non-radical yet reactive oxygen moieties, e.g., singlet states of oxygen (Halliwell 2006). The formation of these short-living radical species is driven by specific reaction conditions among which the temperature and the pH play the key role by influencing the kinetics of each of the simultaneously running reactions. Another important parameter is the reactor exposure to light, especially to the light of shorter wavelengths (UV) (see also Chap. 10, which focuses on UV water treatment). UV light mechanism of action is multimodal. Besides direct organic compound photolysis, ROS are generated under oxygen saturated or oversaturated water conditions. In addition, UV is often used to initiate ROS production in advanced oxidation processes (AOPs) for drinking water supply since UV can provide the right portion of energy for symmetrical bond cleavage. Hydroxyl ($^{\bullet}OH$) and sulfate radicals ($SO_4{}^{\bullet-}$) are species that are directly responsible for transformation of organic pollutants and for water disinfection. This chapter focuses on describing the conditions for radical species generation, on radical reactivity with the target pollutants and radical scavenging by other water components, and on disinfection by-products (DBP). Some of the processes tend to be more employed in drinking water supply (chlorine and ozone), other find its application first in wastewater treatment and groundwater remediation (Fenton reagent and peroxydisulfate).

9.2 Chlorine Species

Chlorine species are traditionally used as drinking water disinfectants, which make water supply safe from microbiological point of view. Presence of organic compounds is generally unfavorable in the treated water (Ramseier 2010), as will be explained further. Therefore, chlorination is not a suitable technology for pollutant degradation.

Most usually, chlorine gas (Cl_2) or sodium hypochlorite (NaOCl) is dosed. In the USA, monochloramine (NH_2Cl) is still in use as residual disinfectant for water distribution systems. High natural organic matter (NOM) and bromide (Br^-) content

Table 9.1 Chlorine species overview

Name	Formula	Cl valence
Chloride/hydrochloric acid	Cl^{-}/HCl	−1
Free chlorine, chlorine atom	$Cl_2/Cl^{\bullet}$	0
Hypochlorite/hypochlorous acid	$OCl^{-}/HOCl^{-}$	+1
Chlorite/chlorous acid	$ClO_2^{-}/HClO_2$	+2
Chlorate/chloric acid	$ClO_3^{-}/HClO_3$	+5
Chlorine dioxide	ClO_2	+4
Monochloramine	NH_2Cl	+1

in the treated water can lead to formation of toxic DBPs. More than 600 DBPs are known today (Richardson et al. 2007). For chlorine species, most important DBPs are:

- NaOCl or Cl_2 – trihalomethanes (THMs), haloacetic acids, haloacetonitriles, chloral hydrates, trihalonitromethanes
- NH_2Cl – nitrosamines, cyanogen halides, iodinated THMs, haloacetaldehyde
- ClO_2 – chlorite, chlorate, organic acids

Chlorine dioxide (ClO_2) offers several advantages in comparison to other chlorine sources as it reacts directly as an electron acceptor rather than via chlorine substitution. Moreover, it is more efficient in protozoa deactivation; its efficiency is pH-independent and no malodorous DBPs are formed. However, in case of ClO_2 some relevant DBPs can also appear. As with other chlorine species, precise dosing with a feedback to matrix components concentrations is absolutely necessary to avoid the DBP formation. An overview of chlorine water species is provided in Table 9.1.

With the exception of ClO_2, which is a free radical, chlorine species act as electron pair transfer oxidants. Only under specific conditions, chlorine atom $Cl^{\bullet}$ (radical) can act in persulfate-based oxidation in turnover of $SO_4^{\bullet-}$ to $^{\bullet}OH$ (Lutze 2013; see also Sect. 9.4).

9.3 Ozone and Hydrogen Peroxide Related AOPs

There are two main oxidants used in AOP: ozone (O_3) and hydrogen peroxide (H_2O_2). Both oxidants react directly with some target pollutants and at the same time they are the source of set of reactive oxygen species. $^{\bullet}OH$ is the active substance produced in solutions of both oxidants.

Ozonation is probably the most common drinking water treatment that provides disinfection, color removal, iron and manganese oxides precipitation, and pollutant control (Lutze 2013). In groundwater and soil remediation, ozonation did not gain such popularity mainly because of the risk connected with gaseous ozone toxicity.

O_3 directly reacts with double bonds or aromatic ring containing compounds, reduced sulfur containing groups, and deprotonated amines (von Gunten 2003). $^{\bullet}OH$ can be formed either in the reaction of O_3 with NOM or, in the pH > 8 range, with OH^- ions. To support $^{\bullet}OH$ formation from O_3, H_2O_2 is sometimes added to make a mix called peroxon (Staehelin and Hoigne 1982) or the pH is adjusted to higher values (Ramseier 2010).

In propagation reactions of $^{\bullet}OH$ and O_3 with NOM and carbonates, other ROS, e.g., perhydroxyl ($HO_2^{\bullet}$) or its deprotonated version superoxide radical anion ($O_2^{\bullet-}$) emerge and make the reaction system even more complex. $HO_2^{\bullet}/O_2^{\bullet-}$ are poorly reactive with organic compounds in water medium, i.e., react very slowly (Gutteridge and Halliwell 1989). This is in contradiction to the reaction mechanisms proposed by some authors; details will be discussed in Sect. 9.5. Superoxide easily undertakes termination reactions, e.g., in the one with ozone in which oxygen is formed. Two superoxide molecules can reconstitute H_2O_2 in dismutation reaction ($2O_2^{\bullet-} + 2H^+ \rightarrow H_2O_2 + O_2$). In termination reactions, O_2 (mainly from $O_2^{\bullet-}$) or H_2O/OH^- (from $^{\bullet}OH$) are formed.

H_2O_2 additions to ozone are also used to avoid by-products formation. In case of ozone, the most relevant DBP is bromate BrO^-, which could be formed if bromide (Br^-) is present in treated water (von Gunten and Hoigne 1994). Avoiding bromate formation is a challenging task, which exceeds single ozone treatment technology. Regarding the presence of ozone recalcitrant pollutants together with bromide and NOM concentration, it might be either more efficient to enhance ozone dosing in some cases or to optimize H_2O_2 addition or to leave ozone AOP and switch over to using UV/H_2O_2 (Lutze 2013; von Gunten and Oliveras 1998). Ozonation is a fairly established technology with solid knowledge foundation coming from water radiochemistry, with many full-scale applications and comprehensive literature, e.g., von Sonntag and von Gunten (2012).

Changing the basic oxidant to H_2O_2, we move from potable water treatment towards wastewater treatment technologies. Diluted H_2O_2 itself is not a sufficiently strong oxidant to treat dissolved TOC or specific organic pollutants. However, various iron species are known to initiate $^{\bullet}OH$ generation from H_2O_2 in homogeneous or heterogeneous Fenton reaction (Eq. 9.1) derived processes, as reviewed by (Navalon et al. 2010, 2011; Pignatello et al. 2007).

$$H_2O_2 + Fe^{2+} \rightarrow Fe^{3+} + OH^- + {}^{\bullet}OH \tag{9.1}$$

Since the presence of iron in the technological unit inevitably leads to the need for sludge management, there are often other methods employed for H_2O_2 one electron reduction (i.e., $^{\bullet}OH$ formation). Gamma radiation, electrolysis, and ultrasound (including cavitation) are examples of the means that can mediate it (Brillas et al. 2009; Ince et al. 2001). Unlike the case of persulfates, heating cannot break the peroxygen bond of H_2O_2.

Subsurface applications of H_2O_2 (so called *in situ* chemical oxidation, ISCO) for contaminated groundwater and soil treatment are a very complex mix of physical and chemical processes that are well reviewed in (Siegrist et al. 2011). In brief, H_2O_2 instantly and exothermically decomposes to oxygen and water ($H_2O_2 \rightarrow H_2O + 1/2O_2$) once in contact with iron and manganese oxides surfaces of aquifer materials. Therefore, organic acids (e.g., citric acid, EDTA) additions are widely studied for H_2O_2 ISCO (Engelmann et al. 2003). This way, via iron and manganese oxide dissolution and chelation by organic acids, the unfavourable exothermic reaction may be slowed down. However, with the addition of organic acids, the question of other (toxic) metals mobilization arises. Even if metals Cu^+, Cr^{2+}, Co^{2+} (Tarr 2003) can substitute iron in $^{\bullet}OH$ formation, their presence has to be monitored to exclude the possibility of mobilisation. Many studies have shown that the mobilisation of heavy metals is a time and space limited process in case of joint applications of chelators and H_2O_2 (Bennedsen et al. 2012).

As pollutants recalcitrant to $^{\bullet}OH$ attack, perhalogenated aliphatic hydrocarbons are reported (Siegrist et al. 2011), especially chlorinated methanes and ethanes. Small kinetic constants were also found for long chain petroleum hydrocarbons.

9.4 Non-consensual Radical Mechanisms

We provide basic information on two non-consensual radical mechanisms: Involvement of superoxide radical anion ($O_2^{\bullet-}$) in tetrachloromethane (CCl_4) degradation and ferryl (FeO^{2+}) specie reactions.

Carbon tetrachloride is an established toxicity model substance (Weber et al. 2003), its transformation mechanism is well described in mammalian hepatic cells. Decades of CCl_4 toxic action research on cellular, tissue, and animal model levels brought postulation of its free radical nature: Liver cell cytochrome P450 initiates the transformation of CCl_4 to trichloromethyl radical ($^{\bullet}CCl_3$) (McCay et al. 1984), which may form trichloromethylperoxyl radicals ($^{\bullet}OOCCl_3$) in the presence of oxygen (Mönig et al. 1983). Both radicals were identified as the species being responsible for the specific toxic effects of CCl_4. Several hundreds of in-vitro and in-vivo studies have been conducted in order to clearly understand the variability of CCl_4-induced toxic effects, including dose-response dependence, preventive or potentiating impact of other compounds (e.g., antioxidants), oxygen partial pressure, or age of model animals (Vulimiri et al. 2011). Chloroform, hexachloroethane, phosgene, and alkanes/alkanals/alkenals belong to CCl_4 metabolism intermediates while carbon monoxide, carbon dioxide, and hydrochloric acid are its metabolic endproducts (Plaa 2000).

In complex scientific coverage of molecular mechanisms in CCl_4-induced toxic effects, no evidence is given for the $O_2^{\bullet-}$ involvement in the initial step of the CCl_4 transformation, even though $O_2^{\bullet-}$ is ubiquitously present in living cells (McCord and Fridovich 1988). Instead, the initial step is attributed to NADPH-cytochrome *P*-450 reductase. Furthermore, hyperbaric oxygen conditions are claimed to prevent the initiation of CCl_4 metabolism via trichloromethyl radical mechanism (Manibusan et al. 2007; Reiner et al. 1972).

In the aquifer, CCl_4 undergoes degradation, which is driven by biotic and abiotic factors (Shao and Butler 2009), whereby oxygen poor conditions are essential. Microbial transformations of CCl_4 are either accomplished by extracellular coenzyme exudates (Hashsham and Freedman 1999) or by extracellular mineral secretion (McCormick and Adriaens 2004). In abiotic degradation pathways, reduced species of iron and sulfur such as pyrite (Kriegman-King and Reinhard 1994), sulfides (Devlin and Müller 1999), adsorbed Fe^{II} (Amonette et al. 2000; Elsner et al. 2004; Kenneke and Weber 2003), or zero-valent iron (ZVI) (Johnson et al. 1998), were identified as reactants for reductive CCl_4 degradation. The half-life of CCl_4 in the aquifer is reported to range from a few days to hundreds of days (Howard 1991). For more information on the degradation of CCl_4 in the aquifer see Pecher et al. (2002) and Elsner and Hofstetter (2011).

Major products of CCl_4 aquifer degradation are chloroform, formate, carbon monoxide, and carbon dioxide. Similarly to CCl_4 mammalian metabolism, the initial step in the aquifer-related CCl_4 degradation—one electron reduction to trichloromethyl radical—is supposed for all identified intermediates and products (Elsner et al. 2004). However, in the degradation mediated by *Pseudomonas stutzeri* KC only traces of chloroform has been reported, with most of the carbon mass being transformed to carbon dioxide (Criddle et al. 1990).

Under oxygen-rich conditions, represented by oxidative water treatment such as ozone and hydroxyl radical based advanced oxidation processes (AOP), CCl_4 is claimed to exhibit chemical stability (von Sonntag 2008).

On the other hand, contrary to the above-mentioned literature, there is a series of studies (Che and Lee 2011; Furman et al. 2009; Howsawkeng et al. 2010; Smith et al. 2004; Stoin et al. 2015; Teel and Watts 2002; Watts et al. 2005) that describe CCl_4 degradation proceeding under peroxide-based conditions. Similar CCl_4 degradation reactions are believed to occur in peroxydisulfate systems, which was reported by (Xu et al. 2014a, b). Moreover, $O_2^{\bullet-}$ was identified as an oxygen species initiating CCl_4 transformation in both oxidative systems. Furthermore, a technique employing CCl_4 was established to prove the $O_2^{\bullet-}$ presence (Corbin III 2008; Watts 2011; US EPA 2014).

We conclude that the CCl_4 reactivity with $O_2^{\bullet-}$ in oxidative systems remains controversial. Both peroxide- and peroxydisulfate-based systems are a very complex interplay of several simultaneous reactions. Since gaseous oxygen is one of the reaction products in oxidative systems, it is extremely uneasy to conduct CCl_4 degradability experiments in a methodologically correct way. We suggest that it is the CCl_4 volatilisation loss rather than its degradation that might be the process leading to the CCl_4 decrease in the reaction systems reported in the previous paragraph. The CCl_4 degradation in oxidative systems is still questionable, so is its initiation by $O_2^{\bullet-}$.

The other non-consensual species, which is often discussed in studies of Fenton-like reaction systems, is ferryl ion FeO^{2+}. Reaction mechanisms in AOP are perceived as radical ($^{\bullet}OH$) based by most of the scientific community. However, there is an alternative to $^{\bullet}OH$—FeO^{2+}. High valent (4+, 5+, and 6+) iron-oxo species chemistry is thoroughly described in Chap. 8. Four-valent iron-oxo complexes are proposed to participate in enzymatic reactions of living cell systems. Ferryl ion is proven to be a part of non-heme iron metalloenzymes (Groves 2006; Krebs et al. 2007), which provide substrate hydroxylation. Some authors propose its equivalent presence in H_2O_2 systems with Fe^{2+} activation (Lee and Sedlak 2009) and its presence is also supposed in zero-valent iron corrosion systems, where H_2O_2 is generated as an intermediate species (Keenan and Sedlak 2008). Calculations of Bossmann et al. (1998) have shown higher thermodynamic favourability for the formation of FeO^{2+} rather than of $^{\bullet}OH$ under the assumption that the Fe^{2+} ion is present as $Fe(H_2O)_5(OH)^+$ at circumneutral pH and H_2O_2 must first be incorporated into this complex to form $Fe(OH)(H_2O_2)(H_2O)4^+$. Let's conclude with Barbusiński (2009) it is possible that both $^{\bullet}OH$ and FeO^{2+} can coexist in Fenton chemistry and, depending on the operating parameters, one of them predominates.

9.5 Persulfates Chemistry

Persulfates arouse enormous research interest. Over the course of the last 2 years about ten review papers have been published on sulfate radical-based advanced oxidation techniques (Boczkaj and Fernandes 2017; Brienza and Katsoyiannis 2017; Ghanbari and Moradi 2017; Hu and Long 2016; Ike et al. 2018; Liu et al. 2018; Matzek and Carter 2016; Oh et al. 2016; Wacławek et al. 2017; Wang and Wang 2018; Wang et al. 2016; Xiao et al. 2018; Ye et al. 2017; B.-T. Zhang et al. 2014).

There could be several different reasons why persulfates generate interest in recent scientific publication that has also been reflected in the number of citations. One of the reasons could be related to the simultaneous development of the heterogeneous catalysis field, which is directly interconnected with the persulfates one (Oh et al. 2016). Another reason could be that some time ago a cheaper method was developed for peroxydisulfate large scale production, which in combination with its cheaper and safer transport due to the solid state (powder) consequently makes it one of the most often used oxidants in ISCO (Ike et al. 2018; Wacławek et al. 2017). Another explanation could be the fact that the sulfate radicals (generated in the PS system) have different oxidation mechanism in comparison to the hydroxyl ones (Boczkaj and Fernandes 2017) and possess a very high redox potential, which consequently makes SO_4^- radical more suitable for degradation of several substances that are immune to standard AOP techniques, e.g., perfluorooctanoic acid (PFOA) (Wacławek et al. 2017). Sulfate radical oxidation mechanism was broadly described in a recent work of Ye et al. (2017) as shown in Fig. 9.1.

Persulfates are a group of two oxidants namely peroxydisulfate (PDS) and peroxymonosulfate (PMS); the main difference between them is that in the PDS molecule, the peroxide group bridges two sulfur atoms, whereas PMS is a type of an S-inorganic hydroperoxide. Both of these oxidants are not very reactive when non-activated, i.e., when they are not generating radical species. The main aim of the catalytic activation is to weaken the O–O bond of the persulfates (which can result in homolytic or heterolytic cleavage of the peroxide bond) (Wacławek et al.

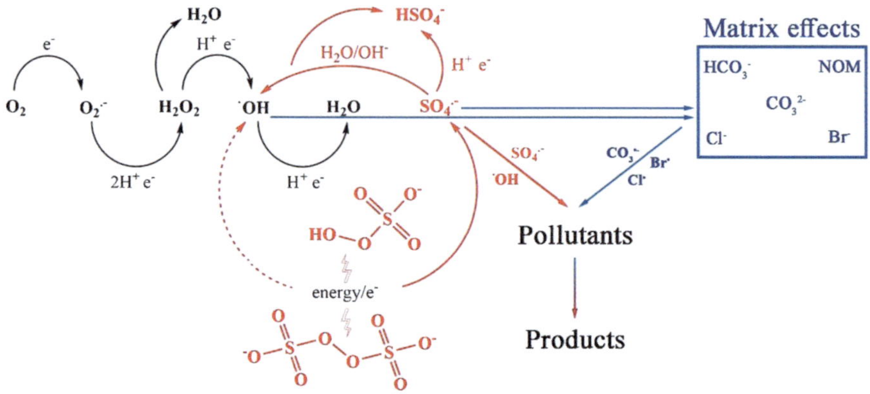

Fig. 9.1 Black arrows—scheme of oxygen molecule reduction to water; red arrows—radical formation and behaviour in a persulfates system; blue arrows—water matrix effects on the persulfates chemistry (dashed arrow—only valid for homolysis of the PMS peroxide bond). This figure is reprinted with a kind permission from Elsevier published paper (Wacławek et al. 2017)

2017), similarly to the Fenton chemistry discussed in the previous sections. In order to activate persulfates, several approaches that are homogeneous (catalyst is in the same phase as PS, e.g., dissolved cation) and heterogeneous (catalyst is in a different phase than PS, e.g., solid metal particles) can be adopted.

The homogeneous (as well as homolytic) activation of persulfates was critically reviewed and discussed by almost all the above-mentioned reviews (Ghanbari and Moradi 2017; Matzek and Carter 2016; Wacławek et al. 2017; Wang and Wang 2018; T. Zhang et al. 2014) and book (Siegrist et al. 2011). Because of this fact, only small part of this chapter will be devoted to these types of reaction. Persulfates can be activated by UV (Yang et al. 2010), heat (Ghauch et al. 2012; Yang et al. 2010), ultrasounds (Wei et al. 2017), and by homogeneous (B.-T. Zhang et al. 2015) and heterogeneous (Hu and Long 2016; Oh et al. 2016) approaches, whereas an example of a chemical element that can be used as a persulfates activator either as homogeneous and heterogeneous catalyst is iron (Fränzle et al. 2010; Fu et al. 2014; Hrabák et al. 2016; Kang et al. 2018; Wacławek et al. 2015).

A very good comparison of various homogeneous activators of PDS was provided by Anipsitakis and Dionysiou who determined Ag(I) as the best activator for PDS compared to other commonly available transition metals (Anipsitakis and Dionysiou 2004); however, there have been more reports published on persulfates catalysed by iron in its various valence states. Iron overall was identified as one of the best catalytic materials for persulfates activation, especially for PDS; as for PMS, divalent cobalt was determined to be the most efficient catalyst according to Anipsitakis and Dionysiou (2003) and Fernandez et al. (2004). The different reasons to explain the dominance of iron may include its price, biocompatibility, and capability of activating PS efficiently.

However, there are also several problems with the homogeneous iron activation of PS, i.e., the selection of an appropriate dose and the type of the iron catalyst since, when it is applied in excess, scavenging of sulfate radicals may become problematic (Zhao et al. 2014).

Moreover, Fe(II) can be quickly oxidized to Fe(III); although there are several researchers claiming that the Fe(III)/PS process can be efficient, the activation mechanism is still being questioned and investigated. For example, an interesting theory proposed by Ike et al. (2017) explains that a possible reaction considering the E^0 of PDS/Fe may involve Fe(III) oxidation to higher valence iron, i.e., ferrates. However, the main problem is the low solubility of Fe(III) in the absence of suitable complexation. Literature data indicates that, under standard conditions, the concentration of Fe(III) in solution can be less than 1.8×10^{-4} mmol/L. This is due to the strong tendency of Fe(III) to hydrolyse and form hydroxide complexes (that have a low solubility in H_2O).

During the last eight years these problems have been thoroughly investigated and several solutions have been found, e.g.,:

- chelation of iron for effective control of the amount of Fe(II) available to react with PS (Rastogi et al. 2009)
- regeneration of Fe(II) by electrolysis. This technique may be very effective and may be competitive where sustainably generated electricity is abundant and cheap (Wacławek et al. 2016; Yuan et al. 2014)
- heterogeneous catalysis of persulfates that (in case of the zero-valent iron) provides slow release of Fe(II) (Oh et al. 2016; Xiao et al. 2018)

Especially the last point has aroused significant interest lately since the heterogeneous catalyst can often be easily separated from the reactant solution and have other interesting properties (Oh et al. 2016; Xiao et al. 2018). Moreover, carbonaceous catalysis of PS (Chen and Carroll 2016; Duan et al. 2015; Lee et al. 2015) was found very popular recently, however, this topic still remains controversial (Lee et al. 2017). Several oxidants can also be heterogeneously activated by subsurface minerals (Ahmad et al. 2010), which makes their application for, e.g., groundwater remediation more facile.

As mentioned before, iron is one of those elements that can be incorporated to the heterogeneous catalysis of PS. Adopting the use of solid iron particles where the release of iron species being responsible for the activation of PDS occurred smoothly without the risk of sulfate radical quenching was reported by Naim and Ghauch (2016). On the other hand, Ayoub and Ghauch (2014) demonstrated that the activation of PDS can be sustained better in solution especially while using bimetallic and trimetallic iron-based particles, which make the process more efficient in long-term application. Such great efficiency of heterogeneous activation by solid iron has led to investigating this material also in smaller dimensions, i.e., nanoscale zero-valent iron (nZVI). nZVI is not a new material in the environmental chemistry field and has already been applied to many contaminated sites *in situ*, where it can

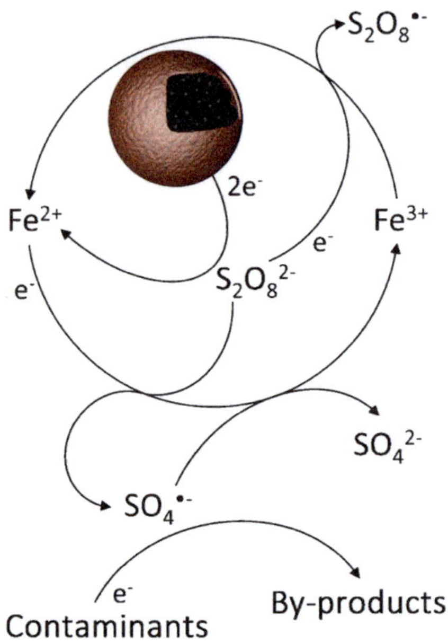

Fig. 9.2 Scheme presenting mechanism of peroxydisulfate activation with nZVI. (Adapted from Kim et al. 2018)

not only adsorb and reduce contaminants but also enhance bioremediation processes (Krol et al. 2013; Němeček et al. 2016). It has been recently found that nZVI can be successfully applied as a heterogeneous catalyst for peroxydisulfate (Kang et al. 2018; Kim et al. 2018) and peroxymonosulfate (Tan et al. 2018; Wang et al. 2017). Especially Kim et al. (2018) focused on the mechanism of the reaction (Fig. 9.2).

For example, it was determined that the $SO_4^{\bullet-}$ yield per mole of PDS was more than two times higher in the PDS/nZVI system in comparison to the PDS/Fe(II) one (Kim et al. 2018). It could be therefore assumed that the radicals were produced more efficiently in the heterogeneous system because aqueous Fe(II) was supplied more slowly, preventing scavenging of $SO_4^{\bullet-}$ by excess Fe(II).

Therefore, a relatively low oxidant dose could be used when nZVI rather than Fe (II) is used as an activator.

Moreover, it should be noted that various coatings of nZVI can alter these processes, making it a real challenge worth solving (Diao et al. 2016). The influence of inorganic coating of nZVI was the subject of a recent study (Rayaroth et al. 2017) wherein the sulfidation of nZVI was investigated leading to the increased applicability of nZVI (efficiency of degradation was extended to alkaline pH).

To the best of our knowledge, the only oxidation state of iron that has not been reported in combination with persulfates is the iron in high oxidation states (i.e., ferrates). In the light of the preliminary experiments that have not been published yet (Wacławek, Hrabák, Filip, Černík, unpublished data), it is believed that ferrates can have a synergistic effect with persulfates (Fig. 9.3).

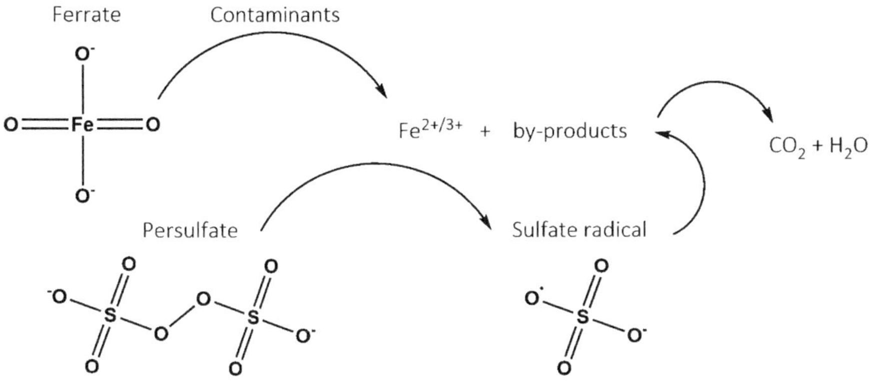

Fig. 9.3 Scheme of possible synergistic effect between ferrates and persulfates

However, this process will strongly depend on the solution pH as well as on many other parameters; therefore it should be thoroughly investigated in the future.

Overall, investigating the influence of the pH on the persulfate decomposition can be considered as another serious challenge. The oxidation of substances by persulfate depends strongly on the concentration of hydronium and hydroxyl ions. Especially PMS is affected by the pH since it reacts differently when the molecule is fully protonated or deprotonated (i.e., H_2SO_5, HSO_5^-, or SO_5^{2-}). Further, a very interesting topic that has not been carefully investigated yet is the rate of the formation of radicals in various pH values. Generally, PS reactions at acidic pH most frequently improve the degradation kinetics, however, the improved removal rates were also observed at neutral pH (between 6 and 8) and slightly basic pH (between 9 and 10, which can be often associated with interchanging of the radical species—sulfate to hydroxyl) depending on the contaminant and activation type (Matzek and Carter 2016).

Another challenging concept involves the PDS activation with hydrogen peroxide, which was studied by several authors (Zhao et al. 2013). However, its exact mechanism remains unknown, but there are several researchers that believe the activation process depends only on the heat generated from the hydrogen peroxide exothermic reactions.

It should be noted that because of several undesirable characteristics, like post-contamination with sulfate ions and reaction with chlorine/bromine species with further chlorination/bromination of the water matrix (both of those processes can lead to the increase in the toxicity), sulfate radical-based systems are not considered the best candidates for drinking water treatment. However, it is possible that a new technology in combination with the membrane techniques will be developed to serve the purpose in the future.

Acknowledgement The authors appreciate kind support of laboratory of Instrumental Analytical Chemistry at University Duisburg-Essen, namely of Prof. Torsten Schmidt, in elucidation of carbon tetrachloride degradability.

References

Ahmad M, Teel AL, Watts RJ (2010) Persulfate activation by subsurface minerals. J Contam Hydrol 115(1–4):34–45. https://doi.org/10.1016/j.jconhyd.2010.04.002

Aieta EM, Berg JD (1986) A review of chlorine dioxide in drinking water treatment. J Am Water Works Assoc 78(6):62–72

Amonette JE, Workman DJ, Kennedy DW, Fruchter JS, Gorby YA (2000) Dechlorination of carbon tetrachloride by Fe(II) associated with goethite. Environ Sci Technol 34 (21):4606–4613. https://doi.org/10.1021/es9913582

Anipsitakis GP, Dionysiou DD (2003) Degradation of organic contaminants in water with sulfate radicals generated by the conjunction of peroxymonosulfate with cobalt. Environ Sci Technol 37(20):4790–4797. https://doi.org/10.1021/es0263792

Anipsitakis GP, Dionysiou DD (2004) Radical generation by the interaction of transition metals with common oxidants. Environ Sci Technol 38(13):3705–3712. https://doi.org/10.1021/es035121o

Ayoub G, Ghauch A (2014) Assessment of bimetallic and trimetallic iron-based systems for persulfate activation: application to sulfamethoxazole degradation. Chem Eng J 256:280–292. https://doi.org/10.1016/j.cej.2014.07.002

Barbusiński K (2009) Fenton reaction – controversy concerning the chemistry. Ecol Chem Eng S 16 (3):347–358

Bennedsen LR, Krischker A, Jørgensen TH, Søgaard EG (2012) Mobilization of metals during treatment of contaminated soils by modified Fenton's reagent using different chelating agents. J Hazard Mater 199–200:128–134. https://doi.org/10.1016/j.jhazmat.2011.10.068

Boczkaj G, Fernandes A (2017) Wastewater treatment by means of advanced oxidation processes at basic pH conditions: a review. Chem Eng J 320:608–633. https://doi.org/10.1016/j.cej.2017.03.084

Bossmann SH, Oliveros E, Göb S, Siegwart S, Dahlen EP, Payawan L, Straub M, Wörner M, Braun AM (1998) New evidence against hydroxyl radicals as reactive intermediates in the thermal and photochemically enhanced Fenton reactions. J Phys Chem A 102(28):5542–5550. https://doi.org/10.1021/jp980129j

Brienza M, Katsoyiannis IA (2017) Sulfate radical technologies as tertiary treatment for the removal of emerging contaminants from wastewater. Sustainability 9(9):1604. https://doi.org/10.3390/su9091604

Brillas E, Sirés I, Oturan MA (2009) Electro-Fenton process and related electrochemical technologies based on Fenton's reaction chemistry. Chem Rev 109(12):6570–6631. https://doi.org/10.1021/cr900136g

Che H, Lee W (2011) Selective redox degradation of chlorinated aliphatic compounds by Fenton reaction in pyrite suspension. Chemosphere 82(8):1103–1108. https://doi.org/10.1016/j.chemosphere.2010.12.002

Chen H, Carroll KC (2016) Metal-free catalysis of persulfate activation and organic-pollutant degradation by nitrogen-doped graphene and aminated graphene. Environ Pollut 215:96–102. https://doi.org/10.1016/j.envpol.2016.04.088

Corbin JF III (2008) Mechanisms of base, mineral, and soil activation of persulfate for groundwater treatment. Dissertation, Washington State University

Criddle CS, DeWitt JT, Grbić-Galić D, McCarty PL (1990) Transformation of carbon tetrachloride by *Pseudomonas* sp. strain KC under denitrification conditions. Appl Environ Microbiol 56 (11):3240–3246

Devlin JF, Müller D (1999) Field and laboratory studies of carbon tetrachloride transformation in a sandy aquifer under sulfate reducing conditions. Environ Sci Technol 33(7):1021–1027. https://doi.org/10.1021/es9806884

Diao Z-H, Xu W-R, Jiang D, Kong L-J, Sun Y-X, Hu Y-X, Hao Q-W, Chen H (2016) Bentonite-supported nanoscale zero-valent iron/persulfate system for the simultaneous removal of Cr (VI) and phenol from aqueous solutions. Chem Eng J 302:213–222. https://doi.org/10.1016/j.cej.2016.05.062

Duan X, Sun H, Kang J, Wang Y, Indrawirawan S, Wang S (2015) Insights into heterogeneous catalysis of persulfate activation on dimensional-structured nanocarbons. ACS Catal 5 (8):4629–4636. https://doi.org/10.1021/acscatal.5b00774

Elsner M, Hofstetter TB (2011) Current perspectives on the mechanisms of chlorohydrocarbon degradation in subsurface environments: insight from kinetics, product formation, probe molecules, and isotope fractionation. In: Aquatic redox chemistry, ACS Symposium Series, vol 1071. American Chemical Society, Washington, DC, pp 407–439. https://doi.org/10.1021/bk-2011-1071.ch019

Elsner M, Haderlein SB, Kellerhals T, Luzi S, Zwank L, Angst W, Schwarzenbach RP (2004) Mechanisms and products of surface-mediated reductive dehalogenation of carbon tetrachloride by Fe(II) on goethite. Environ Sci Technol 38(7):2058–2066. https://doi.org/10.1021/es034741m

Engelmann MD, Bobier RT, Hiatt T, Cheng IF (2003) Variability of the Fenton reaction characteristics of the EDTA, DTPA, and citrate complexes of iron. Biometals 16(4):519–527. https://doi.org/10.1023/A:1023480617038

Fernandez J, Maruthamuthu P, Renken A, Kiwi J (2004) Bleaching and photobleaching of Orange II within seconds by the oxone/Co^{2+} reagent in Fenton-like processes. Appl Catal B 49 (3):207–215. https://doi.org/10.1016/j.apcatb.2003.12.018

Fränzle S, Silbernagel H, Uchlier L, Liepelt G (2010) Environmental heterogeneous catalysis and water purification by activated interfaces: a survey of different ways of surface activation and demonstration of a novel, simple and efficient procedure. Ecol Chem Eng S 17(1):25–36

Fu F, Dionysiou DD, Liu H (2014) The use of zero-valent iron for groundwater remediation and wastewater treatment: a review. J Hazard Mater 267:194–205. https://doi.org/10.1016/j.jhazmat.2013.12.062

Furman O, Laine DF, Blumenfeld A, Teel AL, Shimizu K, Cheng IF, Watts RJ (2009) Enhanced reactivity of superoxide in water−solid matrices. Environ Sci Technol 43(5):1528–1533. https://doi.org/10.1021/es802505s

Ghanbari F, Moradi M (2017) Application of peroxymonosulfate and its activation methods for degradation of environmental organic pollutants: review. Chem Eng J 310:41–62. https://doi.org/10.1016/j.cej.2016.10.064

Ghauch A, Tuqan AM, Kibbi N, Geryes S (2012) Methylene blue discoloration by heated persulfate in aqueous solution. Chem Eng J 213:259–271. https://doi.org/10.1016/j.cej.2012.09.122

Groves JT (2006) High-valent iron in chemical and biological oxidations. J Inorg Biochem 100 (4):434–447. https://doi.org/10.1016/j.jinorgbio.2006.01.012

Gutteridge JMC, Halliwell B (1989) 1 Iron toxicity and oxygen radicals. Baillière's Clin Haematol 2(2):195–256. https://doi.org/10.1016/S0950-3536(89)80017-4

Halliwell B (2006) Reactive species and antioxidants. Redox biology is a fundamental theme of aerobic life. Plant Physiol 141(2):312–322. https://doi.org/10.1104/pp.106.077073

Halliwell B, Gutteridge JMC (1984) Oxygen toxicity, oxygen radicals, transition metals and disease. Biochem J 219(1):1–14. https://doi.org/10.1042/bj2190001

Hashsham SA, Freedman DL (1999) Enhanced biotransformation of carbon tetrachloride by *Acetobacterium woodii* upon addition of hydroxocobalamin and fructose. Appl Environ Microbiol 65(10):4537–4542

Hayyan M, Hashim MA, AlNashef IM (2016) Superoxide ion: generation and chemical implications. Chem Rev 116(5):3029–3085. https://doi.org/10.1021/acs.chemrev.5b00407

Howard PH (1991) Handbook of environmental fate and exposure data: for organic chemicals, vol. III pesticides. CRC Press/Routledge, Boca Raton/New York

Howsawkeng J, Teel AL, Hess TF, Crawford RL, Watts RJ (2010) Simultaneous abiotic reduction–biotic oxidation in a microbial-MnO_2-catalyzed Fenton-like system. Sci Total Environ 409 (2):439–445. https://doi.org/10.1016/j.scitotenv.2010.10.009

Hrabák P, Homolková M, Wacławek S, Černík M (2016) Chemical degradation of PCDD/F in contaminated sediment. Ecol Chem Eng S 23(3):473–482. https://doi.org/10.1515/eces-2016-0034

Hu P, Long M (2016) Cobalt-catalyzed sulfate radical-based advanced oxidation: a review on heterogeneous catalysts and applications. Appl Catal B 181:103–117. https://doi.org/10.1016/j.apcatb.2015.07.024

Huber MM, Korhonen S, Ternes TA, von Gunten U (2005) Oxidation of pharmaceuticals during water treatment with chlorine dioxide. Water Res 39(15):3607–3617. https://doi.org/10.1016/j.watres.2005.05.040

Ike IA, Foster SL, Shinn SR, Watson ST, Orbell JD, Greenlee LF, Duke MC (2017) Advanced oxidation of orange G using phosphonic acid stabilised zerovalent iron. J Environ Chem Eng 5 (4):4014–4023. https://doi.org/10.1016/J.JECE.2017.07.069

Ike IA, Linden KG, Orbell JD, Duke M (2018) Critical review of the science and sustainability of persulphate advanced oxidation processes. Chem Eng J 338:651–669. https://doi.org/10.1016/j.cej.2018.01.034

Ince NH, Tezcanli G, Belen RK, Apikyan İG (2001) Ultrasound as a catalyzer of aqueous reaction systems: the state of the art and environmental applications. Appl Catal B 29(3):167–176. https://doi.org/10.1016/S0926-3373(00)00224-1

IUPAC (1997) Compendium of chemical terminology, 2nd edn. (the "Gold Book"). Compiled by A. D. McNaught and A. Wilkinson. Blackwell Scientific Publications, Oxford. XML on-line corrected version: http://goldbook.iupac.org (2006–) created by M. Nic, J. Jirat, B. Kosata. Updates compiled by A. Jenkins. ISBN:0-9678550-9-8. https://doi.org/10.1351/goldbook

Johnson TL, Fish W, Gorby YA, Tratnyek PG (1998) Degradation of carbon tetrachloride by iron metal: complexation effects on the oxide surface. J Contam Hydrol 29(4):379–398. https://doi.org/10.1016/S0169-7722(97)00063-6

Kang Y-G, Yoon H, Lee W, Kim E-j, Chang Y-S (2018) Comparative study of peroxide oxidants activated by nZVI: removal of 1,4-Dioxane and arsenic(III) in contaminated waters. Chem Eng J 334:2511–2519. https://doi.org/10.1016/J.CEJ.2017.11.076

Keenan CR, Sedlak DL (2008) Factors affecting the yield of oxidants from the reaction of nanoparticulate zero-valent iron and oxygen. Environ Sci Technol 42(4):1262–1267. https://doi.org/10.1021/es7025664

Kenneke JF, Weber EJ (2003) Reductive dehalogenation of halomethanes in iron- and sulfate-reducing sediments. 1. Reactivity pattern analysis. Environ Sci Technol 37(4):713–720. https://doi.org/10.1021/es0205941

Kim C, Ahn J-Y, Kim TY, Shin WS, Hwang I (2018) Activation of persulfate by nanosized zero-valent iron (NZVI): mechanisms and transformation products of NZVI. Environ Sci Technol 52 (6):3625–3633. https://doi.org/10.1021/acs.est.7b05847

Krebs C, Galonić Fujimori D, Walsh CT, Bollinger JM Jr (2007) Non-heme Fe(IV)–oxo intermediates. Acc Chem Res 40(7):484–492. https://doi.org/10.1021/ar700066p

Kriegman-King MR, Reinhard M (1994) Transformation of carbon tetrachloride by pyrite in aqueous solution. Environ Sci Technol 28(4):692–700. https://doi.org/10.1021/es00053a025

Krol MM, Oleniuk AJ, Kocur CM, Sleep BE, Bennett P, Xiong Z, O'Carroll DM (2013) A field-validated model for in situ transport of polymer-stabilized nZVI and implications for subsurface injection. Environ Sci Technol 47(13):7332–7340. https://doi.org/10.1021/es3041412

Lee C, Sedlak DL (2009) A novel homogeneous Fenton-like system with Fe(III)–phosphotungstate for oxidation of organic compounds at neutral pH values. J Mol Catal A Chem 311:1–2):1–6. https://doi.org/10.1016/j.molcata.2009.07.001

Lee H, Lee H-J, Jeong J, Lee J, Park N-B, Lee C (2015) Activation of persulfates by carbon nanotubes: oxidation of organic compounds by nonradical mechanism. Chem Eng J 266:28–33. https://doi.org/10.1016/j.cej.2014.12.065

Lee H, Lee C, Kim J-H (2017) Response to comment on "activation of persulfate by graphitized nanodiamonds for removal of organic compounds". Environ Sci Technol 51(9):5353–5354. https://doi.org/10.1021/acs.est.7b01642

Liu C, Wu B, Chen X (2018) Sulfate radical-based oxidation for sludge treatment: a review. Chem Eng J 335:865–875. https://doi.org/10.1016/j.cej.2017.10.162

Lutze H (2013) Sulfate radical based oxidation in water treatment. Dissertation, Universität Duisburg-Essen

Manibusan MK, Odin M, Eastmond DA (2007) Postulated carbon tetrachloride mode of action: a review. J Environ Sci Health Part C 25(3):185–209. https://doi.org/10.1080/10590500701569398

Matzek LW, Carter KE (2016) Activated persulfate for organic chemical degradation: a review. Chemosphere 151:178–188. https://doi.org/10.1016/j.chemosphere.2016.02.055

McCay PB, Lai EK, Poyer JL, DuBose CM, Janzen EG (1984) Oxygen- and carbon-centered free radical formation during carbon tetrachloride metabolism. Observation of lipid radicals in vivo and in vitro. J Biol Chem 259(4):2135–2143

McCord JM, Fridovich I (1988) Superoxide dismutase: the first twenty years (1968–1988). Free Radic Biol Med 5(5–6):363–369. https://doi.org/10.1016/0891-5849(88)90109-8

McCormick ML, Adriaens P (2004) Carbon tetrachloride transformation on the surface of nanoscale biogenic magnetite particles. Environ Sci Technol 38(4):1045–1053. https://doi.org/10.1021/es030487m

Mönig J, Bahnemann D, Asmus K-D (1983) One electron reduction of CCl_4 in oxygenated aqueous solutions: a $CCl_3O_2\bullet$-free radical mediated formation of Cl^- and CO_2. Chem Biol Interact 47(1):15–27. https://doi.org/10.1016/0009-2797(83)90144-8

Naim S, Ghauch A (2016) Ranitidine abatement in chemically activated persulfate systems: assessment of industrial iron waste for sustainable applications. Chem Eng J 288:276–288. https://doi.org/10.1016/j.cej.2015.11.101

Navalon S, Alvaro M, Garcia H (2010) Heterogeneous Fenton catalysts based on clays, silicas and zeolites. Appl Catal B 99:1–2):1–26. https://doi.org/10.1016/j.apcatb.2010.07.006

Navalon S, Dhakshinamoorthy A, Alvaro M, Garcia H (2011) Heterogeneous Fenton catalysts based on activated carbon and related materials. ChemSusChem 4(12):1712–1730. https://doi.org/10.1002/cssc.201100216

Němeček J, Pokorný P, Lhotský O, Knytl V, Najmanová P, Steinová J, Černík M, Filipová A, Filip J, Cajthaml T (2016) Combined nano-biotechnology for in-situ remediation of mixed contamination of groundwater by hexavalent chromium and chlorinated solvents. Sci Total Environ 563–564:822–834. https://doi.org/10.1016/j.scitotenv.2016.01.019

Oh W-D, Dong Z, Lim T-T (2016) Generation of sulfate radical through heterogeneous catalysis for organic contaminants removal: current development, challenges and prospects. Appl Catal B 194:169–201. https://doi.org/10.1016/j.apcatb.2016.04.003

Pecher K, Haderlein SB, Schwarzenbach RP (2002) Reduction of polyhalogenated methanes by surface-bound Fe(ii) in aqueous suspensions of iron oxides. Environ Sci Technol 36(8):1734–1741. https://doi.org/10.1021/es011191o

Pignatello JJ, Oliveros E, MacKay A (2007) Advanced oxidation processes for organic contaminant destruction based on the Fenton reaction and related chemistry. Crit Rev Environ Sci Technol 37(3):273–275. https://doi.org/10.1080/10643380601163809

Plaa GL (2000) Chlorinated methanes and liver injury: highlights of the past 50 years. Annu Rev Pharmacol Toxicol 40(1):43–65. https://doi.org/10.1146/annurev.pharmtox.40.1.43

Qian SY, Buettner GR (1999) Iron and dioxygen chemistry is an important route to initiation of biological free radical oxidations: an electron paramagnetic resonance spin trapping study. Free Radic Biol Med 26(11–12):1447–1456. https://doi.org/10.1016/S0891-5849(99)00002-7

Ramseier MK (2010) Assimilable organic carbon formation and disinfection during oxidative drinking water treatment. Doctoral thesis, ETH Zurich. https://doi.org/10.3929/ethz-a-006371326

Rastogi A, Al-Abed SR, Dionysiou DD (2009) Effect of inorganic, synthetic and naturally occurring chelating agents on Fe(II) mediated advanced oxidation of chlorophenols. Water Res 43(3):684–694. https://doi.org/10.1016/j.watres.2008.10.045

Rayaroth MP, Lee C-S, Aravind UK, Aravindakumar CT, Chang Y-S (2017) Oxidative degradation of benzoic acid using Fe^0- and sulfidized Fe^0-activated persulfate: a comparative study. Chem Eng J 315:426–436. https://doi.org/10.1016/j.cej.2017.01.031

Reiner O, Athanassopoulos S, Hellmer KH, Murray RE, Uehleke H (1972) Bildung von Chloroform aus Tetrachlorkohlenstoff in Lebermikrosomen, Lipidperoxidation und Zerstörung von Cytochrom P-450. Arch Toxikol 29(3):219–233. https://doi.org/10.1007/BF00315600

Richardson SD, Plewa MJ, Wagner ED, Schoeny R, DeMarini DM (2007) Occurrence, genotoxicity, and carcinogenicity of regulated and emerging disinfection by-products in drinking water: a review and roadmap for research. Mutat Res Rev Mutat Res 636(1–3):178–242. https://doi.org/10.1016/j.mrrev.2007.09.001

Shao H, Butler EC (2009) The relative importance of abiotic and biotic transformation of carbon tetrachloride in anaerobic soils and sediments. Soil Sediment Contam 18(4):455–469. https://doi.org/10.1080/15320380902962346

Siegrist RL, Crimi M, Simpkin TJ (eds) (2011) In situ chemical oxidation for groundwater remediation, 1st edn. Springer, New York. https://doi.org/10.1007/978-1-4419-7826-4

Smith BA, Teel AL, Watts RJ (2004) Identification of the reactive oxygen species responsible for carbon tetrachloride degradation in modified Fenton's systems. Environ Sci Technol 38 (20):5465–5469. https://doi.org/10.1021/es0352754

Staehelin J, Hoigne J (1982) Decomposition of ozone in water: rate of initiation by hydroxide ions and hydrogen peroxide. Environ Sci Technol 16(10):676–681. https://doi.org/10.1021/es00104a009

Stoin U, Mojon A, Sasson Y (2015) Fast and complete *in situ* mineralization of contaminated soils using a novel method for superoxide generation. RSC Adv 5(9):6571–6577. https://doi.org/10.1039/C4RA08015G

Tan C, Dong Y, Fu D, Gao N, Ma J, Liu X (2018) Chloramphenicol removal by zero valent iron activated peroxymonosulfate system: kinetics and mechanism of radical generation. Chem Eng J 334:1006–1015. https://doi.org/10.1016/j.cej.2017.10.020

Tarr MA (ed) (2003) Chemical degradation methods for wastes and pollutants: environmental and industrial applications, 1st edn. CRC Press, Boca Raton

Teel AL, Watts RJ (2002) Degradation of carbon tetrachloride by modified Fenton's reagent. J Hazard Mater 94(2):179–189. https://doi.org/10.1016/S0304-3894(02)00068-7

US EPA (2014) Extramural research competitions. Final Report | Fenton-like reductions for the enhanced desorption and degradation of biorefractory contaminants. Reports and assessments. http://cfpub.epa.gov/ncer_abstracts/INDEX.cfm/fuseaction/display.abstractDetail/abstract/774/report/F. Accessed 19 Oct 2014

von Gunten U (2003) Ozonation of drinking water: Part II. Disinfection and by-product formation in presence of bromide, iodide or chlorine. Water Res 37(7):1469–1487. https://doi.org/10.1016/S0043-1354(02)00458-X

von Gunten U, Hoigne J (1994) Bromate formation during ozonization of bromide-containing waters: interaction of ozone and hydroxyl radical reactions. Environ Sci Technol 28 (7):1234–1242. https://doi.org/10.1021/es00056a009

von Gunten U, Oliveras Y (1998) Advanced oxidation of bromide-containing waters: bromate formation mechanisms. Environ Sci Technol 32(1):63–70. https://doi.org/10.1021/es970477j

von Sonntag C (2008) Advanced oxidation processes: mechanistic aspects. Water Sci Technol 58 (5):1015–1021. https://doi.org/10.2166/wst.2008.467

von Sonntag C, von Gunten U (2012) Chemistry of ozone in water and wastewater treatment: from basic principles to applications. IWA Publishing, London

Vulimiri SV, Berger A, Sonawane B (2011) The potential of metabolomic approaches for investigating mode(s) of action of xenobiotics: case study with carbon tetrachloride. Mutat Res Genet Toxicol Environ Mutagen 722(2):147–153. https://doi.org/10.1016/j.mrgentox.2010.02.013

Wacławek S, Nosek J, Cádrová L, Antoš V, Černík M (2015) Use of various zero valent irons for degradation of chlorinated ethenes and ethanes. Ecol Chem Eng S 22(4):577–587. https://doi.org/10.1515/eces-2015-0034

Wacławek S, Antoš V, Hrabák P, Černík M, Elliott D (2016) Remediation of hexachlorocyclohexanes by electrochemically activated persulfates. Environ Sci Pollut Res 23 (1):765–773. https://doi.org/10.1007/s11356-015-5312-y

Wacławek S, Lutze HV, Grübel K, Padil VVT, Černík M, Dionysiou DD (2017) Chemistry of persulfates in water and wastewater treatment: a review. Chem Eng J 330:44–62. https://doi.org/10.1016/j.cej.2017.07.132

Wang J, Wang S (2018) Activation of persulfate (PS) and peroxymonosulfate (PMS) and application for the degradation of emerging contaminants. Chem Eng J 334:1502–1517. https://doi.org/10.1016/j.cej.2017.11.059

Wang Y, Ao Z, Sun H, Duan X, Wang S (2016) Activation of peroxymonosulfate by carbonaceous oxygen groups: experimental and density functional theory calculations. Appl Catal B 198:295–302. https://doi.org/10.1016/j.apcatb.2016.05.075

Wang Z, Ai L, Huang Y, Zhang J, Li S, Chen J, Yang F (2017) Degradation of azo dye with activated peroxygens: when zero-valent iron meets chloride. RSC Adv 7(49):30941–30948. https://doi.org/10.1039/c7ra03872k

Watts RJ (2011) Final report. Enhanced reactant-contaminant contact through the use of persulfate in situ chemical oxidation (ISCO). SERDP Project ER-1489. Washington State University

Watts RJ, Howsawkeng J, Teel AL (2005) Destruction of a carbon tetrachloride dense nonaqueous phase liquid by modified Fenton's reagent. J Environ Eng 131(7):1114–1119. https://doi.org/10.1061/(ASCE)0733-9372(2005)131:7(1114)

Weber LWD, Boll M, Stampfl A (2003) Hepatotoxicity and mechanism of action of haloalkanes: carbon tetrachloride as a toxicological model. Crit Rev Toxicol 33(2):105–136. https://doi.org/10.1080/713611034

Wei Z, Villamena FA, Weavers LK (2017) Kinetics and mechanism of ultrasonic activation of persulfate: an in situ EPR spin trapping study. Environ Sci Technol 51(6):3410–3417. https://doi.org/10.1021/acs.est.6b05392

Xiao R, Luo Z, Wei Z, Luo S, Spinney R, Yang W, Dionysiou DD (2018) Activation of peroxymonosulfate/persulfate by nanomaterials for sulfate radical-based advanced oxidation technologies. Curr Opin Chem Eng 19:51–58. https://doi.org/10.1016/j.coche.2017.12.005

Xu M, Du H, Gu X, Lu S, Qiu Z, Sui Q (2014a) Generation and intensity of active oxygen species in thermally activated persulfate systems for the degradation of trichloroethylene. RSC Adv 4(76):40511–40517. https://doi.org/10.1039/C4RA04942J

Xu M, Gu X, Lu S, Qiu Z, Sui Q (2014b) Role of reactive oxygen species for 1,1,1-trichloroethane degradation in a thermally activated persulfate system. Ind Eng Chem Res 53(3):1056–1063. https://doi.org/10.1021/ie403689d

Yang S, Wang P, Yang X, Shan L, Zhang W, Shao X, Niu R (2010) Degradation efficiencies of azo dye Acid Orange 7 by the interaction of heat, UV and anions with common oxidants: persulfate, peroxymonosulfate and hydrogen peroxide. J Hazard Mater 179(1–3):552–558. https://doi.org/10.1016/j.jhazmat.2010.03.039

Ye T, Wei Z, Spinney R, Tang C-J, Luo S, Xiao R, Dionysiou DD (2017) Chemical structure-based predictive model for the oxidation of trace organic contaminants by sulfate radical. Water Res 116:106–115. https://doi.org/10.1016/j.watres.2017.03.015

Yuan S, Liao P, Alshawabkeh AN (2014) Electrolytic manipulation of persulfate reactivity by iron electrodes for trichloroethylene degradation in groundwater. Environ Sci Technol 48(1):656–663. https://doi.org/10.1021/es404535q

Zhang T, Chen Y, Wang Y, Le Roux J, Yang Y, Croué J-P (2014) Efficient peroxydisulfate activation process not relying on sulfate radical generation for water pollutant degradation. Environ Sci Technol 48(10):5868–5875. https://doi.org/10.1021/es501218f

Zhang B-T, Zhang Y, Teng Y, Fan M (2015) Sulfate radical and its application in decontamination technologies. Crit Rev Environ Sci Technol 45(16):1756–1800. https://doi.org/10.1080/10643389.2014.970681

Zhao D, Liao X, Yan X, Huling SG, Chai T, Tao H (2013) Effect and mechanism of persulfate activated by different methods for PAHs removal in soil. J Hazard Mater 254–255:228–235. https://doi.org/10.1016/j.jhazmat.2013.03.056

Zhao L, Hou H, Fujii A, Hosomi M, Li F (2014) Degradation of 1,4-dioxane in water with heat- and Fe^{2+}-activated persulfate oxidation. Environ Sci Pollut Res 21(12):7457–7465. https://doi.org/10.1007/s11356-014-2668-3

Chapter 10
Photo-oxidation Technologies for Advanced Water Treatment

Rakesh Kumar Sharma, Bhavya Arora, Sriparna Dutta, and Manoj B. Gawande

Abstract Providing a sustainable source of clean and fresh water represents the most serious global challenge facing humanity today. Major factors that contribute towards impairing the quality of water are industrial and agricultural runoffs containing toxic contaminants including dyes, pesticides, and other organic moeities. They have identified to be highly problematic since they enter the food chain, which leads to a large number of disorders. Such a critical issue indeed calls for urgent action. The need to address this challenge has also been reflected in the 2030 Agenda for Sustainable Development, which ranks water quality among the 17 primary goals. Photo-oxidation technologies that rely on the power of absorbed ultraviolet components of solar radiation to oxidize or reduce different substances in diverse environmental strata have emerged as one of the most promising and economically viable techniques for advanced water cleanup. This chapter provides a detailed account of the progress that has been made in this field ranging from the fundamental concept behind the UV radiation mediated processes to all the engineered photooxidative technologies that are widely employed. A special emphasis is put on various potential and commercial photocatalytic reactors that have been designed with the specific aim of degrading pollutants. In addition, a special section has been devoted to the recently developed nano-photocatalytic systems that offer excellent prospects for managing water pollution. A few case studies illustrate practical implementation of UV-oxidation processes by leading companies such as Merck & Co., BASF, and GlaxoSmithKline.

Keywords Photooxidation · UV irradiation · Degradation · Contaminants · Water treatment · PhotoReactors

R. K. Sharma (✉) · B. Arora · S. Dutta
Department of Chemistry, Green Chemistry Network Centre, University of Delhi, Delhi, India

M. B. Gawande (✉)
Regional Centre of Advanced Technologies and Materials, Palacký University Olomouc, Olomouc, Czech Republic
e-mail: manoj.gawande@upol.cz

J. Filip et al. (eds.), *Advanced Nano-Bio Technologies for Water and Soil Treatment*, Applied Environmental Science and Engineering for a Sustainable Future,
https://doi.org/10.1007/978-3-030-29840-1_10

10.1 Introduction

It is a universally accepted fact that *"water is the elixir of life"* as almost all forms of living organisms depend on it in order to survive. However, with the incessantly growing population and rapid modernization, lack of an adequate supply of clean drinking water has become an issue of serious concern across the globe, as poor water quality not only threatens the health of people and ecosystems but also limits economic productivity and development opportunities.

Although several water purification techniques such as ion exchange (Raghu and Basha 2007), reverse osmosis (Dialynas et al. 2008), electrolysis (Jin et al. 2003), precipitation (Yan et al. 2017), ultrafiltration (Fan et al. 2014), and adsorption (Liang et al. 2017) have been developed to reduce toxic effluents (Fig. 10.1) from contaminated water, they unfortunately suffer from a number of drawbacks such as high capital cost, generation of secondary waste product that cannot be treated again and dumped as such, sludge formation, etc. Therefore, the critical need of the hour is to increase the efficacy of current wastewater treatment techniques that will provide

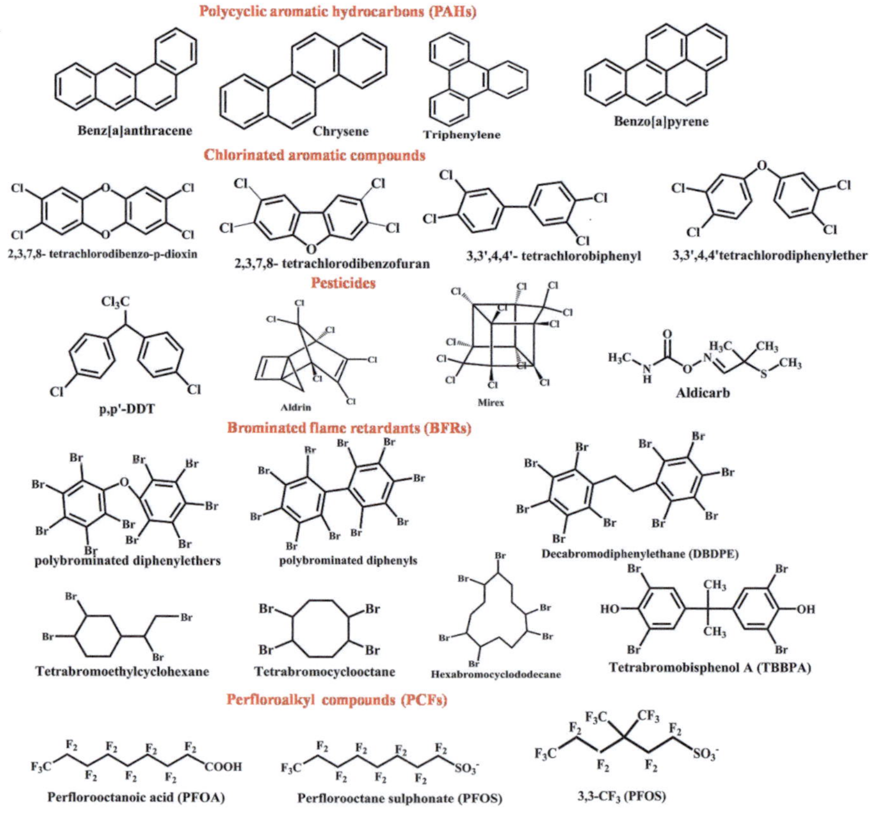

Fig. 10.1 Various pollutants present in water bodies

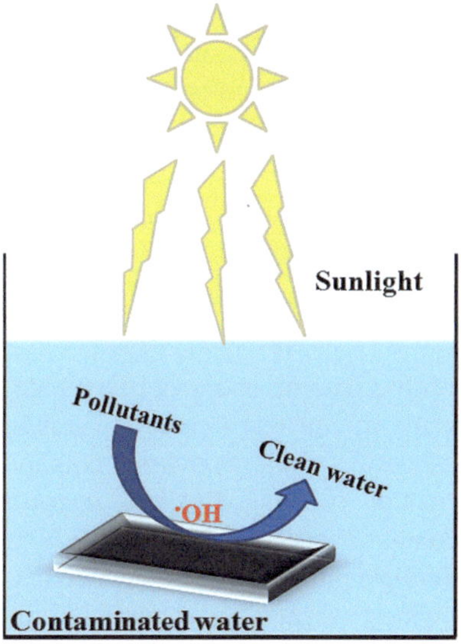

Fig. 10.2 Schematic description showing photo-oxidation technique (Gupta et al. 2015)

high efficiency, multiple functionality, and high flexibility in system size. Photo-oxidation technology possesses all the salient features that may offer leapfrogging opportunities in water treatment.

This methodology involves the irradiation of contaminated water under UV light/sunlight in the presence of oxidant/catalyst, which generates reactive oxidizing species that are able to completely decompose organic pollutants into environmentally benign nontoxic molecules such as H_2O, CO_2, and inorganic salts. So, it is the absorbed solar radiation that provides the primary driving force for the various chemical, physical, and biological processes that oxidize or reduce substances in the environment as shown in Fig. 10.2. Examples of oxidants include hydrogen peroxide (H_2O_2), persulfate (PS), peroxymonosulfate (PMS), and ozone that have been reported to create receptive oxidizing species (Sun et al. 2016). Thus, this procedure utilizes a variety of bright segment of retained sun-based radiation, which results in immediate and circuitous photoreactions in nature. In comparison with other treatment processes such as adsorption, bioremediation, incineration, etc., photo-oxidation has developed as an efficient technique that reduces the risk of sludge generation and eliminates the need for further secondary processes in water treatment. Some of the advantages of the photo-oxidation methodologies are listed below (UMEX GmbH):

- Irreversible removal of organic pollutants.
- Significant reduction in overall chemical oxygen demand (COD).
- No requirement of off-gas treatment.
- Reduction in the multiple stages of treatment.
- No generation of sludge or emissions of volatile organic compounds (VOCs).
- Production of innocuous, stable, and mineralized products, e.g., H_2O, CO_2, etc., produced.

- No increase of salinity.
- Easy process control.
- Effectiveness in batch reactions.

10.2 Fundamental Concept Behind the Photocatalyst UV Irradiation Mediated Decontamination

With a specific end goal to emulate the nature, a noteworthy advancement has been made by our researchers towards use of photochemical response to dispose of the dangerous pollutants from contaminated water, which shall be discussed in the ensuing sections. But it is important to understand what actually happens when contaminants are exposed to UV radiation. A radical reaction is directly initiated during a photocatalyst-mediated reaction due to the activation energy supplied by the highly energetic UV radiation to the catalyst surface and the water molecules in the effluent. Different types of semiconductor oxides have been used as catalysts for this particular purpose. Upon illumination with radiation of energy greater than its optical band gap, the semiconductor molecules transfer an electron to the conduction band leaving behind a hole in the valence band. The electron transfer takes place either from the conduction band to an acceptor in the solution or from a donor in solution to the valence band at the solid–liquid interface. These processes compete with the recombination of electron and hole to produce thermal energy, which is not useful for decontamination. In the absence of suitable electron and hole scavengers, the stored energy is dissipated within a few nanoseconds by recombination. The valence band holes are powerful oxidants (1.0 to 3.5 V) while conduction band electrons are good reductants (0.5 to −1.5 eV). This eventually leads to photocatalytic redox processes ultimately leading to the destruction of the organic molecules.

10.2.1 Reactive Oxidizing Species

The effectiveness of the photo-oxidation process, (i.e., the complete oxidation of pollutants into carbon dioxide and water) depends on the oxidation potential of the oxidizing agent employed for this purpose. The oxidation potentials of some important oxidants are given below in Table 10.1 (Barrera-Díaz 2014; Hassaan and El Nemr 2017).

Legrini et al. (1993) reported that fluorine has higher oxidation potential than the hydroxyl radical, but because of its adverse effects on humans and environment, it is undesirable to use fluorine in water treatment. Therefore, hydroxyl radical stands out to be the most promising reactive species for oxidation processes (Fig. 10.3) (Buthiyappan et al. 2016).

Table 10.1 Relative oxidation potential of the different oxidizing species

Oxidizing agent	Symbol	Oxidation potential E^0 (V)
Fluorine	F_2	3.03
Hydroxyl radical	$^{\bullet}OH$	2.80
Atomic oxygen	O	2.42
Ozone	O_3	2.07
Hydrogen peroxide	H_2O_2	1.78
Perhydroxyl radical	$^{\bullet}O_2H$	1.70
Permanganate ion	MnO_4^-	1.67
Hypochlorous acid	HClO	1.49
Chlorine	Cl_2	1.36
Chlorine dioxide	ClO_2	1.27
Oxygen	O_2	1.23

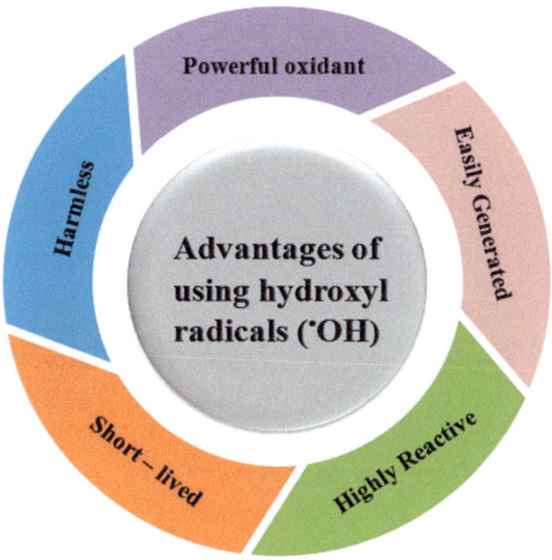

Fig. 10.3 Advantages of using hydroxyl radicals as oxidizing species

Fenton (1894) first discovered the hydroxyl radical by the oxidation of tartaric acid via hydrogen peroxide. The breakdown of the hydrogen peroxide to hydroxyl radicals is accomplished via UV radiation having wavelength range between 190 nm and 280 nm. For this reaction, the hydrogen peroxide molecule must absorb a photon (Deng and Zhao 2015).

$$H_2O_2 + h\gamma \rightarrow 2^{\bullet}OH \tag{10.1}$$

It is also possible to generate hydroxyl radicals using ozone (O_3). However, this is a two-step reaction and two photons are needed to produce hydroxyl radical (Deng and Zhao 2015).

$$O_3 + h\gamma + H_2O \rightarrow H_2O_2 + O_2 \tag{10.2}$$

$$H_2O_2 + O_2 \rightarrow 2^{\bullet}OH \tag{10.3}$$

10.2.1.1 Processes by Which Hydroxyl Radicals Are Generated and Their Subsequent Use in Advanced Water Treatment

It is well-known that hydroxyl radicals have a short lifetime, they are just produced through various techniques such as via the mixing of oxidizing operators (e.g., H_2O_2 and O_3), light (e.g., bright light or ultrasound), and salts (e.g., Fe^{2+}). The practices for the generation of hydroxyl radicals are summarized below.

H_2O_2/UV Process

One of the most effective techniques in treatment of wastewaters containing toxic organic pollutants is the combination of hydrogen peroxide with ultraviolet photolysis (H_2O_2/UV). This process involves two types of reactions simultaneously—photolysis and oxidation.

The degradation of water contaminants depends on the molecular structure and the wavelength of irradiated UV light. Generally, in order to meet these objectives, two types of mercury-containing UV lamps are used:

(a) **Low-pressure (LP) UV lamps, emitting irradiation with a wavelength of 253.7 nm.**

Braun et al. (1991) reported that the low-pressure mercury light is a monochromatic UV source that has a solid outflow of unreversed resonance line at 253 nm and 184 nm alongside other significantly weaker lines. UV radiation likewise gives the vitality to start the decay of ozone, which prompts the arrangement of two hydroxyl radicals (OH). The mercury outflow at 253.7 nm delivered specifically by low-weight circular segments has been accounted for degradation of chemicals in water. Numerous analyses have been conducted to decompose chlorinated hydrocarbons such as 85% tetrachloroethene, 55% trichloroethene, 45% 1,1,1-trichloroethene with a low-weight mercury light at 253.7 nm (Lee and Lee 2005).

(b) **Medium-pressure (MP) UV lamps, emitting over a wavelength range of 200 to 300 nm.**

The MP UV lamps are loaded with mercury vapor and work at a pressure of ca 1 atm. With the increase of pressure in the discharge tube, two major changes occur:

- Due to the growing number of collision with the energetic electrons, the gas temperature increases.
- The high temperature becomes localized at the center of the discharge and a temperature gradient is developed towards the walls, which are much cooler.

A broad range of wavelengths in the UV region is discharged by medium- and high-pressure lights and as they possess high light intensity, they have the ability to penetrate deeper. Therefore, the completion of reaction is done in a short span of time.

The prime reason behind the combination of UV and H_2O_2 for formation of hydroxyl radicals is the cleavage of H_2O_2 with UV light. This produces a quantum yield of two hydroxyl radicals per unit of radiation absorbed (Glaze et al. 1987) as shown in Eq. 10.4.

$$H_2O_2 + h\gamma \rightarrow 2^{\bullet}OH \tag{10.4}$$

Legrini et al. (1993) reported the sequence of the chemical reaction between hydroxyl radicals and organic matter present in drinking matter. Hydroxyl radicals react with organic compounds to produce organic radicals, which then further react with the dissolved oxygen to produce hydroperoxyl radicals. Then, these formed radicals initiate the oxidation reaction to form innocuous smaller weight fragments. This technique can be compatibly used for the removal of hazardous waste from groundwater, chemical process water, drinking water and industrial wastewater.

Various organic pollutants are efficiently degraded by the UV/H_2O_2 process. In literature, several successful applications of UV/H_2O_2 have been reported. For instance, Rivas et al. (1998) used this process for phenol oxidation while Zhang et al. (2008) investigated the treatment of contaminated waters.

Furthermore, this UV/H_2O_2 technology has also been employed for the remediation of ground water containing mixtures of hazardous aliphatic compound trichloroethylene (TCE) present in the concentration range of about 2000–10,000 (μg/L). The contaminated water was irradiated with medium pressure Hg arc of 3000 W of electric power followed by the addition of 50 mg/L of hydrogen peroxide. Under these conditions, significant elimination of TCE was achieved from 3700–4000 (μg/L) to 0.7 (μg/L) within 50 s of the irradiation (Sharma and Celin 2005). In order to investigate the effectiveness of this process, water samples were collected from paper pulp bleaching industries, distillery, and tomato processing plants. The extent of waste removal could be achieved after one hour of the treatment, which resulted in 25% reduction in COD. Although, this process has been used widely for wastewater treatment due to lower cost and ready availability of hydrogen peroxide but it suffers from a number of drawbacks, which are illustrated below (Stasinakis 2008).

- H_2O_2 has poor UV light absorption characteristics, because of which a lot of UV light becomes absorbed by water matrix and thus most of the light input gets wasted.
- It cannot utilize solar light as the source of UV light because UV energy required for the photolysis of the oxidizer is not available in the solar spectrum (Niaounakis and Halvadakis 2006).
- The UV light can also be absorbed by the interfering compounds and turbidity in the wastewater as a result; the efficiency of the system is thus reduced.

- An excessive concentration of H_2O_2 would act as a radical scavenger reducing the rate of oxidation, which is shown below:

$$H_2O_2 + HO^{\bullet} \rightarrow HO_2\bullet + H_2O \quad (10.5)$$

$$H_2O_2 + HO_2\bullet \rightarrow HO^{\bullet} + H_2O + O_2 \quad (10.6)$$

$$2HO^{\bullet} \rightarrow H_2O_2 \quad (10.7)$$

$$2HO_2\bullet \rightarrow H_2O_2 + O_2 \quad (10.8)$$

$$HO^{\bullet} + HO_2\bullet \rightarrow H_2O + O_2 \quad (10.9)$$

A too low H_2O_2 dosage leads to insufficient formation of hydroxyl radicals, thus decreasing the oxidation rate.

Photolytic Ozonation (UV/O_3)

Recently, a list of best accessible technologies was released by the United States Environmental Protection Agency for enhanced removal of contaminants for water treatment and ozone (O_3) was classified as one the most powerful oxidants and a potent disinfectant. In fact, the use of ozone has been found to be highly beneficial for the improvement of water quality as it can degrade microbes, remove micropollutants, nonprotonated amines, taste, and odor (von Gunten 2003; Camel and Bermond 1998). Theoretically, it has been considered that both organic and inorganic pollutants are oxidized by ozone. However, it acts as an electrophile in the reaction and is highly selective towards the pollutants (Chokshi and Ruparelia 2015). In water, ozone is wobbly, so it decomposes via a series of complex reactions shown below (Mishra et al. 2017):

$$O_3 + {}^{-}OH \rightarrow {}^{\bullet}O_2H \quad (10.10)$$

or

$$O_3 + {}^{-}OH \rightarrow O_2^{-} \quad (10.11)$$

$$O_2 + O_3 \rightarrow O_2 + O_3^{-} \quad (10.12)$$

$$O_3^{-} \leftrightarrow {}^{\bullet}O_3H \quad (10.13)$$

$${}^{\bullet}O_3H \rightarrow {}^{\bullet}OH + O_2 \quad (10.14)$$

$${}^{\bullet}OH + O_3 \rightarrow O_4H \quad (10.15)$$

$${}^{\bullet}O_4H \rightarrow O_2 + {}^{\bullet}O_2H \quad (10.16)$$

$${}^{\bullet}O_2H \rightarrow O_2^{-} \quad (10.17)$$

Table 10.2 The effective treatment of various pollutants via ozonation

Category	Specific pollutant
Microbial (cyanobacteria) product	Microcystin-LR
Pesticide	Carbofuran, dinoseb
Solvent	Vinyl chloride, dichloroethenes, nitrilotriacetate (NTA)
Ligands	Ethylenediaminetetraacetate (EDTA), diclofenac, carbamazepine, benzafibrate
Pharmaceuticals	Diclofenac, ibuprofen, sulfamethoxazole, roxithromycin, iopromide, 17α-ethynylestradiol
Inorganic micropollutants	Fe(II), CN^-, Mn(II), H_2S, NO_2^-

Initially, in water the concentration of ozone decreases rapidly and first-order kinetics is followed, whereas second-order kinetics is followed in the second phase of ozone decrease through oxidation (von Gunten 2003; Stettler et al. 1998; Exner 1988). Being an extremely strong oxidant, ozone ($E^0 = 2.07$ V) reacts rapidly with most of organic compounds. The oxidation of organic pollutants is enhanced by ozone when electron-donating groups like $-CH_3$, $-O$, $-OCH_3$ are present, whereas it gets reduced as a result of the presence of electron-withdrawing groups like $-Cl$, $-NO_2$ (von Gunten 2003). There is a decrease in the oxidation potential with protonation of species due to the decrease in the nucleophilicity. The decolorization using ozone as an oxidant is fast because ozone molecule attacks the unsaturated groups present in chromophores. On the other hand, the saturated compounds exhibit low reactivity because they get accumulated during the ozonation process. The presence of sulfide and amino groups in the organic compounds shows high ozone reactivity. The half-life of dissolved ozone varies from seconds to hours depending upon the pH of water, alkalinity, and waste organic matter content.

This technique is extensively used in European countries, e.g., Switzerland and Germany in wastewater treatment plants (Badia-Fabregat et al. 2017). There are various pollutants to which ozonation can be effectively applied as illustrated in Table 10.2 (Mishra et al. 2017).

However, some of the factors that may affect the ozonation process are listed below:

- Dosage of ozone
- pH of the solution
- Concentration of waste content

The removal efficiency of micropollutants is enhanced at higher dosage of ozone with low bromide concentration and with increase in the pH (Bourgin et al. 2017). The reaction of O_3 and $^{\bullet}OH$ active compounds such as diclofenac, sulfamethoxazole, carbamazepine, trimethoprim, hydrochlorothiazide, phenazone, tramadol, metoprolol, etc., is independent of variation in the pH change (Bourgin et al. 2017).

Limitations of This Technique (Mishra et al. 2017)

- Compounds like benzenes, geosmin, and 2-methylisoborneol (MIB), trihalomethanes (THMs), etc., do not become efficiently oxidized by ozone
- Removal of ammonia by this process is slow as it exhibits low oxidation potential
- *Cryptosporidium parvum* cyst requires high amount of ozone and detrimental by-products are formed during this process in drinking water
- If bromide is present in wastewater, it reacts with O_3 and $^{\bullet}OH$ to form bromate, which is a potential carcinogen
- Complete mineralization of the pollutant into carbon dioxide, which makes ozonation hardly possible
- High production cost
- Relatively low solubility and stability in water
- Selectively reacts with organic compounds at acidic pH

By introducing the UV radiation ($\lambda = 254$ nm) in the ozonation process, the removal efficiency of contaminants in water is enhanced. This process is also known as photolytic ozonation.

Advantages of the UV/O_3 Technique (Krishnan et al. 2017)

- The efficiency of the combined UV/O_3 process is typically higher than the additive efficiencies of UV alone and ozone.
- Compared with the H_2O_2/UV process, the combined O_3/UV process is more efficient in generating hydroxyl radicals using LP UV lamps.
- This process is more stochiometrically efficient than other processes in generating OH radicals.
- This process is capable of degradation of stubborn organic compounds that resist degradation.

In this photolytic ozonation method, ozone is energized and combines with water to generate H_2O_2 as an intermediate, which further decomposes to form OH radicals. The ability of the ozone molecules is enhanced more effectively because of the highly active and nonselective hydroxyl radicals acting on many targets. The reactions involved in photolysis of ozone are exemplified in the equation below (Hassaan and El Nemr 2017):

$$O_3 + H_2O + h\gamma \rightarrow O_2 + H_2O_2 \text{ at } (h\gamma : \lambda < 300 \text{ nm}) \tag{10.18}$$

$$H_2O_2 + h\gamma \rightarrow 2^{\bullet}OH \tag{10.19}$$

$$2O_3 + H_2O_2 \rightarrow 2^{\bullet}OH + 3O_2 \tag{10.20}$$

The combination of UV and O_3 process has been found to be more effective as compared to either O_3 or UV individually. For the success of this ozone/UV process, the key parameters that affect this system are (Krishnan et al. 2017):

Ozone Dosage

For the effective degradation of contaminants, optimum dosage of ozone is necessary. This can be accomplished by the effective transfer of ozone gas, which helps to maintain the high dissolved ozone rate into the solution.

UV Illumination Level

For proper UV irradiation and mixing of ozone in the system, an effective O_3/UV system with pressurized injection secondary mix UV/O_3 reactor needs to be designed. Such system generates microbubbles with the constant renewal of gas to liquid—mixing zone. This enhances the gas solubility for the better utilization of UV irradiation.

pH of the System

With the increase in the pH (>8) of the system, ozone becomes readily converted to hydroxyl radicals, which increase the oxidation rate of contaminants such as pesticides and cyanide.

In order to achieve the maximum degradation efficiency, the equilibrium between the pH level, UV photolysis, and ozone dosage must be considered in system design.

In fact, UV/O_3 technology has been extensively used to manage contamination in wastewater obtained from textiles, petroleum refineries, electroplating waste, electronic chip manufacturing, etc., because it not only leads to the efficient degradation of pollutants but also successfully reduces the total organic carbon (TOC) and COD. Several pollutants have been degraded using photolytic ozonation technique (Table 10.3) (Mishra et al. 2017). In a study performed for the removal of Reactive Blue 19 dye from wastewater, Emami et al. (2010) found that, as compared to ozonation (O_3) process, the photolytic ozonation (UV/O_3) process was much more efficient in COD reduction.

Photo-Fenton Chemistry

It was H. J. H. Fenton, a French scientist who perceived that when Fe^{2+} reacts with H_2O_2, OH radicals are produced for oxidizing tartaric acid (Fenton 1894). Haber and

Table 10.3 Listing of various pollutants degraded through photolytic ozonation

Categories	Specific pollutant
Dyes	Rhodamine B-19 (RB-19)
	Direct Blue-86 (DB-86)
	Mordant Violet-40 (MV-40)
	Direct Yellow 50
Phenols	4-Chlorophenol
Pharmaceuticals	Ketoprofen
	Caffeine
Pesticides	Linuron
Other organic chemicals	Nitrosopyrrolidine
	Bisphenol A

Weiss (1934) also proposed that there is a formation of $^{\bullet}OH$ due to the Fenton reaction, where iron salt works as a catalyst. Nowadays, Fenton's process is an attractive technique for oxidation as it can generate $^{\bullet}OH$ without the use of special apparatus or chemicals under ambient temperature and pressure. Besides, it is an effortless technique that is environmentally safe and involves the use of cheap precursors like iron salt, H_2O_2. The generation of $^{\bullet}OH$ by the reaction of hydrogen peroxide with iron salt is shown below in the equation:

$$Fe^{2+} + H_2O_2 \rightarrow Fe^{3+} + OH^- + HO^{\bullet} \quad (10.21)$$

Although Fenton's reagent was revealed about 100 years ago, its application as an oxidizing agent for degradation of toxic organics was not applied until the late 1960s. The main reasons for the huge popularity and widespread applicability of Fenton oxidation processes are (Bokare and Choi 2014):

- The breakdown of the organic pollutants into nontoxic CO_2 can occur easily because of the high efficiency of mineralization associated with the Fenton process.
- Owing to the rapid reaction between iron and H_2O_2, generation of hydroxyl radicals is completed in the shortest reaction time.
- Oxidizing radicals are generated at ambient pressure and temperature, and, thus, the requirement of complex reactor facilities can be avoided.

A variety of industrial waste containing toxic organic compounds such as 4-chlorophenol, pentachlorobenzene, and chlorophenoxy herbicides has been efficiently treated by Fenton's process.

Homogeneous Fenton Reaction

Recently, the mechanism of Fenton process has been broadly exploited in several wastewater treatments as homogeneous Fenton (as Fe^{2+}/H_2O_2) and Fenton-like systems (if the reaction between H_2O_2 and other cations such as Fe^{3+}, Co^{2+}, Mn^{2+}, and Cu^{2+} takes place, it is called a Fenton-like reaction). The suggested mechanism involves a sequence of seven reactions shown below (Mishra et al. 2017):

$$Fe^{2+} + H_2O_2 \rightarrow Fe^{3+} + OH^- + {}^{\bullet}OH \quad (10.22)$$

$$Fe^{3+} + H_2O_2 \rightarrow Fe^{2+} + {}^{\bullet}O_2H + H^+ \quad (10.23)$$

$${}^{\bullet}OH + H_2O_2 \rightarrow {}^{\bullet}O_2H + H_2O \quad (10.24)$$

$${}^{\bullet}OH + Fe^{2+} \rightarrow Fe^{3+} + OH^- \quad (10.25)$$

$$Fe^{3+} + {}^{\bullet}O_2H \rightarrow Fe^{2+} + O_2H^+ \quad (10.26)$$

$$Fe^{2+} + {}^{\bullet}O_2H + H^+ \rightarrow Fe^{3+} + H_2O_2 \quad (10.27)$$

$$^{\bullet}O_2H + {}^{\bullet}O_2H \rightarrow H_2O_2 + O_2 \tag{10.28}$$

The organic compound (RH/R) reacts with the OH radicals to produce ($R^{\bullet}$) or ($^{\bullet}ROH$) by abstraction of hydrogen and addition of hydroxyl, respectively, as described in Eqs. 10.29 and 10.30 (Walling and Kato 1971):

$$RH + {}^{\bullet}OH \rightarrow H_2O + R^{\bullet} \rightarrow \text{Further oxidation} \tag{10.29}$$

$$R + {}^{\bullet}OH \rightarrow {}^{\bullet}ROH \rightarrow \text{Further oxidation} \tag{10.30}$$

Heterogeneous Fenton Reaction

One of the biggest challenges that stimulated the use of Fenton's process in heterogeneous catalysis is the removal of dissolved iron from the treated water. The heterogeneous Fenton-like reaction is the reaction between H_2O_2 and Fe in a solid matrix such as an Fe-oxide, Fe-bearing zeolite, Fe-bearing silica, pillared clay, etc. The heterogeneous Fenton-like process has several advantages over the homogeneous process. These include (Walling and Kato 1971):

- Enabling reactions to occur at neutral pH
- Acidification and neutralization steps are not required
- Reduces the possibility of sludge formation
- Solid catalysts are easy to handle
- Recycling and regeneration can be easily done

Among different Fe-bearing solid species, iron oxides have been extensively used owing to the following advantages they offer:

- Exceptional properties such as nontoxicity, high thermal stability, abundant availability, high crystallinity, chemical inertness, and low cost.
- Most significantly, it is their superparamagnetic nature that aids in facile separation of the catalyst from the reaction mixture.
- Besides, they possess excellent stability in the pH range of solutions commonly used in oxidation reactions of organic compounds in water.

Photo-Fenton Reaction

Further research has been done to modify the Fenton process and it has been observed that the degradation rate of organic pollutants is strongly accelerated with the UV–VIS light irradiation at wavelengths greater than 300 nm (from 300 nm to 650 nm) (Muruganandham et al. 2014). The UV irradiation-mediated Fenton process is called a photo-Fenton system. Due to the introduction of light, there is also some supplement to the Fenton mechanism (Nguyen and Ngo 2017).

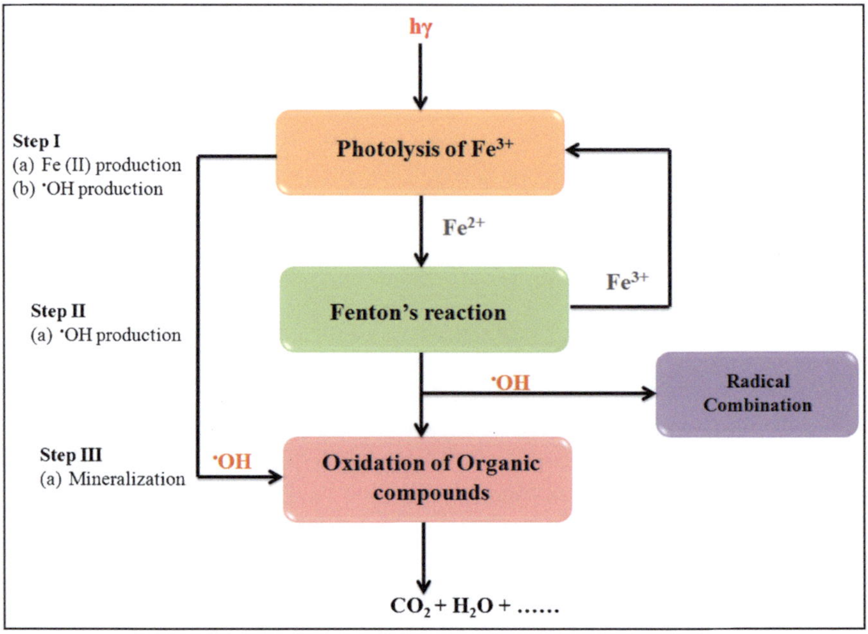

Fig. 10.4 Schematic depiction of the response pathway of photo-Fenton reaction

$$Fe^{III}(OH)^{2+} + h\nu \rightarrow {}^{\bullet}OH + Fe^{2+} \tag{10.31}$$

$$H_2O_2 + h\nu \rightarrow 2^{\bullet}OH \tag{10.32}$$

The hydroxyl radicals formed in the photo-Fenton process are highly reactive oxidants and initiate a radical chain oxidative decomposition of organic substances (RH) in water, which may lead to a total mineralization of the organic pollutants:

$$RH + {}^{\bullet}OH \rightarrow H_2O + {}^{\bullet}R \tag{10.33}$$

$${}^{\bullet}R + H_2O_2 \rightarrow ROH + {}^{\bullet}OH \tag{10.34}$$

$${}^{\bullet}R + O_2 \rightarrow ROO^{\bullet} \tag{10.35}$$

The generated organic radical reacts instantaneously with dissolved oxygen to yield a peroxyl radical, thus initiating subsequent oxidation by chain reactions. As a result, the photo-Fenton process has been reported to be more efficient than Fenton treatment. The schematic design of the response pathway of photo-Fenton reaction has been depicted in Fig. 10.4 (Huang 2012).

There are a number of examples that highlight the importance of the photo-Fenton oxidation process in the treatment of water contaminated with organic pollutants. Bleaching industries containing dyes deploy this technique using 250 mM of H_2O_2 and 5 mM Fe^{2+} under optimal conditions. The obtained results

Table 10.4 The illustrative works related to degradation of various pollutants using photo-Fenton/heterogeneous process

Pollutants	Photo-Fenton/heterogeneous catalyst	Removal efficiency
Bisphenol A (BPA)	Photo-Fenton/magnetite and ethylenediamine-N,N′-disuccinic acid	BPA = 70% in 11 h
Micropollutants from municipal wastewater	Photo-Fenton	Micropollutants = 40%
Paracetamol	Photo-Fenton oxidation with zeolite as catalyst	Paracetamol removal = ~99%, TOC removal = 60% in 5 h

reveal 96% decolorization and 89% dechlorination of wastewater. In another study, wastewater samples collected from pulp industries containing lignin sugars were decomposed by photo-Fenton process. Some other descriptive works related to degradation of various pollutants using photo-Fenton/heterogeneous process are summarized in Table 10.4 (Mishra et al. 2017).

This indicates that the photo-Fenton process has been successful in the removal of drugs, dyes and other pollutants from wastewater.

TiO_2/UV Process

In comparison with other wastewater treatments, heterogeneous photocatalysis is one of the excellent practices that includes the use of light irradiation source and a light-absorbing photocatalyst. They are relevant from two perspectives: (a) it is established as an emerging green technology for wastewater treatment and (b) an ultimate solution for elimination of pollutants from natural aquatic water bodies. This process includes redox reactions (oxidation and reduction) simultaneously. There are a number of compounds that utilize light irradiation and undergo photolysis to catalyze redox reactions. This technique involves TiO_2 as an efficient heterogeneous photocatalyst as it has good spectral overlap with near-UV lamps or solar radiation and high quantum efficiencies. Titanium peroxide semiconductor absorbs UV light and produces hydroxyl radicals. Usually, these compounds possess a band structure with a filled valence band and empty conduction band. When the photocatalyst is under UV illumination, the electrons are excited from the valence band to the conduction band only if the photon energy exceeds the energy gap of the semiconductor (Fig. 10.5) (Ge et al. 2016; Dong et al. 2015). Once the electron is excited, it produces holes in the valence band and electrons in the conduction band. The electrons in the conduction band interact with surface adsorbed molecular oxygen (O_2) in order to yield superoxide radical anions, while the holes in the valence band interact with water to produce hydroxyl radical (Chen et al. 2005):

The organic contaminants present in wastewater decompose by the reductive cleavage through the conduction band electrons as well as by the reaction with the valence bond holes, hydroxyl and peroxide radicals in order to generate CO_2 and

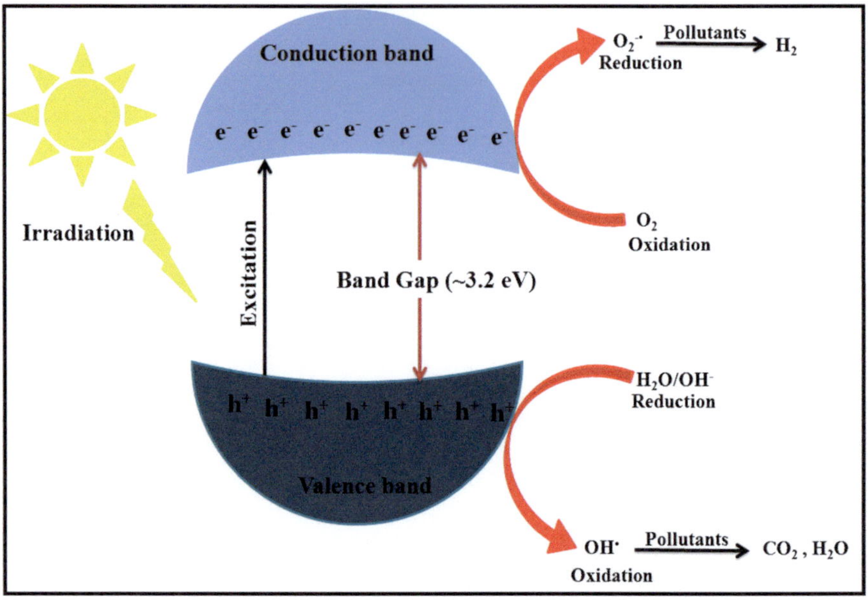

Fig. 10.5 Schematic diagram showing the mechanism of TiO_2 when irradiated with UV light

H_2O. Up to date, the TiO_2/UV light process is believed to be appropriate for wastewater treatment at ambient conditions as it shows no mass transfer limitations and effective use of solar irradiation. Moreover, TiO_2 is extensively used as a photocatalyst owing to its high photostability, nontoxicity, chemical and biological inertness, cost-effectiveness, and ready availability. Since the band gap energy of TiO_2 is ~3.2 eV, a large array of organic pollutants such as herbicides, dyes, pesticides, phenolic compounds, tetracycline, sulfamethazine, etc., can undergo decomposition to form harmless compounds such as CO_2 and H_2O (Chatterjee and Dasgupta 2005). This technique involves in situ generation of active oxidizing species under solar irradiation in the presence of a catalyst, titanium dioxide, as shown below in Eqs. 10.36, 10.37, 10.38, 10.39 and 10.40. The active oxidizing species as generated thereafter degrade the organic pollutants (Eqs. 10.41, 10.42 and 10.43) (Akpan and Hameed 2009; Fujishima et al. 2008):

$$TiO_2 + h\gamma\ (UV) \rightarrow TiO_2\ (e^-{}_{CB} + h^+{}_{VB}) \tag{10.36}$$

$$TiO_2\ (h^+{}_{VB}) + H_2O \rightarrow TiO_2 + H^+ + {}^{\bullet}OH \tag{10.37}$$

$$TiO_2\ (h^+{}_{VB}) + {}^{-}OH \rightarrow TiO_2 + {}^{\bullet}OH \tag{10.38}$$

$$TiO_2\ (e^-{}_{CB}) + O_2 \rightarrow TiO_2 + O_2{}^{-\bullet} \tag{10.39}$$

$$O_2{}^{-\bullet} + H^+ \rightarrow HO_2{}^{\bullet} \tag{10.40}$$

Table 10.5 Bandgap energies for some common semiconductor materials at 300 K (Thiruvenkatachari et al. 2008)

Semiconductors	Band gap energy (eV)
Diamond	5.4
WO_3	2.76
CdS	2.42
Si	1.17
ZnS	3.6
Ge	0.744
ZnO	3.436
Fe_2O_3	2.3
TiO_2	3.03
PbS	0.286
PbSe	0.165
SnO_2	3.54
ZrO_2	3.87
CdSe	1.7
Cu_2O	2.172

$$\text{Organic pollutants} + {}^{\bullet}\text{OH} \rightarrow \text{Degraded products} \quad (10.41)$$

$$\text{Organic pollutants} + \text{h}^{+}{}_{\text{VB}} \rightarrow \text{Oxidation products} \quad (10.42)$$

$$\text{Organic pollutants} + \text{e}^{-}{}_{\text{CB}} \rightarrow \text{Reduction products} \quad (10.43)$$

Okomoto and Sakthivel observed that TiO_2 has greater photocatalytic activity as compared to CdS, α-Fe_2O_3, ZrO_2, WO_3, and SnO_2 (Table 10.5) (Thiruvenkatachari et al. 2008). The light absorbed by the TiO_2 is below the visible range of light spectrum. Hence, the UV radiation of wavelength less than or equal to 384 nm is required having maximum absorbance at 340 nm.

Factors That Affect TiO_2/UV Light Processes

There are number of operational parameters on which the photodegradation rates and efficiency of a photocatalyst depends described below (Fig. 10.6) (Kumar and Pandey 2017).

1. **Effect of catalyst amount**: With the increase in amount of catalyst, the degradation efficiency of photocatalyst increases. This is because of the increase in the number of active sites on the photocatalytic surface that accelerates the generation of hydroxyl radicals. But, beyond an optimum amount, if the catalyst amount increases the solution becomes turbid and there is a decrease in the degradation rate, which is due to blocking of UV radiation for the reaction.
2. **Presence of ionic species**: There are number of inorganic ions present in wastewater system such as magnesium, iron, zinc, copper, bicarbonate, phosphate, nitrate, sulfate, and chloride that can affect the photocatalytic degradation efficiency of organic pollutants as these ions become adsorbed on the surface of TiO_2 nanocomposites. The presence of these ions in wastewater diminishes the colloidal stability, increases mass transfer, and decreases the contact between the pollutant and the photocatalyst.

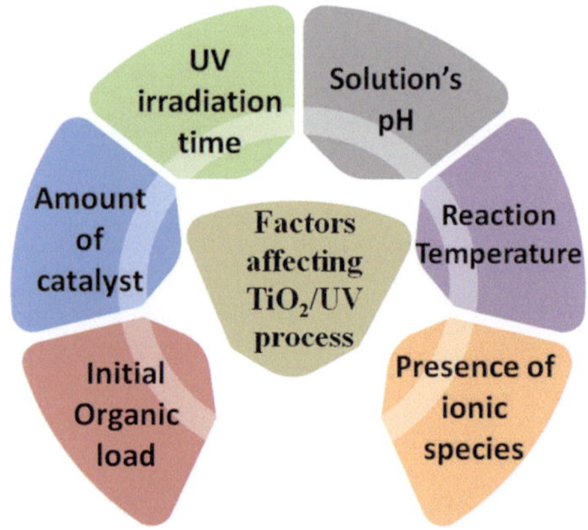

Fig. 10.6 Diagram representing the factors affecting TiO_2/UV process

3. **Initial organic load**: The presence of high concentration of pollutants in wastewater leads to the saturation of TiO_2 surface, which results in the reduction of photocatalyst activity.
4. **Effect of reaction temperature**: In general, the photocatalytic activity of catalyst increases with the increase in the reaction temperature (>80 °C). This is because high temperature favors the recombination of charge carriers and avoids the adsorption of organic compounds on TiO_2 surface. On the other hand, the reaction temperature below 80 °C promotes the adsorption and a further decrease in the temperature to 0 °C leads to an increase in apparent activation energy. Hence, for significant photodegradation of organic pollutants, temperature 20–80 °C has been considered as an optimum temperature range.
5. **UV irradiation time**: The intensity of UV light and the time of irradiation both significantly affect photodegradation of organic pollutants. It has been observed from the results that at lower light intensity, the rate would increase linearly with the increase in the intensity of light and follows first-order kinetics; at higher intensities of light, the rate is independent of light intensity.
6. **Effect of pH**: One of the main factors, which affects the photodegradation of wastewater contaminant, is the pH of the solution. The acidic and alkaline conditions can protonate as well as deprotonate the surface of titania, respectively, which will consequently either increase or decrease the rate of reaction depending on the nature of the pollutant.

In order to examine the photocatalytic efficiency of titanium dioxide, it was used for the treatment of diluted (1/100) olive mill wastewater. After a 24 h treatment, it was observed that the COD, colouration at 330 nm and level of phenols could decrease to 22%, 57% and 94% respectively (El Hajjouji et al. 2008).

Figure 10.7 displays some of the disadvantages of the TiO_2/UV process (Dong et al. 2015; Bagheri et al. 2017). As photocatalytic degradation mainly occurs on the surface of TiO_2, TiO_2 has poor affinity towards the organic contaminants because of which the adsorption of pollutants onto the surface of TiO_2 is slow, which leads to the reduction of photocatalytic degradation rates. Furthermore, owing to the

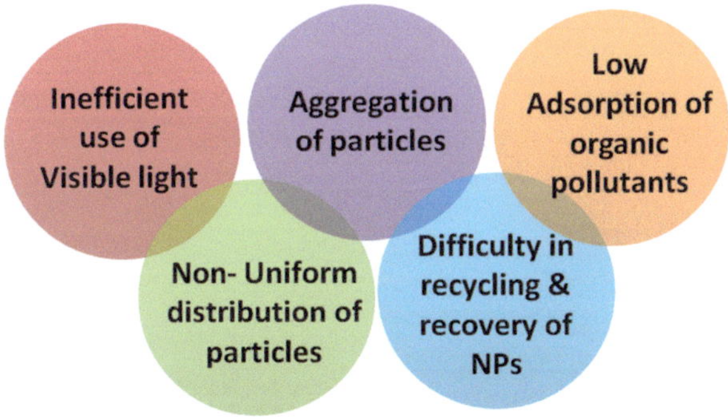

Fig. 10.7 Disadvantages of the TiO_2/UV process

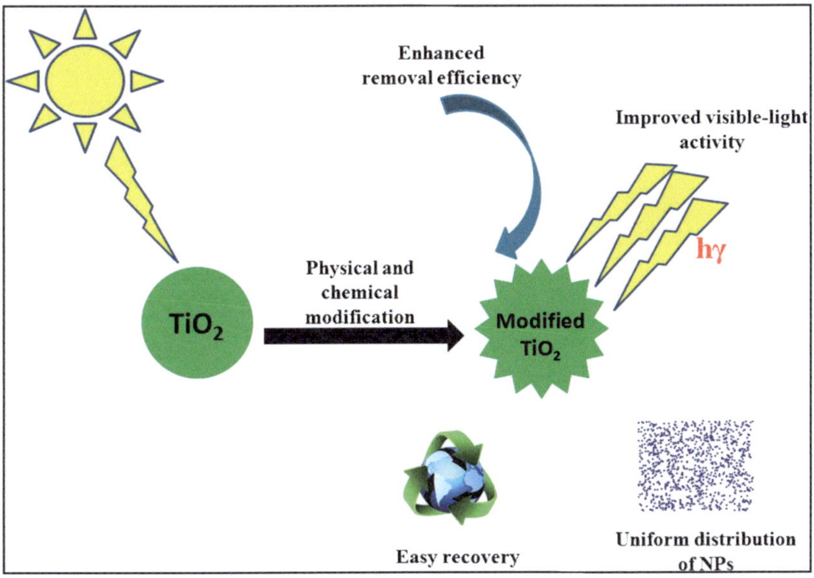

Fig. 10.8 Schematic representation of the advantages of modified TiO_2 NPs

instability of the nanosized particles, the TiO_2 nanoparticles (NPs) may aggregate during the photocatalytic degradation. This aggregation of TiO_2 nanocomposites may obstruct the incident light on active centers, which consequently results in the reduction of catalytic activity. Besides this, one of the main practical challenges is to recover nanosized TiO_2 particles from the treated wastewater. To overcome these limitations, the lattice of TiO_2 is doped with other metal ions (Sahoo et al. 2005). Such doping results in an improved visible-light activity and improved removal efficiency as demonstrated in Fig. 10.8.

In a recent study, for the removal of Acid Red 88, Behnajady et al. (2008) found that silver-doped TiO_2 was more efficient than the undoped TiO_2. The advantages of the doped TiO_2-based photocatalyst and the respective improvements in the mechanism have been summarized in Table 10.6.

Table 10.6 Advantages of doped TiO_2-based photocatalyst and the respective improvements in mechanism (Dong et al. 2015)

Advantages	Techniques	Improvements
Enhancing the visible-light photocatalytic activity of TiO_2 particles	Doping of metals	Narrowing band gap
		Retarding electron–hole recombination
	Doping of nonmetals	Enhancing adsorption of contaminants
		Narrowing band gap
		Enhancing adsorption of contaminants
	Co-doping technique	Enhancing conductivity of TiO_2
	Surface organic modification (dye sensitization/ organic coating)	Improving interfacial charge transfer
		Synergistic effect of two elements co-doping
		Enhancing visible-light absorption
		Improving electron–hole separation
Enhancing adsorption of organic pollutants on TiO_2 particles	Surface organic modification	Hydrogen bonding, n-п and п-п interactions result in stronger adsorption
	Doping of carbon-based nanoparticles	Providing high surface area, −good conductivity and higher visible-light absorption intensity
		Suppression of electron–hole recombination (due to the high electrical conductivity)
		Band gap narrowing resulting from the presence of Ti-O-C bonds
Stabilization of TiO_2 particles	Stabilization by support structures	Immobilizing the TiO_2 photocatalyst to preventing aggregation
	Stabilization by surface modification	Preventing the particle aggregation and also enable the homogeneous dispersion of TiO_2 NPs
Separation of TiO_2 particles	Immobilization on support structures	Immobilizing the TiO_2 photocatalyst for easy recovery
	Magnetic separation	This type of composite photocatalyst comprises of a TiO_2 shell and a magnetic core, making them recoverable due to their magnetic properties

10.3 Use of Photoreactors in Water Treatment

In recent decades, the photochemical oxidation of pollutants in wastewater has become an attractive technology for the water treatment. For the specific concentration range of hazardous contaminants, this method is suggested because artificial source requires high delivery of energy. The key parameters that play a crucial role in water treatment efficiencies are (1) selection of light source, (2) oxidation system, and (3) determination of optimal conditions. Sufficient UV penetration into the radiated liquid is crucial for the efficient degradation of pollutants in waste water. The significant interaction between the pollutant and photocatalyst is crucial for the

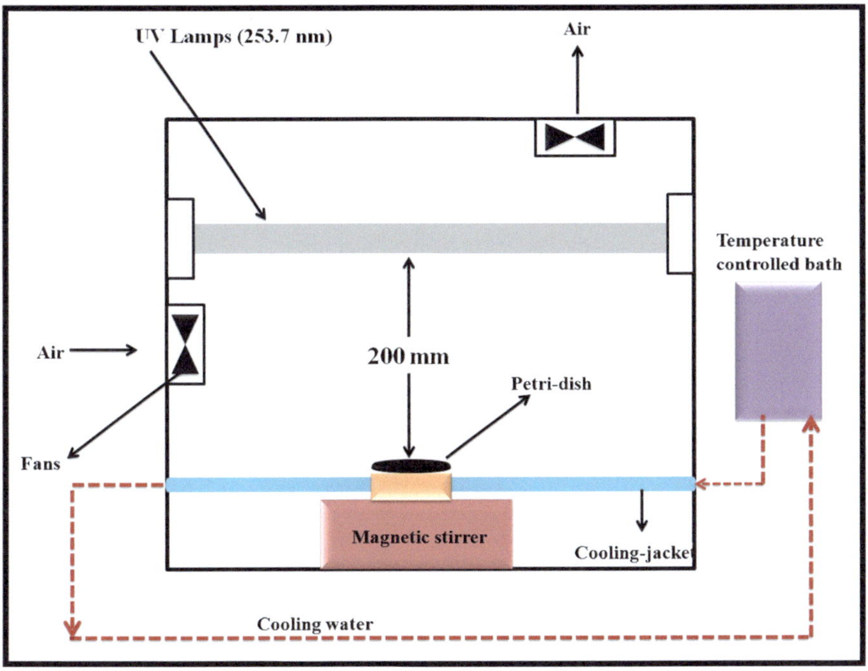

Fig. 10.9 A schematic diagram of lab-scale treatment batch photoreactor involving UV/H_2O_2 process

high mass transfer rates. Another important parameter for practical applicability during this process is the high oxygen uptake at gas-liquid interface. Therefore, designing a suitable reactor for effective treatment of wastewater has remained a challenging task for the researchers in today's global scenario. The ensuing section throws light on some of the photoreactors that have been employed for water treatment.

10.3.1 UV/H_2O_2 Photoreactor

A batch photoreactor was designed by Jelena Mitrović and her colleagues for the decolorization of textile azo dye (Reactive Orange 16) via UV/H_2O_2 process. As demonstrated in Fig. 10.9, all photochemical reactions were conducted in a batch photoreactor by using low-pressure mercury vapor lamps with maximum emission at 253.7 nm (28 W, UV-C, Philips, Holland) as a light source (Mitrović et al. 2012). At the top of photoreactor, ten parallel UV lamps were fixed. For the efficient light reflection, the inner surface of the photoreactor was composed of highly polished stainless steel. The photoreactor also contained air cooling system with electrical fans in order to vent the heat outside from the reactor and avert the lamps from overheating.

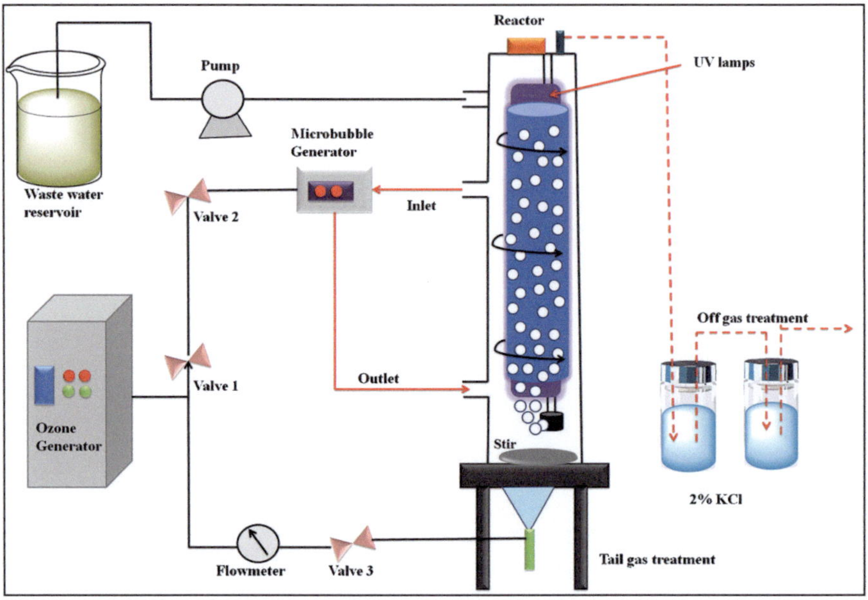

Fig. 10.10 Schematic diagram of microbubble-ozonation/UV combined process experimental setup

In addition, for the controlled release of UV light, the lamps were placed at a distance of 220 mm above from the working solution. The UV radiation intensity was measured by a UV radiometer solar meter model 8.0 (UVC) (Solartech, USA). Further, to study the effect of the important parameters on the decolorization rate of dye, some of the experiments were carried out by using the lowest UV light intensity.

10.3.2 Photoreactor for Photolytic Ozonation

Zheng et al. (2015) designed a combined microbubble ozonation/UV irradiation process for the advanced treatment of acryclic fibre in wastewater. In Fig. 10.10, the experimental setup is displayed. The reactor comprises of transparent rigid Plexiglas with an inner diameter of 80 mm, a height of 1200 mm, and an effective volume of 6 L. In order to form microbubbles in the reactor, a TCRI microbubble generator (Japan) was employed and operated under the pressure of 0.4 MPa, which produced bubbles having a mean diameter less than 45 μm. At the center of the reactor, a low-pressure mercury vapor UV lamp was equipped with a quartz cell (Zheng et al. 2015). To study the effect of irradiation intensity of UV light on reaction sample, different UV lamps such as 10 W (15 mm diameter × 212 mm height), 14 W (15 mm diameter × 303 mm height), 18 W (15 mm diameter × 356 mm height), 23 W (15 mm diameter × 436 mm height), and 28 W (15 mm diameter × 550 mm height) produced by Cnlight Co., Ltd., Guangdong, China, were used.

A 40 μm cylindrical microporous titanium plate, which was planted at the bottom of the reactor, played a crucial role in generating ozone macrobubbles (with a mean bubble diameter of approximately 1 mm). Moreover, an ozone generator (CF-YG5, Shanmei Shuimei Co. Beijing) was attached with the reactor, which produced ozone gas at the rate of 5 g O_3/h by using dehumidified air as gas source (flow rate of air of 0.5 L/min). The reaction process began by injecting acrylic fiber manufacturing wastewater (3 L) with a peristaltic pump. The wastewater persistently circulated between the microbubble generator and the reactor. The ozone gas that came out from the reactor was absorbed with 2% KCl solution. At preselected regular time intervals, the samples were withdrawn and instantly purged by N_2 gas in order to eliminate the residual ozone. Throughout the experiment, the reaction temperature was maintained at 20 °C.

10.3.3 Photo-Fenton Reactor

The photo-assisted Fenton process adds UV radiation to the system resulting in the formation of more hydroxyl free radicals. Recently, studies have been carried out to compare the Fenton and photo-assisted Fenton process. In general, various reactor geometries have been used among which cylindrical reactor with UV lamps placed in the center has gained the highest popularity. For the testing of photodegradation of organic pollutants, a glass reactor with an external loop, which had a capacity of 1.6 L, was utilized and operated in batch mode as shown in Fig. 10.11 (Bañuelos et al. 2014). A medium-pressure Hg lamp (type EQ 1023 Z4) surrounded by a cooling pipe made of quartz glass was located in the center. In this photo-Fenton reactor, jet-loop reactor and UV reactor were connected by a polyvinyl chloride (PVC) tube. The obtained results showed that the degradation efficiency of pollutants got improved with continuous aeration.

10.3.4 Photocatalytic Reactor

A vertical circulating photocatalytic reactor was designed by Shahrezaei and colleagues for the photocatalytic degradation of aniline using TiO_2 NPs. A 400 W mercury UV lamp of 22 cm body length and 16 cm arc length was used as a UV source and placed inside a quartz tube which is totally immersed in the reactor (Fig. 10.12) (Shahrezaei et al. 2012).

In order to attain homogeneity of the nanocomposites along the quartz tube, a pump was used to provide a tunable circulating stream and fed wastewater from the top of the reactor. In this process, continuous airflow was delivered to the reactor at a constant flow rate (3 L/min) with the aid of a lab-scale air compressor. It was not anticipated that through the short reactor a single pass of polluted water would give sufficient degradation.

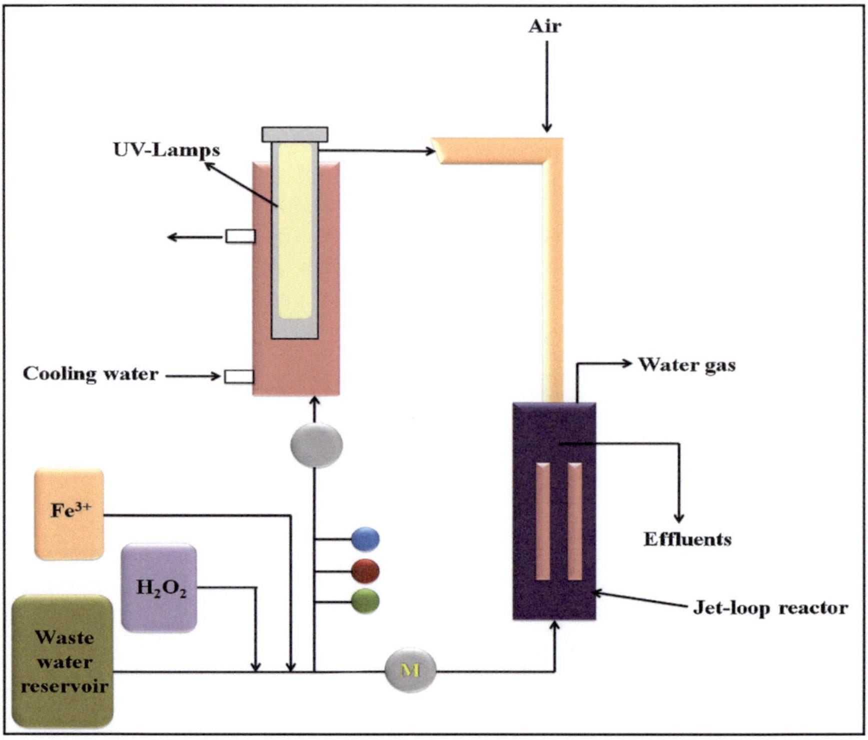

Fig. 10.11 Schematic depiction of photo-Fenton reactor (Bañuelos et al. 2014)

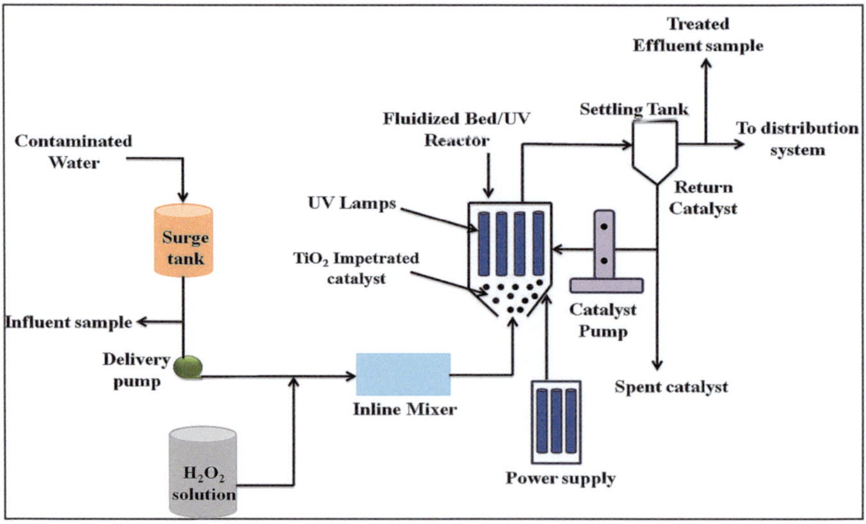

Fig. 10.12 Schematic description of degradation of aniline by using photocatalytic reactor

10.4 Case Studies

Nowadays, photo-oxidation technologies are also extensively employed by renowned pharmaceutical companies for the remediation of wastewater. A few examples that have been discussed below elucidate potential and practical implementation of UV-oxidation processes by Merck & Co., BASF, and GlaxoSmithKline.

10.4.1 Merck & Co.

Merck & Co. is one of the world's largest American pharmaceutical companies established in 1891 as the US subsidiary of the German company. The pharmaceutical manufacturing site of Merck KGaA in Altdorf, Switzerland, synthesizes an antihypertensive containing a phenolic structure.

The production of this antihypertensive drug has led to water contamination. To deal with this problem, the wastewater was initially collected and then transported to incineration plant for treatment. Afterwards, by eliminating all solvents and all aromatic structural compounds this wastewater was successfully treated (Sörensen et al. 2015). In this way, toxicity was completely removed. In order to gain the best quality water, the water was then passed from a recuperation system, which was integrated into the UV plant. The whole system is fully incorporated into Merck's factory management and operated with minimum effort. The data from the treatment at Merck Altdorf, Switzerland, is tabulated in Table 10.7. Utilization of this system has proven to be an essential step for improvement of the environmental credential as well as economic gains for Merck.

10.4.2 BASF

On April 6, 1865, BASF fully named as Badische Anilin und Soda Fabrik or in English Baden Aniline and Soda Factory started its journey in Mannheim, a city located in Germany. The headquarters of this company is situated in Ludwigshafen, Germany and it operates in six integrated production sites. Today, this company is the largest chemical producer in the world. They deal in different segments such as chemicals, plastics, performance products, functional solutions, agricultural solutions, and oil and gas.

BASF has been involved in synthesizing chelating agents for detergents and cleaners, pulp and paper and agriculture, but this synthesis has resulted in the contamination of water and this wastewater has been found to contain EDTA.

The effluent of the EDTA-manufacturing line at BASF, Ludwigshafen showed a completely bioavailable wastewater; however, it was found that even after the

Table 10.7 Data of treatment at Merck Altdorf, Switzerland

Parameter	Data	Result
Flow rate	37 m^3/d	
TOC	≈50,000 mg/L	≈5000 mg/L
Phenols	15,000–24,000 mg/L	<5 mg/L
Dichloromethane (DCM)	≈5000 mg/L	<1 mg/L
Bioavailability	<5%	≈95%
Operating expenses (OPEX)	≈20 €/m^3	

Table 10.8 Data of treatment at BASF Ludwigshafen, Germany

Parameter	Data	Result
Flow rate	240–280 m^3/d	
COD	≈3000 mg/L	No significant impact
EDTA	400–600 mg/L	≈50% reduction rate
Bioavailability of target	0%	≈90% of photo-transformed molecules
OPEX	≈0.5 €/m^3	

extraction step, EDTA still could not be removed completely. Therefore, this company directed its effort towards developing a highly selective treatment for the elimination of this single substance only. Finally, they developed a highly selective UV treatment that was able to reduce the concentration of EDTA without any detectable change in the matrix and this was selected as the pretreatment methodology. The prime advantage of this strategy is that it reduces the operational costs (shown in Table 10.8) (Sörensen et al. 2015). Similar UV-mediated water treatment technologies have been applied for the selective removal of active pharmaceutical ingredient (API) from industrial wastewater at many pharmaceutical manufacturing sites of leading pharmaceutical companies.

10.4.3 GlaxoSmithKline

GlaxoSmithKline (GSK), a British pharmaceutical company situated at Brentford, London, is the world's sixth largest pharmaceutical company. This company manufactures drugs such as Advair, Avodart, Flovent, Augmentin, Lovaza, and Lamictal that help people to fight against various diseases such as asthma, cancer, infections, diabetes, and mental disorders. Historically, medicines discovered or developed at GSK were sold as generic drugs, which include amoxicillin and amoxicillin-clavulanate, etc. The largest production site of antibiotics is amoxicillin manufacturing site of GSK in Singapore. It is well-known that biological treatment methodologies are not capable of treating water that is rich in pharmaceutical waste. As a result, for the effective removal of the several toxic substances from the wastewater, GSK has been searching for a more viable process. In this regard, it is worth mentioning that this

Table 10.9 Data of treatment at GSK Singapore

Parameter	Data	Result
Flow rate	54–100 m^3/d	
COD	58,000–70,000 mg/L	≈40–50% reduction rate
Phenol-based organics from amoxicillin process	≈1000–5000 mg/L	<0.5 mg/L
Bioavailability	≈0%	≈70–90%
OPEX	≈20 €/m^3	

company has employed photo-oxidation technology for the removal of toxicity, elimination of phenol-based active structures, and for increasing the bioavailability of strong wastewater. This installed treatment system has increased the capacity of wastewater from 54 m^3/d to 100 m^3/d. GSK even won an award (Singapore Environment Achievement Award) for all its environmental activities including the installation of the photo-oxidation technology. The data from the treatment of wastewater at GSK Singapore in summarized in Table 10.9 (Sörensen et al. 2015).

10.5 Recent Advancements: Nanotechnology Coupled with Photo-Oxidation Process

The advent of nanotechnology has provided multiple opportunities to expand the range of next-generation water treatment techniques. Significant research has been conducted in nanotechnology, which reveals an improved performance over other conventional techniques. In the last decades, nanotechnology has quickly changed from an academic pursuit to commercial reality. This technique brings the dimension of catalytic material down to nanoscale range (<100 nm) in the form of NPs or nanosize porous supports with controlled shapes and sizes. In fact, nanomaterials possess astonishing properties such as high surface area, photosensitivity, high catalytic activity, antimicrobial activity, electrochemical; optical; and magnetic properties, and tunable pore size and surface that make a proficient use of these in wastewater treatment. It is difficult to manage water with high pollutant concentration using photo-oxidation methodology alone; however, this problem can be resolved by combining it with nanotechnology (Das et al. 2017).

As previously discussed, TiO_2 photocatalyst under UV illumination effectively produces reactive oxygen species, superoxide, and hydroxyl radicals. However the effectiveness of the photocatalytic process for treatment of wastewater increases, when bulk TiO_2 is replaced by nano-TiO_2 owing to its outstanding advantages such as large surface area (i.e., more surface reactive sites), lower volume e^-/h^+ recombination, and faster interfacial charge transfer.

Chalasani and Vasudevan (2013) reported the efficient photocatalytic degradation of endocrine-disrupting chemicals present in water bodies via magnetically retrievable

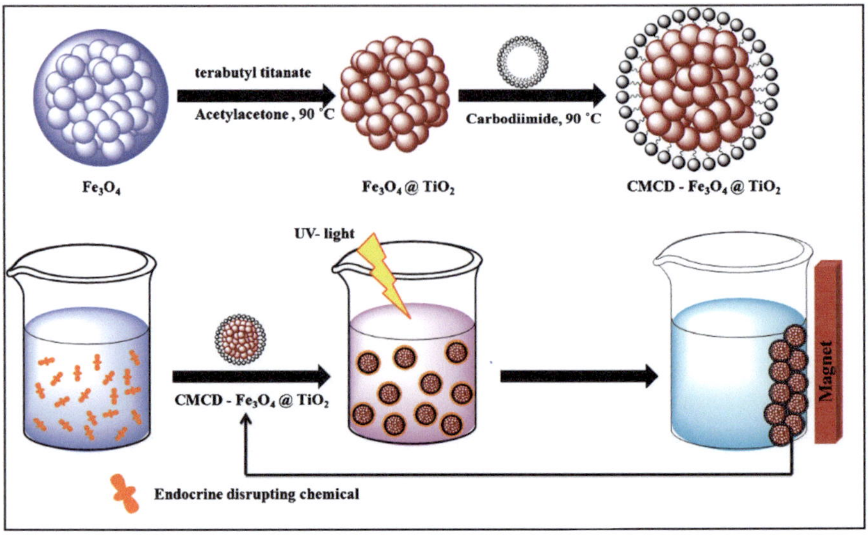

Fig. 10.13 Pictorially summarized the whole cycle from the capture to the photodestruction of the endocrine-disrupting chemical (CMCD—carboxymethyl-β-cyclodextrin)

cyclodextrin-functionalized Fe_3O_4@TiO_2 NPs. These magnetic core–shell nanoparticles were prepared by anchoring cyclodextrin cavities onto TiO_2 shell. As β–cyclodextrin is amphiphilic in nature, it possesses both hydrophobic as well as hydrophilic cavities. The hydrophilic cyclodextrin cavity consists of exposed hydroxyl groups, which are coordinatively linked to the TiO_2 shell via a carboxymethyl linkage and they are responsible for the aqueous dispersibility of the NPs.

The hydrophobic cavity present in anchored cyclodextrin plays a crucial role in capturing the nonpolar organic pollutants such as endocrine-disrupting chemicals, BPA, dibutyl phthalate, etc. It forms 1:1 inclusion complex with these pollutants. The trapping of the organic pollutants by cyclodextrin brings them in the closer proximity to the surface of TiO_2 shell, which allows their significant photodegradation and complete mineralization, compared with the bare Fe_3O_4@TiO_2 NPs. Once the photocatalytic degradation is complete, the important step is the facile separation of the catalyst from the reaction mixture, not just for reusability purpose but also to elude adverse biological effects of semiconductor NPs even in the absence of light. As depicted in Fig. 10.13, a fascinating feature of these "capture and destroy" NPs is the presence of superparamagnetic iron oxide core, which allows the facile magnetic separation of these NPs with the aid of external magnetic force and allows the reuse with minimal loss in its photocatalytic activity.

The enhanced photocatalytic activity for the degradation of rhodamine B using core–shell ellipsoid-like $BiVO_4$@g-C_3N_4 photocatalyst using visible light illumination has been studied by Sui and colleagues. By using different amounts of g-C_3N_4, these novel nanocomposites are successfully prepared via simple hydrothermal and ultrasound chemisorption methods. The entire sequence for the synthesis of promising novel core–shell photocatalyst is depicted in Fig. 10.14. The photodegradation

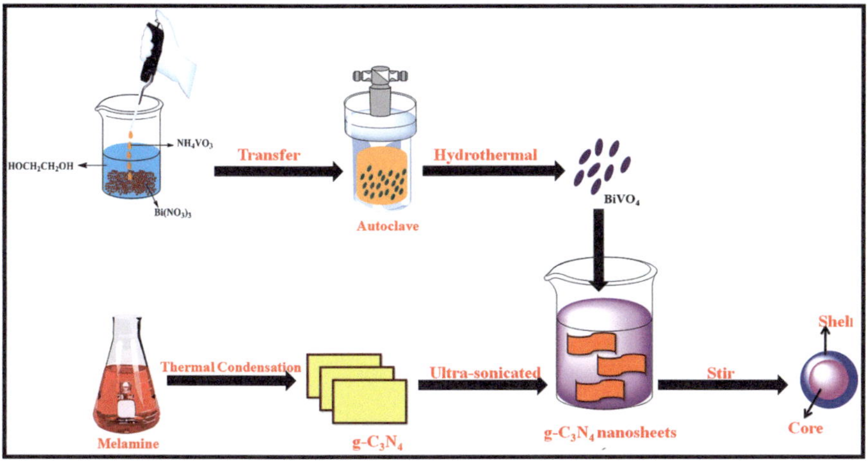

Fig. 10.14 Pictorial illustration for the $BiVO_4$@g-C_3N_4 preparation process

efficiency was found to be nine times greater than pristine $BiVO_4$ when 7% g-C_3N_4 content was used (Zhang et al. 2017). The thin g-C_3N_4 plays a crucial role in increasing the extent of degradation by reducing the charge carrier transfer distance, which further suppresses the recombination of e^-/h^+ pairs. In addition, the stable core–shell structure not only enlarges the contact area but also strengthens the interaction of chemical bonds between the $BiVO_4$ core and g-C_3N_4 nanosheet, which results in more exposed active sites. Furthermore, even after the five photocatalytic cycles the structure and activity of photocatalyst remained the same.

Recently, Sedghi has synthesized a recyclable graphene oxide supported TiO_2 photocatalyst for the efficient elimination of methylene blue from the wastewater. The design of this nanocomposite employs a modular synthesis technique as shown in Fig. 10.15. This method is anticipated to allow simple tuning of the ratio of particle loading on the graphene oxide (GO) surface (Nabid et al. 2014). In order to determine the photocatalytic ability of the photocatalyst, a model dye (methylene blue) is taken. The observed results reveal the capability of photocatalyst to degrade the aqueous solution containing 3 mg/L methylene blue over ten simultaneous cycles with the little or no catalytic loss. Typically, this catalyst could be recovered simply by getting an external magnet in the vicinity of the reaction vessel. With the aim to employ these nanocomposites for treatment of real water samples, rapid degradation of pharmaceutical compounds—caffeine and carbamazepine—via commercial photocatalyst P25 was conducted. The enhanced activity is shown by the nanocomposites rather than by commercial photocatalyst P25 by a factor of 1.2 with the additional advantage of excellent recoverability, reusability, and easy production. Moreover, it was found that after multiple trials, the catalyst was be able to gain its initial efficiency through simple UV irradiation. The illustrative works related to degradation of various pollutants using synergistic effect of photo-oxidation and nanotechnologies are summarized in Table 10.10.

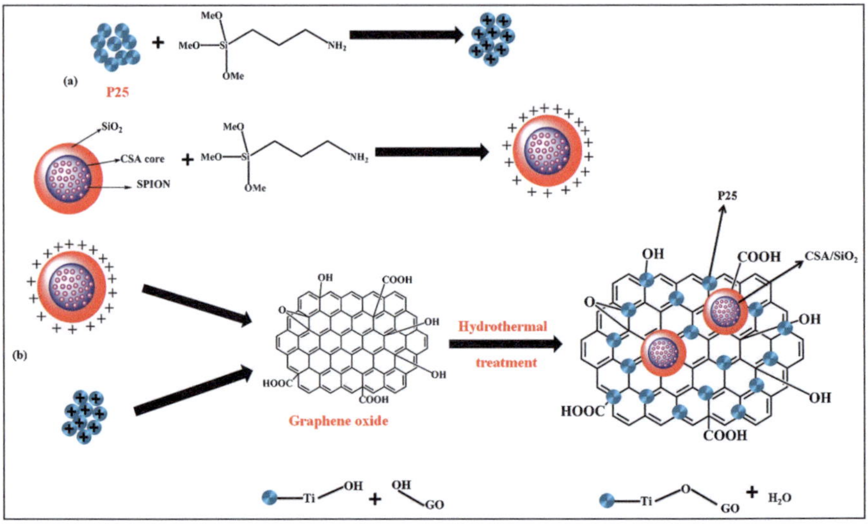

Fig. 10.15 Schematic description of recyclable graphene oxide-supported titanium dioxide nanocomposite as a photocatalyst

Table 10.10 Degradation of various pollutants using synergistic effect of photo-oxidation and nanotechnologies

Catalyst	Pollutants	Removal efficiency (%)	Time (min)	References
RGO/g-C_3N_4 $BiVO_4$	Tetracycline hydrochloride	72.5	150	Jiang et al. (2017)
N-doped graphene quantum dots-$BiVO_4$/g-C_3N_4	Tetracycline	91.5	30	Yan et al. (2016)
	Oxytetracycline	66.7	120	
	Ciprofloxacin	72.4	120	
Ln-MOFs	Rhodamine B	99	12	Xia et al. (2017)
ZnO/$ZnFe_2O_4$ magnetic heterostructures	Rhodamine B	100	240	Xu et al. (2017)
	Methylene blue			
Fe(II) metal–organic framework	Bis(p-nitrophenyl) phosphate	>90	90	Li ct al. (2018)
	Paraoxon-ethyl	100	120	
	Tris(4-nitrophenyl) phosphate	100	300	
Graphene nanosheets on anastase–TiO_2	Methyl orange	~90	60	Liu et al. (2018)
Graphene-TiO_2 nanotube	Malachite green	~80	75	Perera et al. (2012)
Fenton-like catalyst, metal–organic framework (CUS-MIL-100(Fe))	Sulfamethazine	100	180	Tang and Wang (2018)
NH_2 – MOFXII@30	Amoxicillin	100	90	Abazari and Mahjoub (2018)
7 Sm_2O_3 – ZnO				
Core–shell Fe_3O_4@TiO_2	Organic pollutants in steel mill wastewater	96	90	Salamat et al. (2017)
NPs				

10.6 Conclusion and Future Scope

The immense enthusiasm of the academic as well as industrial researchers towards utilization of photo-oxidation techniques in wastewater treatment is reflected in the increase in the critical number of publications over the past decades. The major center of attention of this procedure is the hydroxyl radical, which nonselectively attacks the pollutants present in waste water once created. However, a photocatalytic (TiO_2/UV) process, photolytic ozonation, a H_2O_2/UV process, and Fenton's responses have been considered and widely utilized for the removal of hard-headed natural chemicals. Their treatment proficiency trusts upon different process parameters that have been thoroughly investigated. In order to develop a suitable method for wastewater treatment, it is relevant to conduct experimental studies. There are various factors that influence the reaction conditions such as initial concentration of the target compounds, the amount of oxidation agents and catalysts, the light intensity, the irradiation time, and the nature of the wastewater's solution (pH, the presence of solids, and other ions). The main challenge to address in the future could be the development of proficient and minimal-effort materials to advance adequate treatment, the utilization of sustainable power sources, the reception of systems for forms reconciliation, and focusing on new classes of contaminants.

References

Abazari R, Mahjoub AR (2018) Amine-functionalized Al-MOF$^{\#}$@$_y^x$$Sm_2O_3$–ZnO: a visible light-driven nanocomposite with excellent photocatalytic activity for the photo-degradation of amoxicillin. Inorg Chem 57(5):2529–2545. https://doi.org/10.1021/acs.inorgchem.7b02880

Akpan UG, Hameed BH (2009) Parameters affecting the photocatalytic degradation of dyes using TiO_2-based photocatalysts: a review. J Hazard Mater 170(2–3):520–529. https://doi.org/10.1016/j.jhazmat.2009.05.039

Badia-Fabregat M, Oller I, Malato S (2017) Overview on pilot-scale treatments and new and innovative technologies for hospital effluent. In: Verlicchi P (ed) Hospital wastewaters: characteristics, management, treatment and environmental risks. Springer, Cham, pp 209–230. https://doi.org/10.1007/698_2017_23

Bagheri S, TermehYousefi A, Do T-O (2017) Photocatalytic pathway toward degradation of environmental pharmaceutical pollutants: structure, kinetics and mechanism approach. Cat Sci Technol 7(20):4548–4569. https://doi.org/10.1039/C7CY00468K

Bañuelos JA, Rodríguez FJ, Manríquez J, Bustos E, Rodríguez A, Godínez LA (2014) A review on arrangement and reactors for Fenton-based water treatment processes. In: Peralta-Hernández JM, Rodrigo-Rodrigo MA, Martínez-Huitle CA (eds) Evaluation of electrochemical reactors as a new way to environmental protection. Research Signpost, Kerala

Barrera-Díaz C, Cañizares P, Fernández FJ, Natividad R, Rodrigo MA (2014) Electrochemical advanced oxidation processes: an overview of the current applications to actual industrial effluents. J Mex Chem Soc 58(3):256–275

Behnajady MA, Modirshahla N, Shokri M, Rad B (2008) Enhancement of photocatalytic activity of TiO_2 nanoparticles by silver doping: Photodeposition versus liquid impregnation methods. Global NEST J 10(1):1–7. https://doi.org/10.30955/gnj.000485

Bokare AD, Choi W (2014) Review of iron-free Fenton-like systems for activating H_2O_2 in advanced oxidation processes. J Hazard Mater 275:121–135. https://doi.org/10.1016/j.jhazmat.2014.04.054

Bourgin M, Borowska E, Helbing J, Hollender J, Kaiser H-P, Kienle C, McArdell CS, Simon E, von Gunten U (2017) Effect of operational and water quality parameters on conventional ozonation and the advanced oxidation process O_3/H_2O_2: kinetics of micropollutant abatement, transformation product and bromate formation in a surface water. Water Res 122:234–245. https://doi.org/10.1016/j.watres.2017.05.018

Braun AM, Maurette M-T, Oliveros E (1991) Photochemical technology. Wiley, Chichester

Buthiyappan A, Aziz ARA, Daud WMAW (2016) Recent advances and prospects of catalytic advanced oxidation process in treating textile effluents. Rev Chem Eng 32(1):1–47. https://doi.org/10.1515/revce-2015-0034

Camel V, Bermond A (1998) The use of ozone and associated oxidation processes in drinking water treatment. Water Res 32(11):3208–3222. https://doi.org/10.1016/S0043-1354(98)00130-4

Chalasani R, Vasudevan S (2013) Cyclodextrin-functionalized Fe_3O_4@TiO_2: reusable, magnetic nanoparticles for photocatalytic degradation of endocrine-disrupting chemicals in water supplies. ACS Nano 7(5):4093–4104. https://doi.org/10.1021/nn400287k

Chatterjee D, Dasgupta S (2005) Visible light induced photocatalytic degradation of organic pollutants. J Photochem Photobiol C 6(2–3):186–205. https://doi.org/10.1016/j.jphotochemrev.2005.09.001

Chen Y, Crittenden JC, Hackney S, Sutter L, Hand DW (2005) Preparation of a novel TiO_2-based p−n junction nanotube photocatalyst. Environ Sci Technol 39(5):1201–1208. https://doi.org/10.1021/es049252g

Chokshi NP, Ruparelia JP (2015) Photocatalytic ozonation for treatment of wastewater. In: 2nd international conference on Multidisciplinary Research and Practice, Ahmedabad, Gujarat, India, 2015

Das R, Vecitis CD, Schulze A, Cao B, Ismail AF, Lu X, Chen J, Ramakrishna S (2017) Recent advances in nanomaterials for water protection and monitoring. Chem Soc Rev 46 (22):6946–7020. https://doi.org/10.1039/C6CS00921B

Deng Y, Zhao R (2015) Advanced oxidation processes (AOPs) in wastewater treatment. Curr Pollut Rep 1(3):167–176. https://doi.org/10.1007/s40726-015-0015-z

Dialynas E, Mantzavinos D, Diamadopoulos E (2008) Advanced treatment of the reverse osmosis concentrate produced during reclamation of municipal wastewater. Water Res 42 (18):4603–4608. https://doi.org/10.1016/j.watres.2008.08.008

Dong H, Zeng G, Tang L, Fan C, Zhang C, He X, He Y (2015) An overview on limitations of TiO_2-based particles for photocatalytic degradation of organic pollutants and the corresponding countermeasures. Water Res 79:128–146. https://doi.org/10.1016/j.watres.2015.04.038

El Hajjouji H, Barje F, Pinelli E, Bailly J-R, Richard C, Winterton P, Revel J-C, Hafidi M (2008) Photochemical UV/TiO_2 treatment of olive mill wastewater (OMW). Bioresour Technol 99 (15):7264–7269. https://doi.org/10.1016/j.biortech.2007.12.054

Emami F, Tehrani-Bagha AR, Gharanjig K, Menger FM (2010) Kinetic study of the factors controlling Fenton-promoted destruction of a non-biodegradable dye. Desalination 257 (1–3):124–128. https://doi.org/10.1016/j.desal.2010.02.035

Exner O (1988) Correlation analysis of chemical data. Springer, New York

Fan X, Tao Y, Wang L, Zhang X, Lei Y, Wang Z, Noguchi H (2014) Performance of an integrated process combining ozonation with ceramic membrane ultra-filtration for advanced treatment of drinking water. Desalination 335(1):47–54. https://doi.org/10.1016/j.desal.2013.12.014

Fenton HJH (1894) LXXIII.—oxidation of tartaric acid in presence of iron. J Chem Soc Trans 65:899–910. https://doi.org/10.1039/CT8946500899

Fujishima A, Zhang X, Tryk DA (2008) TiO_2 photocatalysis and related surface phenomena. Surf Sci Rep 63(12):515–582. https://doi.org/10.1016/j.surfrep.2008.10.001

Ge M, Cao C, Huang J, Li S, Chen Z, Zhang K-Q, Al-Deyab SS, Lai Y (2016) A review of one-dimensional TiO_2 nanostructured materials for environmental and energy applications. J Mater Chem A 4(18):6772–6801. https://doi.org/10.1039/C5TA09323F

Glaze WH, Kang J-W, Chapin DH (1987) The chemistry of water treatment processes involving ozone, hydrogen peroxide and ultraviolet radiation. Ozone Sci Eng 9(4):335–352. https://doi.org/10.1080/01919518708552148

Gupta A, Saurav JR, Bhattacharya S (2015) Solar light based degradation of organic pollutants using ZnO nanobrushes for water filtration. RSC Adv 5(87):71472–71481. https://doi.org/10.1039/C5RA10456D

Haber F, Weiss J (1934) The catalytic decomposition of hydrogen peroxide by iron salts. Proc R Soc A 147(861):332–351. https://doi.org/10.1098/rspa.1934.0221

Hassaan MA, El Nemr A (2017) Advanced oxidation processes for textile wastewater treatment. Int J Photochem Photobiol 1(1):27–35. https://doi.org/10.11648/j.ijpp.20170101.15

Huang W (2012) Homogeneous and heterogeneous Fenton and photo-Fenton processes: impact of iron complexing agent ethylenediamine-N,N′-disuccinic acid (EDDS). Dissertation, Université Blaise Pascal – Clermont-Ferrand II

Jiang D, Xiao P, Shao L, Li D, Chen M (2017) RGO-promoted all-solid-state g-C_3N_4/$BiVO_4$ Z-scheme heterostructure with enhanced photocatalytic activity toward the degradation of antibiotics. Ind Eng Chem Res 56(31):8823–8832. https://doi.org/10.1021/acs.iecr.7b01840

Jin Y-z, Zhang Y-f, Li W (2003) Micro-electrolysis technology for industrial wastewater treatment. J Environ Sci 15(3):334–338

Krishnan S, Rawindran H, Sinnathambi CM, Lim JW (2017) Comparison of various advanced oxidation processes used in remediation of industrial wastewater laden with recalcitrant pollutants. IOP Conf Ser Mater Sci Eng 206:012089. https://doi.org/10.1088/1757-899X/206/1/012089

Kumar A, Pandey G (2017) A review on the factors affecting the photocatalytic degradation of hazardous materials. Mater Sci Eng Int J 1(3):106–114. https://doi.org/10.15406/mseij.2017.01.00018

Lee KY, Lee J-Y (2005) Photochemical destruction of tetrachloroethylene and trichloroethylene from the exhaust of an air stripper. J Environ Eng 131(10):1441–1446. https://doi.org/10.1061/(ASCE)0733-9372(2005)131:10(1441)

Legrini O, Oliveros E, Braun AM (1993) Photochemical processes for water treatment. Chem Rev 93(2):671–698. https://doi.org/10.1021/cr00018a003

Li W-J, Li Y, Ning D, Liu Q, Chang L, Ruan WJ (2018) An Fe(ii) metal–organic framework as a visible responsive photo-Fenton catalyst for the degradation of organophosphates. New J Chem 42(1):29–33. https://doi.org/10.1039/C7NJ03815A

Liang X, Lu Y, Li Z, Yang C, Niu C, Su X (2017) Bentonite/carbon composite as highly recyclable adsorbents for alkaline wastewater treatment and organic dye removal. Microporous Mesoporous Mater 241:107–114. https://doi.org/10.1016/j.micromeso.2016.12.016

Liu H, Chen Z, Zhang L, Zhu D, Zhang Q, Luo Y, Shao X (2018) Graphene grown on anatase–TiO_2 nanosheets: enhanced photocatalytic activity on basis of a well-controlled interface. J Phys Chem C 122(11):6388–6396. https://doi.org/10.1021/acs.jpcc.7b12305

Mishra NS, Reddy R, Kuila A, Rani A, Mukherjee P, Nawaz A, Pichiah S (2017) A review on advanced oxidation processes for effective water treatment. Curr World Environ 12(3):470–490. https://doi.org/10.12944/CWE.12.3.02

Mitrović J, Radović M, Bojić D, Anđelković T, Purenović M, Bojić A (2012) Decolorization of textile azo dye reactive Orange 16 by the UV/H_2O_2 process. J Serb Chem Soc 77(4):465–481. https://doi.org/10.2298/JSC110216187M

Muruganandham M, Suri RPS, Jafari S, Sillanpää M, Lee G-J, Wu JJ, Swaminathan M (2014) Recent developments in homogeneous advanced oxidation processes for water and wastewater treatment. Int J Photoenergy 2014:821674. https://doi.org/10.1155/2014/821674

Nabid MR, Sedghi R, Behbahani M, Arvan B, Heravi MM, Oskooie HA (2014) Application of poly 1,8-diaminonaphthalene/multiwalled carbon nanotubes-COOH hybrid material as an efficient

sorbent for trace determination of cadmium and lead ions in water samples. J Mol Recognit 27 (7):421–428. https://doi.org/10.1002/jmr.2361
Nguyen XS, Ngo KD (2017) The role of metal-doped into magnetite catalysts for the photo-Fenton degradation of organic pollutants. J Surf Eng Mater Adv Technol 8(1):1–14. https://doi.org/10.4236/jsemat.2018.81001
Niaounakis M, Halvadakis CP (2006) Olive processing waste management. Literature review and patent survey, Waste management series, vol 5, 2nd edn. Elsevier, Burlington
Perera SD, Mariano RG, Vu K, Nour N, Seitz O, Chabal Y, Balkus KJ (2012) Hydrothermal synthesis of graphene-TiO_2 nanotube composites with enhanced photocatalytic activity. ACS Catal 2(6):949–956. https://doi.org/10.1021/cs200621c
Raghu S, Basha CA (2007) Chemical or electrochemical techniques, followed by ion exchange, for recycle of textile dye wastewater. J Hazard Mater 149(2):324–330. https://doi.org/10.1016/j.jhazmat.2007.03.087
Rivas FJ, Kolaczkowski ST, Beltrán FJ, McLurgh DB (1998) Development of a model for the wet air oxidation of phenol based on a free radical mechanism. Chem Eng Sci 53(14):2575–2586. https://doi.org/10.1016/S0009-2509(98)00060-8
Sahoo C, Gupta AK, Pal A (2005) Photocatalytic degradation of Methyl Red dye in aqueous solutions under UV irradiation using Ag^+ doped TiO_2. Desalination 181(1–3):91–100. https://doi.org/10.1016/j.desal.2005.02.014
Salamat S, Younesi H, Bahramifar N (2017) Synthesis of magnetic core–shell Fe_3O_4@TiO_2 nanoparticles from electric arc furnace dust for photocatalytic degradation of steel mill wastewater. RSC Adv 7(31):19391–19405. https://doi.org/10.1039/C7RA01238A
Shahrezaei F, Mansouri Y, Zinatizadeh AAL, Akhbari A (2012) Photocatalytic degradation of aniline using TiO_2 nanoparticles in a vertical circulating photocatalytic reactor. Int J Photoenergy 2012:430638. https://doi.org/10.1155/2012/430638
Sharma RK, Celin SM (2005) Photodegradation of organic pollutants. In: Livingston JV (ed) Trends in water pollution research. Nova Science Publishers Inc, New York, pp 221–236
Sörensen M, Zegenhagen F, Weckenmann J (2015) State of the art wastewater treatment in pharmaceutical and chemical industry by advanced oxidation. Pharm Ind 77(4):594–607
Stasinakis AS (2008) Use of selected advanced oxidation processes (AOPs) for wastewater treatment – a mini review. Global NEST J 10(3):376–385. https://doi.org/10.30955/gnj.000598
Stettler R, Courbat R, von Gunten U, Kaiser H-P, Walther JL, Gaille P, Jordan R, Ramseier S, Revelly P (1998) Utilisation de l'ozone pour le traitement des eaux potables en Suisse. GWA 78 (11):876–890
Sun H, Yang X, Zhao L, Xu T, Lian J (2016) One-pot hydrothermal synthesis of octahedral CoFe/$CoFe_2O_4$ submicron composite as heterogeneous catalysts with enhanced peroxymonosulfate activity. J Mater Chem A 4(24):9455–9465. https://doi.org/10.1039/C6TA02126C
Tang J, Wang J (2018) Metal organic framework with coordinatively unsaturated sites as efficient Fenton-like catalyst for enhanced degradation of sulfamethazine. Environ Sci Technol 52 (9):5367–5377. https://doi.org/10.1021/acs.est.8b00092
Thiruvenkatachari R, Vigneswaran S, Moon IS (2008) A review on UV/TiO_2 photocatalytic oxidation process (Journal Review). Korean J Chem Eng 25(1):64–72. https://doi.org/10.1007/s11814-008-0011-8
UMEX GmbH. Photooxidation. https://www.umex-gmbh.de/en/products/abox-uv-technologie/photooxidation/. Accessed 13/07/2019
von Gunten U (2003) Ozonation of drinking water: Part I. oxidation kinetics and product formation. Water Res 37(7):1443–1467. https://doi.org/10.1016/S0043-1354(02)00457-8
Walling C, Kato S (1971) Oxidation of alcohols by Fenton's reagent. Effect of copper ion. J Am Chem Soc 93(17):4275–4281. https://doi.org/10.1021/ja00746a031
Xia Q, Yu X, Zhao H, Wang S, Wang H, Guo Z, Xing H (2017) Syntheses of novel lanthanide metal–organic frameworks for highly efficient visible-light-driven dye degradation. Cryst Growth Des 17(8):4189–4195. https://doi.org/10.1021/acs.cgd.7b00504

Xu Y, Wu S, Li X, Huang Y, Wang Z, Han Y, Wu J, Meng H, Zhang X (2017) Synthesis, characterization, and photocatalytic degradation properties of $ZnO/ZnFe_2O_4$ magnetic heterostructures. New J Chem 41(24):15433–15438. https://doi.org/10.1039/C7NJ03373G

Yan M, Zhu F, Gu W, Sun L, Shi W, Hua Y (2016) Construction of nitrogen-doped graphene quantum dots-$BiVO_4$/g-C_3N_4 Z-scheme photocatalyst and enhanced photocatalytic degradation of antibiotics under visible light. RSC Adv 6(66):61162–61174. https://doi.org/10.1039/C6RA07589D

Yan X, Chai L, Li Q, Ye L, Yang B, Wang Q (2017) Abiological granular sludge formation benefit for heavy metal wastewater treatment using sulfide precipitation. Clean Soil Air Water 45 (4):1500730. https://doi.org/10.1002/clen.201500730

Zhang J, Zhuang J, Gao L, Zhang Y, Gu N, Feng J, Yang D, Zhu J, Yan X (2008) Decomposing phenol by the hidden talent of ferromagnetic nanoparticles. Chemosphere 73(9):1524–1528. https://doi.org/10.1016/j.chemosphere.2008.05.050

Zhang Z, Wang M, Cui W, Sui H (2017) Synthesis and characterization of a core–shell $BiVO_4$@g-C_3N_4 photo-catalyst with enhanced photocatalytic activity under visible light irradiation. RSC Adv 7(14):8167–8177. https://doi.org/10.1039/C6RA27766G

Zheng T, Zhang T, Wang Q, Tian Y, Shi Z, Smale N, Xu B (2015) Advanced treatment of acrylic fiber manufacturing wastewater with a combined microbubble-ozonation/ultraviolet irradiation process. RSC Adv 5(95):77601–77609. https://doi.org/10.1039/C5RA14575A

Chapter 11
The Use of Nanomaterials in Electro-Fenton and Photoelectro-Fenton Processes

Ignasi Sirés and Enric Brillas

Abstract Nowadays, electro-Fenton and photoelectro-Fenton are considered two of the best electrochemical advanced oxidation processes to ensure fast degradation of organic pollutants and inactivation of toxic pathogens. Despite their apparent simplicity, two critical factors must be carefully controlled in order to enhance the production of $^{\bullet}OH$ from homogeneous Fenton's reaction: the H_2O_2 electrogeneration from cathodic O_2 reduction, and the characteristics of metal species employed to catalyze the quick decomposition of H_2O_2. In conventional electrochemical Fenton-based systems, the cathodes are made of either raw carbon powder or massive carbon pieces since these materials are known to promote the two-electron reduction of O_2 to yield H_2O_2, whereas Fe(II) or Fe(III) salts are added to allow homogeneous catalysis at optimum pH ~3. On the basis of this concept, good progress has been made when working with nanomaterials because the increased specific surface area provides: (i) more active cathode surfaces to produce H_2O_2; (ii) better heterogeneous catalysts, which may react to a larger extent with H_2O_2, within a wider pH range; (iii) a larger absorption of photons, promoting photoreduction and photooxidation reactions; and (iv) a larger area to adsorb pollutants that can be further degraded. As will be discussed in this chapter, most of the work has addressed the use of pristine, functionalized, and decorated nanocarbons as cathodes; the synthesis of nanocatalysts containing iron since this is the best metal for Fenton's reaction; and the preparation of nanostructured anodes.

Keywords Carbon felt · Carbon nanotubes · Electrochemical reactor · Gas diffusion · Graphene · Heterogeneous catalysis · Homogeneous catalysis · Hydrogen peroxide · Electro-Fenton · Nanomaterials · Photocatalysis · Photoelectro-Fenton · Surface functionalization · Water treatment

---

I. Sirés (✉) · E. Brillas
Laboratori d'Electroquímica dels Materials i del Medi Ambient, Departament de Química Física, Facultat de Química, Universitat de Barcelona, Barcelona, Spain
e-mail: i.sires@ub.edu

J. Filip et al. (eds.), *Advanced Nano-Bio Technologies for Water and Soil Treatment*, Applied Environmental Science and Engineering for a Sustainable Future,
https://doi.org/10.1007/978-3-030-29840-1_11

11.1 Introduction

The growing imbalance between water demand and water resources throughout the world, along with the scarcity of fresh water to sustain all human activities, has stimulated the development of a large variety of physical, chemical, photochemical, and electrochemical methods that show great efficiency in treating urban and industrial wastewater, aiming at its further reuse. The electro-Fenton (EF) process, initially called electrogenerated Fenton's reagent, was first used for wastewater treatment in the mid '80s (Brillas et al. 2009). EF is considered an electrochemical advanced oxidation process (EAOP) since it consists in the generation of homogeneous hydroxyl radical ($^{\bullet}OH$). This is achieved upon the occurrence of Fenton's reaction between Fe^{2+} and electrogenerated H_2O_2 (Brillas et al. 2009; Oturan and Aaron 2014; Sirés et al. 2014). $^{\bullet}OH$ is the second strongest oxidizing species known after fluorine, showing a large ability to mineralize most organic pollutants in water. The most characteristic feature of EF is that H_2O_2 is continuously dosed to the reaction medium from the two-electron reduction of O_2. This gas can be either directly fed as pure O_2 or air into the wastewater or pumped through a gas-diffusion device, further accepting the electrons typically supplied to a carbonaceous cathode. Thanks to the production of H_2O_2 on site, expensive and dangerous steps like industrial synthesis, transportation, storage, and handling can be avoided. Conventional homogeneous EF involves the addition of Fe^{2+} to the wastewater, although Fe^{3+} can be used alternatively because it is reduced to the former ion at the cathode surface (Oturan and Aaron 2014; Martínez-Huitle et al. 2015). The continuous Fe^{2+} and H_2O_2 (re)generation represents a significant advantage over the classical chemical Fenton process. In addition, the use of an undivided cell can accelerate the decontamination because of the production of heterogeneous M($^{\bullet}OH$) from water discharge at the surface of the anode M. On the other hand, one important drawback of EF when applied to organics removal is the production of Fe(III)-carboxylate complexes as final byproducts, which are quite refractory to $^{\bullet}OH$ attack. This phenomenon also occurs in the dark Fenton. To overcome this problem, the photoelectro-Fenton (PEF) process was developed in the mid '90s (Brillas et al. 2009). It consists in the simultaneous irradiation of the solution with artificial UV light or, more recently, sunlight, thus promoting the photolysis of intermediates like the Fe(III)-carboxylate species. As a result, a larger mineralization can be usually attained. The application of EF and PEF, along with the related processes, to the treatment of synthetic and real wastewater has been discussed in several authoritative reviews (Brillas et al. 2009; Oturan and Aaron 2014; Sirés et al. 2014; Martínez-Huitle et al. 2015; Moreira et al. 2017). Recently, the fundamentals, reactions, and applications of these EAOPs have been summarized in a comprehensive book edited by Zhou et al. (2018).

Other drawbacks of homogeneous EF and PEF include the limited pH range, i.e., ~3, to avoid the catalyst loss by precipitation, and the increase of soluble iron content in the treated effluent (Martínez-Huitle et al. 2015). Over the last decade, many efforts have been made to overcome such limitations, thus re-enforcing the EF and

PEF viability. Among the most valuable, nanoengineering has allowed the development of new nanomaterials as alternative cathodes, anodes, and catalysts. This chapter reviews the characteristics and applications of these materials, mainly prepared as: (i) cathodes to enhance the H_2O_2 electrogeneration and/or to promote heterogeneous Fenton's reaction, (ii) anodes to favor the electrocatalytic and photoelectrocatalytic production of $^{\bullet}OH$, and (iii) Fe-based catalysts to extend the pH range. Nanomaterials for hybrid treatments involving EF or PEF in combination with adsorption are described as well.

11.2 Nanomaterials as Cathodes

Many nanomaterials have been utilized to enhance the H_2O_2 production at the cathode of an electrolytic cell from the two-electron reduction of O_2 gas by reaction in Eq. 11.1. In some cases, they can simultaneously act as catalysts that favor the heterogeneous Fenton's reaction, mimicking the conventional Fenton's reaction (Eq. 11.2) with $^{\bullet}OH$ generation (Brillas et al. 2009; Oturan and Aaron 2014; Sirés et al. 2014; Martínez-Huitle et al. 2015; Moreira et al. 2017). The use of these modified cathodes is described in this section.

$$O_{2(g)} + 2H^+ + 2e^- \rightarrow H_2O_2 \tag{11.1}$$

$$Fe^{2+} + H_2O_2 \rightarrow Fe^{3+} + {}^{\bullet}OH + OH^- \tag{11.2}$$

11.2.1 Carbon-Based Nanomaterials

Carbonaceous materials are appealing targets to be used as catalysts for H_2O_2 electrogeneration since they are abundant, cheap, durable, and show good faradaic efficiency in the reaction in Eq. 11.1 (Čolić et al. 2018). A recent theoretical study by Chai et al. (2017) has reported that the reduction of O_2 molecules takes place by approaching the hydrogen sites on the carbon surface to form the hydroperoxide ion (HO_2^-), whose subsequent protonation yields H_2O_2. The two-electron oxygen reduction reaction (ORR) is then very effective in acidic media, which agrees with the optimum pH of conventional Fenton's reaction (Eq. 11.2), whereas in alkaline medium, the four-electron ORR to form H_2O prevails. The use of carbon-based nanomaterials allows increasing the electroactive surface area of the cathode, and consequently higher O_2 mass transport rates are achieved.

Carbon-based nanomaterials may exhibit outstanding electronic, photonic, electrocatalytic, chemical, and mechanical properties, depending on their nanoscale structure. Two main groups of materials can be distinguished: nanosized and nanostructured carbons (Khataee and Hasanzadeh 2017). The former is

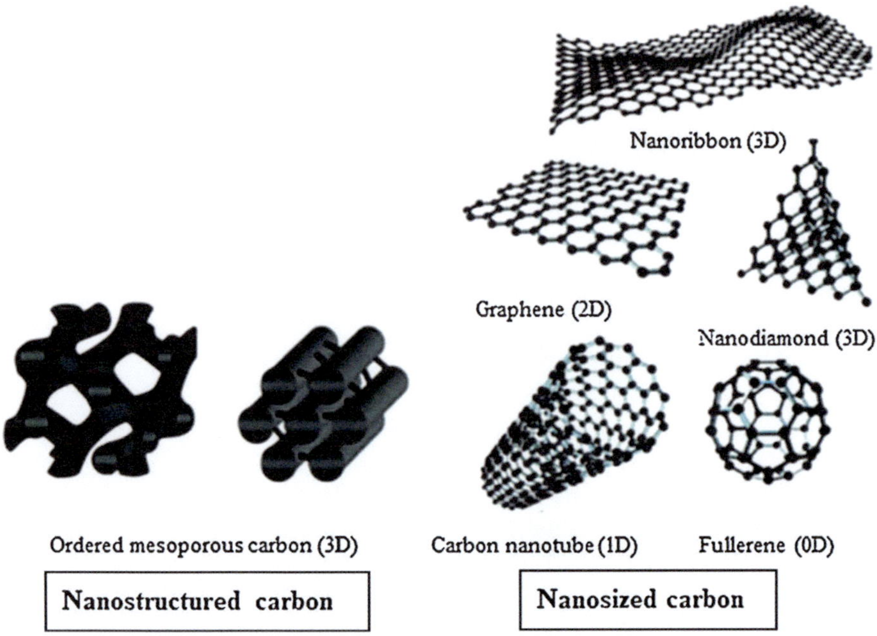

Fig. 11.1 Schematic illustration of some carbon-based nanomaterials. (Adapted from Khataee and Hasanzadeh (2017), Copyright 2017, with permission from Springer Nature)

characterized by nanometric shell size and thickness; and among others includes carbon nanotubes (CNTs), graphene (Gr) and its derivatives (graphene oxide (GO) and reduced graphene oxide (rGO)), nanofibers, nanodiamonds, nanocoils, nanoribbon, and fullerene. The latter group is composed of ordered mesoporous carbons and carbon fibers, which are constructed through various methodologies. Some examples of nanocarbons are shown in Fig. 11.1.

CNTs are ubiquitous carbon-based materials that allow the electrocatalysis of the reaction in Eq. 11.1. They have a hexagonal 1D structure with sp^2 hybridization (see Fig. 11.1), being formed upon the rolling up of a graphite layer that yields a nanoscale tube form (about 1 nm in diameter). The occurrence of two or more coaxial CNTs, with expanding diameters and tube separation of around 0.34 nm, gives rise to multi-walled CNTs (MWCNTs). CNTs sponge (Wang et al. 2014), classical CNTs (Khataee and Hasanzadeh 2017; Tian et al. 2016a), or MWCNTs (Roth et al. 2016) deposited onto gas-diffusion electrodes (GDEs), and graphite mixed with MWCNTs (Chu et al. 2013; Pajootan et al. 2014; Babaei-Sati and Parsa 2017) have been utilized to enhance the H_2O_2 generation, hence the removal of several organics by EF. For example, the Khataee's group reported that the amount of H_2O_2 produced using CNTs-based GDE was up to three-fold higher than using conventional GDE with activated carbon (Khataee and Hasanzadeh 2017). Wang et al. (2014) degraded 120 mL of 50 mg/L dimethyl phthalate in 0.1 M Na_2SO_4 with 0.5 mM Fe^{2+} at pH 3.0 by EF using an undivided cell with a Pt anode and either a

4 cm^2 CNTs sponge or graphite-based GDE as the cathode, at a cathodic potential (E_{cath}) of −0.50 V/saturated calomel electrode (SCE). After 120 min of electrolysis, the pollutant removal was 96% with 75% of total organic carbon (TOC) abatement using the sponge. These values are much higher than the 48% and 35% obtained for graphite-based GDE, respectively. A superior degradation performance upon the incorporation of CNTs in the GDE has also been described during the EF treatment of the dye Rhodamine B (Tian et al. 2016a). Excellent removal has also been reported for Acid Red 14 with a MWCNTs-based GDE (Roth et al. 2016) and for *m*-cresol (Chu et al. 2013), Basic Blue 41 (Pajootan et al. 2014), and Acid Red 14 and Acid Blue 92 (Babaei-Sati and Parsa 2017) with graphite-MWCNTs.

Graphene is composed of a 2D planar sheet with monoatomic thickness, made of sp^2 carbon atoms densely organized into a honeycomb structure (see Fig. 11.1). As in the case with CNTs, multilayers are also feasible. Graphene has been applied to the EF process as a pristine nanomaterial (Mousset et al. 2016b; Chen et al. 2016a), coated onto carbon felt (Le et al. 2015, 2017; Yang et al. 2017), carbon cloth (Mousset et al. 2016a), carbon-cloth GDEs (Garcia-Rodriguez et al. 2018), or carbon fiber (Mousset et al. 2017a), and mixed with graphite and PTFE (Zhang et al. 2018). Table 11.1 collects the main results obtained for several target organic pollutants using these cathodes. All the trials were carried out at bench scale using undivided cells at the optimum pH for Fenton's reaction (Eq. 11.2). Three-electrode cells with a constant E_{cath} between the cathode and the reference electrode (SCE) or two-electrode cells with constant current (I) or current density (j) were utilized.

In most cases, O_2 was bubbled into the solutions to ensure their saturation, aiming at enhancing the largest H_2O_2 production by the reaction in Eq. 11.1. In setups equipped with carbon cloth GDE (Garcia-Rodriguez et al. 2018), O_2 was pumped through the dry surface to produce H_2O_2 at the wet surface, which is in contact with the solution. Table 11.1 shows that very short electrolysis time was needed to degrade the model pollutants, whereas much longer time (up to 480 min) was necessary for achieving significant mineralization degrees because of the greater recalcitrance of byproducts to the attack of $^{\bullet}OH$. In general, worse results were found with raw cathodes without Gr. Figure 11.2 schematizes the preparation of a Gr/carbon-cloth cathode, involving the electrochemical exfoliation of a graphite rod to obtain graphene, which was mixed with Nafion® (PTFE suspension) and then ultrasonicated to obtain the graphene ink to coat the carbon cloth. Figure 11.2 also highlights the greater H_2O_2 electrogeneration with the Gr/carbon-cloth cathode compared with the uncoated cloth. This led to the greatest degradation (92%) and mineralization (57%) degrees for a 1.4 mM phenol solution using the coated cathode (see Table 11.1), only reaching 72% and 41% in the absence of Gr. Therefore, the larger H_2O_2 production favored a greater accumulation of $^{\bullet}OH$, thus resulting in the faster oxidation of organics.

Table 11.1 Selected results obtained for the EF treatment of several organic pollutants using undivided cells with cathodes based on graphene (Gr)

Cathode	Substrate	Experimental remarks	Best performance	Ref.
Gr	Phenol	150 mL of 1 mM substrate in 0.05 M K_2SO_4, 0.1 mM Fe^{2+}, pH 3.0, Pt anode, $E_{cath} = -0.60$ V/SCE	75% substrate decay (180 min), 49% TOC reduction (480 min)	Mousset et al. (2016b)
	Methylene Blue	100 mL of 11 mg/L dye in 0.1 mM Na_2SO_4, 11.1 mM Fe^{2+}, pH 3.0, Pt anode, $E_{cath} = -1.0$ V/SCE	100% color removal (160 min)	Chen et al. (2016a)
Gr/carbon felt	Acid Orange 7	30 mL of 0.1 mM dye in 0.05 M Na_2SO_4, 0.2 mM Fe^{2+}, pH 3.0, Pt anode, $I = 40$ mA	94% TOC removal (480 min)	Le et al. (2015)
	Orange II	100 mL of 50 mg/L dye in 0.05 M Na_2SO_4, 0.4 mM Fe^{2+}, pH 3.0, DSA® anode, $E_{cath} = -0.90$ V/SCE[e]	100% color and 79% TOC decays (60 min)	Yang et al. (2017)
Gr/carbon cloth[a]	Phenol	80 mL of 1.4 mM substrate in 0.05 M K_2SO_4, 0.1 mM Fe^{2+}, pH 3.0, Pt anode, $j = 1.25$ mA/cm^2	92% substrate decay (180 min), 57% TOC reduction (480 min)	Mousset et al. (2016a)
Gr/carbon cloth (GDE)[b]	Electronic wastewater	400 mL of wastewater with 53.5 mg/L TOC, 0.2 mM Fe^{2+}, pH 3.0, BDD anode. $j = 29$ mA/cm^2 [f]	92% TOC decay (180 min)	Garcia-Rodriguez et al. (2018)
Gr/carbon fiber[c]	Phenol	400 mL of 0.33 mM substrate in 0.05 M K_2SO_4, 0.1 mM Fe^{2+}, pH 3.0, Pt anode, $j = 1.25$ mA/cm^2	99% substrate decay (120 min), 98% TOC reduction (360 min)	Mousset et al. (2017a)
Gr@graphite[d]	Rhodamine B	50 mL of 20 mg/L dye in 0.05 M Na_2SO_4, 0.3 M Fe^{2+}, pH 3.0, DSA® anode, $j = 20$ mA/cm^2 [e]	100% color and 79% COD decays (60 min)[g]	Zhang et al. (2018)

Cathode area: [a]15, [b]20, [c]126, and [d]1.5 cm^2. [e]DSA®—dimensionally stable anode. [f]BDD—boron-doped diamond. [g]COD—chemical oxygen demand

11.2.2 Chemically Modified Carbonaceous Nanomaterials

The ability of carbonaceous cathodes to electrogenerate H_2O_2 from the reaction in Eq. 11.1 can be enhanced by chemically modifying their surface. Zhou et al. (2014) modified a graphite felt (GF) with pure ethanol, yielding a material (GF-A) with carbon nanoparticles of average diameter ca. 500 nm on the surface of the filaments. A second kind of modification employing 90:10 (v/v) ethanol/hydrazine hydrate mixtures yielded oxygen- and nitrogen-containing functional groups (GF-B), resulting in weaker hydrophobicity compared with GF and GF-A, thus favoring the H_2O_2 generation. Its accumulation increased in the sequence: GF (67 mg/L) < GF-A (110 mg/L) < GF-B (176 mg/L) upon electrolysis of 130 mL of an O_2-saturated 0.05 M Na_2SO_4 solution at pH 3.0 using a three-electrode cell at

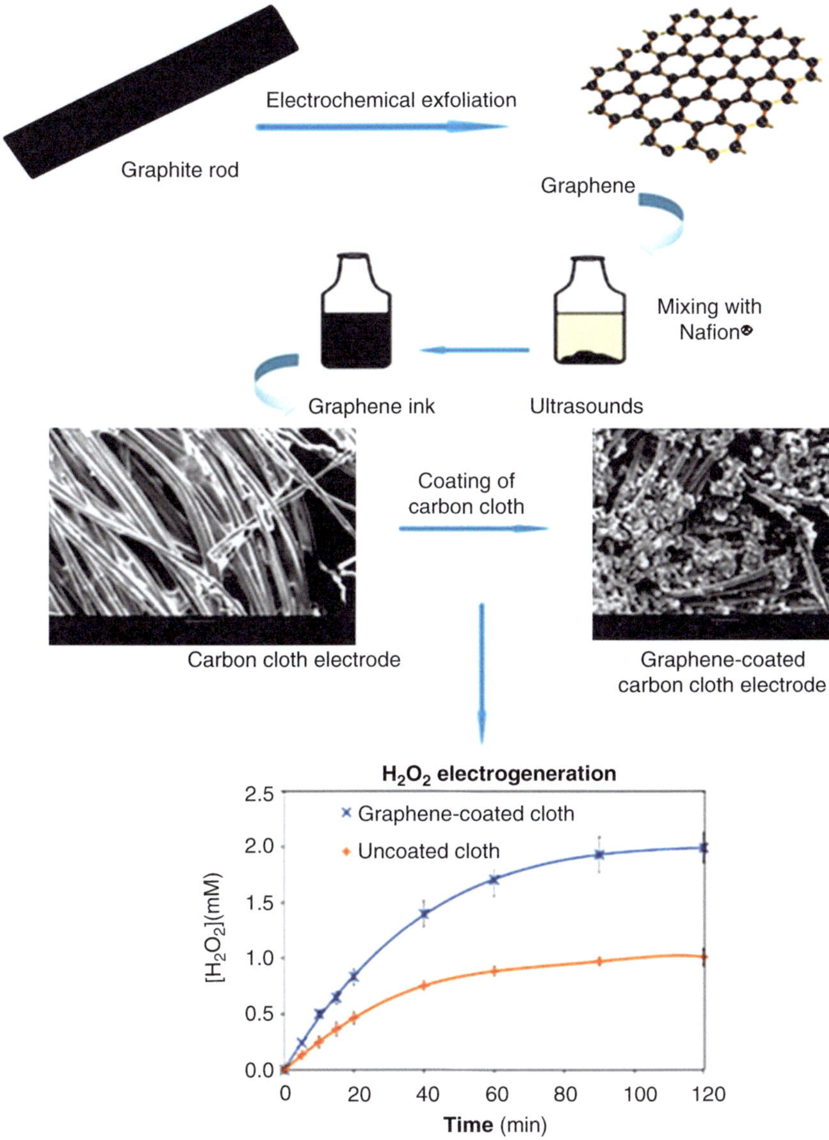

Fig. 11.2 Steps for the preparation of graphene-coated carbon cloth cathode. (Adapted from Khataee and Hasanzadeh (2017), Copyright 2017, with permission from Springer Nature). The graph represents the concentration of H_2O_2 accumulated in 80 mL of 0.05 M K_2SO_4 at pH 3.0 using an undivided cell with a Pt anode and uncoated or graphene-coated carbon cloth cathode, all with 15 cm^2 area, at current density of 1.25 mA/cm^2. (Adapted from Mousset et al. (2016a), Copyright 2016, with permission from Elsevier)

E_{cath} = −0.65 V/SCE for 120 min. The same trend was obtained when the performance of the EF process was assessed using 50 mg/L *p*-nitrophenol under the above conditions using 0.2 mM Fe^{3+} as catalyst, reaching 67%, 83%, and 99% substrate decay at 60 min and 22%, 32%, and 51% mineralization at 120 min with GF, GF-A, and GF-B, respectively. The superiority of GF-B containing N-doped carbon nanoparticles was related to the higher current recorded, which promoted the H_2O_2 production and the reduction of Fe^{3+} to Fe^{2+}, thereby increasing the amount of $^{\bullet}OH$ generated.

N-doping then favors the occurrence of the two-electron ORR to yield H_2O_2 since it increases the chemically active sites, the O_2 chemisorption, and the hydrophilicity of the carbon surface (Khataee and Hasanzadeh 2017). This has been confirmed for the other modified nanocarbons like N-CNTs (Zhang et al. 2008) and an N-Gr@CNTs composite GDE (Liu et al. 2016), which were used to treat 250 mL of Methyl Orange and 100 mL of 50 mg/L dimethyl phthalate solutions in 0.05 M Na_2SO_4 with 0.2–0.5 mM Fe^{2+} at pH 3.0 by EF using three-electrode undivided cells with a Pt anode at E_{cath} of −0.85 and −0.50 V/SCE, respectively. In the former system, 100% decolorization was obtained after 60 min using N-CNTs, a value higher than 82% found with an unmodified CNTs cathode. In the other case, the highest TOC removal of 52% at 180 min with the N-Gr@CNTs GDE was superior to 26%, 12%, 7.4%, and 38% reached with Gr@CNTs GDE, CNTs-based GDE, Gr-based GDE, and graphite-based GDE as the cathode, respectively. However, a contradictory interpretation of the behavior of the N-doped carbonaceous cathodes has been proposed in the recent work of Yang et al. (2018b). These authors prepared GF cathodes coated with Gr and N-Gr. Figure 11.3a and b evidence that the N-Gr particles deposited onto the graphite fibers were more uniform than the Gr ones. The degradation of 100 mL of 50 mg/L phenol in 0.05 M Na_2SO_4 at pH 3.0 with a DSA® anode at E_{cath} = −0.90 V/SCE revealed the removal of 99% and 78% phenol using GF with N-Gr and Gr within 50 min, respectively. This was ascribed to the formation of $^{\bullet}OH$ from the H_2O_2 reduction via the in situ metal-free processes illustrated in Fig. 11.3c, which was enhanced with the N-Gr-coated GF, as detected by ESR spectroscopy. Figure 11.3d shows phenol removal efficiency of 99% using the N-Gr-GF cathode at pH 3.0, quite similar to the 97% obtained using EF with Gr and 0.40 mM Fe^{2+}. At pH 7.0, phenol disappearance slowed down under the latter conditions, attaining 41%. When using N-Gr-GF without an iron catalyst, removal efficiency of up to 92% was achieved. These surprising results should be confirmed in future works, because such a large formation of $^{\bullet}OH$ from H_2O_2 reduction at GF coated with N-Gr is difficult to justify considering the state-of-the-art of EAOPs.

On the other hand, Quan and coauthors have recently developed hierarchically porous carbon (HPC) as a novel structure with a high electrocatalytic activity for the ORR to form H_2O_2. These materials are derived from the carbonization of metal–organic frameworks (MOFs), giving rise to abundant micro-, meso-, or even macropores with transference of C-sp^2 to C-sp^3 that can act as active sites for O_2 adsorption, eventually improving the kinetics of ORR (Liu et al. 2015b). F-doped HPC cathodes have been reported as good electrocatalysts for H_2O_2 generation (Zhao et al. 2018b). Liu et al. (2015a) prepared an HPC electrode from Zn^{2+} and

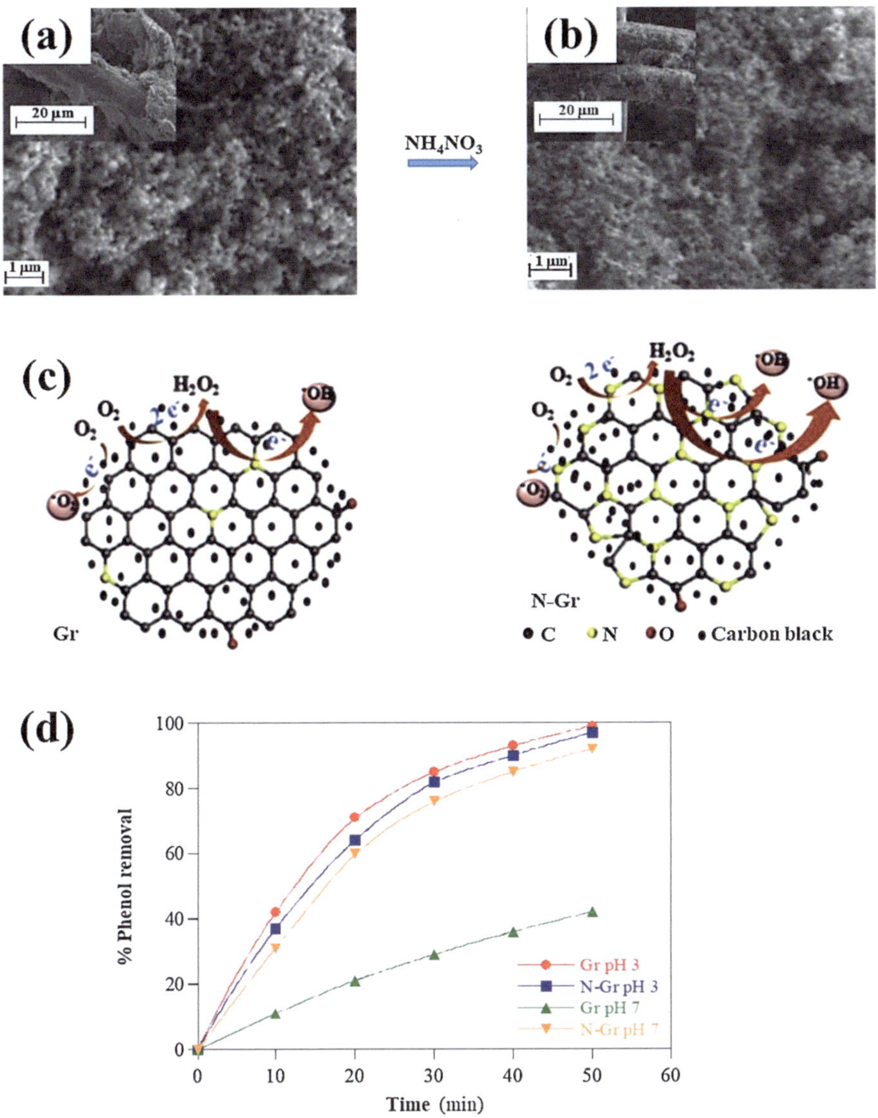

Fig. 11.3 SEM images of raw graphite-felt cathode modified with (**a**) graphene (Gr) and (**b**) N-doped graphene (N-Gr). (**c**) Schematic illustration of the in situ metal-free reactions occurring in the above cathodes. (**d**) Percentage of phenol removal during the treatment of 100 mL of 50 mg/L phenol in 0.05 M Na_2SO_4 using a cell with a DSA® anode and each of the above cathodes at $E_{cath} = -0.90$ V/SCE. The EF trials with Gr were made using 0.40 mM Fe^{2+}. No Fe^{2+} was added using N-Gr. (Adapted from Yang et al. (2018b), Copyright 2018, with permission from Elsevier)

1,4-benzenedicarboxylic acid as precursors, carrying out the carbonization under H_2 atmosphere. This electrode was used as the cathode of a three-electrode undivided cell with a Pt anode, yielding 97% of substrate decay and 91% TOC abatement after

180 and 240 min, respectively, upon EF treatment of 50 mg/L of perfluorooctanoate in 0.05 M Na_2SO_4 with 1.0 mM Fe^{2+} at pH 2.0 and $E_{cath} = -0.40$ V/SCE.

11.2.3 Non-Ferrous Metal-Modified Carbon Nanomaterials

For a given cathode material, the H_2O_2 production can be maximized by optimizing main experimental parameters like the electrolyte composition, pH, temperature, cell configuration, and applied current or E_{cath} (Yang et al. 2018a). Recently, the modification of carbonaceous cathodes with nonferrous metals has been addressed aiming at improving their electroactivity, i.e., trying to cause the ORR at less negative potential in order to reach higher current values at lower overvoltage. In this scenario, several authors have reported a large H_2O_2 enhancement by using Au-Pd (Pizzutilo et al. 2017) and Pt-Hg (Siahrostami et al. 2013) nanoparticles immobilized onto glassy carbon (GC). However, only few papers have described the use of bimetallic nanomaterials for the EF treatment of wastewater. For example, Félix-Navarro et al. (2013) deposited bimetallic Pt-Pd nanoparticles on MWCNTs, which were sprayed onto a reticulated vitreous carbon (RVC) GDE. This cathode allowed the accumulation of 71 mM H_2O_2 in 20 mL of 0.5 M H_2SO_4 after 20 min at $E_{cath} = -0.50$ V vs. Ag/AgCl. This was much greater than 2.2 mM obtained with a similar cathode without nanoparticles, as a result of the great electroactive area of Pt-Pd nanoparticles. Under these conditions, nitrobenzene was rapidly and completely degraded by EF with 0.1 mM Fe^{2+}.

Other authors have used metallic oxide nanoparticles, such as TaO_2 deposited onto GDE (Carneiro et al. 2016), $WO_{2.72}$ mixed with Vulcan carbon to prepare GDEs (Paz et al. 2018), and $Ce_xA_{1-x}O_2$ (A = Zr, Cu, or Ni) immobilized onto carbon felt (Li et al. 2017), to enhance the electroactivity regarding the H_2O_2 production. The use of $WO_{2.72}$@Vulcan GDE and Pt as the cathode and anode in a three-electrode cell to degrade 350 mL of a 0.260 mM Orange II solution in 0.1 M K_2SO_4 with 0.50 mM Fe^{2+} at pH 3.0 by EF yielded 100% decolorization after 120 min at $E_{cath} = -0.70$ V vs Ag/AgCl. This potential was selected because it led to the maximum H_2O_2 accumulation (i.e., 480 mg/L) in the absence of a catalyst (Paz et al. 2018). The crucial role of $^{\bullet}OH$ formed from Fenton's reaction (Eq. 11.2) was confirmed from the poorer color loss, ca. 20%, achieved under analogous conditions but without Fe^{2+}. In that so-called electrochemical oxidation with electrogenerated H_2O_2 (EO-H_2O_2) process, heterogeneous Pt($^{\bullet}OH$) is formed from water discharge from the reaction in Eq. 11.3 (Brillas et al. 2009; Oturan and Aaron 2014; Sirés et al. 2014; Martínez-Huitle et al. 2015; Moreira et al. 2017). In EF, both $^{\bullet}OH$ and Pt($^{\bullet}OH$) contribute to the dye oxidation, with predominance of the former radical.

$$Pt + H_2O \rightarrow Pt(^{\bullet}OH) + H^+ + e^- \quad (11.3)$$

Co and Co-based materials also possess excellent characteristics for H_2O_2 electrogeneration in EF and PEF. Barros et al. prepared a GDE modified with 5% Co-phthalocyanine for the EF treatment of 400 mL of solutions containing 100 mg/L of the food dyes Amaranth (Barros et al. 2014b) and Tartrazine (Barros et al. 2014a), with 0.1 M K_2SO_4 and 0.15 mM Fe^{2+} at pH 2.5, using a three-electrode undivided cell with a Pt anode at $E_{cath} = -0.70$ V vs. Ag/AgCl. After 90 min, 79% and 98% decolorization, with 67% and 75% TOC removal, related to 370 and 219 kWh per kg TOC removed, were obtained, respectively. Ridruejo et al. (2018) showed the viability of a CoS_2/MWCNTs GDE cathode to degrade 150 mL of 0.112 mM of the anaesthetic tetracaine in 0.05 M Na_2SO_4 at pH 3.0 in a two-electrode cell with a BDD anode at $j = 100$ mA/cm^2. The authors found the complete drug disappearance at 120 and 60 min of EO-H_2O_2 and PEF with a 6 W UVA lamp, respectively, with TOC reductions of 43% and 60% at 180 min. The superiority of PEF was explained by the generation of $^{\bullet}OH$ from Fenton's reaction (Eq. 11.2) and the photoreduction of photoactive $Fe(OH)^{2+}$ via the reaction in Eq. 11.4, along with the photodecomposition of final Fe(III)-carboxylate species ($Fe(OOCR)^{2+}$) via the reaction in Eq. 11.5:

$$Fe(OH)^{2+} + h\nu \rightarrow Fe^{2+} + {}^{\bullet}OH \tag{11.4}$$

$$Fe(OOCR)^{2+} + h\nu \rightarrow Fe^{2+} + CO_2 + R^{\bullet} \tag{11.5}$$

What is also worth highlighting, Liang et al. (2016, 2017) proposed a heterogeneous EF-like treatment to decolorize 100 mL of 50 mg/L of Methyl Orange in 0.05 M Na_2SO_4 at pH 3–9 in a two-electrode undivided cell, with a Ti/IrO_2-RuO_2 anode (DSA®) and a cathode composed of GDE or GF coated with Co, at $j = 5$ mA/cm^2. A large H_2O_2 accumulation (about 595 mg/L) was obtained in the background electrolytes after 120 min of electrolysis, whereas 100% and 86% decolorization of the dye solution was attained after 180 min at pH 3.0 and pH 9.0, respectively, using coatings with 1.0–1.5 wt.% Co. The cathodes presented a great stability, with <4% of loss of color removal after ten cycles. The authors also showed the formation of $^{\bullet}OH$, which was ascribed to the Fenton-like reaction (Eq. 11.6) between Co^{2+} and H_2O_2 at the cathode surface. The Co^{3+} ion formed was further reduced to Co^{2+}.

$$Co^{2+} + H_2O_2 \rightarrow Co^{3+} + {}^{\bullet}OH + OH^- \tag{11.6}$$

11.2.4 Fe-Loaded Carbon Nanomaterials

During the twenty-first century, many research efforts have been made to develop heterogeneous electro-Fenton (hetero-EF) and photoelectro-Fenton (hetero-PEF) processes using carbonaceous materials loaded with nanoparticles of Fe (Li et al. 2011; Bañuelos et al. 2015), Fe-Fe_2O_3 (Li et al. 2009; Ai et al. 2007; Ding et al.

2017), Fe_2O_3 (Wang et al. 2013; Peng et al. 2015; Sklari et al. 2015), FeOOH (Zhang et al. 2012, 2015a), Fe_3O_4 (Chen et al. 2016b; Tian et al. 2016b; Zhang et al. 2017), γ-Fe_2O_3/Fe_3O_4 (Plakas et al. 2016), Fe-Cu (Garrido-Ramírez et al. 2016; Zhao et al. 2016, 2018a), CoFe hydroxide (Ganiyu et al. 2017), and Fe_3S_4/Fe_7S_8 (Choe et al. 2018). In these processes, no soluble iron catalyst is added to the solution since Fenton's reaction (Eq. 11.2) occurs between Fe(II) adhered to the cathode surface and H_2O_2 produced at the same surface. Furthermore, leaching of iron ions from the catalyst can contribute to conventional homogeneous Fenton's reaction. The main aim when using heterogeneous catalysts is to make the application of water treatments without pH regulation feasible, i.e., at natural pH, which is usually circumneutral. These materials minimize the formation of iron hydroxide, with the consequent reduction of costs associated to sludge management. A potential drawback of such catalysts is their gradual solubilization upon the use and the reuse, which limits their lifetime, especially under acidic conditions. Therefore, it is crucial to test their stability in consecutive degradation cycles. Selected degradation and mineralization results for some organic pollutants treated by hetero-EF and hetero-PEF are summarized in Table 11.2. A good performance can be observed for both processes within the pH range 2–7, although it is slightly better at pH 3.0, which is in agreement with the optimum conditions of Fenton's reaction (Eq. 11.2). The data corroborate the viability of the heterogeneous treatments at neutral pH, showing acceptable effectiveness. A clear superiority of hetero-PEF over hetero-EF, leading to faster removal of the target pollutant and TOC, can also be observed under comparable conditions, which can be accounted for by the additional generation of $^{\bullet}OH$ from photoreaction (Eq. 11.4) and the photolysis of intermediates such as final Fe(III)-carboxylate complexes via the reaction in Eq. 11.5.

As an example, Fig. 11.4a shows that magnetite (Fe_3O_4) presents a relatively uniform and spherical shape, with an average diameter of 40–50 nm, whereas Fig. 11.4b confirms its successful loading on the activated carbon acting as GDE (Zhang et al. 2017). Figure 11.4c illustrates the hetero-EF mechanism proposed to destroy tetracycline upon the attack of different reactive oxygen species (ROS) including: (i) the superoxide radical ($O_2^{\bullet-}$) produced from O_2 via the reaction in Eq. 11.7, (ii) H_2O_2 formed from O_2 via the reaction in Eq. 11.1, and (iii) $^{\bullet}OH$ produced from heterogeneous Fenton's reaction (Eq. 11.8), with the subsequent reduction of Fe^{III} to Fe^{II} via the reaction in Eq. 11.9

$$O_{2(g)} + e^- \rightarrow O_2^{\bullet -} \tag{11.7}$$

$$\equiv Fe^{II} + H_2O_2 \rightarrow \equiv Fe^{III} + {}^{\bullet}OH + OH^- \tag{11.8}$$

$$\equiv Fe^{III} + e^- \rightarrow \equiv Fe^{II} \tag{11.9}$$

where $\equiv$ means the Fe_3O_4 surface. On the other hand, Fig. 11.4d schematizes the generation of $^{\bullet}OH$ to attack Acid Orange 7, using a CoFe hydroxide/carbon felt cathode (Ganiyu et al. 2017). $^{\bullet}OH$ is formed homogeneously from the reactions in Eqs. 11.2 and 11.6 in acidic medium, owing to Fe^{2+} and Co^{2+} leaching from the

Table 11.2 Selected results obtained for the hetero-EF and hetero-PEF treatment of several organic pollutants using undivided cells with cathodes based on Fe-loaded carbon nanomaterials

Cathode	Pollutant	Experimental remarks	Best performance	Ref.
Hetero-EF treatment				
Fe/activated carbon	Methyl Orange	25 mL of 10 mg/L dye in 0.05 M Na_2SO_4, pH 3.0, Pt anode, $E_{cath} = -0.70$ V vs. Ag/AgCl	96% dye and 88% TOC decays (30 min)	Bañuelos et al. (2015)
$Fe@Fe_2O_3$/ACF[a]	Atrazine	30 mg/L herbicide in 0.05 M Na_2SO_4, pH 3.0, BDD anode, $I = 30$ mA	100% substrate decay (60 min), 87% TOC decay (240 min)	Ding et al. (2017)
Fe_2O_3/carbon aerogel[b]	Imidacloprid	100 mL of 200 mg/L insecticide in 0.1 M Na_2SO_4, pH 6.9, Pt anode, $j = 10$ mA/cm^2	69% insecticide decay (150 min)	Peng et al. (2015)
FeOOH/AC/GF	Aramanth[d]	300 mL of 80 mg/L dye in 0.2 M Na_2SO_4, pH 4.0, Pt anode, $E_{cath} = -0.64$ V/SCE	99% dye decay (210 min), 53% TOC removal (360 min)	Zhang et al. (2012)
Fe_3O_4/GDE	Tetracycline	100 mL of 50 mg/L drug in 0.05 M N_2SO_4, pH 3.0, Pt anode, $E_{cath} = -0.80$ V/SCE	100% drug, 65% COD, and 57% TOC reductions (120 min)	Zhang et al. (2017)
Fe-cu allophane/GC	Phenol	100 mL of 0.5 mM substrate in 0.05 M Na_2SO_4, pH 3.0 and 5.5, Pt anode, $E_{cath} = -0.60$ V vs. Ag/AgCl	100% substrate decay at 120 min (pH 3.0) and 240 min (pH 5.5)	Garrido-Ramírez et al. (2016)
CoFe hydroxide/carbon felt[c]	Acid Orange 7	300 mL of 0.1 mM dye in 0.05 M Na_2SO_4, pH 2.0–7.1, Pt anode, $j = 4.2$ mA/cm^2	100% color decay (10 min), 87% TOC dye (120 min) at pH 3.0	Ganiyu et al. (2017)
Hetero-PEF treatment				
Fe/activated carbon	Methyl Orange	25 mL of 10 mg/L dye in 0.05 M Na_2SO_4, pH 3.0, Pt anode, 75 mW/ cm^2 UVA, $E_{cath} = -0.70$ V vs. Ag/AgCl	100% dye and 98% TOC decays (30 min)	Bañuelos et al. (2015)
Fe_2O_3/carbon aerogel[b]	Imidacloprid	100 mL of 200 mg/L insecticide in 0.1 M Na_2SO_4, pH 6.9, Pt anode, 500 W Xe lamp, $j = 10$ mA/cm^2	95% substrate removal (150 min), 98% TOC decay (480 min)	Peng et al. (2015)

[a]ACF—activated carbon fiber. Cathode area: [b]3, and [c]4.5 cm^2. [d]Divided cell with 300 mL of catholyte and 100 mL of anolyte

cathode. In contrast, this radical originates from the heterogeneous reactions in Eqs. 11.10 and 11.11, with regeneration of the Fe^{II}/Co^{II}-OH catalyst and no metal release, at neutral pH.

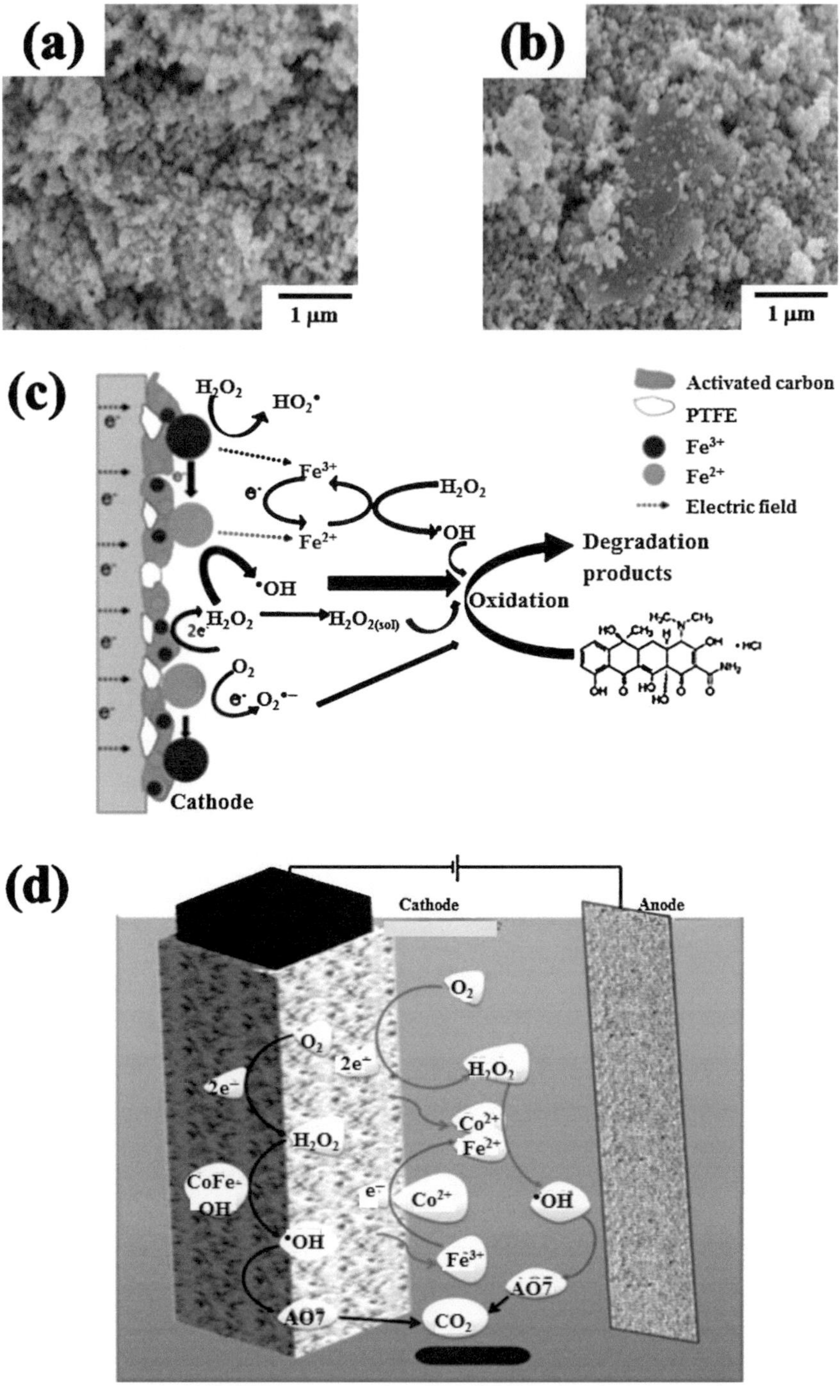

Fig. 11.4 SEM images of (**a**) pure Fe_3O_4 and (**b**) fresh Fe_3O_4/GDE. (**c**) Schematic illustration of the hetero-EF mechanism with the latter cathode. (Adapted from Zhang et al. (2017), Copyright 2017, with permission from Elsevier). (**d**) Schematic illustration of Acid Orange 7 (AO7) degradation in the hetero-EF system with a Pt anode and a CoFe–LDH/CF cathode). (Adapted from Ganiyu et al. (2017), Copyright 2017, with permission from The Royal Society of Chemistry)

$$\equiv Fe^{III}/Co^{III}\text{-}OH + H_2O_2 \rightarrow \equiv Fe^{II}/Co^{II}\text{-}OH + HO_2^{\bullet} + H^+ \quad (11.10)$$

$$\equiv Fe^{II}/Co^{II}\text{-}OH + H_2O_2 \rightarrow Fe^{III}/Co^{III}\text{-}OH + {}^{\bullet}OH + OH^- \quad (11.11)$$

11.3 Nanomaterials as Anodes

The EF and PEF treatments are usually carried out in cells equipped with conventional anodes like plates, rods/bars, or meshes. Alternatively, they can be performed with nanostructured anodes that can act as photoanodes or more efficient electrocatalysts. In the former case, hybrid processes with photoelectrocatalysis (PEC) have been devised. This section is focused on the preparation and application of such materials.

11.3.1 TiO_2-Based Photoanodes

The PEC method involves the use of a photoanode, commonly a nanocrystalline TiO_2-based material, in order to cause the light-induced oxidation of organic pollutants in aqueous media. This semiconductor has low cost and toxicity, and possesses a wide band gap of 3.2 eV (in the anatase form) (Garcia-Segura and Brillas 2017). Deposited as a thin film, it can absorb UV photons ($\lambda < 380$ nm) to promote an electron from the valence band to the conduction band (e^-_{CB}), with generation of a positively charged vacancy or hole (h^+_{VB}) via the reaction in Eq. 11.12. Organics can then be oxidized by: (i) the hole, (ii) heterogeneous $^{\bullet}OH$ formed from the reaction in Eq. 11.13 between h^+_{VB} and adsorbed water, and (iii) different ROS initiated from the reduction of O_2 by e^-_{CB} according to the reactions in Eqs. 11.14–11.17 (Sirés et al. 2014; Garcia-Segura and Brillas 2017).

$$TiO_2 + h\nu \rightarrow e^-_{CB} + h^+_{VB} \quad (11.12)$$

$$h^+_{VB} + H_2O \rightarrow {\bullet}OH + H^+ \quad (11.13)$$

$$e^-_{CB} + O_2 \rightarrow O_2^{\bullet -} \quad (11.14)$$

$$O_2^{\bullet -} + H^+ \rightarrow HO_2^{\bullet} \quad (11.15)$$

$$2HO_2^{\bullet} \rightarrow H_2O_2 + O_2 \quad (11.16)$$

$$H_2O_2 + O_2^{\bullet -} \rightarrow {}^{\bullet}OH + OH^- + O_2 \quad (11.17)$$

The recombination of the e^-_{CB}/h^+_{VB} pair is the major drawback since it leads to significant efficiency loss. But this can be overcome by applying either an optimum constant I or a constant bias anodic potential (E_{anod}) to the illuminated photoanode (Brillas et al. 2009; Sirés et al. 2014; Garcia-Segura and Brillas 2017). This ensures the continuous extraction of photoinduced electrons, which are conveyed to the cathode through the external electrical circuit.

When a photoanode is combined with an appropriate carbonaceous cathode that electrogenerates H_2O_2 in the presence of Fe^{2+} as catalyst, the hybrid PEC/EF process takes place (Peralta-Hernández et al. 2006; Xie and Li 2006; Li et al. 2007; Esquivel et al. 2009; Ramírez et al. 2010; Ding et al. 2012, 2014; Lin et al. 2013; Almeida et al. 2014; Mousset et al. 2017b). In this case, only the photoanode is exposed to UV radiation. In contrast, if also the solution is illuminated, the process is the so-called PEC/PEF (Mousset et al. 2017b; Almeida et al. 2015). For example, Fig. 11.5a schematizes the formation of the main oxidant $^{\bullet}OH$ at a Bi_2WO_6 anode in PEC from the reaction in Eq. 11.13, which is combined with its formation via the reaction in Eq. 11.8 in hetero-PEF using an Fe@Fe_2O_3/activated carbon fiber (ACF) cathode. Both radicals are employed to destroy the dye Rhodamine B (Ding et al. 2012).

Good performance of PEC/EF and PEC/PEF processes for the removal of several organics can be observed in Table 11.3. Apart from TiO_2, other effective photocatalysts like Bi_2WO_6 (Ding et al. 2012) and Pt/TiO_2 NTs (Almeida et al. 2014; Mousset et al. 2017b) have been tested. Table 11.3 also evidences, as expected, greater efficiency of PEC/PEF when compared to PEC/EF for the treatment of a phenol solution under comparable conditions, which results from the acceleration of the oxidation process thanks to the contribution of the reactions in Eqs. 11.4 and 11.5 (Mousset et al. 2017b). In most of these works, classical anodes were comparatively tested to confirm the better performance of the hybrid treatments. In the case of hybrid PEC/EF, for example, Rhodamine B solutions were mineralized up to 94% in a cell equipped with a Bi_2WO_6 photoanode under 300 W tungsten halogen lamp irradiation and an Fe@ACF cathode (see Table 11.3). In contrast, only 78% and 14% mineralization were found by EF with Pt/Fe@Fe_2O_3 cell and PEC with Bi_2WO_6/Pt cell, respectively, under analogous conditions. The outperformance of PEC/EF was explained by: (i) better separation of e^-_{CB}/h^+_{VB} pairs, with accumulation of a larger amount of oxidant holes at the photoanode surface, and (ii) the additional injection of photoinduced electrons to the Fe@Fe_2O_3/ACF cathode, enhancing the H_2O_2 production and yielding greater quantities of $^{\bullet}OH$. On the other hand, Fig. 11.5b illustrates the better performance of PEC/PEF with Pt/TiO_2 NTs anode under 80 W UVA illumination and GDE as cathode, as compared to EF with a Pt/GDE cell and PEC with a Pt/TiO_2 NTs photoanode and GDE as cathode (without added Fe^{2+}) (Almeida et al. 2015). As can be seen, the TOC of an Orange G solution was reduced by 97%, 87%, and 80%, respectively, after 200 mA h/L of specific electrical consumption. The combined oxidation by

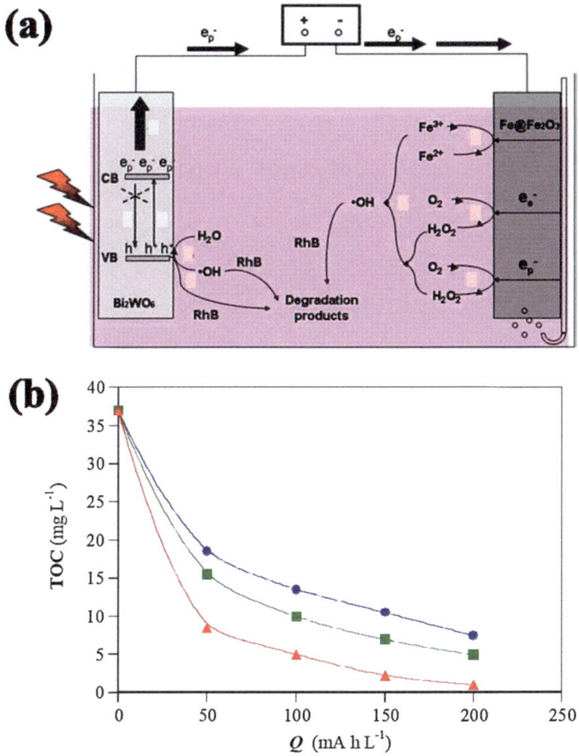

Fig. 11.5 (**a**) Schematic view of the main reactions to destroy Rhodamine B (RhB) by PEC with a Bi_2WO_6 photoanode combined with hetero-EF with a Fe@Fe_2O_3/ACF cathode. (Adapted from Ding et al. (2012), Copyright 2012, with permission of Elsevier). (**b**) Total organic carbon decay vs. applied electric charge for the (●) PEC, (■) EF, and (▲) PEC/PEF treatments of 500 mL of 85.4 mg/L Orange G solutions in 0.05 M Na_2SO_4 at pH 3.0. Anode in EF: Pt. Anode in PEC and PEC/PEF: Pt/TiO_2 NTs. Cathode: GDE. Applied current: 50 mA. Fe^{2+} concentration: 0.50 mM. Irradiation: 80 W UVA lamp. (Adapted from Almeida et al. (2015), Copyright 2015, with permission of Elsevier))

•OH formed from the reactions in Eqs. 11.2, 11.4, and 11.13, along with the photolytic action of UVA photons, accounts for the faster removal of organics by PEC/PEF.

As closely related to the aforementioned processes, the Khataee's group explored the characteristics of a PEF/photocatalysis treatment (Khataee et al. 2010, 2012, 2013, 2014; Zarei et al. 2010; Khataee and Zarei 2011), which can be considered as an alternative to the PEC/PEF system. Cubic undivided cells of 1.0–3.0 L capacity were equipped with the following elements: (i) a Pt anode, (ii) a CNTs-PTFE GDE as cathode fed with an O_2 flow, (iii) a 6 W UVA, UVB, or UVC lamp within a quartz tube, and (iv) four glass or ceramic plates coated with TiO_2 (Khataee et al. 2010, 2012, 2014; Zarei et al. 2010), N-doped TiO_2 (Khataee et al. 2013), or ZnO (Khataee and Zarei 2011) nanoparticles, which were placed covering the four inner cell walls.

Table 11.3 Selected results obtained for the PEC/EF and PEC/PEF treatment of several organics using undivided cells with nanostructured photoanodes

Photoanode, cathode	Substrate	Experimental remarks	Best performance	Ref.
PEC/EF treatment				
TiO_2/Ti, GF[a]	2,4-Dichlorophenol	50 mL of 15 mg/L substrate in 0.02 M Na_2SO_4, pH 3.0, 8 W UVA lamp, $I = 3.1$ mA	93% substrate and 78% TOC removal (60 min)	Li et al. (2007)
TiO_2/SS[b], graphite	Orange G	200 mL of 64 mg/L dye in 0.01 M Na_2SO_4, pH = 3.0, 330 μW/cm^2 UVA lamp, $E_{anod} = 1.0$ V/SCE	64% TOC removal (210 min)	Lin et al. (2013)
Bi_2WO_6, Fe@Fe_2O_3/ACF	Rhodamine B	100 mL of 10 μM dye in 0.05 M Na_2SO_4, pH 6.2, 300 W tungsten halogen lamp, $I = 0.3$ mA	94% TOC decay (240 min)	Ding et al. (2012)
Pt/TiO_2NTs, GDE	Acid Red 29	500 mL of 85.4 mg/L dye in 0.05 M Na_2SO_4, 0.50 mM Fe^{2+}, pH = 3.0, 80 W UVA lamp, $E_{anod} = 2.0$ V vs. Ag/AgCl	100% color removal (7 min), 98% TOC decay (200 mA h/L^{-1})	Almeida et al. (2014)
TiO_2/FTO[c], carbon felt	Phenol	200 mL of 1.4 mM substrate in 0.05 M K_2SO_4, 0.1 mM Fe^{2+}, pH = 3.0, 6 W UVA lamp, $I = 50$ mA	67% substrate decay (300 min), 76% TOC removal (480 min)	Mousset et al. (2017b)
PEC/PEF treatment				
TiO_2/FTO[c], carbon felt	Phenol	200 mL of 1.4 mM substrate in 0.05 M K_2SO_4, 0.1 mM Fe^{2+}, pH = 3.0, 6 W UVA lamp, $I = 50$ mA	100% substrate decay (300 min), 97% TOC removal (480 min)	Mousset et al. (2017b)
Pt/TiO_2NTs, GDE	Orange G	500 mL of 85.4 mg/L dye in 0.05 M Na_2SO_4, 0.50 mM Fe^{2+}, pH = 3.0, 80 W UVA lamp, $I = 50$ mA	100% color removal (30 min), 97% TOC decay (200 mA h/L)	Almeida et al. (2015)

[a]Second Fe anode at $I = 0.1$ mA as Fe^{2+} source. [b]SS—stainless steel, which acts as the source of an Fe^{2+} catalyst. [c]FTO—fluorine-doped tin oxide

Solutions of the dyes Acid Red 17 (Khataee et al. 2010), Basic Red 46 (Zarei et al. 2010), Acid Yellow 36 (Khataee et al. 2012), Direct Red 23 (Khataee et al. 2013), and Direct Yellow 12 (Khataee and Zarei 2011), as well as phenol (Khataee et al. 2014), in 0.05 M Na_2SO_4 with 0.1–0.2 mM Fe^{3+} at pH 3.0 were comparatively treated by the individual processes to show the benefits of the hybrid PEF/photocatalysis treatment. For example, for 2 L of 50 mg/L Direct Yellow 12 with 0.1 mM Fe^{3+} treated for 90 min at 100 mA under a 6 W UVC light irradiation (Khataee and Zarei 2011), the decolorization ability decreased as follows: PEF/photocatalysis (93%) > PEF (69%) > EF (56%), being much lower (39%) for photocatalysis alone (ZnO irradiated by UVC). The high mineralization power of the

PEF/photocatalysis method was shown by large TOC reduction (97%) achieved at 360 min. This can be attributed to the destruction of organics by: (i) heterogeneous Pt(•OH) formed from the reaction in Eq. 11.3 and heterogeneous •OH from photogenerated holes on ZnO by the reaction in Eq. 11.13, (ii) homogeneous •OH formed from Fenton's reaction (Eq. 11.2), from photolysis of $Fe(OH)^{2+}$ species via the reaction in Eq. 11.4, and from UVC photolysis of electrogenerated H_2O_2, (iii) photoinduced holes produced on ZnO, similarly to the reaction in Eq. 11.12, and (iv) photodecomposition of intermediates under UVC radiation.

11.3.2 Electrocatalytic Anodes

Several authors have improved the EF method by synthesizing anodes with greater electrocatalytic ability as compared to conventional ones, thereby enhancing the production of heterogeneous •OH from water discharge, similarly to the reaction in Eq. 11.3. Examples of some stable, nanostructured anodes for such a purpose are β-PbO_2 (Sirés et al. 2010), porous Ni/BDD/Ta (Li et al. 2018), and ZnO-TiO_2/GF (El-Kacemi et al. 2017). The excellent performance of 3D β-PbO_2 deposits obtained from carbon/polyvinyl-ester composite (Sirés et al. 2010) for the EF process has been well proven for the treatment of 275 mL of an O_2-saturated solution with 0.25 mM Methyl Orange in 0.05 M Na_2SO_4 with 0.2 mM Fe^{2+} at pH 3.0, using an undivided glass cell with a carbon-felt cathode at $I = 60$ mA. Total decolorization was achieved after 60 min of electrolysis, a time shorter than 240 min needed for the analogous EO with Ni cathode in the absence of Fe^{2+}, at 300 mA. This corroborates much greater oxidation power of •OH formed from Fenton's reaction (Eq. 11.2) than PbO_2(•OH) in EF. Further work of these authors showed the applicability of PbO_2 deposits onto RVC to EO (Recio et al. 2011; Ramírez et al. 2016), being feasible to incorporate TiNTs (Ramírez et al. 2016), which could be extended in the future to EF and PEF. Similarly, for 60 mg/L Methylene Blue in 0.05 M Na_2SO_4 at pH 3.0, the use of a porous Ni/BDD/Ta anode allowed the overall loss of color after 80 min of the EF treatment with 0.5 mM Fe^{2+} at 120 mA. This was faster than the comparable EO process, which required 240 min (Li et al. 2018). On the other hand, a nano-structured ZnO-TiO_2/GF anode was coupled to a GF cathode to treat 250 mL of an O_2-saturated solution with 0.12 mM of the dye Amido Black 10, 0.06 M Na_2SO_4, and 0.1 mM Fe^{2+} at pH 3.0 by EF at $I = 100$ mA (El-Kacemi et al. 2017). Under these conditions, the dye solution was completely decolorized in 60 min, whereas 91% mineralization was reached after 360 min of the treatment, showing good effectiveness of this anode for dye destruction.

11.4 Suspended Nanocatalysts

The synthesis of nanocatalysts to be suspended in solution, with the ability to form heterogeneous $^{\bullet}OH$ upon the reaction with electrogenerated H_2O_2, has widened the applicability of Fenton-based EAOPs for wastewater treatment through the development of hetero-EF and hetero-PEF. Similarly to what has been discussed in the case of Fe-loaded carbon nanomaterials used as cathodes, the main advantage of suspended nanocatalysts is that they allow operating within a larger pH range, and, therefore, neutralization of the final effluent may be avoided (Ganiyu et al. 2018; Poza-Nogueiras et al. 2018). This allows the minimization of sludge formation and prevents an excessive accumulation of iron ions in the treated effluent, which are common handicaps in the homogeneous processes. Moreover, the nanocatalyst is easy to handle, safe to store, it can be efficiently recovered and has the possibility of being reused (Ganiyu et al. 2018). As main drawbacks, suspended nanocatalysts may lose some of their active sites, thus reducing their catalytic ability, and they can undergo partial solubilization. This means that, on many occasions, there exists the conjunction of both homogeneous and heterogeneous processes during the treatment, even if this is disregarded by many authors.

A large number of nanomaterials has been synthesized and tested for hetero-EF, including: (i) Fe-based nanoparticles, like Fe_3O_4 (He et al. 2014; Es'haghzade et al. 2017), Pd/Fe_3O_4 (Luo et al. 2014; Huang et al. 2017), Fe molybdophosphate (Baiju et al. 2018), and zero-valent iron (Babuponnusami and Muthukumar 2012); (ii) Fe-carbon nanoparticles such as Fe-C/PTFE (Zhang et al. 2015b, c) and Fe_3O_4-CNTs (Shen et al. 2014); (iii) mineral-like nanocatalysts such as martite (Khataee et al. 2017); (iv) MOFs, only used in non-electrochemical systems so far, as described in detail in recent reviews (Dias and Petit 2015; Cheng et al. 2018); (v) supported Fe catalysts, such as Fe–zeolite (Rostamizadeh et al. 2018), Fe_3O_4–chitosan (Rezgui et al. 2018), Fe–silica (Jinisha et al. 2018), and Fe_2O_3–kaolin (Özcan et al. 2017); and (vi) non-Fe-based, like Cu/C (Xu et al. 2013). Table 11.4 collects selected results obtained in these works.

Successful degradations can be observed within the pH range 2–7, which confirms the viability of nanomaterials to operate at neutral pH—the condition required for the treatment of wastewater under real conditions.

In the case of Fe-based nanocatalysts, the proposed heterogeneous mechanism involves the oxidation of organics with heterogeneous $^{\bullet}OH$ produced via the reaction in Eq. 11.8, where H_2O_2 is generated at a suitable carbonaceous cathode (Ganiyu et al. 2018; Poza-Nogueiras et al. 2018). The regeneration of Fe^{II} at the catalyst surface can occur either by the reaction in Eq. 11.9, when the material comes into contact with the cathode upon stirring or recirculation, or by the heterogeneous Fenton-like reaction in Eq. 11.18.

$$\equiv Fe^{III} + H_2O_2 \rightarrow \equiv Fe^{II} + HO_2^{\bullet} + H^+ \qquad (11.18)$$

Table 11.4 Selected results obtained for the hetero-EF treatment of several organic pollutants using undivided cells with suspended nanostructured catalysts in solution

Nanocatalyst	Substrate	Experimental remarks	Best performance	Ref.
Fe_3O_4	Reactive Blue 19	200 mL of 100 mg/L dye in 0.05 M Na_2SO_4, 1 g/L catalyst, pH = 3.0, Fe/ACF cell, I = 120 mA	87% TOC removal (120 min)	He et al. (2014)
	Acid Red 14	250 mL of 50 mg/L substrate in 4 g/L Na_2SO_4, 0.6 g/L catalyst, pH = 7, graphite/graphite cell, I = 180 mA	92% (pH = 3), 83% (pH = 7), and 85% (pH = 9) color removal (120 min)	Es'haghzade et al. (2017)
Pd/Fe_3O_4	Humic acid + Cr(VI)	200 mL of 100 mg/L humic acid +20 mg/L Cr(VI), 1 g/L catalyst, pH = 3.0, BDD/Pt cell, I = 40 mA	90% TOC decay (480 min), total removal Cr (VI) (120 min)	Huang et al. (2017)
Zero-valent iron	Phenol	1 L of 200 mg/L substrate, 0.5 g/L catalyst, 500 mg/L H_2O_2, pH = 6.2, SS/SS cell, I = 60 mA	88% substrate decay (60 min)	Babuponnusami and Muthukumar (2012)
Fe-C/PTFE	2,4-Dichlorophenol	150 mL of 120 mg/L substrate in 0.05 M Na_2SO_4, 6 g/L catalyst, pH = 6.7, Ti/IrO_2-RuO_2/GDE cell, I = 100 mA	97% substrate and 34% TOC removals (120 min)	Zhang et al. (2015b)
Fe_3O_4/CNTs	Methylene Blue	150 mL of 0.24 mM dye in 0.10 M Na_2SO_4, 10 g/L catalyst, Ti/SnO_2/graphite cell, E_{cell} = 5 V [a]	70% (pH = 3), 73% (pH = 7), and 44% (pH = 9) color decay (30 min)	Shen et al. (2014)
Martite	Paraquat	100 mL of 20 mg/L herbicide in 0.05 M Na_2SO_4, 1 g/L catalyst, pH = 6, Pt/graphite cell, I = 300 mA	86% herbicide decay (150 min)	Khataee et al. (2017)
Fe–zeolite	Reactive Red 120	200 mL of 10 mg/L herbicide in 0.05 M Na_2SO_4, 1 g/L catalyst, pH = 3, graphite/graphite cell, I = 100 mA	94% color removal (30 min)	Rostamizadeh et al. (2018)
Fe_3O_4–chitosan	Chlordimeform	30 mL of 37.5 mg/L insecticide in 0.05 M Na_2SO_4, 0.5 g/L	100% insecticide decay (30 min),	Rezgui et al. (2018)

(continued)

Table 11.4 (continued)

Nanocatalyst	Substrate	Experimental remarks	Best performance	Ref.
		catalyst, pH = 3, Pt/carbon felt cell, $I = 100$ mA	80% TOC decay (360 min)	
Fe–silica	Rhodamine B	750 mL of 10 mg/L drug in 20 mg/ L Na_2SO_4, 15 mg/L catalyst, pH = 2, graphite/graphite cell, $E_{cell} = 8$ V	98% color and 35% TOC removals (180 min)	Jinisha et al. (2018)
Fe_2O_3–kaolin	Enoxacin	175 mL of 0.25 mM drug in 0.05 M Na_2SO_4, 0.3 g/L catalyst, pH = 5.1, Pt/carbon felt cell, $I = 100$ mA	100% drug removal (15 min), 99% TOC decay (360 min)	Özcan et al. (2017)
Cu/C	Phenol	200 mL of 2 mg/L substrate in 0.01 M Na_2SO_4, 1 g/L catalyst, 10 mg/L Fe^{2+}, pH = 3, Pt/Pt cell, $I = 50$ mA	80% (pH = 3) and 50% (pH = 5 and 7) substrate removals (180 min)	Xu et al. (2013)

[a] E_{cell}—potential difference between anode and cathode

Some authors have suggested more complex reaction mechanisms for hetero-EF when H_2O_2 is not cathodically produced. Figure 11.6a schematizes the steps proposed for the removal of humic acids (HAs) and Cr(VI) in a BDD/Pt cell using Pd/Fe_3O_4 as catalyst (Huang et al. 2017). These include: (i) H_2 and O_2 gas generation at the cathode and anode, respectively, (ii) diffusion of O_2 to the Pd/Fe_3O_4 nanoparticles to be reduced to the oxidant $O_2^{\bullet -}$, (iii) H_2O_2 formation from the reaction between H_2 and O_2 at the catalyst surface, (iv) generation of heterogeneous $^{\bullet}OH$ from heterogeneous Fenton's reaction (Eq. 11.8) between Fe^{II} at the Pd/Fe_3O_4 surface and produced H_2O_2, with subsequent Fe^{II} regeneration via the reaction in Eq. 11.9, and (v) $^{\bullet}OH$ formation at the BDD surface from a reaction similar to the one in Eq. 11.3. HAs are then oxidized by $O_2^{\bullet -}$ and $^{\bullet}OH$. According to step 6 shown in Fig. 11.6a, Cr(VI) is reduced to Cr(III) at the cathode, and by atomic H at the nanocatalyst surface, whereas as shown in step 7, Cr(III) can be removed upon the deposition of chromite ($FeCr_2O_4$). Similarly, for the treatment of phenol solutions with Cu/C nanoparticles (Xu et al. 2013), the H_2O_2 production from the heterogeneous reaction of H_2 and O_2 gases at the catalyst surface has been suggested. Hence, the oxidant $^{\bullet}OH$ is formed homogeneously from Fenton's reaction (Eq. 11.2) thanks to the addition of Fe^{2+} ion to the solution.

Several authors have also compared the homogeneous EF and hetero-EF treatments to show the excellent performance of the latter method. Figure 11.6b illustrates the normalized TOC abatement from a 0.25 mM enoxacin solution at pH 3.0 by homogeneous EF with 0.3 mM Fe^{2+} and hetero-EF with 0.1 g of Fe_2O_3–kaolin

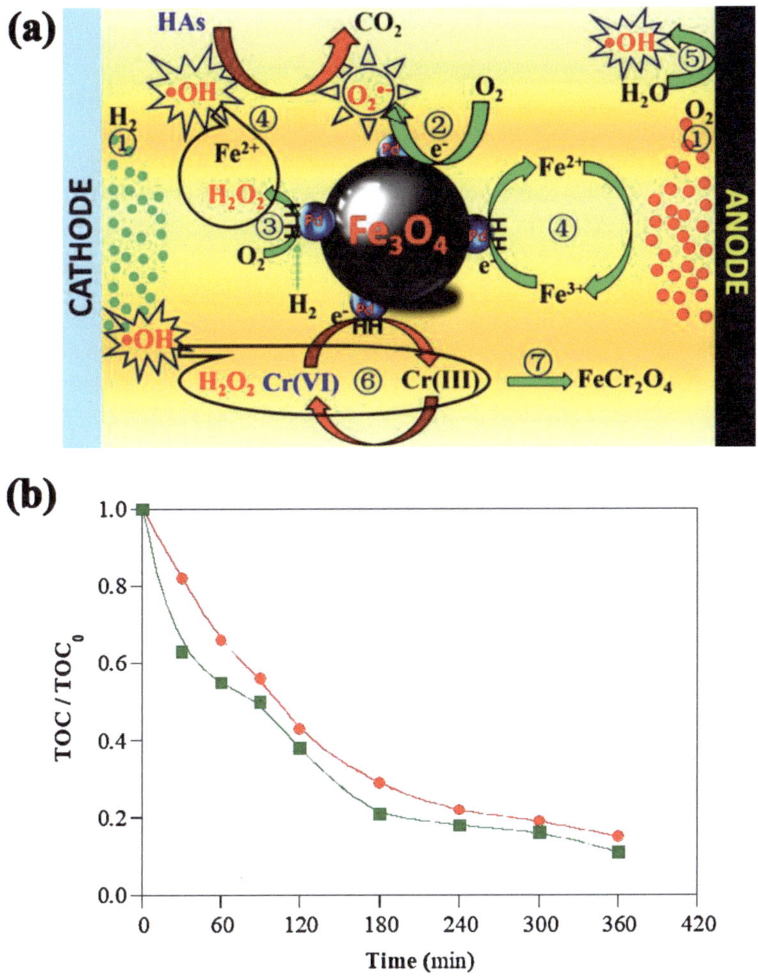

Fig. 11.6 (**a**) Schematic mechanisms for the removal of humic acid (HAs) and Cr(VI) by hetero-EF process with a BDD/Pt cell using Pd/Fe_3O_4 as catalyst. (Adapted from Huang et al. (2017), Copyright 2017, with permission from Elsevier). (**b**) Normalized TOC removal with electrolysis time for the treatment of 175 mL of 0.25 mM of the antibiotic enoxacin in 0.05 M Na_2SO_4 at pH 3.0 by (●) homogeneous EF with 0.30 mM Fe^{2+} and (■) hetero-EF with 0.1 g of Fe_2O_3–kaolin using a Pt/carbon felt cell. Applied current: 60 mA. (Adapted from Özcan et al. (2017), Copyright 2017, with permission from Elsevier)

using a Pt/carbon felt cell at 60 mA. After 360 min of electrolysis, 85% and 89% mineralization, respectively, was achieved, indicating the high effectiveness of the nanocatalyst, even superior to that of Fe^{2+} for $^{\bullet}OH$ generation. This should further encourage the research into novel nanomaterials with a higher catalytic activity and great stability for practical wastewater treatment, aiming to implement hetero-EF at industrial scale.

11.5 Nanomaterials in Hybrid Processes

Some other hybrid processes have been described to upgrade the oxidation ability of the EF and PEF methods (Brillas and Sirés 2018). However, the use of nanomaterials is rather limited to date, mainly being reduced to different kinds of carbon materials as adsorbents. Recently, Zhao et al. (2017) proposed a hybrid method involving electrosorption/EF to treat 100 mL of an O_2-saturated solution with 50 mg/L dimethyl phthalate in 0.05 M Na_2SO_4 at natural pH 6 using an undivided cell with a Pt anode and an Fe-doped carbon aerogel with Gr sheets as cathode at $I = 135$ mA. It was found that about 90% of the substrate was adsorbed on the cathode surface, whereas 98% of it was removed after 150 min of hetero-EF. The authors suggested a degradation mechanism involving the destruction of physisorbed dimethyl phthalate with heterogeneous $^{\bullet}OH$ formed from the reaction in Eq. 11.8. Wang et al. (2018) synthesized nanomaterials like $La_{1-x}Nd_xFeO_3$@activated carbon as anodes with a high adsorption ability. In only 10 min of EF treatment of Methyl Orange at pH 2.0 using GDE as cathode at $I = 60$ min, 99% color and 97% chemical oxygen demand (COD) was found, attaining 17% and 36% removal of such by direct adsorption at the same time. Direct removal of the adsorbed dye by heterogeneous $^{\bullet}OH$ formed from water oxidation at the anode and homogeneous $^{\bullet}OH$ from Fenton's reaction (Eq. 11.2) between the anodically leached Fe^{2+} and the produced H_2O_2 was proposed as the main degradation mechanism.

11.6 Conclusions

This chapter has reviewed the main current developments in the synthesis, modification, use, and reuse of key nanomaterials that enhance the performance of EF and PEF processes for the removal of organic pollutants from water. Since Fenton's reaction relies on the decomposition of H_2O_2 upon the reaction with Fe(II) to produce large amounts of $^{\bullet}OH$ in the bulk solution, efforts are particularly concentrated on new Fe-based nanocatalysts. We believe that research should especially focus on nanomaterials used as suspended catalysts since catalysts supported on the carbonaceous cathode surface tend to be detrimental for the two-electron ORR, and they may affect the material's stability and diminish the exposed carbon surface area for H_2O_2 electrogeneration. Investigation into potential Fe-based catalysts should pay special attention to their robustness, reducing the iron leaching to ensure a minimum secondary contamination and a greater reusability with reproducible performance. The use of MOFs in heterogeneous electrochemical Fenton-based EAOPs is forecast as a relevant topic in the near future. Novel cathodes, particularly those based on non-metal and non-ferrous nanocatalysts, may become the material of choice to electrogenerate H_2O_2 on site, and new electrode configurations might contribute to a significant improvement of such in situ electrosynthesis. The preparation of all these mentioned catalysts should, hopefully, be based on green synthesis principles, as shown in the case of 3D PbO_2 electrodeposited using biocompatible methanesulfonic acid. Regarding the use of nanomaterials in hybrid processes, adsorption with nanomaterials like Gr followed by EF or PEF post-treatments of the resulting effluents is an interesting way worth exploring, which can be combined with washing nanoadsorbents by means of Fenton-based EAOPs, as recently addressed by some authors. Finally, it is worth mentioning that the tests of these new materials and reactors should be carried out with real wastewater matrices in order to correctly assess their viability when compared to robust separation technologies like adsorption, (electro)coagulation, or membrane separation. Reusability tests in those complex matrices are thus needed to assess the impact of the catalyst degradation on the treatment costs as well as the reproducibility upon successive usage.

References

Ai Z, Mei T, Liu J, Li J, Jia F, Zhang L, Qiu J (2007) Fe@Fe_2O_3 core–shell nanowires as an iron reagent. 3. Their combination with CNTs as an effective oxygen-fed gas diffusion electrode in a neutral electro-Fenton system. J Phys Chem C 111(40):14799–14803. https://doi.org/10.1021/jp073617c

Almeida LC, Silva BF, Zanoni MVB (2014) Combined photoelectrocatalytic/electro-Fenton process using a Pt/TiO_2NTs photoanode for enhanced degradation of an azo dye: a mechanistic study. J Electroanal Chem 734:43–52. https://doi.org/10.1016/j.jelechem.2014.09.035

Almeida LC, Silva BF, Zanoni MVB (2015) Photoelectrocatalytic/photoelectro-Fenton coupling system using a nanostructured photoanode for the oxidation of a textile dye: kinetics study and oxidation pathway. Chemosphere 136:63–71. https://doi.org/10.1016/j.chemosphere.2015.04.042

Babaei-Sati R, Parsa JB (2017) Electrogeneration of H_2O_2 using graphite cathode modified with electrochemically synthesized polypyrrole/MWCNT nanocomposite for electro-Fenton process. J Ind Eng Chem 52:270–276. https://doi.org/10.1016/j.jiec.2017.03.056

Babuponnusami A, Muthukumar K (2012) Removal of phenol by heterogenous photo electro Fenton-like process using nano-zero valent iron. Sep Purif Technol 98:130–135. https://doi.org/10.1016/j.seppur.2012.04.034

Baiju A, Gandhimathi R, Ramesh ST, Nidheesh PV (2018) Combined heterogeneous electro-Fenton and biological process for the treatment of stabilized landfill leachate. J Environ Manag 210:328–337. https://doi.org/10.1016/j.jenvman.2018.01.019

Bañuelos JA, García-Rodríguez O, Rodríguez-Valadez FJ, Godínez LA (2015) Electrochemically prepared iron-modified activated carbon electrodes for their application in electro-Fenton and photoelectro-Fenton processes. J Electrochem Soc 162(9):E154–E159. https://doi.org/10.1149/2.0581509jes

Barros WRP, Alves SA, Franco PC, Steter JR, Rocha RS, Lanza MRV (2014a) Electrochemical degradation of tartrazine dye in aqueous solution using a modified gas diffusion electrode. J Electrochem Soc 161(9):H438–H442. https://doi.org/10.1149/2.015409jes

Barros WRP, Franco PC, Steter JR, Rocha RS, Lanza MRV (2014b) Electro-Fenton degradation of the food dye amaranth using a gas diffusion electrode modified with cobalt (II) phthalocyanine. J Electroanal Chem 722–723:46–53. https://doi.org/10.1016/j.jelechem.2014.03.027

Brillas E, Sirés I (2018) Chapter 11 - Hybrid and sequential chemical and electrochemical processes for water decontamination. In: Martínez-Huitle CA, Rodrigo MA, Scialdone O (eds)

Electrochemical water and wastewater treatment. Butterworth-Heinemann, Oxford, pp 267–304. https://doi.org/10.1016/B978-0-12-813160-2.00011-0

Brillas E, Sirés I, Oturan MA (2009) Electro-Fenton process and related electrochemical technologies based on Fenton's reaction chemistry. Chem Rev 109(12):6570–6631. https://doi.org/10.1021/cr900136g

Carneiro JF, Rocha RS, Hammer P, Bertazzoli R, Lanza MRV (2016) Hydrogen peroxide electrogeneration in gas diffusion electrode nanostructured with Ta_2O_5. Appl Catal A 517:161–167. https://doi.org/10.1016/j.apcata.2016.03.013

Chai G-L, Hou Z, Ikeda T, Terakura K (2017) Two-electron oxygen reduction on carbon materials catalysts: mechanisms and active sites. J Phys Chem C 121(27):14524–14533. https://doi.org/10.1021/acs.jpcc.7b04959

Chen C-Y, Tang C, Wang H-F, Chen C-M, Zhang X, Huang X, Zhang Q (2016a) Oxygen reduction reaction on graphene in an electro-Fenton system: in situ generation of H_2O_2 for the oxidation of organic compounds. ChemSusChem 9(10):1194–1199. https://doi.org/10.1002/cssc.201600030

Chen W, Yang X, Huang J, Zhu Y, Zhou Y, Yao Y, Li C (2016b) Iron oxide containing graphene/carbon nanotube based carbon aerogel as an efficient E-Fenton cathode for the degradation of methyl blue. Electrochim Acta 200:75–83. https://doi.org/10.1016/j.electacta.2016.03.044

Cheng M, Lai C, Liu Y, Zeng G, Huang D, Zhang C, Qin L, Hu L, Zhou C, Xiong W (2018) Metal-organic frameworks for highly efficient heterogeneous Fenton-like catalysis. Coord Chem Rev 368:80–92. https://doi.org/10.1016/j.ccr.2018.04.012

Choe YJ, Byun JY, Kim SH, Kim J (2018) Fe_3S_4/Fe_7S_8-promoted degradation of phenol via heterogeneous, catalytic H_2O_2 scission mediated by S-modified surface Fe^{2+} species. Appl Catal B 233:272–280. https://doi.org/10.1016/j.apcatb.2018.03.110

Chu Y, Zhang D, Liu L, Qian Y, Li L (2013) Electrochemical degradation of *m*-cresol using porous carbon-nanotube-containing cathode and Ti/SnO_2–Sb_2O_5–IrO_2 anode: kinetics, byproducts and biodegradability. J Hazard Mater 252–253:306–312. https://doi.org/10.1016/j.jhazmat.2013.03.018

Čolić V, Yang S, Révay Z, Stephens IEL, Chorkendorff I (2018) Carbon catalysts for electrochemical hydrogen peroxide production in acidic media. Electrochim Acta 272:192–202. https://doi.org/10.1016/j.electacta.2018.03.170

Dias EM, Petit C (2015) Towards the use of metal–organic frameworks for water reuse: a review of the recent advances in the field of organic pollutants removal and degradation and the next steps in the field. J Mater Chem A 3(45):22484–22506. https://doi.org/10.1039/C5TA05440K

Ding X, Ai Z, Zhang L (2012) Design of a visible light driven photo-electrochemical/electro-Fenton coupling oxidation system for wastewater treatment. J Hazard Mater 239–240:233–240. https://doi.org/10.1016/j.jhazmat.2012.08.070

Ding X, Ai Z, Zhang L (2014) A dual-cell wastewater treatment system with combining anodic visible light driven photoelectro-catalytic oxidation and cathodic electro-Fenton oxidation. Sep Purif Technol 125:103–110. https://doi.org/10.1016/j.seppur.2014.01.046

Ding X, Wang S, Shen W, Mu Y, Wang L, Chen H, Zhang L (2017) Fe@Fe_2O_3 promoted electrochemical mineralization of atrazine via a triazinon ring opening mechanism. Water Res 112:9–18. https://doi.org/10.1016/j.watres.2017.01.024

El-Kacemi S, Zazou H, Oturan N, Dietze M, Hamdani M, Es-Souni M, Oturan MA (2017) Nanostructured ZnO-TiO_2 thin film oxide as anode material in electrooxidation of organic pollutants. Application to the removal of dye Amido black 10B from water. Environ Sci Pollut Res 24(2):1442–1449. https://doi.org/10.1007/s11356-016-7920-6

Es'haghzade Z, Pajootan E, Bahrami H, Arami M (2017) Facile synthesis of Fe_3O_4 nanoparticles via aqueous based electro chemical route for heterogeneous electro-Fenton removal of azo dyes. J Taiwan Inst Chem Eng 71:91–105. https://doi.org/10.1016/j.jtice.2016.11.015

Esquivel K, Arriaga LG, Rodríguez FJ, Martínez L, Godínez LA (2009) Development of a TiO_2 modified optical fiber electrode and its incorporation into a photoelectrochemical reactor for wastewater treatment. Water Res 43(14):3593–3603. https://doi.org/10.1016/j.watres.2009.05.035

Félix-Navarro RM, Beltrán-Gastélum M, Salazar-Gastélum MI, Silva-Carrillo C, Reynoso-Soto EA, Pérez-Sicairos S, Lin SW, Paraguay-Delgado F, Alonso-Núñez G (2013) Pt–Pd bimetallic nanoparticles on MWCNTs: catalyst for hydrogen peroxide electrosynthesis. J Nanopart Res 15:1802. https://doi.org/10.1007/s11051-013-1802-3

Ganiyu SO, Le TXH, Bechelany M, Esposito G, van Hullebusch ED, Oturan MA, Cretin M (2017) A hierarchical CoFe-layered double hydroxide modified carbon-felt cathode for heterogeneous electro-Fenton process. J Mater Chem A 5(7):3655–3666. https://doi.org/10.1039/C6TA09100H

Ganiyu SO, Zhou M, Martínez-Huitle CA (2018) Heterogeneous electro-Fenton and photoelectro-Fenton processes: a critical review of fundamental principles and application for water/wastewater treatment. Appl Catal B 235:103–129. https://doi.org/10.1016/j.apcatb.2018.04.044

Garcia-Rodriguez O, Lee YY, Olvera-Vargas H, Deng F, Wang Z, Lefebvre O (2018) Mineralization of electronic wastewater by electro-Fenton with an enhanced graphene-based gas diffusion cathode. Electrochim Acta 276:12–20. https://doi.org/10.1016/j.electacta.2018.04.076

Garcia-Segura S, Brillas E (2017) Applied photoelectrocatalysis on the degradation of organic pollutants in wastewaters. J Photochem Photobiol C 31:1–35. https://doi.org/10.1016/j.jphotochemrev.2017.01.005

Garrido-Ramírez EG, Marco JF, Escalona N, Ureta-Zañartu MS (2016) Preparation and characterization of bimetallic Fe–Cu allophane nanoclays and their activity in the phenol oxidation by heterogeneous electro-Fenton reaction. Microporous Mesoporous Mater 225:303–311. https://doi.org/10.1016/j.micromeso.2016.01.013

He Z, Gao C, Qian M, Shi Y, Chen J, Song S (2014) Electro-Fenton process catalyzed by Fe_3O_4 magnetic nanoparticles for degradation of C.I. reactive blue 19 in aqueous solution: operating conditions, influence, and mechanism. Ind Eng Chem Res 53(9):3435–3447. https://doi.org/10.1021/ie403947b

Huang B, Qi C, Yang Z, Guo Q, Chen W, Zeng G, Lei C (2017) Pd/Fe_3O_4 nanocatalysts for highly effective and simultaneous removal of humic acids and Cr(VI) by electro-Fenton with H_2O_2 *in situ* electro-generated on the catalyst surface. J Catal 352:337–350. https://doi.org/10.1016/j.jcat.2017.06.004

Jinisha R, Gandhimathi R, Ramesh ST, Nidheesh PV, Velmathi S (2018) Removal of rhodamine B dye from aqueous solution by electro-Fenton process using iron-doped mesoporous silica as a heterogeneous catalyst. Chemosphere 200:446–454. https://doi.org/10.1016/j.chemosphere.2018.02.117

Khataee A, Hasanzadeh A (2017) Modified cathodes with carbon-based nanomaterials for electro-Fenton process. In: Zhou M, Oturan MA, Sirés I (eds) Electro-Fenton process: new trends and scale-up. Springer, Singapore, pp 111–143. https://doi.org/10.1007/698_2017_74

Khataee AR, Zarei M (2011) Photocatalysis of a dye solution using immobilized ZnO nanoparticles combined with photoelectrochemical process. Desalination 273(2–3):453–460. https://doi.org/10.1016/j.desal.2011.01.066

Khataee AR, Zarei M, Asl SK (2010) Photocatalytic treatment of a dye solution using immobilized TiO_2 nanoparticles combined with photoelectro-Fenton process: optimization of operational parameters. J Electroanal Chem 648(2):143–150. https://doi.org/10.1016/j.jelechem.2010.07.017

Khataee AR, Safarpour M, Zarei M, Aber S (2012) Combined heterogeneous and homogeneous photodegradation of a dye using immobilized TiO_2 nanophotocatalyst and modified graphite electrode with carbon nanotubes. J Mol Catal A Chem 363–364:58–68. https://doi.org/10.1016/j.molcata.2012.05.016

Khataee A, Marandizadeh H, Vahid B, Zarei M, Joo SW (2013) Combination of photocatalytic and photoelectro-Fenton/citrate processes for dye degradation using immobilized N-doped TiO_2 nanoparticles and a cathode with carbon nanotubes: central composite design optimization. Chem Eng Process 73:103–110. https://doi.org/10.1016/j.cep.2013.07.007

Khataee AR, Fathinia M, Zarei M, Izadkhah B, Joo SW (2014) Modeling and optimization of photocatalytic/photoassisted-electro-Fenton like degradation of phenol using a neural network coupled with genetic algorithm. J Ind Eng Chem 20(4):1852–1860. https://doi.org/10.1016/j.jiec.2013.08.042

Khataee A, Sajjadi S, Hasanzadeh A, Vahid B, Joo SW (2017) One-step preparation of nanostructured martite catalyst and graphite electrode by glow discharge plasma for heterogeneous electro-Fenton like process. J Environ Manag 199:31–45. https://doi.org/10.1016/j.jenvman.2017.04.095

Le TXH, Bechelany M, Lacour S, Oturan N, Oturan MA, Cretin M (2015) High removal efficiency of dye pollutants by electron-Fenton process using a graphene based cathode. Carbon 94:1003–1011. https://doi.org/10.1016/j.carbon.2015.07.086

Le TXH, Bechelany M, Cretin M (2017) Advances in carbon felt material for electro-Fenton process. In: Zhou M, Oturan MA, Sirés I (eds) Electro-Fenton process: new trends and scale-up. Springer, Singapore, pp 145–173. https://doi.org/10.1007/698_2017_55

Li XZ, Zhao BX, Wang P (2007) Degradation of 2,4-dichlorophenol in aqueous solution by a hybrid oxidation process. J Hazard Mater 147(1–2):281–287. https://doi.org/10.1016/j.jhazmat.2006.12.077

Li J, Ai Z, Zhang L (2009) Design of a neutral electro-Fenton system with Fe@Fe_2O_3/ACF composite cathode for wastewater treatment. J Hazard Mater 164(1):18–25. https://doi.org/10.1016/j.jhazmat.2008.07.109

Li H, Lei H, Chen K, Yao C, Zhang X, Leng Q, Wang W (2011) A nano-Fe^0/ACF cathode applied to neutral electro-Fenton degradation of Orange II. J Chem Technol Biotechnol 86(3):398–405. https://doi.org/10.1002/jctb.2530

Li Y, Han J, Mi X, Mi X, Li Y, Zhang S, Zhan S (2017) Modified carbon felt made using $Ce_xA_{1-x}O_2$ composites as a cathode in electro-Fenton system to degrade ciprofloxacin. RSC Adv 7 (43):27065–27078. https://doi.org/10.1039/C7RA03302H

Li X, Li H, Li M, Li C, Sun D, Lei Y, Yang B (2018) Preparation of a porous boron-doped diamond/Ta electrode for the electrocatalytic degradation of organic pollutants. Carbon 129:543–551. https://doi.org/10.1016/j.carbon.2017.12.052

Liang L, An Y, Zhou M, Yu F, Liu M, Ren G (2016) Novel rolling-made gas-diffusion electrode loading trace transition metal for efficient heterogeneous electro-Fenton-like. J Environ Chem Eng 4(4):4400–4408. https://doi.org/10.1016/j.jece.2016.10.006

Liang L, Yu F, An Y, Liu M, Zhou M (2017) Preparation of transition metal composite graphite felt cathode for efficient heterogeneous electro-Fenton process. Environ Sci Pollut Res 24 (2):1122–1132. https://doi.org/10.1007/s11356-016-7389-3

Lin W-C, Chen C-H, Tang H-Y, Hsiao Y-C, Pan JR, Hu C-C, Huang C (2013) Electrochemical photocatalytic degradation of dye solution with a TiO_2-coated stainless steel electrode prepared by electrophoretic deposition. Appl Catal B 140–141:32–41. https://doi.org/10.1016/j.apcatb.2013.03.032

Liu Y, Chen S, Quan X, Yu H, Zhao H, Zhang Y (2015a) Efficient mineralization of perfluorooctanoate by electro-Fenton with H_2O_2 electro-generated on hierarchically porous carbon. Environ Sci Technol 49(22):13528–13533. https://doi.org/10.1021/acs.est.5b03147

Liu Y, Quan X, Fan X, Wang H, Chen S (2015b) High-yield electrosynthesis of hydrogen peroxide from oxygen reduction by hierarchically porous carbon. Angew Chem Int Ed 127 (23):6941–6945. https://doi.org/10.1002/ange.201502396

Liu T, Wang K, Song S, Brouzgou A, Tsiakaras P, Wang Y (2016) New electro-Fenton gas diffusion cathode based on nitrogen-doped graphene@carbon nanotube composite materials. Electrochim Acta 194:228–238. https://doi.org/10.1016/j.electacta.2015.12.185

Luo M, Yuan S, Tong M, Liao P, Xie W, Xu X (2014) An integrated catalyst of Pd supported on magnetic Fe_3O_4 nanoparticles: simultaneous production of H_2O_2 and Fe^{2+} for efficient electro-Fenton degradation of organic contaminants. Water Res 48:190–199. https://doi.org/10.1016/j.watres.2013.09.029

Martínez-Huitle CA, Rodrigo MA, Sirés I, Scialdone O (2015) Single and coupled electrochemical processes and reactors for the abatement of organic water pollutants: a critical review. Chem Rev 115(24):13362–13407. https://doi.org/10.1021/acs.chemrev.5b00361

Moreira FC, Boaventura RAR, Brillas E, Vilar VJP (2017) Electrochemical advanced oxidation processes: a review on their application to synthetic and real wastewaters. Appl Catal B 202:217–261. https://doi.org/10.1016/j.apcatb.2016.08.037

Mousset E, Ko ZT, Syafiq M, Wang Z, Lefebvre O (2016a) Electrocatalytic activity enhancement of a graphene ink-coated carbon cloth cathode for oxidative treatment. Electrochim Acta 222:1628–1641. https://doi.org/10.1016/j.electacta.2016.11.151

Mousset E, Wang Z, Hammaker J, Lefebvre O (2016b) Physico-chemical properties of pristine graphene and its performance as electrode material for electro-Fenton treatment of wastewater. Electrochim Acta 214:217–230. https://doi.org/10.1016/j.electacta.2016.08.002

Mousset E, Wang Z, Hammaker J, Lefebvre O (2017a) Electrocatalytic phenol degradation by a novel nanostructured carbon fiber brush cathode coated with graphene ink. Electrochim Acta 258:607–617. https://doi.org/10.1016/j.electacta.2017.11.104

Mousset E, Weiqi VH, Kai BFY, Koh JS, Tng JW, Wang Z, Lefebvre O (2017b) A new 3D-printed photoelectrocatalytic reactor combining the benefits of a transparent electrode and the Fenton reaction for advanced wastewater treatment. J Mater Chem A 5(47):24951–24964. https://doi.org/10.1039/C7TA08182K

Oturan MA, Aaron J-J (2014) Advanced oxidation processes in water/wastewater treatment: principles and applications. A review. Crit Rev Environ Sci Technol 44(23):2577–2641. https://doi.org/10.1080/10643389.2013.829765

Özcan A, Özcan AA, Demirci Y, Şener E (2017) Preparation of Fe_2O_3 modified kaolin and application in heterogeneous electro-catalytic oxidation of enoxacin. Appl Catal B 200:361–371. https://doi.org/10.1016/j.apcatb.2016.07.018

Pajootan E, Arami M, Rahimdokht M (2014) Discoloration of wastewater in a continuous electro-Fenton process modified graphite electrode with multi-walled carbon nanotubes/surfactant. Sep Purif Technol 130:34–44. https://doi.org/10.1016/j.seppur.2014.04.025

Paz EC, Aveiro LR, Pinheiro VS, Souza FM, Lima VB, Silva FL, Hammer P, Lanza MRV, Santos MC (2018) Evaluation of H_2O_2 electrogeneration and decolorization of Orange II azo dye using tungsten oxide nanoparticle-modified carbon. Appl Catal B 232:436–445. https://doi.org/10.1016/j.apcatb.2018.03.082

Peng Q, Zhao H, Qian L, Wang Y, Zhao G (2015) Design of a neutral photo-electro-Fenton system with 3D-ordered macroporous Fe_2O_3/carbon aerogel cathode: high activity and low energy consumption. Appl Catal B 174–175:157–166. https://doi.org/10.1016/j.apcatb.2015.02.031

Peralta-Hernández JM, Meas-Vong Y, Rodríguez FJ, Chapman TW, Maldonado MI, Godínez LA (2006) In situ electrochemical and photo-electrochemical generation of the Fenton reagent: a potentially important new water treatment technology. Water Res 40(9):1754–1762. https://doi.org/10.1016/j.watres.2006.03.004

Pizzutilo E, Kasian O, Choi CH, Cherevko S, Hutchings GJ, Mayrhofer KJJ, Freakley SJ (2017) Electrocatalytic synthesis of hydrogen peroxide on Au-Pd nanoparticles:from fundamentals to continuous production. Chem Phys Lett 683:436–442. https://doi.org/10.1016/j.cplett.2017.01.071

Plakas KV, Sklari SD, Yiankakis DA, Sideropoulos GT, Zaspalis VT, Karabelas AJ (2016) Removal of organic micropollutants from drinking water by a novel electro-Fenton filter: pilot-scale studies. Water Res 91:183–194. https://doi.org/10.1016/j.watres.2016.01.013

Poza-Nogueiras V, Rosales E, Pazos M, Sanromán MÁ (2018) Current advances and trends in electro-Fenton process using heterogeneous catalysts – a review. Chemosphere 201:399–416. https://doi.org/10.1016/j.chemosphere.2018.03.002

Ramírez J, Godínez LA, Méndez M, Meas Y, Rodríguez FJ (2010) Heterogeneous photo-electro-Fenton process using different iron supporting materials. J Appl Electrochem 40(10):1729–1736. https://doi.org/10.1007/s10800-010-0157-z

Ramírez G, Recio FJ, Herrasti P, Ponce-de-León C, Sirés I (2016) Effect of RVC porosity on the performance of PbO_2 composite coatings with titanate nanotubes for the electrochemical oxidation of azo dyes. Electrochim Acta 204:9–17. https://doi.org/10.1016/j.electacta.2016.04.054

Recio FJ, Herrasti P, Sirés I, Kulak AN, Bavykin DV, Ponce-de-León C, Walsh FC (2011) The preparation of PbO_2 coatings on reticulated vitreous carbon for the electro-oxidation of organic pollutants. Electrochim Acta 56(14):5158–5165. https://doi.org/10.1016/j.electacta.2011.03.054

Rezgui S, Amrane A, Fourcade F, Assadi A, Monser L, Adhoum N (2018) Electro-Fenton catalyzed with magnetic chitosan beads for the removal of Chlordimeform insecticide. Appl Catal B 226:346–359. https://doi.org/10.1016/j.apcatb.2017.12.061

Ridruejo C, Alcaide F, Álvarez G, Brillas E, Sirés I (2018) On-site H_2O_2 electrogeneration at a CoS_2-based air-diffusion cathode for the electrochemical degradation of organic pollutants. J Electroanal Chem 808:364–371. https://doi.org/10.1016/j.jelechem.2017.09.010

Rostamizadeh M, Jafarizad A, Gharibian S (2018) High efficient decolorization of reactive red 120 azo dye over reusable Fe-ZSM-5 nanocatalyst in electro-Fenton reaction. Sep Purif Technol 192:340–347. https://doi.org/10.1016/j.seppur.2017.10.041

Roth H, Gendel Y, Buzatu P, David O, Wessling M (2016) Tubular carbon nanotube-based gas diffusion electrode removes persistent organic pollutants by a cyclic adsorption – electro-Fenton process. J Hazard Mater 307:1–6. https://doi.org/10.1016/j.jhazmat.2015.12.066

Shen L, Yan P, Guo X, Wei H, Zheng X (2014) Three-dimensional electro-Fenton degradation of methylene blue based on the composite particle electrodes of carbon nanotubes and nano-Fe_3O_4. Arab J Sci Eng 39(9):6659–6664. https://doi.org/10.1007/s13369-014-1184-6

Siahrostami S, Verdaguer-Casadevall A, Karamad M, Deiana D, Malacrida P, Wickman B, Escudero-Escribano M, Paoli EA, Frydendal R, Hansen TW, Chorkendorff I, Stephens IEL, Rossmeisl J (2013) Enabling direct H_2O_2 production through rational electrocatalyst design. Nat Mater 12:1137–1143. https://doi.org/10.1038/nmat3795

Sirés I, Low CTJ, Ponce-de-León C, Walsh FC (2010) The deposition of nanostructured β-PbO_2 coatings from aqueous methanesulfonic acid for the electrochemical oxidation of organic pollutants. Electrochem Commun 12(1):70–74. https://doi.org/10.1016/j.elecom.2009.10.038

Sirés I, Brillas E, Oturan MA, Rodrigo MA, Panizza M (2014) Electrochemical advanced oxidation processes: today and tomorrow. A review. Environ Sci Pollut Res 21(14):8336–8367. https://doi.org/10.1007/s11356-014-2783-1

Sklari SD, Plakas KV, Petsi PN, Zaspalis VT, Karabelas AJ (2015) Toward the development of a novel electro-Fenton system for eliminating toxic organic substances from water. Part 2. Preparation, characterization, and evaluation of iron-impregnated carbon felts as cathodic electrodes. Ind Eng Chem Res 54(7):2059–2073. https://doi.org/10.1021/ie5048779

Tian J, Olajuyin AM, Mu T, Yang M, Xing J (2016a) Efficient degradation of rhodamine B using modified graphite felt gas diffusion electrode by electro-Fenton process. Environ Sci Pollut Res 23(12):11574–11583. https://doi.org/10.1007/s11356-016-6360-7

Tian J, Zhao J, Olajuyin AM, Sharshar MM, Mu T, Yang M, Xing J (2016b) Effective degradation of rhodamine B by electro-Fenton process, using ferromagnetic nanoparticles loaded on modified graphite felt electrode as reusable catalyst: in neutral pH condition and without external aeration. Environ Sci Pollut Res 23(15):15471–15482. https://doi.org/10.1007/s11356-016-6721-2

Wang Y, Zhao G, Chai S, Zhao H, Wang Y (2013) Three-dimensional homogeneous ferrite-carbon aerogel: one pot fabrication and enhanced electro-Fenton reactivity. ACS Appl Mater Interfaces 5(3):842–852. https://doi.org/10.1021/am302437a

Wang Y, Liu Y, Liu T, Song S, Gui X, Liu H, Tsiakaras P (2014) Dimethyl phthalate degradation at novel and efficient electro-Fenton cathode. Appl Catal B 156–157:1–7. https://doi.org/10.1016/j.apcatb.2014.02.041

Wang Q, Zhou S, Xiao S, Wei F, Zhao X, Qu J, Wang H (2018) Novel perovskite-based composites, $La_{1-x}Nd_xFeO_3$@activated carbon, as efficient catalysts for the degradation of organic pollutants by heterogeneous electro-Fenton reactions. RSC Adv 8(27):14775–14786. https://doi.org/10.1039/C8RA00244D

Xie YB, Li XZ (2006) Interactive oxidation of photoelectrocatalysis and electro-Fenton for azo dye degradation using TiO_2–Ti mesh and reticulated vitreous carbon electrodes. Mater Chem Phys 95(1):39–50. https://doi.org/10.1016/j.matchemphys.2005.05.048

Xu X, Liao P, Yuan S, Tong M, Luo M, Xie W (2013) Cu-catalytic generation of reactive oxidizing species from H_2 and O_2 produced by water electrolysis for electro-Fenton degradation of organic contaminants. Chem Eng J 233:117–123. https://doi.org/10.1016/j.cej.2013.08.046

Yang W, Zhou M, Cai J, Liang L, Ren G, Jiang L (2017) Ultrahigh yield of hydrogen peroxide on graphite felt cathode modified with electrochemically exfoliated graphene. J Mater Chem A 5 (17):8070–8080. https://doi.org/10.1039/C7TA01534H

Yang S, Verdaguer-Casadevall A, Arnarson L, Silvioli L, Čolić V, Frydendal R, Rossmeisl J, Chorkendorff I, Stephens IEL (2018a) Toward the decentralized electrochemical production of H_2O_2: a focus on the catalysis. ACS Catal 8(5):4064–4081. https://doi.org/10.1021/acscatal.8b00217

Yang W, Zhou M, Liang L (2018b) Highly efficient in-situ metal-free electrochemical advanced oxidation process using graphite felt modified with N-doped graphene. Chem Eng J 338:700–708. https://doi.org/10.1016/j.cej.2018.01.013

Zarei M, Khataee AR, Ordikhani-Seyedlar R, Fathinia M (2010) Photoelectro-Fenton combined with photocatalytic process for degradation of an azo dye using supported TiO_2 nanoparticles and carbon nanotube cathode: neural network modeling. Electrochim Acta 55(24):7259–7265. https://doi.org/10.1016/j.electacta.2010.07.050

Zhang X, Fu J, Zhang Y, Lei L (2008) A nitrogen functionalized carbon nanotube cathode for highly efficient electrocatalytic generation of H_2O_2 in electro-Fenton system. Sep Purif Technol 64(1):116–123. https://doi.org/10.1016/j.seppur.2008.07.020

Zhang G, Wang S, Yang F (2012) Efficient adsorption and combined heterogeneous/homogeneous Fenton oxidation of amaranth using supported nano-FeOOH as cathodic catalysts. J Phys Chem C 116(5):3623–3634. https://doi.org/10.1021/jp210167b

Zhang G, Zhou Y, Yang F (2015a) FeOOH-catalyzed heterogeneous electro-Fenton system upon anthraquinone@graphene nanohybrid cathode in a divided electrolytic cell: catholyte-regulated catalytic oxidation performance and mechanism. J Electrochem Soc 162(6):H357–H365. https://doi.org/10.1149/2.0691506jes

Zhang C, Zhou M, Ren G, Yu X, Ma L, Yang J, Yu F (2015b) Heterogeneous electro-Fenton using modified iron–carbon as catalyst for 2,4-dichlorophenol degradation: influence factors, mechanism and degradation pathway. Water Res 70:414–424. https://doi.org/10.1016/j.watres.2014.12.022

Zhang C, Zhou M, Yu X, Ma L, Yu F (2015c) Modified iron-carbon as heterogeneous electro-Fenton catalyst for organic pollutant degradation in near neutral pH condition: characterization, degradation activity and stability. Electrochim Acta 160:254–262. https://doi.org/10.1016/j.electacta.2015.01.092

Zhang Y, Gao M, Wang S-G, Zhou W, Sang Y, Wang X-H (2017) Integrated electro-Fenton process enabled by a rotating Fe_3O_4/gas diffusion cathode for simultaneous generation and activation of H_2O_2. Electrochim Acta 231:694–704. https://doi.org/10.1016/j.electacta.2017.02.091

Zhang Z, Meng H, Wang Y, Shi L, Wang X, Chai S (2018) Fabrication of graphene@graphite-based gas diffusion electrode for improving H_2O_2 generation in electro-Fenton process. Electrochim Acta 260:112–120. https://doi.org/10.1016/j.electacta.2017.11.048

Zhao H, Qian L, Guan X, Wu D, Zhao G (2016) Continuous bulk FeCuC aerogel with ultradispersed metal nanoparticles: an efficient 3D heterogeneous electro-Fenton cathode over a wide range of pH 3−9. Environ Sci Technol 50(10):5225–5233. https://doi.org/10.1021/acs.est.6b00265

Zhao H, Wang Q, Chen Y, Tian Q, Zhao G (2017) Efficient removal of dimethyl phthalate with activated iron-doped carbon aerogel through an integrated adsorption and electro-Fenton oxidation process. Carbon 124:111–122. https://doi.org/10.1016/j.carbon.2017.08.034

Zhao H, Qian L, Chen Y, Wang Q, Zhao G (2018a) Selective catalytic two-electron O_2 reduction for onsite efficient oxidation reaction in heterogeneous electro-Fenton process. Chem Eng J 332:486–498. https://doi.org/10.1016/j.cej.2017.09.093

Zhao K, Su Y, Quan X, Liu Y, Chen S, Yu H (2018b) Enhanced H_2O_2 production by selective electrochemical reduction of O_2 fluorine-doped hierarchically porous carbon. J Catal 357:118–126. https://doi.org/10.1016/j.jcat.2017.11.008

Zhou L, Zhou M, Hu Z, Bi Z, Groenen Serrano K (2014) Chemically modified graphite felt as an efficient cathode in electro-Fenton for *p*-nitrophenol degradation. Electrochim Acta 140:376–383. https://doi.org/10.1016/j.electacta.2014.04.090

Zhou M, Oturan MA, Sirés I (eds) (2018) Electro-Fenton process: new trends and scale-up. The handbook of environmental chemistry, vol 61, 1st edn. Springer, Singapore. https://doi.org/10.1007/978-981-10-6406-7

Chapter 12
Field Study III: Evidence Gained from Site Studies for the Performance of Ferrate(VI) in Water and Wastewater Treatment

Jia-Qian Jiang

Abstract The work presented in this chapter was to validate whether ferrate(VI) can be used as an alternative to the existing coagulant (e.g., ferric chloride) for both drinking water and domestic sewage treatment via a series of pilot-plant trials. For drinking water treatment, a ferrate(VI) dose of 0.1 mg/L can achieve 93% and 97% particle removal (in terms of particle counting) after the filtration for the raw water and for the ozonized water, respectively, which is satisfactory to the treated water quality demand for the particle removal. Moreover, ferrate(VI) can remove 10% metformin, benzotriazole and acesulfame from raw water, but $FeCl_3$ with ozonation cannot. When treating domestic sewage at pilot-scale trials, ferrate(VI) demonstrated encouraging performance as well; at a very low dose range, 0.1–0.2 mg $Fe^{(VI)}$/L, ferrate(VI) achieved better performance in comparison with high-dosed ferric sulfate. This will reduce chemical demand and sludge production, and, therefore, it results in a low operating cost and substantial cost saving in treating sewage.

Keywords BOD removal · Coagulation · COD removal · Drinking water treatment · Ferrate(VI) · Micropollutant reduction · Particle removal · Phosphorus removal · Sewage treatment

12.1 Introduction

Ferrate(VI) is a very strong oxidant. Under acidic conditions, the redox potential of ferrate(VI) ions is 2.2 V, which is well compatible with that of ozone (2.0 V) (Jiang and Lloyd 2002). Exploration of the use of ferrate(VI) for water and wastewater treatment has been addressed (Jiang 2007, 2014; Jiang and Lloyd 2002; Sharma 2002; Sharma et al. 2015). The studies revealed that ferrate(VI) can disinfect microorganisms, partially degrade and/or oxidize organic and inorganic impurities,

J.-Q. Jiang (✉)
Department of Civil Engineering & Environmental Management, SCEBE, Glasgow Caledonian University, Glasgow, Scotland, UK
e-mail: jiaqian.jiang@gcu.ac.uk

J. Filip et al. (eds.), *Advanced Nano-Bio Technologies for Water and Soil Treatment*, Applied Environmental Science and Engineering for a Sustainable Future,
https://doi.org/10.1007/978-3-030-29840-1_12

remove suspended/colloidal particulate materials, and reduce phosphate concentrations significantly in sewage treatment. Most recently, researches have reported using ferrate(VI) to treat emerging micropollutants in water purification processes (Lee et al. 2005; Jiang et al. 2005). However, the implementation of ferrate (VI) technology in practice represents a challenge due to the instability of a ferrate (VI) solution and a high production cost of solid ferrate(VI) products. The research has been directed at generation and application of ferrate(VI) in situ (Jiang et al. 2009). Practical advantages of ferrate(VI) over the existing water and wastewater treatment methods can only be demonstrated when water industry is able to implement the technology into full-scale application. To do so, a series of pilot-scale trials using ferrate(VI) for water and wastewater treatment are needed to establish the database of the comparative treatment performance and to assess the operating cost against the existing technologies.

This chapter reviews the field work in drinking water and wastewater treatment carried out by this author's team to identify the optimal operating conditions for using ferrate(VI) as an alternative to currently used chemicals in the treatment of drinking water and sewage.

12.2 Materials and Methods

12.2.1 Pilot-Scale Trials of Using Ferrate(VI) Coagulation before Filtration in Drinking Water Treatment Processes

The pilot plant was designed and set up by Lake Constance Water Supply according to the parameters shown in Table 12.1. The water flew through a microsieve filter (15 μm), and then into the customized ozone mixer followed by seven contact tanks. Next, ferrate(VI) and $FeCl_3$ were pumped into two flowing waters to be treated by peristaltic pumps separately with the required volume dosage. Water/coagulant mixtures were directed into two separate chambers in which suitable flocculation occurred before the flow entered two parallel filter columns with similar flow conditions. The filter columns are made of steel tube

Table 12.1 Design parameters of pilot plant filters

Filter parameter	Unit	Details
Total height	m	3.6
Filter area	m^2	0.283
Average flow rate	L/h	~1700
Average flow velocity	m/h	~6
Running time	h	40–100
Filter media		40 cm EVERZIT N (0.8–1.6 mm); 60 cm sand (0.4–0.7 mm); ~18 cm supporting material

Table 12.2 Pilot plant operating conditions (Fe dose = 0.1 mg/L)

Parameters	Details
Initial/final flow rate (L/h)	1500/1000
Running time (h)	5–7
Online measurement instrument	Particle counter; flow rate, pH, and conductivity
Final water sampling time	After 4 h of dosing coagulant
Ozone dosing (mg/L)	~1.2 (dose); ~0.7 (at ozone mixer outlet)
Residual ozone concentration before sand filters (mg/L)	0.05–0.08

running vertically having the design parameters mentioned in Table 12.1. The operating conditions of filters can be seen in Table 12.2.

Analysis of various water quality parameters and residual ozone concentration followed the standard methods (Eaton et al. 1995). Analysis of micropollutants, metformin, benzotriazole, and acesulfame was carried out using an Agilent 1100 LC system (Agilent, Waldbronn, Germany) equipped with an API 4000 triple quadrupole mass spectrometer with electrospray ionization (Applied Biosystems, Darmstadt, Germany). The column was Ultra Aqueous C18 (250 mm × 4.6 mm) from Restek (Bad Homburg, Germany). Water (eluent A) and acetonitrile/water (95/5 vol%/vol%, eluent B) with 0.1 vol% formic acid were used as mobile phase with a flow rate of 0.75 mL/min. The column was brought to a constant temperature 25 °C. 100 μL of the sample was injected directly without any further sample pretreatment. The eluent program started with 5% eluent B, increased linearly within 6 min to 80% eluent B and increased linearly from 6 to 12 min to 95% eluent B. After the analytic run the eluent was set back to 5% eluent B from 12 to 18 min. The LC column was coupled to the mass spectrometer directly into the ion source, which was heated to 650 °C inside the ionization section with nitrogen gas flows of 40 psi for curtain gas and 60 psi for the ion source gases 1 and 2, respectively. The ion spray voltage was set to 5.5 kV. The mass spectrometer was operated in the positive mode. The detection of metformin was performed with three multiple reaction monitoring transitions: from m/z 130 to m/z 71 at a collision energy of 19 V, from m/z 130 to m/z 60 at a collision energy of 29 V, and from m/z 130 to m/z 85 at a collision energy of 25 V.

12.2.2 *Pilot-Scale Trials of Dosing Ferrate(VI) into Crude Sewage for Wastewater Treatment*

A pilot-scale reactor system consists of two major components: (1) the ferrate (VI) production component including an alkali tank, a chemical transfer pump, an electrochemical reactor, a power supply, a ferrate(VI) product storage tank, a chemical dosing pump, and a control box; and (2) the treatment part including a submerged pump situated in the influent channel, hose, artificial channel where the ferrate(VI) was dosed, and the exit hose that goes back to the main wastewater flow channel.

Ferrate(VI) production time was 30 min per each preparation. The resulting ferrate(VI) was measured using an established spectroscopy method where the absorbance of the ferrate(VI) solution was measured at 505 nm. The absorbance was converted to the concentration using an absorption coefficient of 1100 1/(M cm). The ferrate(VI) dosing flow rate was determined based on the desired dose and the measured ferrate(VI) concentration. The samples after ferrate (VI) dosing and mixing were collected and analyzed for the concentrations of suspended solids (SS), chemical oxygen demand (COD), biochemical oxygen demand (BOD), total phosphorus (P), pH, and residual Fe.

12.3 Results and Discussion

12.3.1 Pilot-Scale Drinking Water Treatment Performance

The tested lake water was of better quality, so the required coagulant dose was low (0.1 mg Fe/L). For the given operating conditions (Table 12.2), particle removal percentage after filtration was 93% for raw water and 97% for the ozonized water (Fig. 12.1). As can be seen in Fig. 12.1, there were larger numbers of 1 μm particles

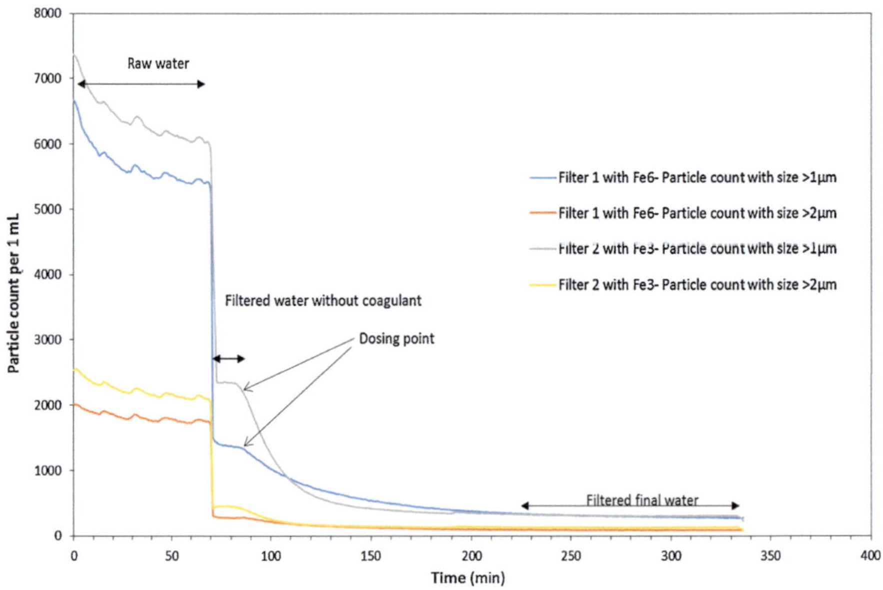

Fig. 12.1 Particle removal by coagulation at 0.1 mg Fe/L and pilot plant filtration from raw water (Filter 1—ferrate, Filter 2—$FeCl_3$). (Reproduced from Jiang et al. 2018)

Table 12.3 Comparative performance of ferrate(VI) and $FeCl_3$

		Raw water		Ozone water	
	Unit	Ferrate(VI)	$FeCl_3$	Ferrate(VI)	$FeCl_3$
Fe dosage	Mg/L	0.1			
Turbidity	%	~80	~80	~90	~90
UV-254	No change				
DOC	No change				
Residual Fe	μg/L	~16	~9	~15	~12
Particle removal	%	~93	~94	~98	~98
Bromate formation	μg/L	0	0	~11	~11
Benzotriazole removal	%	10	0	10	0
Acesulfame removal	%	10	0	10	0
Metformin removal	%	10	0	10	0
X-ray contrast medium removal	%	100	100	100	100

than those of 2 μm. For both raw water and ozonated water, two filters had different performance; Filter 1 achieved slightly better performance than Filter 2. However, after dosing the coagulants, such differences were not shown.

The field trials were carried out at the pilot plant where the operating conditions followed the main plants. Therefore, the ferrate(VI) dose used was very low, 0.1 mg Fe/L, in order to equally compare with the performance of ferric chloride (0.1 mg Fe/L) and ozonation (4 mg O_3/L) running at the main plant. Table 12.3 shows the comparative performance of ferrate(VI) and $FeCl_3$ at 0.1 mg/L dosage in pilot-scale experiments. Both performed similarly in the removal of particles, UV absorbers, and dissolved organic carbon (DOC) under the given conditions. However, ferrate (VI) can achieve 10% reduction of metformin, benzotriazole, and acesulfame, but $FeCl_3$ with ozonation cannot remove any of these compounds. Moreover, the ferrate (VI)-treated water did not generate bromate, but the ozonated water did although the resulting bromate concentration was 11 μg/L.

12.3.2 *Crude Sewage Treatment Performance in the Pilot Plant*

During the pilot-plant study, properties of the crude sewage were tested. Concentrations of the target quality parameters varied from 242 to 730 mg/L for the SS, 523 to 1125 mg/L for the COD, 235 to 441 mg/L for the BOD, and 11.3 to 18.5 mg/L for the phosphate as total P.

The comparative performance of ferrate(VI) and ferric sulfate can be seen in Table 12.4. With a low dose (0.03 mg Fe/L), ferrate(VI) can achieve similar or better performance than a high dose of ferric sulfate (37 mg Fe/L).

Table 12.4 Comparative performance[a] of crude sewage treatment with ferric sulfate and ferrate (VI)

Chemical and dose	Average percentage removal for the given doses of given chemical (%)			
	SS	P	COD	BOD
Ferrate(VI) (0.03 mg Fe/L)	79	56	50	30
Ferric sulfate (37 mg Fe/L)	78	59	54	43

aCrude sewage properties: [SS] = 730 mg/L; [P] = 18.5 mg/L; [COD] = 1125 mg/L; [BOD] = 388 mg/L

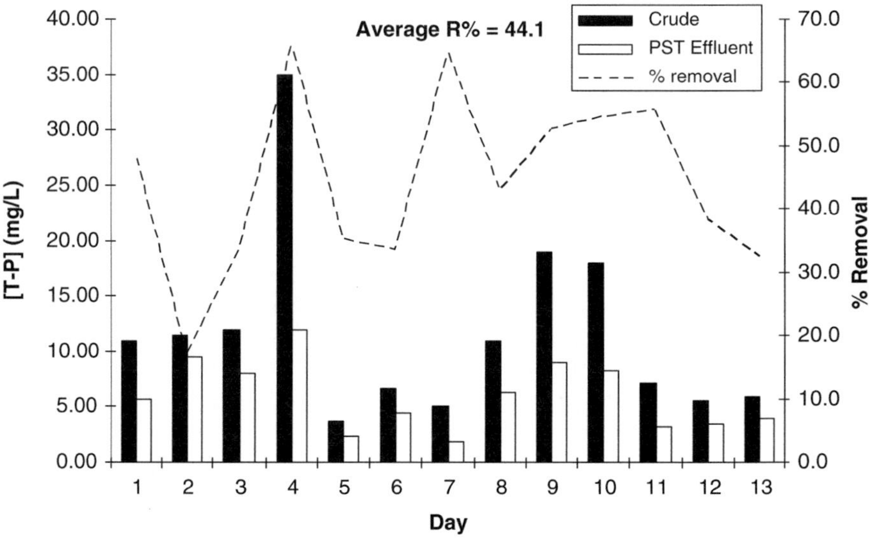

Fig. 12.2 Total phosphorus concentration in crude sewage and the effluent after pre-sedimentation tank (PST) and percentage removals of P. (Reproduced from Jiang et al. 2018)

Figs. 12.2, 12.3, 12.4 and 12.5 show concentrations of P, COD, BOD, and SS from various samples during the given test running period as well as the relevant percentage removals. For the above-stated low dose (0.16 mg $Fe^{(VI)}$/L), ferrate (VI) achieved, in average removals, 64% of SS, 44% of phosphate, 46% of COD, and 40% of BOD when the pH was above 9. The comparative performance of ferrate (VI) and ferric sulfate shows that the ferrate(VI) dose of 0.16 mg $Fe^{(VI)}$/L can achieve similar results to those of ferric sulfate at a relatively high dose (25 mg $Fe^{(III)}$/L). Much smaller dose demand of ferrate(VI) can significantly reduce the chemical requirement and sludge production, so it could result in cutting the operating cost in the treatment of sewage.

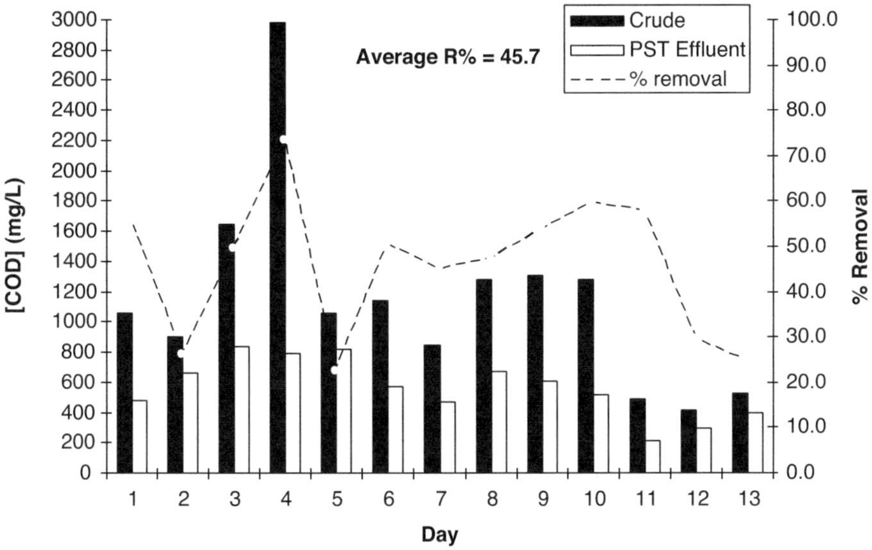

Fig. 12.3 COD concentration in crude sewage and the effluent after pre-sedimentation tank (PST) and percentage removals of COD. (Reproduced from Jiang et al. 2018)

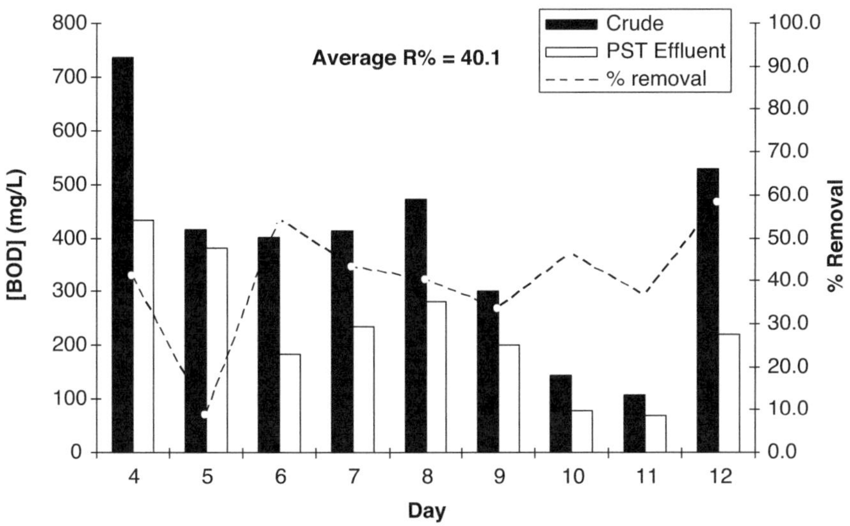

Fig. 12.4 BOD concentration in crude sewage and the effluent after pre-sedimentation tank (PST) and percentage removals of BOD. (Reproduced from Jiang et al. 2018)

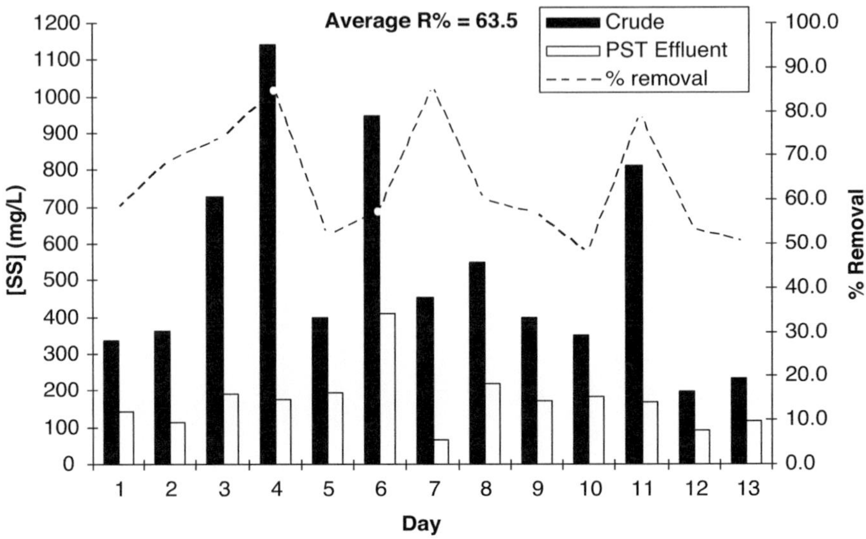

Fig. 12.5 Suspended solids concentration in crude sewage and the effluent after pre-sedimentation tank (PST) and percentage removals of SS. (Reproduced from Jiang et al. 2018)

12.4 Conclusions

Pilot-scale field tests with low ferrate(VI) doses for drinking water treatment achieved average particle removal percentage of 93% for the raw water and 97% for the ozonized water in terms of particle counting data. No pH neutralization was required after dosing ferrate(VI). In comparison with using ozonation and $FeCl_3$ coagulation, ferrate(VI) has the additional benefits: it can remove 10% metformin, benzotriazole, and acesulfame, whereas $FeCl_3$ with ozonation cannot. Additionally, the ferrate(VI)-treated water did not generate bromate, while the ozonated water did.

As for the sewage treatment, pilot-scale tests demonstrated that with very low dose range, 0.1–0.2 mg $Fe^{(VI)}$/L, ferrate(VI) achieved removal targets of phosphorus, COD, BOD, and suspended solids from the crude sewage, while the dose demand of ferric sulfate (25 mg Fe/L) was much higher than that of ferrate(VI); these will reduce the chemical demand and sludge production and therefore generate substantial cost saving in treating sewage. Depending on individual circumstances, the ferrate(VI) technology could be implemented in wastewater treatment practice.

Acknowledgement This author acknowledges the funding and staff support from Southern Water in England and Zweckverband Bodensee-Wasserversorgung Water Board of Germany to run the field trials. Thanks also go to author's team members including PhD students and postdoc researchers who assisted staffs when conducting the field tests. The views of this chapter may not necessarily represent those of the stated water companies.

References

Eaton AD, Clesceri LS, Greenberg AE, Franson MAH (eds) (1995) Standard methods for the examination of water and wastewater, 19th edn. American Public Health Association, American Water Works Association, Water Environment Federation, Washington

Jiang JQ (2007) Research progress in the use of ferrate(VI) for the environmental remediation. J Hazard Mater 146(3):617–623. https://doi.org/10.1016/j.jhazmat.2007.04.075

Jiang J-Q (2014) Advances in the development and application of ferrate(VI) for water and wastewater treatment. J Chem Technol Biotechnol 89(2):165–177. https://doi.org/10.1002/jctb.4214

Jiang J-Q, Lloyd B (2002) Progress in the development and use of ferrate(VI) salt as an oxidant and coagulant for water and wastewater treatment. Water Res 36(6):1397–1408. https://doi.org/10.1016/S0043-1354(01)00358-X

Jiang JQ, Yin Q, Zhou JL, Pearce P (2005) Occurrence and treatment trials of endocrine disrupting chemicals (EDCs) in wastewaters. Chemosphere 61(4):544–550. https://doi.org/10.1016/j.chemosphere.2005.02.029

Jiang J-Q, Stanford C, Alsheyab M (2009) The online generation and application of ferrate(VI) for sewage treatment—a pilot scale trial. Sep Purif Technol 68(2):227–231. https://doi.org/10.1016/j.seppur.2009.05.007

Jiang J-Q, Stanford C, Petri M (2018) Practical application of ferrate(VI) for water and wastewater treatment – site study's approach. Water Energy Nexus 1(1):42–46. https://doi.org/10.1016/j.wen.2018.05.001

Lee Y, Yoon J, von Gunten U (2005) Kinetics of the oxidation of phenols and phenolic endocrine disruptors during water treatment with ferrate (Fe(VI)). Environ Sci Technol 39(22):8978–8984. https://doi.org/10.1021/es051198w

Sharma VK (2002) Potassium ferrate(VI): an environmentally friendly oxidant. Adv Environ Res 6 (2):143–156. https://doi.org/10.1016/S1093-0191(01)00119-8

Sharma VK, Zboril R, Varma RS (2015) Ferrates: greener oxidants with multimodal action in water treatment technologies. Acc Chem Res 48(2):182–191. https://doi.org/10.1021/ar5004219

Chapter 13
Field Study IV: Arsenic Removal from Groundwater by Ferrate with the Concurrent Disinfecting Effect: Semi-Pilot On-site Application

Monika Heřmánková, Roman Vokáč, Jan Slunský, and Jan Filip

Abstract The effect of ferrates on arsenic removal from groundwater was initially observed using laboratory tests. Two different sources of groundwater containing an elevated concentration of As (ca 100 μg/L) were used for the semi-pilot scale testing, both possessing the potential for drinking water production. Groundwater treated by ferrates under laboratory conditions proved to meet the requirements for As limits in drinking water, i.e., 10 μg/L (according to Czech legislation—Public notice No. 252/2004 Sb.). A prototype of a portable technological unit for an on-site ferrate application (Fe(IV), Fe(V)) has been constructed to prove on a semi-pilot scale that ferrates are applicable for production of drinking water. Water flow of 100 L/h, two sources of groundwater containing arsenic in concentrations ten times exceeding the limit for drinking water, and ferrates (in the form of commercial ferrate, dosed in different amounts—5 mg/L, 10 mg/L, 15 mg/L, and 20 mg/L) were used. The quality of the treated water from both sites was in compliance with the requirements for drinking water. The minimal dose of commercial ferrates to reach the As limit for potable water was set to 10 mg/L for the first water source and 15 mg/L for the second one. Elimination of microbial organisms was achieved even with a lower dose of commercial ferrates—10 mg/L. The main advantages of the tested

M. Heřmánková (✉)
DEKONTA, a.s., Stehelčeves, Czech Republic

AECOM CZ, s.r.o., Praha, Czech Republic
e-mail: hermankova@dekonta.cz

R. Vokáč
AECOM CZ, s.r.o., Praha, Czech Republic

J. Slunský
NANO IRON, s.r.o., Židlochovice, Czech Republic

J. Filip
Regional Centre of Advanced Technologies and Materials, Palacký University Olomouc, Olomouc, Czech Republic

J. Filip et al. (eds.), *Advanced Nano-Bio Technologies for Water and Soil Treatment*, Applied Environmental Science and Engineering for a Sustainable Future,
https://doi.org/10.1007/978-3-030-29840-1_13

technology over the existing ones are minimal reagent consumption, minimal sludge production with As (hazardous waste), and disinfecting effect.

Keywords Ferrates · Arsenic removal · Drinking water treatment · Groundwater · Disinfecting effect · Disinfecting agents

13.1 Introduction

Ferrates have been described in the literature for the past two decades as a new type of substance enabling versatile employment in environmental applications. Laboratory tests on the level of fundamental research proved their significant potential applicable for water remediation (e.g., removal of metals, triazine pesticides, endocrine disruptors, cyanides, chemical warfare agents, etc., disinfecting effects (Filip et al. 2011; Zboril et al. 2012; Prucek et al. 2013, 2015; Yates et al. 2014). Ferrates show strong oxidation effect; their final products after the reaction—ferric oxides and oxyhydroxides—are considered nontoxic, and, moreover, they can act as coagulants. Chemical properties of ferrates such as fast reaction with contaminants or low reagent consumption and the resulting lower waste footprint represent an advantage for their application compared with other commonly used technologies.

High-quality resources of drinking water have been disappearing in the Czech Republic, hence the need to begin deploying resources of lower quality. On account of that, research into the disinfecting effect of ferrates has been included within the treatment scheme for arsenic removal. From a microbiological point of view, the remediated water met the requirements for drinking water. Such a promising effect may have been caused by the combination of several factors: disinfecting effect of ferrates reacting in water environment, sorption of microorganisms on the formed nanoparticles of iron oxide/oxyhydroxide, and ultrafiltration with the pore size of 0.45 μm.

13.2 Characterization of Ferrates Used for the Testing

For the purpose of laboratory tests aimed at arsenic removal from groundwater, we used commercial ferrates. In this case, these were commercial ferrates from NANO IRON company (Czech Republic) containing about 33 wt.% of K_2FeO_4 in aqueous solution (UV–VIS measurement done by spectrophotometry at 510 nm and subsequent calculation according to Bielski and Thomas (1987). Detailed characterization of the commercial ferrate is provided in Tables 13.1 and 13.2.

The commercial ferrate is formed by coarse-grained aggregates of a black-violet or almost black color containing crystals of ferrates Fe(V) and Fe(VI). When it is dissolved in water, ferrate Fe(V) disproportionates to Fe(VI) and Fe(III) in the ratio about 2:1. Ferrates in aqueous solution exist mainly in the form of Fe(VI). Fe(VI) is unstable in aqueous solution, after 24 h the concentration reaches only about 10% of

Table 13.1 Concentration of ferrate in the commercial product (solid product)—characterized by Mössbauer and atomic absorption spectroscopy

ENVIFER batch 019, solid product	
Compound	Concentration (%)
$KFe(III)O_2$	29 ± 3
$K_3Fe(V)O_4$	50 ± 5
$K_2Fe(VI)O_4$	6 ± 2
K_2O	<3

The difference of up to 100% is due to the content of oxide impurities (approx. 5–10%), the accumulation of measurement errors and the content of unidentified components with different stoichiometry

Table 13.2 Concentration of ferrate in the commercial product (aqueous solution)—characterized by UV–VIS spectrometry, 510 nm

ENVIFER batch 019, aqueous solution 0.5 g/L		
Time after dissolving	Absorbance	Concentration of $K_2Fe(VI)O_4$ (%)
1 min	0.98	33.35
24 h	0.1	3.40

With regard to practical use, checking the current content of ferrates in the aqueous solution in the form of Fe(VI) using UV–VIS spectrometry is sufficient and accessible to most of the laboratories

the original concentration in the solution (with the commercial ferrate, it is only 3.4%; see Table 13.2). In industrial practice, ferrates could be processed in a solid form on plants enabling dry storage and dosing. Another option being considered is dosing ferrates in the form of tablets (Czölderová et al. 2018).

13.3 The Principle of the Method for Arsenic Separation by Ferrates

Reaction of ferrates with water leads to the formation of nanoparticles in the form of amorphous γ-Fe_2O_3 and γ-FeOOH (Kolařík et al. 2018). Arsenic is adsorbed, and partly incorporated, within 3–4 min into the structure of the nanoparticles. Subsequently, the microflakes formed during the process containing Fe hydroxides and As have to be thoroughly separated from the water (Prucek et al. 2013).

Initially, we carried out a series of laboratory tests with two different sources of groundwater having a naturally elevated arsenic concentration (ca 100 µg/L), both of which could be used for drinking water production. Following the application of ferrates, the groundwater met the requirements for arsenic limits in drinking water (ca 10 µg/L) in both of the tested sources. On the basis of the experience with the laboratory tests, the prototype of a mobile device for a ferrate application (Fe (IV) and Fe(V)) on-site with the flow of 100 L/h has been designed. This device has been constructed to prove on a semi-pilot scale that ferrates are applicable for production of drinking water.

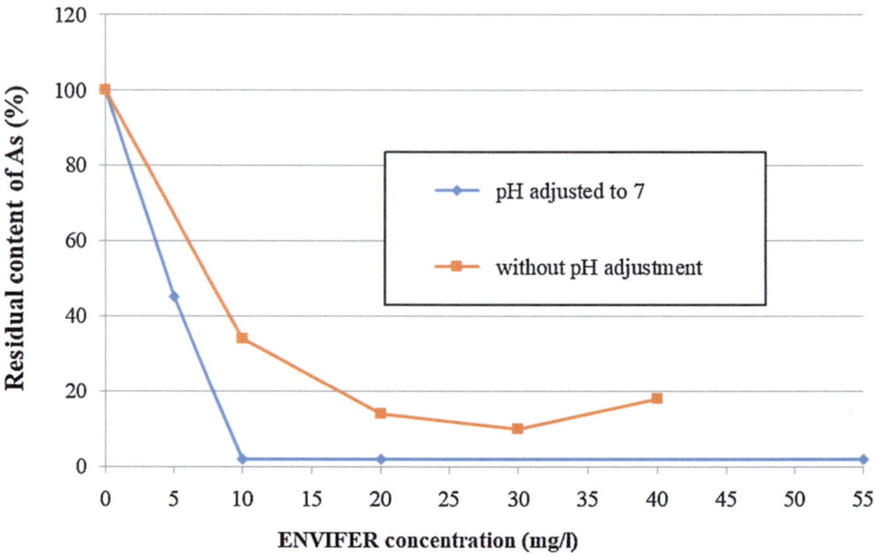

Fig. 13.1 pH optimization for dosing of commercial ferrate to get a maximal As removal in water KLU (see below)—a laboratory test

13.4 Optimal Conditions for Arsenic Separation Set under Laboratory Conditions for the Target Groundwater

Ferrates exhibit alkali reaction with water. Therefore, the pH rises as the amount of ferrate increases. pH values, oxidation–reduction potential (ORP), and temperature were continuously measured during the reaction of groundwater with ferrate. Laboratory tests confirmed the optimal pH value for As removal to be 7.0 (see Fig. 13.1). Further decrease in the pH value leads to lower ferrate stability.

Maximal oxidation–reduction potential E_h measured during the reaction reached 850 mV. This value confirms strong oxidizing environment. Optimal reaction time was set to 20 min, including adding the ferrate, adjusting the pH, and adding a polymeric flocculant. It was followed by 10-min ageing enabling possible sedimentation and by another 15 min of three-phase filtration.

13.4.1 Laboratory-Verified Conditions for as Removal from Groundwater Using Ferrates

- *pH*: after adding the ferrate, adjust the pH to 7 with HCl
- *Redox potential*: ORP E_h maximum 850 mV; max measured ORP on AgCl electrode 630 mV

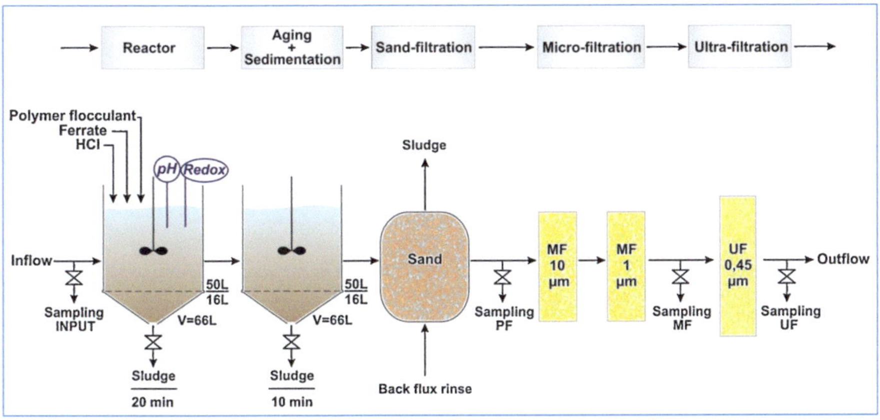

Fig. 13.2 Scheme of technological unit for arsenic removal from groundwater by ferrates. (Capacity approx. 100 L/h)

- *Larger particles formation support*: application of polymeric flocculant solution increases efficiency in separation of the Fe-As complex by filtration
- *Reaction time*: 20 min (1 min ferrate addition, 5–7 min acid addition, 12–15 min flocculant addition)
- *Microflakes separation time*: 10 min sedimentation (negligible sludge formation), 15 min three-step filtration (sand filtration + microfiltration + ultrafiltration)

13.5 Description of a Mobile Semi-Pilot Plant for Ferrate Application

On the basis of the laboratory tests results, a prototype of a mobile technological unit for ferrate (Fe(IV), Fe(V)) application has been designed. This mobile device is aimed at on-site application operating at the flow rate of 100 L/h (see Fig. 13.2).

The aim was to verify the applicability of ferrate for drinking water production on a semi-pilot scale.

The whole line is placed in a transportable container that can be fixed to a tow bar of a car (see Fig. 13.3). The size of the container is 2 m × 2 m × 3 m (width, height, length). The container runs on electrical power 380 V.

The line configuration (see Fig. 13.4)

- mixing reactor with ferrate dosing, diluted HCl, and a solution of polymeric flocculant
- tank for ageing of the reaction and possible sedimentation
- sand filtration
- microfiltration—10 µm and 1 µm
- ultrafiltration—0.45 µm

Fig. 13.3 Transport of mobile technological unit to a site

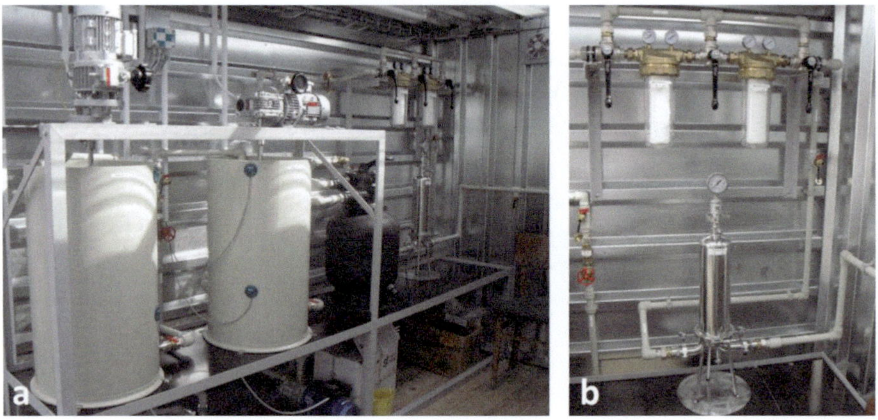

Fig. 13.4 (**a**) The overall view of the semi-pilot plant—mixing reactor, tank for ageing, sand filtration (the black container), two-phase microfiltration (in the top right-hand corner), ultrafiltration (the silver container below the microfiltration). (**b**) Detailed picture of microfiltration (at the top) and ultrafiltration (silver container below)

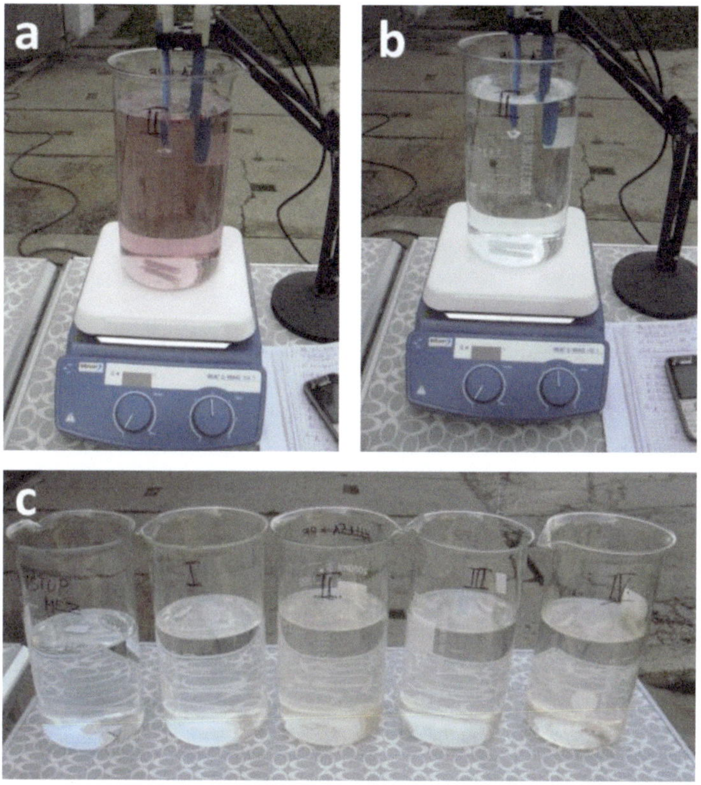

Fig. 13.5 (**a**) Sample of water right after ferrates were added. Commercial ferrate dose—10 mg/L. (**b**) The same sample after 30 min of ferrate addition. (**c**) Verifying on-site tests: progress in the change of the color monitored from the beginning—(INPUT, see "VSTUP MEZ") until the end of the process (left to right). The numbers refer to different doses of the commercial ferrate—5 mg/L (I), 10 mg/L (II), 15 mg/L (III), and 20 mg/L (IV)

13.6 Test Results of the Mobile Line on Site

13.6.1 Arsenic Separation

The proposed technology was tested using actual groundwater with a naturally elevated concentration of As in two different localities being 12 km far away from each other (unspecified localities in the Czech Republic labeled MEZ and KLU). The water from MEZ contained about 90 μg/L As, the water from KLU contained about 100 μg/L, while the limit for concentration of As in drinking water is 10 μg/L. The dose and timing of the reagents was adjusted on site by technological laboratories to suit the conditions of the water being currently treated (see Fig. 13.5).

Subsequently, the target water was treated on a semi-pilot plant using different doses of commercial ferrates (5 mg/L, 10 mg/L, 15 mg/L, 20 mg/L). The operating

Table 13.3 Arsenic removal using ferrates, semi-pilot plant, duplicate testing of each dose of commercial ferrate

As separation Semi-pilot tests	INPUT As (μg/L)	OUTPUT As (μg/L)							
		ENVIFER 5 mg/L		ENVIFER 10 mg/L		ENVIFER 15 mg/L		ENVIFER 20 mg/L	
		A	B	A	B	A	B	A	B
Water MEZ	89	11	17	<5	<5	<5	<5	<5	<5
Water KLU	102	–	–	13	17	12	<5	<5	5

Table 13.4 The effect of ferrates on the water in KLU locality after using 15 mg/L of commercial ferrate, duplicate test (A, B)

Sample	As μg/L	Fe mg/L	K mg/L	pH –
Detection limit	0.0050	0.0020	0.0150	1
A – INPUT	102	0.0035	2.5	7.44
A – After PF	69	1.26	9.03	6.59
A – After MF	24.9	0.252	8.73	7.35
A – After UF	12.5	0.0814	8.75	7.22
B – INPUT	102	0.0035	2.5	7.44
B – After PF	62.5	1.3	9.26	7.39
B – After MF	43.2	0.784	9.37	7.37
B – After UF	<0.005	0.0746	9.85	7.25

50 L Water KLU + 750 mg Commercial Ferrate +0.5 mL Flocculant +4.0 mL 10% HCl

volume of the mixing reactor for a semi-continuous process is 50 L; the reaction lasted for ca 30 min including the time for emptying the reactor. 100 L of raw water could be processed within 1 h. The pH values, the oxidation–reduction potential, and temperature were measured continually. The maximum value of E_h during the reaction was 850 mV. Each dose of ferrate was verified in duplicate on the semi-pilot plant. In order to evaluate the effectivity of the As separation on the semi-pilot plant, four different samples were collected—raw water (input), water after sand filter (PF), water after microfiltration 1 μm (MF), and water after ultrafiltration (UF), i.e., the treated water. The chemical analysis of the samples was carried out in an accredited commercial laboratory (ALS). The final concentrations of As after applying the target doses of commercial ferrates in each locality are shown in Table 13.3.

Arsenic concentrations in the treated water from both localities were below the set limit for drinking water (max. 10 μg/L). The optimum dose of a commercial ferrate for MEZ water was 10 mg/L, for KLU water 15 mg/L. It is highly likely that the 12.5 μg/L value achieved in the first test (A) was due to the previous test with a lower dose of a commercial ferrate (10 mg/L), which was proved to be insufficient for arsenic removal from KLU locality (Filip et al. 2017).

The agents used for arsenic removal from groundwater obviously affect the quality of the remediated water. Ferrates (K_2FeO_4) raise the concentration of K, Fe, and the pH. Table 13.4 shows the example of such changes. Specifically, herein we present the results of a duplicate test on KLU water after applying a 15 mg/L dose

Table 13.5 Comparison of the effect of arsenic removal using particular filtration steps for the test water employing the optimum dose of commercial ferrate

	MEZ water (commercial ferrate 10 mg/L)		KLU water (commercial ferrate 15 mg/L)	
	As concentration in water (%)	Ratio of As removed by a particular filtration phase (%)	As concentration in water (%)	Ratio of As removed by a particular filtration phase (%)
INPUT	100	0	100	0
Sand filtration	49.5	50.5	61.3	38.7
Microfiltration	11.9	37.6	42.4	18.9
Ultrafiltration	0	11.9	0	42.4

of a commercial ferrate. Such a dose guarantees that the limit for arsenic concentration in drinking water will be met.

If we focus on the test B performance and its results, i.e., the output after ultrafiltration, it is clear that the final concentrations—Fe 0.0746 mg/L (limit 0.2 mg/L), K 9.85 mg/L (limit is not set; potassium is an essential element in humans and is seldom, if ever, found in drinking water at levels that could be a concern for healthy humans and the recommended daily requirement is greater than 3000 mg), and pH 7.25 (limit for pH 6–8)—comply with the requirements for drinking water. Adding ferrates has only a minimum impact on the original water content.

The effect of arsenic separation using different filtration steps for the optimal dose of commercial ferrates in KLU water and MEZ water is different—see Table 13.5.

With MEZ water, the highest proportion of arsenic was already removed during the sand filtration phase (50%); with KLU water the highest proportion of arsenic was removed only during the last filtration phase—ultrafiltration (42.4%). It is clear that the water chemistry has a considerable impact on the reaction of the water with the ferrate and on forming Fe microflakes for arsenic separation. As a result, the operating costs of the treatment will be different for the two localities even if the initial concentration of arsenic is almost the same in each one. It has been confirmed repeatedly that proposing a technology and reagent doses only according to laboratory tests based on spiked water is inaccurate. Results that are to be reliable must be achieved with real water that has been remediated. The results of the semi-pilot unit tests showed that the proposed technology is sufficiently robust and it reliably achieves the removal of Fe-As microflakes regardless of their size, so that the arsenic concentration in the remediated water complies with the limit for drinking water.

13.6.2 Microbiological Analysis

The results of the microbiological analysis of KLU water carried out on a semi-pilot unit are presented in Table 13.6. We monitored the microbiological rejuvenation of the input and output water from the testing line after ultrafiltration for each tested dose of the commercial ferrate.

Table 13.6 Results of the microbiological analysis of KLU water—input and output water from the tested technology for each dose of the commercial ferrate (sampling point—after UF)

Bacterial microflora	Unit	KLU INPUT	KLU ENVIFER 10 mg/L without flocculant	KLU ENVIFER 10 mg/L	KLU ENVIFER 15 mg/L	KLU ENVIFER 20 mg/L
Clostridium perfringens	CFU/ 100 mL	0	0	0	0	0
Coliform bacteria	CFU/ 100 mL	100	0	0	0	0
Enterococci	CFU/ 100 mL	57	0	0	0	0
Escherichia coli	CFU/ 100 mL	0	0	0	0	0
Microorganisms cultivated at 22 °C	CFU/ mL	3300	18	0	3	0
Microorganisms cultivated at 36 °C	CFU/ mL	2900	10	0	0	0

Table 13.6 clearly shows that the water coming out from the tested technology complies with the requirements for drinking water from the microbiological point of view. The disinfecting effect of ferrates has already been described in the literature. Nevertheless, such results have been achieved owing to the combination of different factors—disinfecting effect caused by ferrates reacting in water, sorption of microorganisms into the formed iron hydroxides microflakes, the impact of ultrafiltration with the pore size of 0.45 μm, and disinfecting the line before the test. However, the reliability of the disinfecting effect of the tested technology has yet to be confirmed under operating conditions in the long run.

13.6.3 Economic Point of View

The price of commercial ferrates for industrial applications is currently (2018) about EUR 80 per kilo. As the dose of ferrate ensuring the effectiveness of the arsenic removal from the tested groundwaters is 10–15 mg/L, i.e., 10–15 g/m^3, the price of the amount sufficient to treat 1 m^3 of water could range between 1.0–1.1 EUR/m^3.

In 2017 the following ferrate-based products were available on the market for the purpose of laboratory tests (see Table 13.7).

Since the production of ferrates is demanding, batches from individual producers may vary. Therefore, it is recommended that the quality of the particular batch should be checked in the following steps:

1. Verify the presence of Fe in a ferrate aqueous solution by fundamental analysis.

Table 13.7 List of ferrates available on the market

Producer/Supplier	Price and comment
Santa Cruz Biotechnology, Inc. (USA)	104 EUR/g Purity ≥92% according to the producer
Sigma Aldrich	200 EUR/g Purity ≥95% according to the producer
NANO IRON (CZ)	0.08 EUR/g Content of K_2FeO_4 30–40%

2. Measure the absorbance of the aqueous solution at 510 nm (or 505 nm). In the literature, it is recommended that both of these wavelengths should be used for measuring the concertation of Fe(VI) in aqueous solution. The Fe(VI) spectrum has a broader peak in this area, i.e., both variants are correct. When comparing UV–VIS spectra of ferrates and permanganate, it is clear that measuring at 510 nm is not sufficient for distinguishing a ferrate solution from permanganate (see Fig. 13.6a, b).
3. Ferrates can contain traces of heavy metals (Cr, Ni, V, Cu, etc.). Although, it is convenient to use ferrates at very low concentrations even from economic point of view (dozens of mg/L), it is still recommended that the final content of heavy metals in the remediated water in which ferrates were applied should be checked.
4. When in solutions with demineralized water, ferrates will show greater efficiency. Nevertheless, such a ferrate is unstable and requires fast processing. When in solutions with drinking water, ferrates will be more stable (1–2 h), but their reaction will take more time.

13.6.4 Waste Products

The amount of waste products was evaluated after completing the test in KLU locality. Within this testing, 516 L of water containing 0.1 mg/L of arsenic was processed under different technological conditions, meaning different ENVIFER doses.

Sand filter, microfiltration (2 pcs), and ultrafiltration separated out ferrate-reaction products containing arsenic from the water having been treated. Filtration units from microfiltration and ultrafiltration showed only slightly yellowish color; neither of them showed any fall in pressure (indication of pore clogging), i.e., they were preserved for the following tests. The sand filter was cleaned using a standard method, i.e., by counter-current flushing with five units of drinking water equalling five units of the sand filter content. The filling volume of the sand filter was 35 L, the total volume of the cleaning water was 175 L (5 × 35 L). The cleaning water was collected in a container. After the process of cleaning the filter had been finished, samples were collected for chemical analysis in an accredited laboratory ALS.

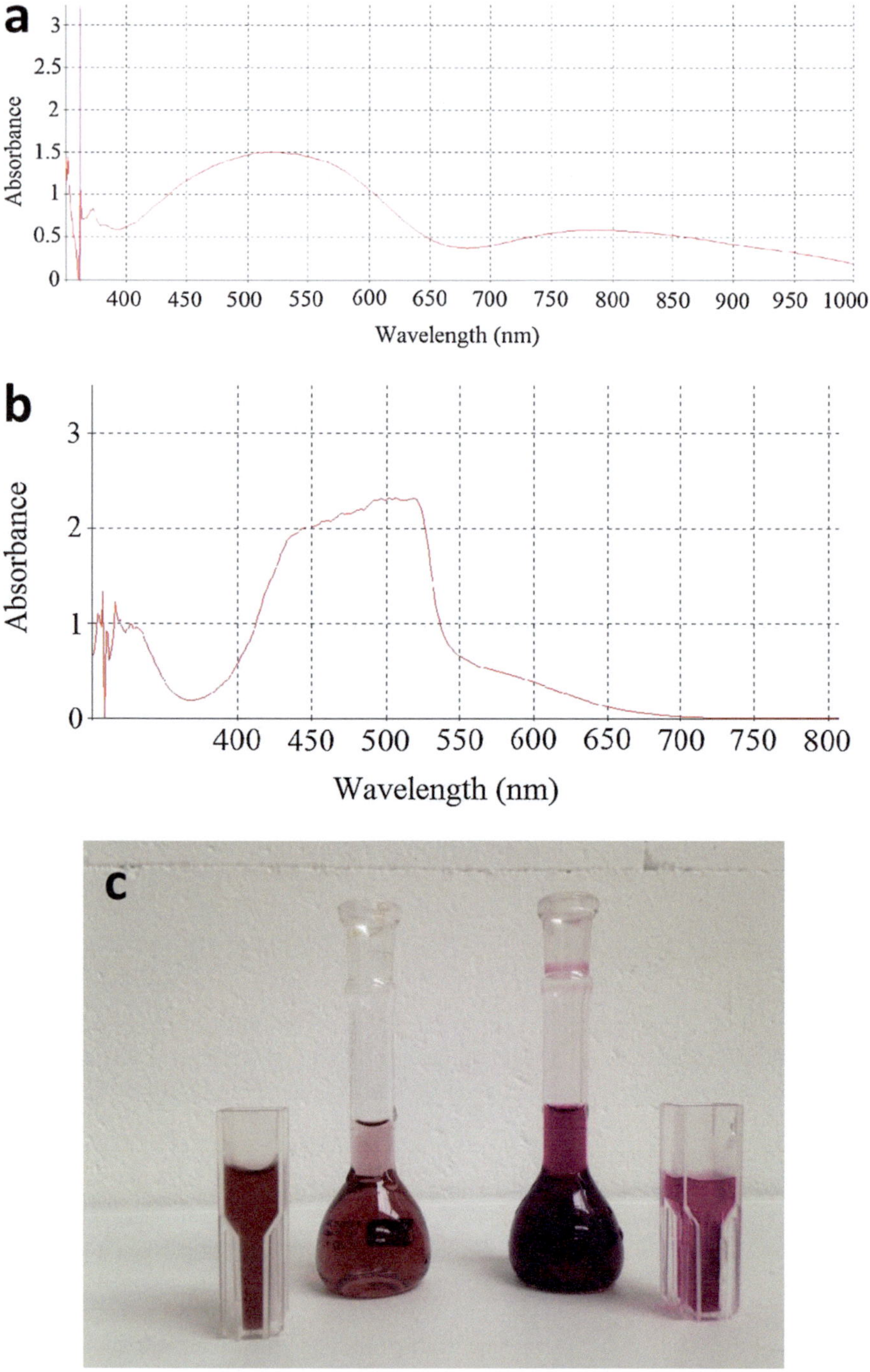

Fig. 13.6 (**a**) UV–VIS spectrum of ENVIFER 036 in aqueous solution at concentration 0.5 g/L. (**b**) UV–VIS spectrum of $KMnO_4$ in aqueous solution at concentration 0.5 g/L. **c** Differently colored aqueous solutions at concentration 0.5 g/L—commercial ferrate (on the left—pinker and lighter) and potassium permanganate (on the right—purpler and darker)

Table 13.8 Results of chemical analysis of the cleaning water from the sand filter

Parameter	SUM (mg/L) Cleaning water from the sand filter
CHSKMn	1.11
NL (105 °C)	10.2
As total	0.119
Fe total	2.74
As dissolved	0.0105
Fe dissolved	0.0125

The results of the laboratory analysis of the cleaning water from the sand filter are presented in Table 13.8.

Table 13.8 suggests the following facts:

- Cleaning water: standard volume of cleaning water equaling five volumes of the sand filter filling is sufficient for flushing the filter. The concentration of the dissolved As (0.0105 mg/L) and the dissolved Fe (0.0125 mg/L) used in the cleaning water meet the requirements for drinking water (Table 13.8), i.e., both arsenic and iron are strongly bonded to the microflakes in the form of undissolved substances remaining in the sludge, and being released into the cleaning water only minimally. After the sludge from the filtration has been removed, the cleaning water does not represent hazardous waste.
- Sludge: any measurable amount of sludge was detected on the sand filter after having treating 516 L of water.
- The used filters from filtration and microfiltration will be disposed of according to the Waste Regulations. If a low number of filters (dozens or hundreds per year) are disposed of, they could be treated as hazardous waste. In case of a larger-scale technology, the possibility of recategorizing the filters as other type of waste should be considered on the basis of evaluating their hazardous properties.

13.6.5 Limitations of the Method

It has been found out that higher concentrations of phosphorus in water can decrease the efficiency of the method employing ferrates for arsenic removal (Kolařík et al. 2018). The cause of such decrease consists in the geochemical relation of the two elements. This fact has been proved with contaminated groundwater (labeled SFA) containing 988 mg/L of phosphorus and 2.5 mg/L of arsenic. The highest efficiency in laboratory tests was achieved with a commercial ferrate dosed at 60 mg/L. The efficiency in arsenic removal was only 30%. 1.73 mg/L of arsenic and 823 mg/L of phosphorus were detected in the output water. Comparing the quality of raw water from SFA locality and raw water from KLU locality is provided in Table 13.9.

The limit for drinking water was not achieved with SFA sample even after employing other methods commonly used in the water industry. Employing

Table 13.9 Phosphorus—effect on arsenic removal from water

ALS laboratory chemical analysis		Phosphorus limits		As removal	
		Site SFA		Site KLU	
Parameter	Unit	SFA INPUT	SFA OUTPUT UF 60 mg/L	KLU INPUT	KLU OUTPUT UF 15 mg/L
Dissolved solids dried at 105 °C	mg/L	4570	–	428	437
Chemical oxygen demand (COD-Mn)	mg/L	4.10	–	0.71	0.84
Chloride	mg/L	114*	–	23.0	31.3
Fluoride	mg/L	23.4	–	<0.200	<0.200
Hydrogen carbonates HCO_3^-	mg/L	615	–	96.2	86.3
Nitrates	mg/L	35.3	–	88.8	88.1
Orthophosphate	**mg/L**	**1590**	**–**	**0.157**	**<0.040**
Sulfate as SO_4^{2-}	mg/L	423	–	53.4	53.5
Electrical conductivity at 25 °C	mS/m	423	–	50.6	51.9
pH value	**–**	**6.54**	**6.56**	**7.44**	**7.25**
Arsenic	**mg/L**	**2.5**	**1.73**	**0.102**	**<0.0050**
Calcium	mg/L	36.1	–	56.8	61.2
Iron	mg/L	<0.0200	0.234	0.0035	0.0746
Magnesium	mg/L	25.9	–	13.8	14.9
Manganese	mg/L	0.184	–	0.00631	0.00780
Phosphorus	**mg/L**	**988**	**823**	**0.098**	**<0.040**
Potassium	mg/L	27.2	–	2.50	9.85
Sodium	mg/L	1450	–	15.0	16.4

The bold entries mark the key parameters influencing the efficiency of ferrates in arsenic removal

commercial sorbent GEH, we achieved as low arsenic concentration in the remediated water as with ferrates—1.7 mg/L. By means of coagulation tests, we achieved the lowest concentration of As in the output water for coagulant $FeCl_3$ and the dose of Fe 190 mg/L (1.2 mg/L and pH 9.5); Al (PAX-18)-based coagulant achieved only minimal efficiency (As in the output water 1.9 mg/L).

13.7 Conclusion

- The efficiency of the proposed technology for arsenic separation employing ferrates was proved on a mobile technological unit with a flow of 100 L/h with two different sources of groundwater ten times exceeding the allowed limit of arsenic concentration in drinking water. In both cases the final concentration of arsenic in the remediated water was lower than the detection limit of the analytical method (<5 μg/L). The optimal dose of ferrate (ENVIFER from NANO IRON, Czech Republic) for MEZ water was 10 mg/L, for KLU water was 15 mg/L.
- It has been found that the different chemistry of the tested water in MEZ locality and KLU locality had a major impact on the reaction of ferrates with arsenic, although the initial concentration of arsenic was the same. With MEZ water, the most part of arsenic removal was achieved during the first filtration stage; with KLU water, it was during the last stage. The operating costs for the treatment will then differ. As a result, proposing a method and the reagent doses only on the basis of laboratory tests with distilled water cannot prove efficient.
- The water remediated with the proposed technology meets the limits for drinking water from the microbiological point of view.
- When working with commercial ferrates, it is recommended that Fe(VI) content as well as presence of heavy metals in the tested products should be checked.
- The presence of phosphates in higher concentrations could be a limitation in removing arsenic from water.

References

Bielski BHJ, Thomas MJ (1987) Studies of hypervalent iron in aqueous solutions. 1. Radiation-induced reduction of iron(VI) to iron(V) by CO_2^-. J Am Chem Soc 109(25):7761–7764. https://doi.org/10.1021/ja00259a026

Czölderová M, Behúl M, Filip J, Zajíček P, Grabic R, Vojs-Staňová A, Gál M, Kerekeš M, Híveš J, Ryba J, Rybanská M, Brandeburová P, Mackuľak T (2018) 3D printed polyvinyl alcohol ferrate (VI) capsules: Effective means for the removal of pharmaceuticals and illicit drugs from wastewater. Chem Eng J 349:269–275. https://doi.org/10.1016/j.cej.2018.05.089

Filip J, Yngard RA, Siskova K, Marusak Z, Ettler V, Sajdl P, Sharma VK, Zboril R (2011) Mechanisms and efficiency of the simultaneous removal of metals and cyanides by using ferrate(VI): Crucial roles of nanocrystalline iron(III) oxyhydroxides and metal carbonates. Chem Eur J 17(36):10097–10105. https://doi.org/10.1002/chem.201100711

Filip J, Kolařík J, Zajíček P, Maršálek B, Stavělová M, Matysíková J, Mackuľak T, Slunský J (2017) Experiences with ferrate (FeV and FeIV) application: Results from pilot-scale treatment of drinking surface and waste waters. In: 10th world congress on chemical engineering, Barcelona, Spain

Kolařík J, Prucek R, Tuček J, Filip J, Sharma VK, Zbořil R (2018) Impact of inorganic ions and natural organic matter on arsenates removal by ferrate(VI): Understanding a complex effect of phosphates ions. Water Res 141:357–365. https://doi.org/10.1016/j.watres.2018.05.024

Prucek R, Tuček J, Kolařík J, Filip J, Marušák Z, Sharma VK, Zbořil R (2013) Ferrate(VI)-induced arsenite and arsenate removal by in situ structural incorporation into magnetic iron(III) oxide nanoparticles. Environ Sci Technol 47(7):3283–3292. https://doi.org/10.1021/es3042719

Prucek R, Tuček J, Kolařík J, Hušková I, Filip J, Varma RS, Sharma VK, Zbořil R (2015) Ferrate (VI)-prompted removal of metals in aqueous media: Mechanistic delineation of enhanced efficiency via metal entrenchment in magnetic oxides. Environ Sci Technol 49(4):2319–2327. https://doi.org/10.1021/es5048683

Yates BJ, Zboril R, Sharma VK (2014) Engineering aspects of ferrate in water and wastewater treatment – a review. J Environ Sci Health, Part A 49(14):1603–1614. https://doi.org/10.1080/10934529.2014.950924

Zboril R, Andrle M, Oplustil F, Machala L, Tucek J, Filip J, Marusak Z, Sharma VK (2012) Treatment of chemical warfare agents by zero-valent iron nanoparticles and ferrate(VI)/(III) composite. J Hazard Mater 211–212:126–130. https://doi.org/10.1016/j.jhazmat.2011.10.094

Chapter 14
Field Study V: Combined Oxidation Technology Using Ferrates (Fe^{IV-VI}) and Hydrogen Peroxide for Rapid and Effective Remediation of Contaminated Water—Comprehensive Practically Focused Study

Petr Lacina and Michal Hegedüs

Abstract Recently, water-soluble compounds of iron in high oxidation states (Fe^{IV-VI}), also known as ferrates, have gained a lot of attention due to their strong oxidation properties. They can potentially be used for degradation/removal of various compounds from contaminated water. To date, the majority of published papers have concerned only laboratory-scale use of ferrates in model solutions. Large-scale applications of ferrates to remediation have proved so far to be rather complicated as the obtained results failed to meet expectations. Therefore, there is an ongoing need to consider the suitability of their large-scale use. Further combination with other oxidizing agents may provide favorable results. The presented research focuses on laboratory experiments using real groundwater followed by a pilot field application realized as an ex situ experiment and subsequently as in situ remediation. Ferrates were combined with hydrogen peroxide in order to enhance their removal efficiency during this pilot remediation test. Such combination proved to be highly effective achieving 60–80% removal of persistent organic contaminants during the first 24 h.

Keywords Ferrate · Hydrogen peroxide · Iron compounds · Water treatment · Groundwater · Contaminated water · Wastewater · Oxidation technologies · Radical oxidation

---

P. Lacina (✉) · M. Hegedüs
GEOtest, a.s, Brno, Czech Republic
e-mail: lacina@geotest.cz

J. Filip et al. (eds.), *Advanced Nano-Bio Technologies for Water and Soil Treatment*, Applied Environmental Science and Engineering for a Sustainable Future,
https://doi.org/10.1007/978-3-030-29840-1_14

14.1 Introduction

Over the last two decades, water-soluble compounds of iron in high oxidation states (especially Fe^{V} and Fe^{VI}), also known as ferrates (Sharma et al. 2005; Yates et al. 2014b), have been presented as a promising material for wastewater treatment with an effort to enhance their efficiency. Not only do they possess relatively strong oxidizing properties with the capability of degrading a wide range of organic and inorganic contaminants, but they also provide an environmentally friendly solution for water treatment processes, which is a number-one priority for newly introduced materials on the market. Depending on the reaction conditions, Fe^{VI} (or Fe^{V}) species are reduced to Fe^{III} or Fe^{II}, which in an aqueous environment precipitate out in a form of various polyhydroxy complexes or hydroxides. These compounds are stable forms of iron commonly found in nature (Yates et al. 2014b) and do not represent any potential danger to the environment (Jiang and Lloyd 2002; Sharma 2002, 2011; Sharma et al. 2005; Prucek et al. 2013).

Strong oxidizing properties of ferrates are given by an unusually high oxidation state of iron. In an aqueous environment, ferrates are unstable and are subjected to a rapid reduction to stable forms of iron (see Eqs. 14.1 and 14.2 below). They act as strong electron acceptors, removing electrons from their surroundings. It has been shown that Fe^{VI} can be a stronger oxidant than ozone under specific conditions (Jiang 2014; Jiang et al. 2016). Its redox potential varies from 0.72 V in alkaline medium (Eq. 14.1) to 2.20 V in acidic medium (Eq. 14.2) (Sharma et al. 2005; Tiwari et al. 2006; Prucek et al. 2013; Jiang 2014; Sharma et al. 2014).

$$FeO_4{}^{2-} + 4H_2O + 3e^- \rightarrow Fe(OH)_3 + 5OH^- \quad (E^0 = +0.72\ V) \tag{14.1}$$

$$FeO_4{}^{2-} + 8H^+ + 3e^- \rightarrow Fe^{3+} + 4H_2O \quad (E^0 = +2.20\ V) \tag{14.2}$$

The oxidation power of ferrates can be thus altered by adjusting the pH value or by the addition of other oxidizing agents. Ferrates are believed to be an effective remediation material for a wide range of organic (Prucek et al. 2013; Sharma et al. 2014; Yates et al. 2014a; Zhou and Jiang 2015) and inorganic (Filip et al. 2011; Sharma 2011; Machala et al. 2015) contaminants but also as a disinfectant agent (Jiang and Wang 2003; Sharma et al. 2005; Jiang 2014; Sharma et al. 2014; Yates et al. 2014b; Talaiekhozani et al. 2016). On top of that, hydrolysis products of Fe^{III} species in aqueous environments may serve as effective adsorbents of by-products of oxidation processes as well as an effective coagulant and flocculant (Bartzatt et al. 1992; Jiang and Lloyd 2002; Lee et al. 2003; Lee et al. 2009; Filip et al. 2011; Yates et al. 2014b; Talaiekhozani et al. 2016).

However, large-scale utilization of ferrates seems to be less effective than it was previously expected. The pilot studies show big deviances from the results obtained in laboratories. The main drawbacks of their application are as follows:

- they show strong oxidation properties only in acidic environment (lowering the pH value will lead to an increase in the salt content in groundwater)
- under these conditions, they are highly unstable (high amount of ferrates is needed for full remediation)
- in strongly polluted waters, they deteriorate quickly failing to show significant removal efficiency of contaminants

Thus, their efficiency may strongly vary depending on the conditions given by the treated groundwater/wastewater. Their inappropriate application to real waters may end up in a failure. Therefore, there is an ongoing need to consider the suitability of their use or consider their possible combination with other agents (e.g., hydrogen peroxide) to achieve favorable results. It is worth mentioning that a practical application of ferrates should also be preceded by laboratory experiments conducted on particular waters in order to verify the suitability of the proposed process.

The presented research reports on a comprehensive practically oriented study on applications of ferrates in combination with hydrogen peroxide in order to enhance their removal efficiency in remediation of a contaminated real water. On the basis of previously obtained results from laboratory experiments, a field pilot application was realized as an ex situ experiment and subsequently as in situ large-scale remediation in areas of two different industrial factories in the Czech Republic.

14.2 Methods

14.2.1 Materials and Chemicals

Commercial ferrates, ENVIFER, were obtained from NANO IRON, s.r.o. (Czech Republic) in a powder form. It contained iron in high oxidation states as $K_3Fe^{V}O_4$ a $K_2Fe^{VI}O_4$ (total content of Fe 18.4 mass%; content of Fe^{V} 57 ± 3 mol%; content of Fe^{VI} < 3 mol%; content of K_2O from production 19 ± 3 mass%). After the product was dissolved in water, Fe^{V} was disproportionate to Fe^{VI} and Fe^{III} (Kokarovtseva et al. 1972; Kolář et al. 2015). The Fe^{VI}/Fe^{V} ratio in water was roughly 0.81. One gram of ENVIFER dissolved in water, thus creating approximately 85 mg of Fe^{VI}. Other used chemicals were as follows: technical hydrogen peroxide 35% (OQEMA, s.r.o., Czech Republic) and sulfuric acid (Chromservis s.r.o., Czech Republic).

14.2.2 *Laboratory Tests*

Laboratory-scale tests were carried out on real groundwater from two different localities in the Czech Republic, where a pilot field ex situ and in situ application of ferrates was planned. During these experiments, the removal of selected contaminants was studied. Water from locality A was strongly contaminated with chlorinated ethenes (CEs)—total value of CEs (ΣCEs) ranged from 60 to 70 mg/L. Water from locality B was contaminated mainly with benzene, toluene, ethylbenzene, and xylenes (BTEX). The aim of these experiments was to verify and compare the efficiency of ferrates in combination with hydrogen peroxide under various conditions in order to find the most effective method for the pilot application.

During the batch tests, 1000 mL reagent bottles were used. Two parallel sets, each containing six reagent bottles, were prepared. The bottles were labeled sp. 1–6. The first set (labeled A) was prepared with water from locality A, the second set (labeled B) was prepared with water from locality B. Both sets included various combinations of reagents for possible applications. Each of the bottles represented one type of a sample prepared as follows:

sp. 1 Ferrates (0.5 g/L) + pH adjustment $\approx$ 3 + H_2O_2 (5 mL/L),
sp. 2 Ferrates (0.5 g/L) + H_2O_2 (5 mL/L),
sp. 3 Ferrates (0.5 g/L) + pH adjustment $\approx$ 3,
sp. 4 Only a dose of ferrates (0.5 g/L),
sp. 5 Only a dose of H_2O_2 (5 mL/L),
sp. 6 Blank

The dose of commercial ferrates was set to 0.5 g/L, which corresponds to the doses of Fe(VI) commonly used for laboratory tests (Prucek et al. 2013). Ferrates were dosed in a powder form and applied directly to the reagent bottle with contaminated groundwater. The bottles were closed and their contents were vigorously mixed with a magnetic stirrer for 30 minutes. Subsequently, the following steps in the selected samples were taken: pH adjustment by 50% H_2SO_4 (sp. 1 and 3) and application of 5 mL hydrogen peroxide (sp. 1, 2, and 5). The closed bottles with the prepared water samples were put on an orbital shaker and were shaken intensively at regular 6-hour intervals for 5 min. Due to the effervescence of the samples (mainly sp. 1 and 2), the lids of the bottles were opened at regular intervals to release accumulated gases. 100 mL of each water sample was collected for analysis after 24, 48, 96, and 168 h and each sample was filtered through a filter paper to remove sludge before analytical sample treatment. The content of contaminants in the samples was analyzed with headspace gas chromatography/mass spectrometry in accredited laboratories. All samples were processed within 24 h of the collection.

14.2.3 Pilot Field Ex Situ Application

Ex situ pilot validation of the obtained laboratory results was carried out in the area of Czech chemical factory, where the groundwater was strongly contaminated with a wide range of organic contaminants. The most important contaminants were as follows: aromatic hydrocarbons (benzene, toluene, ethylbenzene, xylenes) and chlorinated aromatic hydrocarbons (chlorobenzene, o-dichlorobenzene, m-dichlorobenzene, p-dichlorobenzene). During the initial monitoring process physicochemical analysis of raw groundwater was carried out and the key parameters were determined (Table 14.1). The list of detected contaminants and their concentration levels before the tests are given in Table 14.2.

The pilot application to the target locality as a container test (i.e., ex situ treatment) was carried out in July 2014 using a 1000 liters intermediate bulk container (IBC) made of high-density polyethylene. The container was filled with 900 L contaminated groundwater pumped out from a borehole located in the area of chemical factory surrounding. The whole pilot test consisted of three application steps:

1. 350 g of commercial ferrates ($\approx$ 0.4 g/L $\approx$ 34 mg Fe^{VI}/L) was applied into the container and the content was then mixed for 15 min.
2. After reduction of all dosed ferrates (visible detection by changing color of water from violet to orange), the first dose of hydrogen peroxide with a volume of 4.5 L ($\approx$ 5 mL/L) was applied. The content of the container was subsequently mixed at 30-min intervals.
3. Four hours after the first application of hydrogen peroxide, the second dose of the same amount 4.5 L H_2O_2 ($\approx$5 mL/L) was applied.

Table 14.1 Qualitative parameters of the raw groundwater

Parameter	Value	Unit
pH	6.29	
Conductivity	4270	µS/cm (20 °C)
Acid-neutralizing capacity ($ACN_{4.5}$)	10.09	mmol/L
Chemical oxygen demand (COD_{Cr})	1090	mg/L
Biochemical oxygen demand (BOD_5)	551	mg/L
Total hardness	8.05	mmol/L
NH^{4+}	37.8	mg/L
Total inorganic nitrogen	10.66	mg/L
Sulfates	855	mg/L
Chlorides	504	mg/L
Nitrates	<3.0	mg/L
Nitrites	<0.10	mg/L
Phosphates	2.92	mg/L
Mineralization	3537	mg/L
Halogenated organic compounds (AOX)	6.42	mg/L

Table 14.2 The list of contaminants detected in groundwater and their concentration level before the start and at the end of the pilot test

Contaminant	Before the start of the test c (μg/L)	At the end of the test c (μg/L)
Benzene	340 ± 85	170 ± 42
Toluene	363,000 ± 90,700	103,000 ± 25,700
∑xylenes	480 ± 120	130 ± 32
1,2,4-trichlorobenzene	62 ± 15	18 ± 4
o-dichlorobenzene	19,000 ± 4700	5800 ± 1400
m-dichlorobenzene	3700 ± 920	970 ± 240
p-dichlorobenzene	4900 ± 1200	1300 ± 320
∑chlorinated ethenes	520 ± 130	<2
Ethylbenzene	140 ± 35	44 ± 11
Chlorobenzene	2900 ± 720	800 ± 200
Naphthalene	3600 ± 900	1200 ± 300
Phenol	3.0 ± 0.7	<2
∑cresols	50 ± 12	220 ± 55
∑dichlorophenols	9 ± 2.2	8.1 ± 2.0
∑chlorophenols	12 ± 3.0	<2
Aniline	730 ± 180	320 ± 80
N-ethylaniline	4.6 ± 1.1	4.2 ± 1.0
2,4,6-trimethylaniline	130 ± 32	<2
Nitrobenzene	2100 ± 520	2300 ± 520

Table 14.3 The time schedule of the pilot test with substantial operations

Time (h)	Operation
0	Filling the container (900 L)
0.75	Application of the ferrate – ENVIFER (350 g)
1.25	Application of 1st dose of H_2O_2 (4.5 L)
5	Application of 2nd dose of H_2O_2 (4.5 L)
24	Termination of the test

The test was terminated after another 19 h. The entire reaction process thus lasted 24 h. Table 14.3 lists a concise time schedule of the pilot test with particular operations. Throughout the test also physicochemical parameters, notably pH, redox potential, and the concentration of dissolved oxygen, were observed and measured by a multiparameter probe Aquaread AP-2000 (Aquaread Ltd., Great Britain).

Simultaneously with the IBC container test, also a blank experiment was realized using 50 L barrel filled with the same groundwater. No oxidizing agents were applied to this barrel, only mechanical and measuring operations (i.e., mixing and measurements of the selected physicochemical parameters) according to the same timetable as in the case of the IBC container (Table 14.3) test were carried out.

14.2.4 Pilot Field In Situ Application

For an in situ pilot validation of results obtained from the laboratory tests, an area of Czech industrial factory aimed at metal production was chosen. Groundwater at this locality was contaminated mainly with CEs. For the pilot verification of results employing the in situ approach, the borehole in strongly contaminated area was chosen. An average level of ΣCEs in this borehole ranged from tens to hundreds mg/L. The in situ pilot application was carried out in three application rounds from February to May 2014. The application procedure was as follows:

1st application round (February 27, 2014)

In the first step, 1500 g of commercial ferrates was dissolved in 700 L of distilled water and the concentrated ferrate solution was applied directly to the application borehole. The water column in the borehole was subsequently mixed by a submersible pump. In the second step, approximately 1 h after the first injection, the pH in the borehole was adjusted to ca value 3 by ca 0.5 L of 50% H_2SO_4. After this, 15 L of 30% H_2O_2 was applied and mixed by submersible pump.

2nd and 3rd application round (March 27 and May 13, 2014)

During these next two application rounds, only 15 L of 30% H_2O_2 was applied.

14.3 Results and Discussion

14.3.1 Results of Laboratory Tests

Groundwater from two different localities (denoted as A and B) was used for the laboratory tests. Both of these localities were selected also for the subsequent pilot test. During the laboratory tests, changes in the concentration of priority contaminants, depending on the reaction time and the type of the prepared sample, were monitored:

- total concentration of CEs (ΣCEs) in the case of locality B (Fig. 14.1)
- total concentration of benzene, toluene, ethylbenzene, and xylenes (ΣBTEX) in the case of locality B (Fig. 14.2)

The zero point on "Reaction time" axis represent initial values, i.e., values of raw water before the start of the test. The graphs in Fig. 14.3 show the overall removal efficiency of the target contaminants after 24 h for each of the prepared samples.

The results obtained from the laboratory tests showed that neither ferrates nor hydrogen peroxide exhibit significant efficiency in the removal of the targeted contaminants, in comparison with the blank (sp. 4, 5, and 6). The reason for the low removal efficiency is high pollution of the tested groundwater, where oxidative effects of ferrates were consumed by easily oxidizable ballast substances present in

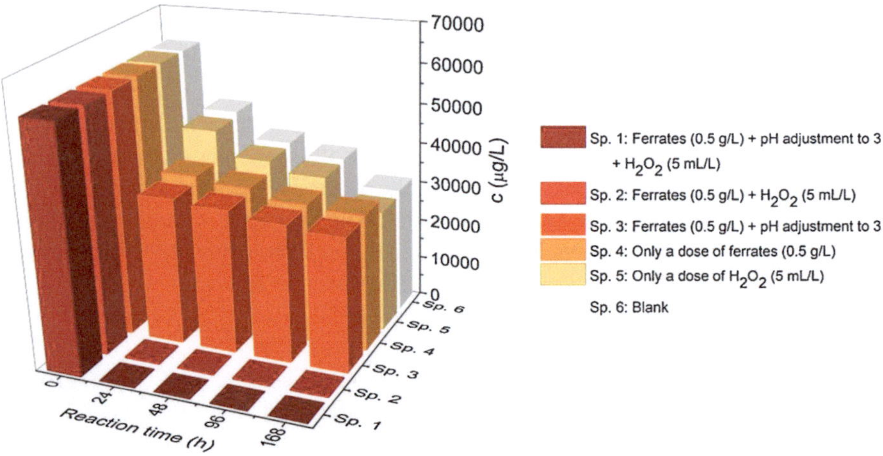

Fig. 14.1 The concentration change of ΣCEs during the laboratory tests (set A)

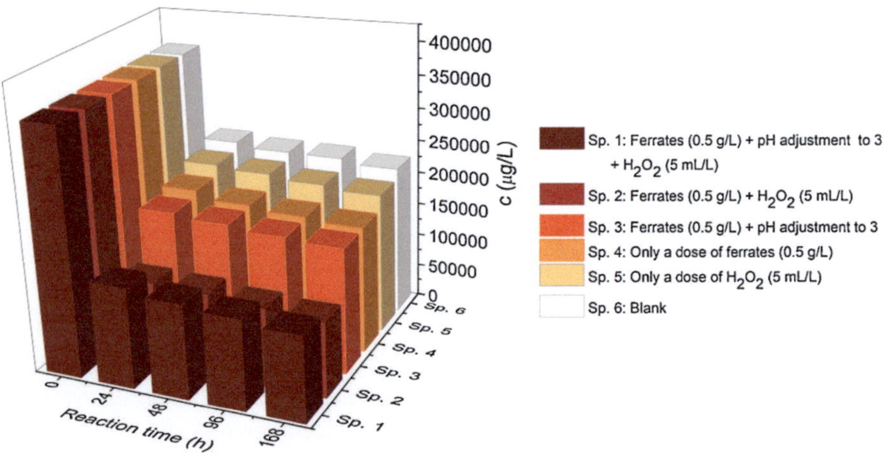

Fig. 14.2 The concentration change of ΣBTEX during the laboratory tests (set B)

the water, and the ferrates degraded too rapidly to be able to degrade the target stable contaminants. Conversely, in samples where only hydrogen peroxide was applied, slow and long decomposition of peroxide without a significant influence on the selected contaminants has been observed. pH adjustment to lower values caused higher degradation of ferrates without a significant effect on the target contaminants—only slight increase in the removal efficiency was observed (sp. 3).

However, the combination of these two agents (sp. 1 and sp. 2) caused a significant increase in the removal efficiency. At the end of the tests on water from locality A, almost 100% efficiency in the removal of the monitored contaminants (CEs) was observed when the pH was not previously adjusted. 60–75% removal efficiency in the selected contaminants (BTEX) was reached in the tests on

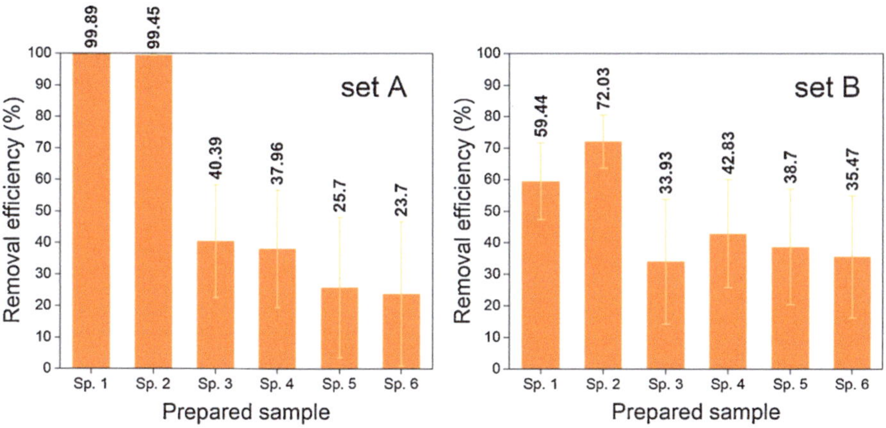

Fig. 14.3 Removal efficiency of target contaminants (ΣCEs of set A and ΣBTEX of set B) after 24-hour reaction time

groundwater from locality B—the highest efficiency in this case was reached without any previous pH adjustment. Both of these reactions (only for sp. 1 and sp. 2) were accompanied by slight fizzing, which was observed until complete decomposition of the applied hydrogen peroxide, detected by starch-iodide papers. The presence of hydrogen peroxide was detected for a maximum of 12–24 h for sp. 1 and sp. 2. In the case of sp. 5, hydrogen peroxide was detected even 96 h after its application. Significantly faster decomposition of hydrogen peroxide in aqueous medium was observed in samples where ferric sludge from gradual ferrate reduction was formed. Such a rapid decrease in the concentration of the target contaminants has been observed only in advanced oxidation processes including radical oxidations by OH radicals (Che et al. 2011; Hrabák 2012; Hara 2012). Considering these facts, it is highly probable that radical oxidation took place in sp. 1 and sp. 2. Characterization of the ferric sludge formed by ferrate reduction gave findings that the sludge comprised of nanostructures, which most likely catalyzed the radical decomposition of hydrogen peroxide. The formed OH radicals then reacted with the target contaminants. This nanoparticle formation after the ferrate addition into natural water was proved by Goodwill et al. (2015). Simultaneously, it was found that the radical reactions were able to run repeatedly after an addition of hydrogen peroxide to the ferric sludge in the water, even without adjusting pH.

14.3.2 Results of Pilot Ex Situ Application

The main aim of the pilot field application was to verify the results from the laboratory experiments in practice. The pilot test was performed as a 24-hour container test carried out directly in the selected area. All substantial operations done during the test are listed in Table 14.3. The design and the implementation of

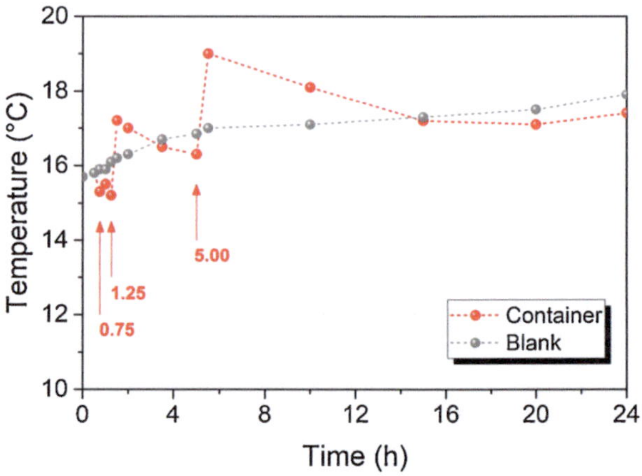

Fig. 14.4 The temperature of water in the reaction container and of the blank sample during the pilot test; the time axis includes times of the applications (corresponding to Table 14.3)

all the operations were based on the findings revealed during the laboratory tests. Raw groundwater pumped into the container was turbid, strongly smelling of organic solvents, and had a gray color. After the first step, i.e., an application of commercial ferrates, color of the raw water rapidly changed to pale orange (within a few minutes) as a consequence of Fe^{3+} ions formation from ferrates reduction. Moreover, this process was associated with the ferric sludge formation (hydrated forms of Fe-oxides) in the form of flakes, which removed the water turbidity and supported sedimentation. After the second step, i.e., hydrogen peroxide application, a gentle fizzing in the whole water volume was observed. Simultaneously, a slight increase in water temperature (ca 2–3 °C) was observed (Fig. 14.4). The effervescence intensity had a decreasing trend during the reaction time along with the temperature of water (Fig. 14.4) and it was no longer observed at the end of the test. Furthermore, the presence of hydrogen peroxide was not detected at the end of the test. After the test terminated, no turbidity was observed, the water was slightly orange, odorless, and the sludge was deposited on the bottom of the container.

During the test, selected physicochemical parameters (pH, redox potential, dissolved oxygen) were monitored (Fig. 14.5). Expected increase in the pH caused by the reaction of ferrates with water and by presence of K_2O (creating KOH) was observed after the first application step (ferrate application in time 0.75 h). However, due to strong buffering capacity of water, the pH increase was lower than previously expected. Because of this, the pH values were within the optimal range for the formation of ferric sludge flakes, which worked as an effective flocculant. A slight decrease in the pH value occurred after the second application step, i.e., the first dose of hydrogen peroxide (in time 1.25 h), because of its weak acidic effects in an aqueous medium. However, during the reaction time, the pH value returned back due to decomposition of the hydrogen peroxide and buffering capacity of the water. After the third step, i.e., the second dose of hydrogen peroxide (in time 5.00 h), no significant change in the pH values was observed, probably due to degradation products, which contributed to the buffering capacity of the water. In the case of the redox potential (expressed as E_h), significant growth was observed immediately

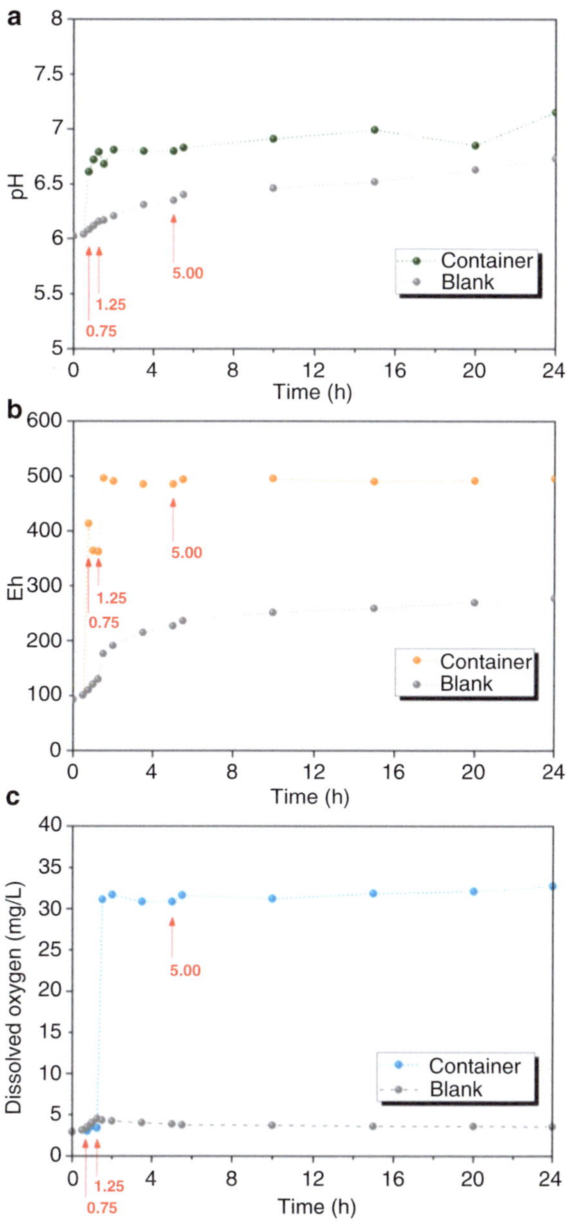

Fig. 14.5 Evolution of physicochemical parameters in the reaction container and in the blank sample during the pilot test: (**a**) pH, (**b**) redox potential (E_h), (**c**) concentration of dissolved oxygen; the time axis includes times of the applications (corresponding to Table 14.3)

after the first application step (ferrate application in time 0.75 h) and also after the second application step (first dose of hydrogen peroxide in time 1.25 h). Rapid increase in the dissolved oxygen concentration after the second application step was caused by decomposition of dosed hydrogen peroxide. Significant increase in both

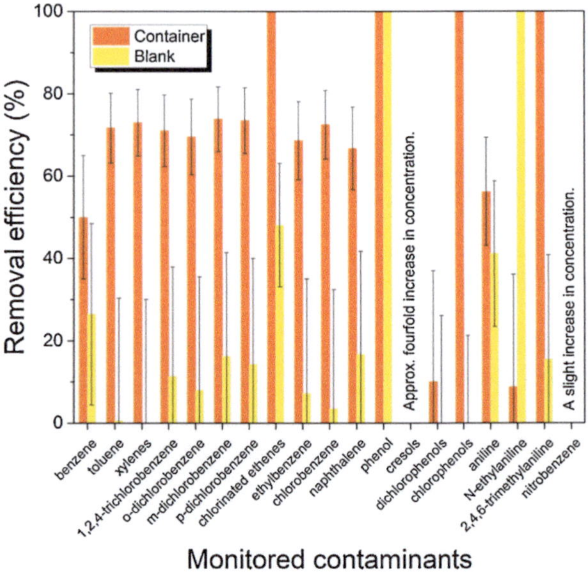

Fig. 14.6 Total removal efficiency of monitored contaminants at the end of the pilot test; compared to the blank

redox potential and dissolved oxygen showed the formation of strong oxidizing conditions in the water medium.

Figure 14.5 also shows physicochemical parameters of the blank sample compared with the pilot test. In the case of the blank experiment all measurements were carried out parallel to the measurements carried out during the container test.

The liquid samples for the analysis were prepared in two rounds: (1) just before the start of the test and (2) immediately after its termination. After they were collected, all the samples were transported quickly to accredited laboratories in order to perform sample treatment and analysis. The results of the analysis are listed in Table 14.2. On the basis of the obtained results, the removal efficiency of the pilot test was determined. The overall removal efficiency in the monitored contaminants at the end of the test, i.e., after 24 hours, is listed in Fig. 14.6 including the removal efficiency in the blank sample. All differences between the initial and the final concentrations in the case of the blank sample are results of mixing, aeration, volatilization, sample handling, etc. The obtained results revealed that there was a significant decrease in the concentration of almost all monitored contaminants during the 24-hour pilot test, compared with the blank sample. The removal efficiency in the majority of monitored contaminants ranged between 60 and 80%. Complete elimination was observed with phenol, chlorophenol, N-ethylaniline, and chlorinated ethenes. An increase in the concentration of cresols and nitrobenzene at the end of the test was observed. This is the consequence of radical oxidation reaction (in presence of OH radicals) of selected aromatic compounds in which aromatic alcohols (cresols) can be formed. According to the input data, high concentration of toluene may be primarily responsible for the observed increase. A slight increase in the concentration of nitrobenzene may have resulted from the oxidation of aniline. Zero removal efficiency in Fig. 14.6 means that no changes in the concentrations of selected contaminants were observed during the pilot test, especially in the case of blank: xylenes, cresols, chlorophenols, dichlorophenols, nitrobenzene.

Based on the obtained results of the pilot-field application, the combination of the two oxidizing agents seems to be a promising way for quick removal of a wide range of organic contaminants. It can not only be used for groundwater remediation but also for wastewater and other types of water purification by ex situ methods. Although ferrates used independently in model solutions exhibit strong oxidative effects, they rapidly deteriorate without noticeable effects on chemically stable contaminants in highly polluted real waters. The efficiency of the reaction can be increased by addition of hydrogen peroxide to the formed ferric sludge. The properties of the two reagents can therefore be combined—efficiency of ferrates alone (oxidation, coagulation) and subsequently efficiency of radical oxidation, which is initiated after the addition of hydrogen peroxide to ferric sludge. In this specific case, the formed sludge acts most likely as a catalyst for radical degradation of hydrogen peroxide. Therefore, the removal efficiency is comparable to the Fenton oxidation. On top of that, ferrates utilization does not lead to increased salinity of water and no other substances are introduced to an aqueous environment (as sulfates, acids, etc.). From the viewpoint of the environmental protection, the proposed combination represents no risk—ferrates are reduced to Fe^{3+}/Fe^{2+} and precipitate in the form of polyhydroxy complexes and hydroxides, which commonly occur in nature, and hydrogen peroxide is degraded to O_2 and H_2O.

14.3.3 Results of Pilot In Situ Application

The purpose of this type of ferrate application was to verify the results obtained from the laboratory experiment employing the method of in situ remediation. It is necessary to emphasize that the pH was adjusted only before the first application round; in another application round, only H_2O_2 was added into the application borehole. The reason for this step was to verify if the remaining ferric sludge, which was detected in the borehole even during the second and the third application round, would react with a new dose of hydrogen peroxide even without pH adjustment. After all applications of hydrogen peroxide, the reaction in the borehole was relatively tempestuous, but it lasted only a few hours. Between all application rounds, regular monitoring was carried out. In the following graphs (Fig. 14.7a–c) physicochemical parameters are shown. It is obvious that rapid increase in the E_h and dissolved oxygen (DO) was observed after the application in all cases, as a consequence of H_2O_2 decomposition. Decrease in the pH was also observed in all cases. In the first application round it was the consequence of target pH adjustment. In other application rounds, when the pH was not adjusted, it was the result of acidic nature of H_2O_2 itself.

According to the graph showing CE contamination development (Fig. 14.7d), it is obvious that a rapid decrease was observed after each application. It made no

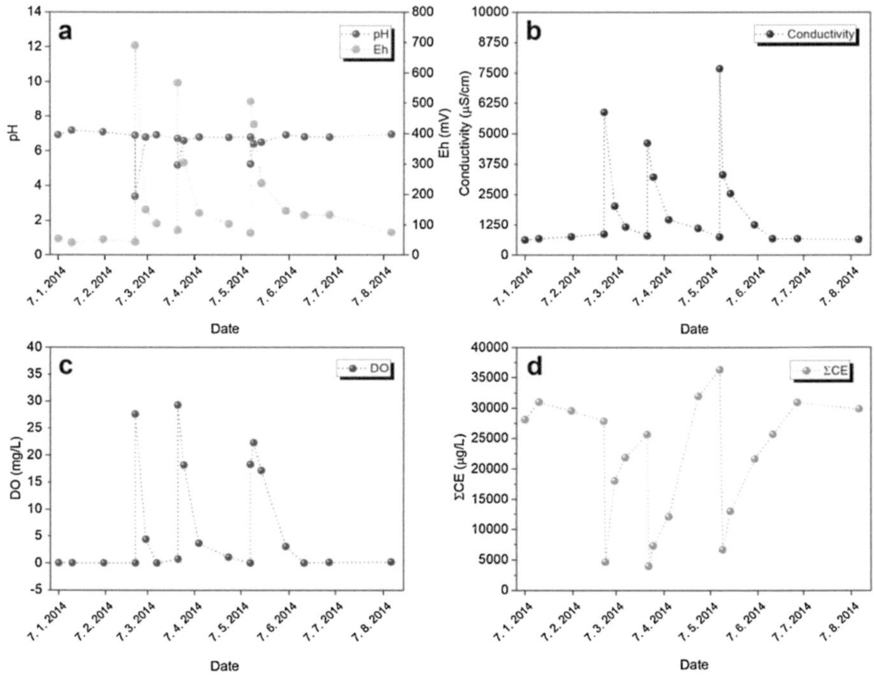

Fig. 14.7 (**a–c**) Evolution of physicochemical parameters of pilot in situ application and (**d**) CEs concentration development in time

significant difference whether the pH was adjusted or not. On the basis of the obtained results, also in the case of in situ application, it was proved that a radical reaction with ferric sludge from ferrate reduction ran even without pH adjustment. However, a rapid increase in the contamination was observed during in situ remediation as a consequence of contamination subsidy from contaminated soil surroundings. This represents a significant difference between in situ and ex situ application. Therefore, in the case of in situ application it is necessary to apply hydrogen peroxide repeatedly at short intervals. The only condition is the presence of ferric sludge from ferrate reduction in the borehole. A major advantage of this approach is the use of hydrogen peroxide solely. Compared to the Fenton oxidation, there is no need for pH adjustment and repeated addition of ferrous sulfate, so there is no salinization of the rock environment.

14.4 Conclusions

The use of ferrates alone during the remediation of heavily contaminated groundwater and wastewater does not seem to provide expected results. Therefore, it is inevitable to reexamine other options for their use (e.g., further purification, coagulation, flocculation, disinfection, etc.) or their use in a combination with other agents, which will assist in achieving the expected effectiveness in remediation practice. One of the options is their combination with hydrogen peroxide, where nanostructured ferric/ferrous sludge is formed by the reduction of ferrates. This sludge can act as an effective catalyst for radical decomposition of hydrogen peroxide. The aim of this study was to test the combination of ferrates with hydrogen peroxide in laboratory conditions and then to realize a field pilot application. The proposed combination showed a relatively high efficiency during remediation of even heavily contaminated water. Ferrates were finally reduced to naturally occurring forms of iron and the final products of the hydrogen peroxide degradation were oxygen and water. Thus, no other contaminants were discharged into the aqueous environment. No increase in salinity of the water was observed either. The reaction proceeded rapidly even without pH value adjustment or other parameters of the reaction being changed. In addition to this, the formed iron sludge acted as an effective flocculant. The presence of this sludge in purified water is the necessary condition for the successful remediation by subsequent hydrogen peroxide application. Comparing ex situ and in situ applications, the removal efficiency is high in both cases, but significant difference consists in the rapid run of the reaction. In the case of water remediation by an ex situ method, this is an advantage as the effective remediation during short time is required. Conversely, in the case of water remediation by an in situ method, this is a disadvantage because the contamination returns rapidly to the initial level as a consequence of contamination of the soil surrounding. In this case, rather long-term remediation reactions are required.

References

Bartzatt R, Cano M, Johnson L, Nagel D (1992) Removal of toxic metals and nonmetals from contaminated water. J Toxicol Environ Health 35(4):205–210. https://doi.org/10.1080/15287399209531611

Che H, Bae S, Lee W (2011) Degradation of trichloroethylene by Fenton reaction in pyrite suspension. J Hazard Mater 185(2–3):1355–1361. https://doi.org/10.1016/j.jhazmat.2010.10.055

Filip J, Yngard RA, Siskova K, Marusak Z, Ettler V, Sajdl P, Sharma VK, Zboril R (2011) Mechanisms and efficiency of the simultaneous removal of metals and cyanides by using ferrate(VI): crucial roles of nanocrystalline iron(III) oxyhydroxides and metal carbonates. Chem Eur J 17(36):10097–10105. https://doi.org/10.1002/chem.201100711

Goodwill JE, Jiang Y, Reckhow DA, Gikonyo J, Tobiason JE (2015) Characterization of particles from ferrate preoxidation. Environ Sci Technol 49(8):4955–4962. https://doi.org/10.1021/acs.est.5b00225

Hara J (2012) Chemical degradation of chlorinated organic pollutants for in situ remediation and evaluation of natural attenuation. In: Puzyn T, Mostrag-Szlichtyng A (eds) Organic pollutants ten years after the Stockholm convention – environmental and analytical update. InTech, Rijeka, pp 345–364. https://doi.org/10.5772/31775

Hrabák P (2012) Kritické studium sanačního využití modifikovaného Fentonova činidla (A critical study of remedial utilization of modified Fenton reagent). Dissertation, Technical University of Liberec

Jiang JQ (2014) Advances in the development and application of ferrate(VI) for water and wastewater treatment. J Chem Technol Biotechnol 89(2):165–177. https://doi.org/10.1002/jctb.4214

Jiang J-Q, Lloyd B (2002) Progress in the development and use of ferrate(VI) salt as an oxidant and coagulant for water and wastewater treatment. Water Res 36(6):1397–1408. https://doi.org/10.1016/S0043-1354(01)00358-X

Jiang J-Q, Wang S (2003) Inactivation of Escherichia coli with ferrate and sodium hypochlorite: a study on the disinfection performance and constant. In: Vogelpohl A (ed) Proceedings of the 3rd international conference on oxidation technologies for water and wastewater treatment, Goslar, 18–22 May 2003, CUTEC-Series Publication No. 57. Papierflieger Verlag, Clausthal-Zellerfeld, pp 403–411

Jiang JQ, Durai HBP, Petri M, Grummt T, Winzenbacher R (2016) Drinking water treatment by ferrate(VI) and toxicity assessment of the treated water. Desalin Water Treat 57 (54):26369–26375. https://doi.org/10.1080/19443994.2016.1203825

Kokarovtseva IG, Belyaev IN, Semenyakova LV (1972) Oxygen compounds of iron(VI, V, IV). Russ Chem Rev 41(11):929–937. https://doi.org/10.1070/RC1972v041n11ABEH002104

Kolář M, Kolařík J, Jančula D, Slunský J, Medřík I, Filip J, Maršálek B, Zbořil R (2015) Feráty: vlastnosti a přehled možných environmentálních aplikací (ferrates: properties and the overview of possible environmental applications). In: Burkhard J, Petráková Kánská K (eds) Sanační technologie XVIII, Uherské Hradiště, 19–21 May 2015, pp 152–154

Lee Y, Um I-h, Yoon J (2003) Arsenic(III) oxidation by iron(VI) (ferrate) and subsequent removal of arsenic(V) by iron(III) coagulation. Environ Sci Technol 37(24):5750–5756. https://doi.org/10.1021/es034203+

Lee Y, Zimmermann SG, Kieu AT, von Gunten U (2009) Ferrate (Fe(VI)) application for municipal wastewater treatment: a novel process for simultaneous micropollutant oxidation and phosphate removal. Environ Sci Technol 43(10):3831–3838. https://doi.org/10.1021/es803588k

Machala L, Filip J, Prucek R, Tucek J, Frydrych J, Sharma VK, Zboril R (2015) Potassium ferrite ($KFeO_2$): synthesis, decomposition, and application for removal metals. Sci Adv Mater 7 (3):579–587. https://doi.org/10.1166/sam.2015.2143

Prucek R, Tuček J, Kolařík J, Filip J, Marušák Z, Sharma VK, Zboril R (2013) Ferrate(VI)-induced arsenite and arsenate removal by in situ structural incorporation into magnetic iron(III) oxide nanoparticles. Environ Sci Technol 47(7):3283–3292. https://doi.org/10.1021/es3042719

Sharma VK (2002) Potassium ferrate(VI): an environmentally friendly oxidant. Adv Environ Res 6 (2):143–156. https://doi.org/10.1016/S1093-0191(01)00119-8

Sharma VK (2011) Oxidation of inorganic contaminants by ferrates (VI, V, and IV)-kinetics and mechanisms: a review. J Environ Manag 92(4):1051–1073. https://doi.org/10.1016/j.jenvman.2010.11.026

Sharma VK, Kazama F, Jiangyong H, Ray AK (2005) Ferrates (iron(VI) and iron(V)): environmentally friendly oxidants and disinfectants. J Water Health 3(1):45–58. https://doi.org/10.2166/wh.2005.0005

Sharma VK, Homonnay Z, Siskova K, Machala L, Zboril R (2014) Mössbauer investigation of the reaction of ferrate(VI) with sulfamethoxazole and aniline in alkaline medium. Hyperfine Interact 224(1–3):7–13. https://doi.org/10.1007/s10751-013-0819-4

Talaiekhozani A, Bagheri M, Talaei MR, Jaafarzadeh N (2016) An overview on production and applications of ferrate(VI). Jundishapur J Health Sci 8(3):e34904

Tiwari D, Kim H-U, Lee S-M, Yang J-K, Kim H (2006) Ferrate(VI) for waste water treatment : oxidation of cyanide in aqueous medium. Environ Eng Res 11(6):318–324. https://doi.org/10.4491/eer.2006.11.6.318

Yates BJ, Darlington R, Zboril R, Sharma VK (2014a) High-valent iron-based oxidants to treat perfluorooctanesulfonate and perfluorooctanoic acid in water. Environ Chem Lett 12 (3):413–417. https://doi.org/10.1007/s10311-014-0463-5

Yates BJ, Zboril R, Sharma VK (2014b) Engineering aspects of ferrate in water and wastewater treatment – a review. J Environ Sci Health Part A 49(14):1603–1614. https://doi.org/10.1080/10934529.2014.950924

Zhou Z, Jiang J-Q (2015) Reaction kinetics and oxidation products formation in the degradation of ciprofloxacin and ibuprofen by ferrate(VI). Chemosphere 119:S95–S100. https://doi.org/10.1016/j.chemosphere.2014.04.006

Part III
Biotechnologies for Water Treatment

Chapter 15
Biotechnologies for Water Treatment

Dietmar Schlosser

Abstract Too low concentrations of organic pollutants as well as too high loads of toxic contaminants and/or other extreme conditions of (waste)waters may impede their efficient treatment by conventional biological technologies. Micropollutants typically occur in only minute amounts in the aquatic environment, which is not in favor of their productive biodegradation and may lead to infinitely low degradation rates. Conventional bioprocesses could even collapse in response to too toxic or otherwise extreme conditions of the waters to be treated. Advanced biological water treatment processes aim to overcome the performance limits of conventional biological processes, and make use of isolated enzymes and whole organisms. Current approaches involve enzyme immobilization to yield biocatalytically active nanomaterials, and bioelectrochemical systems employing immobilized microorganisms for wastewater treatment along with energy production. Such applications are mostly still at the experimental stage, and issues related to process upscaling, long-term stability, and cost efficiency still need to be solved. Membrane bioreactors (MBRs) retain the biocatalytically active sludge in the system. They can be augmented with additional degrader organisms or used in combination with physicochemical processes such as activated carbon adsorption and ozonation. MBRs have already successfully been tested at the pilot and sometimes even at full scale. Constructed wetlands also have also the potential to remove micropollutants from urban wastewater. They may be considered as an alternative for wastewater treatment in small and scattered communities, and for the final step in the treatment of special wastewaters.

Keywords Constructed wetlands · Bioelectrochemical systems · Immobilized enzymes · Membrane bioreactors · Micropollutants · Wastewater treatment

---

D. Schlosser (✉)
Department of Environmental Microbiology, Helmholtz Centre for Environmental Research – UFZ, Leipzig, Germany
e-mail: dietmar.schlosser@ufz.de

J. Filip et al. (eds.), *Advanced Nano-Bio Technologies for Water and Soil Treatment*, Applied Environmental Science and Engineering for a Sustainable Future,
https://doi.org/10.1007/978-3-030-29840-1_15

15.1 Biotechnologies for Water Treatment

Under certain circumstances, the performance limits of current biological (waste) water treatment technologies are exceeded. With regard to organic chemicals polluting waters and the accompanying conditions determining the activity and survival of degrader organisms, two major groups of scenarios are of particular importance as they necessitate development and implementation of more efficient, advanced biological treatment technologies.

One of these groups of scenarios relates to nowadays widely manifested concerns regarding the occurrence of a multitude of overwhelmingly nonregulated, primarily synthetic organic chemicals with quite diverse structures, origins, and uses, which are usually referred to as emerging contaminants or micropollutants. These chemicals range from quite hydrophobic to very hydrophilic (polar and ionic) structures, and arise from, e.g., urban, industrial, and agricultural activities. Accordingly, micropollutants are emitted from both point and nonpoint sources, and are found in, e.g., pharmaceutical production plants, daily household products, landfills, municipal sewage, sewage sludge, hospital wastewater, and the natural aquatic environment (Ahmed et al. 2017; Hochstrat et al. 2015; Solé and Schlosser 2015). Micropollutants and their metabolites can enter the water cycle because of frequently insufficient degradation and retention in conventional wastewater treatment plants (WWTPs), which were not originally designed for their removal (Ahmed et al. 2017; Cecconet et al. 2017; Hochstrat et al. 2015; Kümmerer 2011; Lapworth et al. 2012; Silva et al. 2012). Besides surface and groundwater, they have been occasionally detected even in drinking water (Kümmerer 2009; Kümmerer 2011; Lapworth et al. 2012). Environmentally relevant concentrations of micropollutants and also their transformation products may pose considerable ecological and human health risks due to, e.g., interferences with the endocrine system of vertebrates; their deleterious effects on growth, reproduction, and development in wildlife; the development of hazardous microbial resistance mechanisms; and accumulation in soil, plants, and animals (Ahmed et al. 2017; Hochstrat et al. 2015; Table 15.1).

There is neither a general definition nor a complete list of micropollutants. Their most common characteristic is a very low concentration in the aquatic environment in the ng/L to the μg/L range (Ahmed et al. 2017; Kümmerer 2011; Murray et al. 2010). Recent classifications based on the respective field of application list industrial chemicals, pesticides, pharmaceuticals and personal care products (PPCPs), with pharmaceutically active compounds (PhACs) being a subcategory of PPCPs (Cecconet et al. 2017; Murray et al. 2010). PhACs include, e.g., analgesics, lipid regulators, antibiotics, diuretics, nonsteroidal anti-inflammatory drugs, stimulant drugs, antiseptics, beta blockers, and antimicrobials, whereas cosmetics, sunscreen agents, food supplements, and fragrances belong to personal care products (PCPs) (Ahmed et al. 2017). Following the aforementioned classification approach, frequently detected micropollutants such as bisphenol A (BPA) and nonylphenol (NP) represent industrial chemicals, whereas carbamazepine (CBZ), diclofenac (DF), 17α-ethinylestradiol (EE2), sulfamethoxazol (SMX), and triclosan (TCS) are

Table 15.1 Examples of prominent micropollutants frequently found in waters, and related characteristics. Adapted with permission from Hochstrat et al. (2015) – Table 1.1 (© IWA Publishing 2015), and Solé and Schlosser (2015) – Table 13.1 (© Wiley-VCH Verlag GmbH & Co. KGaA, Boschstr. 12, 69469 Weinheim, Germany)

Compound(s)	**Sources & uses**	**Contaminated environmental compartments / spread**	**(Potential) hazardous effects**
Bisphenol A (BPA)	Used in polycarbonate polymer and epoxy resin production; also a degradation product of polybrominated flame retardants	Leaching from plastics including epoxy resin-coated water pipes; contaminant in WWTP effluents and receiving waters, and seepage from landfills	Endocrine disrupting chemical (EDC); adverse effects on growth, reproduction, and development in wildlife
Carbamazepine (CBZ)	Antiepileptic	Contaminant in WWTP effluents (communal and hospital wastewaters) and receiving waters due to insufficient removal in WWTPs, groundwater/bank filtrate contamination due to a particularly high mobility and persistence	Several adverse effects including blurred vision, drowsiness, headache, and birth defects
Diclofenac (DF)	Anti-inflammatory	Contaminant in WWTP effluents (communal and hospital wastewaters) and receiving waters due to incomplete removal in WWTPs, considerably mobile thus (potentially) causing groundwater/bank filtrate contamination	Renal insufficiency in carcass-eating wildlife (vultures) resulting in reduced/missing elimination of corpse pathogens; harmful to freshwater fish species
N,N-′-Dimethylsulfamide (DMS)	Formed by microbial transformation in soil from the pesticides (fungicides) tolylfluanid and dichlofluanid	Contaminates surface and groundwater due to its high mobility, is not removed by riverbank filtration and hence can enter drinking water treatment processes	Formation of the carcinogenic N-nitrosodimethylamine (NDMA) from DMS during ozonation (used for disinfection of drinking water)

(continued)

Table 15.1 (continued)

Compound(s)	Sources & uses	Contaminated environmental compartments / spread	(Potential) hazardous effects
Endosulfan	Worldwide use in agriculture to control insect pests	Environmentally ubiquitous (air, water, wildlife), subject to long-range atmospheric transport, semi-volatile, persistent to degradation, bioaccumulation in fish	High acute toxicity to insects and mammals, acts as EDC, delays sexual maturity in male children, carcinogenic effects under debate
17α-Ethinylestradiol (EE2)	Synthetic estrogen in oral contraceptives	Contaminant in WWTP effluents and receiving waters due to incomplete removal in WWTPs, moderate sorption to WWTP sludge and aquatic sediments	EDC, can potentially impact the sustainability of aquatic wildlife populations
p-Nonylphenol (NP) isomers with variously branched side chains, e.g., 4-(1-ethyl-1,3-dimethylpentyl) phenol	Release of various nonylphenol isomers with branched side chains resulting from incomplete degradation in WWTPs of nonylphenol ethoxylate surfactants, which are used in various cleaning agents and contaminate wastewaters	Presence of nonylphenols and other long-chain alkylphenols in WWTP effluents and receiving waters due to their persistence in WWTPs, strong sorption to WWTP sludge and aquatic sediments, may contaminate soils when WWTP sludge is used as a fertilizer	EDCs, can potentially impact the sustainability of aquatic wildlife populations, cause sperm count reduction in male offspring of mammals when exposed during pregnancy and lactation
Pentabromodiphenyl ether (pentaBDE) congeners, e.g., 2,2′,4,4′,5-pentabromodiphenyl ether	Polybrominated diphenyl ethers (PBDEs) used as flame retardants, with pentaBDE commonly used in polyurethane foam; released into the environment, e.g., from pentaBDE-containing products and from emissions related to their production	Elevated concentrations in soil, water, sediments, sludge, air-suspended particulate matter, food, and organisms; strong biomagnification in carnivores and humans, highly persistent	May affect the liver, thyroid (EDC effects), and neurobehavioral development of animals

(continued)

Table 15.1 (continued)

Compound(s)	Sources & uses	Contaminated environmental compartments / spread	(Potential) hazardous effects
Phthalates, e.g., di (2-ethylhexyl) phthalate (DEHP)	Used to soften plastics (plasticizer), in particular polyvinyl chloride; phased out of many products in Europe, the USA, and Canada for health concerns	Easy release from plastics (accelerated release by breaking and aging of plastics), strong sorption to WWTP sludge and aquatic sediments especially under anoxic conditions	EDCs; adverse effects on growth, reproduction, and development in wildlife
Polycyclic musk fragrances, e.g., 1,3,4,6,7,8-hexahydro-4,6,6,7,8,8-hexamethylcyclopenta [*g*]-2-benzopyran (HHCB, Galaxolide™)	Personal care product (PCP) ingredients (cleaning agents, air fresheners, and other hygiene/household products)	Presence in WWTP effluents and receiving waters due to its persistence in WWTPs, strong sorption to WWTP sludge and aquatic sediments, may contaminate soils when WWTP sludge is used as a fertilizer	Inhibition of MXR transporters in aquatic organisms, suspected EDCs
Sulfamethoxazole (SMX)	Sulfonamide bacteriostatic antibiotic, mostly used in synergistic combination with the bacteriostatic antibiotic trimethoprim (abbreviations SMX-TMP or SMZ-TMP)	Contaminant in WWTP effluents (communal and hospital wastewaters) and receiving waters due to incomplete removal in WWTPs, high mobility resulting in groundwater/bank filtrate contamination	Concerns related to a possibly accelerated development of bacterial antibiotic resistance
Triclosan (TCS)	Antiseptic in many consumer products (e.g., toothpastes, soaps, shampoos)	Contaminant in WWTP effluents and receiving waters due to incomplete removal in WWTPs; strong sorption to WWTP sludge, aquatic sediments and soil	Suspected EDC, toxic to aquatic bacteria and algae, concerns regarding the potential development of bacterial cross-resistances

PPCPs (Table 15.1). Among them, the antimicrobial TCS may be considered as a hygienic PCP, whereas the other compounds represent PhACs (Hochstrat et al. 2015; Hofmann and Schlosser 2016). Another classification approach, which is based on the biological effect mechanism of certain micropollutants, refers to the compounds causing interferences with the endocrine system of vertebrates as endocrine disrupting chemicals (EDCs). Among the aforementioned micropollutants, EE2, BPA, NP, and (possibly) also TCS exhibit endocrine activities, and, therefore, they are also referred to as EDCs (Ahmed et al. 2017; Hofmann and Schlosser 2016; Jahangiri et al. 2017; Table 15.1).

As a precondition for the function of any biological treatment technology, both anthropogenic and naturally occurring chemicals can only be biologically degraded under favorable conditions (Kolvenbach et al. 2014). New catabolic pathways, which are directed against the existing chemicals and the new ones in order to detoxify and/or utilize them as growth substrates, are known to be constantly evolving especially in bacteria (Kolvenbach et al. 2014). However, typically very low environmental concentrations of micropollutants make them only poor growth substrates for microbes, which is not in favor of the evolution of productive microbial degradation pathways typically found in bacteria (Harms et al. 2011; Hochstrat et al. 2015). Moreover, too low concentrations of potentially toxic pollutants may not adversely affect organisms and thus might not drive the evolution of degradative pathways aimed at detoxification (Kolvenbach et al. 2014). Furthermore, micropollutants usually occur in a mixture, whereas pollutant-degrading microbes such as bacteria are often more or less compound-specific (Harms et al. 2011). Degradation rates of those polluting chemicals, which are susceptible to potentially existing, sufficiently promiscuous enzymes, may remain infinitely low at too low environmental concentrations (Kolvenbach et al. 2014). Depending on the nature of a chemical, these obstacles may become even more complicated by physicochemical constraints related to bioavailability and bioaccessibility (Kolvenbach et al. 2014). Altogether, the aforementioned limitations largely contribute to the insufficient performance of conventional biological (waste)water treatment technologies towards micropollutants.

Besides too low pollutant concentrations, also too high loads of toxic contaminants and/or other extreme conditions of waters represent a second group of scenarios that also may hamper or even stop the efficient treatment by conventional biological technologies. The examples include acidic olive oil mill wastewaters contaminated with recalcitrant and toxic phenolic and lipid compounds; and colored and frequently alkaline effluents from the textile and dyestuff industries, which contain highly concentrated mixtures of structurally diverse, recalcitrant, often toxic, and sometimes potentially carcinogenic dyes and pigments together with high loads of different inorganic additives. Furthermore, highly toxic molasses-based wastewaters polluted with melanoidin-type high molecular weight compounds. and toxic pulp and paper bleach plant effluents of the alkaline extraction stage (containing phenolic wastes) and of chlorine-mediated bleaching processes (polluted with various chloroaromatic and chloroaliphatic compounds, and further colored substances) are also well-known to be inhibitory to conventional biological water treatment (Arora and Sharma 2010; Asgher et al. 2008; D'Souza et al. 2006; Harms et al. 2011; Majeau et al. 2010; Pérez et al. 2016; Pointing 2001; Pophali et al. 2003; Saparrat et al. 2010; Wesenberg et al. 2003).

Current advanced biological (waste)water treatment processes, which aim to overcome the performance limits of conventional biological processes, make use of either isolated enzymes or whole organisms. Immobilization of suitable, isolated enzymes yields biocatalytically active nanomaterials with an encouraging perspective for application in wastewater treatment despite certain issues such as upscaling the process, long-term stability, and cost efficiency, which still needs to be solved. Moreover most of the related applications are still at the experimental stage (Arca-Ramos et al. 2016; Ba et al. 2013; Fernández-Fernández et al. 2013). Other approaches, also mostly still experimental, such as bioelectrochemical systems (BESs) employ specific surface-immobilized microorganisms for wastewater treatment along with energy production based on the microbes' ability to catalyze redox reactions on or near the electrodes (Cecconet et al. 2017). Membrane bioreactors (MBRs), which retain the biocatalytically active sludge on the membrane surfaces and may also be augmented with additional organisms (e.g., fungi) (Ahmed et al. 2017), have been demonstrated to be highly effective in the removal of EDCs and PCPs already at pilot and full-scale (Cecconet et al. 2017). Biological activated carbon processes were reported to be suitable for the removal of PhACs and pesticides (Ahmed et al. 2017; Cecconet et al. 2017). Hybrid systems based on the combinations of biological and physicochemical processes have been developed in order to enhance the efficiency of water treatment processes. For instance, MBR performance can be improved by combination with activated carbon adsorption (Cecconet et al. 2017), and ozonation followed by biological activated carbon was reported to be highly efficient on PhACs and pesticides (Ahmed et al. 2017). Some of such hybrid technologies have already been successfully tested for PhAC removal at pilot and sometimes even at full-scale (MBR followed by ozonation and powdered activated carbon, respectively) (Alrhmoun et al. 2015). Constructed wetlands, which require more space compared to other biological water treatment systems, have also the potential to remove common PhACs from urban wastewater. Their implementation has been considered to be an alternative for wastewater treatment in small and scattered communities and for the final step in the treatment of special wastewaters from, e.g., healthcare or hospital facilities (Verlicchi and Zambello 2014). The following chapters will focus on particularly promising state-of-the-art biological treatment processes in detail, and hereby build a bridge between the tailored design of biocatalytically active nanomaterials on the basis of oxidative (laccases, peroxidases, tyrosinases) and hydrolytic enzymes (e.g., esterases, lipases) and the assessment of the efficiency of advanced biological processes in field studies.

Acknowledgement This work was supported by the Helmholtz Association of German Research Centres and contributes to the Chemicals in the Environment (CITE) Research Programme conducted at the Helmholtz Centre for Environmental Research – UFZ.

References

Ahmed MB, Zhou JL, Ngo HH, Guo W, Thomaidis NS, Xu J (2017) Progress in the biological and chemical treatment technologies for emerging contaminant removal from wastewater: a critical review. J Hazard Mater 323:274–298. https://doi.org/10.1016/j.jhazmat.2016.04.045

Alrhmoun M, Dagot C, Gonzales Ospina A, Jiang JQ, Klepiszewski K, Lyko S, Venditti S (2015) 6. Occurrence and removal of pharmaceuticals by advanced treatment of hospital wastewater. noPILLS report (http://wwwno-pillseu/conference/BS_NoPills_Final%20Report_long_ENpdf). June 2015 edn. EMSCHERGENOSSENSCHAFT, Essen, pp 82–95

Arca-Ramos A, Ammann EM, Gasser CA, Nastold P, Eibes G, Feijoo G, Lema JM, Moreira MT, Corvini PF-X (2016) Assessing the use of nanoimmobilized laccases to remove micropollutants from wastewater. Environ Sci Pollut Res 23(4):3217–3228. https://doi.org/10.1007/s11356-015-5564-6

Arora DS, Sharma RK (2010) Ligninolytic fungal laccases and their biotechnological applications. Appl Biochem Biotechnol 160(6):1760–1788. https://doi.org/10.1007/s12010-009-8676-y

Asgher M, Bhatti HN, Ashraf M, Legge RL (2008) Recent developments in biodegradation of industrial pollutants by white rot fungi and their enzyme system. Biodegradation 19 (6):771–783. https://doi.org/10.1007/s10532-008-9185-3

Ba S, Arsenault A, Hassani T, Jones JP, Cabana H (2013) Laccase immobilization and insolubilization: from fundamentals to applications for the elimination of emerging contaminants in wastewater treatment. Crit Rev Biotechnol 33(4):404–418. https://doi.org/10.3109/07388551.2012.725390

Cecconet D, Molognoni D, Callegari A, Capodaglio AG (2017) Biological combination processes for efficient removal of pharmaceutically active compounds from wastewater: a review and future perspectives. J Environ Chem Eng 5(4):3590–3603. https://doi.org/10.1016/j.jece.2017.07.020

D'Souza DT, Tiwari R, Sah AK, Raghukumar C (2006) Enhanced production of laccase by a marine fungus during treatment of colored effluents and synthetic dyes. Enzym Microb Technol 38(3–4):504–511. https://doi.org/10.1016/j.enzmictec.2005.07.005

Fernández-Fernández M, Sanromán MÁ, Moldes D (2013) Recent developments and applications of immobilized laccase. Biotechnol Adv 31(8):1808–1825. https://doi.org/10.1016/j.biotechadv.2012.02.013

Harms H, Schlosser D, Wick LY (2011) Untapped potential: exploiting fungi in bioremediation of hazardous chemicals. Nat Rev Microbiol 9(3):177–192. https://doi.org/10.1038/nrmicro2519

Hochstrat R, Schlosser D, Corvini P, Wintgens T (2015) Introduction. In: Hochstrat R, Wintgens T, Corvini P (eds) Immobilised biocatalysts for bioremediation of groundwater and wastewater. IWA Publishing, London, pp 1–14

Hofmann U, Schlosser D (2016) Biochemical and physicochemical processes contributing to the removal of endocrine-disrupting chemicals and pharmaceuticals by the aquatic ascomycete *Phoma* sp. UHH 5-1-03. Appl Microbiol Biotechnol 100(5):2381–2399. https://doi.org/10.1007/s00253-015-7113-0

Jahangiri E, Seiwert B, Reemtsma T, Schlosser D (2017) Laccase- and electrochemically mediated conversion of triclosan: metabolite formation and influence on antibacterial activity. Chemosphere 168:549–558. https://doi.org/10.1016/j.chemosphere.2016.11.030

Kolvenbach BA, Helbling DE, Kohler H-PE, Corvini PF-X (2014) Emerging chemicals and the evolution of biodegradation capacities and pathways in bacteria. Curr Opin Biotechnol 27:8–14. https://doi.org/10.1016/j.copbio.2013.08.017

Kümmerer K (2009) Antibiotics in the aquatic environment – a review – part I. Chemosphere 75 (4):417–434. https://doi.org/10.1016/j.chemosphere.2008.11.086

Kümmerer K (2011) 04 - Emerging contaminants. In: Wilderer P (ed) Treatise on water science, vol 3. Elsevier, Oxford, pp 69–87. https://doi.org/10.1016/B978-0-444-53199-5.00052-X

Lapworth DJ, Baran N, Stuart ME, Ward RS (2012) Emerging organic contaminants in groundwater: a review of sources, fate and occurrence. Environ Pollut 163:287–303. https://doi.org/10.1016/j.envpol.2011.12.034

Majeau J-A, Brar SK, Tyagi RD (2010) Laccases for removal of recalcitrant and emerging pollutants. Bioresour Technol 101(7):2331–2350. https://doi.org/10.1016/j.biortech.2009.10.087

Murray KE, Thomas SM, Bodour AA (2010) Prioritizing research for trace pollutants and emerging contaminants in the freshwater environment. Environ Pollut 158(12):3462–3471. https://doi.org/10.1016/j.envpol.2010.08.009

Pérez A, Poznyak T, Chairez I (2016) Effect of inorganic additives in the textile dyes removal by ozonation. In: Kumbasar EPA, Körlü AE (eds) Textile wastewater treatment. IntechOpen, Rijeka, pp 55–73. https://doi.org/10.5772/62286

Pointing S (2001) Feasibility of bioremediation by white-rot fungi. Appl Microbiol Biotechnol 57 (1–2):20–33. https://doi.org/10.1007/s002530100745

Pophali GR, Kaul SN, Mathur S (2003) Influence of hydraulic shock loads and TDS on the performance of large-scale CETPs treating textile effluents in India. Water Res 37 (2):353–361. https://doi.org/10.1016/S0043-1354(02)00268-3

Saparrat MCN, Jurado M, Díaz R, Romera IG, Martínez MJ (2010) Transformation of the water soluble fraction from "alpeorujo" by *Coriolopsis rigida*: the role of laccase in the process and its impact on *Azospirillum brasiliense* survival. Chemosphere 78(1):72–76. https://doi.org/10.1016/j.chemosphere.2009.09.050

Silva CP, Otero M, Esteves V (2012) Processes for the elimination of estrogenic steroid hormones from water: a review. Environ Pollut 165:38–58. https://doi.org/10.1016/j.envpol.2012.02.002

Solé M, Schlosser D (2015) Xenobiotics from human impacts. In: Krauss G-J, Nies DH (eds) Ecological biochemistry: environmental and interspecies interactions. Wiley-VCH, Weinheim, pp 259–275

Verlicchi P, Zambello E (2014) How efficient are constructed wetlands in removing pharmaceuticals from untreated and treated urban wastewaters? A review. Sci Total Environ 470–471:1281–1306. https://doi.org/10.1016/j.scitotenv.2013.10.085

Wesenberg D, Kyriakides I, Agathos SN (2003) White-rot fungi and their enzymes for the treatment of industrial dye effluents. Biotechnol Adv 22(1–2):161–187. https://doi.org/10.1016/j.biotechadv.2003.08.011

Chapter 16
Enzyme-Based Nanomaterials in Bioremediation

Monika Čvančarová, Patrick Shahgaldian, and Philippe F.-X. Corvini

Abstract The combination of enzymes and nanomaterials represents an innovative technology that can be used in the field of environmental bioremediation. Various biologically active molecules can be effectively immobilized onto nanomaterials, which are considered very interesting matrices because of their unique physicochemical properties. Nanoparticles are carriers that have been intensively investigated because of their potential to produce nanobiocatalysts. Currently, the development of novel hybrid nanocomposites such as nanographene, nanotubes, nanofibers, and nanogels is significantly increasing. In the first part of this chapter, importance and advantages of enzyme immobilization, properties of free enzymes, and parameters influencing the immobilization processes are discussed. Description of different types of nanomaterials and the corresponding immobilization techniques such as adsorption, covalent binding, entrapment, encapsulation, and cross-linking is also involved. The second part of this chapter is about degradation of various pollutants including phenolic compounds, synthetic dyes, pharmaceuticals, and PAHs by nanobiocatalysts. Many studies have been published over the past 5 years, which indicates an increasing interest in this application. The most important parameters and results of the published papers have been reviewed and organized in a comprehensive table. It is obvious that the utilization of nanobiocatalysts for remediating contaminated wastewater and soil is still under development. However, enzyme-based nanomaterials found an interesting application in the field of biosensors, which is described at the end of this chapter. Nanobiosensors are rapidly developing and more and more often used for detecting and monitoring various pollutants in the environment.

M. Čvančarová (✉) · P. F.-X. Corvini
Institute for Ecopreneurship, School of Life Sciences, University of Applied Sciences and Arts Northwestern Switzerland, Muttenz, Switzerland
e-mail: monika.cvancarova@fhnw.ch

P. Shahgaldian
Institute for Chemistry and Bioanalytics, School of Life Sciences, University of Applied Sciences and Arts Northwestern Switzerland, Muttenz, Switzerland

J. Filip et al. (eds.), *Advanced Nano-Bio Technologies for Water and Soil Treatment*, Applied Environmental Science and Engineering for a Sustainable Future,
https://doi.org/10.1007/978-3-030-29840-1_16

Keywords Biodegradation · Bioremediation · Enzyme · Immobilization · Laccase · Micropollutant · Nanobiocatalyst · Nanobiosensor · Nanocomposite · Nanomaterial · Nanoparticle · Peroxidase

16.1 Introduction

Water and soil contamination rates are increasing, therefore the need for developing effective remediation techniques is becoming more and more urgent. Nanotechnologies, which have been extensively studied in recent years, have the potential to provide satisfactory solutions for many types of environmental pollution (Rizwan et al. 2014). However, the release of nanoparticles into the environment causes certain health and environmental concerns (Handy et al. 2008). Nanomaterials exhibit unique physicochemical properties and represent novel and interesting matrices for the immobilization of various biologically active molecules. Immobilized enzymes have several applications in various industrial sectors (Ansari and Husain 2012). However, in the field of bioremediation, nanobiocatalysts represent a new approach, and the high number of recently published research papers indicates an increasing interest in this area.

This chapter includes a list of the enzymes that are typically used for bioremediation purposes. The main focus of this chapter is on the basic parameters influencing enzyme immobilization, the different types of nanomaterials used for this purpose, and the corresponding immobilization techniques. Advantages and disadvantages of the nanomaterials as immobilization matrix are discussed as well. Current development of biosensors for pollutant detection is also described, and a recent progress in the application of various nanobiocatalysts for bioremediation purposes is evaluated.

16.2 Free Enzymes Used for Bioremediation

Enzymes are specific biocatalysts that accelerate the conversion of substrates into products by providing favorable conditions that lower the activation energy of the reaction. A large number of enzymes from bacteria, fungi, and plants have been reported to be involved in the biodegradation of toxic organic pollutants (Karigar and Rao 2011).

The main detoxification enzymes (oxygenases, laccases, peroxidases) belong to the group of oxidoreductases (EC 1). These enzymes catalyze the transfer of electrons from a donor to an acceptor and the contaminants are often oxidized into less harmful compounds. Monooxygenases are involved in the process of desulfurization, dehalogenation, denitrification, ammonification, hydroxylation, biotransformation, and biodegradation of various aromatic and aliphatic compounds (Arora et al. 2010). However, large-scale applications of monooxygenases are hampered by their price and the price of their cofactors. Laccases represent another interesting group of oxidoreductases that exhibit a great bioremediation potential (Viswanath et al. 2014). They are produced intra- and extracellularly and can be relatively simply isolated and purified from different organisms that include mainly bacteria (Muthukumarasamy et al. 2015) and white-rot fungi (Shraddha et al. 2011). Laccases catalyze the oxidation of a wide range of phenolic and aromatic compounds. They can decolorize industrial textile effluents (Zucca et al. 2015) and initiate depolymerization reactions (Dashtban et al. 2010). It has been reported that laccases are involved in the degradation of highly recalcitrant environmental pollutants such as pharmaceuticals (Yang et al. 2017) and polycyclic aromatic hydrocarbons (Kadri et al. 2017). Peroxidases are oxidoreductases that oxidize organic compounds using hydrogen peroxide. Peroxidases are produced by a variety of different organisms and the most investigated types include lignin peroxidase, manganese-dependent peroxidase, and versatile peroxidase. Peroxidases have the potential for bioremediation of wastewater and soil contaminated with phenols, cresols, textile dyes, endocrine disruptive chemicals as well as several pesticides and herbicides (Bansal and Kanwar 2013).

Hydrolytic enzymes participate in the degradation of pollutants by breaking major chemical bonds. Lipases hydrolyze triacylglycerols into glycerol and free fatty acids and are used for effective removal of oil spills and reduction of hydrocarbons (Cammarota and Freire 2006). Further hydrolytic enzymes such as amylases, celluloses, and proteases are involved in the degradation of macromolecules in the activated sludge process (Guo and Xu 2011).

Further details on the role of microbial enzymes in the biodegradation of pollutants have been thoroughly reviewed elsewhere (Karigar and Rao 2011; Eibes et al. 2015; Kües 2015).

16.3 Why Immobilization?

Enzymatic degradation represents an eco-friendly remediation technique. However, it faces major hurdles which still hinder its sustainable and low cost applications. Soluble enzymes often exhibit a limited stability in contaminated environmental matrices. The immobilization of enzymes on solid supports in order to enhance the stability and improve the activity has been demostrated.

The immobilization process usually increases the enzyme stability against high temperature, pH variations, organic solvents and detergents during both storage and process operation. Enzyme activity is maintained or even significantly enhanced upon immobilization, which could be caused by conformational changes (Secundo 2013). Furthermore, the immobilized enzymes can be retained, separated, and applied in bioreactors operated in a continuous mode. Immobilization may reduce operational costs not only by improving the catalyst properties but also by enabling efficient recycling and controlling the process. Immobilization increases the area of application and facilitates large-scale implementation of the enzymatic processes (Rao et al. 2014).

16.4 Parameters Influencing Immobilization

When planning enzyme immobilization some parameters should be carefully considered. Especially physical properties of the selected carrier, i.e., the particle size, shape, porosity, and surface area, are very important. The type of the carrier consisting of organic or inorganic material with hydrophobic or hydrophilic properties and carrying surface charge and functionalization should be considered before starting the immobilization procedure. Mechanical and chemical stability of the carrier is important when selecting the type of reactors, reaction media, and reaction conditions (Cao 2006; Hanefeld et al. 2009).

Catalytic properties are linked to the enzymes and involve the turnover rate, selectivity, substrate specificity and stability at various pH values, extreme temperatures, and in different aqueous and organic media. Some other parameters, such as the size of the enzyme, its isoeletric point, conformational flexibility, surface functional groups, surface charge, the presence of hydrophobic and hydrophilic domains, should be taken into account when planning the immobilization. Catalytic functions are usually designed to reach high productivity, long-term stability, and broad applicability of the catalysts (Cao 2006; Hanefeld et al. 2009).

The third group of important parameters are specific factors related to the reaction system. Composition and viscosity of the reaction medium, reaction thermodynamics, mass transfer limitation, enzyme inhibition or precipitation, non-specific solute-support interactions represent only a few examples (Hanefeld et al. 2009).

Immobilization procedure should lead to the production of robust catalysts. Proper selection of the carrier, enzyme, and reaction parameters is crucial to achieve success. The resulting biocatalytic activity is influenced by the enzyme density on the carrier, enzyme orientation and conformation (Ding et al. 2015).

Several analytical methods have been developed to control the immobilization process and track the post-immobilization changes in the enzymes. These techniques are commonly based on surface analysis and involve, e.g., scanning electron microscopy, surface plasmon resonance, circular dichroism, and Förster resonance energy transfer (Mohamad et al. 2015). Time-of-flight secondary ion mass spectroscopy (TOF-SIMS) represents one of the most powerful analytical tools that can provide detailed surface characterization including the composition, structure, orientation, and spatial distribution of the molecules on the surface (Kim et al. 2015). Conformational analysis of the immobilized enzymes is very important and helps to develop and improve the enzyme immobilization strategies (Secundo 2013).

16.5 Basic Immobilization Techniques

Selection of the appropriate technique is crucial for effective immobilization of enzymes. The most common techniques involve adsorption, covalent binding, entrapment, encapsulation, and cross-linking. The prepared nanobiocatalysts are either carrier-bond or carrier-free enzymes. Detailed methodology has been published in several reviews (Brady and Jordaan 2009; Hanefeld et al. 2009) and a brief summary of each technique is summarized in this chapter.

Immobilization on a support depends on the properties of both the enzyme and the carrier material. A typical and straightforward method is physisorption (i.e., physical adsorption). It is a reversible process commonly based on hydrophobic or ionic interactions, van der Waals forces, and hydrogen bonding. This method is very simple and requires only mixing the enzyme and the carrier material under appropriate pH conditions and ionic strength. Because of weak interactions between the enzyme and the carrier, several factors like temperature fluctuations, changes in the pH, and ionic concentrations may cause desorption of the enzyme (Ahmad and Sardar 2015). Spontaneous desorption is the main disadvantage. It can be controlled, however, and utilized for regeneration and reloading the support with fresh enzyme. The cost of the enzyme is often a primary factor in the overall cost of the immobilized catalysts (Mohamad et al. 2015).

Chemisorption (i.e., covalent binding) is a widely used strategy for enzyme immobilization. It is mostly irreversible once covalent bonds have been formed between the enzyme and the carrier. Side chain functional groups of the enzyme like amino, thiol, carboxylic, imidazole, and phenolic are involved in the binding; however, it is important to verify that the groups recruited for immobilization are not essential for the catalytic activity of the enzyme. The surface of the support is commonly activated by electrophilic groups, e.g., epoxide, which react with nucleophiles on the protein. Covalent immobilization prevents enzyme release from the carrier surface into the reaction system, which results in higher stability. Conformational changes and orientation of the enzyme after the attachment to the support may cause an increase or decrease in the specific activity. Compared to the physical adsorption, covalent binding is irreversible and if the enzyme activity is decayed, it cannot be easily desorbed and the support cannot be simply regenerated (Mohamad et al. 2015).

Entrapment involves inclusion of an enzyme in a polymer network (gel lattice) such as an organic polymer or a silica sol–gel, or a membrane device such as a hollow fiber or a microcapsule. Entrapment requires the synthesis of the polymeric network in the presence of the enzyme (Homaei et al. 2013). Encapsulation of an enzyme is the formation of a membrane-like physical barrier around an enzyme preparation. Encapsulated enzymes are enclosed within the internal capsule phase by a semipermeable membrane coating or a porous polymeric network structure (Koyani and Vazquez-Duhalt 2016). Entrapment and encapsulation restrict the movement of the enzyme and only the substrate and products can pass through. By adequate selection of the polymer material and modification of the immobilization process, it is possible to create an ideal microenvironment for the enzyme with optimal pH and polarity. The entrapped and encapsulated enzyme is either free molecule or additionally covalently attached to the support. Entrapment and encapsulation of the enzyme in polymer network can improve the mechanical stability, prevent enzyme leaching, and protect the enzyme against external environment. The main disadvantage of these methods is mass transfer limitations. Large substrates cannot reach the active sites of the entrapped or encapsulated enzymes because of small pores in the polymer network. Low loading capacity and possible deactivation during immobilization represent another drawback of these techniques (Hwang and Gu 2013; Ahmad and Sardar 2015).

Carrier-free immobilized enzymes do not require a support and are prepared by direct cross-linking of different enzyme preparations. Several modifications such as cross-linked dissolved enzymes (CLEs), cross-linked enzyme aggregates (CLEAs), cross-linked enzyme crystals (CLECs), and cross-linked spray-dried enzymes (CLSDs) have been developed to date (Cao et al. 2003). The most common material is CLEAs. Preparation of CLEAs is relatively simple and consists of two steps. Soluble enzymes are precipitated into aggregates by water-miscible organic solvents, salts, or nonionic polymers. The process does not require highly pure enzymes. Then the aggregates are cross-linked by bi- or multifunctional reagents. Glutaraldehyde is a typically used agent because of high reactivity and low cost. CLEAs are easy to recover and reuse and have higher operational stability. By precise control of the process, CLEAs can be prepared as uniformly sized cross-linked aggregates (Nguyen and Yang 2014). Production of CLEAs is a relatively simple and cheap process. Carrier-free immobilization has several drawbacks though. Enzymes frequently undergo conformational changes that can cause the enzyme activity loss. Production of large and stable enzyme aggregates without denaturation can be problematic since the process of cross-linking is difficult to control and can result in mass transfer limitations of large substrates (Sheldon and van Pelt 2013).

Many variations based on combinations of these basic immobilization techniques have been developed. Further physicochemical factors, advantages and drawbacks of each technique should be considered when planning immobilization of enzymes; they are reviewed by Hanefeld et al. (2009), Hwang and Gu (2013), and Zucca and Sanjust (2014), respectively.

16.6 Nanomaterials

A broad variety of carrier materials are available for enzyme immobilization (Fig. 16.1) (Datta et al. 2013). Nanoparticles made of silica, magnetite, and titanium are the most common materials employed for biocatalysis. More recently, novel nanomaterials such as nanotubes, nanofibres, nanogels, and graphene oxide nanosheets have been tested. Nanomaterials exhibit unique physicochemical properties. Especially large specific surface area and effective enzyme loading are the key factors that determine the biocatalyst efficiency. High mechanical strength and minimal diffusional problems represent further advantages offered by nanomaterials (Cipolatti et al. 2014). The production of nanomaterials is usually a cheap process. Several reviews on the immobilization of various enzymes onto different types of nanomaterials have been published earlier (Gupta et al. 2011; Cipolatti et al. 2014;

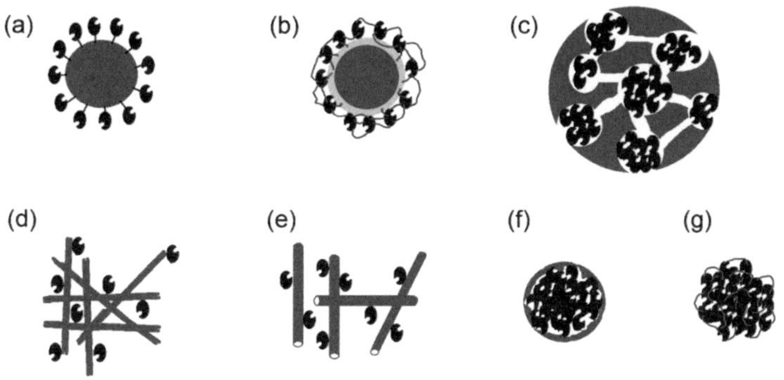

Fig. 16.1 Basic nanostructures for enzyme immobilization. (**a**) Nanoparticle with covalently bound enzymes. (**b**) Nanoparticle with cross-linked enzymes. (**c**) Mesoporous nanosphere. (**d**) Nanofibers. (**e**) Nanotubes. (**f**) Carrier-free enzyme nanogel. (**g**) Cross-linked enzyme aggregate. (Adapted from Wang (2006))

Min and Yoo 2014; Wang et al. 2015). In this chapter only recent papers that deal with the immobilization of enzymes for bioremediation application are discussed.

16.6.1 Nanoparticles

Spherical nanoparticles are the most studied enzyme carriers that offer many advantages due to their uniform size and spherical shape; however, handling of dry powders of nanoparticles presents certain health and environmental concerns (Handy et al. 2008). Retention of the nanoparticles and separation of the reaction medium are the main obstacles to their large-scale implementation. Sophisticated instrumentation for such hybrid membrane bioreactors with high-pressure pumps and proper regulation may help to solve this problem (Gasser et al. 2014). Magnetic nanoparticles have an additional advantage; they can be separated from the reaction media by an external magnetic field. Enzyme–nanoparticles conjugates have good Brownian motion mobility in a homogeneous reaction environment (Jia et al. 2003). Many types of materials, e.g., silica (Patel et al. 2014), magnetite (Mohamed et al. 2017), TiO_2 (Ahmad and Sardar 2014), gold, silver (Kim et al. 2017; Petkova et al. 2012), polymer (Tay et al. 2016), and core–shell $Fe_3O_4@SiO_2$ nanoparticles (Xia et al. 2017) have been successfully applied for the immobilization of various enzymes with pollutant-degradation potential.

The surface of nanoparticles is commonly modified by the introduction of specific functional groups or coated by specific layers to confer the requested properties that are essential for successful application. Immobilization of enzymes on modified nanoparticles is a broad topic and only selected examples are described in the following text. Laccase immobilization on functionalized magnetite nanoparticles was optimized by Fortes et al. (2017) and the preparation of chitosan-coated magnetite nanoparticles with covalently bound laccase was investigated by Kalkan et al. (2012). Koyani and Vazquez-Duhalt (2016) tested chitosan nanoparticles with encapsulated laccase in real bioremediation conditions. In soil, compost, and wastewater, the nano-encapsulated preparation showed higher stability against microbial degradation and thus longer operational activity compared to the free enzyme.

Nanomembranes prepared from nanoparticles with immobilized enzymes represent a new hybrid nanocomposite and are attracting more and more attention. Aghababaie et al. (2016) covalently immobilized a lipase enzyme on the surface of $Fe_3O_4@SiO_2$ nanoparticles/ultrafiltration membrane, which greatly improved the relative enzymatic activity and loading capacity in comparison to the unmodified ultrafiltration membrane. Hou et al. (2014b) compared the immobilization of laccase onto titanium nanoparticles and titanium-functionalized polyethersulfone membranes. These biocatalytic membranes showed greater immobilization efficiency, displayed good enzyme stability along with higher tolerance to a wider range of pH values and vigorous filtration conditions applied during the water treatment. Benefits of both nanoparticles and membrane filtration system have been successfully combined. Membrane bioreactor with magnetic biocatalytic membrane was tested by Gebreyohannes et al. (2015). The authors immobilized polygalacturonase and xylanase and observed that the enzymes prevented membrane fouling, which may be useful in industrial production, environmental remediation, or bio-energy generation.

16.6.2 *Novel Nanomaterials*

Recently a growing interest in nanographene, nanotubes, nanofibers, nanogels, and mesoporous nanospheres has been reported. Specific hybrid nanocomposites can be synthetized by functionalization and combination of various nanostructures.

16.6.2.1 Nanographene

Graphene is made of a single layer of carbon atoms that are bonded together in a repeating pattern of hexagons. Immobilization of enzymes onto graphene oxide nanosheets can be performed via physical adsorption, covalent attachment, and additional cross-linking; and show high thermal and solvent stability (Hermanová et al. 2015). Graphene oxide nanosheets have been used for successful immobilization of chloroperoxidase (Ding et al. 2017). Laccase of *Aspergillus oryzae* was immobilized on graphene nanosheets by Skoronski et al. (2017). The enzyme quickly lost its activity after the second reaction cycle when immobilized via physical adsorption, while the covalent binding technique retained around 80% of the activity after 6 cycles. Self-assembled free-standing graphene oxide membranes can be constructed by individual graphene oxide sheets through layer-by-layer stacking, resulting in excellent mechanical and optical performances (Chen et al. 2009).

16.6.2.2 Nanotubes

Amongst various nanomaterials, carbon nanotubes possess unique structural, mechanical, thermal, and biocompatibility properties. They have generated interest among researchers due to their potential for biotechnological applications (Feng and Ji 2011). The efficiency of nanotubes can be improved by surface functionalization (Verma et al. 2013). Kinetic, thermodynamic, and stability studies of laccase immobilized on multi-walled carbon nanotubes were performed by Tavares et al. (2015). Immobilization of various lipases on multi-walled carbon nanotubes was achieved by Badgujar et al. (2015). Chao et al. (2013) investigated halloysite nanotubes and laccase immobilized through dopamine self-polymerization process. Entrapped laccase could retain more than 90% of the initial activity after five repeated uses and exhibited a rapid degradation rate and high degradation efficiency with respect to the removal of phenolic compounds.

16.6.2.3 Nanofibers

Nanofibres are one-dimensional materials that offer a number of attractive features. Among them, electrospun nanofibers seem to have high benefits. Due to their reduced thickness and high porosity, mass transfer limitation is alleviated. They provide high surface area and are exceptionally long and uniform in diameter. Composites made from nanofibers are easier to recover and reuse than nanoparticles or carbon nanotubes. They have the benefit of being easy to produce and handle. Surface of nanofibers can be modified to optimize the enzyme loading and activity (Wang et al. 2009).

Covalent immobilization of enzyme on the surface of electrospun polymer nanofibers was investigated by Kim et al. (2005). This new approach of enzyme coating on nanofibers, which yields high activity and stability, creates a useful new biocatalytic system with potential applications in bioconversion, bioremediation, and biosensing The collection of randomly arrayed nanofibers usually forms a non-woven mesh or membrane (Gopal et al. 2006). Sulaiman et al. (2017) applied cellulose nanofiber from kenaf bast fibers for enzyme immobilization and preparation of ultrafiltration membrane, which showed good enzymatic performance and ability to be reused. Various organic and inorganic nanomaterials can be used to make hybrid composite nanofibers with additional physical properties and mechanical stability. Laccase was efficiently immobilized on amidoxime polyacrylonitrile/montmorillonite composite nanofibers (Feng et al. 2016) and Li et al. (2017b) reported a convenient method for preparing co-immobilized enzyme and magnetic nanoparticles using metal coordinated hydrogel nanofibers. Electrospun nanofibrous membranes filled with carbon nanotubes have been characterized by Wan et al. (2008).

16.6.2.4 Nanogels

Nanogels are composed of cross-linked three-dimensional polymer chain networks that are formed via covalent linkages or self-assembly processes. Two types of nanogels can be synthetized. Support-free nanogel is the enzyme itself surrounded with a porous polymeric network of a few nanometers in thickness. Support-based immobilized enzyme has nanogel as a major component. Enzymes are immobilized mainly via encapsulation and entrapment. Nanogels as enzyme support are designed to have accessible and reactive functional groups on their surfaces that enable covalent binding of enzyme to nanogel (Chauhan 2014).

Zhang et al. (2015) fabricated a novel biocatalyst by coating polyethylenimine onto the native laccase followed by cross-linking with glutaraldehyde. The immobilized enzyme exhibited significantly higher decolorization efficiency in the degradation of a representative azo dye, Acid orange 7. Jia et al. (2013) studied laccase nanogel that was prepared by encapsulation in polyacrylamide as the result of N-acryloxysuccinimide modification and in situ polymerization.

16.6.2.5 Nanoflowers

Nanoflowers are innovative 3D structures for enzyme immobilization (Ge et al. 2012). Biocatalytic nanoflowers were investigated by Li et al. (2017a), who synthesized a 3D flower-like structured self-assembly hybrid nanocomposite with copper phosphate, laccase, graphite oxide, and carbon nanotubes. The prepared nanocomposite exhibited very high enzyme loading and improved laccase activity. Efficient removal of organic dye and micropollutant was achieved by the hybrid nanoflowers.

16.6.2.6 Mesoporous Nanosphere

Mesoporous nanomaterials possess high surface area, narrow pore size distribution in the nanometer range, and defined pore geometry. Mesoporous materials are mechanically stable. Enzymes can be immobilized by physical adsorption; however, it is a reversible process. To prevent the enzymes continuously leaching from the material, their surface can be functionalized. Mesoporous nanoparticles have been synthesized and tested for enzyme immobilization by Ibrahim et al. (2016) and Kalantari et al. (2017). Specific hybrid composite can be also prepared by combination of mesoporous nanospheres with various nanomaterials. Composite of Fe_3O_4 magnetic nanoparticles with mesoporous silica nanospheres was used by

Zhu et al. (2007) for laccase immobilization. Laccase sorbed onto carbon-based mesoporous magnetic nanocomposite was applied by Liu et al. (2012) for successful removal of phenol and *p*-chlorophenol.

The interest in the synthesis of various hybrid nanocomposites is rapidly increasing and several specific techniques have been published in recent years. Song et al. (2017) reported a novel strategy for enzyme immobilization based on DNA strand displacement on modified magnetic nanoparticles using alkaline phosphatase and horseradish peroxidase. Woo et al. (2015) introduced a nanosized magnetite impregnated mesocellular foam composite with Cu ligand as an enzyme carrier. Cao et al. (2016) prepared and successfully used magnetic cellulose nanocrystals as an enzyme support for *Pseudomonas cepacia* lipase immobilization via precipitation followed by a cross-linking process.

16.7 Application of Nanobiocatalyst for Remediation

Nanobiocatalysts may be used for remediation of contaminated wastewater or soil. Many studies have been published over the past 5 years (Table 16.1). It is obvious that field application of nanobiocatalysts is still in its infancy. Most of the work is focused on synthesis and evaluation of new nanocomposites. Remediation efficiency of the nanobiocatalysts was tested with various pollutants including phenolic compounds, synthetic dyes, pharmaceuticals, and PAHs. The treatment was mostly performed in artificially spiked standard solutions in batch systems, which often does not reflect real conditions. Removal of the pollutants by biocatalyst-containing nanocomposites is actually the result of many concurrent processes, which include mainly enzymatic degradation, physical adsorption, and sometimes polymerization. Removal efficiency is dependent especially on the pH, the concentration of the pollutant, and the presence of redox mediators, as well as matrix effects. The biocatalyst nanocomposites usually showed high enzyme loading and enhanced catalytic activity and stability compared to the corresponding free enzymes. The pollutant removal efficiency was mostly >80%; however, as it was mentioned above, most of the research was performed in buffered conditions.

Several research groups tested remediation processes using nanobiocatalysts in continuous flow mode. To retain the immobilized enzymes and separate them from the treated water, membrane hybrid reactors are commonly operated. A laboratory scale perfusion basket reactor was used by Kumar et al. (2014) in order to study the continuous decolorization of dyes by a laccase enzyme from *Trametes versicolor* immobilized onto amino-functionalized magnetic nanoparticles. The efficient decolorization (>90%) of Remazol brilliant blue R and slight decrease in nanoparticle activity were measured over a 10-hour period of continuous operation. Zhang et al. (2015) synthesized a novel biocatalyst by coating polyethylenimine onto the native laccase followed by cross-linking with glutaraldehyde. The carrier-free nanogels exhibited enhanced stability, high catalytic activity, and favorable properties for membrane separation. The cyclic decolorization of Acid orange 7 demonstrated that

Table 16.1 Enzymes immobilized on various nanomaterials and their applications for bioremediation purposes

Nanomaterial/nanocomposite	Source of laccase	Immobilization technique	Pollutant	Matrix	Scale	Reference
Amino-functionalized silica nanoparticles	*Coriolopsis polyzona*	Covalent	Bisphenol A	Standard solution	Batch	Galliker et al. (2010)
Electrospun nanofibrous membranes	*Trametes versicolor*	Encapsulation	Polycyclic aromatic hydrocarbons	Shoal soil	Batch	Dai et al. (2011)
Electrospun nanofibrous membranes	*Trametes versicolor*	Encapsulation	Polycyclic aromatic hydrocarbons	Standard solution	Biocatalytic membrane reactor	Dai et al. (2013)
Polyacrylonitrile nanofibrous membrane	Laccase	Covalent	2,4,6-trichlorophenol	Standard solution	Biocatalytic membrane reactor	Xu et al. (2013)
CLEAs	*Trametes versicolor* and mushroom tyrosinase	Cross-linking	Acetaminophen	Municipal and hospital wastewater	Batch	Ba et al. (2014)
Amino-functionalized silica nanoparticles	*Thielavia* genus	Cross-linking	Bisphenol A	Municipal and industrial wastewater	Hybrid membrane reactor - pilot	Gasser et al. (2014)
TiO_2 sol–gel coated PVDF membrane	*Trametes versicolor*	Covalent	Bisphenol A	Standard solution	Biocatalytic membrane reactor	Hou et al. (2014a)
CLEAs and amino-functionalized magnetic iron nanoparticles	*Trametes versicolor*	Cross-linking	Remazol brilliant blue R, Malachite green and Reactive black 5 dyes	Standard solution	Batch and perfusion basket reactor	Kumar et al. (2014)
Polyacrylonitrile/montmorillonite/graphene oxide nanofibrous membrane	*Trametes versicolor*	Physical adsorption	Catechol	Standard solution	Biocatalytic membrane reactor	Wang et al. (2014a)
Biotitania-coated magnetic iron nanoparticles	*Thielavia* sp.	Entrapment	Sulfamethoxazole, carbamazepine, bisphenol A, 17aethinylestradiol, diclofenac and triclosan	Standard solution	Batch	Ardao et al. (2015)

Chitosan-coated magnetic iron nanoparticles	*Trametes versicolor*	Cross-linking	Direct blue 78, Acid blue 225, Reactive red 195, Acid blue 74 and Phenol red	Standard solution	Batch	Jořenek and Zajoncová (2015)
ZnO nanowires in macroporous SiO_2 composite	*Trametes versicolor*	Physical adsorption	Remazol brilliant blue B and Acid blue 25 dyes	Standard solution	Batch	Li et al. (2015)
Fullerene, multi-walled carbon nanotubes, oxidized-multi-walled carbon nanotubes, graphene oxide	*Trametes versicolor*	Physical adsorption	Bisphenol A, catechol	Standard solution	Batch	Pang et al. (2015)
Carrier-free nanogel	*Trametes versicolor*	Cross-linking	Acid orange 7	Standard solution	Hybrid membrane reactor	Zhang et al. (2015)
Amino-functionalized silica nanoparticles	*Trametes versicolor* and *Myceliophthora thermophila*	Cross-linking	Bisphenol A, diclofenac	Municipal wastewater	Batch	Arca-Ramos et al. (2016)
Electrospun fibrous membranes modified by multi-walled carbon nanotubes	*Trametes versicolor*	Encapsulation	Bisphenol A, triclosan, 2,4-dichlorophenol	Standard solution	Batch	Dai et al. (2016)
TiO_2 nanoparticles and TiO_2 sol–gel coated PVDF membrane	*Trametes versicolor*	Covalent	Carbamazepine	Standard solution	Hybrid membrane reactor and bio-catalytic membrane reactor	Ji et al. (2016a)
Functionalized graphene oxide nanosheets	*Trametes versicolor*	Covalent	Anthracene and pinacyanol chloride	Standard solution	Batch	Patila et al. (2016)
$MnFe_2O_4$/calcium alginate nanocomposite	Glucose oxidase and laccase	Entrapment	Methylene blue, Indigo and Acid red 14 dyes	Standard solution	Batch	Shojaat et al. (2016)
Chitosan-coated magnetic iron nanoparticles	*Trametes versicolor*	Covalent	Chlorpyrifos	Standard solution	Batch	Das et al. (2017)

(continued)

Table 16.1 (continued)

Nanomaterial/nanocomposite	Source of laccase	Immobilization technique	Pollutant	Matrix	Scale	Reference
EDTA-Cu(II) chelating magnetic iron nanoparticles	*Aspergillus oryzae* (Novozym ® 51,003)	Chelation with copper(II) ions	Indigo carmine and Congo red dyes	Standard solution	Batch	Fernandes et al. (2017)
Amino-functionalized silica nanoparticles	*Ganoderma cupreum* AG-1	Covalent	Reactive violet 1 dye	Standard solution	Batch	Gahlout et al. (2017)
Amino-functionalized TiO_2 nanoparticles	*Pleurotus ostreatus*	Covalent	Bisphenol A and carbamazepine	Standard solution	Hybrid membrane reactor	Ji et al. (2017)
Polyethylenimine/ polyethersulfone electrospun nanofibrous membrane	*Rhus vernificera*	Cross-linking	Bisphenol A	Standard solution	Bio-catalytic membrane reactor	Koloti et al. (2018)
Nanoflowers with graphene oxide and carbon nanotubes	*Trametes versicolor*	Encapsulation	Crystal violet and Neutral red dyes	Standard solution	Batch	Li et al. (2017b)
Amine-functionalized Fe_3O_4@C nanoparticles	*Trametes versicolor*	Chelation with copper(II) ions	*o*-phenylenediamine	Standard solution	Batch	Lin et al. (2017)
ZnO nanotubes and MnO_2 nanoparticles	*Trametes versicolor*	Chelation with copper(II) ions	Alizarin red S dye	Standard solution	Batch	Rani et al. (2017)
Dopamine-coated magnetic iron nanoparticles	*Trametes versicolor*	Entrapment	4-chlorophenol	Standard solution	Batch	Zhang et al. (2017)
Silica-coated magnetic iron nanoparticles	Horseradish	Covalent	2,4-dichlorophenol	Standard solution	Batch	Chang and Tang (2014)
Magnetic iron nanoparticles/ graphene oxide nanocomposite	Horseradish	Covalent	2-chlorophenol, 4-chlorophenol and 2,4-dichlorophenol	Standard solution	Batch	Chang et al. (2015)
TiO_2 hollow nanofibers	Horseradish	Encapsulation	2,4-dichlorophenol	Standard solution	Batch	Ji et al. (2016b)

Silica-coated magnetic iron nanoparticles	Soybean	Covalent	Ferulic acid	Standard solution	Batch	Silva et al. (2016)
Chitosan nanoparticles	*Bjerkandera adusta*	Encapsulation	Ten phenolic compounds, including pesticides, halogenated compounds, endocrine disruptors and antibacterials	Standard solution	Batch	Alarcón-Payán et al. (2017)
Silica-coated TiO_2 nanocomposite	Ginger	Covalent	Acid yellow 42	Standard solution	Batch	Ali et al. (2017)
CLEAs	Horseradish	Cross-linking	Methyl orange, Indigo, Rhodamine B, Rhodamine 6G, and Pararosaniline dyes	Standard solution	Packed bed reactor	Bilal et al. (2017)
ZnO nanowires/macroporous SiO_2 composite	Horseradish	Cross-linking	Acid blue 113 and Acid black 10 BX	Standard solution	Batch	Sun et al. (2017)
Functionalized nanoporous activated carbon	Lipase from *Pseudomonas otitidis*	Physical adsorption	Oil	Wastewater from oil refinery	Packed column reactor	Saranya et al. (2014)
Functionalized multi-walled carbon nanotubes	Protocatechuate 3,4-dioxygenase from *Pseudomonas* sp.	Covalent	3,4-dihydroxybenzoic acid	Standard solution	Batch	Das et al. (2016)
Chitosan-coated magnetic iron nanoparticles	Endoglucanase, cellobiohydrolase, β-glucosidase from *Aspergillus niger*	Covalent	Biomass degradation, anthocyanins recycling	Fruit and vegetable waste	Batch	Yuan et al. (2016)

hybrid membrane reactors based on the laccase-coated nanogels could be an attractive strategy for water treatment. Ji et al. (2017) immobilized crude enzyme extracts from *Pleurotus ostreatus* onto functionalized TiO_2 nanoparticles. The nanobiocatalysts were used to treat water contaminated with two micropollutants, namely bisphenol A and carbamazepine, in a hybrid membrane reactor. The biocatalyst gave performance comparable to the purified commercial enzyme. Crude enzyme extracts have significant potential for cost-effective applications. Bilal et al. (2017) prepared horseradish peroxidase cross-linked enzyme aggregates. Biocatalytic efficiency of these aggregates was investigated for bioremediation purposes using a newly developed packed-bed reactor system. Successful decolorization, as well as removal of different synthetic dyes, was achieved in the reactor. After seven consecutive dye degradation cycles, 60% of the initial enzymatic activity was retained by the aggregates. The newly developed horseradish peroxidase cross-linked enzyme aggregates have promising potential for the removal of synthetic dyes.

Biocatalytic membranes represent a new approach for continuous treatment of contaminated effluents. Membranes can be designed by combining various nanostructures of the membrane with immobilized enzymes. Production of biocatalytic membranes becomes popular because of their promising application in continuous flow reactors that may be used for remediation of real wastewater. Hou et al. (2014a) immobilized laccase on TiO_2 sol–gel coated 0.1 and 0.45 μm polyvinylidene fluoride (PVDF) membranes. The results showed that both the pore size of the membrane support and the number of coating cycles had a significant impact on the biocatalytic membrane performance. The 0.1 μm thick membranes exhibited higher activity recovery and better stability regarding bisphenol A degradation performances. Substantial improvement in bisphenol A removal efficiency and stability were obtained under moderate flow-rate conditions, and the biodegradation process showed negligible fouling impact on the coated 0.1 μm membrane. Ji et al. (2016a) immobilized laccase of *T. versicolor* onto TiO_2 nanoparticles and TiO_2 sol–gel coated PVDF membrane and tested the degradation of carbamazepine in two different reactors. Over 65% removal of carbamazepine was achieved in the membrane hybrid reactor containing biocatalytic TiO_2 suspension within 96 h, while the biocatalytic membrane reactor removed only 40% of carbamazepine during the same period under identical operational conditions. The difference may be attributed to the relatively short contact time between the attached enzymes and the substrates in the biocatalytic membrane. In the study reported by Xu et al. (2013), a laccase enzyme was immobilized on polyacrylonitrile nanofibrous membranes and successfully applied for the removal of 2,4,6-trichlorophenol from water. Nanomembranes with fiber diameters from 200 nm to 300 nm were fabricated via electrospinning and provided a large surface area for enzyme immobilization and catalytic reactions. Koloti et al. (2018) covalently bound laccase from *Rhus vernificera* on hyperbranched polyethylenimine/polyethersulfone electrospun nanofibrous membranes. Recyclability study indicated that the laccase-modified membranes maintained a high bisphenol A removal level, up to 79% even after four filtration cycles. The laccase-modified membranes also maintained a constant permeate flux

(7.07 ± 5.54 L/m^2 h) throughout the filtration process. Polyacrylonitrile/montmorillonite/graphene oxide composite nanofibers were evaluated by Wang et al. (2014a) in a homemade nanofibrous membrane reactor. These membranes with immobilized laccase from *Trametes versicolor* were successfully applied for the removal of catechol from an aqueous solution. The addition of graphene oxide in the nanomembrane composite significantly improved its operational and storage stability. The treatment method was simple, low cost, and produced no secondary pollution, making it a good candidate for future industrial and remediation applications. Four types of electrospun fibrous membranes with laccase catalytic activity were fabricated by emulsion electrospinning by Dai et al. (2013). The membranes were employed in biocatalytic membrane reactor for removal of polycyclic aromatic hydrocarbons from water. The rapid adsorption of polycyclic aromatic hydrocarbons onto the membranes significantly improved their degradation efficiencies by a laccase.

A few attempts were made to remediate real contaminated samples. Ba et al. (2014) prepared laccase and tyrosinase cross-linked enzyme aggregates. In batch mode, the aggregates transformed more than 80% to nearly 100% of acetaminophen from municipal wastewater and more than 90% from hospital wastewater. Arca-Ramos et al. (2016) treated secondary effluent of municipal wastewater by laccase immobilized onto fumed silica nanoparticles. Compared to soluble laccases, immobilized enzymes led to much slower rates of bisphenol A transformation. For instance, after 24 h the percentages of bisphenol A removal by free laccases or immobilized enzymes reached 67.8 ± 5.2 and $27.0 \pm 3.9\%$, respectively. The transformation of diclofenac could not be achieved with the tested biocatalysts under real municipal wastewater conditions. Dai et al. (2011) investigated laccase-carrying electrospun nanofibrous membranes for adsorption and degradation of polycyclic aromatic hydrocarbons in shoal soils. The removal efficiencies for phenanthrene, fluoranthene, benz[*a*]anthracene, and benzo[*a*]pyrene after 6 h were greater than 95.1%, 93.2%, 79.1%, and 72.5%, respectively. Saranya et al. (2014) treated wastewater from oil refinery using lipase immobilized on functionalized nanoporous activated carbon. The immobilized lipase showed high efficiency for the hydrolysis of oil in batch and continuous mode and high operational stability of up to 50 cycles of use. The authors concluded that the prepared biocatalysts have great potential for remediation of lipid-containing wastewater from the industrial sectors. Yuan et al. (2016) introduced novel magnetic molecularly imprinted immobilized cellulases. The biocatalysts were used for the effective degradation of fruit and vegetable waste and for synergistic recovery of anthocyanins.

Recently published papers demonstrated good stability and performance of newly synthetized biocatalytic nanocomposites and nanomembranes. However, the efficiency of the nanostructures was mainly tested in pure water or buffer solutions. It is obvious from the few studies mentioned above that real matrix has a significant impact on the efficiency of these bioremediation processes. Further studies are essential to evaluating the feasibility of utilizing such biocatalytic nanocomposite systems in real contaminated matrices. Another weakness of the works performed so far is the scale of the tested processes. Almost all experiments using hybrid membrane reactor or biocatalytic membrane were operated in a lab-scale and batch mode. This underlines the fact that bioremediation technology with nanobiocatalysts is still under development. Only one pilot experiment has been performed so far; it will be described in the following chapters of this book (Gasser et al. 2014).

16.8 Nanobiosensors

Enzymes immobilized onto nanomaterials can be used for designing biosensors that can be applied for the detection of various micropollutants in the environment. A biosensor is a device for the selective detection of analytes that combines a biological component such as an enzyme and a physicochemical detector associated with electronics and a signal processing system. Biosensors can be considered a complementary tool to classical analytical methods that include gas and liquid chromatography. Most biosensors are based on an electrochemical principle that employs redox reactions in order to quantify the amount of an analyte. Electrochemical biosensors can be categorized as potentiometric, amperometric, and conductometric. A potentiometric biosensor measures oxidation or reduction potential of an electrochemical reaction using an indicator electrode and a reference electrode. An amperometric biosensor is based on the measurement of current as a function of time resulting from the oxidation and reduction of an electroactive species in a biochemical reaction that mainly depends on the concentration of an analyte with a fixed potential. A conductometric biosensor measures changes in the ionic strength, and thus the conductivity of the solution between two electrodes as a result of an enzymatic reaction (Perumal and Hashim 2014). Common biosensors are based on cells or enzymes bound to the surface of a membrane or an electrode. These biosensors have been tested for the detection of heavy metals, phenolic compounds like pesticides, herbicides, and pharmaceuticals in wastewater samples and soil and many of them have been reviewed by Nigam and Shukla (2015) and Rebollar-Pérez et al. (2015).

A large number of nanobiosensors have been developed and tested over the past 5 years (Table 16.2). They have been constructed of diverse nanomaterials ranging from nanoparticles, nanotubes, nanorods, nanowires, graphene nanosheets, and various hybrid nanocomposites. Nanomaterials incorporated in the biosensors represent a new approach with unique features, which were summarized by Malik et al. (2013) and Hammond et al. (2016). High selectivity, low detection limits, simplicity, relatively low cost, and reproducibility represent the most promising properties of nanobiosensors. Nanobiosensors are nowadays under development, and especially laccase, tyrosinase, and acetylcholinesterase have been tested for the detection of pesticides and phenolic compounds. The interest in enzyme-based biosensors has significantly increased thanks to the properties, which can extend the range of applications utilizing nanobiocatalysts.

Table 16.2 Enzyme-based nanobiosensors exploited for detection of various pollutants

Nanomaterial	Enzyme	Pollutant	Reference
Multi-walled carbon nanotubes/cobalt phthalocyanine/silk fibroin composite	Mushroom tyrosinase	Bisphenol A	Yin et al. (2010)
Nanographene and multi-walled carbon nanotubes	Mushroom tyrosinase	Bisphenol A	Wu et al. (2012)
Chitosan-carbon coated nickel nanoparticles	Mushroom tyrosinase	Catechol	Yang et al. (2012)
Liposome bioreactor and chitosan nanocomposite	Mushroom tyrosinase	Phenolic compounds	Guan et al. (2013)
Self-assembled laccase/multi-walled carbon nanotubes	Laccase	Catechol and hydroquinone	Qu et al. (2013)
Nanostructured Au film	Horseradish peroxidase	4-chlorophenol	Qiu et al. (2013)
Osmium tetroxide and multi-walled carbon nanotubes	Laccase from *Trametes versicolor*	Pyrocatechol	Das et al. (2014)
Electrospun copper/carbon composite nanofibers	Laccase	Catechol	Fu et al. (2014)
Mesoporous carbon/Co_3O_4 nanocomposite	Mushroom tyrosinase	Phenolic compounds	Wang et al. (2014b)
Mesoporous carbon nitride	Mushroom tyrosinase	Catechol and phenol	Zhou et al. (2014)
Gold nanoparticles on zein ultrafine fibers	Laccase	Catechol	Chen et al. (2015)
ZnO nanoparticles	Urease type from Jack beans	Urea	Eghbali et al. (2015)
Sol–gel TiO_2/multi-walled carbon nanotubes/polycationic polymer/Nafion composite	Mushroom tyrosinase	Bisphenol A	Kochana et al. (2015)
Graphene oxide/palladium–copper alloyed nanocages	Laccase	Catechol	Mei et al. (2015)
Graphene-gold nanoparticle composite	Mushroom tyrosinase	Bisphenol A	Pan et al. (2015)
Multi-walled carbon nanotubes	Acetylcholinesterase from electric eel	Organophosphorus pesticides	Yu et al. (2015)
Multi-walled carbon nanotubes	Mushroom tyrosinase	Bisphenol A	Zehani et al. (2015)
Graphene oxide-glycol chitosan nanohybrid	Laccase from *Trametes versicolor*	Phenolic compounds	Boujakhrout et al. (2016)
Core–shell magnetic iron nanoparticles	Acetylcholinesterase	Organophosphorus pesticides	Cancar et al. (2016)
Nanostructured polyaniline-Nafion ®/Au/Al_2O_3	Urease type from Jack beans	Mercury and lead ions	Do and Lin (2016)
Silver nanoparticles	Acetylcholinesterase from electric eel	Organophosphorus pesticides	Zheng et al. (2016)

16.9 Conclusions and Future Prospects for Nanobiocatalysts

Specific properties of nanomaterials give an opportunity to develop unique nanobiocatalysts that differ from the traditional immobilized enzymes. Proper selection of the nanomaterial, enzyme, and immobilization technique is always crucial for successful preparation of stable and active biocatalysts. Nanobiocatalysts have already been introduced in various branches of industries and medicine. A new approach of nanobiocatalysts utilization is their application for remediation of contaminated water and soil. In recent years, an enormous number of research papers have been published on this topic. The application of nanobiocatalysts for wastewater and soil treatment is still under development; however, the increasing interest of researchers is pushing this technology closer to being applied in real conditions. Most of the nanobiocatalysts have been tested in standard solutions and showed enhanced stability and catalytic activity compared to the free form enzymes. The results pointed to the remediation potential they have. Although a few attempts have already been made, further studies will have to be performed to evaluate this technology in real contaminated matrix. All remediation experiments except one were performed in a lab-scale unit. The lack of pilot scale studies is definitively limiting broader implementation of nanobiocatalysts for bioremediation of real wastewater or soil. Another significant hindrance is the price of enzymes. Enzyme-based nanomaterials found an interesting application in the field of biosensors. Nanobiosensors have high selectivity, fast response, and low detection limits. Due to their simplicity and relative low cost, they are rapidly developing and are more often used for detection and monitoring various pollutants in the environment.

References

Aghababaie M, Beheshti M, Razmjou A, Bordbar A-K (2016) Covalent immobilization of *Candida rugosa* lipase on a novel functionalized $Fe_3O_4@SiO_2$ dip-coated nanocomposite membrane. Food Bioprod Process 100:351–360. https://doi.org/10.1016/j.fbp.2016.07.016

Ahmad R, Sardar M (2014) Immobilization of cellulase on TiO_2 nanoparticles by physical and covalent methods: a comparative study. Indian J Biochem Biophys 51(4):314–320

Ahmad R, Sardar M (2015) Enzyme immobilization: an overview on nanoparticles as immobilization matrix. Biochem Anal Biochem 4(2):1000178. https://doi.org/10.4172/2161-1009.1000178

Alarcón-Payán DA, Koyani RD, Vazquez-Duhalt R (2017) Chitosan-based biocatalytic nanoparticles for pollutant removal from wastewater. Enzym Microb Technol 100:71–78. https://doi.org/10.1016/j.enzmictec.2017.02.008

Ali M, Husain Q, Alam N, Ahmad M (2017) Enhanced catalytic activity and stability of ginger peroxidase immobilized on amino-functionalized silica-coated titanium dioxide nanocomposite: a cost-effective tool for bioremediation. Water Air Soil Pollut 228(1):22. https://doi.org/10.1007/s11270-016-3205-4

Ansari SA, Husain Q (2012) Potential applications of enzymes immobilized on/in nano materials: a review. Biotechnol Adv 30(3):512–523. https://doi.org/10.1016/j.biotechadv.2011.09.005

Arca-Ramos A, Ammann EM, Gasser CA, Nastold P, Eibes G, Feijoo G, Lema JM, Moreira MT, Corvini PF-X (2016) Assessing the use of nanoimmobilized laccases to remove micropollutants from wastewater. Environ Sci Pollut Res 23(4):3217–3228. https://doi.org/10.1007/s11356-015-5564-6

Ardao I, Magnin D, Agathos SN (2015) Bioinspired production of magnetic laccase-biotitania particles for the removal of endocrine disrupting chemicals. Biotechnol Bioeng 112 (10):1986–1996. https://doi.org/10.1002/bit.25612

Arora PK, Srivastava A, Singh VP (2010) Application of monooxygenases in dehalogenation, desulphurization, denitrification and hydroxylation of aromatic compounds. J Bioremed Biodegr 1(3):1000112. https://doi.org/10.4172/2155-6199.1000112

Ba S, Haroune L, Cruz-Morató C, Jacquet C, Touahar IE, Bellenger J-P, Legault CY, Jones JP, Cabana H (2014) Synthesis and characterization of combined cross-linked laccase and tyrosinase aggregates transforming acetaminophen as a model phenolic compound in wastewaters. Sci Total Environ 487(1):748–755. https://doi.org/10.1016/j.scitotenv.2013.10.004

Badgujar KC, Sasaki T, Bhanage BM (2015) Synthesis of lipase nano-bio-conjugates as an efficient biocatalyst: characterization and activity–stability studies with potential biocatalytic applications. RSC Adv 5(68):55238–55251. https://doi.org/10.1039/C5RA10032A

Bansal N, Kanwar SS (2013) Peroxidase(s) in environment protection. Sci World J 2013:714639. https://doi.org/10.1155/2013/714639

Bilal M, Iqbal HMN, Hu H, Wang W, Zhang X (2017) Development of horseradish peroxidase-based cross-linked enzyme aggregates and their environmental exploitation for bioremediation purposes. J Environ Manag 188:137–143. https://doi.org/10.1016/j.jenvman.2016.12.015

Boujakhrout A, Jimenez-Falcao S, Martínez-Ruiz P, Sánchez A, Díez P, Pingarrón JM, Villalonga R (2016) Novel reduced graphene oxide–glycol chitosan nanohybrid for the assembly of an amperometric enzyme biosensor for phenols. Analyst 141(13):4162–4169. https://doi.org/10.1039/C5AN02640G

Brady D, Jordaan J (2009) Advances in enzyme immobilisation. Biotechnol Lett 31 (11):1639–1650. https://doi.org/10.1007/s10529-009-0076-4

Cammarota MC, Freire DMG (2006) A review on hydrolytic enzymes in the treatment of wastewater with high oil and grease content. Bioresour Technol 97(17):2195–2210. https://doi.org/10.1016/j.biortech.2006.02.030

Cancar HD, Soylemez S, Akpinar Y, Kesik M, Göker S, Gunbas G, Volkan M, Toppare L (2016) A novel acetylcholinesterase biosensor: core–shell magnetic nanoparticles incorporating a conjugated polymer for the detection of organophosphorus pesticides. ACS Appl Mater Interfaces 8 (12):8058–8067. https://doi.org/10.1021/acsami.5b12383

Cao L (2006) Introduction: immobilized enzymes: past, present and prospects. In: Carrier-bound immobilized enzymes. https://doi.org/10.1002/3527607668.ch1

Cao L, van Langen L, Sheldon RA (2003) Immobilised enzymes: carrier-bound or carrier-free? Curr Opin Biotechnol 14(4):387–394. https://doi.org/10.1016/S0958-1669(03)00096-X

Cao S-L, Huang Y-M, Li X-H, Xu P, Wu H, Li N, Lou W-Y, Zong M-H (2016) Preparation and characterization of immobilized lipase from pseudomonas cepacia onto magnetic cellulose nanocrystals. Sci Rep 6(1):20420. https://doi.org/10.1038/srep20420

Chang Q, Tang H (2014) Immobilization of horseradish peroxidase on NH_2-modified magnetic Fe_3O_4/SiO_2 particles and its application in removal of 2,4-dichlorophenol. Molecules 19 (10):15768–15782. https://doi.org/10.3390/molecules191015768

Chang Q, Jiang G, Tang H, Li N, Huang J, Wu L (2015) Enzymatic removal of chlorophenols using horseradish peroxidase immobilized on superparamagnetic Fe_3O_4/graphene oxide nanocomposite. Chin J Catal 36(7):961–968. https://doi.org/10.1016/S1872-2067(15)60856-7

Chao C, Liu J, Wang J, Zhang Y, Zhang B, Zhang Y, Xiang X, Chen R (2013) Surface modification of halloysite nanotubes with dopamine for enzyme immobilization. ACS Appl Mater Interfaces 5(21):10559–10564. https://doi.org/10.1021/am4022973

Chauhan GS (2014) Evaluation of nanogels as supports for enzyme immobilization. Polym Int 63 (11):1889–1894. https://doi.org/10.1002/pi.4734

Chen C, Yang Q-H, Yang Y, Lv W, Wen Y, Hou P-X, Wang M, Cheng H-M (2009) Self-assembled free-standing graphite oxide membrane. Adv Mater 21(29):3007–3011. https://doi.org/10.1002/adma.200803726

Chen X, Li D, Li G, Luo L, Ullah N, Wei Q, Huang F (2015) Facile fabrication of gold nanoparticle on zein ultrafine fibers and their application for catechol biosensor. Appl Surf Sci 328:444–452. https://doi.org/10.1016/j.apsusc.2014.12.070

Cipolatti EP, Silva MJA, Klein M, Feddern V, Feltes MMC, Oliveira JV, Ninow JL, de Oliveira D (2014) Current status and trends in enzymatic nanoimmobilization. J Mol Catal B Enzym 99:56–67. https://doi.org/10.1016/j.molcatb.2013.10.019

Dai Y, Yin L, Niu J (2011) Laccase-carrying electrospun fibrous membranes for adsorption and degradation of PAHs in shoal soils. Environ Sci Technol 45(24):10611–10618. https://doi.org/10.1021/es203286e

Dai Y, Niu J, Yin L, Xu J, Xu J (2013) Laccase-carrying electrospun fibrous membrane for the removal of polycyclic aromatic hydrocarbons from contaminated water. Sep Purif Technol 104:1–8. https://doi.org/10.1016/j.seppur.2012.11.013

Dai Y, Yao J, Song Y, Wang S, Yuan Y (2016) Enhanced adsorption and degradation of phenolic pollutants in water by carbon nanotube modified laccase-carrying electrospun fibrous membranes. Environ Sci Nano 3(4):857–868. https://doi.org/10.1039/C6EN00148C

Das P, Barbora L, Das M, Goswami P (2014) Highly sensitive and stable laccase based amperometric biosensor developed on nano-composite matrix for detecting pyrocatechol in environmental samples. Sensors Actuators B Chem 192:737–744. https://doi.org/10.1016/j.snb.2013.11.021

Das R, Hamid SBA, Annuar MSM (2016) Highly efficient and stable novel nanoBiohybrid catalyst to avert 3,4-dihydroxybenzoic acid pollutant in water. Sci Rep 6(1):33572. https://doi.org/10.1038/srep33572

Das A, Singh J, Yogalakshmi KN (2017) Laccase immobilized magnetic iron nanoparticles: fabrication and its performance evaluation in chlorpyrifos degradation. Int Biodeterior Biodegrad 117:183–189. https://doi.org/10.1016/j.ibiod.2017.01.007

Dashtban M, Schraft H, Syed TA, Qin W (2010) Fungal biodegradation and enzymatic modification of lignin. Int J Biochem Mol Biol 1(1):36–50

Datta S, Christena LR, Rajaram YRS (2013) Enzyme immobilization: an overview on techniques and support materials. 3 Biotech 3(1):1–9. https://doi.org/10.1007/s13205-012-0071-7

Ding S, Cargill AA, Medintz IL, Claussen JC (2015) Increasing the activity of immobilized enzymes with nanoparticle conjugation. Curr Opin Biotechnol 34:242–250. https://doi.org/10.1016/j.copbio.2015.04.005

Ding Y, Cui R, Hu M, Li S, Zhai Q, Jiang Y (2017) Well-oriented bioarchitecture for immobilization of chloroperoxidase on graphene oxide nanosheets by site-specific interactions and its catalytic performance. J Mater Sci 52(17):10001–10012. https://doi.org/10.1007/s10853-017-1202-7

Do J-S, Lin K-H (2016) Kinetics of urease inhibition-based amperometric biosensors for mercury and lead ions detection. J Taiwan Inst Chem Eng 63:25–32. https://doi.org/10.1016/j.jtice.2016.03.011

Eghbali M, Farahbakhsh A, Rohani A, Pour AN (2015) Urea biosensor based on immobilization of urease on ZnO nanoparticles. Orient J Chem 31(2):1237–1242. https://doi.org/10.13005/ojc/310284

Eibes G, Arca-Ramos A, Feijoo G, Lema JM, Moreira MT (2015) Enzymatic technologies for remediation of hydrophobic organic pollutants in soil. Appl Microbiol Biotechnol 99 (21):8815–8829. https://doi.org/10.1007/s00253-015-6872-y

Feng W, Ji P (2011) Enzymes immobilized on carbon nanotubes. Biotechnol Adv 29(6):889–895. https://doi.org/10.1016/j.biotechadv.2011.07.007

Feng Q, Wu D, Huan S, Li M, Li X (2016) Study on the preparation of the AOPAN/MMT composite nanofibers and their application for laccase immobilization. J Eng Fibers Fabr 11 (3):45–54

Fernandes RA, Daniel-da-Silva AL, Tavares APM, Xavier AMRB (2017) EDTA-Cu (II) chelating magnetic nanoparticles as a support for laccase immobilization. Chem Eng Sci 158:599–605. https://doi.org/10.1016/j.ces.2016.11.011

Fortes CCS, Daniel-da-Silva AL, Xavier AMRB, Tavares APM (2017) Optimization of enzyme immobilization on functionalized magnetic nanoparticles for laccase biocatalytic reactions. Chem Eng Process Process Intensif 117:1–8. https://doi.org/10.1016/j.cep.2017.03.009

Fu J, Qiao H, Li D, Luo L, Chen K, Wei Q (2014) Laccase biosensor based on electrospun copper/ carbon composite nanofibers for catechol detection. Sensors 14(2):3543–3556. https://doi.org/ 10.3390/s140203543

Gahlout M, Rudakiya DM, Gupte S, Gupte A (2017) Laccase-conjugated amino-functionalized nanosilica for efficient degradation of Reactive Violet 1 dye. Int Nano Lett 7(3):195–208. https://doi.org/10.1007/s40089-017-0215-1

Galliker P, Hommes G, Schlosser D, Corvini PF-X, Shahgaldian P (2010) Laccase-modified silica nanoparticles efficiently catalyze the transformation of phenolic compounds. J Colloid Interface Sci 349(1):98–105. https://doi.org/10.1016/j.jcis.2010.05.031

Gasser CA, Yu L, Svojitka J, Wintgens T, Ammann EM, Shahgaldian P, Corvini PF-X, Hommes G (2014) Advanced enzymatic elimination of phenolic contaminants in wastewater: a nano approach at field scale. Appl Microbiol Biotechnol 98(7):3305–3316. https://doi.org/10.1007/ s00253-013-5414-8

Ge J, Lei J, Zare RN (2012) Protein–inorganic hybrid nanoflowers. Nat Nanotechnol 7:428–432. https://doi.org/10.1038/nnano.2012.80

Gebreyohannes AY, Bilad MR, Verbiest T, Courtin CM, Dornez E, Giorno L, Curcio E, Vankelecom IFJ (2015) Nanoscale tuning of enzyme localization for enhanced reactor performance in a novel magnetic-responsive biocatalytic membrane reactor. J Membr Sci 487:209–220. https://doi.org/10.1016/j.memsci.2015.03.069

Gopal R, Kaur S, Ma Z, Chan C, Ramakrishna S, Matsuura T (2006) Electrospun nanofibrous filtration membrane. J Membr Sci 281(1–2):581–586. https://doi.org/10.1016/j.memsci.2006. 04.026

Guan H, Liu X, Wang W (2013) Encapsulation of tyrosinase within liposome bioreactors for developing an amperometric phenolic compounds biosensor. J Solid State Electrochem 17 (11):2887–2893. https://doi.org/10.1007/s10008-013-2181-5

Guo J-s, Xu Y-f (2011) Review of enzymatic sludge hydrolysis. J Bioremed Biodegr 2(5):1000130. https://doi.org/10.4172/2155-6199.1000130

Gupta MN, Kaloti M, Kapoor M, Solanki K (2011) Nanomaterials as Metrices for enzyme immobilization. Artif Cells, Blood Substitutes, Biotechnol 39(2):98–109. https://doi.org/10. 3109/10731199.2010.516259

Hammond JL, Formisano N, Estrela P, Carrara S, Tkac J (2016) Electrochemical biosensors and nanobiosensors. Essays Biochem 60(1):69–80. https://doi.org/10.1042/EBC20150008

Handy RD, Owen R, Valsami-Jones E (2008) The ecotoxicology of nanoparticles and nanomaterials: current status, knowledge gaps, challenges, and future needs. Ecotoxicology 17(5):315–325. https://doi.org/10.1007/s10646-008-0206-0

Hanefeld U, Gardossi L, Magner E (2009) Understanding enzyme immobilisation. Chem Soc Rev 38(2):453–468. https://doi.org/10.1039/B711564B

Hermanová S, Zarevúcká M, Bouša D, Pumera M, Sofer Z (2015) Graphene oxide immobilized enzymes show high thermal and solvent stability. Nanoscale 7(13):5852–5858. https://doi.org/ 10.1039/C5NR00438A

Homaei AA, Sariri R, Vianello F, Stevanato R (2013) Enzyme immobilization: an update. J Chem Biol 6(4):185–205. https://doi.org/10.1007/s12154-013-0102-9

Hou J, Dong G, Ye Y, Chen V (2014a) Enzymatic degradation of bisphenol-a with immobilized laccase on TiO_2 sol–gel coated PVDF membrane. J Membr Sci 469:19–30. https://doi.org/10.1016/j.memsci.2014.06.027

Hou J, Dong G, Ye Y, Chen V (2014b) Laccase immobilization on titania nanoparticles and titania-functionalized membranes. J Membr Sci 452:229–240. https://doi.org/10.1016/j.memsci.2013.10.019

Hwang ET, Gu MB (2013) Enzyme stabilization by nano/microsized hybrid materials. Eng Life Sci 13(1):49–61. https://doi.org/10.1002/elsc.201100225

Ibrahim ASS, Al-Salamah AA, El-Toni AM, Almaary KS, El-Tayeb MA, Elbadawi YB, Antranikian G (2016) Enhancement of alkaline protease activity and stability via covalent immobilization onto hollow Core-mesoporous shell silica nanospheres. Int J Mol Sci 17 (2):184. https://doi.org/10.3390/ijms17020184

Ji C, Hou J, Wang K, Zhang Y, Chen V (2016a) Biocatalytic degradation of carbamazepine with immobilized laccase-mediator membrane hybrid reactor. J Membr Sci 502:11–20. https://doi.org/10.1016/j.memsci.2015.12.043

Ji X, Su Z, Xu M, Ma G, Zhang S (2016b) TiO_2-horseradish peroxidase hybrid catalyst based on hollow nanofibers for simultaneous photochemical-enzymatic degradation of 2,4-dichlorophenol. ACS Sustain Chem Eng 4(7):3634–3640. https://doi.org/10.1021/acssuschemeng.6b00075

Ji C, Nguyen LN, Hou J, Hai FI, Chen V (2017) Direct immobilization of laccase on titania nanoparticles from crude enzyme extracts of *P. ostreatus* culture for micro-pollutant degradation. Sep Purif Technol 178:215–223. https://doi.org/10.1016/j.seppur.2017.01.043

Jia H, Zhu G, Wang P (2003) Catalytic behaviors of enzymes attached to nanoparticles: the effect of particle mobility. Biotechnol Bioeng 84(4):406–414. https://doi.org/10.1002/bit.10781

Jia H, Zhong C, Huang F, Wang C, Jia L, Zhou H, Wei P (2013) The preparation and characterization of a laccase nanogel and its application in naphthoquinone synthesis. ChemPlusChem 78 (5):451–458. https://doi.org/10.1002/cplu.201300066

Jořenek M, Zajoncová L (2015) Immobilization of laccase on magnetic carriers and its use in decolorization of dyes. Chem Biochem Eng Q 29(3):457–466. https://doi.org/10.15255/CABEQ.2014.2079

Kadri T, Rouissi T, Brar SK, Cledon M, Sarma S, Verma M (2017) Biodegradation of polycyclic aromatic hydrocarbons (PAHs) by fungal enzymes: a review. J Environ Sci 51:52–74. https://doi.org/10.1016/j.jes.2016.08.023

Kalantari M, Yu M, Yang Y, Strounina E, Gu Z, Huang X, Zhang J, Song H, Yu C (2017) Tailoring mesoporous-silica nanoparticles for robust immobilization of lipase and biocatalysis. Nano Res 10(2):605–617. https://doi.org/10.1007/s12274-016-1320-6

Kalkan NA, Aksoy S, Aksoy EA, Hasirci N (2012) Preparation of chitosan-coated magnetite nanoparticles and application for immobilization of laccase. J Appl Polym Sci 123 (2):707–716. https://doi.org/10.1002/app.34504

Karigar CS, Rao SS (2011) Role of microbial enzymes in the bioremediation of pollutants: a review. Enzyme Res 2011:805187. https://doi.org/10.4061/2011/805187

Kim BC, Nair S, Kim J, Kwak JH, Grate JW, Kim SH, Gu MB (2005) Preparation of biocatalytic nanofibres with high activity and stability via enzyme aggregate coating on polymer nanofibres. Nanotechnology 16(7):S382–S388. https://doi.org/10.1088/0957-4484/16/7/011

Kim Y-P, Shon HK, Shin SK, Lee TG (2015) Probing nanoparticles and nanoparticle-conjugated biomolecules using time-of-flight secondary ion mass spectrometry. Mass Spectrom Rev 34 (2):237–247. https://doi.org/10.1002/mas.21437

Kim GB, Lee JO, Kim Y-P (2017) Graying the self-assembly of gold nanoparticles for improved enzyme activity assays. Sensors Actuators B Chem 246:271–277. https://doi.org/10.1016/j.snb.2017.02.067

Kochana J, Wapiennik K, Kozak J, Knihnicki P, Pollap A, Woźniakiewicz M, Nowak J, Kościelniak P (2015) Tyrosinase-based biosensor for determination of bisphenol A in a flow-batch system. Talanta 144:163–170. https://doi.org/10.1016/j.talanta.2015.05.078

Koloti LE, Gule NP, Arotiba OA, Malinga SP (2018) Lacasse-immobilized dendritic nanofibrous membranes as a novel approach towards the removal of bisphenol A. Environ Technol 39 (3):392–404. https://doi.org/10.1080/09593330.2017.1301570

Koyani RD, Vazquez-Duhalt R (2016) Laccase encapsulation in chitosan nanoparticles enhances the protein stability against microbial degradation. Environ Sci Pollut Res 23(18):18850–18857. https://doi.org/10.1007/s11356-016-7072-8

Kües U (2015) Fungal enzymes for environmental management. Curr Opin Biotechnol 33:268–278. https://doi.org/10.1016/j.copbio.2015.03.006

Kumar VV, Sivanesan S, Cabana H (2014) Magnetic cross-linked laccase aggregates — bioremediation tool for decolorization of distinct classes of recalcitrant dyes. Sci Total Environ 487:830–839. https://doi.org/10.1016/j.scitotenv.2014.04.009

Li W-X, Sun H-Y, Zhang R-F (2015) Immobilization of laccase on a novel ZnO/SiO_2 nanocomposited support for dye decolorization. IOP Conf Ser: Mater Sci Eng 87:012033. https://doi.org/10.1088/1757-899X/87/1/012033

Li H, Hou J, Duan L, Ji C, Zhang Y, Chen V (2017a) Graphene oxide-enzyme hybrid nanoflowers for efficient water soluble dye removal. J Hazard Mater 338:93–101. https://doi.org/10.1016/j.jhazmat.2017.05.014

Li C, Jiang S, Zhao X, Liang H (2017b) Co-immobilization of enzymes and magnetic nanoparticles by metal-nucleotide hydrogelnanofibers for improving stability and recycling. Molecules 22 (1):179. https://doi.org/10.3390/molecules22010179

Lin J, Wen Q, Chen S, Le X, Zhou X, Huang L (2017) Synthesis of amine-functionalized Fe_3O_4@C nanoparticles for laccase immobilization. Int J Biol Macromol 96:377–383. https://doi.org/10.1016/j.ijbiomac.2016.12.051

Liu Y, Zeng Z, Zeng G, Tang L, Pang Y, Li Z, Liu C, Lei X, Wu M, Ren P, Liu Z, Chen M, Xie G (2012) Immobilization of laccase on magnetic bimodal mesoporous carbon and the application in the removal of phenolic compounds. Bioresour Technol 115:21–26. https://doi.org/10.1016/j.biortech.2011.11.015

Malik P, Katyal V, Malik V, Asatkar A, Inwati G, Mukherjee TK (2013) Nanobiosensors: concepts and variations. ISRN Nanomater 2013:327435. https://doi.org/10.1155/2013/327435

Mei L-P, Feng J-J, Wu L, Zhou J-Y, Chen J-R, Wang A-J (2015) Novel phenol biosensor based on laccase immobilized on reduced graphene oxide supported palladium–copper alloyed nanocages. Biosens Bioelectron 74:347–352. https://doi.org/10.1016/j.bios.2015.06.060

Min K, Yoo YJ (2014) Recent progress in nanobiocatalysis for enzyme immobilization and its application. Biotechnol Bioprocess Eng 19(4):553–567. https://doi.org/10.1007/s12257-014-0173-7

Mohamad NR, Marzuki NHC, Buang NA, Huyop F, Wahab RA (2015) An overview of technologies for immobilization of enzymes and surface analysis techniques for immobilized enzymes. Biotechnol Biotechnol Equip 29(2):205–220. https://doi.org/10.1080/13102818.2015.1008192

Mohamed SA, Al-Harbi MH, Almulaiky YQ, Ibrahim IH, El-Shishtawy RM (2017) Immobilization of horseradish peroxidase on Fe_3O_4 magnetic nanoparticles. Electron J Biotechnol 27:84–90. https://doi.org/10.1016/j.ejbt.2017.03.010

Muthukumarasamy NP, Jackson B, Raj AJ, Sevanan M (2015) Production of extracellular laccase from *Bacillus subtilis* MTCC 2414 using Agroresidues as a potential substrate. Biochem Res Int 2015:765190. https://doi.org/10.1155/2015/765190

Nguyen LT, Yang K-L (2014) Uniform cross-linked cellulase aggregates prepared in millifluidic reactors. J Colloid Interface Sci 428:146–151. https://doi.org/10.1016/j.jcis.2014.04.033

Nigam VK, Shukla P (2015) Enzyme based biosensors for detection of environmental pollutants-a review. J Microbiol Biotechnol 25(11):1773–1781. https://doi.org/10.4014/jmb.1504.04010

Pan D, Gu Y, Lan H, Sun Y, Gao H (2015) Functional graphene-gold nano-composite fabricated electrochemical biosensor for direct and rapid detection of bisphenol A. Anal Chim Acta 853:297–302. https://doi.org/10.1016/j.aca.2014.11.004

Pang R, Li M, Zhang C (2015) Degradation of phenolic compounds by laccase immobilized on carbon nanomaterials: diffusional limitation investigation. Talanta 131:38–45. https://doi.org/10.1016/j.talanta.2014.07.045

Patel SKS, Kalia VC, Choi J-H, Haw J-R, Kim I-W, Lee JK (2014) Immobilization of laccase on SiO_2 nanocarriers improves its stability and reusability. J Microbiol Biotechnol 24(5):639–647. https://doi.org/10.4014/jmb.1401.01025

Patila M, Kouloumpis A, Gournis D, Rudolf P, Stamatis H (2016) Laccase-functionalized graphene oxide assemblies as efficient nanobiocatalysts for oxidation reactions. Sensors 16(3):287. https://doi.org/10.3390/s16030287

Perumal V, Hashim U (2014) Advances in biosensors: principle, architecture and applications. J Appl Biomed 12(1):1–15. https://doi.org/10.1016/j.jab.2013.02.001

Petkova GA, Záruba K, Žvátora P, Král V (2012) Gold and silver nanoparticles for biomolecule immobilization and enzymatic catalysis. Nanoscale Res Lett 7:287. https://doi.org/10.1186/1556-276X-7-287

Qiu C, Chen T, Wang X, Li Y, Ma H (2013) Application of horseradish peroxidase modified nanostructured Au thin films for the amperometric detection of 4-chlorophenol. Colloids Surf B Biointerfaces 103:129–135. https://doi.org/10.1016/j.colsurfb.2012.10.017

Qu J, Lou T, Kang S, Du X (2013) Simultaneous determination of catechol and hydroquinone using a self-assembled laccase biosensor based on nanofilm. Sens Lett 11(9):1567–1572. https://doi.org/10.1166/sl.2013.3017

Rani M, Shanker U, Chaurasia AK (2017) Catalytic potential of laccase immobilized on transition metal oxides nanomaterials: degradation of alizarin red S dye. J Environ Chem Eng 5 (3):2730–2739. https://doi.org/10.1016/j.jece.2017.05.026

Rao MA, Scelza R, Acevedo F, Diez MC, Gianfreda L (2014) Enzymes as useful tools for environmental purposes. Chemosphere 107:145–162. https://doi.org/10.1016/j.chemosphere.2013.12.059

Rebollar-Pérez G, Campos-Terán J, Ornelas-Soto N, Méndez-Albores A, Torres E (2015) Biosensors based on oxidative enzymes for detection of environmental pollutants. Biocatalysis 1 (1):118–129. https://doi.org/10.1515/boca-2015-0010

Rizwan M, Singh M, Mitra CK, Morve RK (2014) Ecofriendly application of nanomaterials: Nanobioremediation. J Nanopart 2014:431787. https://doi.org/10.1155/2014/431787

Saranya P, Ramani K, Sekaran G (2014) Biocatalytic approach on the treatment of edible oil refinery wastewater. RSC Adv 4(21):10680–10692. https://doi.org/10.1039/C3RA43668C

Secundo F (2013) Conformational changes of enzymes upon immobilisation. Chem Soc Rev 42 (15):6250–6261. https://doi.org/10.1039/C3CS35495D

Sheldon RA, van Pelt S (2013) Enzyme immobilisation in biocatalysis: why, what and how. Chem Soc Rev 42(15):6223–6235. https://doi.org/10.1039/C3CS60075K

Shojaat R, Saadatjoo N, Karimi A, Aber S (2016) Simultaneous adsorption–degradation of organic dyes using $MnFe_2O_4$/calcium alginate nano-composites coupled with GOx and laccase. J Environ Chem Eng 4(2):1722–1730. https://doi.org/10.1016/j.jece.2016.02.029

Shraddha SR, Sehgal S, Kamthania M, Kumar A (2011) Laccase: microbial sources, production, purification, and potential biotechnological applications. Enzyme Res 2011:217861. https://doi.org/10.4061/2011/217861

Silva MC, Torres JA, Nogueira FGE, Tavares TS, Corrêa AD, Oliveira LCA, Ramalho TC (2016) Immobilization of soybean peroxidase on silica-coated magnetic particles: a magnetically recoverable biocatalyst for pollutant removal. RSC Adv 6(87):83856–83863. https://doi.org/10.1039/C6RA17167B

Skoronski E, Hoefling Souza D, Ely C, Broilo F, Fernandes M, Fúrigo Júnior A, Gomes Ghislandi M (2017) Immobilization of laccase from *Aspergillus oryzae* on graphene nanosheets. Int J Biol Macromol 99:121–127. https://doi.org/10.1016/j.ijbiomac.2017.02.076

Song J, Su P, Yang Y, Yang Y (2017) Efficient immobilization of enzymes onto magnetic nanoparticles by DNA strand displacement: a stable and high-performance biocatalyst. New J Chem 41(14):6089–6097. https://doi.org/10.1039/C7NJ00284J

Sulaiman S, Cieh NL, Mokhtar MN, Naim MN, Kamal SMM (2017) Covalent immobilization of cyclodextrin glucanotransferase on kenaf cellulose nanofiber and its application in ultrafiltration membrane system. Process Biochem 55:85–95. https://doi.org/10.1016/j.procbio.2017.01.025

Sun H, Jin X, Long N, Zhang R (2017) Improved biodegradation of synthetic azo dye by horseradish peroxidase cross-linked on nano-composite support. Int J Biol Macromol 95:1049–1055. https://doi.org/10.1016/j.ijbiomac.2016.10.093

Tavares APM, Silva CG, Dražić G, Silva AMT, Loureiro JM, Faria JL (2015) Laccase immobilization over multi-walled carbon nanotubes: kinetic, thermodynamic and stability studies. J Colloid Interface Sci 454:52–60. https://doi.org/10.1016/j.jcis.2015.04.054

Tay T, Köse E, Keçili R, Say R (2016) Design and preparation of nano-lignin peroxidase (NanoLiP) by protein block copolymerization approach. Polymers 8(6):223. https://doi.org/10.3390/polym8060223

Verma ML, Naebe M, Barrow CJ, Puri M (2013) Enzyme immobilisation on amino-functionalised multi-walled carbon nanotubes: structural and biocatalytic characterisation. PLoS One 8(9): e73642. https://doi.org/10.1371/journal.pone.0073642

Viswanath B, Rajesh B, Janardhan A, Kumar AP, Narasimha G (2014) Fungal laccases and their applications in bioremediation. Enzyme Res 2014:163242. https://doi.org/10.1155/2014/163242

Wan L-S, Ke B-B, Xu Z-K (2008) Electrospun nanofibrous membranes filled with carbon nanotubes for redox enzyme immobilization. Enzym Microb Technol 42(4):332–339. https://doi.org/10.1016/j.enzmictec.2007.10.014

Wang P (2006) Nanoscale biocatalyst systems. Curr Opin Biotechnol 17(6):574–579. https://doi.org/10.1016/j.copbio.2006.10.009

Wang Z-G, Wan L-S, Liu Z-M, Huang X-J, Xu Z-K (2009) Enzyme immobilization on electrospun polymer nanofibers: an overview. J Mol Catal B Enzym 56(4):189–195. https://doi.org/10.1016/j.molcatb.2008.05.005

Wang Q, Cui J, Li G, Zhang J, Li D, Huang F, Wei Q (2014a) Laccase immobilized on a PAN/adsorbents composite nanofibrous membrane for catechol treatment by a biocatalysis/adsorption process. Molecules 19(3):3376–3388. https://doi.org/10.3390/molecules19033376

Wang X, Lu X, Wu L, Chen J (2014b) Direct electrochemical tyrosinase biosensor based on mesoporous carbon and Co_3O_4 nanorods for the rapid detection of phenolic pollutants. ChemElectroChem 1(4):808–816. https://doi.org/10.1002/celc.201300208

Wang M, Qi W, Su R, He Z (2015) Advances in carrier-bound and carrier-free immobilized nanobiocatalysts. Chem Eng Sci 135:21–32. https://doi.org/10.1016/j.ces.2015.03.051

Woo E-J, Kwon H-S, Lee C-H (2015) Preparation of nano-magnetite impregnated mesocellular foam composite with a Cu ligand for His-tagged enzyme immobilization. Chem Eng J 274:1–8. https://doi.org/10.1016/j.cej.2015.03.123

Wu L, Deng D, Jin J, Lu X, Chen J (2012) Nanographene-based tyrosinase biosensor for rapid detection of bisphenol A. Biosens Bioelectron 35(1):193–199. https://doi.org/10.1016/j.bios.2012.02.045

Xia G-H, Liu W, Jiang X-P, Wang X-Y, Zhang Y-W, Guo J (2017) Surface modification of Fe_3O_4@SiO_2 magnetic nanoparticles for immobilization of lipase. J Nanosci Nanotechnol 17(1):370–376. https://doi.org/10.1166/jnn.2017.10964

Xu R, Chi C, Li F, Zhang B (2013) Laccase–polyacrylonitrile nanofibrous membrane: highly immobilized, stable, reusable, and efficacious for 2,4,6-trichlorophenol removal. ACS Appl Mater Interfaces 5(23):12554–12560. https://doi.org/10.1021/am403849q

Yang L, Xiong H, Zhang X, Wang S (2012) A novel tyrosinase biosensor based on chitosan-carbon-coated nickel nanocomposite film. Bioelectrochemistry 84:44–48. https://doi.org/10.1016/j.bioelechem.2011.11.001

Yang J, Li W, Ng TB, Deng X, Lin J, Ye X (2017) Laccases: production, expression regulation, and applications in pharmaceutical biodegradation. Front Microbiol 8:832. https://doi.org/10.3389/fmicb.2017.00832

Yin H, Zhou Y, Xu J, Ai S, Cui L, Zhu L (2010) Amperometric biosensor based on tyrosinase immobilized onto multiwalled carbon nanotubes-cobalt phthalocyanine-silk fibroin film and its application to determine bisphenol A. Anal Chim Acta 659(1–2):144–150. https://doi.org/10.1016/j.aca.2009.11.051

Yu G, Wu W, Zhao Q, Wei X, Lu Q (2015) Efficient immobilization of acetylcholinesterase onto amino functionalized carbon nanotubes for the fabrication of high sensitive organophosphorus pesticides biosensors. Biosens Bioelectron 68:288–294. https://doi.org/10.1016/j.bios.2015.01.005

Yuan B, X-q Y, L-w X, Y-n F, Jiang J-h (2016) A novel recycling system for nano-magnetic molecular imprinting immobilised cellulases: synergistic recovery of anthocyanin from fruit and vegetable waste. Bioresour Technol 222:14–23. https://doi.org/10.1016/j.biortech.2016.09.088

Zehani N, Fortgang P, Lachgar MS, Baraket A, Arab M, Dzyadevych SV, Kherrat R, Jaffrezic-Renault N (2015) Highly sensitive electrochemical biosensor for bisphenol A detection based on a diazonium-functionalized boron-doped diamond electrode modified with a multi-walled carbon nanotube-tyrosinase hybrid film. Biosens Bioelectron 74:830–835. https://doi.org/10.1016/j.bios.2015.07.051

Zhang X, Hua M, Lv L, Pan B (2015) Ionic polymer-coated laccase with high activity and enhanced stability: application in the decolourisation of water containing AO7. Sci Rep 5:8253. https://doi.org/10.1038/srep08253

Zhang D, Deng M, Cao H, Zhang S, Zhao H (2017) Laccase immobilized on magnetic nanoparticles by dopamine polymerization for 4-chlorophenol removal. Green Energy Environ 2(4):393–400. https://doi.org/10.1016/j.gee.2017.04.001

Zheng Q, Yu Y, Fan K, Ji F, Wu J, Ying Y (2016) A nano-silver enzyme electrode for organophosphorus pesticide detection. Anal Bioanal Chem 408(21):5819–5827. https://doi.org/10.1007/s00216-016-9694-6

Zhou Y, Tang L, Zeng G, Chen J, Cai Y, Zhang Y, Yang G, Liu Y, Zhang C, Tang W (2014) Mesoporous carbon nitride based biosensor for highly sensitive and selective analysis of phenol and catechol in compost bioremediation. Biosens Bioelectron 61:519–525. https://doi.org/10.1016/j.bios.2014.05.063

Zhu Y, Kaskel S, Shi J, Wage T, van Pée K-H (2007) Immobilization of *Trametes versicolor* laccase on magnetically separable mesoporous silica spheres. Chem Mater 19(26):6408–6413. https://doi.org/10.1021/cm071265g

Zucca P, Sanjust E (2014) Inorganic materials as supports for covalent enzyme immobilization: methods and mechanisms. Molecules 19(9):14139–14194. https://doi.org/10.3390/molecules190914139

Zucca P, Cocco G, Sollai F, Sanjust E (2015) Fungal laccases as tools for biodegradation of industrial dyes. Biocatalysis 1(1):82–108. https://doi.org/10.1515/boca-2015-0007

Chapter 17
Bioelectrochemical Processes for the Treatment of Oil-Contaminated Water and Sediments

Matteo Daghio and Andrea Franzetti

Abstract Non-halogenated and halogenated hydrocarbons are widespread toxic contaminants. Several physicochemical and biological strategies have been developed for the remediation of contaminated environments. The main goal of biological approaches is to stimulate the microbial activity by overcoming limiting factors such as the presence of nutrients and/or the presence of electron acceptors/donors. Bioelectrochemical systems (BESs) are innovative devices that can be exploited for the bioremediation of polluted environments. In a BES an electrode can be used by several groups of bacteria as solid electron acceptor (anode) or donor (cathode) and the electrons that flow from the anode to the cathode produce an electrical current. BESs have been successfully applied to remove both non-halogenated and halogenated hydrocarbons in lab-scale studies. Several advantages can be obtained using electrodes for the stimulation of the microbial degradation: operational costs can be potentially decreased, the electrical current generated can be used for the real-time monitoring of the degradation, the potential of the electrode (i.e., electron acceptor/donor) can be controlled and adapted to the conditions. This chapter discusses the key aspects linked to the application of bioelectrochemical technologies for the removal of non-halogenated and halogenated hydrocarbons.

Keywords Bioremediation · Bioelectrochemical systems · Chlorinated solvents · Hydrocarbons

M. Daghio (✉)
Department of Earth and Environmental Sciences, University of Milano-Bicocca, Milan, Italy

Department of Agriculture, Food, Environment and Forestry, University of Florence, Florence, Italy
e-mail: matteo.daghio@unifi.it

A. Franzetti
Department of Earth and Environmental Sciences, University of Milano-Bicocca, Milan, Italy

J. Filip et al. (eds.), *Advanced Nano-Bio Technologies for Water and Soil Treatment*, Applied Environmental Science and Engineering for a Sustainable Future,
https://doi.org/10.1007/978-3-030-29840-1_17

17.1 Introduction

The most common approach for bioremediation of contaminated environments involves stimulation of the microbial activity by overcoming the factors that are limiting the metabolism of microorganisms (Alvarez and Illman 2005). The stimulation is usually achieved by supplying electron acceptors/donors to sustain the oxidation/reduction of the contaminants (Alvarez and Illman 2005; Gkorezis et al. 2016).

In hydrocarbon-contaminated environments, thermodynamically favorable electron acceptors (e.g., O_2) are quickly depleted, and, therefore, they need continuous supplying to sustain the process (Farhadian et al. 2008). Despite the relatively high redox potential of oxygen (O_2/H_2O + 820 mV vs SHE—standard hydrogen electrode, all the potentials reported in this chapter are relative to SHE unless stated otherwise) and the high efficiency of oxygenases for the aerobic biodegradation of petroleum hydrocarbons (Baldwin et al. 2009), some disadvantages are linked to the use of aerobic bioremediation strategies (Daghio et al. 2017). Oxygen solubility in water is low, and it can be consumed by unwanted side reactions (e.g., by oxidation of reduced Fe^{2+} and Mn^{2+}) (Broden et al. 1997; Tuxen et al. 2006). Several microorganisms showed the ability to degrade petroleum hydrocarbons in anaerobic environments, which is, however, usually a slower process compared to aerobic biodegradation (Weelink et al. 2010). The drawbacks linked to the aerobic biodegradation can be overcome by the use of anaerobic strategies. The addition of chelators to solubilize Fe^{3+} for the microbial respiration or the addition of soluble electron shuttles (e.g., humic substances) to favor the electron transfer to solid electron acceptors (e.g., Fe^{3+} and Mn^{2+}) has been suggested (Lovley et al. 1994, 1996a, b). Alternatively, anaerobic metabolism can be stimulated by the addition of nitrate or sulfate (Mihelcic and Luthy 1988; Weiner et al. 1998). The main disadvantage linked to the above-mentioned strategies (both aerobic and anaerobic) is that the reagents are rapidly consumed and can migrate away from the contaminated area. The cost of the bioremediation can therefore increase because of the need of continuous replenishment of reagents to sustain the microbial activity (Daghio et al. 2017; Zhang et al. 2010).

Similarly to the strategies described above, which can be used for the stimulation of the degradation of non-halogenated hydrocarbons, halogenated hydrocarbons can be remediated by stimulating the reductive dehalogenation (de Bruin et al. 1992; Seshadri et al. 2005). Hydrogen (H_2) is the typical electron donor for the dehalogenation (Aulenta et al. 2006; He et al. 2002) and it can be supplied directly or by passive dissolution (Fang et al. 2002; Ma et al. 2003). Alternatively, H_2 can be indirectly supplied by using organic substrates (e.g., butyric acid, ethanol, or lactic acid) that can be fermented (Aulenta et al. 2005; Fennell et al. 1997). However, as mentioned above for water-soluble electron acceptors, also water-soluble electron donors have to be continuously supplied, which can result in high costs (Daghio et al. 2017). In addition, fermentation products (e.g., volatile fatty acids) can accumulate, which results in deterioration of the water quality

(Aulenta et al. 2007b), and controlling H_2 delivery is crucial to prevent electrons diverting to side processes (e.g., methanogenesis, homoacetogenesis, and sulfate reduction) (Daghio et al. 2017).

The use of bioelectrochemical technologies, in which electrodes are used as solid electron acceptors/donors by several groups of microorganisms, has been extensively reviewed and suggested to overcome some of the limitations of the classical bioremediation approaches (Daghio et al. 2017; Li and Yu 2015; Wang et al. 2015); therefore, they will be deeply discussed in this chapter.

17.2 Introduction to Bioelectrochemical Systems

Bioelectrochemical systems (BESs) are devices in which the ability of specific microorganisms to catalyze redox reactions on or near electrodes is exploited (Logan et al. 2006; Rosenbaum et al. 2011). This technology was initially developed to convert chemical energy into electrical energy during wastewater treatment (Logan et al. 2006). The main advantage was the opportunity to recover energy from the waste, instead of consuming it to supply the oxygen needed for the aerobic oxidation of the organic substances. In natural environments, several microorganisms are able to oxidize organic substrates in anaerobic conditions using solid electron acceptors (e.g., iron oxides) (Shi et al. 2016; Weber et al. 2006). In BESs the electrons obtained from the oxidation of the organic substrate are transferred to an electrode (anode), which is a solid electron acceptor, and flow through a circuit. The electrons reach another electrode (cathode) where the reduction of a final electron acceptor takes place (Logan et al. 2006) (Fig. 17.1). By convention, an electric current flows into the circuit in a direction that is the opposite to the direction of the electrons.

Several reactors' configurations can be used. The traditional H-shaped two-chamber reactor was widely used for the early studies in which BESs were investigated. In this configuration, two bottles (one for the anodic chamber and one for the cathodic chamber) are connected with a tube and separated with a cation exchange membrane (CEM). The membrane is used to allow the exchange of protons between the two chambers and to avoid the diffusion of the organic substrate and of the final electron acceptor (Logan et al. 2006).

BESs can be divided in two main groups: microbial fuel cells (MFCs) and microbial electrolysis cells (MECs). In a MFC, a net energy gain is achieved because the redox potential of the oxidation reaction at the anode is lower than the redox potential of the reduction reaction at the cathode. In a MEC, an energy input is used to drive the reaction (Daghio et al. 2017).

MECs can be operated at a fixed potential (Rozendal et al. 2006) with the advantage that the favorable conditions to drive the desired reaction, or for a certain biocatalyst, can be created. Alternatively, a MEC can be operated by setting a fixed current (Andersen et al. 2013) to control the reaction rates.

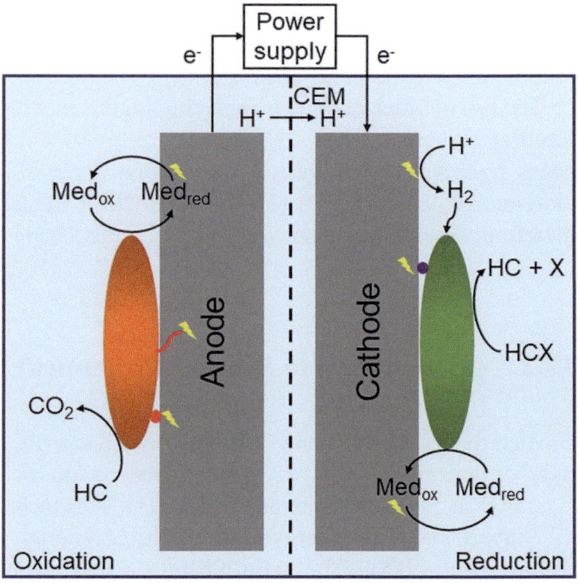

Fig. 17.1 Scheme of a BES. An electron donor (e.g., non-halogenated hydrocarbons—HC) is oxidized in the anodic compartment and an electron acceptor (e.g., halogenated hydrocarbons — HCX) is reduced in the cathodic compartment. The flash symbols represent the steps in which electrons are transferred to the anode or from the cathode. The anodic compartment and the cathodic compartment are separated by a cation exchange membrane (CEM) for the transfer of H^+ ions

The first works on BESs focused on the anodic biological oxidation of organic compounds making use of expensive catalysts (e.g., platinum) to catalyze a chemical cathodic reaction (e.g., O_2 reduction to H_2O). Later, the development of biocathodes (cathodes that exploit biocatalysts) opened up new possibilities for the application of BESs (Rosenbaum et al. 2011).

Several groups of microorganisms are able to transfer electrons to/from solid electron acceptors/donors via different extracellular electron transfer (EET) mechanisms as widely reviewed by Lovley (2012), Rosenbaum et al. (2011), and Shi et al. (2016). Three main mechanisms used by bacteria to transfer electrons to electrodes have been described so far. Short-range electron transfer has been extensively studied in *Geobacter sulfurreducens*. *G. sulfurreducens* is able to perform EET by redox active proteins (*c*-type cytochromes) located on the outer membrane. In particular, OmcZ seems to play a key role in the process (Lovley 2012). Another possible strategy to perform EET is the long-range transport by soluble redox mediators (i.e., the electron transfer shuttles). The electron transfer shuttles are reduced at the outer cell surface and they diffuse to the anode where they are oxidized and the electrons are donated to the electrode (Lovley 2012). The shuttles can be exogenous, but several microorganisms are able to use endogenous molecules (e.g., flavins in *Shewanella oneidensis* or phenazines in *Pseudomonas aeruginosa*) (Lovley 2012; Marsili et al. 2008; Rabaey et al. 2005). Exogenous shuttles can be natural molecules such as humic acids, which have been largely proved to act as electron transfer mediators when both minerals and electrodes are used for the microbial respiration (Milliken and May 2007; Van der Zee and Cervantes 2009). Artificial exogenous shuttles (e.g., neutral red, ferricyanide, methyl viologen—MV, riboflavin, anthraquinone-2-sulfonate—AQS, or antraquinone-2,6-disulfonate—AQDS) have also been used in some studies (Adelaja et al. 2015; Aulenta et al. 2007a, 2008a, 2010a; Leitão et al. 2016; Lin et al. 2014). Their use is limited to laboratory experiments since these molecules can be toxic and, if used during bioremediation of contaminated environments, they can have disadvantages similar to the ones reported for the soluble electron acceptors, e.g., diffusion away from the reaction area or occurrence of side reactions (Daghio et al. 2017). However, recent studies suggested that the limits linked to the diffusion of soluble electron shuttles (e.g., AQDS) could be overcome by the immobilization of the mediator on the electrode surface (Leitão et al. 2017). A third mechanism of EET involves long-range electron transport through the biofilm via electrically conductive pili called nanowires (Lovley 2011). Experimental evidences suggest that in *G. sulfurreducens* the electrons transported along the pili are transferred to the electrode by extracellular *c*-type cytochromes that accumulate near the anode surface (Inoue et al. 2011; Lovley and Nevin 2011). EET by outer membrane *c*-type cytochromes and mediators can also be involved in the electrons uptake from the cathode (Rosenbaum et al. 2011). Furthermore, the cathodic electron uptake can involve hydrogenases, which can accept the electrons from the electrode. The electrons can then be used for the H_2 production or for the reduction of the final electron acceptor (Rosenbaum et al. 2011).

17.3 Use of Bioelectrochemical Systems for Bioremediation of Oil-Contaminated Environments

Several authors have reported on the possibility of using electrodes as electron acceptors for the microbial respiration during the bioremediation of environments polluted by a wide range of organic contaminants (Daghio et al. 2017). Refinery wastewater and diesel-contaminated groundwater were used in the first studies, which showed the possibility of coupling the current production with the hydrocarbons removal (Morris et al. 2009; Morris and Jin 2007). Later, a high number of studies reported the degradation of total petroleum hydrocarbons (TPH) and diesel in laboratory-scale BES reactors (Chandrasekhar and Venkata Mohan 2012; Cheng et al. 2017; Cruz Viggi et al. 2015; Li et al. 2014, 2015; Lu et al. 2014a, b; Morris and Jin 2012; Venkata Mohan and Chandrasekhar 2011; Wang et al. 2012; Zhang et al. 2015). Other studies showed the degradation of BTEX compounds (benzene, toluene, ethylbenzene, *o*-xylene, *m*-xylene, *p*-xylene) such as toluene (Espinoza Tofalos et al. 2018; Daghio et al. 2016; Lin et al. 2014; Zhang et al. 2010) and benzene (Adelaja et al. 2015; Rakoczy et al. 2013; Wei et al. 2015; Wu et al. 2013; Zhang et al. 2010), which were removed by both pure cultures and mixed cultures using the anode of a BES. The bioelectrochemical degradation of monoaromatic hydrocarbons was also assessed using a mixture containing all the BTEX compounds. The authors showed that the removal of toluene, *m*-xylene, and *p*-xylene was improved by using a MEC at different applied external voltages

(0.8 V, 1.0 V, and 1.2 V) compared to the open circuit control (Daghio et al. 2018a). Similarly, also phenol (Friman et al. 2013; Huang et al. 2011; Palma et al. 2018; Zhang et al. 2017) and recalcitrant polycyclic aromatic hydrocarbons—PAHs (Adelaja et al. 2014, 2015; Hamdan et al. 2017; Yan et al. 2012; Yu et al. 2017; Zhang et al. 2010) were successfully removed in BESs. Anodic oxidation of the halogenated hydrocarbon 1,2-dichloroethane (1,2-DCA) was also shown (Pham et al. 2009).

During the stimulation of the bioremediation by anodic oxidation of the contaminant, the electrode is the final electron acceptor for the microbial respiration. It is thus possible to hypothesize that with higher anodic potentials the energy gain for the microorganisms is higher, and, therefore, the biodegradation rate can be increased. Several studies have tried to correlate the anodic potential with the microbial activity using easily degradable substrates such as acetate. The data, however, gave controversial results since a positive correlation was observed in some cases, while in other studies the opposite was observed (Aelterman et al. 2008; Wagner et al. 2010). The data on the influence of the anodic potential when hydrocarbons are the electron donors are lacking. During the bioelectrochemical toluene removal in BES reactors inoculated with a marine sediment, and with artificial seawater as growth medium, two anodic potentials were tested (i.e., +200 mV and +500 mV) without any appreciable effect on the biodegradation rate (Daghio et al. 2016). When the anode was polarized at +500 mV, an unidentified redox active moiety with a midpoint potential of about +400 mV was detected and its involvement in the EET was hypothesized (Daghio et al. 2016). More information on the influence of the anode potential is, therefore, necessary, but it is reasonable that the optimal anodic potential is highly dependent on specific in situ conditions and on the microbial communities involved in the biodegradation.

Oxygen evolution using high-potential electrodes is an alternative strategy to use electrodes for the stimulation of the oxidation of petroleum hydrocarbons in contaminated environments (Daghio et al. 2017). Indeed, oxygen is not only an important electron acceptor for the microbial respiration, but also a key reagent involved in the initial attack of hydrocarbons in aerobic environments through the addition of hydroxyl groups catalyzed by oxygenases (Gkorezis et al. 2016; Weelink et al. 2010). A small amount of oxygen can then be useful to trigger the degradation of hydrocarbon contaminants. Examples of stimulation of the biodegradation by oxygen evolution with electrodes have been reported for *cis*-dichloroethene (*cis*-DCE) removal, with an electrode polarized at +1500 mV (Aulenta et al. 2013), and for TPH removal by applying an external voltage (2 V) on dimensionally stable anodes—DSA; i.e., Ti mesh covered with mixed metal oxides, primarily consisting of Ir and Ru (Bellagamba et al. 2017).

When the reduction of the contaminant is the reaction that has to be stimulated (e.g., for the dehalogenation of halogenated hydrocarbons) the cathodic process can be exploited. Several studies have successfully showed dehalogenation of 2,6-dichlorophenol (2,6-DCP) (Skadberg et al. 1999), trichloroethene (TCE) (Aulenta et al. 2008b), and tetrachloroethene (PCE) (Lohner and Tiehm 2009) through H_2 production using electrodes. However, a direct electron uptake from the cathode is a better strategy to avoid the side reactions (e.g., methanogenesis) that divert the electrons, thus decreasing the efficiency of the process. In compliance with this line, other works showed the possibility of performing the TCE dehalogenation with the redox mediators (e.g., MV or the humic acid analogue AQDS) (Aulenta et al. 2007a, 2008a, 2010a) and without the redox mediators suggesting that TCE removal is also possible by a direct electron uptake from the electrode (Aulenta et al. 2009, 2010b, 2011; Verdini et al. 2015). Bioelectrochemical dehalogenation with a cathode as a sole electron donor has been reported also for PCE, which was reduced to *cis*-DCE by a pure culture of *Geobacter lovleyi* (Strycharz et al. 2008).

A key parameter that has to be controlled for the stimulation of the reductive dechlorination in order to minimize the occurrence of the side reactions is the cathode potential. TCE dehalogenation via H_2 evolution was studied at cathodic potentials between −800 mV and −500 mV. The highest dechlorination rate was obtained at the more negative potentials with a fast increase in the range from −650 mV to −625 mV. In the same potential range, however, the efficiency of the process was decreased by diverting the electrons toward methane production (Aulenta et al. 2008b). Later, the electron uptake from an electrode poised at −450 mV was observed during the TCE dechlorination. A redox active moiety potentially involved in the EET (midpoint potential around −400 mV) was detected in the supernatant (Aulenta et al. 2009). A subsequent study, performed with a continuous flow reactor, showed an improvement of the dehalogenation rate (from 15.5 ± 1.2 μmol e^-/L to 58 ± 1 μmol e^-/L) linked to the enhancement of the H_2 production by decreasing the cathode potential from −250 mV to −450 mV (Aulenta et al. 2011). However, at lower potential a drastic drop in the coulombic efficiency (from nearly 100% at −250 mV to less than 1% at −750 mV) was observed because of methanogenesis (Aulenta et al. 2011), confirming the conclusions of the previous studies (Aulenta et al. 2008b). A further demonstration of this trend has been reported with 1,2-DCA. The dehalogenation rate increased from −300 mV to −900 mV, but the higher coulombic efficiency (about 70%) was obtained at the higher cathodic potential (−300 mV) (Leitão et al. 2015). Methane production is not the only side reaction that can decrease the efficiency of the process during the bioelectrochemical dehalogenation. At potentials below −550 mV also nitrate reduction and sulfate reduction can have negative effects on the coulombic efficiency (Lai et al. 2015). Reductive dechlorination can also be affected by mass transport (e.g., groundwater flow) at different cathodic potentials (Verdini et al. 2015). TCE dehalogenation was studied at different flow velocities (from 0.3 m/day to 1.7 m/day) by varying the cathode potential (from −450 mV to −250 mV). At more reducing potentials the dechlorination rate increased with the flow velocity because the process was influenced by mass transport of H_2 generated at the cathode. At higher potentials the dehalogenation was sustained by direct EET that was not affected by the flow rate (Verdini et al. 2015).

A simple configuration suggested for the application of BES-based technologies during the bioremediation of aquatic environments involves the placement of an anode in anaerobic benthic sediments (Bond et al. 2002) (Fig. 17.2a). The electrode can thus be used as electron acceptor for the anodic oxidation of the contaminant.

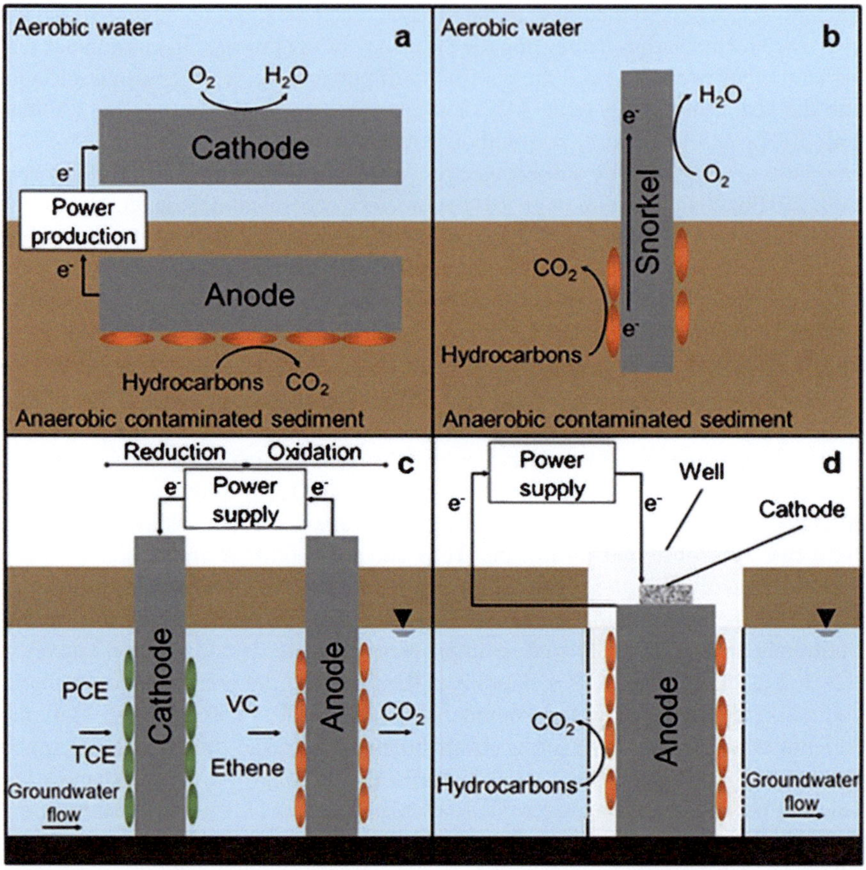

Fig. 17.2 Possible configurations of a BES for bioremediation of sediments (**a** and **b**) and groundwater (**c** and **d**) contaminated with halogenated and non-halogenated hydrocarbons. (**a**) Benthic BES, (**b**) bioelectrochemical snorkel, (**c**) sequential reductive–oxidative treatment, (**d**) bioelectric well

The electrons transferred to the anode flow to a cathode, which is placed in the overlying aerobic water column, where oxygen is the final electron acceptor (Lovley and Nevin 2011). By placing a borehole anode in a contaminated aquifer and the cathode at the ground surface, a similar design may also be used for the bioremediation of groundwater environments (Williams et al. 2010).

Another possibility is the use of the "bioelectrochemical snorkels" which consist of conductive rods buried in the contaminated sediment. The rods pass throughout both the anaerobic zone (i.e., anodic area) and the aerobic zone (i.e., cathodic area) and act as both electron collector (i.e., anode) and cathode (Cruz Viggi et al. 2015; Matturro et al. 2017) (Fig. 17.2b).

One of the first designs proposed for in situ applications of a BES during the bioremediation of contaminated aquifers was inspired by the "lasagna process", which is an electrochemical technology developed for the electrokinetic transport coupled to the in situ treatment (Ho et al. 1995), and it involves the placement of a BES-based biobarrier that intercepts the contaminated plume (Daghio et al. 2017). For the treatment of halogenated hydrocarbons (e.g., TCE or PCE) it could be feasible to place a cathodic biobarrier upstream to an anodic biobarrier. This setup allows the reductive dehalogenation in the cathodic area, followed by the anodic oxidation of the reduced compounds (Fig. 17.2c). The possibility to apply a sequential reductive–oxidative treatment has been suggested by several lab-scale studies (Aulenta et al. 2011; Lai et al. 2015, 2017; Lohner and Tiehm 2009). However, a crucial parameter that has to be considered with this architecture is the spacing between the electrodes since an excessive energy input may be required with a distance between the electrodes higher than a few meters (Daghio et al. 2017).

An alternative configuration suggested for the bioremediation of hydrocarbon-contaminated groundwater is the so-called "bioelectric well" (Palma et al. 2018) (Fig. 17.2d). A concentric anode made of granular graphite is placed around a concentric cathode in order to minimize the spacing between the electrodes and the resulting energy loss. The two electrodes can be kept physically separated by a thin layer of plastic material (e.g., polyethylene mesh) that, however, has to ensure hydraulic connection. This system could be easily installed in wells already placed in situ (Palma et al. 2018), potentially decreasing the installation cost for the treatment system.

17.4 Microorganisms Involved in Bioremediation with Bioelectrochemical Systems

The characterization of the microorganisms able to use electrodes as electron acceptors/donors during the oxidation/reduction of organic contaminants is pivotal for the comprehension of the mechanisms involved, and thus for the optimization of the technology. Unfortunately, not all of the published studies have reported information about the taxonomy of the microorganisms inoculated or enriched in the reactors.

Pure cultures of *Geobacter metallireducens* and *S. oneidensis* were able to degrade toluene and phenanthrene in a BES (Adelaja et al. 2014; Zhang et al. 2010). However, iron reducers were not the only microorganisms that were successfully inoculated in BES reactors. *P. aeruginosa* was able to remove phenanthrene faster than *S. oneidensis* (Adelaja et al. 2014). *Cupriavidus basilensis* degraded phenol with an electrode poised at +325 mV as electron acceptor and an oxidation peak at +140 mV was observed (Friman et al. 2013).

In one of the first reports that showed the possibility of coupling diesel degradation with current production, facultative anaerobes (i.e., NO_3^- reducing bacteria) were enriched in a single chamber MFC reactor (Morris et al. 2009). In another experiment, during the PAHs degradation, microorganisms belonging to Chloroflexi and to Nitrospira were detected (Yan et al. 2012) while a complex community was described during the TPH degradation with a BES in contaminated soil (Lu et al. 2014a). The community was highly enriched in Proteobacteria (up to 94% on the anodes) and the most abundant classes within this phylum were Betaproteobacteria (main genera *Achromobacter*, *Alcaligenes*, *Bordetella*, *Comamonas*, and *Pusillimonas*) and Gammaproteobacteria (main genera *Enterobacter*, *Fulvimonas*, *Pseudomonas*, *Pseudoxanthomonas*, and *Stenotrophomonas*) and included microorganisms already described for their electrochemical activity (e.g., *Comamonas* and *Pseudomonas*), but also bacteria that were not reported as exoelectrogens (e.g., *Bordetella*) (Lu et al. 2014a). Iron reducers represented only a minor fraction of the microorganisms enriched in the above-reported studies. However, microorganisms of the genus *Geobacter* were enriched on the anodes of the bioelectrochemical reactors during the phenol removal (Palma et al. 2018; Zhang et al. 2017). The presence of *Geobacter* was also observed in microbial communities enriched in bioelectrochemical reactors during the degradation of recalcitrant hydrocarbons such as BTEX compounds (Daghio et al. 2018a) and PAHs (Yu et al. 2017). While in the first case, the members of the family Geobacteraceae accounted for a minor fraction of the microorganisms in anodic communities dominated by the family Desulfobulbaceae (Daghio et al. 2018a), in the other report the genus *Geobacter* represented nearly 39% of the microorganisms enriched on the electrodes of sediment MFCs in which removal of anthracene, phenanthrene, and pyrene was observed (Yu et al. 2017).

It is intriguing that iron reducers, which are often enriched in the anodic communities with easily degradable electron donors (e.g., acetate) (Kiely et al. 2011), seem to have a marginal role in several studies during the degradation of recalcitrant substrates (e.g., hydrocarbons). Some authors have hypothesized that when readily biodegradable carbon sources are provided, the selection of the microbial community is mainly driven by the ability to use the electrodes as electron acceptors (Daghio et al. 2017). Conversely, with complex and recalcitrant substrates, the ability to use the electron donor represents a selective pressure that can lead to the selection of microorganisms with less efficient EET mechanisms (Daghio et al. 2017).

The occurrence of complex interactions within the microbial communities has been described during bioelectrochemical hydrocarbon degradation in sulfur-rich environments. Benzene removal in a BES reactor filled with sulfide-rich groundwater led to the selection of putative aerobic degraders probably because of the diffusion of a small amount of oxygen into the reactor. The involvement of aerobic hydrocarbon degraders in benzene removal was suggested by carbon and hydrogen isotope fractionation patterns, which indicated that the benzene activation occurred by a reaction catalyzed by a monooxygenase (Rakoczy et al. 2013). However, the occurrence of the sulfide removal and the sulfate formation, together with the presence of several groups of microorganisms linked to the sulfur cycle, led to the hypothesis that benzene might have been degraded in the bulk and that the electron transfer to the anode was performed via sulfide-sulfate (Rakoczy et al. 2013). A confirmation of this hypothesis came from the characterization of the microbial communities able to remove toluene and TPH in BES reactors and microcosms that simulate marine environments (Cruz Viggi et al. 2015; Daghio et al. 2016; Matturro et al. 2017). In particular, microorganisms of the family Desulfobulbaceae seem to be the key players since their presence has been reported in several studies (Daghio et al. 2016, 2018a, b; Matturro et al. 2017). Members of the family Desulfobulbaceae, called cable bacteria, have been extensively described to couple sulfide oxidation with reduction of spatially distant oxygen by delivering electrons over long distances (Müller et al. 2016; Pfeffer et al. 2012; Schauer et al. 2014), and they might be also involved in electron transfer processes during the hydrocarbon removal in BES. Hydrocarbons can be oxidized via sulfate reduction (with formation of sulfide) and the bacteria of the family Desulfobulbaceae could oxidize sulfide to elemental sulfur on the anode surface and transfer the electrons to the anode (Daghio et al. 2018a). Elemental sulfur can be reduced to sulfide by members of the Desulfuromonadaceae forming a cycle (Bond et al. 2002; Daghio et al. 2016, 2018a; Dutta et al. 2009) or being oxidized back to sulfate (Zhang et al. 2014) regenerating the electron acceptor for sulfate reducers. The effect of the sulfate reduction during the removal of PAHs (i.e., naphthalene, 2-methylnaphthalene, and phenanthrene) in marine sediments was assessed (Hamdan et al. 2017). The performances of the reactors where molybdate was added as an inhibitor of the sulfate reduction were compared with the results obtained in the reactors without the inhibition of the sulfate-reducing bacteria. Slightly higher hydrocarbon removal was observed in the reactors where the sulfate reduction was inhibited. The microbial community in the MFCs where molybdate was added differed from the community enriched when the inhibitor was not supplied. Indeed, with molybdate, the enrichment of *Desulfuromonas* was observed on the anodes, as well as iron reducers of the genus *Geoalkalibacter*. A decrease in sulfate concentration, however, was also observed in the molybdate-supplied reactors (i.e., the sulfate concentration dropped from 1130 mg/kg of dry sediment to 590 mg/kg of dry sediment after 28 weeks); therefore, the authors hypothesized a partial inhibition of the sulfate reduction (Hamdan et al. 2017). Further experiments might help in a more detailed description of the effects the sulfur cycle can have on the bioelectrochemical processes in sulfate/sulfide-rich environments.

Pure cultures of *G. lovleyi* were inoculated in bioelectrochemical reactors for the reductive dehalogenation of PCE and TCE to *cis*-DCE that was not further reduced (Aulenta et al. 2009; Strycharz et al. 2008). *Dehalococcoides* spp. are the only microorganisms able to completely reduce TCE to ethene (Maymó-Gatell et al. 2001; West et al. 2008) and they were widely used for the study of the reductive dehalogenation in BESs (Daghio et al. 2017). Fluorescence in situ hybridization (FISH) analysis of the microbial communities both on the cathode and in the bulk of reactors used for bioelectrochemical TCE and 1,2-DCA removal showed the presence of active *Dehalococcoides* proving their importance during the reductive dehalogenation in BESs (Aulenta et al. 2009, 2010b; Leitão et al. 2015, 2016). In particular, a strong enrichment of *Dehalococcoides mccartyi* was observed in a bioelectrochemical reactor during 1,2-DCA dehalogenation with an electrode poised at −300 mV in the presence of AQDS (Leitão et al. 2016). The presence of

Dehalococcoides, however, can be influenced by the cathode potential. Catalyzed reporter deposition-FISH (CARD-FISH) analysis was performed on the cathode effluent of a bioelectrochemical reactor run at different potentials (from −750 mV to −250 mV). Members of the Chloroflexi represented from 65% to 100% of the total bacteria at all the tested potentials. However, at potentials from −750 mV to −550 mV, when the dehalogenation was probably sustained by H_2 evolution, the genus *Dehalococcoides* accounted for almost all the Chloroflexi, but it strongly decreased at higher potentials (from −450 mV to −250 mV), when the process was probably driven by a direct electron uptake from the cathode, and was partially outcompeted by an unidentified member of the Chloroflexi (Di Battista et al. 2012).

17.5 Advantages and Disadvantages of Bioelectrochemical Processes

Several potential advantages could be linked to the use of novel BES-based approaches for the bioremediation of contaminated environments. The main benefit that can be achieved when using a BES to remove contaminants is the decrease in operational costs. Since no information about field-scale plants is currently available, a complete estimation of the costs associated with the application of bioelectrochemical technologies for the bioremediation is not possible yet. Electrodes can serve as virtually inexhaustible electron donors/acceptors because a stable electron flow can be maintained for long operational periods without the need of continuous supply of water-soluble electron donors/acceptors (Morris and Jin 2007). In a recent review, a 2.4 kWh energy input was calculated for the complete PCE dehalogenation to ethane with a dehalogenation rate of ~10 (mol PCE)/day (i.e., 1.66 kg/day) (Daghio et al. 2017). This estimation, however, does not consider any possible unwanted side reactions that may occur leading to a lower efficiency of the process, and, thus, a higher energy input might be required in field operations. In terms of installation costs it is reasonable to presume that the main cost will be associated to the electrode material, which could be variable from € 30–50 for 1 m^2 up to € 500 for 1 m^2 for DSA (Daghio et al. 2017). However, it is important to point out that this evaluation has to be based on considering the lifetime of the components that has never been tested in real applications.

The choice of the electrode material is crucial. Appropriate electrode materials have to be biocompatible and stable, ensuring high electrical conductivity and a low cost/life ratio (Daghio et al. 2017). For the stimulation of the oxidation with anodes, some authors used stainless steel electrodes or even DSA when high potential anodes were used for oxygen evolution. However, carbon materials (e.g., graphite or activated carbon) are used in the majority of studies (Daghio et al. 2017). Carbon materials can adsorb the contaminants and colocalize the electron donor, the electron acceptor, and the catalyst (i.e., microorganisms) in the same highly metabolically active area (Zhang et al. 2010). Adsorption usually decreases the bioavailability of the contaminant. However, studies with ^{14}C-labeled toluene and benzene and isotopic analysis demonstrated that the biodegradation of hydrocarbons is not negatively affected by adsorption on the electrodes (Rakoczy et al. 2013; Zhang et al. 2010).

Another important advantage that can be linked to the use of bioelectrochemical technologies is the possibility of adjusting the energy level adapting it to the environmental conditions. The optimal circumstances for the reaction can thus be achieved (Daghio et al. 2017). The importance of this aspect has been discussed in relation to the bioelectrochemical stimulation of the reductive dehalogenation. Indeed, in this process, the electron's delivery rate is pivotal to maintain the H_2 concentration and to avoid the unwanted side reactions (e.g., methanogenesis) (Aulenta et al. 2009, 2010b, 2011). Another aspect linked to the control of the process is the selectivity of the biological treatments that, in some conditions, could prevent the formation of unwanted end products (Daghio et al. 2017). As an example, it is possible to consider the removal of nitrobenzene by bioelectrochemical reduction. Almost 99% of the nitrobenzene was reduced to aniline (more degradable compared to nitrobenzene) without production of toxic nitrosobenzene, which was detected in abiotic conditions (Wang et al. 2011).

Some authors also suggested the possibility of using pre-colonized electrodes to facilitate the bioaugmentation (Venkidusamy et al. 2016), which could be limited by the environmental conditions and competition with indigenous bacteria (Gkorezis et al. 2016).

The use of BESs during the bioremediation of contaminated environments could be also beneficial for the process monitoring. The generated current can be correlated to the microbial activity (Daghio et al. 2017; Tront et al. 2008a; Williams et al. 2010) or it could be used to power sensors (Lovley 2006; Shantaram et al. 2005). Several studies have investigated the correlation between the electrical signal produced in a BES and the degradation of the organic substrates, and have demonstrated that the current production can be correlated to the chemical oxygen demand (COD) and to the biochemical oxygen demand (BOD) by Monod-type kinetics (Franzetti et al. 2017; Kim et al. 2003; Kumlanghan et al. 2007; Min and Logan 2004). The electrical current is proportional to the charge exchanged per time unit [1 A (Ampere) = 1 C (Coulomb) /1 s (second); C = n F where n is the number of electrons and F is the Faraday constant (i.e., 9.64853×10^4 C/mol)]. It should be therefore proportional to the rate at which the electrons are transferred to/from the electrode by the oxidation/reduction reaction. The electrical signal produced in a BES can thus be exploited for the real-time monitoring of the microbial activity and of the pollutants' concentration during the bioremediation of the contaminated sites (Daghio et al. 2017).

The relation between current production and substrate concentration has been studied in MFCs inoculated with pure cultures of *G. sulfurreducens* and of *S. oneidensis* where easily degradable substrates (i.e., acetate and lactate, respectively) were provided (Tront et al. 2008a, b). Acetate concentration (in the range 0–2.3 mM) and lactate concentration (in the range 0–41 mM) correlated with the current (Tront et al. 2008a, b), suggesting that the process is governed by a first-order kinetics.

The possibility of using the electrical signal produced in a BES for monitoring the bioremediation process has also been explored in the field. An MFC-based biosensor was successfully applied in a field study to assess the rate of U(VI) bioreduction providing acetate as the electron donor (Williams et al. 2010). Similar devices could be easily installed in existing monitoring well networks with the main benefit of decreasing both the number of analyses and the time required to assess the microbial activity (Tront et al. 2008a).

For the development of BES-based biosensors, it is, however, important to consider that the oxidation/reduction of chemical compounds other than the contaminant could lead to the presence of a background signal (Daghio et al. 2017). The placement of control sensors outside the contaminated areas or the development of selective enzymatic biosensors could be possible strategies to overcome issues due to the generation of background currents (Daghio et al. 2017).

One of the main drawbacks linked to the use of BES-based technologies to stimulate the bioremediation in anaerobic environments is the lower efficiency of the degradation, which leads to a slower biodegradation compared to the aerobic metabolisms (Weelink et al. 2010). The degradation rate at the anode can also be limited by the kinetic of the cathodic reaction when an MFC configuration is used (Logan 2009; Wei et al. 2011).

The scale-up of the technology is challenging and the data from field-scale plants are not currently available (Daghio et al. 2017). One of the most important aspects that has to be considered when applying a BES to the bioremediation in real plants is the radius of influence, which can greatly affect the process. Attempts to evaluate the radius of influence have been made during TPH degradation with anodes serving as electron acceptors (Lu et al. 2014a, b; Wang et al. 2012). In a biodegradation study performed over 120 days with cylindrical anodes (radius 7.5 cm), the authors predicted the maximum radius of influence of 90 cm, assuming a linear correlation between the distance of the electrode and the stimulation of TPH removal (Lu et al. 2014b). Water content, mass transport, and the type of soil as well as the design of the treatment plant (e.g., the type of electrode) can potentially affect the radius of influence; thus, the site-specific conditions have to be considered (Daghio et al. 2017).

Future scale-up of this innovative technology will lead to a more detailed comprehension of the advantages and the limitations linked to the use of BESs for bioremediation, which is now mainly limited to small-scale reactors.

17.6 Possible Opportunities for Future Development

Several studies proved that a broad range of halogenated and non-halogenated hydrocarbons can be successfully reduced/oxidized in BESs and that this technology could be effective in the stimulation of the microbial metabolism in contaminated environments. From the analysis of the literature reported in this chapter it is clear that the main challenge that has to be faced in the near future is the progressive scale-up of this innovative technology, which was also previously suggested (Daghio et al. 2017). Pilot-scale experiments will allow the investigation of critical factors, such as the extension of the radius of influence, a more comprehensive estimation of installation and operational costs, and the estimation of the lifetime of the components. The data from the field plants are also essential for the evaluation of the advantages and disadvantages compared to the common bioremediation approaches.

From lab-scale experiments, benefits could be obtained by a better description of the microbial communities enriched in order to shed light on the complex interactions that may occur between different populations and to better identify the key players. Understanding the role of different microorganisms can be useful to identify the parameters that can be modified in order to improve the technology and gain higher efficiencies.

New possibilities to assist the bioremediation using BESs can also be explored. An example can be the use of bioelectrochemical technologies for the removal of metabolites that could be inhibitory for the microbial activity if accumulated (Daghio et al. 2017). For example, toxic sulfide, which accumulates during hydrocarbons' degradation via sulfate reduction, can be removed by anodic oxidation to elemental sulfur and oxidized back to sulfate (Daghio et al. 2018b; Dutta et al. 2008; Zhang et al. 2014). It has also been observed that, using sediment MFCs, microorganisms can grow by using plant root exudates with production of electricity and reduction of greenhouse emissions (i.e., CH_4) from wetlands (Arends et al. 2014; Kaku et al. 2008; De Schamphelaire et al. 2008). The interaction between the plants and the bacteria during the hydrocarbon biodegradation is a well-known process (Gkorezis et al. 2016). It is therefore possible to test combined phytoremediation-BES treatments (Daghio et al. 2017) as already reported for the bioremediation of pyrene and benzo[*a*]pyrene using a macrophyte (*Acorus calamus*) in combination with an MFC (Yan et al. 2015).

References

Adelaja O, Keshavarz T, Kyazze G (2014) Enhanced biodegradation of phenanthrene using different inoculum types in a microbial fuel cell. Eng Life Sci 14(2):218–228. https://doi.org/10.1002/elsc.201300089

Adelaja O, Keshavarz T, Kyazze G (2015) The effect of salinity, redox mediators and temperature on anaerobic biodegradation of petroleum hydrocarbons in microbial fuel cells. J Hazard Mater 283:211–217. https://doi.org/10.1016/j.jhazmat.2014.08.066

Aelterman P, Freguia S, Keller J, Verstraete W, Rabaey K (2008) The anode potential regulates bacterial activity in microbial fuel cells. Appl Microbiol Biotechnol 78(3):409–418. https://doi.org/10.1007/s00253-007-1327-8

Alvarez PJ, Illman WA (2005) Bioremediation and natural attenuation: process fundamentals and mathematical models. Wiley, Hoboken

Andersen SJ, Pikaar I, Freguia S, Lovell BC, Rabaey K, Rozendal RA (2013) Dynamically adaptive control system for bioanodes in serially stacked bioelectrochemical systems. Environ Sci Technol 47(10):5488–5494. https://doi.org/10.1021/es400239k

Arends JBA, Speeckaert J, Blondeel E, De Vrieze J, Boeckx P, Verstraete W, Rabaey K, Boon N (2014) Greenhouse gas emissions from rice microcosms amended with a plant microbial fuel cell. Appl Microbiol Biotechnol 98(7):3205–3217. https://doi.org/10.1007/s00253-013-5328-5

Aulenta F, Gossett JM, Petrangeli Papini M, Rossetti S, Majone M (2005) Comparative study of methanol, butyrate, and hydrogen as electron donors for long-term dechlorination of tetrachloroethene in mixed anerobic cultures. Biotechnol Bioeng 91(6):743–753. https://doi.org/10.1002/bit.20569

Aulenta F, Di Tomassi C, Cupo C, Petrangeli Papini M (2006) Influence of hydrogen on the reductive dechlorination of tetrachloroethene (PCE) to ethene in a methanogenic biofilm reactor: role of mass transport phenomena. J Chem Technol Biotechnol 81(9):1520–1529. https://doi.org/10.1002/jctb.1562

Aulenta F, Catervi A, Majone M, Panero S, Reale P, Rossetti S (2007a) Electron transfer from a solid-state electrode assisted by methyl viologen sustains efficient microbial reductive dechlorination of TCE. Environ Sci Technol 41(7):2554–2559. https://doi.org/10.1021/es0624321

Aulenta F, Pera A, Rossetti S, Petrangeli Papini M, Majone M (2007b) Relevance of side reactions in anaerobic reductive dechlorination microcosms amended with different electron donors. Water Res 41(1):27–38. https://doi.org/10.1016/j.watres.2006.09.019

Aulenta F, Canosa A, Majone M, Panero S, Reale P, Rossetti S (2008a) Trichloroethene dechlorination and H_2 evolution are alternative biological pathways of electric charge utilization by a dechlorinating culture in a bioelectrochemical system. Environ Sci Technol 42(16):6185–6190. https://doi.org/10.1021/es800265b

Aulenta F, Reale P, Catervi A, Panero S, Majone M (2008b) Kinetics of trichloroethene dechlorination and methane formation by a mixed anaerobic culture in a bio-electrochemical system. Electrochim Acta 53(16):5300–5305. https://doi.org/10.1016/j.electacta.2008.02.084

Aulenta F, Canosa A, Reale P, Rossetti S, Panero S, Majone M (2009) Microbial reductive dechlorination of trichloroethene to ethene with electrodes serving as electron donors without the external addition of redox mediators. Biotechnol Bioeng 103(1):85–91. https://doi.org/10.1002/bit.22234

Aulenta F, Di Maio V, Ferri T, Majone M (2010a) The humic acid analogue antraquinone-2,6-disulfonate (AQDS) serves as an electron shuttle in the electricity-driven microbial dechlorination of trichloroethene to cis-dichloroethene. Bioresour Technol 101(24):9728–9733. https://doi.org/10.1016/j.biortech.2010.07.090

Aulenta F, Reale P, Canosa A, Rossetti S, Panero S, Majone M (2010b) Characterization of an electro-active biocathode capable of dechlorinating trichloroethene and cis-dichloroethene to ethene. Biosens Bioelectron 25(7):1796–1802. https://doi.org/10.1016/j.bios.2009.12.033

Aulenta F, Tocca L, Verdini R, Reale P, Majone M (2011) Dechlorination of trichloroethene in a continuous-flow bioelectrochemical reactor: effect of cathode potential on rate, selectivity, and electron transfer mechanisms. Environ Sci Technol 45(19):8444–8451. https://doi.org/10.1021/es202262y

Aulenta F, Verdini R, Zeppilli M, Zanaroli G, Fava F, Rossetti S, Majone M (2013) Electrochemical stimulation of microbial cis-dichloroethene (cis-DCE) oxidation by an ethene-assimilating culture. New Biotechnol 30(6):749–755. https://doi.org/10.1016/j.nbt.2013.04.003

Baldwin BR, Nakatsu CH, Nebe J, Wickham GS, Parks C, Nies L (2009) Enumeration of aromatic oxygenase genes to evaluate biodegradation during multi-phase extraction at a gasoline-contaminated site. J Hazard Mater 163(2–3):524–530. https://doi.org/10.1016/j.jhazmat.2008.07.002

Bellagamba M, Cruz Viggi C, Ademollo N, Rossetti S, Aulenta F (2017) Electrolysis-driven bioremediation of crude oil-contaminated marine sediments. New Biotechnol 38:84–90. https://doi.org/10.1016/j.nbt.2016.03.003

Bond DR, Holmes DE, Tender LM, Lovley DR (2002) Electrode-reducing microorganisms that harvest energy from marine sediments. Science 295(5554):483–485. https://doi.org/10.1126/science.1066771

Broden RC, Goin RT, Kao C-M (1997) Control of BTEX migration using a biologically enhanced permeable barrier. Groundwater Monit Rem 17(1):70–80. https://doi.org/10.1111/j.1745-6592.1997.tb01186.x

Chandrasekhar K, Venkata Mohan S (2012) Bio-electrochemical remediation of real field petroleum sludge as an electron donor with simultaneous power generation facilitates biotransformation of PAH: effect of substrate concentration. Bioresour Technol 110:517–525. https://doi.org/10.1016/j.biortech.2012.01.128

Cheng Y, Wang L, Faustorilla V, Megharaj M, Naidu R, Chen Z (2017) Integrated electrochemical treatment systems for facilitating the bioremediation of oil spill contaminated soil. Chemosphere 175:294–299. https://doi.org/10.1016/j.chemosphere.2017.02.079

Cruz Viggi C, Presta E, Bellagamba M, Kaciulis S, Balijepalli SK, Zanaroli G, Petrangeli Papini M, Rossetti S, Aulenta F (2015) The "Oil-Spill Snorkel": an innovative bioelectrochemical approach to accelerate hydrocarbons biodegradation in marine sediments. Front Microbiol 6:881. https://doi.org/10.3389/fmicb.2015.00881

Daghio M, Vaiopoulou E, Patil SA, Suárez-Suárez A, Head IM, Franzetti A, Rabaey K (2016) Anodes stimulate anaerobic toluene degradation via sulfur cycling in marine sediments. Appl Environ Microbiol 82(1):297–307. https://doi.org/10.1128/AEM.02250-15

Daghio M, Aulenta F, Vaiopoulou E, Franzetti A, Arends JBA, Sherry A, Suárez-Suárez A, Head IM, Bestetti G, Rabaey K (2017) Electrobioremediation of oil spills. Water Res 114:351–370. https://doi.org/10.1016/j.watres.2017.02.030

Daghio M, Espinoza Tofalos A, Leoni B, Cristiani P, Papacchini M, Jalilnejad E, Bestetti G, Franzetti A (2018a) Bioelectrochemical BTEX removal at different voltages: assessment of the degradation and characterization of the microbial communities. J Hazard Mater 341:120–127. https://doi.org/10.1016/j.jhazmat.2017.07.054

Daghio M, Vaiopoulou E, Aulenta F, Sherry A, Head I, Franzetti A, Rabaey K (2018b) Anode potential selection for sulfide removal in contaminated marine sediments. J Hazard Mater 360:498–503. https://doi.org/10.1016/j.jhazmat.2018.08.016

de Bruin WP, Kotterman MJ, Posthumus MA, Schraa G, Zehnder AJ (1992) Complete biological reductive transformation of tetrachloroethene to ethane. Appl Environ Microbiol 58 (6):1996–2000

De Schamphelaire L, Van den Bossche L, Dang SH, Höfte M, Boon N, Rabaey K, Verstraete W (2008) Microbial fuel cells generating electricity from rhizodeposits of rice plants. Environ Sci Technol 42(8):3053–3058. https://doi.org/10.1021/es071938w

Di Battista A, Verdini R, Rossetti S, Pietrangeli B, Majone M, Aulenta F (2012) CARD-FISH analysis of a TCE-dechlorinating biocathode operated at different set potentials. New Biotechnol 30(1):33–38. https://doi.org/10.1016/j.nbt.2012.06.002

Dutta PK, Rabaey K, Yuan Z, Keller J (2008) Spontaneous electrochemical removal of aqueous sulfide. Water Res 42(20):4965–4975. https://doi.org/10.1016/j.watres.2008.09.007

Dutta PK, Keller J, Yuan Z, Rozendal RA, Rabaey K (2009) Role of sulfur during acetate oxidation in biological anodes. Environ Sci Technol 43(10):3839–3845. https://doi.org/10.1021/es803682k

Espinoza Tofalos A, Daghio M, González M, Papacchini M, Franzetti A, Seeger M (2018) Toluene degradation by Cupriavidus metallidurans CH34 in nitrate-reducing conditions and in Bioelectrochemical Systems. FEMS Microbiol Lett 365(12):fny119. https://doi.org/10.1093/femsle/fny119

Fang Y, Hozalski RM, Clapp LW, Novak PJ, Semmens MJ (2002) Passive dissolution of hydrogen gas into groundwater using hollow-fiber membranes. Water Res 36(14):3533–3542. https://doi.org/10.1016/S0043-1354(02)00046-5

Farhadian M, Vachelard C, Duchez D, Larroche C (2008) In situ bioremediation of monoaromatic pollutants in groundwater: a review. Bioresour Technol 99(13):5296–5308. https://doi.org/10.1016/j.biortech.2007.10.025

Fennell DE, Gossett JM, Zinder SH (1997) Comparison of butyric acid, ethanol, lactic acid, and propionic acid as hydrogen donors for the reductive dechlorination of tetrachloroethene. Environ Sci Technol 31(3):918–926. https://doi.org/10.1021/es960756r

Franzetti A, Daghio M, Parenti P, Truppi T, Bestetti G, Trasatti SP, Cristiani P (2017) Monod kinetics degradation of low concentration residual organics in membraneless microbial fuel cells. J Electrochem Soc 164(3):H3091–H3096. https://doi.org/10.1149/2.0141703jes

Friman H, Schechter A, Nitzan Y, Cahan R (2013) Phenol degradation in bio-electrochemical cells. Int Biodeterior Biodegrad 84:155–160. https://doi.org/10.1016/j.ibiod.2012.04.019

Gkorezis P, Daghio M, Franzetti A, Van Hamme JD, Sillen W, Vangronsveld J (2016) The interaction between plants and bacteria in the remediation of petroleum hydrocarbons: an environmental perspective. Front Microbiol 7:1836. https://doi.org/10.3389/fmicb.2016.01836

Hamdan HZ, Salam DA, Hari AR, Semerjian L, Saikaly P (2017) Assessment of the performance of SMFCs in the bioremediation of PAHs in contaminated marine sediments under different redox conditions and analysis of the associated microbial communities. Sci Total Environ 575:1453–1461. https://doi.org/10.1016/j.scitotenv.2016.09.232

He J, Sung Y, Dollhopf ME, Fathepure BZ, Tiedje JM, Löffler FE (2002) Acetate versus hydrogen as direct electron donors to stimulate the microbial reductive dechlorination process at chloroethene-contaminated sites. Environ Sci Technol 36(18):3945–3952. https://doi.org/10.1021/es025528d

Ho SV, Sheridan PW, Athmer CJ, Heitkamp MA, Brackin JM, Weber D, Brodsky PH (1995) Integrated *in situ* soil remediation technology: the lasagna process. Environ Sci Technol 29 (10):2528–2534. https://doi.org/10.1021/es00010a011

Huang D-Y, Zhou S-G, Chen Q, Zhao B, Yuan Y, Zhuang L (2011) Enhanced anaerobic degradation of organic pollutants in a soil microbial fuel cell. Chem Eng J 172(2–3):647–653. https://doi.org/10.1016/j.cej.2011.06.024

Inoue K, Leang C, Franks AE, Woodard TL, Nevin KP, Lovley DR (2011) Specific localization of the *c*-type cytochrome OmcZ at the anode surface in current-producing biofilms of *Geobacter sulfurreducens*. Environ Microbiol Rep 3(2):211–217. https://doi.org/10.1111/j.1758-2229.2010.00210.x

Kaku N, Yonezawa N, Kodama Y, Watanabe K (2008) Plant/microbe cooperation for electricity generation in a rice paddy field. Appl Microbiol Biotechnol 79(1):43–49. https://doi.org/10.1007/s00253-008-1410-9

Kiely PD, Regan JM, Logan BE (2011) The electric picnic: synergistic requirements for exoelectrogenic microbial communities. Curr Opin Biotechnol 22(3):378–385. https://doi.org/10.1016/j.copbio.2011.03.003

Kim BH, Chang IS, Gil GC, Park HS, Kim HJ (2003) Novel BOD (biological oxygen demand) sensor using mediator-less microbial fuel cell. Biotechnol Lett 25(7):541–545. https://doi.org/10.1023/A:1022891231369

Kumlanghan A, Liu J, Thavarungkul P, Kanatharana P, Mattiasson B (2007) Microbial fuel cell-based biosensor for fast analysis of biodegradable organic matter. Biosens Bioelectron 22 (12):2939–2944. https://doi.org/10.1016/j.bios.2006.12.014

Lai A, Verdini R, Aulenta F, Majone M (2015) Influence of nitrate and sulfate reduction in the bioelectrochemically assisted dechlorination of cis-DCE. Chemosphere 125:147–154. https://doi.org/10.1016/j.chemosphere.2014.12.023

Lai A, Aulenta F, Mingazzini M, Palumbo MT, Petrangeli Papini M, Verdini R, Majone M (2017) Bioelectrochemical approach for reductive and oxidative dechlorination of chlorinated aliphatic hydrocarbons (CAHs). Chemosphere 169:351–360. https://doi.org/10.1016/j.chemosphere.2016.11.072

Leitão P, Rossetti S, Nouws HPA, Danko AS, Majone M, Aulenta F (2015) Bioelectrochemically-assisted reductive dechlorination of 1,2-dichloroethane by a *Dehalococcoides*-enriched microbial culture. Bioresour Technol 195:78–82. https://doi.org/10.1016/j.biortech.2015.06.027

Leitão P, Rossetti S, Danko AS, Nouws H, Aulenta F (2016) Enrichment of *Dehalococcoides mccartyi* spp. from a municipal activated sludge during AQDS-mediated bioelectrochemical dechlorination of 1,2-dichloroethane to ethene. Bioresour Technol 214:426–431. https://doi.org/10.1016/j.biortech.2016.04.129

Leitão P, Nouws H, Danko AS, Aulenta F (2017) Bioelectrochemical dechlorination of 1,2-DCA with an AQDS-functionalized cathode serving as electron donor. Fuel Cells 17(5):612–617. https://doi.org/10.1002/fuce.201700045

Li W-W, Yu H-Q (2015) Stimulating sediment bioremediation with benthic microbial fuel cells. Biotechnol Adv 33(1):1–12. https://doi.org/10.1016/j.biotechadv.2014.12.011

Li X, Wang X, Zhang Y, Cheng L, Liu J, Li F, Gao B, Zhou Q (2014) Extended petroleum hydrocarbon bioremediation in saline soil using Pt-free multianodes microbial fuel cells. RSC Adv 4(104):59803–59808. https://doi.org/10.1039/C4RA10673C

Li X, Wang X, Ren ZJ, Zhang Y, Li N, Zhou Q (2015) Sand amendment enhances bioelectrochemical remediation of petroleum hydrocarbon contaminated soil. Chemosphere 141:62–70. https://doi.org/10.1016/j.chemosphere.2015.06.025

Lin C-W, Wu C-H, Chiu Y-H, Tsai S-L (2014) Effects of different mediators on electricity generation and microbial structure of a toluene powered microbial fuel cell. Fuel 125:30–35. https://doi.org/10.1016/j.fuel.2014.02.018

Logan BE (2009) Exoelectrogenic bacteria that power microbial fuel cells. Nat Rev Microbiol 7:375–381. https://doi.org/10.1038/nrmicro2113

Logan BE, Hamelers B, Rozendal R, Schröder U, Keller J, Freguia S, Aelterman P, Verstraete W, Rabaey K (2006) Microbial fuel cells: methodology and technology. Environ Sci Technol 40 (17):5181–5192. https://doi.org/10.1021/es0605016

Lohner ST, Tiehm A (2009) Application of electrolysis to stimulate microbial reductive PCE dechlorination and oxidative VC biodegradation. Environ Sci Technol 43(18):7098–7104. https://doi.org/10.1021/es900835d

Lovley DR (2006) Bug juice: harvesting electricity with microorganisms. Nat Rev Microbiol 4:497–508. https://doi.org/10.1038/nrmicro1442

Lovley DR (2011) Live wires: direct extracellular electron exchange for bioenergy and the bioremediation of energy-related contamination. Energy Environ Sci 4(12):4896–4906. https://doi.org/10.1039/C1EE02229F

Lovley DR (2012) Electromicrobiology. Annu Rev Microbiol 66:391–409. https://doi.org/10.1146/annurev-micro-092611-150104

Lovley DR, Nevin KP (2011) A shift in the current: new applications and concepts for microbe-electrode electron exchange. Curr Opin Biotechnol 22(3):441–448. https://doi.org/10.1016/j.copbio.2011.01.009

Lovley DR, Woodward JC, Chapelle FH (1994) Stimulated anoxic biodegradation of aromatic hydrocarbons using Fe(III) ligands. Nature 370:128–131. https://doi.org/10.1038/370128a0

Lovley DR, Coates JD, Blunt-Harris EL, Phillips EJP, Woodward JC (1996a) Humic substances as electron acceptors for microbial respiration. Nature 382:445–448. https://doi.org/10.1038/382445a0

Lovley DR, Woodward JC, Chapelle FH (1996b) Rapid anaerobic benzene oxidation with a variety of chelated Fe(III) forms. Appl Environ Microbiol 62(1):288–291

Lu L, Huggins T, Jin S, Zuo Y, Ren ZJ (2014a) Microbial metabolism and community structure in response to bioelectrochemically enhanced remediation of petroleum hydrocarbon-contaminated soil. Environ Sci Technol 48(7):4021–4029. https://doi.org/10.1021/es4057906

Lu L, Yazdi H, Jin S, Zuo Y, Fallgren PH, Ren ZJ (2014b) Enhanced bioremediation of hydrocarbon-contaminated soil using pilot-scale bioelectrochemical systems. J Hazard Mater 274:8–15. https://doi.org/10.1016/j.jhazmat.2014.03.060

Ma X, Novak PJ, Clapp LW, Semmens MJ, Hozalski RM (2003) Evaluation of polyethylene hollow-fiber membranes for hydrogen delivery to support reductive dechlorination in a soil column. Water Res 37(12):2905–2918. https://doi.org/10.1016/S0043-1354(03)00111-8

Marsili E, Baron DB, Shikhare ID, Coursolle D, Gralnick JA, Bond DR (2008) *Shewanella* secretes flavins that mediate extracellular electron transfer. Proc Natl Acad Sci U S A 105 (10):3968–3973. https://doi.org/10.1073/pnas.0710525105

Matturro B, Cruz Viggi C, Aulenta F, Rossetti S (2017) Cable bacteria and the bioelectrochemical snorkel: the natural and engineered facets playing a role in hydrocarbons degradation in marine sediments. Front Microbiol 8:952. https://doi.org/10.3389/fmicb.2017.00952

Maymó-Gatell X, Nijenhuis I, Zinder SH (2001) Reductive dechlorination of *cis*-1,2-dichloroethene and vinyl chloride by "*Dehalococcoides ethenogenes*". Environ Sci Technol 35(3):516–521. https://doi.org/10.1021/es001285i

Mihelcic JR, Luthy RG (1988) Degradation of polycyclic aromatic hydrocarbon compounds under various redox conditions in soil-water systems. Appl Environ Microbiol 54(5):1182–1187

Milliken CE, May HD (2007) Sustained generation of electricity by the spore-forming, Gram-positive, *Desulfitobacterium hafniense* strain DCB2. Appl Microbiol Biotechnol 73 (5):1180–1189. https://doi.org/10.1007/s00253-006-0564-6

Min B, Logan BE (2004) Continuous electricity generation from domestic wastewater and organic substrates in a flat plate microbial fuel cell. Environ Sci Technol 38(21):5809–5814. https://doi.org/10.1021/es0491026

Morris JM, Jin S (2007) Feasibility of using microbial fuel cell technology for bioremediation of hydrocarbons in groundwater. J Environ Sci Health, Part A 43(1):18–23. https://doi.org/10.1080/10934520701750389

Morris JM, Jin S (2012) Enhanced biodegradation of hydrocarbon-contaminated sediments using microbial fuel cells. J Hazard Mater 213–214:474–477. https://doi.org/10.1016/j.jhazmat.2012.02.029

Morris JM, Jin S, Crimi B, Pruden A (2009) Microbial fuel cell in enhancing anaerobic biodegradation of diesel. Chem Eng J 146(2):161–167. https://doi.org/10.1016/j.cej.2008.05.028

Müller H, Bosch J, Griebler C, Damgaard LR, Nielsen LP, Lueders T, Meckenstock RU (2016) Long-distance electron transfer by cable bacteria in aquifer sediments. ISME J 10:2010–2019. https://doi.org/10.1038/ismej.2015.250

Palma E, Daghio M, Franzetti A, Petrangeli Papini M, Aulenta F (2018) The bioelectric well: a novel approach for *in situ* treatment of hydrocarbon-contaminated groundwater. Microb Biotechnol 11(1):112–118. https://doi.org/10.1111/1751-7915.12760

Pfeffer C, Larsen S, Song J, Dong M, Besenbacher F, Meyer RL, Kjeldsen KU, Schreiber L, Gorby YA, El-Naggar MY, Leung KM, Schramm A, Risgaard-Petersen N, Nielsen LP (2012) Filamentous bacteria transport electrons over centimetre distances. Nature 491:218–221. https://doi.org/10.1038/nature11586

Pham H, Boon N, Marzorati M, Verstraete W (2009) Enhanced removal of 1,2-dichloroethane by anodophilic microbial consortia. Water Res 43(11):2936–2946. https://doi.org/10.1016/j.watres.2009.04.004

Rabaey K, Boon N, Höfte M, Verstraete W (2005) Microbial phenazine production enhances electron transfer in biofuel cells. Environ Sci Technol 39(9):3401–3408. https://doi.org/10.1021/es048563o

Rakoczy J, Feisthauer S, Wasmund K, Bombach P, Neu TR, Vogt C, Richnow HH (2013) Benzene and sulfide removal from groundwater treated in a microbial fuel cell. Biotechnol Bioeng 110 (12):3104–3113. https://doi.org/10.1002/bit.24979

Rosenbaum M, Aulenta F, Villano M, Angenent LT (2011) Cathodes as electron donors for microbial metabolism: which extracellular electron transfer mechanisms are involved? Bioresour Technol 102(1):324–333. https://doi.org/10.1016/j.biortech.2010.07.008

Rozendal RA, Hamelers HVM, Euverink GJW, Metz SJ, Buisman CJN (2006) Principle and perspectives of hydrogen production through biocatalyzed electrolysis. Int J Hydrog Energy 31(12):1632–1640. https://doi.org/10.1016/j.ijhydene.2005.12.006

Schauer R, Risgaard-Petersen N, Kjeldsen KU, Tataru Bjerg JJ, Jørgensen BB, Schramm A, Nielsen LP (2014) Succession of cable bacteria and electric currents in marine sediment. ISME J 8:1314–1322. https://doi.org/10.1038/ismej.2013.239

Seshadri R, Adrian L, Fouts DE, Eisen JA, Phillippy AM, Methe BA, Ward NL, Nelson WC, Deboy RT, Khouri HM, Kolonay JF, Dodson RJ, Daugherty SC, Brinkac LM, Sullivan SA, Madupu R, Nelson KE, Kang KH, Impraim M, Tran K, Robinson JM, Forberger HA, Fraser CM, Zinder SH, Heidelberg JF (2005) Genome sequence of the PCE-dechlorinating bacterium *Dehalococcoides ethenogenes*. Science 307(5706):105–108. https://doi.org/10.1126/science.1102226

Shantaram A, Beyenal H, Veluchamy RRA, Lewandowski Z (2005) Wireless sensors powered by microbial fuel cells. Environ Sci Technol 39(13):5037–5042. https://doi.org/10.1021/es0480668

Shi L, Dong H, Reguera G, Beyenal H, Lu A, Liu J, Yu H-Q, Fredrickson JK (2016) Extracellular electron transfer mechanisms between microorganisms and minerals. Nat Rev Microbiol 14:651–662. https://doi.org/10.1038/nrmicro.2016.93

Skadberg B, Geoly-Horn SL, Sangamalli V, Flora JRV (1999) Influence of pH, current and copper on the biological dechlorination of 2,6-dichlorophenol in an electrochemical cell. Water Res 33 (9):1997–2010. https://doi.org/10.1016/S0043-1354(98)00431-X

Strycharz SM, Woodard TL, Johnson JP, Nevin KP, Sanford RA, Löffler FE, Lovley DR (2008) Graphite electrode as a sole electron donor for reductive dechlorination of tetrachlorethene by *Geobacter lovleyi*. Appl Environ Microbiol 74(19):5943–5947. https://doi.org/10.1128/AEM.00961-08

Tront JM, Fortner JD, Plötze M, Hughes JB, Puzrin AM (2008a) Microbial fuel cell biosensor for *in situ* assessment of microbial activity. Biosens Bioelectron 24(4):586–590. https://doi.org/10.1016/j.bios.2008.06.006

Tront JM, Fortner JD, Plötze M, Hughes JB, Puzrin AM (2008b) Microbial fuel cell technology for measurement of microbial respiration of lactate as an example of bioremediation amendment. Biotechnol Lett 30(8):1385–1390. https://doi.org/10.1007/s10529-008-9707-4

Tuxen N, Reitzel LA, Albrechtsen H-J, Bjerg PL (2006) Oxygen-enhanced biodegradation of phenoxy acids in ground water at contaminated sites. Groundwater 44(2):256–265. https://doi.org/10.1111/j.1745-6584.2005.00104.x

Van der Zee FP, Cervantes FJ (2009) Impact and application of electron shuttles on the redox (bio) transformation of contaminants: a review. Biotechnol Adv 27(3):256–277. https://doi.org/10.1016/j.biotechadv.2009.01.004

Venkata Mohan S, Chandrasekhar K (2011) Self-induced bio-potential and graphite electron accepting conditions enhances petroleum sludge degradation in bio-electrochemical system with simultaneous power generation. Bioresour Technol 102(20):9532–9541. https://doi.org/10.1016/j.biortech.2011.07.038

Venkidusamy K, Megharaj M, Marzorati M, Lockington R, Naidu R (2016) Enhanced removal of petroleum hydrocarbons using a bioelectrochemical remediation system with pre-cultured anodes. Sci Total Environ 539:61–69. https://doi.org/10.1016/j.scitotenv.2015.08.098

Verdini R, Aulenta F, de Tora F, Lai A, Majone M (2015) Relative contribution of set cathode potential and external mass transport on TCE dechlorination in a continuous-flow bioelectrochemical reactor. Chemosphere 136:72–78. https://doi.org/10.1016/j.chemosphere.2015.03.092

Wagner RC, Call DF, Logan BE (2010) Optimal set anode potentials vary in bioelectrochemical systems. Environ Sci Technol 44(16):6036–6041. https://doi.org/10.1021/es101013e

Wang A-J, Cheng H-Y, Liang B, Ren N-Q, Cui D, Lin N, Kim BH, Rabaey K (2011) Efficient reduction of nitrobenzene to aniline with a biocatalyzed cathode. Environ Sci Technol 45 (23):10186–10193. https://doi.org/10.1021/es202356w

Wang X, Cai Z, Zhou Q, Zhang Z, Chen C (2012) Bioelectrochemical stimulation of petroleum hydrocarbon degradation in saline soil using U-tube microbial fuel cells. Biotechnol Bioeng 109 (2):426–433. https://doi.org/10.1002/bit.23351

Wang H, Luo H, Fallgren PH, Jin S, Ren ZJ (2015) Bioelectrochemical system platform for sustainable environmental remediation and energy generation. Biotechnol Adv 33 (3–4):317–334. https://doi.org/10.1016/j.biotechadv.2015.04.003

Weber KA, Achenbach LA, Coates JD (2006) Microorganisms pumping iron: anaerobic microbial iron oxidation and reduction. Nat Rev Microbiol 4:752–764. https://doi.org/10.1038/nrmicro1490

Weelink SAB, van Eekert MHA, Stams AJM (2010) Degradation of BTEX by anaerobic bacteria: physiology and application. Rev Environ Sci Biotechnol 9(4):359–385. https://doi.org/10.1007/s11157-010-9219-2

Wei J, Liang P, Huang X (2011) Recent progress in electrodes for microbial fuel cells. Bioresour Technol 102(20):9335–9344. https://doi.org/10.1016/j.biortech.2011.07.019

Wei M, Harnisch F, Vogt C, Ahlheim J, Neu TR, Richnow HH (2015) Harvesting electricity from benzene and ammonium-contaminated groundwater using a microbial fuel cell with an aerated cathode. RSC Adv 5(7):5321–5330. https://doi.org/10.1039/C4RA12144A

Weiner JM, Lauck TS, Lovley DR (1998) Enhanced anaerobic benzene degradation with the addition of sulfate. Biorem J 2(3–4):159–173. https://doi.org/10.1080/10889869809380374

West KA, Johnson DR, Hu P, DeSantis TZ, Brodie EL, Lee PKH, Feil H, Andersen GL, Zinder SH, Alvarez-Cohen L (2008) Comparative genomics of "*Dehalococcoides ethenogenes*" 195 and an enrichment culture containing unsequenced "*Dehalococcoides*" strains. Appl Environ Microbiol 74(11):3533–3540. https://doi.org/10.1128/AEM.01835-07

Williams KH, Nevin KP, Franks A, Englert A, Long PE, Lovley DR (2010) Electrode-based approach for monitoring in situ microbial activity during subsurface bioremediation. Environ Sci Technol 44(1):47–54. https://doi.org/10.1021/es9017464

Wu C-H, Lai C-Y, Lin C-W, Kao M-H (2013) Generation of power by microbial fuel cell with ferricyanide in biodegradation of benzene. Clean: Soil, Air, Water 41(4):390–395. https://doi.org/10.1002/clen.201200198

Yan Z, Song N, Cai H, Tay J-H, Jiang H (2012) Enhanced degradation of phenanthrene and pyrene in freshwater sediments by combined employment of sediment microbial fuel cell and amorphous ferric hydroxide. J Hazard Mater 199–200:217–225. https://doi.org/10.1016/j.jhazmat.2011.10.087

Yan Z, Jiang H, Cai H, Zhou Y, Krumholz LR (2015) Complex interactions between the macrophyte *Acorus calamus* and microbial fuel cells during pyrene and benzo[*a*]pyrene degradation in sediments. Sci Rep 5:10709. https://doi.org/10.1038/srep10709

Yu B, Tian J, Feng L (2017) Remediation of PAH polluted soils using a soil microbial fuel cell: influence of electrode interval and role of microbial community. J Hazard Mater 336:110–118. https://doi.org/10.1016/j.jhazmat.2017.04.066

Zhang T, Gannon SM, Nevin KP, Franks AE, Lovley DR (2010) Stimulating the anaerobic degradation of aromatic hydrocarbons in contaminated sediments by providing an electrode as the electron acceptor. Environ Microbiol 12(4):1011–1020. https://doi.org/10.1111/j.1462-2920.2009.02145.x

Zhang T, Bain TS, Barlett MA, Dar SA, Snoeyenbos-West OL, Nevin KP, Lovley DR (2014) Sulfur oxidation to sulfate coupled with electron transfer to electrodes by *Desulfuromonas* strain TZ1. Microbiology 160:123–129. https://doi.org/10.1099/mic.0.069930-0

Zhang Y, Wang X, Li X, Cheng L, Wan L, Zhou Q (2015) Horizontal arrangement of anodes of microbial fuel cells enhances remediation of petroleum hydrocarbon-contaminated soil. Environ Sci Pollut Res 22(3):2335–2341. https://doi.org/10.1007/s11356-014-3539-7

Zhang D, Li Z, Zhang C, Zhou X, Xiao Z, Awata T, Katayama A (2017) Phenol-degrading anode biofilm with high coulombic efficiency in graphite electrodes microbial fuel cell. J Biosci Bioeng 123(3):364–369. https://doi.org/10.1016/j.jbiosc.2016.10.010

Chapter 18
Field Study VI: The Effect of Loading Strategies on Removal Efficiencies of a Hybrid Constructed Wetland Treating Mixed Domestic and Agro-Industrial Wastewaters

Michal Šereš, Tereza Hnátková, Petr Maršík, Tomáš Vaněk, Petr Soudek, and Jan Vymazal

Abstract The treatment of domestic wastewater using constructed wetlands (CWs) is common in the Czech Republic; however, the treatment of agro-industrial wastewater is still at the beginning of its application. A hybrid CW consisting of two horizontal filters (HF1 and HF2), one vertical filter (VF), and three stabilization ponds (SP1–SP3) was put into operation in 2012 in Chrámce, Ústí region, Czech Republic. The hybrid system treats mixed household and agro-industrial wastewater mainly from sheep farms and wine and fruit juice production factories. The filters are planted with *Phragmites australis*, *Phalaris arundinacea*, *Iris pseudacorus*, *Iris sibirica*, *Glyceria maxima*, and *Lythrum salicaria*. In a fed-batch operation, the inflow values vary on the basis of wine processing seasons. In high season, it reaches 17,012 mg L^{-1} of chemical oxygen demand (COD), 1806 mg L^{-1} of biochemical oxygen demand (BOD_5), and 43,723 mg L^{-1} of total suspended solids (TSS) on average. Despite such high inlet concentrations, the removal efficiency (RE) for the

M. Šereš
DEKONTA a.s., Stehelčeves, Czech Republic

Institute for Environmental Studies, Faculty of Science, Charles University, Prague, Czech Republic

T. Hnátková (✉)
DEKONTA a.s., Stehelčeves, Czech Republic

Department of Applied Ecology, Faculty of Environmental Sciences, Czech University of Life Sciences Prague, Prague, Czech Republic
e-mail: hnatkova@dekonta.cz

P. Maršík · T. Vaněk · P. Soudek
Laboratory of Plant Biotechnologies, Institute of Experimental Botany, Academy of Sciences of the Czech Republic, Prague, Czech Republic

J. Vymazal
Department of Applied Ecology, Faculty of Environmental Sciences, Czech University of Life Sciences Prague, Prague, Czech Republic

J. Filip et al. (eds.), *Advanced Nano-Bio Technologies for Water and Soil Treatment*, Applied Environmental Science and Engineering for a Sustainable Future,
https://doi.org/10.1007/978-3-030-29840-1_18

three selected parameters reached up to 99% for all the parameters. In this work, the removal of four pharmaceuticals (diclofenac, ibuprofen, ketoprofen, and naproxen) in the hybrid system was studied as well. The CW performed high RE for all of the four pharmaceuticals when the RE of diclofenac, ibuprofen, ketoprofen, and naproxen was 81.1%, 93.2%, 96.7%, and 78.3%, respectively.

Keywords Agro-industrial wastewater · Hybrid constructed wetlands · Macrophytes · Pharmaceuticals · NSAIDs

18.1 Introduction

Constructed wetlands (CWs) is an efficient technology for treatment of wide range of wastewaters, which has been used for more than 50 years (Vymazal 2011a). Domestic and municipal wastewater is usually treated using CWs (Brix and Arias 2005; Kadlec and Wallace 2009), but other types of wastewater such as agro-industrial wastewater (Cronk 1996; Knight et al. 2000; Sultana et al. 2015) can be effectively treated by CWs as well. Constructed wetlands are also used for treating diffuse agricultural pollution (Kasak et al. 2018; Mendes et al. 2018) and they can be used for pesticide mitigation from non-point sources as well (Vymazal and Březinová 2015). In the Czech Republic, the last reported number of CWs for municipal wastewater treatment was 250 (Vymazal 2011b); however, we estimate that the number has increased up to 400 constructed wetlands recently. Municipal and agro-industrial wastewater might contain many different pharmaceuticals and personal care products (PPCPs) some of which can be considered endocrine disruptors (EDs). Degradation of such compounds in constructed wetlands was also tested in many studies (e.g., Ávila and García 2015; Hakk et al. 2018).

18.1.1 *Hybrid Constructed Wetlands*

Many different types of constructed wetlands such as free water surface CWs, subsurface flow CWs with upward flow, etc., were developed and used during the 50-year history of the CWs application. However, currently the most common ones are wetlands with subsurface horizontal (also called horizontal filters or horizontal flow constructed wetlands—HF CWs) (Vymazal 2011c) and subsurface vertical flow (also called vertical filters or vertical flow constructed wetlands—VF CWs) (Brix and Arias 2005). The difference of VF lies in filtration medium fractions distribution, distribution and drainage pipes placement, and most importantly in the water flow direction. VFs are commonly fed intermittently with large batches of wastewater (Vymazal 2016). The combination of HF and VF in the so-called hybrid systems is currently becoming more and more popular (Vymazal 2013). The main reason for using the hybrid system is to exploit the advantages of each filter in order to fulfill the requirements for the quality of the discharged water and especially to fulfill nitrogen concentration limits (Tunçsiper 2009). HF CWs are anaerobic and provide the necessary conditions to attain denitrification, while the intermittently fed VF CWs are aerobic, thus enabling nitrification (Vymazal and Kröpfelová 2015). The removal of total nitrogen in hybrid CWs may amount to 85% (Kantawanichkul et al. 2001). The hybrid CW Kotenčice, Czech Republic, was built in 2011 in Central Bohemia and consists of various HF beds with the total area of 911 m^2 and VF beds with the total area of 300 m^2 being alternately connected (Hudcová et al. 2013). This hybrid CW has been designed to test the optimum combination of the HF and VF beds. The results from this CW shows that nitrogen was reduced up to 79% in the summer period and up to 62% in the winter period.

18.1.2 CWs for Treatment of Agro-Industrial Wastewaters

Constructed wetlands are used, among others, for treating agricultural wastewaters, especially wastewaters from animal feedlots (Knight et al. 2000) or drainage waters (Xue et al. 1999; Poe et al. 2003). The problem of feedlot wastewater is that the organics content characterized by biochemical oxygen demand (BOD_5) is usually one order of magnitude higher in domestic waters (Cronk 1996). Healy et al. (2007) reported BOD_5 concentrations reaching up to 7130 mg L^{-1} (the average of seven observed CWs treating concentrated animal waste). Dairy, swine, and cattle wastewater respectively are also rich in total suspended solids (TSS). Knight et al. (2000) reports the TSS concentrations reaching up to 1111 mg L^{-1}, 128 mg L^{-1}, and 291 mg L^{-1}, respectively. Total nitrogen concentrations from pig farms, poultry farms, and dairies averaged 407 mg L^{-1}, 89 mg L^{-1}, and 103 mg L^{-1}, respectively. Agricultural wastewater also includes food processing. Such wastewaters can be characterized by high loads of organics. Especially chemical oxygen demand (COD) loads can reach up to 7406 mg L^{-1} during the wine making season; however, even during the rest of the year it can be as high as 1721 mg L^{-1} (Grismer et al. 2003). Some authors report even higher concentrations—up to 15,400 mg L^{-1} (Grismer et al. 2003) or 25,400 mg L^{-1} (Šereš et al. 2017)—in winemaking facilities. The results of the review done by Sultana et al. (2015) show that the use of hybrid systems is desirable for treating such waters because VF CWs are able to treat extremely high loads of pollution.

18.1.3 Emerging Pollutants and CWs

Due to the increased consumption of pharmaceutical products, there is a rising concern regarding the harmful effects of such substances on the environment (Reif et al. 2008). The concerns are also connected with the overuse of veterinary pharmaceuticals and other active substances belonging to the so-called pharmaceutical and personal care products (PPCPs). The persistence of pharmaceutical residues and their metabolites in the environment is one of the most significant problems that can lead to a cumulative effect on the nontarget organisms in the aquatic environment (Han et al. 2010). One of the most frequently used groups of drugs that can be often found in water environments are so-called nonsteroidal anti-inflammatory drugs (NSAIDs) (Wang et al. 2011). Some studies already examined the elimination capacity of constructed wetlands for most common NSAIDs—ibuprofen, diclofenac, naproxen, and ketoprofen (Hijosa-Valsero et al. 2010). In the groundwater analysis performed in the Czech Republic, these four substances were the most abundant pharmaceuticals in the collected samples (Helenkár et al. 2010; Kotowska et al. 2014). Marsik et al. (2017) confirm significant concentrations of the NSAIDs in the Elbe river and its tributaries. In this work, the efficiency of the hybrid constructed wetland Chrámce, Ústí region, in the removal of basic chemical parameters (COD, BOD_5, TSS, N, and P) and NSAIDs from mixed domestic and agro-industrial wastewater was evaluated.

18.2 Materials and Methods

18.2.1 Experimental Plant Description

This study was performed in a hybrid constructed wetland Chrámce, Ústí region, Czech Republic. This CW receives sewage water from a residential building (negligible part of the total flow); wastewaters from fruit, fruit juice, and wine processing (the total area of orchards and vineyards is 195 ha and 1.9 ha, respectively); and wastewater from a sheep farm (150 sheep in total). The total calculated capacity of the CW is 50 population equivalent (PE) and the average calculated flow through the system is 0.09 L s^{-1}, which represents 7.8 m^3 d^{-1}. The calculated theoretical retention time is 10.5 days and the tracers tests that were performed according to Kadlec and Wallace (2009) using a KBr indicator confirmed the mean retention time to be 10 days. The CW has been operating since 2012. The inflow is not continuous, but the CW is batch fed by vacuum trucks.

The system consists of eight treatment stages (see Fig. 18.1): a reservoir for wastewater (3 × 10 m^3), a three-chamber septic tank (14.1 m^3), a series of three wetland filters—horizontal filter 1 (HF1), vertical filter (VF), and horizontal filter 2 (HF2). Tertiary treatment of wastewater is performed in three small stabilization ponds with littoral zones connected with a shallow meandering stream (SP1–SP3). The treated water is discharged into the existing pond. More detailed characteristics are summarized in Table 18.1.

Filters are planted with reed canary grass (*Phalaris arundinacea*) and common reed (*Phragmites australis*). Meandering streams and littoral zones are planted with sweet manna grass (*Glyceria maxima*), purple loosestrife (*Lythrum salicaria*), yellow iris (*Iris pseudacorus*), Siberian iris (*Iris sibirica*), sweet flag (*Acorus calamus*), beaked sedge (*Carex rostrata*), and broadleaf cattail (*Typha latifolia*).

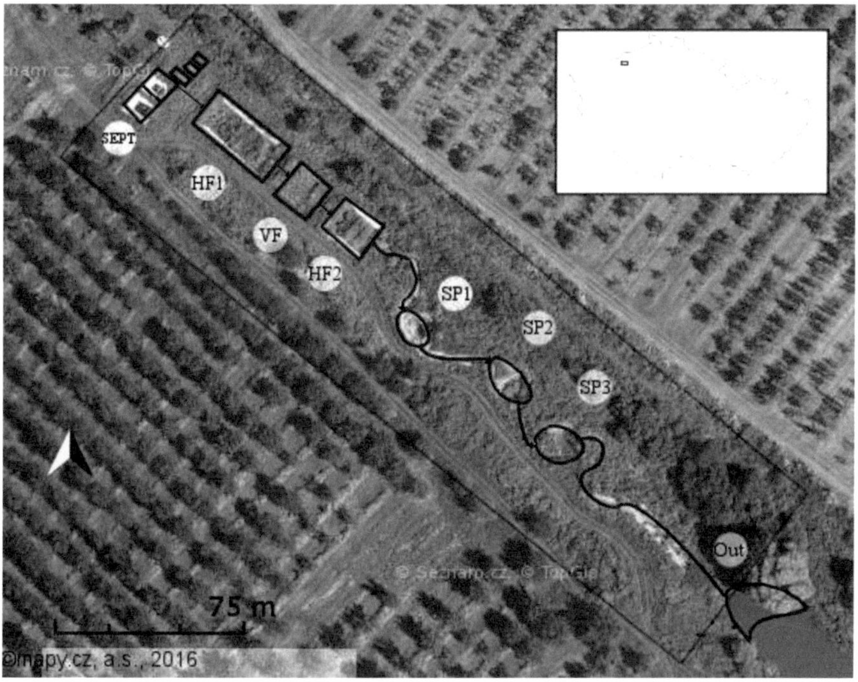

Fig. 18.1 Map and layout of the hybrid constructed wetland in Chrámce

Table 18.1 Major design parameters of constructed wetland

Parameter	Effective area (m^2)	Depth (m)	Filtration material*	Volume (m^3)	HLR (cm/d)
HF1	133	0.7	a, b, c	97.2	3.8
VF	49	1.3	a, b, c, d	50.4	10.2
HF2	70	0.9	a, b, c, e	48.6	7.1

Adapted from Šereš et al. (2017)

*(a) Washed gravel 2–4 mm (protective material); (b) washed gravel 4–8 mm, porosity: 0.45, conductivity 16 cm/s; (c) washed gravel 8–16 mm, porosity: 0.44, conductivity 94 cm/s; (d) washed gravel 32–64 mm, porosity: 0.46, conductivity 350 cm/s; (e) slag 8–16 mm, porosity: 0.51, conductivity 47 cm/s

18.2.2 CW Performance Evaluation

Sampling is performed at ten sampling points—two samples in septic tank (1st and 3rd chamber), two samples in HF1 (middle and outlet), other six samples are taken at the outflow from the outlets at each resting stage (VF, HF2, SP1–SP3, and final discharge point). For the treatment efficiency evaluation, the following parameters regulated by Czech laws were monitored: COD, BOD_5, TP, TN, NH_4^+, TSS. The samples were analyzed according to the standard methods (APHA 1998). Detailed sampling was performed on a weekly basis during the periods September–October 2013 (n = 10 samples) during the high production season. Sampling then continued during the period April–October 2014 (n = 17 samples) and February–October 2015 (n = 17 samples).

18.2.3 NSAIDs Removal Evaluation

For the evaluation of NSAIDs removal in the constructed wetland, an in vitro analytical method was developed and an analysis of these antiphlogistics in a real system was performed thereafter. Metabolic potential of cell cultures of *Melilotus officinalis* and *Rheum palmatum* in vitro was verified in *Phragmites australis* cells, tissues, and whole plants, and finally tested in the CW. Some measurements of NSAIDs concentrations in this CW were reported by Marsik et al. (2015).

The sampling and preparation procedure of the samples was done according to Marsik et al. (2015). The samples were collected in the period June–September 2014 and February–June 2015 (15 samples in total) from the same sampling points as mentioned above. Wastewater samples were analyzed using UPLC-MS/MS. For the analysis, the samples were first acidified with acetic acid at pH 2.5 and then they were filtered through a nylon membrane filter (0.45 μm). For quantitative analysis, system Q-Trap 4000 (AB Sciex, USA) was used. The sample (25 mL) was applied on solid phase extraction columns (Strata C8 (55 μm, 70 A), 500 mg/3 mL, Phenomenex, USA) preconditioned with 5 mL of methanol and 10 mL of deionized water. The columns were then washed with acidified water (acetic acid, pH 2) and eluted with 5 mL of methanol. The samples were then evaporated under stream of nitrogen to dryness and stored at −80 °C. Before the analysis, the samples were dissolved in 1 mL of methanol and then dissolved 100× for analysis.

18.3 Results and Discussion

18.3.1 Performance of the Hybrid Constructed Wetland

In 2013 the monitoring was focused mainly on the period of wine harvesting and processing. In Table 18.2, mean concentrations of COD, BOD_5, and TSS are summarized. The results show that the discharge limits (see in Table 18.3) were fulfilled at the outflow from the CW filters; however, the quality of the water decreased in the stabilization ponds when the COD increased from 40 mg L^{-1} at the outlet from HF2 to 181 mg L^{-1} at the outlet from SP3. As for the TSS, the C^{out} increased from average 3 mg L^{-1} at the outlet from HF2 to 1047 mg L^{-1}at the outlet from SP3. This increase was caused mainly by the growth of algae and decay of plants in the stabilization ponds (see Fig. 18.2). Shepherd et al. (2001) reported very

Table 18.2 CW performance after 1 year of operation (September–November 2013) in high production season

	CW filters[a]			Stabilization ponds[b]	
Parameter	C^{in} (mg/L)	C^{out} (mg/L)	RE (%)	C^{out} (mg/L)	RE[c] (%)
COD	17,012 ± 11,862	40 ± 22	96.3 ± 5	181 ± 135	83.6 ± 22.3
BOD_5	1806 ± 1179	1 ± 0.5	99.7 ± 0.3	14 ± 15	98.5 ± 1.8
TSS	43,723 ± 30,797	3 ± 1	99.6 ± 0.5	1047 ± 1121	20.1 ± 100

[a]CW filters (HF1-VF-HF2); [b]Stabilization ponds (SP1–SP3); [c]Total value of RE after flowing through SP

Table 18.3 Discharge limits set for the constructed wetland Chrámce

Parameter	Mean limit (mg/L)	Max limit (mg/L)	Removal limit (%)
COD	110	170	70–75%
BOD_5	30	50	80–85%
TSS	40	60	90–95%

Fig. 18.2 Algae bloom observed stabilization ponds (2013)

high COD concentrations, rising up to 45,500 mg L^{-1}, in winery wastewaters and the CW treating such wastewaters still provided great removal efficiency (RE), up to 99%.

The quality of agro-industrial wastewaters varies during the year. Such variations can be observed even here if we compare the inlet concentrations of all the parameters in the high season of 2013 (Table 18.2) and average of the off-seasons in years

Table 18.4 CW performance after 2 and 3 years of operation (2014–2015)

	Root filters[a]			Stabilization ponds[b]	
Parameter	C^{in} (mg/L)	C^{out} (mg/L)	RE (%)	C^{out} (mg/L)	RE[c] (%)
COD	477 ± 442	65 ± 101	73 ± 32	42 ± 77	77 ± 42
BOD_5	226 ± 228	14 ± 23	86 ± 19	3 ± 1	89 ± 21
TSS	240 ± 416	3 ± 1	76 ± 47	5 ± 6	76 ± 46
N_{total}	33 ± 21	4 ± 4	86 ± 10	3 ± 3	87 ± 15
P_{total}	10 ± 14	2 ± 7	75 ± 44	0.1 ± 0.1	97 ± 3
Ammonia	25 ± 17	0.1 ± 0.1	98 ± 4	0.1 ± 0.1	98 ± 4

[a]CW filters (HF1-VF-HF2); [b]Stabilization ponds (SP1–SP3); [c]Total value of RE after flowing through SP

2014 and 2015 (Table 18.4). We can observe for COD 35× higher C^{in}, for BOD_5 8× higher C^{in}, and for TSS even 200× higher C^{in}.

The observation shows that during the study period 2014 and 2015 the average elimination efficiency of the hybrid systems fulfilled the legislative requirements for most of the observed parameters. The results are also comparable with other hybrid constructed wetlands (Vymazal 2010, 2013) and previous studies done on this CW (Šereš et al. 2017). We observed relatively high inlet concentrations of COD and BOD_5; however, the outflow concentrations were very low already at the outflow from the CW filters (65 and 15 mg L^{-1} respectively) and decreased even more at the outflow from the stabilization ponds (42 and 3 mg L^{-1} respectively). In comparison with the year 2013, we can observe a decrease in the removal efficiency for COD, BOD_5, and TSS, yet we can observe lower inlet concentrations. As some authors explains, the percentage efficiency depends on the inflow concentration, and, thus, lower C^{in} can cause lower RE (Ghermandi et al. 2007). We can also observe that COD and BOD were reduced from 52 and 56% reduced in HF1 (see Fig. 18.3). The only parameter that in average outlet concentrations did not fulfill the legislative limit was the TSS removal efficiency. Mean C^{in} of TSS were reduced by only 76% in both stages (wetland filters and stabilization ponds). Nevertheless, the outflow concentrations were far below the discharge limits (3–5 mg L^{-1}). The higher outlet concentrations of the suspended solids might be caused by the presence of planktonic algae in the stabilization ponds (Hijosa-Valsero et al. 2012) or by filtration material decomposition (Ghermandi et al. 2007).

As for the nitrogen removal, we can observe that total nitrogen was eliminated by 87%. From Fig. 18.3, we can see that almost 65% of N_{total} that was formed mainly by ammonia N was eliminated in the first horizontal filter and almost 86% of N_{total} was eliminated after the flow through the whole hybrid system. The stabilization ponds made an insignificant contribution to the nitrogen removal. Very high nitrogen removal is common (Vymazal 2013) compared to single HF or VF CWs. Vymazal and Kröpfelová (2015) report similar overall RE of N_{total} (79.9%) in their three-stage CW (VF-VF-HF).

Ammonia N was removed from wastewater mostly in HF1 (61%) and the overall removal reached 98%. Although hybrid CWs are efficient in the removal of total

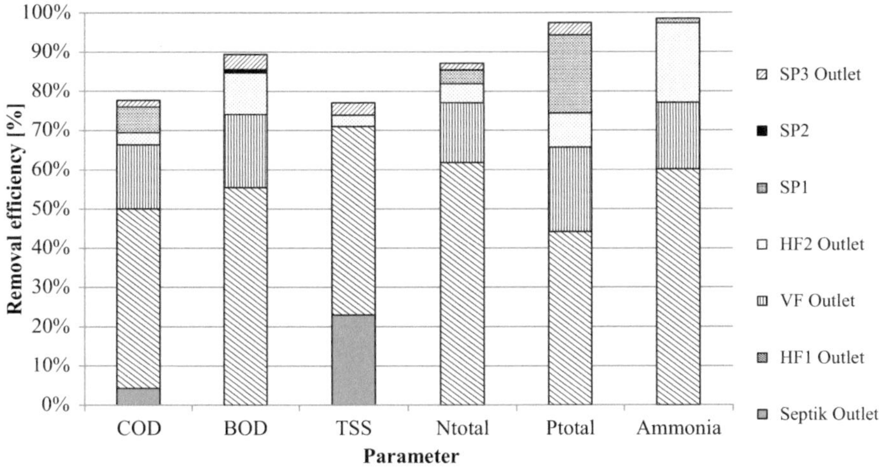

Fig. 18.3 Mean removal efficiencies for the monitored chemical parameters in each stage of the CW

nitrogen, the elimination of ammonia is usually promoted in aerobic conditions of vertical filters. Such high ammonia removal might be explained by higher exploitation of anammox processes in the CW filter environment (Tao et al. 2012). For removal of phosphorus in CWs, mostly physio-chemical processes are applied (Abe et al. 2010), but macrophytes could also be important in phosphorus storage in the short term (Wang et al. 2018). Phosphorus removal of instream CWs usually reaches up to 62% (Kasak et al. 2018); however, Zhu et al. (2012) report the P_{total} removal reaching up to 95% from the livestock wastewater containing 17.99 mg L^{-1}. In our study, we observed the P removal reaching 97% with the average inlet concentrations of 10 mg L^{-1}.

18.3.2 Elimination of NSAIDs in Hybrid CW Chrámce

In this study, the elimination of four selected NSAIDs in hybrid intermittently fed CW was evaluated. In Fig. 18.4, the concentrations observed in each treatment step of the CW are summarized for each NSAID. These results show that ibuprofen, diclofenac, ketoprofen, and naproxen are efficiently degraded and accumulated after flowing through the whole wetland system and all the analyzed pharmaceuticals at the outflow from the stabilization ponds were on average close to the detection limits. The average removal of diclofenac was 81% (see Fig. 18.4) that corresponds with the findings of Hijosa-Valsero et al. (2010) who report the average removal of diclofenac in wetland treatment systems 65–87% while conventional activated sludge wastewater treatment plant removes diclofenac only by 44% (Ávila and García 2015). The average inlet concentration of diclofenac in our study was 389.6 ng L^{-1} and the C^{out} was in average 21.3 ng L^{-1}. Vystavna et al. (2017)

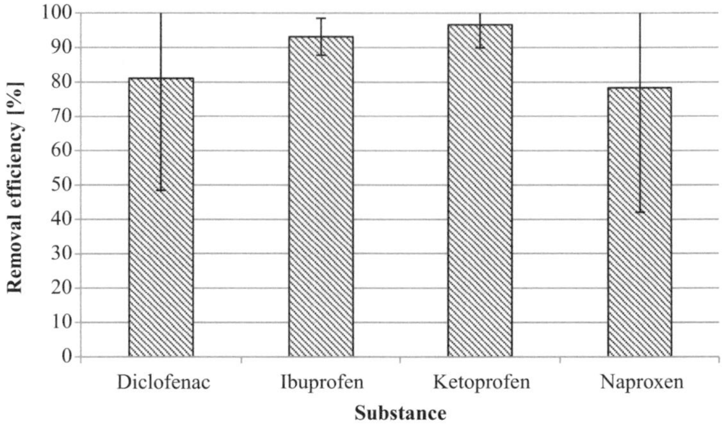

Fig. 18.4 Removal efficiency of four selected NSAIDs in hybrid CW

reported the inlet concentrations of diclofenac to the four-stage CW treating hospital wastewaters to be 1862 ng L^{-1}. Removal efficiency of this CW for diclofenac was higher than 50%. The mean inlet concentration 361.4 ng L^{-1} of ibuprofen was in our system removed by 93.2% from which 40% corresponds to the vertical filter. Ávila et al. (2015) report the removal of ibuprofen reaching 100% also mainly in the VF. Matamoros and Salvadó (2012) report the removal efficiency of ibuprofen, ketoprofen, and naproxen to be 93%, 98%, and 88%, respectively, which is similar to our study. In our study we also observed the highest RE for ketoprofen (97%) and the lowest for naproxen (78%). In the work of Vystavna et al. (2017), the highest removal efficiency was reached for naproxen (>80%) and only above 50% for other observed substances, which conflicted with our findings. However, in their work, the inlet concentration of naproxen was only 0.7 ng L^{-1}.

From Fig. 18.5, we can observe that concentrations of all four substances were reduced mainly after the VF, which is in line with the observations of Ávila and García (2015) who propose that aerobic pathways are more efficient in degradation of PPCPs. Nevertheless, the stabilization ponds played a major role in reaching the RE reported above. The removal of the selected pharmaceuticals was probably performed by combination of the processes. It is expected that some portion of pharmaceuticals might be assimilated by plants; however, Zhang et al. (2013) reported that the naproxen uptake by the wetland plants reached only 4% of the total substance mass; therefore, other processes such as biodegradation must be used here. Also Matamoros et al. (2008) suggest that ibuprofen is removed by degradation rather than by sorption. Biodegradability of all the observed substances is proven also by Yu et al. (2006). In the graph below, we can observe that the concentration of all the substances slightly increased in the first horizontal filter. As we can observe, the removal of NSAIDs did not occur in HF1; therefore, concentrating of substances in this filter might have taken place in the long term and the biodegradation occurred after the VF.

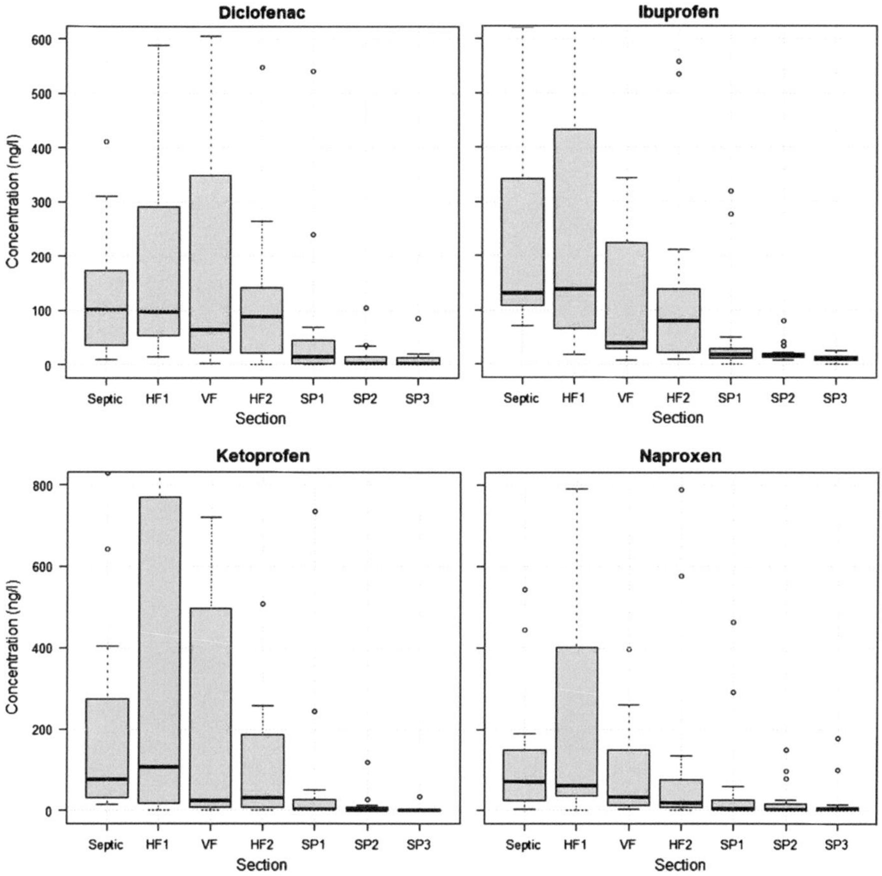

Fig. 18.5 Boxplot diagram of four NSAIDs (diclofenac, ibuprofen, ketoprofen, naproxen) in each CW section

18.4 Conclusions

The hybrid constructed wetland in Chrámce treating mixed household and agro-industrial wastewaters coming mainly from sheep farm and wine and juice production facility performed high treatment efficiencies. The main conclusions are:

- The removal efficiency for COD, BOD_5, and TSS in high wine processing season is 96.3%, 99.7%, and 99.6, respectively, in wetland filters but overall removal efficiency is only 83.6%, 98.5%, and 20.1%, respectively, due to algae growth and decay of plants in the stabilization ponds.
- The average RE of the system in seasons 2014–2015 of COD, BOD_5, and TSS was 77%, 89%, and 76% and the mean outlet concentrations easily fulfilled the discharge limits.

- Nitrogen was effectively removed in the hybrid CW because of the combination of three subsurface wetlands with various feeding strategy. The removal of N_{total} reached 87%. The ammonia removal reached 97% and more work should be done in the future to discover if anammox might take place in the ammonia removal.
- Monitoring four widely used human pharmaceuticals from the group of nonsteroidal anti-inflammatory drugs showed very high removal efficiency of the hybrid CW. The average removal efficiency for diclofenac, ibuprofen, ketoprofen, and naproxen, was 81.1%, 93.2%, 96.7%, and 78.3%, respectively.

Acknowledgements The study was partly supported also by the Technology Agency of the Czech Republic, grant no. TA01020573 "Biotechnological system for treatment of wastewaters from agriculture and their recycling".

References

Abe K, Komada M, Ookuma A (2010) Purification performance of the FWS constructed wetland in biotope area over three years. In: Gilkes R, Prakongkep N (eds) 19th World Congress of Soil Science; Soil Solutions for a Changing World, Brisbane, Australia, 2010. International Union of Soil Sciences, pp 18–21

APHA (1998) Standard methods for the examination of water and wastewater, 20th edn. American Public Health Association, Washington, DC

Ávila C, García J (2015) Chapter 6 – Pharmaceuticals and Personal Care Products (PPCPs) in the environment and their removal from wastewater through constructed wetlands. In: Zeng EY (ed) Comprehensive analytical chemistry, vol 67. Elsevier, pp 195–244. https://doi.org/10.1016/B978-0-444-63299-9.00006-5

Ávila C, Bayona JM, Martín I, Salas JJ, García J (2015) Emerging organic contaminant removal in a full-scale hybrid constructed wetland system for wastewater treatment and reuse. Ecol Eng 80:108–116. https://doi.org/10.1016/J.ECOLENG.2014.07.056

Brix H, Arias CA (2005) The use of vertical flow constructed wetlands for on-site treatment of domestic wastewater: new Danish guidelines. Ecol Eng 25(5):491–500. https://doi.org/10.1016/j.ecoleng.2005.07.009

Cronk JK (1996) Constructed wetlands to treat wastewater from dairy and swine operations: a review. Agric Ecosyst Environ 58(2–3):97–114. https://doi.org/10.1016/0167-8809(96)01024-9

Ghermandi A, Bixio D, Thoeye C (2007) The role of free water surface constructed wetlands as polishing step in municipal wastewater reclamation and reuse. Sci Total Environ 380(1–3):247–258. https://doi.org/10.1016/j.scitotenv.2006.12.038

Grismer ME, Carr MA, Shepherd HL (2003) Evaluation of constructed wetland treatment performance for winery wastewater. Water Environ Res 75(5):412–421. https://doi.org/10.2175/106143003X141213

Hakk H, Sikora L, Casey FXM (2018) Fate of estrone in laboratory-scale constructed wetlands. Ecol Eng 111:60–68. https://doi.org/10.1016/j.ecoleng.2017.11.005

Han S, Choi K, Kim J, Ji K, Kim S, Ahn B, Yun J, Choi K, Khim JS, Zhang X, Giesy JP (2010) Endocrine disruption and consequences of chronic exposure to ibuprofen in Japanese medaka (*Oryzias latipes*) and freshwater cladocerans *Daphnia magna* and *Moina macrocopa*. Aquat Toxicol 98(3):256–264. https://doi.org/10.1016/j.aquatox.2010.02.013

Healy MG, Rodgers M, Mulqueen J (2007) Treatment of dairy wastewater using constructed wetlands and intermittent sand filters. Bioresour Technol 98(12):2268–2281. https://doi.org/10.1016/j.biortech.2006.07.036

Helenkár A, Sebők Á, Záray G, Molnár-Perl I, Vasanits-Zsigrai A (2010) The role of the acquisition methods in the analysis of the non-steroidal anti-inflammatory drugs in Danube River by gas chromatography – mass spectrometry. Talanta 82(2):600–607. https://doi.org/10.1016/j.talanta.2010.05.014

Hijosa-Valsero M, Matamoros V, Martín-Villacorta J, Bécares E, Bayona JM (2010) Assessment of full-scale natural systems for the removal of PPCPs from wastewater in small communities. Water Res 44(5):1429–1439. https://doi.org/10.1016/j.watres.2009.10.032

Hijosa-Valsero M, Sidrach-Cardona R, Bécares E (2012) Comparison of interannual removal variation of various constructed wetland types. Sci Total Environ 430:174–183. https://doi.org/10.1016/j.scitotenv.2012.04.072

Hudcová T, Vymazal J, Kriška Dunajský M (2013) Reconstruction of a constructed wetland with horizontal subsurface flow after 18 years of operation. Water Sci Technol 68(5):1195–1202. https://doi.org/10.2166/wst.2013.374

Kadlec RH, Wallace SD (2009) Treatment wetlands, 2nd edn. CRC Press, Boca Raton

Kantawanichkul S, Neamkam P, Shutes RBE (2001) Nitrogen removal in a combined system: vertical vegetated bed over horizontal flow sand bed. Water Sci Technol 44(11–12):137–142. https://doi.org/10.2166/wst.2001.0820

Kasak K, Kill K, Pärn J, Mander Ü (2018) Efficiency of a newly established in-stream constructed wetland treating diffuse agricultural pollution. Ecol Eng 119:1–7. https://doi.org/10.1016/j.ecoleng.2018.05.015

Knight RL, Payne VWE Jr, Borer RE, Clarke RA Jr, Pries JH (2000) Constructed wetlands for livestock wastewater management. Ecol Eng 15(1–2):41–55. https://doi.org/10.1016/S0925-8574(99)00034-8

Kotowska U, Kapelewska J, Sturgulewska J (2014) Determination of phenols and pharmaceuticals in municipal wastewaters from Polish treatment plants by ultrasound-assisted emulsification-microextraction followed by GC-MS. Environ Sci Pollut Res 21(1):660–673. https://doi.org/10.1007/s11356-013-1904-6

Marsik P, Soudek P, Hudcova T, Syrovatka J, Vanek T (2015) Pharmaceuticals in domestic and agricultural waste waters – problem and its solution. Water Pract Technol 10(3):564–572. https://doi.org/10.2166/wpt.2015.065

Marsik P, Rezek J, Židková M, Kramulová B, Tauchen J, Vaněk T (2017) Non-steroidal anti-inflammatory drugs in the watercourses of Elbe basin in Czech Republic. Chemosphere 171:97–105. https://doi.org/10.1016/j.chemosphere.2016.12.055

Matamoros V, Salvadó V (2012) Evaluation of the seasonal performance of a water reclamation pond-constructed wetland system for removing emerging contaminants. Chemosphere 86(2):111–117. https://doi.org/10.1016/j.chemosphere.2011.09.020

Matamoros V, Caselles-Osorio A, García J, Bayona JM (2008) Behaviour of pharmaceutical products and biodegradation intermediates in horizontal subsurface flow constructed wetland. A microcosm experiment. Sci Total Environ 394(1):171–176. https://doi.org/10.1016/j.scitotenv.2008.01.029

Mendes LRD, Tonderski K, Iversen BV, Kjaergaard C (2018) Phosphorus retention in surface-flow constructed wetlands targeting agricultural drainage water. Ecol Eng 120:94–103. https://doi.org/10.1016/j.ecoleng.2018.05.022

Poe AC, Piehler MF, Thompson SP, Paerl HW (2003) Denitrification in a constructed wetland receiving agricultural runoff. Wetlands 23(4):817–826. https://doi.org/10.1672/0277-5212(2003)023[0817:DIACWR]2.0.CO;2

Reif R, Suárez S, Omil F, Lema JM (2008) Fate of pharmaceuticals and cosmetic ingredients during the operation of a MBR treating sewage. Desalination 221(1–3):511–517. https://doi.org/10.1016/j.desal.2007.01.111

Šereš M, Hnátková T, Vymazal J, Vaněk T (2017) Removal efficiency of constructed wetland for treatment of agricultural wastewaters. Chem J Mol 12(1):45–52. https://doi.org/10.19261/cjm.2017.403

Shepherd HL, Grismer ME, Tchobanoglous G (2001) Treatment of high-strength winery wastewater using a subsurface-flow constructed wetland. Water Environ Res 73(4):394–403. https://doi.org/10.2175/106143001X139434

Sultana M-Y, Akratos CS, Vayenas DV, Pavlou S (2015) Constructed wetlands in the treatment of agro-industrial wastewater: a review. Hem Ind 69(2):127–142. https://doi.org/10.2298/HEMIND150121018S

Tao W, He Y, Wang Z, Smith R, Shayya W, Pei Y (2012) Effects of pH and temperature on coupling nitritation and anammox in biofilters treating dairy wastewater. Ecol Eng 47:76–82. https://doi.org/10.1016/J.ECOLENG.2012.06.035

Tunçsiper B (2009) Nitrogen removal in a combined vertical and horizontal subsurface-flow constructed wetland system. Desalination 247(1–3):466–475. https://doi.org/10.1016/j.desal.2009.03.003

Vymazal J (2010) Constructed wetlands for wastewater treatment. Water 2(3):530–549. https://doi.org/10.3390/w2030530

Vymazal J (2011a) Constructed wetlands for wastewater treatment: five decades of experience. Environ Sci Technol 45(1):61–69. https://doi.org/10.1021/es101403q

Vymazal J (2011b) Constructed wetlands in the Czech Republic: 20 years of experience. In: Vymazal J (ed) Water and nutrient management in natural and constructed wetlands. Springer Netherlands, Dordrecht, pp 169–178. https://doi.org/10.1007/978-90-481-9585-5_13

Vymazal J (2011c) Long-term performance of constructed wetlands with horizontal sub-surface flow: ten case studies from the Czech Republic. Ecol Eng 37(1):54–63. https://doi.org/10.1016/j.ecoleng.2009.11.028

Vymazal J (2013) The use of hybrid constructed wetlands for wastewater treatment with special attention to nitrogen removal: a review of a recent development. Water Res 47(14):4795–4811. https://doi.org/10.1016/j.watres.2013.05.029

Vymazal J (2016) Constructed wetlands for water quality regulation. In: Finlayson CM, Everard M, Irvine K et al (eds) The wetland book: I: structure and function, management and methods. Springer Netherlands, Dordrecht, pp 1–8. https://doi.org/10.1007/978-94-007-6172-8_234-2

Vymazal J, Březinová T (2015) The use of constructed wetlands for removal of pesticides from agricultural runoff and drainage: a review. Environ Int 75:11–20. https://doi.org/10.1016/j.envint.2014.10.026

Vymazal J, Kröpfelová L (2015) Multistage hybrid constructed wetland for enhanced removal of nitrogen. Ecol Eng 84:202–208. https://doi.org/10.1016/j.ecoleng.2015.09.017

Vystavna Y, Frkova Z, Marchand L, Vergeles Y, Stolberg F (2017) Removal efficiency of pharmaceuticals in a full scale constructed wetland in East Ukraine. Ecol Eng 108:50–58. https://doi.org/10.1016/j.ecoleng.2017.08.009

Wang C, Shi H, Adams CD, Gamagedara S, Stayton I, Timmons T, Ma Y (2011) Investigation of pharmaceuticals in Missouri natural and drinking water using high performance liquid chromatography-tandem mass spectrometry. Water Res 45(4):1818–1828. https://doi.org/10.1016/j.watres.2010.11.043

Wang M, Zhang D, Dong J, Tan SK (2018) Application of constructed wetlands for treating agricultural runoff and agro-industrial wastewater: a review. Hydrobiologia 805(1):1–31. https://doi.org/10.1007/s10750-017-3315-z

Xue Y, Kovacic DA, David MB, Gentry LE, Mulvaney RL, Lindau CW (1999) In situ measurements of denitrification in constructed wetlands. J Environ Qual 28(1):263–269. https://doi.org/10.2134/jeq1999.00472425002800010032x

Yu JT, Bouwer EJ, Coelhan M (2006) Occurrence and biodegradability studies of selected pharmaceuticals and personal care products in sewage effluent. Agric Water Manag 86(1–2):72–80. https://doi.org/10.1016/j.agwat.2006.06.015

Zhang DQ, Hua T, Gersberg RM, Zhu J, Ng WJ, Tan SK (2013) Carbamazepine and naproxen: fate in wetland mesocosms planted with *Scirpus validus*. Chemosphere 91(1):14–21. https://doi.org/10.1016/j.chemosphere.2012.11.018

Zhu D, Sun C, Zhang H, Wu Z, Jia B, Zhang Y (2012) Roles of vegetation, flow type and filled depth on livestock wastewater treatment through multi-level mineralized refuse-based constructed wetlands. Ecol Eng 39:7–15. https://doi.org/10.1016/j.ecoleng.2011.11.002

Chapter 19
Field Study VII: Field Study of Three Different Injectable Oxygen Sources to Enhance Mono-Aromatic Solvents In Situ Biodegradation

Ondřej Lhotský

Abstract The present field study describes a pilot remediation test of calcium peroxide, modified calcium peroxide, and gelatinous hydrogen peroxide as oxygen sources to enhance mono-aromatic solvents in situ biodegradation. The field study was carried out in well permeable aquifer contaminated mainly by benzene, toluene, and chlorobenzene. Three different Oxygen Release Compounds (ORCs) were injected near three different monitoring wells situated in the center of the contamination plume. The injections were performed via direct push top-down method. Each oxygen source tested was injected into two direct push probes in close vicinity of respective monitoring well. Membrane interface probe (MIP) investigation was utilized for targeting of the oxygen sources injections into most contaminated layers in the subsurface. Although there was a high heterogeneity of the conditions prevailing in the respective injection areas, it can be stated that the injection of all the materials enhanced the in situ biodegradation and led to significant drops in the concentrations of the contaminants. Therefore, in situ biodegradation without bioaugmentation can be considered a feasible method for the clean-up of the tested site.

Keywords Biodegradation · ORC · Direct-push · MIP

19.1 Introduction

Mono-aromatic solvents (e.g., BTEX and monochlorobenzene) are widespread groundwater and soil contaminants (Kao et al. 2006; Nijenhuis et al. 2007) and their presence in the environment often poses a threat to both human health and

O. Lhotský (✉)
DEKONTA a.s., Prague, Czech Republic

Institute for Environmental Studies, Faculty of Science, Charles University, Prague, Czech Republic
e-mail: lhotsky@dekonta.cz

J. Filip et al. (eds.), *Advanced Nano-Bio Technologies for Water and Soil Treatment*, Applied Environmental Science and Engineering for a Sustainable Future,
https://doi.org/10.1007/978-3-030-29840-1_19

ecosystems. Therefore, it is generally required to remove these contaminants from the underground environment.

Among the possible remediation strategies that can be implemented, bioremediation is one of the most frequently used because of its lower equipment demand and labor and energy requirements compared with classical physicochemical techniques. This concept results in lower costs and increased sustainability of the remediation works (US EPA 2012). Comprehensive characterization of site-specific biodegradation processes is essential to verify whether in situ biodegradation is feasible and to determine whether these processes could be efficient enough or can be enhanced in order to replace conventional clean-up technologies (Bombach et al. 2010).

This work is focused on the assessment of the possibilities to enhance in situ biodegradation of mono-aromatic solvents via injection of three different injectable oxygen sources on a typical environmental burden site in the Czech Republic.

19.2 Materials and Methods

19.2.1 The Locality

The studied locality is situated on the premises of an old chemical and pharmaceutical plant where chemical production began almost 100 years ago. The entire site is highly contaminated mainly by mono-aromatic solvents including monochlorobenzene (**MCB**), benzene (**B**), and toluene (**T**) as well as other contaminants. The field study was carried out on one of the most contaminated spots in the area. The aquifer is developed in Quaternary sandy gravels with a clayey admixture with saturated thickness of approximately 4 m. The hydraulic conductivity of the aquifer ranges from 1.4×10^{-3} to 2.2×10^{-4} m/s, based on the Mallet-Pacquant method (Chery and de Marsily 2007). The base of the aquifer consists of clays with a sand admixture with a hydraulic conductivity below 1.9×10^{-7} m/s. The groundwater is characterized by elevated mineralization (specific conductance ranging from 127 to 1053 mS/m), slightly acidic to neutral pH (6.5–7.0), oxidation–reduction potentials ranging from −45 to +45 mV, dissolved oxygen below 0.2 mg/L, and a temperature of approximately 14.2 °C. The concentrations of the contaminants just before the test varied as follows—**B** 6.7–57.2 μg/L, **T** 41.3–3245 μg/L, and **MCB** 451–2840 μg/L. Old monitoring wells installed approximately 15 years prior to the test were utilized for monitoring the process. The wells had long screen interval along the whole aquifer thickness.

19.2.2 Site Investigation

A membrane interface probe (MIP) (Geoprobe Systems®, USA) was used for the investigation of the contamination close to the particular monitoring wells. One MIP borehole was performed per one monitoring well. The MIP provides semi-quantitative data on the presence of volatile organic compounds (VOCs) in the soil and it allows delineation of the VOCs source (see Chap. 31 for details). Dual tube soil sampling (Geoprobe Systems®, USA) was later utilized to collect four discrete soil samples in accordance with the standard guide (ASTM 1998) in the nearby soil monitoring probe. The material was also collected for in situ microcosm experiments.

19.2.3 Remediation Pilot Test Setup

Three different oxygen release compounds were injected near three different monitoring wells (approximately 25 m from one another) situated in the center of the contamination plume. The injections were performed via direct push top-down method into the subsurface using Geoprobe 7822 DT drilling rig together with Geoprobe Pressure-Activated Injection Probe (Geoprobe Systems®, USA) and Wanner G10 diaphragm pump (Wanner Engineering, Inc., USA). ORCs tested were as follows: **calcium peroxide** (purity of 75%; Ixper 75c, Solvay, Belgium) as a typical compound used as oxygen source; **modified calcium peroxide** – 1:1 mixture of calcium hydroxide oxide and calcium peroxide (ORC Advanced® and ORC® Primer, Regenesis, USA) that should provide oxygen release in longer term; and **gelatinous hydrogen peroxide** (OXYGEL, Biorem, Belgium) which slowly releases hydrogen peroxide from the gel matrix. The oxygen release materials were injected directly into the probes (injection flowrate and pressure 30 L/min, 6–20 bar, respectively). Four to five injection horizons per probe were performed on the basis of the MIP profile from the nearby MIP probe. An example of the targeted injection scheme based on the MIP profile is shown in Fig. 19.1.

The materials, regardless of their actual chemical structure, were injected at the amount capable of releasing 17 kg of O_2 per borehole. In case of calcium peroxide and modified calcium peroxide, this means that 100 kg of material diluted in 500 L of water taken from the aerated bioreactor used within the pump-and-treat technology installed further on site was injected per one injection point. Gelatinous hydrogen peroxide contains 10.6% of releasable oxygen and therefore 160 kg of the product mixed with 140 L of water taken from the aerated bioreactor was injected per one injection point.

Mixing water from the aerated bioreactor contained significant numbers of bacteria capable of aerobic biodegradation, although these bacteria were probably significantly inhibited by the high pH of the suspensions (and presence of hydrogen peroxide), some might have survived and later helped to enhance biodegradation of target contaminants. The scheme of the field study layout is displayed in Fig. 19.2. The distance between the injection points and the monitoring well was approximately 1.5 m. MIP and soil monitoring probes were approximately 0.75 m far from the monitoring well.

MIP and soil monitoring probes were performed prior to the injection and the holes were sealed by cement afterwards.

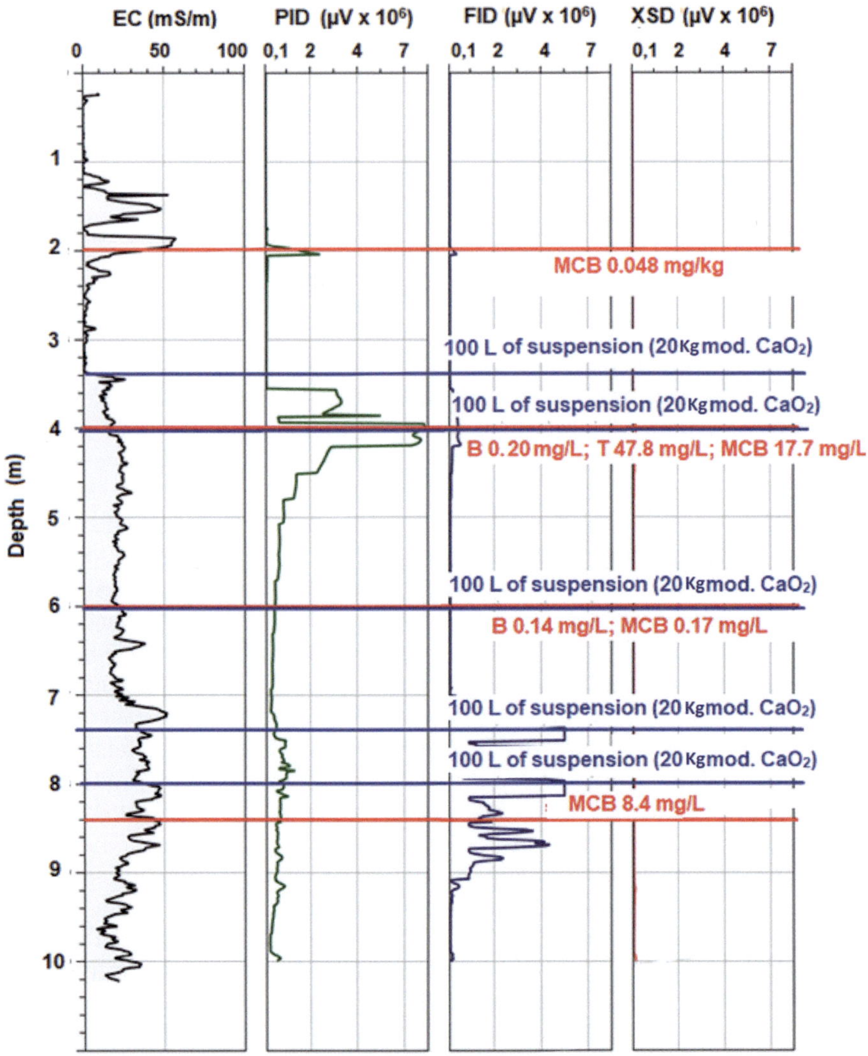

Fig. 19.1 Modified CaO_2 injection scheme based on the MIP profile. Red text indicates concentration of contaminants analyzed in discrete soil samples taken from the nearby soil monitoring probe. Blue text indicates volume of the suspension injected per horizon and the respective amount of the oxygen source

19.2.4 Monitoring

The monitoring wells were monitored prior to and after the injection of the oxygen sources for physicochemical parameters. Biological and chemical monitoring started just before the injection of the oxygen sources.

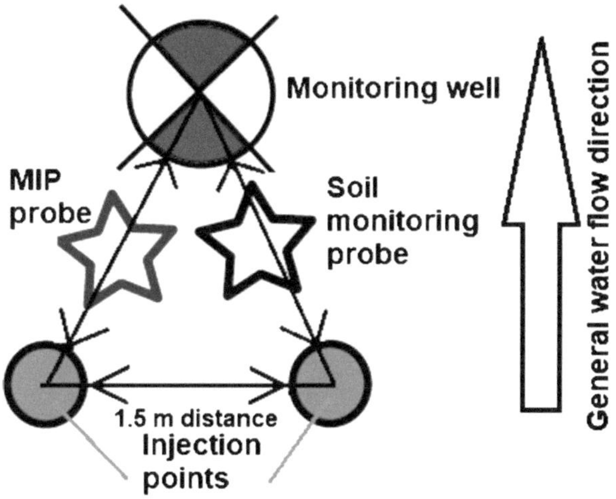

Fig. 19.2 General field study layout

19.2.5 Groundwater Samples

Gigant (Eijkelkamp, The Netherlands) submersible sampling pump placed close to the bottom of the monitoring well was used in order to take samples for phospholipid fatty acid analysis (PLFA) and chemical parameters analysis as well as for determination of physicochemical parameters. Approximately three borehole volumes of groundwater were pumped out from the well in accordance with standard procedure prior to sampling (ISO 2009). The physicochemical parameters were recorded using a Multimeter WTW-3430 (WTW, Germany) during sampling and the final stable values were recorded.

19.2.6 In Situ Microcosms for Measurement of the Microbiological Populations in Soil

Perforated 50 mL centrifuge plastic tubes filled with the 0.125 mm to 4 mm fraction of the soil collected during the direct push investigation of the locality and wrapped in polyethylene protective mesh were used as the microcosms. The tubes were installed in the lower part of the monitoring wells and collected when needed for monitoring and sent to the laboratory for PLFA in order to monitor development of microbial populations.

19.2.7 Analytical Methods

VOCs were analyzed in a commercial laboratory using standard methods. PLFA were analyzed by the Institute of Microbiology of the Czech Academy of Sciences. Detailed description of the utilized analytical methods can be found in Lhotský et al. (2017a, b).

19.3 Results and Discussion

19.3.1 Calcium Peroxide

The initial concentrations of the contaminants in the respective monitoring well just few hours before the start of injections were as follows—**B** 32.4 μg/L; **T** 87.8 μg/L; **MCB** 1386 μg/L. The development of the monitored parameters during the field study is shown in Fig. 19.3.

19.3.1.1 Physicochemical Parameters During the Remediation Pilot Test

As described above, calcium peroxide was injected into two injection points in total amount of 200 kg in 1000 L of water from the bioreactor. After the injections it was found out that calcium peroxide penetrated in to the respective monitoring well causing a significant increase in the pH up to pH 11 one day after the injection. Ten days after the injection the pH dropped to the neutral range and it remained more or less stagnant throughout the rest of the test. The E_h rose for one day after the injection up to +225 mV and from that moment on it was gradually decreasing to +75 mV 147 days after the injection. Oxygen concentrations also rose significantly but the maxima were reached 10 days after the injection reaching up to 7.73 mg/L. Later on, it decreased again and reached values similar to those prior to the test 56 days after the injection. CaO_2 injection also caused a gradual increase in the groundwater temperature (21 days after the injection, data not shown) up to +1.1 °C compared with the situation before the injection. Later on, the temperature decreased gradually. One day after the injection a significant increase in conductivity (data not shown) was also observed; however, it decreased again to values similar to those prior to the injections.

19.3.1.2 Contaminant Concentrations Throughout the Test

The concentrations of the contaminants are shown in C/C_0 scale so that it can be easily compared with the data obtained with other oxygen sources. The concentrations of **T** and **MCB** significantly rose for 21 days after the injection because of the pressure pulse caused by the injection while the concentration of **B** decreased. This was probably caused by the fact that both **T** and **MCB** were partly bound to the soil

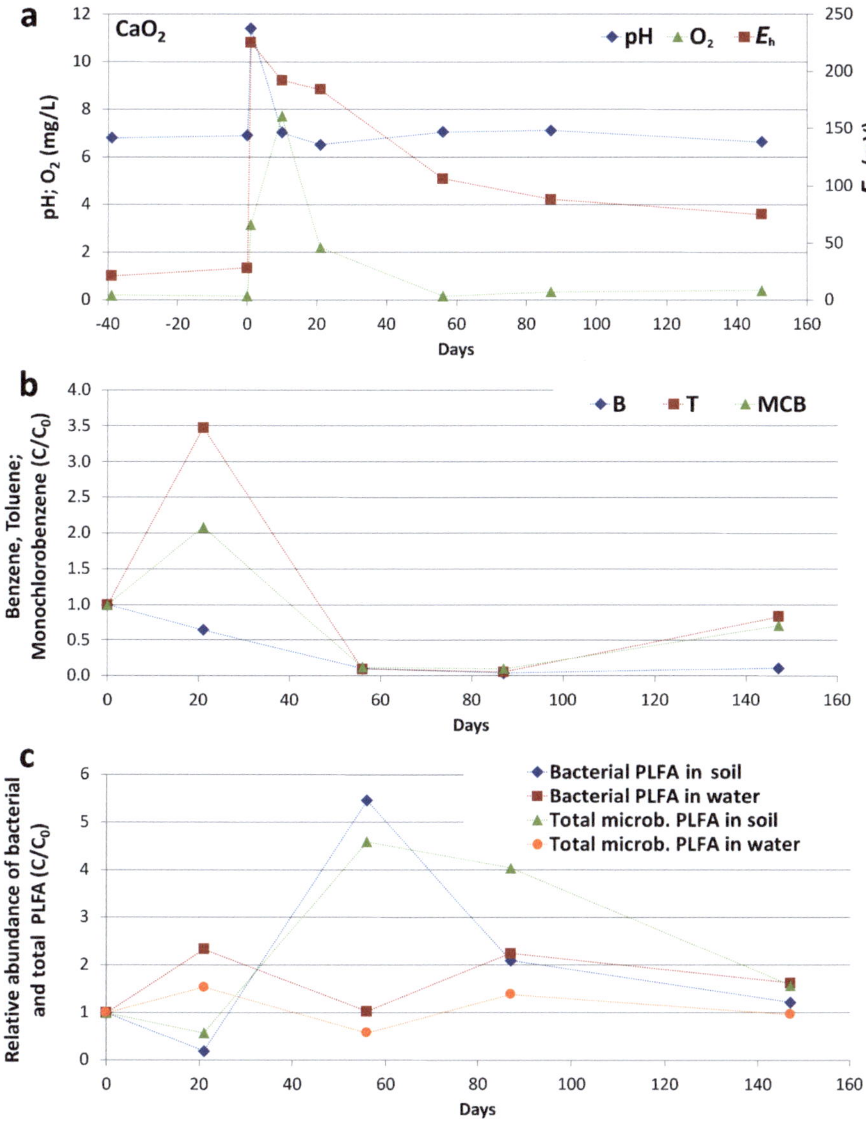

Fig. 19.3 Calcium peroxide - The course of (**a**) physicochemical parameters, (**b**) main contaminant concentrations, and (**c**) PLFA measured in the water and in soil of in situ microcosms during the pilot test

matrix (see Fig. 19.3) while **B** was mainly present dissolved in the groundwater. 56 days after the injection the sum of the contaminants concentration decreased below 12% of their initial value and remained low the whole period of 87 days after the injection. 147 days after the injection **MCB** and **T** concentration rose again to 71% and 83% of their initial concentrations while **B** remained low. On the basis of the results of the tests, it can be assumed that the increase in concentrations was caused mainly by the influx of the contaminated water from the surrounding untreated areas rather than by a rebound effect.

19.3.1.3 Microbial Biomass

The concentration of the respective PLFA is shown as C/C_0 so that it can be easily compared with the data obtained with other oxygen sources. It can be seen that the water samples provide different results compared with those from the in situ microcosms. In the case of the water samples, variation of the results was significant while in the case of soil, it was clear that the initial high pH caused by the presence of CaO_2 in the monitoring well had negative impact on the bacteria present in the well even 21 days after the injection. In the following days the bacterial biomass in the microcosms grew substantially to reach the maximum 56 days after the injection; later on, it gradually decreased although even 147 days after the injection it was still elevated in comparison with the initial values.

19.3.2 Modified Calcium Peroxide

The initial concentrations of contaminants in the respective monitoring well just few hours before the start of the injections were as follows—**B** 57.2 μg/L; **T** 3245 μg/L; **MCB** 2840 μg/L. The development of the monitored parameters during the field study is shown in Fig. 19.4.

19.3.2.1 Physicochemical Parameters During the Remediation Pilot Test

Modified CaO_2 application results are shown in Fig. 19.4. This material did not penetrate into the monitoring well and its impact on the pH values was rather low and gradual as the pH reached 7.14, 87 days after the injection. The effects on the E_h were also less pronounced and more gradual compared with common CaO_2, the maximum of +134 mV was achieved 21 days after the injection and it remained elevated even 87 days after the injection. In the case of the O_2 concentration, the effects were similar and a gradual increase to values around 0.4 mg/L was observed. Afterwards, the O_2 concentration remained stagnant until 87 days after the injection. 147 days after the injection the O_2 concentration dropped to values comparable with the ones measured prior to the injection. In the case of groundwater temperature, a slight gradual increase (data not shown) up to +2.4 °C was observed 10 days after the injection. Later on, the temperature decreased gradually. No significant effect of the injection on conductivity (data not shown) was observed.

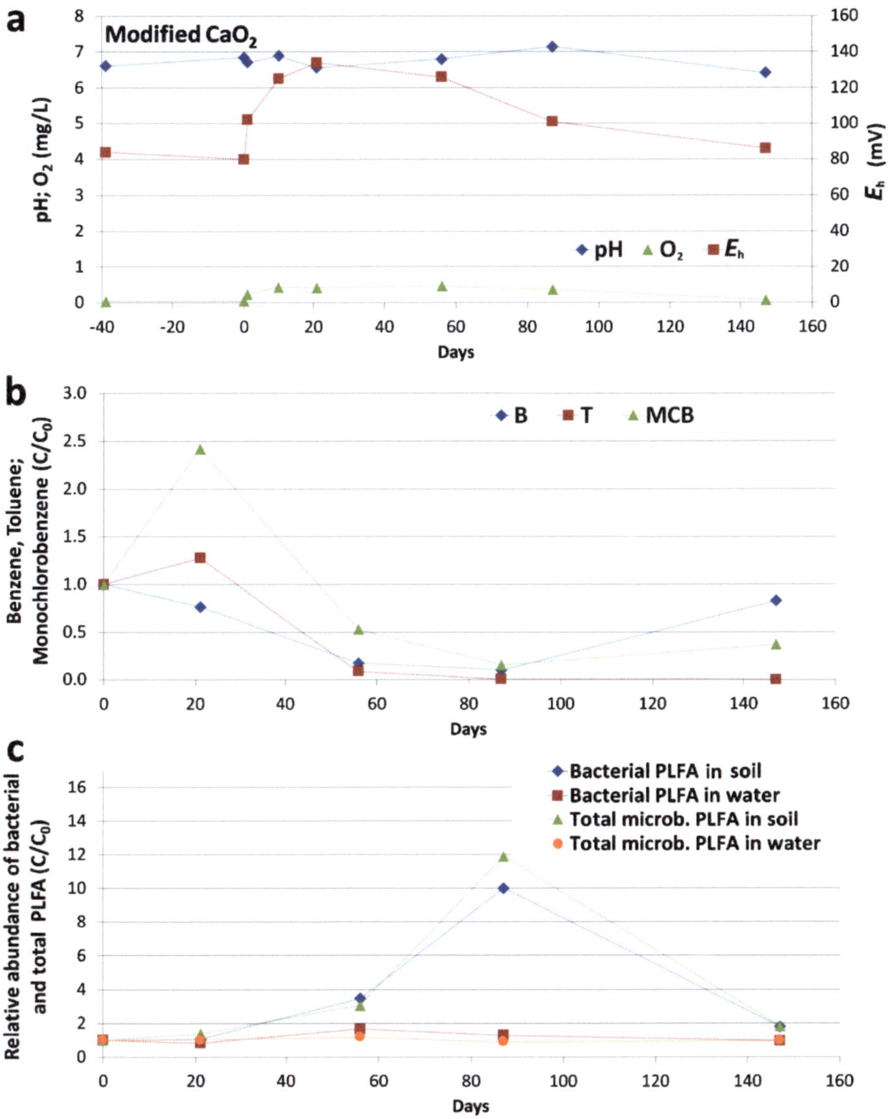

Fig. 19.4 Modified calcium peroxide - The course of (**a**) physicochemical parameters, (**b**) main contaminant concentrations, and (**c**) PLFA measured in the water and in soil of in situ microcosms during the pilot test

19.3.2.2 Contaminant Concentrations Throughout the Test

Similarly to the injection of common CaO_2, the concentrations of **T** and **MCB** rose 21 days after the injection because of the pressure pulse caused by the injection while the **B** concentration decreased slightly. 56 days after the injection the concentration of **B** and **T** decreased below 18% of their initial values while **MCB** decreased only to

53%. 87 days after the injection the contaminants concentrations were below 16% of their initial value. 147 days after the injection **MCB** and **B** concentration rose again to 37% and 83% of their initial concentration, respectively, while **T** almost disappeared.

19.3.2.3 Microbial Biomass

In the case of the water samples, the PLFA concentrations remained more or less stagnant. In the case of the soil from the in situ microcosms, a gradual increase of bacterial biomass was observed to reach their maxima 87 days after the injection. At this time, approximately ten times higher abundance of bacterial biomass compared with the initial values was observed. 147 days after the injection the bacterial biomass was still almost two times higher compared with the initial values.

19.3.3 Gelatinous Hydrogen Peroxide

The initial concentrations of the contaminants in the respective monitoring well just few hours before the start of the injections were as follows—**B** 6.7 μg/L; **T** 41.3 μg/L; **MCB** 451 μg/L. The course of the monitored parameters during the field study is shown in the graphs below (Fig. 19.5). Although pure hydrogen peroxide gel did not penetrate into the monitoring well, low H_2O_2 concentrations were detectable in the well (up to 0.5 mg/L) 21 days after the injection.

19.3.3.1 Physicochemical Parameters During the Remediation Pilot Test

H_2O_2 gel injection caused very slight increase in the pH a day after the injection. Later on, it was more or less stagnant. On the other hand, the effect on the E_h was significant as it increased from −45 mV to +166 mV per day. Later on, it gradually decreased to levels around 0 mV 147 days after the injection. The O_2 concentration rose to 2.2 mg/L one day after the injection and decreased quickly to the original background values after 21 days. The groundwater temperature (data not shown) rose 1 day after the injection to +3.7 °C and remained elevated even 10 days after the injection. Afterward, the temperature decreased gradually. No significant effect of the injection on the conductivity (data not shown) was observed.

19.3.3.2 Contaminant Concentrations Throughout the Test

Unlike the case of the solid oxygen sources, injection contaminants concentration decreased 21 days after the injection. We assume the contaminant mobilization

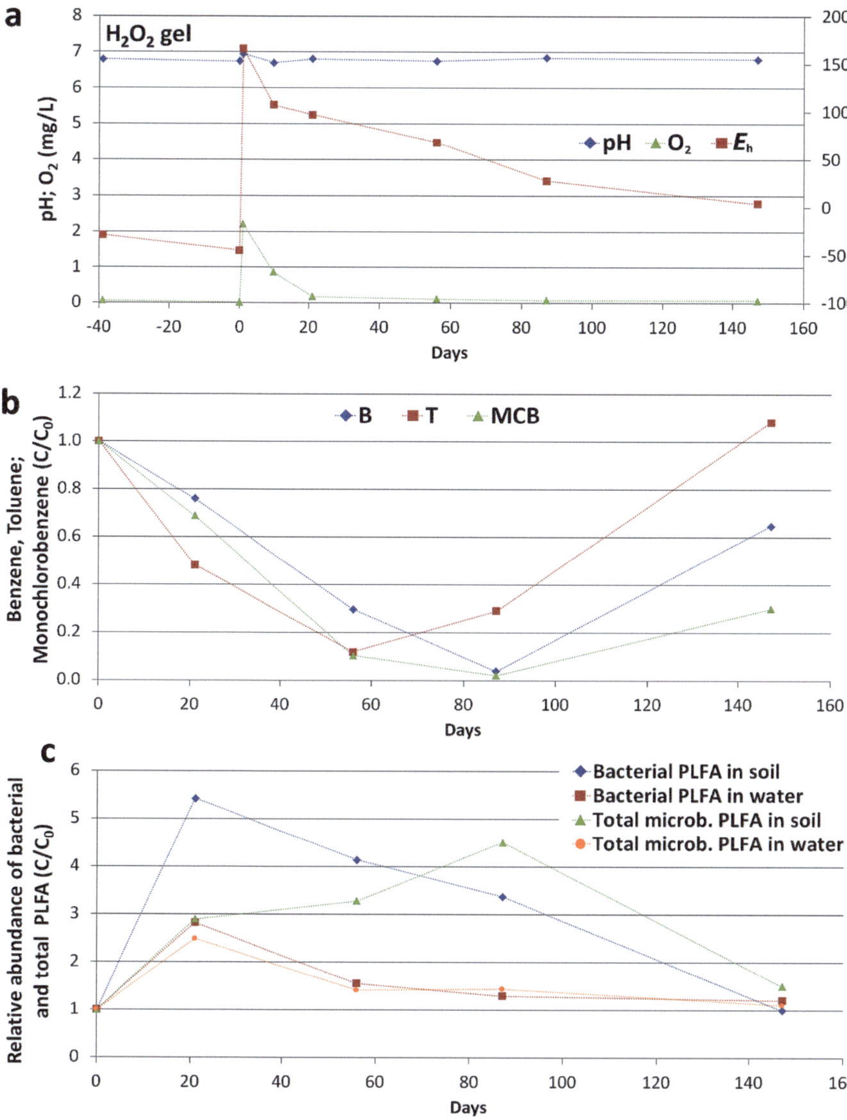

Fig. 19.5 Gelatinous hydrogen peroxide - The course of (**a**) physicochemical parameters, (**b**) main contaminant concentrations, and (**c**) PLFA measured in the water and in soil of in situ microcosms during the pilot test

caused by the pressure pulse was eliminated mainly by a quick chemical oxidation (Fenton's reaction) of the contaminants that occurred in situ because of high concentration of dissolved iron in the groundwater. Later on, bacteria probably contributed to the subsequent decrease of the contamination when the **B** concentration decreased to 30% of its initial value while the **T** and **MCB** decreased to less than 12%. 87 days after the injection, the **B** and **MCB** amounts were below 4% while the **T** concentration started to rise. 147 days after the injection the **MCB** and **B** concentration rose again to 30% and 65% of their initial concentrations while the **T** concentration was higher than the initial values.

19.3.3.3 Microbial Biomass

Both water and soil samples reached their maxima of the PLFA concentrations 21 days after the injection. The abundance of the bacterial biomass in the soil was 5.5 times higher compared with the initial values while in the water it was 2.5 times higher. Subsequently, a gradual decrease of the bacterial biomass was observed to reach more or less similar levels like the initial ones 147 days after the injection.

19.4 Conclusion

Although it is difficult to compare the tested oxygen sources as the conditions prevailing in the respective injection areas were different, it can be stated that the injection of all the materials enhanced the in situ biodegradation and led to significant drops in the concentrations of the contaminants. In all the cases, the bacterial biomass more substantially increased in soil microcosms compared with the bacteria present in the aqueous phase. Therefore, in situ biodegradation without bioaugmentation can be considered a feasible method for the clean-up of the tested sites.

Some general remarks on the performance of the different materials can be also drafted. The longevity of the oxygen sources tested on this particular site was as follows—modified calcium peroxide > calcium peroxide > gelatinous hydrogen peroxide. While gelatinous hydrogen peroxide was probably mainly consumed on the Fenton's ISCO reactions and it released oxygen for only limited amount of time, modified calcium peroxide was capable of releasing oxygen for more than 87 days. Standard calcium peroxide provided oxygen release for about 1 month. Quick consumption of the oxygen sources might be caused by high iron concentrations in the groundwater. In the case of calcium peroxide, high pH of the material can inhibit bacterial growth for several days or weeks after an injection. High H_2O_2 concentrations caused mainly by gelatinous hydrogen peroxide injection can also inhibit bacteria temporary.

References

ASTM (1998) Standard guide for direct push soil sampling for environmental site characterizations. ASTM D6282–98. ASTM International, West Conshohocken. https://doi.org/10.1520/D6282-98

Bombach P, Richnow HH, Kästner M, Fischer A (2010) Current approaches for the assessment of in situ biodegradation. Appl Microbiol Biotechnol 86(3):839–852. https://doi.org/10.1007/s00253-010-2461-2

Chery L, de Marsily G (eds) (2007) Aquifer systems management: Darcy's legacy in a world of impending water shortage. Selected papers on hydrogeology 10, 1st edn. CRC Press, London

ISO (2009) Water quality—sampling—part 11: guidance on sampling of groundwaters. ISO 5667-11:2009

Kao CM, Huang WY, Chang LJ, Chen TY, Chien HY, Hou F (2006) Application of monitored natural attenuation to remediate a petroleum-hydrocarbon spill site. Water Sci Technol 53 (2):321–328. https://doi.org/10.2166/wst.2006.066

Lhotský O, Krákorová E, Linhartová L, Křesinová Z, Steinová J, Dvořák L, Rodsand T, Filipová A, Kroupová K, Wimmerová L, Kukačka J, Cajthaml T (2017a) Assessment of biodegradation potential at a site contaminated by a mixture of BTEX, chlorinated pollutants and pharmaceuticals using passive sampling methods – case study. Sci Total Environ 607–608:1451–1465. https://doi.org/10.1016/j.scitotenv.2017.06.193

Lhotský O, Krákorová E, Mašín P, Žebrák R, Linhartová L, Křesinová Z, Kašlík J, Steinová J, Rødsand T, Filipová A, Petrů K, Kroupová K, Cajthaml T (2017b) Pharmaceuticals, benzene, toluene and chlorobenzene removal from contaminated groundwater by combined UV/H2O2 photo-oxidation and aeration. Water Res 120:245–255. https://doi.org/10.1016/j.watres.2017.04.076

Nijenhuis I, Stelzer N, Kästner M, Richnow H-H (2007) Sensitive detection of anaerobic monochlorobenzene degradation using stable isotope tracers. Environ Sci Technol 41 (11):3836–3842. https://doi.org/10.1021/es0621896

US EPA (2012) A citizen's guide to bioremediation. In: A citizen's guide to cleanup technologies, 2nd ed.; US EPA: EPA 542-F-12-003

Chapter 20
Nano-Bioremediation: Nanoscale Zero-Valent Iron for Inorganic and Organic Contamination

Jaroslav Semerád, Martin Pivokonský, and Tomáš Cajthaml

Abstract The present chapter is describing two pilot-scale experiments and showing the potential of the combined technology (i.e., nanoremediation and bioremediation) for in situ heavy metal removal and dechlorination of chlorinated solvents. Successful applications of the nano-bioremediation approach at two different sites in the Czech Republic contaminated with hexavalent chromium are summarized and evaluated in this chapter. High effectivity of nZVI and the further microbial bioaugmentation step by whey addition resulted in reduction and geofixation of hexavalent chromium in both localities. In addition, a high removal of chlorinated ethenes was observed in the locality 2 after the whey application. The successive abiotic and biotic processes involving the chemical reduction and further microbial bioreduction of chromium and dechlorination of chlorinated ethenes were evaluated by monitoring of physicochemical and biological parameters. The results of both applications of the combined remediation technology clearly prove feasibility and high efficiency of this approach for chromium and chlorinated ethenes removal.

Keywords Nano-bioremediation · Bioreduction · Geofixation · Ecotoxicity · Hexavalent chromium

20.1 Introduction—Two Field Studies

Since zero-valent iron nanoparticles have been invented, the number of published laboratory studies has reached several hundreds. The outcomes of these studies have proved the effectiveness of the above-mentioned nanoparticles and their composites for degradation and elimination of organic and inorganic contamination. In the light

J. Semerád (✉) · T. Cajthaml
Institute of Microbiology of the Czech Academy of Sciences, Prague, Czech Republic

Institute for Environmental Studies, Faculty of Science, Charles University, Prague, Czech Republic
e-mail: jaroslav.semerad@biomed.cas.cz

M. Pivokonský
Institute of Hydrodynamics of the Czech Academy of Sciences, Prague, Czech Republic

J. Filip et al. (eds.), *Advanced Nano-Bio Technologies for Water and Soil Treatment*, Applied Environmental Science and Engineering for a Sustainable Future,
https://doi.org/10.1007/978-3-030-29840-1_20

of this knowledge, the number of successful full-scale applications has been increasing for the past 15 years and the degradation effectivity of nZVI has also been confirmed by in situ applications (Mueller et al. 2012). Recent studies from the contaminated sites where nZVI was applied also reported its stimulation effect on selected microbial species and its potential for the combination of nanoremediation and bioremediation (Němeček et al. 2014, 2015, 2016). This chapter is summarizing two pilot-scale experiments done in the Czech Republic during which nZVI and whey as a microbial substrate were successfully applied for geofixation of Cr (VI) and dechlorination of chlorinated ethenes.

20.2 Locality 1

20.2.1 Sampling and Site Monitoring

The first test site is situated at the Kortan site in the north of the Czech Republic where a leather processing plant was set up in the past. Up to the 1990s, the factory used potassium dichromate for leather tanning whence the contamination with Cr (VI) comes. The current levels of Cr(VI) in groundwater are close to 3 mg/L. The aquifer is surrounded by a mixture of Quaternary sands and gravels with clayey admixture. The saturated zone of the aquifer is approximately 5 m thick; the groundwater table fluctuates at 4.5–5.5 m below the surface of the ground and the flow velocity staggers from 0.2 to 2 m/day. Finally, the groundwater discharges to a nearby river located approximately 500 m from the site. The physicochemical parameters of the groundwater reach values of pH 5.4, total organic carbon ⍰1.5 mg/L, the total dissolved solids ⍰0.3 g/L, and a high oxidation-reduction potential (450 to 550 mV).

The site monitoring was carried out by 19 sampling rounds during both phases, whereas groundwater samples of each round were collected from seven different wells. Before the pilot test, nine wells were drilled: three nZVI injection wells (PV-209, PV-213, and PV-215), three downgradient (PV-214, PV-216, PV-217) and one upgradient monitoring well (PV-212), which served as a reference. Besides the seven main wells, another two downgradient monitoring wells (PV-201 and PV-210) were sampled for additional analysis. The positions of the monitoring and the injection wells are displayed in Fig. 20.1.

20.2.2 Pilot Application of nZVI and Subsequential Whey Injection

The pilot test at the Kortan site consisted of two phases—abiotic and biotic. In summer 2012, during the first (abiotic) phase of the pilot test, 120 kg of nZVI

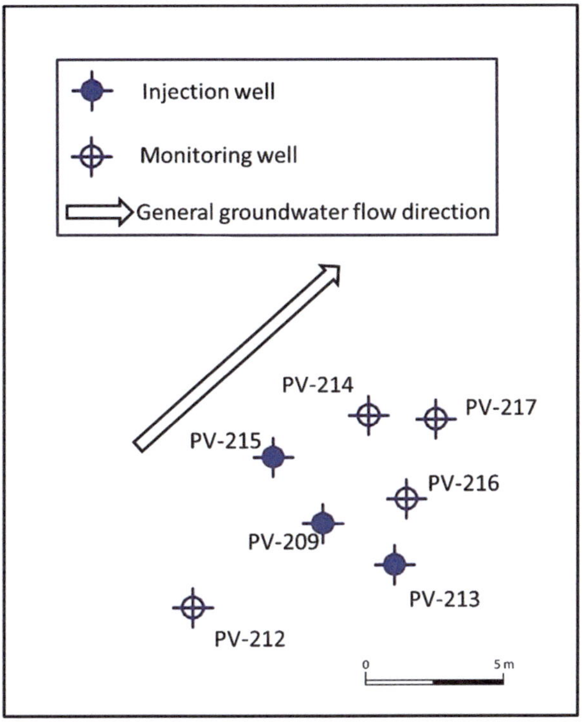

Fig. 20.1 Positions of monitoring and injection wells in locality 1. (Adapted from Němeček et al. 2014)

(NANOFER 25, NANO IRON, s.r.o., Czech Republic) in tap water (2 g/L) was applied into three injection wells (PV-209, PV-213, PV-215). The injection wells were situated in one line with a spacing of 2.8 m to form the barrier configuration. Nine months after the injection of nZVI, the biotic phase was set up. During the biotic phase, to support indigenous bacteria, 5 m^3 of cheese whey was injected into the wells as the microbial substrate. The substrate was mixed with groundwater in a volumetric ratio of 1:50 abstracted from the downgradient wells and then injected back into the upgradient injection wells. Afterward this circulation carried on for 50 days. At the end of the circulation process, the natural groundwater flow was restored. The whole pilot test consisting of both the abiotic and the biotic phase lasted approximately 21 months.

20.2.3 Evolution of Physicochemical Parameters and Concentration of Contaminants

The measurements of physicochemical parameters such as pH and oxidation-reduction potential (ORP) were carried out directly on the site electrochemically.

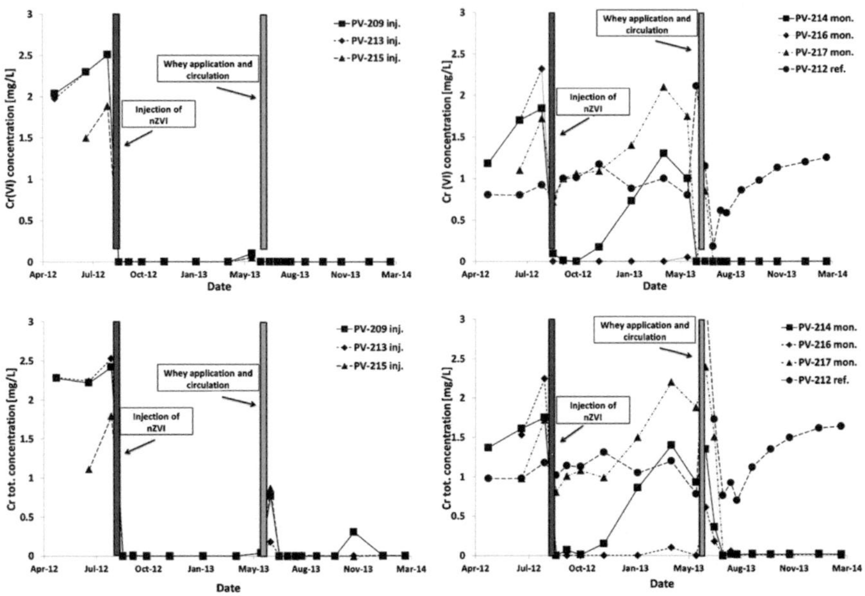

Fig. 20.2 The evolution of chromium contamination (Cr(VI) and Cr(total)) in groundwater during whole pilot-scale test. (Adapted from Němeček et al. 2015)

Other inorganic parameters were determined after several preparation steps in a laboratory. Additionally, the evolution in the Cr(VI) concentration dissolved in groundwater is displayed in Fig. 20.2.

From the results of the thorough site monitoring, it is clear that the immediate decrease of ORP was closely connected with the nZVI application. The fast decrease of ORP (down to −484 mV within one day in injection wells and − 330 mV in monitoring wells) and the reduction of dissolved Cr(VI) concentrations to the detection limits in all injection wells and the two nearest monitoring wells (PV-214 and PV-216) were observed. Additionally, in the distance of 7 m from the nZVI barrier (PV-217), the effect of the nZVI injection was less pronounced. The decrease in the concentrations of Cr(VI) was attributed to the reduction of Cr (VI) to Cr(III) by nZVI followed by the geofixation of reduced Cr(III) into the soil in the system of insoluble Fe(III)-Cr(III)-oxyhydroxides. Therefore an additional step was required, so the abiotic phase was followed by the biotic phase. After the abiotic phase, which was started by the injection of nZVI and lasted nine months, whey was applied according to the previously described circulation system in order to support the resident microbial species. The application of the organic substrate resulted in a gradual and substantial grow of bacteria followed by a further repeated decrease in the Cr(VI) concentrations in all the injection and the monitoring wells, see Fig. 20.2.

20.2.4 Microbial and Ecotoxicological Assessment of Both Pilot Test Phases

In both studies, Němeček et al. (2014 and 2015), the authors also provided detailed microbial and ecotoxicological assessment of the whole 21-month period of the pilot test duration. Using standard approaches such as microbial cultivation tests and inhibition of bioluminescence of *Vibrio fischeri*, the authors assessed the possible ecotoxicological impact of nZVI. The impact on autochthonous microbial population was also monitored by phospholipid fatty acid analysis (PLFA) and 16S rRNA 454 pyrosequencing. Overall, the authors did not observe any negative effect on the microbial population in the groundwater after the nZVI application and even growth stimulation of Gram-positive bacteria was observed in the soil samples using PLFA.

Notwithstanding the acute toxicity of nZVI detected during the in vitro tests (Semerád et al. 2018), the lack of negative or less pronounced toxic effect during in situ application corresponds with the findings of several other authors (Semerád and Cajthaml 2016). Bacterial stimulation was probably caused by the elimination of highly toxic Cr(VI) from the aquifer and thus by the decrease of overall toxicity during the first abiotic phase. Additionally, in the second phase, after nine months from nZVI application the microbial growth was stimulated by the addition of cheese whey and the increase of anaerobic bacteria contributed to the biotic reduction of residual Cr(VI). The direct or indirect effect of biotic reduction was clearly long-term and persisted even after ten months from the whey injection even when the substrate was completely depleted.

20.3 Locality 2

20.3.1 Sampling and Site Monitoring

The second locality is polluted by Cr(VI) and chlorinated ethenes originating from historical degreasing and chromium coating activities. The initial levels of contamination in the groundwater reached values from 4.4 to 57 mg/L for Cr(VI) and from 400 to 6526 μg/L for the sum of chlorinated ethenes at the beginning of the pilot test. Similarly to locality 1, the aquifer is situated in Quaternary sands and gravels with silty admixture and is covered by a layer of clay and clayey loam with a thickness of 5 m. The saturated zone of the aquifer is approximately 4 m thick and the hydraulic conductivity of the aquifer is 7.6×10^{-4} m/s. The groundwater discharges into the neighboring river at the distance of 430 m from the contaminated site. Physico-chemical parameters of the groundwater reached values of pH 6.9–7.0, total organic carbon ⍰1.0 to 5.4 mg/L, the total dissolved solids from 0.9 to 1.2 g/L, and oxidation-reduction potential ranging from +130 to −490 mV. Samples of the groundwater were collected during the whole pilot application from seven different monitoring

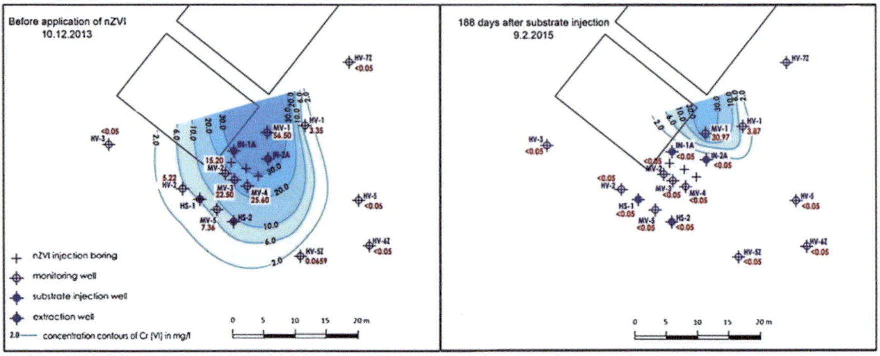

Fig. 20.3 Distribution of injection and monitoring wells and 2D model of Cr(VI) contamination of groundwater before and after the pilot test in locality 2. (Adapted from Němeček et al. 2016)

wells representing two downgradient lines (first line: MV-2, MV-3, MV-4 and the second line: MV-5, HV-2, HS-1, HS-2) and one upgradient well, which served as a control to provide background values (MV-1). In addition, well HV-8 contaminated only by chlorinated ethenes and located cross-gradient from the contamination source was sampled. The detailed distribution of wells in locality 2 is presented in Fig. 20.3.

Groundwater samples were collected from the monitoring wells and used for further analyses of various physical, chemical, and inorganic parameters, including microbial populations using PLFA. In addition, two types of passive samplers were emplaced in each monitoring well for more detailed evaluation of microbial populations and for solid phase analyses (oxidation state of Cr via X-ray photoelectron spectroscopy). The last type of sample was the sediment from five different monitoring wells (HV-8, MV-1, MV-3, MV-4, and MV-5), which was collected at the end of the pilot test and used for microbial community analysis.

20.3.2 *Application of Combined Nano-Biotechnology*

The pilot test in locality 2 included abiotic and biotic phases similarly to the test in locality 1 (Němeček et al. 2016). The abiotic phases consisted of two applications of nZVI at four months interval and lasted six months and a half. nZVI (NANO IRON, s.r.o., Czech Republic) was injected in a water suspension using direct push technology. During the first injection, surface-passivated nZVI (NANOFER STAR) was injected in a concentration of 1 g/L. For the second injection, the author used the same nZVI as in locality 1, i.e., NANOFER 25S in a concentration of 2 g/L. The injection boreholes were located perpendicularly to the groundwater flow, see Fig. 20.3.

The abiotic phase was followed by a biotic phase, which started approximately two months and a half after the second injection of nZVI. Cheese whey was used as the substrate owing to its successful application in locality 1 in order to support microbial populations. Whey was applied using a circulation system; the substrate was mixed with groundwater from downgradient monitoring wells (HS-1 and HS-2) and then injected back to the injection wells (IN-1 A and IN-2 A) at a rate of 0.5–0.7 L/s. During the biotic phase, 8.3 m^3 of whey was applied in total in a dosing ratio of 1/50 (substrate/groundwater). After the application period (five weeks) the circulation of groundwater continued for nine days and therefore the natural groundwater flow was reestablished and the locality was monitored for another five months.

20.3.3 Monitoring of Physicochemical Parameters and Levels of Contamination

Similarly to the pilot test in locality 1, the drop in the oxidation-reduction (to −400 mV) potential was observed after the first injection of nZVI. A less significant decrease in E_h was also observed after the second nZVI injection and the substrate application. Finally, the oxidation-reduction potential was stabilized at 50 mV and 100 mV in the first and second monitoring line, respectively. The drop of the E_h supported by the analysis of inorganics documents the course of reducing conditions in the monitoring wells during the whole remediation experiment including the biotic phase.

The first injection of nZVI resulted in a decrease of the Cr(VI) concentration (below the detection limits) in the first monitoring line, but the concentrations started to rebound within several weeks. After the second injection, the level of chromium contamination reached only 10–15% of the initial concentrations. The reductive effect of nZVI in the second monitoring line was less pronounced. However, the final Cr(VI) removal was achieved by the second phase. After addition of the microbial substrate, the concentration of Cr(VI) decreased below the analytical limits of detection in a very short time and lasted for the whole monitoring period, even after 188 days from the whey application, shown in Fig. 20.3. In addition, the laboratory studies showed that the concentration of total chromium in the soil samples corresponded to the initial concentration of Cr(VI), which documents fixation of Cr(III) in the soil matrix.

Transformation of chlorinated ethenes by nZVI even after the second injection was not observed in the monitoring wells. The authors attributed the limited degradation efficiency during the first phase to the competition with Cr(VI), which is a thermodynamically more favored reducible compound. However, the addition of the organic substrate resulted in a rapid decrease of parent trichloroethene (TCE) below the limit of detection together with a temporal increase and a further decrease of its transformation products—less chlorinated and dechlorinated ethenes. The concentration of parent TCE remained below the detection limit for the rest of the monitoring period (six months) in the first monitoring line. In the second monitoring line, 36–97% removal of chlorinated ethenes was achieved at the end of the pilot test.

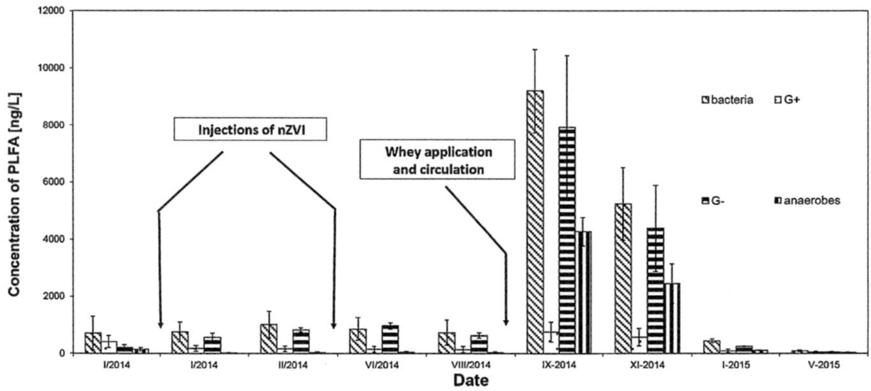

Fig. 20.4 The estimation of biomass via PLFA. (Adapted from Němeček et al. 2016)

20.3.4 Microbes and Their Role in the Whole Nano-Bioremediation Process

The results of the microbial analysis done by the authors indicate that the degradation of the pollutants to a level below the toxicity threshold created preferable environmental conditions for autochthonous microflora in the site. On the basis of the results of qPCR and PLFA, the injection of the substrate resulted in an increase of microbial biomass including microbial degraders. Despite the same results obtained from both types of the analyses, the massive increase (approximately 10× estimated by PLFA) in the biomass was only temporal. After depletion of the substrate, the biomass expeditiously decreased within two months from the whey application (see Fig. 20.4). Additionally, a stable reducing condition persisted even after the complete consumption of the substrate and the simultaneous decrease in microbial biomass. The possible explanation proposed by the authors is that the persistent reducing conditions were partially caused by regenerated iron formed by microbial reduction of Fe(III) to Fe(II) during the application of whey and the increase in biomass. A solid-state analysis of sample MV-4 (passive samplers) by ^{57}Fe Mössbauer spectroscopy supported this hypothesis. At the end of the experiment, the results showed that 21% of present Fe is in oxidation state Fe(II).

Moreover, microbial community analysis of sediment samples at the end of the pilot test proved the occurrence of reducing bacterial genera, such as *Geobacter* and *Rhodoferax*, which are able to reduce iron and sulfur during catabolic processes. These reduced forms could have served as reducing agents in Cr(VI) fixation even after the substrate depletion. The Illumina sequencing data also showed the presence of halorespiring bacteria, which were possibly involved in degradation of chlorinated ethenes. In addition, some of the found phylotypes could (besides chloroethene respiration) have reduced also the metals (Fe(III) and Cr(VI)). Another interesting genus found in the sediment samples was *Dehalococcoides*, which is a typical representative of chlororespiring bacteria. After the whey circulation period, an increase in *vcr*A gene copies was recorded, which greatly corresponded to an increase in the detected degradation products of chlorinated ethenes and to a decreasing chlorine number.

20.4 Conclusion

The results of the two presented pilot tests clearly demonstrate the potential of nZVI for nano-bioremediation application of sites with Cr(VI) contamination via geofixation of the metal into the soil. The decrease in contamination after the nZVI application and the closely connected improvement of the living conditions for microbial species (e.g., lower toxicity, ORP) facilitated the implementation of the next bioremediation step. The subsequent injection of the inexpensive organic substrate caused an increase in the microbial biomass, which was consequently able to continue with reduction of chromium or transformation of TCE even after complete oxidation of nZVI. Both studies document the feasibility and high efficiency of the combination of nanotechnology and biotechnology in remediation of chromium-contaminated sites with a clearly long-term effect.

References

Mueller NC, Braun J, Bruns J, Černík M, Rissing P, Rickerby D, Nowack B (2012) Application of nanoscale zero valent iron (NZVI) for groundwater remediation in Europe. Environ Sci Pollut Res 19(2):550–558. https://doi.org/10.1007/s11356-011-0576-3

Němeček J, Lhotský O, Cajthaml T (2014) Nanoscale zero-valent iron application for in situ reduction of hexavalent chromium and its effects on indigenous microorganism populations. Sci Total Environ 485–486:739–747. https://doi.org/10.1016/j.scitotenv.2013.11.105

Němeček J, Pokorný P, Lacinová L, Černík M, Masopustová Z, Lhotský O, Filipová A, Cajthaml T (2015) Combined abiotic and biotic in-situ reduction of hexavalent chromium in groundwater using nZVI and whey: a remedial pilot test. J Hazard Mater 300:670–679. https://doi.org/10.1016/j.jhazmat.2015.07.056

Němeček J, Pokorný P, Lhotský O, Knytl V, Najmanová P, Steinová J, Černík M, Filipová A, Filip J, Cajthaml T (2016) Combined nano-biotechnology for in-situ remediation of mixed contamination of groundwater by hexavalent chromium and chlorinated solvents. Sci Total Environ 563–564:822–834. https://doi.org/10.1016/j.scitotenv.2016.01.019

Semerád J, Cajthaml T (2016) Ecotoxicity and environmental safety related to nano-scale zerovalent iron remediation applications. Appl Microbiol Biotechnol 100(23):9809–9819. https://doi.org/10.1007/s00253-016-7901-1

Semerád J, Čvančarová M, Filip J, Kašlík J, Zlotá J, Soukupová J, Cajthaml T (2018) Novel assay for the toxicity evaluation of nanoscale zero-valent iron and derived nanomaterials based on lipid peroxidation in bacterial species. Chemosphere 213:568–577. https://doi.org/10.1016/j.chemosphere.2018.09.029

Part IV
Biotechnologies for Soil Treatment

Chapter 21
Biotechnologies for Soil Treatment

Petra Najmanová and Martin Halecký

Abstract Biodegradation performed by indigenous bacteria is one of the major technologies that can effectively remove organic pollution from the environment. The success of the bioremediation process depends on the ability to set up and maintain the optimal conditions for the contaminant biodegradation. Soil excavation followed by ex situ bioremediation is mostly applied to accelerate the site clean-up efficiency of the contaminated soil. An overview of various bioremediation techniques for soil treatment will be provided in this chapter. Basic information on the optimal conditions, the biodegradable contaminants, and the different kinds of bioremediation stimulants with their efficiency will be presented in comparison with the application of allochthonous bacteria. This chapter is especially devoted to the petroleum hydrocarbons, which still represent the main pollutants that are removable by bioremediation in real applications. The topic of bioremediation of persistent organic pollutants will be mentioned only marginally as it will be discussed in detail in the following chapters.

Keywords Soil bioremediation · Petroleum hydrocarbons · Biopiles · Landfarming · Biostimulation · Bioaugmentation

21.1 Introduction

During the last two decades a wide range of research papers on known biodegraders has been published. The research has covered their isolation, description, and formation of genetically modified organisms with improved characteristics.

P. Najmanová (✉)
DEKONTA, a.s., Stehelčeves, Czech Republic

Department of Biotechnology, University of Chemistry and Technology, Prague, Czech Republic
e-mail: najmanova@dekonta.cz

M. Halecký
Department of Biotechnology, University of Chemistry and Technology, Prague, Czech Republic

J. Filip et al. (eds.), *Advanced Nano-Bio Technologies for Water and Soil Treatment*, Applied Environmental Science and Engineering for a Sustainable Future,
https://doi.org/10.1007/978-3-030-29840-1_21

However, their practical use in decontamination technologies is not well known. Some publications deal with relatively low biodegradation of different types of persistent pollutants rather than focusing on the improvement that has been made to the technology. Hardly can we find publications dealing with the solutions to current problems, especially employing alternative approaches to conventional technologies or methods of risk assessment.

Although modern tools using genetic modifications or nanobiotechnologies are exploited, classical remediation techniques still play a major role in real-life applications.

Bioremediation is a method preferred for remediation of petroleum hydrocarbon contaminated soils because it is considered a cost-effective and sustainable technology (Coulon et al. 2012; Jiang et al. 2016). It is important to investigate and understand many factors, such as the soil and the contaminant characteristics, the bioavailability, or the presence of inhibitors to successfully achieve the target limits. All these factors can significantly influence the efficiency of the bioremediation process (Jiang et al. 2016).

Bioremediation is not a new concept and it is being increasingly used as a relatively economical environmental remediation technology. It is an environmentally friendly and cost-effective method of treating soils containing organic chemicals, thus enabling appropriate reuse of the treated soil and minimising disposal of waste soil to landfill whilst providing adequate protection for human health and the environment (EPA 2005).

The ex situ bioremediation treatment of soil is generally used in the biopiles or by landfarming. Landfarming is an above-ground remediation technology for soils that reduces concentrations of petroleum hydrocarbons through biodegradation. This technology usually involves spreading excavated contaminated soils on the ground surface in a thin layer and stimulating the aerobic microbial activity within the soils through aeration and/or the addition of minerals, nutrients, and moisture (US EPA 2017).

From the microbiological point of view, bioremediation is defined as an accelerated process using microorganisms, indigenous or introduced, to degrade and detoxify organic substances to harmless compounds, such as carbon dioxide and water, in a confined and controlled environment. Bioremediation is suitable for treating a variety of organic chemicals, including (i) volatile organic compounds; (ii) benzene, toluene, ethylbenzene, xylene (BTEX) compounds; (iii) phenolic compounds; (iv) polycyclic aromatic hydrocarbons (PAHs), particularly the simpler aromatic compounds; (v) total petroleum hydrocarbons (TPH); (vi) nitroaromatic compounds. Bioremediation exploits microorganisms' ability to degrade or transform organic pollutants to harmless products, yet a spontaneous biodegradation process on a reclaimed site is very slow for the following reasons: (i) absence of a specific microflora able to decompose present hydrocarbons, (ii) low solubility of present pollutants, (iii) non-homogenous distribution of organic pollutants, (iv) deficiency of oxygen and nutrients, etc. To accelerate the natural decontamination process it is convenient to create ideal conditions for the bacterial growth, especially within the following parameters optimizing the conditions: temperature, moisture content, concentration of oxygen and nutrients in the soil.

Different bioremediation approaches should be considered before the site clean-up. Enhancing the indigenous microflora by different kinds of amendments, referred to as biostimulation, can improve the effectivity of bioremediation (Ortíz et al. 2013; Silva et al. 2004). In case of lack of indigenous microflora with biodegradable potential, the addition of specific bacteria able to decompose the target contaminant, so-called bioaugmentation, is supposed to be a convenient solution (Salunkhe et al. 2015; Chen et al. 2014).

21.2 Bioremediation of Petroleum Hydrocarbons

Petroleum hydrocarbons represent the main source of energy. Due to its extensive use in industry, contamination with TPH still poses one of the major environmental problems to which bioremediation seems to be an effective and cheap solution.

The three different approaches (natural attenuation, landfarming, and bioaugmentation) were compared with the treatment of TPH-contaminated soil. The moisture content was 15–20% w/w. The highest TPH removal with 86% efficiency was achieved in the bioaugmentation variant when compared with landfarming where the efficiency was 70% in 90 days at 27 °C (Guarino et al. 2017). In different studies (Martinéz Álvarez et al. 2017; Wang et al. 2016), biostimulation was found to be sufficient for effective contaminant removal. While in some cases biostimulation is more effective than bioaugmentation, in others the addition of TPH-degrading bacteria is the best solution for petroleum hydrocarbon removal (Guarino et al. 2017).

To enhance the bioremediation process an extensive range of amendments can be used. Brown et al. (2017) compared in his study the effect of nutrient addition, bulking and sorption agents, biosurfactants, earthworm enzymes, and soil neutralization. The results showed that most amendments did not prove more effective than the fertilizer addition, only the combination of the nutrients, biochar, and rhamnolipid facilitated the removal of TPH by 23%. The biochar function consisted of hydrophobicity reduction, water capacity holding, and microbial colonization (Brown et al. 2017), and rhamnolipids served as surface active compounds that improved the bioavailability of hydrocarbons (Whang et al. 2008). Nevertheless, the agriculture fertilizer causing the highest TPH degradation efficiency of 45% was chosen as the best amendment for Niger Delta soils landfarming clean-up because of its cost and performance.

In a different study, pea straw addition was evaluated as the best solution for the bioremediation of diesel-contaminated Libyan soil (Koshlaf et al. 2016). Plant residues such as the above-mentioned pea straw represent cheap materials that can promote biodegradation of contaminated soils by providing essential nutrients leading to effective bioremediation (Zhang et al. 2008).

Different additives, such as nutrients, activated sludge from an oil-refining wastewater facility, compost, TPH-degrading bacteria, and fern chips were also used for landfarming system enhancement in the study by Wang et al. (2016). The most efficient amendments were the activated sludge and the compost with 83% and 80% efficiency of TPH removal and the highest first-order TPH degradation rate 0.016 and 0.015 per day, respectively. The fern chip addition and TPH-degrading bacteria had also a positive effect on the TPH removal efficiency with the same decay rate 0.013 per day. The sludge and compost supplement caused an increase in the bacterial diversity, resulting in higher TPH decay rates. Thus, these supplements can substitute TPH-degrading bacteria addition.

Bioremediation works in a wide range of temperatures. Martínez Álvarez et al. (2017) proved the bioremediation process of soil contaminated with hydrocarbons in Antarctica to be efficient at a mean temperature of 6.7 °C. The high density polyethylene geomembranes were used to protect the treated soil (biostimulation and also control piles) against wind and ice. A biostimulation pile, involving nitrogen (as ammonium nitrate) and phosphorus (as monosodium phosphate) addition, was significantly higher than the control pile without nitrogen and phosphorus additions (75.8% vs. 49.5%). However, most microorganisms with biodegradation potential are mesophilic with optimal temperatures ranging from 25 °C to 37 °C (Admassu and Korus 1996). The microbial activity of most bacteria important to petroleum hydrocarbon biodegradation also diminishes at temperatures greater than 45 °C. For practical use, the temperature range from 10 °C to 45 °C was set up by EPA as an effective temperature range for the landfarming system (US EPA 2017).

The ratio of C:N:P should be considered before each bioremediation. The optimal ratio has been reported as 100:10:1 (Dibble and Bartha 1979). The scarcity of essential nutrients such as nitrogen and phosphorus has been recognized as a crucial factor that inhibits the microbial bioremediation performance. The lack of suitable and available nutrients in the contaminated soil required the implementation of engineered bioremediation strategies. Each soil has a different microbial diversity and requirements and the optimal C:N:P ratio should be determined before each bioremediation process.

If we take into account all the above-mentioned conditions, the process of ex situ bioremediation technology for TPH removal is summarized in Fig. 21.1.

21.3 Biodegradation of Persistent Organic Pollutants

Persistent organic pollutants (POPs) are chemicals of global concern because of their potential for long-range transport, persistence in the environment, and ability to bioaccumulate in ecosystems as well as their significant negative effects on both human health and the environment. The most commonly emerged POPs are organochlorine pesticides, such as DDT, industrial chemicals, polychlorinated biphenyls (PCB), or unintentional by-products of many industrial processes, especially polychlorinated dibenzo-*p*-dioxins and dibenzofurans (WHO).

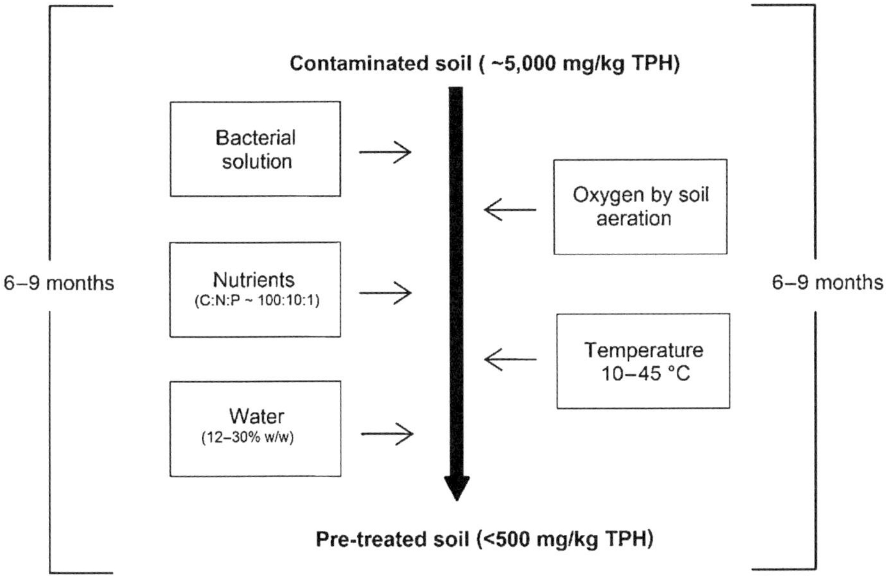

Fig. 21.1 Scheme of ex situ soil bioremediation

As most sites contaminated with POPs, especially pesticides, contain complex mixtures of different compounds, their remediation is a complicated process. Physical, chemical, or biological processes can be used for their elimination from the environment. The conventional methods for the remediation of pesticides in soil, such as landfilling, incineration, or chemical destruction are effective technologies. However, they are also costly and some can pose an environmental risk (Morillo and Villaverde 2017), while bioremediation of pesticides-polluted sites is an emerging technology that is cost-effective and environmentally friendly (Niti et al. 2013).

Bioremediation of soils contaminated with pesticides has drawn considerable research interest mainly because the pesticides pollution is constantly increasing (Morillo and Villaverde 2017). Both biodegradation approaches such as bioaugmentation (Salunkhe et al. 2015; Chen et al. 2014) and biostimulation are studied (Ortíz et al. 2013; Silva et al. 2004).

The biodegradation of herbicide methoxychlor using *Acinetobacter* resulted in 60% efficiency (Fuentes et al. 2014). Application of landfarming for soils heavily contaminated (>5 g/kg) with hexachlorocyclohexane (HCH) isomer was reported by Rubinos et al. (2007). The soil showed a significant decrease in α- and γ-HCH, nearly 90% after 11 months of the treatment while biodegradation of β and δ isomers was insignificant.

From the above-mentioned examples, it is obvious that classic bioremediation technology is not always sufficient and the advanced bioremediation techniques should be used for the degradation of persistent organic compounds. Composting (see Chaps. 23 and 25) seems to be a promising technology for the removal of PAHs

(Covino et al. 2016) or ligninolytic fungi (see Chap. 22) can be a solution for PCB remediation (Čvančarová et al. 2012; Stella et al. 2017).

21.4 Nanobioremediation

Although many technologies are available for the remediation of contaminated sites, the use of a single technology for remediation of recalcitrant contaminants may be expensive and not effective (Singh et al. 2013; Němeček et al. 2016). The combination of two or more different technologies in sequence is a possible solution for cost reduction (He et al. 2009) and the acceleration of site rehabilitation. Nanobioremediation integrates two remediation technologies and the use of nanomaterials, e.g., iron, followed by bioremediation to gain effective, efficient, and sustainable remediation technology (Cecchin et al. 2017).

Singh et al. (2013) evaluated the effect of an integrated technique involving the use of stabilized Pd/Fe^0 bimetallic nanoparticles (CMC-Pd/nFe^0) and *Sphingomonas* sp. strain NM05, on the degradation of γ-HCH in soil. The integration showed synergistic effect on γ-HCH degradation, 99% degradation was achieved within six days of incubation.

21.5 Conclusion

Despite the significant increase in the use of oxidative and reductive technologies over recent decades, biotechnologies still play an irreplaceable role in the field of soil remediation. Even though quick site rehabilitation is usually required, biotechnology will always offer an inexpensive and waste-free solution for a great deal of contaminated sites.

References

Admassu W, Korus RA (1996) Engineering of bioremediation processes: needs and limitations. In: Crawford RL, Crawford DL (eds) Bioremediation: principles and applications. Biotechnology Research. Cambridge University Press, Cambridge, pp 13–34. https://doi.org/10.1017/CBO9780511608414.003

Brown DM, Okoro S, van Gils J, van Spanning R, Bonte M, Hutchings T, Linden O, Egbuche U, Bruun KB, Smith JWN (2017) Comparison of landfarming amendments to improve bioremediation of petroleum hydrocarbons in Niger Delta soils. Sci Total Environ 596–597:284–292. https://doi.org/10.1016/j.scitotenv.2017.04.072

Cecchin I, Reddy KR, Thomé A, Tessaro EF, Schnaid F (2017) Nanobioremediation: integration of nanoparticles and bioremediation for sustainable remediation of chlorinated organic contaminants in soils. Int Biodeterior Biodegrad 119:419–428. https://doi.org/10.1016/j.ibiod.2016.09.027

Chen S, Chang C, Deng Y, An S, Dong YH, Zhou J, Hu M, Zhong G, Zhang L-H (2014) Fenpropathrin biodegradation pathway in *Bacillus* sp. DG-02 and its potential for bioremediation of pyrethroid-contaminated soils. J Agric Food Chem 62(10):2147–2157. https://doi.org/10.1021/jf404908j

Coulon F, Brassington KJ, Bazin R, Linnet PE, Thomas KA, Mitchell TR, Lethbridge G, Smith JWN, Pollard SJT (2012) Effect of fertilizer formulation and bioaugmentation on biodegradation and leaching of crude oils and refined products in soils. Environ Technol 33 (16):1879–1893. https://doi.org/10.1080/09593330.2011.650221

Covino S, Fabianová T, Křesinová Z, Čvančarová M, Burianová E, Filipová A, Voříšková J, Baldrian P, Cajthaml T (2016) Polycyclic aromatic hydrocarbons degradation and microbial community shifts during co-composting of creosote-treated wood. J Hazard Mater 301:17–26. https://doi.org/10.1016/j.jhazmat.2015.08.023

Čvančarová M, Křesinová Z, Filipová A, Covino S, Cajthaml T (2012) Biodegradation of PCBs by ligninolytic fungi and characterization of the degradation products. Chemosphere 88 (11):1317–1323. https://doi.org/10.1016/j.chemosphere.2012.03.107

Dibble JT, Bartha R (1979) Rehabilitation of oil-inundated agricultural land: a case history. Soil Sci 128(1):56–60

EPA (2005) EPA guidelines. Soil bioremediation. Available: https://www.epa.sa.gov.au/files/8372_guide_soil.pdf. Accessed 28 Mar 2018

Fuentes MS, Alvarez A, Saez JM, Benimeli CS, Amoroso MJ (2014) Use of actinobacteria consortia to improve methoxychlor bioremediation in different contaminated matrices. In: Alvarez A, Polti MA (eds) Bioremediation in Latin America: current research and perspectives. Springer, Cham, pp 267–277. https://doi.org/10.1007/978-3-319-05738-5_17

Guarino C, Spada V, Sciarrillo R (2017) Assessment of three approaches of bioremediation (Natural Attenuation, Landfarming and Bioagumentation – Assistited Landfarming) for a petroleum hydrocarbons contaminated soil. Chemosphere 170:10–16. https://doi.org/10.1016/j.chemosphere.2016.11.165

He N, Li P, Zhou Y, Fan S, Ren W (2009) Degradation of pentachlorobiphenyl by a sequential treatment using Pd coated iron and an aerobic bacterium (H1). Chemosphere 76 (11):1491–1497. https://doi.org/10.1016/j.chemosphere.2009.06.046

Jiang Y, Brassington KJ, Prpich G, Paton GI, Semple KT, Pollard SJT, Coulon F (2016) Insights into the biodegradation of weathered hydrocarbons in contaminated soils by bioaugmentation and nutrient stimulation. Chemosphere 161:300–307. https://doi.org/10.1016/j.chemosphere.2016.07.032

Koshlaf E, Shahsavari E, Aburto-Medina A, Taha M, Haleyur N, Makadia TH, Morrison PD, Ball AS (2016) Bioremediation potential of diesel-contaminated Libyan soil. Ecotoxicol Environ Saf 133:297–305. https://doi.org/10.1016/j.ecoenv.2016.07.027

Martínez Álvarez LM, Ruberto LAM, Lo Balbo A, Mac Cormack WP (2017) Bioremediation of hydrocarbon-contaminated soils in cold regions: development of a pre-optimized biostimulation biopile-scale field assay in Antarctica. Sci Total Environ 590–591:194–203. https://doi.org/10.1016/j.scitotenv.2017.02.204

Morillo E, Villaverde J (2017) Advanced technologies for the remediation of pesticide-contaminated soils. Sci Total Environ 586:576–597. https://doi.org/10.1016/j.scitotenv.2017.02.020

Němeček J, Pokorný P, Lhotský O, Knytl V, Najmanová P, Steinová J, Černík M, Filipová A, Filip J, Cajthaml T (2016) Combined nano-biotechnology for in-situ remediation of mixed contamination of groundwater by hexavalent chromium and chlorinated solvents. Sci Total Environ 563–564:822–834. https://doi.org/10.1016/j.scitotenv.2016.01.019

Niti C, Sunita S, Kamlesh K, Rakesh K (2013) Bioremediation: an emerging technology for remediation of pesticides. Res J Chem Environ 17(4):88–105

Ortíz I, Velasco A, Le Borgne S, Revah S (2013) Biodegradation of DDT by stimulation of indigenous microbial populations in soil with cosubstrates. Biodegradation 24(2):215–225. https://doi.org/10.1007/s10532-012-9578-1

Rubinos DA, Villasuso R, Muniategui S, Barral MT, Díaz-Fierros F (2007) Using the landfarming technique to remediate soils contaminated with hexachlorocyclohexane isomers. Water Air Soil Pollut 181(1–4):385–399. https://doi.org/10.1007/s11270-006-9309-5

Salunkhe VP, Sawant IS, Banerjee K, Wadkar PN, Sawant SD (2015) Enhanced dissipation of triazole and multiclass pesticide residues on grapes after foliar application of grapevine-associated *Bacillus* species. J Agric Food Chem 63(50):10736–10746. https://doi.org/10.1021/acs.jafc.5b03429

Silva E, Fialho AM, Sá-Correia I, Burns RG, Shaw LJ (2004) Combined bioaugmentation and biostimulation to cleanup soil contaminated with high concentrations of atrazine. Environ Sci Technol 38(2):632–637. https://doi.org/10.1021/es0300822

Singh R, Manickam N, Mudiam MKR, Murthy RC, Misra V (2013) An integrated (nano-bio) technique for degradation of γ-HCH contaminated soil. J Hazard Mater 258–259:35–41. https://doi.org/10.1016/j.jhazmat.2013.04.016

Stella T, Covino S, Čvančarová M, Filipová A, Petruccioli M, D'Annibale A, Cajthaml T (2017) Bioremediation of long-term PCB-contaminated soil by white-rot fungi. J Hazard Mater 324:701–710. https://doi.org/10.1016/j.jhazmat.2016.11.044

US EPA (2017) How to evaluate alternative cleanup technologies for underground storage tank sites. A guide for corrective action plan reviewers. Chapter V Landfarming. Available: https://www.epa.gov/sites/production/files/2014-03/documents/tum_ch5.pdf. Accessed 22 May 2018

Wang S-Y, Kuo Y-C, Hong A, Chang Y-M, Kao C-M (2016) Bioremediation of diesel and lubricant oil-contaminated soils using enhanced landfarming system. Chemosphere 164:558–567. https://doi.org/10.1016/j.chemosphere.2016.08.128

Whang L-M, Liu P-WG, Ma C-C, Cheng S-S (2008) Application of biosurfactants, rhamnolipid, and surfactin, for enhanced biodegradation of diesel-contaminated water and soil. J Hazard Mater 151(1):155–163. https://doi.org/10.1016/j.jhazmat.2007.05.063

WHO Persistent organic pollutants (POPs). Available: http://www.who.int/foodsafety/areas_work/chemical-risks/pops/en/. Accessed 6 Aug 2018

Zhang K, Hua X-F, Han H-L, Wang J, Miao C-C, Xu Y-Y, Huang Z-D, Zhang H, Yang J-M, Jin W-B, Liu Y-M, Liu Z (2008) Enhanced bioaugmentation of petroleum- and salt-contaminated soil using wheat straw. Chemosphere 73(9):1387–1392. https://doi.org/10.1016/j.chemosphere.2008.08.040

Chapter 22
Mycoremediation of Contaminated Soils

Tatiana Stella

Abstract This chapter deals with the importance of fungi in the decontamination of polluted soils. Because of their physiological and biochemical features, both ligninolytic and non-ligninolytic fungi can degrade and, to some extent, mineralize a wide range of organic and xenobiotic pollutants such as oil-derived products, polycyclic aromatic hydrocarbons, polychlorinated biphenyls, dioxins and furans, pesticides, herbicides, and nitroaromatic explosives, and pharmaceuticals. Thus, fungi are excellent candidates for the design of effective bioremediation technologies. The potentialities of the fungal application in the remediation of soils for the degradation of persistent chlorinated and non-chlorinated organic pollutants are presented in this chapter through a number of studies at laboratory scale as well as at pilot- and full-scale trials. Furthermore, the shortcomings that make this technology still unsuitable for large-scale trials are also analyzed in this chapter.

Keywords Ligninolytic fungi · White-rot fungi · Chlorinated hydrocarbons · Aromatic hydrocarbons · Biodegradation

22.1 Introduction

Intensive industrial processes and agricultural practices have led to either deliberate or accidental release of potentially toxic chemicals into the environment over the last decades. These compounds of either natural or anthropogenic origin cause a great concern due to their adverse effects on ecosystems and human health. They can affect any environmental compartments (air, water, and soil); they can be tightly bound into or onto soil organic matter and clay particles, released into the atmosphere by volatilization, or they could leak into water. Furthermore, living organisms need to face a big hurdle while attempting to degrade most of these compounds. Consequently, they can accumulate in food chains and in the environment affecting

T. Stella (✉)
Department of Earth and Environmental Sciences (DISAT), University of Milano-Bicocca, Milan, Italy
e-mail: tatiana.stella@unimib.it

J. Filip et al. (eds.), *Advanced Nano-Bio Technologies for Water and Soil Treatment*, Applied Environmental Science and Engineering for a Sustainable Future, https://doi.org/10.1007/978-3-030-29840-1_22

also the nutrient cycling. They encompass various organic and inorganic pollutants (Walker et al. 2006): petroleum hydrocarbons, halogenated solvents, chlorinated aromatic hydrocarbons, explosives, dioxins, endocrine disrupting compounds, herbicides, pesticide, heavy metals, and radionuclides. These pollutants are commonly present in the environment in the form of a complex mixture, making the remediation of these co-contaminated sites a challenge to face.

Conventional remediation technologies based on physical and chemical approaches are still the most applied ones for the disposal of hazardous chemicals despite the fact that they are not sustainable in terms of both cost and the environmental impact. Thus, biological treatments are being considered as a valid alternative to traditional remedial strategies.

A specific branch of bioremediation, called mycoremediation, has been arousing real interest in recent years. This technology is based on the use of fungi for the remediation of polluted soils or other solid–liquid matrices.

Fungi are a diverse group of microorganisms ubiquitously present in natural ecosystems and able to thrive in a variety of habitats. They are the major decomposers of complex biomaterials. Saprotrophic fungi in particular are responsible for biomass decomposition.

Among saprotrophic organisms, wood decay fungi evolved an efficient non-specific extracellular enzymatic system for degradation of complex heterogeneous polymers, such as lignocellulose and chitin. Owing to the production of such lignocellulolytic enzymes, they can transform a wide array of organic substrates including organic pollutants that are inaccessible to most of prokaryotes.

Due to these capabilities, ligninolytic fungi (LFs) are considered to play a key role in the biodegradation processes of recalcitrant organic pollutants in a number of contaminated matrices.

22.2 Ligninolytic Fungi and Their Enzymes

Ligninolytic fungi variously secrete one or more of three main extracellular lignin-degrading enzymes, namely lignin peroxidase (LiP, E.C. 1.11.1.14), manganese-dependent peroxidase (MnP, E.C. 1.11.1.13), and laccase (Lac, E.C. 1.10.3.2) (Tuor et al. 1995).

However, novel manganese-independent peroxidase activity and other versatile peroxidase activities have been more recently found in some genera of white-rot fungi (Heinfling et al. 1998; Camarero et al. 1999; Ruiz-Dueñas et al. 2001).

Moreover, other enzymes associated with lignin-degrading enzymes are produced by LFs: glyoxal oxidase (E.C. 1.2.3.5), superoxide dismutase (E.C. 1.15.1.1), glucose oxidase (E.C. 1.1.3.4), aryl alcohol oxidase (E.C. 1.1.3.7), and cellobiose dehydrogenase (E.C. 1.1.99.18). They produce H_2O_2, which is essential for the catalytic cycle of peroxidases (LiP and MnP), or degrade the non-phenolic substructures of lignin by the formation of reactive hydroxyl radicals $^{\bullet}OH$ (Hatakka 2001; Leonowicz et al. 2001; Lundell et al. 2010). All the three major ligninolytic enzymes are encoded by gene families that lead to the production of multiple enzyme isoforms (Thurston 1994; Tuor et al. 1995; Martínez 2002).

Ligninolytic enzyme production occurs mostly during the secondary metabolism and the ligninolytic activities are strongly induced by starvation conditions (i.e., nitrogen, carbohydrate, and sulfur limitation) as well as by the presence of trace metals like Mg^{2+} and Ca^{2+} (Jeffries et al. 1981). Furthermore, growth conditions (i.e., temperature) significantly affect the level of activity of ligninolytic enzymes (Podgornik et al. 2001; Bermek et al. 2004). Additionally, the ligninolytic enzyme production is also affected by the presence/absence of mediators or various chemicals and by specific concentrations of metals Mn^{2+} and/or Cu^{2+} (Scheel et al. 2000; Galhaup et al. 2002).

A new type of H_2O_2-requiring enzyme was also described in the basidiomycete *Agrocybe aegerita* (Ullrich et al. 2004) and, later, in other fungi (Hofrichter and Ullrich 2014). These proteins are heme-thiolate haloperoxidases, classified as fungal unspecific peroxygenases (UPO, EC 1.11.2.1). Their catalytic cycle combines the typical pathways of extracellular peroxidases and intracellular monooxygenases (unspecific cytochrome P450 system, CYP450, EC 1.14.14.1), thus enabling them to catalyze a wide array of reactions (Hofrichter and Ullrich 2014).

22.3 Mycoremediation and White-Rot Fungi

Fungi with contaminant-degrading capabilities are mostly wood-degrading basidiomycetes belonging to Agaricales (i.e., *Stropharia rugosoannulata*, *Agrocybe praecox,* and *Pleurotus ostreatus*) and Polyporales orders (i.e., *Phanerochaete chysosporium*, *Trametes versicolor*, *Bjerkandera adusta*, *Irpex lacteus*, etc.) (Gadd 2001; Singh 2006). In addition, several fungi belonging to the phyla Ascomycota (i.e., *Trichoderma* spp.), Zygomycota (i.e., *Mucor* spp.), and anamorphic ascomycetes (i.e., *Aspergillus* spp., *Penicillium* spp., *Paecilomyces* spp.) have also demonstrated the ability to degrade contaminants (Tortella et al. 2005; Tigini et al. 2009; Carvalho et al. 2009).

Among contaminant-degrading fungi, the group of white-rot fungi (WRFs), which is specialized in lignin breakdown, is particularly promising in the bioremediation of contaminated matrices. Indeed, in addition to their natural substrate, these fungi can degrade and, to some extent mineralize, a wide range of organic and xenobiotic pollutants structurally similar to lignin such as petroleum hydrocarbons, chlorophenols, polycyclic aromatic hydrocarbons (PAHs), polychlorinated biphenyls (PCBs), dioxins and furans, pesticides, herbicides, and nitroaromatic explosives (Pointing 2001; Mougin et al. 2002; Rabinovich et al. 2004; Baldrian 2008; Cerniglia and Sutherland 2006; Cajthaml and Svobodová 2012).

These fungi can easily penetrate into the polluted matrix thanks to their hyphal structures and reach the less bioavailable contaminants by secreting ligninolytic enzymes (i.e., Lac, LiP, and MnP) into the extracellular environment. As already mentioned, these oxidases exhibit very low substrate specificity and can attack the contaminants by non-specific radical-based reactions. WRFs are endowed also with an intracellular enzymatic system involving cytochrome P450 monooxygenase enzymes (Črešnar and Petrič 2011; Stella et al. 2013). This intracellular pathway, which occurs in all eukaryotic organisms, mainly regulates the bioconversion of hormones and the detoxification of drugs and xenobiotics (Bernhardt 2006). In WRFs, CYP450 is implied as well as the ligninolytic system in the degradation of xenobiotics (van den Brink et al. 1998; Stella et al. 2013). Moreover, as mentioned above, WRFs evolved a spatially extensive hyphal growth enabling them to penetrate across air-filled soil pores, air-water interfaces, and even rock matrices (Bornyasz et al. 2005) and act as dispersion vectors of bacteria (Kohlmeier et al. 2005; Bonfante and Anca 2009). Fungi can also tolerate high concentrations of organic contaminants and heavy metals without deleterious effects on their enzyme activities (Baldrian et al. 2000; Baldrian 2003; Tuomela et al. 2005).

Unlike bacteria, fungi cannot assimilate contaminants as a source of carbon and energy; thus, they use the lignocellulosic residues as amendments to support their growth in the co-metabolic process of contaminant degradation (Singh 2006).

Furthermore, fungi are involved in a soil humification process leading to decontamination and also to reuse of the soil for agricultural purposes (Bollag 1992; Michels 1998). Therefore, from the circular economy strategy point of view, the use of these organisms in soil remediation is highly recommended.

Nevertheless, mycoremediation has not been considered yet as a method for consistently recovering contaminated soils or sediments at pilot-scale level, but the number of bench-scale studies is increasing over time in view of their potential applications. Within this frame, a number of laboratory studies and also semi- and pilot-scale studies concerning the mycoremediation of soils contaminated by recalcitrant organic pollutants will be presented in the following sections.

22.4 Chlorinated Aromatic Pollutants

22.4.1 Polychlorinated Biphenyls

Polychlorinated biphenyls (PCBs) are synthetic compounds the structure of which consists of a biphenyl molecule that carries from one to ten chlorine atoms. In the past, chemical, petrochemical, paint and coating industries widely used PCB mixtures because of their thermal and chemical stability, flame resistance, and dielectric properties. Consequently, their use led to an extensive environmental contamination either through accidental releases or inappropriate methods of disposal. Owing to their aforementioned inertness, PCBs are still present in those areas where industrial activities based on their use had been carried out for decades, although their production and application were banned long time ago. PCBs rank among the most hazardous contaminants in the world; thus, they are at the forefront of public health concern. Indeed, the teratogenic, carcinogenic, and endocrine-disrupting effects of these xenobiotics have been widely reported in the literature (Crinnion

2011; Helmfrid et al. 2012; Kramer et al. 2012; El Majidi et al. 2013) especially for coplanar PCBs (not substituted at the *ortho* ring positions), which exhibit dioxin-like toxicity features (US EPA 2004; Van den Berg et al. 2006). Therefore, the cleanup of PCB-contaminated sites has become a global priority.

A large number of LFs belonging to Basidiomycota phylum were tested in laboratory-scale model liquid systems for their ability to degrade technical PCB mixtures or single PCB congeners, namely *Phanerochaete chrysosporium* (Eaton 1985; Thomas et al. 1992; Vyas et al. 1994; Yadav et al. 1995; Kamei et al. 2006a), *T. versicolor* (Zeddel et al. 1993; Vyas et al. 1994; Cloete and Celliers 1999), *Lentinus edodes* (Ruiz-Aguilar et al. 2002), *P. ostreatus* (Kubátová et al. 2001; Čvančarová et al. 2012), *Grifola frondosa* (Seto et al. 1999), *Coriolopsis polyzona* (Vyas et al. 1994; Novotný et al. 1997), *Phlebia brevispora* (Kamei et al. 2006b), *I. lacteus* (Čvančarová et al. 2012), *B. adusta* (Beaudette et al. 1998; Čvančarová et al. 2012), *Pycnoporus cinnabarinus* (Čvančarová et al. 2012), *Phanerochaete magnoliae* (Čvančarová et al. 2012). Generally, all these studies confirmed that the extent of degradation significantly decreases with increasing chlorine content.

Moreover, the mechanisms by which fungi degrade PCBs were widely investigated (Beaudette et al. 1998; Krčmář et al. 1999; Kamei et al. 2006a, 2006b; Takagi et al. 2007; Čvančarová et al. 2012; Stella et al. 2017).

Nevertheless, few studies investigated the ability of WRFs to degrade PCBs in actual contaminated soils (Borazjani et al. 2005; Federici et al. 2012; Stella et al. 2017; Siracusa et al. 2017). Borazjani et al. (2005) tested different treatments for the remediation of soil contaminated by the technical PCB mixtures Aroclor 1242, 1248, and 1260. The highest degradation extent for the higher chlorinated mixture 1260 was observed when kenaf (a lignocellulosic substrate) inoculated with the white-rot fungus *P. chrysosporium* was mixed with soil (3% w/w). Another study demonstrated that 33.6% of Aroclor 1260 depletion and 23.2% of dechlorination was reached in 60 days by a bioaugmentation approach based on the use of maize stalk-immobilized fungus *Lentinus tigrinus* (Federici et al. 2012).

The PCB-degrading abilities of two white-rot fungi, namely *P. ostreatus* and *I. lacteus*, were also assessed in long-term PCB-contaminated soils the contamination of which was mostly caused by the presence of Delor 103 PCB mixture (Stella et al. 2017). Three soil samples, which differed in their PCB contents, were selected for microcosm preparation: bulk soil, topsoil, and rhizosphere soil. The soils were mixed with either *P. ostreatus*- or *I. lacteus*-colonized lignocellulosic substrate (LS) for bioaugmentation treatments to reach a final soil:LS mass ratio of 5:1 (w/w) and incubated for 12 weeks. The greatest PCB depletion (50.5%) was achieved in the *P. ostreatus*-augmented rhizosphere soil, whilst *I. lacteus* led to the degradation of 30.3% of the original amount after 12 weeks of the incubation. On the other hand, for the most contaminated soil (bulk soil), no significant decrease in the PCB concentration was observed within the first 6 weeks of the incubation regardless of the treatment employed. Moderate removals (18.5 and 19.3%) were observed only after 12 weeks in *P. ostreatus*- and *I. lacteus*-augmented microcosms, respectively.

Furthermore, a recent study evaluated the feasibility of exploiting *P. ostreatus* spent mushroom substrate (SMS-a low-cost waste of the industrial production of this fungus) to treat PCB-contaminated soil coming from a former industrial area (Siracusa et al. 2017). The fungal treatment was firstly assessed at mesocosm scale, and then validated at pilot scale. In the latter test, 94.1% of the PCB degradation was observed in dynamic biopiles when the soil was mixed in a ratio of 10% w/w with *P. ostreatus* substrate.

Compared with the PCB biotransformation processes mediated by basidiomycetes, few studies investigated the PCB biodegradation potential of non-ligninolytic fungi. Most of the reports referred to liquid culture laboratory tests to evaluate the degradation capabilities of these fungi towards technical mixture of PCBs: *Aspergillus niger* (Dmochewitz and Ballschmiter 1988), *Aspergillus*, *Penicillium*, *Fusarium* and *Scedosporium* genera isolated from an aged PCB-contaminated soil (Tigini et al. 2009) and *Doratomyces* spp. *Myceliophthora thermophila*, *Phoma eupyrena* and *Thermoascus crustaceus* (Mouhamadou et al. 2013).

With regard to ascomycetes, a fungal consortium was used for the remediation of both PCB-contaminated soil and sediment (Sage et al. 2014). In this study, 18.7 and 33.3% of 15 target PCBs were degraded in soil and sediment mesocosms, respectively, after 6 months of treatment. Afterwards, the fungal strains were re-inoculated. However, no additional PCB depletion was observed within another 2 months of treatment suggesting that the bioavailable fraction of PCB had already been completely removed and the remaining fraction was inaccessible to fungi.

22.4.2 Chlorinated Dioxins and Furans

Polychlorinated dibenzodioxins (PCDDs) are a group of polyhalogenated organic compounds that consist of two benzene rings joined by two oxygen bridges (dibenzo-1,4-dioxin) with one or several chlorine substituents replacing hydrogen atoms. Polychlorinated dibenzofurans (PCDFs) are a group of organopollutants with properties and chemical structures similar to PCDDs. PCDDs and PCDFs are released into the environment as by-products of several industrial processes (i.e., incineration of waste, chemical processes, and manufacturing of pesticides). They can be easily accumulated in humans and wildlife because of their lipophilic properties. Further, they can cause developmental disorders and cancer, as already reported several decades ago (Huff et al. 1980). For this reason, they were listed among the most persistent hazardous organic pollutants in the Stockholm convention.

Several white-rot fungi were assessed for their ability to degrade chlorinated dioxins in model liquid systems. At first, the oxidation of 2,3,7,8-tetrachlorodibenzo-*p*-dioxin (2,3,7,8-tetraCDD) (Bumpus et al. 1985) and 2,7-dichlorodibenzo-*p*-dioxin (2,7-diCDD) (Valli et al. 1992) by the white-rot fungus *P. chrysosporium* under ligninolytic conditions was demonstrated. Thereafter, another strain of *Phanerochaete* (*P. sordida*) was studied for its capability of degrading a mixture of 10 tetra- to octachlorodibenzo-*p*-dioxins and furans (Takada et al. 1996). Moreover, strains belonging to *Phlebia* and *Bjerkandera* genera were also proved to be efficient in metabolizing both polychlorinated dibenzo-*p*-dioxins and furans (Mori and Kondo 2002; Kamei and Kondo 2005; Kamei et al. 2005; Manji and Ishihara 2004). Other WRFs were also studied for their ability to degrade dibenzo-*p*-dioxins as reviewed by Pinedo-Rivilla et al. (2009): several studies reported the capability of *Trametes* sp. CH2, *Irpex* sp. W3, and *Pleurotus pulmonarius* to mediate the hydroxylation and methoxylation of PCDDs (Yamaguchi et al. 2007; Nam et al. 2008).

Due to the promising results achieved in liquid fungal cultures, a recent study focused on the feasibility of treating actual PCDD- and PCDF-contaminated soils by fungi. Anasonye et al. (2015) firstly performed a 30-day screening of six fungal strains to assess their ability to grow on historically contaminated soils. Pine bark was used as lignocellulosic substrate to inoculate the soils. Two strains, namely *Phanerochaete velutina* and *Stropharia rugosoannulata*, were then selected to carry out the treatment of a PCDD- and PCDF-contaminated soil from a sawmill site: glass bioreactors were filled with contaminated soil mixed with fungal inoculum grown on pine bark. Bioreactors were incubated for 90 days and continuously aerated with moist air. Bioaugmentation with *P. velutina* and *S. rugosoannulata* brought about 62% and 64% of PCDD and PCDF degradation, respectively.

On the other hand, a limited number of studies dealt with PCDDs and PCDFs degradation by non-ligninolytic fungi. A dioxin-degrading organism was isolated from the activated sludge of a landfill treatment plant and identified as *Acremonium* sp. (Nakamiya et al. 2002). This fungus (later identified as *Pseudallescheria boydii*) was tested for the degradation of a mixture of highly chlorinated dioxins (2 ng/mL): this fungal culture was able to remove up to 82% of the mixture with respect to the initial concentration in one day under denitrifying conditions. Due to these promising results, *P. boydii* was further used in a slurry bioreactor system for the treatment of dioxin-contaminated soils (Ishii and Furuichi 2007). However, *P. boydii* was not able to remove the dioxins up to the environmental standard level, and, thus, the mycoremediation treatment was coupled with a physicochemical process (ethanol extraction) to reach the required quality standard. Two years after, the same authors demonstrated that this fungus requires glucose as carbon source to grow and degrade 2,3,7,8-tetraCDD through a co-metabolic process (Ishii et al. 2009).

Among ascomycetes, the cyclic ether degrading fungus *Cordyceps sinensis* was proved to degrade not only non-chlorinated dioxins, but also 2,3,7-trichlorodibenzo-*p*-dioxin and octachlorodibenzo-*p*-dioxin (Nakamiya et al. 2005). Chemical analyses revealed that the concentration of chlorinated dioxins halved after 50 days of incubation with the fungus leading to the formation of chlorocatechols as biotransformation products. A few years later, another strain of *Cordyceps* (*C. militaris*) was tested in liquid cultures for di-, tri-, and tetra-CDD degradation activity (Mori et al. 2015). In this case, only 55% of 2,7-diCDD was degraded after 30 days of incubation in glucose-basal salt medium. This fungus was also able to degrade 2,3,7-trichlorodibenzo-*p*-dioxin (2,3,7-triCDD) and 2,4,8-trichlorodibenzo-*p*-furan

(2,4,8-triCDF) at different extents (lower than for dichlorodibenzo-*p*-dioxins), but it was completely unable to transform 1,3,6,8-, 1,3,7,8-, and 1,2,8,9-tetraCDDs during the same incubation period.

22.5 Non-chlorinated Aromatic Pollutants

22.5.1 Polycyclic Aromatic Hydrocarbons

Polycyclic aromatic hydrocarbons (PAHs), unlike the contaminants (PCBs, PCDDs, and PCDFs) described so far, are natural compounds. However, most of them arise from the incomplete combustion of fossil fuels and biomasses associated with different industrial activities. From a chemical viewpoint, PAHs consist of two or more fused benzene rings arranged either in a linear or a cluster mode. They are included among priority pollutants owing to their toxic, mutagenic, and, in some cases, carcinogenic properties (Haritash and Kaushik 2009). Their hydrophobicity, toxicity, and recalcitrance to biodegradation processes increase as the molecular weight increases making PAHs particularly persistent in the environment.

To overcome bacterial degradation limitations, fungal transformation of PAHs has extensively been studied over the last decades. Ligninolytic fungi, mostly WRFs such as *Trametes*, *Pleurotus*, *Phanerochaete*, *Bjerkandera*, *Irpex*, *Phlebia*, *Nematoloma*, and *Lentinus* spp. were demonstrated to efficiently metabolize PAHs in model liquid cultures or in soils (Sack et al. 1997; Pointing 2001; D'Annibale et al. 2005; Valentín et al. 2006; Giubilei et al. 2009; Novotný et al. 2009; Covino et al. 2010a, b; Lladó et al. 2013).

Concerning the mycoremediation of PAH-contaminated soil, one of the first studies dealt with the treatment of a historically aromatic hydrocarbons-contaminated soil (D'Annibale et al. 2005). Two WRFs, namely *P. chrysosporium* NRRL 6361 and *P. pulmonarius* CBS 664.97 were tested for their ability to grow under non-sterile conditions and to degrade various hydrocarbons. In the case of naphthalene, both fungi completely removed this contaminant within 30 days of the incubation. Fungal strains belonging to genera of *Phanerochaete*, *Polyporus*, *Stereum*, *Lentinus*, *Bjerkandera*, *Irpex*, *Pleurotus*, and *Phlebia* were also screened for their ability to degrade a mixture of PAHs (phenanthrene, fluoranthrene, pyrene, and chrysene) in artificially spiked forest and marsh soils (Valentín et al. 2006). Marsh soil test revealed that all fungi, with the exception of *Phlebia* ssp., were capable of degrading the mixture of PAHs to some extent after 30 days of the incubation. In particular, *Pleurotus eryngii*, *B. adusta*, and *I. lacteus* were the most outstanding strains with an average degradation of PAHs being 67, 55, and 53%, respectively. Furthermore, these three fungal strains partially degraded PAHs also under saline conditions (marsh soil test).

A few years later, Covino and colleagues deeply investigated the effect of both lignocellulosic substrate (inoculum carrier) and contaminant bioavailability on the degradation performance of different fungal strains (*L. tigrinus*, *I. lacteus*,

D. squalens, *P. ostreatus*, and *Coprinus comatus*) towards a historically PAH-contaminated soil collected from a wood preservation plant (Covino et al. 2010a, b). *L. tigrinus* was significantly more efficient in PAH removal than *I. lacteus* regardless of the inoculum carrier used (wheat straw, corn cobs, and commercial pellets) (Covino et al. 2010a). Fluorene, phenanthrene, and anthracene were the most degraded ones (degradation extent ranging from 75% to 91%), whereas chrysene was the least degraded compound (percentages of removal ranging from 1 to 9%). The lower degradation of chrysene was attributed to the higher hydrophobicity of this hydrocarbon with respect to other PAHs, thus to its low bioavailability (available fraction of chrysene was only 41.1%). In the second study, *P. ostreatus* immobilized on commercial pellets resulted to be the most effective fungal inoculant being able to partially deplete even the most recalcitrant PAHs such as benzo[*a*] pyrene and chrysene (Covino et al. 2010b).

The fungal bioaugmentation treatment of an aged creosote-contaminated soil was also evaluated by Lladó et al. (2013). In this case, the degradation capabilities of two strains of WRFs (*T. versicolor* and *L. tigrinus*) towards high-molecular-weight polycyclic aromatic hydrocarbons (HMW-PAHs) were compared. The soil, collected after a 180-days-long biopiling treatment at pilot scale, contained mostly 4- and 5-ring PAHs, the former ones being the most degraded in *L. tigrinus* microcosms supplemented with Mn^{2+} ions.

Concerning non-ligninolytic fungi, bioaugmented microcosms with SMS of the basidiomycete *Agaricus bisporus* were set up at laboratory scale to treat a soil contaminated by Pb and PAHs (García-Delgado et al. 2015). These microcosms led to higher PAH removal and ecotoxicity reduction than their respective biostimulated ones (soil mixed only with sterilized SMS). Moreover, the highest PAH degradation (27% with respect to the initial concentration of 757 mg/kg) was achieved in microcosms where *A. bisporus* was re-inoculated.

A fungal strain isolated from a contaminated soil (*Mucor mucedo*) was also tested for its ability to degrade PAHs in two soils collected from a manufactured gas plant (MGP) and from a wastewater-irrigated site (WIS) (Guo et al. 2016). *Mucor*-augmented microcosms led to a removal of PAHs of 72 and 62% in MGP and WIS soils, respectively, after 30 days of incubation with the highest degradation observed for the 3-ring PAHs.

In a recent research study, three fungal strains of *Aspergillus* spp., one of *Rhizomucor* sp., one of *Trichoderma* sp., and eight bacterial strains were isolated from heavy crude oil-contaminated soils and selected for the construction of two consortia to use for the decontamination of a PAH-polluted soil (Zafra et al. 2017). The first consortium (C1), constituted by the aforementioned fungal and bacterial strains, was able to remove only low molecular weight PAHs, whereas the second consortium C2 (C1 with the addition of two genetically modified strains of *Aspergillus niger* expressing LiP and MnP genes from *P. chrysosporium*) degraded also the high-molecular-weight PAHs.

22.6 Pilot- and Field-Scale Mycoremediation

As reported so far, a number of studies at laboratory scale dealing with the mycoremediation of contaminated matrices was carried out. In contrast, there are only few reports concerning the use of fungi at field scale, which demonstrate the effectiveness and economic viability of the fungal technology application at pilot and full scale. For instance, despite its great effectiveness in degrading organic pollutants in laboratory-scale experiments, the model white-rot fungus *P. chrysosporium* has never been so effective in any field-scale tests.

One of the first trials was reported in the 1990s: the white-rot fungus *P. chrysosporium* was used in the treatment of a trinitrotoluene (TNT)-contaminated soil from a former Naval Submarine Base Bangor (Washington, USA) at pilot scale. The initial TNT content of 1844 ppm was reduced up to 41% in 4 months. However, the final concentration was still significantly above the target level and thus the test was considered a failure (US EPA 1993).

A decade later, a field-scale trial for the remediation of crude oil-contaminated soil was performed. In this case, the efficiency of two biological approaches was compared: composting biopiles technology where the soil was aerated and amended with bulking agents vs. fungal bioaugmentation (Li et al. 2004). Three fungal strains characterized by high lipase activities (*Mucor* sp., *Cunninghamella* sp., and *Fusarium* sp.) and previously isolated, were reintroduced into the contaminated matrices either with or without *P. chrysosporium*. The results showed that the introduction of the allochthonous fungus did not improve the efficiency of total petroleum hydrocarbons (TPH) removal process.

A similar outcome was also achieved a few years later in a pilot-scale experiment for the treatment of soil contaminated by TNT, PAHs, PCDDs, and PCDFs (Tuomela et al. 2012). In this case, the litter-decomposing fungus *P. velutina* inoculum was firstly prepared using pine bark as growth substrate. Once the inoculum was prepared, the allochthonous fungus was inoculated into soil and the extent of degradation of the target pollutants was evaluated. *P. velutina* growth was completely inhibited when introduced into highly affected soil (PAHs and TNT concentrations of 5000 mg/kg and > 10 g/kg, respectively) and no degradation occurred. Furthermore, *P. velutina* growth was limited by the presence of molds such as *Trichoderma*. Contrarily, the degradation process significantly enhanced when PAH- and TNT-contaminated soil was diluted with garden compost. However, the role of the litter-decomposing soil fungus in the removal of PAHs and TNT could not be clarified since also the growth of indigenous microorganisms was clearly stimulated under such conditions. These results were also confirmed 2 years later when a field-scale experiment (2 t) for the treatment of a PAH-contaminated soil was carried out (Winquist et al. 2014). At first, a laboratory-scale experiment was performed to compare the extent of PAH degradation in soil mixed with green compost and incubated with or without *P. velutina*. A higher degradation (up to 96% for 4-ring PAHs and 39% for 5- and 6-ring PAHs) was achieved in microcosms inoculated with the fungus than in the biostimulated ones after 3 months. Secondly, a larger-scale trial was accomplished. Also in this case, the percentage of degraded PAHs in soil under field condition was extremely high, but the efficiency of both treatments (*P. velutina*-inoculated and non-inoculated soil) was similar. Therefore, the degradation of PAHs in this test could not be ascribed univocally to the use of the allochthonous fungus.

Concerning this white-rot fungus, *P. velutina* was selected also for another pilot-scale (0.3 t) treatment trial of a TNT-contaminated soil (Anasonye et al. 2015). Pine bark was used as growth substrate and the contaminated soil was diluted with composted green waste (1:20). TNT was efficiently degraded up to 80% in 107 days. The quantification of the fungal Internal Transcribed Spacer (ITS) region DNA copy number by quantitative PCR (qPCR) using specific primers for *P. velutina* proved that the fungus could withstand high concentration of TNT and autochthonous microorganisms. However, the real efficiency of the fungus in TNT removal could not be proved because non-inoculated soil was not prepared as control due to economical and practical issues and thus the degradation ability of *P. velutina* could not be compared with the indigenous microbiota one.

The mycoremediation of a pentachlorophenol (PCP)-contaminated soil was also assessed at field scale (Walter et al. 2005). The soil was collected from a timber treatment plant where PCP was widely used as a wood preservative. In this case, the capability of the white-rot fungus *T. versicolor* in the remediation of this contaminated matrix was evaluated in 500 L soil cells. At first, the contaminated soil (1000 mg/kg) was diluted with non-contaminated soil and mixed with wood chips obtained from cuttings and shreds. Thereafter, such blended pile of material was mixed with a formulation composed of sawdust, cornmeal, and starch previously inoculated with *T. versicolor* to obtain a final ratio of soil:woodchips:fungal inoculum of 40:20:40 or 60:20:20. A significant decrease in PCP concentration was observed within the first 25 weeks of incubation, while the degradation rate remarkably decreased in the following weeks. Nonetheless, more than 90% of the original PCP content was removed in both cells (20 or 40% of fungal inoculum used) after 2.5 years. These results confirmed that the PCP depletion efficiency was similar regardless of the amount of fungal inoculum used.

P. ostreatus was also applied in several field-scale tests. Explosive-contaminated soil from the Naval Weapons Station Yorktown (Virginia, USA), where munitions manufacturing was carried out for decades, was treated using this fungus (Axtell et al. 2000). Two plots (4.59 m^3) of contaminated soil were blended with 2.29 m^3 of a substrate mixture previously colonized by *P. ostreatus*. The concentration of three target explosives was monitored during the treatment which lasted 62 days. The content of these compounds noticeably decreased; however, the degradation extent and rate in *P. ostreatus*-augmented plots was comparable to those achieved in soil amended only with the substrate mixture during the same period. This outcome showed that the addition of amendments enhanced the growth and activity of indigenous microorganisms promoting the degradation of the explosives in soil. As for other fungi, the role of the *P. ostreatus* in this field-scale test was not elucidated.

The use of fungi was evaluated for the remediation of creosote-contaminated soil at a wood-preserving site in West Virginia (Lamar et al. 2002). At first, a preliminary bench-scale test was carried out to assess the degradation capabilities of two WRFs, namely *P. ostreatus* and *I. lacteus*. Afterwards, *P. ostreatus* from a commercial spawn provider was selected to scale up the process and two treatment biocells were set up at pilot scale. The contaminated soil was mixed with a sterile substrate (wood splinters or wood splinters with sawdust) and *P. ostreatus* spawn. The concentration of all the 16 PAHs was reduced in both biocells below the industrial risk-based level in 276 days with the exception of benzo[*a*]pyrene (Lamar et al. 2002).

P. ostreatus-based treatment was evaluated for the remediation of PAH-contaminated soils under field conditions also in the Northern temperate zone (Hestbjerg et al. 2003). The contaminated soils were collected from two different areas of an asphalt factory: a former shipyard and a coal tar storage. Concrete cylinders were filled with both soils and used to test different conditions: non-amended soil (control), soil mixed with autoclaved sawdust substrate, and soil mixed with *P. ostreatus* spawn. In the case of the soil from the coal tar storage, the addition of the oyster mushroom significantly enhanced the degradation rate of 4-ring PAHs compared to the rate in either soil control or biostimulated cylinder, while 5- and 6-ring PAHs were only slightly removed and no significant difference was observed between the two treatments. By contrast, the growth and the activity of *P. ostreatus* were completely prevented when inoculated in the other soil, although the factors that resulted in its inefficacy were not clarified.

22.7 The Potentialities and Drawbacks of Soil Mycoremediation

The potentialities of using both ligninolytic and non-ligninolytic fungi for the degradation of persistent chlorinated and non-chlorinated organopollutants in contaminated soils have been clearly demonstrated through a number of studies at laboratory scale since the 1980s.

However, as it was highlighted in the previous section, the low number of larger-scale trials showed remarkable limitations in the practical application of fungal-based technologies. Consequently, all the attempts to promote soil mycoremediation on the market as a valuable on-site bioremediation option have failed (Šašek 2003; Baldrian 2008).

One of the reasons that limited the use of the fungal-augmentation strategy in the remediation of contaminated soils was the competition between the inoculated fungi and the autochthonous soil microorganisms. Indeed, when a single allochthonous fungal species is introduced into soil, its colonization, growth, and metabolism are strongly affected by the type and activity of other resident soil organisms. Moreover, the inoculation carrier onto which fungi are immobilized prior to their introduction in non-sterile soil (e.g., straw, corn cobs, seeds husks, wood chips or shavings, and bark) represents a source of nutrients also for the autochthonous soil microbiota. Even if the addition of nutrients positively impacts on the bioremediation efficiency (Lang et al. 2000; Steffen et al. 2007; Federici et al. 2012), the growth-inhibition and mycophagy phenomena could still occur as a result (de Boer et al. 2005; Baldrian 2008).

Other constraints for the application of fungal technologies at field scale are the strict growth conditions required by most WRFs. For example, 37 and 30 °C represent the optimum temperatures for *P. chrysosporium* growth and ligninolytic enzymes production, respectively. These conditions are not easily maintained under actual environmental conditions due to daily and seasonal fluctuations (Hestbjerg et al. 2003). Contrary to *P. chrysosporium*, *P. ostreatus* is less affected by temperature changes: the degradation of organic pollutants by this WRF was demonstrated even at very low temperatures (up to 8 °C) (Eggen and Sveum 1999; Lang et al. 2000; Hestbjerg et al. 2001). Furthermore, *Pleurotus* sp. is highly competitive with soil microbiota and thus more successful to colonize the contaminated matrix than other fungi (in der Wiesche et al. 1996; Lang et al. 1997, 1998; Stella et al. 2017).

Another parameter that deeply influences the effectiveness of bioremediation is the bioavailability of contaminants, which depends on soil organic matter content, soil texture and structure, soil aging, and hydrophobicity of the contaminants. For instance, historically contaminated matrices are characterized by a significant non-bioavailable fraction of compounds and the complete biodegradation of recalcitrant pollutants, such as PAHs and PCBs, is hardly reached. Nonetheless, several authors asserted that bioavailability should be the underlying basis for risk assessment and in setting the remediation goals at contaminated sites (Latawiec et al. 2010; Naidu et al. 2015; Stella et al. 2013). This concept is more and more readily accepted by decision makers and regulatory organizations (Naidu et al. 2015). Therefore, the efficiency of certain bioremediation applications at field scale should be evaluated by the estimation of both concentration and bioavailability of the target contaminants. Moreover, the feasibility of applying a mycoremediation strategy, as well as other bio-based remediation technologies, should be evaluated performing accurate compound-specific and ecotoxicological analyses to assess if the oxidized and more water-soluble products exert a higher toxicity than parent molecules themselves. In case polar metabolites are formed, as in the case of oxidized HMW-PAHs, it is of utmost importance to determine the presence of other microorganisms that are able to cooperate with fungi in the complete mineralization of contaminants.

Another important parameter to take into account when a mycoremediation approach is selected for the restoration of contaminated sites is the formulation and the amount of inoculum carrier. Indeed, ligninolytic fungi are not able to grow in soil unless an external substrate is provided. Therefore, the selection and the pre-colonization of the lignocellulosic carrier by the fungus is crucial to support an efficient and long-lasting fungal growth into soil. However, using a large amount of lignocellulose-immobilized fungi is not always economically feasible and would significantly increase the volume of the soil after treatment. A valid alternative to bypass the high costs associated with the production of fungal inoculum is the reuse of SMS, which is a cheap and readily available source of both fungal mycelia and enzymes (Phan and Sabaratnam 2012). The spent residues from mushroom industries are still considered waste to be disposed of, and, thus, the application of commercially edible fungi spawn (i.e., *P. ostreatus*, *A. bisporus*) in bioremediation trials represents a valid alternative to disposal. In this respect, *P. ostreatus* refuse from commercial mushroom production was exploited as a source of viable mycelia to remediate contaminated soils under field conditions as reported in the previous section.

22.8 Conclusions

The potentialities of fungi in the remediation of contaminated matrices were highlighted and the shortcomings that make this technology still unsuitable for large-scale trials were also analyzed in this chapter. A step forward in the application of fungal-based technology for the treatment of polluted environmental matrices is strongly required. In this respect, a more comprehensive knowledge of the ecology and physiology of fungi in their natural environment is needed, with specific regard to the condition when fungi come in contact with the recalcitrant organic pollutants.

With this regard, the recent advances in genomics have triggered a revolution in understanding even the most complex biological systems. Nowadays, indeed, these innovative molecular biology techniques allow gaining insight into the structure and dynamics of microbial community during the fungal colonization of contaminated solid matrices as well as into the interactions between allochthonous and autochthonous species.

Mycoremediation, as all the other bioremediation technologies, cannot be considered as "the" method of choice nowadays, but it still deserves attention because the combination of fungal treatment with bacteria-based approaches or with phytoremediation could overcome the limitations associated with their stand-alone application.

References

Anasonye F, Winquist E, Räsänen M, Kontro J, Björklöf K, Vasilyeva G, Jørgensen KS, Steffen KT, Tuomela M (2015) Bioremediation of TNT contaminated soil with fungi under laboratory and pilot scale conditions. Int Biodeterior Biodegrad 105:7–12. https://doi.org/10.1016/j.ibiod.2015.08.003

Axtell C, Johnston CG, Bumpus JA (2000) Bioremediation of soil contaminated with explosives at the naval weapons station Yorktown. J Soil Contam 9(6):537–548. https://doi.org/10.1080/10588330091134392

Baldrian P (2003) Interactions of heavy metals with white-rot fungi. Enzym Microb Technol 32(1):78–91. https://doi.org/10.1016/S0141-0229(02)00245-4

Baldrian P (2008) Wood-inhabiting ligninolytic basidiomycetes in soils: ecology and constraints for applicability in bioremediation. Fungal Ecol 1(1):4–12. https://doi.org/10.1016/j.funeco.2008.02.001

Baldrian P, in der Wiesche C, Gabriel J, Nerud F, Zadražil F (2000) Influence of cadmium and mercury on activities of ligninolytic enzymes and degradation of polycyclic aromatic hydrocarbons by *Pleurotus ostreatus* in soil. Appl Environ Microbiol 66(6):2471–2478. https://doi.org/10.1128/AEM.66.6.2471-2478.2000

Beaudette LA, Davies S, Fedorak PM, Ward OP, Pickard MA (1998) Comparison of gas chromatography and mineralization experiments for measuring loss of selected polychlorinated biphenyl congeners in cultures of white rot fungi. Appl Environ Microbiol 64(6):2020–2025

Bermek H, Gülseren I, Li K, Jung H, Tamerler C (2004) The effect of fungal morphology on ligninolytic enzyme production by a recently isolated wood-degrading fungus *Trichophyton rubrum* LSK-27. World J Microbiol Biotechnol 20(4):345–349. https://doi.org/10.1023/B:WIBI.0000033055.52660.03

Bernhardt R (2006) Cytochromes P450 as versatile biocatalysts. J Biotechnol 124(1):128–145. https://doi.org/10.1016/j.jbiotec.2006.01.026

Bollag JM (1992) Decontaminating soil with enzymes. Environ Sci Technol 26(10):1876–1881. https://doi.org/10.1021/es00034a002

Bonfante P, Anca I-A (2009) Plants, mycorrhizal fungi, and bacteria: a network of interactions. Annu Rev Microbiol 63:363–383. https://doi.org/10.1146/annurev.micro.091208.073504

Borazjani H, Wiltcher D, Diehl S (2005) Bioremediation of polychlorinated biphenyl and petroleum contaminated soil. In: Lyon WG, Hong JJ, Reddy RK (eds) Proceedings of the first international conference on environmental science and technology. American Science Press, New Orleans, pp 502–507

Bornyasz MA, Graham RC, Allen MF (2005) Ectomycorrhizae in a soil-weathered granitic bedrock regolith: linking matrix resources to plants. Geoderma 126(1–2):141–160. https://doi.org/10.1016/j.geoderma.2004.11.023

Bumpus JA, Tien M, Wright D, Aust SD (1985) Oxidation of persistent environmental pollutants by a white rot fungus. Science 228(4706):1434–1436. https://doi.org/10.1126/science.3925550

Cajthaml T, Svobodová K (2012) Biodegradation of aromatic pollutants by ligninolytic fungal strains. In: Singh SN (ed) Microbial degradation of xenobiotics. Springer, Berlin/Heidelberg, pp 291–316

Camarero S, Sarkar S, Ruiz-Dueñas FJ, Martínez MJ, Martinez ÁT (1999) Description of a versatile peroxidase involved in the natural degradation of lignin that has both manganese peroxidase and lignin peroxidase substrate interaction sites. J Biol Chem 274(15):10324–10330. https://doi.org/10.1074/jbc.274.15.10324

Carvalho MB, Martins I, Leitão MC, Garcia H, Rodrigues C, San Romão V, McLellan I, Hursthouse A, Pereira CS (2009) Screening pentachlorophenol degradation ability by environmental fungal strains belonging to the phyla Ascomycota and Zygomycota. J Ind Microbiol Biotechnol 36(10):1249–1256. https://doi.org/10.1007/s10295-009-0603-2

Cerniglia CE, Sutherland JB (2006) Relative roles of bacteria and fungi in polycyclic aromatic hydrocarbon biodegradation and bioremediation of contaminated soils. In: Gadd GM (ed) Fungi in biogeochemical cycles. British Mycological Society Symposia. Cambridge University Press, Cambridge, pp 182–211. https://doi.org/10.1017/CBO9780511550522.009

Cloete TE, Celliers L (1999) Removal of Aroclor 1254 by the white rot fungus *Coriolus versicolor* in the presence of different concentrations of Mn(IV)oxide. Int Biodeterior Biodegrad 44 (4):243–253. https://doi.org/10.1016/S0964-8305(99)00085-2

Covino S, Čvančarová M, Muzikář M, Svobodová K, D'Annibale A, Petruccioli M, Federici F, Křesinová Z, Cajthaml T (2010a) An efficient PAH-degrading *Lentinus* (*Panus*) *tigrinus* strain: effect of inoculum formulation and pollutant bioavailability in solid matrices. J Hazard Mater 183(1–3):669–676. https://doi.org/10.1016/j.jhazmat.2010.07.078

Covino S, Svobodová K, Čvančarová M, D'Annibale A, Petruccioli M, Federici F, Křesinová Z, Galli E, Cajthaml T (2010b) Inoculum carrier and contaminant bioavailability affect fungal degradation performances of PAH-contaminated solid matrices from a wood preservation plant. Chemosphere 79(8):855–864. https://doi.org/10.1016/j.chemosphere.2010.02.038

Črešnar B, Petrič Š (2011) Cytochrome P450 enzymes in the fungal kingdom. Biochim Biophys Acta, Proteins Proteomics 1814(1):29–35. https://doi.org/10.1016/j.bbapap.2010.06.020

Crinnion WJ (2011) Polychlorinated biphenyls: persistent pollutants with immunological, neurological, and endocrinological consequences. Altern Med Rev 16(1):5–13

Čvančarová M, Křesinová Z, Filipová A, Covino S, Cajthaml T (2012) Biodegradation of PCBs by ligninolytic fungi and characterization of the degradation products. Chemosphere 88 (11):1317–1323. https://doi.org/10.1016/j.chemosphere.2012.03.107

D'Annibale A, Ricci M, Leonardi V, Quaratino D, Mincione E, Petruccioli M (2005) Degradation of aromatic hydrocarbons by white-rot fungi in a historically contaminated soil. Biotechnol Bioeng 90(6):723–731. https://doi.org/10.1002/bit.20461

de Boer W, Folman LB, Summerbell RC, Boddy L (2005) Living in a fungal world: impact of fungi on soil bacterial niche development. FEMS Microbiol Rev 29(4):795–811. https://doi.org/10.1016/j.femsre.2004.11.005

Dmochewitz S, Ballschmiter K (1988) Microbial transformation of technical mixtures of polychlorinated biphenyls (PCB) by the fungus *Aspergillus niger*. Chemosphere 17 (1):111–121. https://doi.org/10.1016/0045-6535(88)90049-5

Eaton DC (1985) Mineralization of polychlorinated biphenyls by *Phanerochaete chrysosporium*: a ligninolytic fungus. Enzym Microb Technol 7(5):194–196. https://doi.org/10.1016/S0141-0229(85)80001-6

Eggen T, Sveum P (1999) Decontamination of aged creosote polluted soil: the influence of temperature, white rot fungus *Pleurotus ostreatus*, and pretreatment. Int Biodeterior Biodegrad 43(3):125–133. https://doi.org/10.1016/S0964-8305(99)00039-6

El Majidi N, Bouchard M, Carrier G (2013) Systematic analysis of the relationship between standardized prenatal exposure to polychlorinated biphenyls and mental and motor development during follow-up of nine children cohorts. Regul Toxicol Pharmacol 66(1):130–146. https://doi.org/10.1016/j.yrtph.2013.03.002

Federici E, Giubilei M, Santi G, Zanaroli G, Negroni A, Fava F, Petruccioli M, D'Annibale A (2012) Bioaugmentation of a historically contaminated soil by polychlorinated biphenyls with *Lentinus tigrinus*. Microb Cell Factories 11:35. https://doi.org/10.1186/1475-2859-11-35

Gadd GM (ed) (2001) Fungi in bioremediation. British Mycological Society Symposia. Cambridge University Press, Cambridge. https://doi.org/10.1017/CBO9780511541780

Galhaup C, Goller S, Peterbauer CK, Strauss J, Haltrich D (2002) Characterization of the major laccase isoenzyme from *Trametes pubescens* and regulation of its synthesis by metal ions. Microbiology 148(7):2159–2169. https://doi.org/10.1099/00221287-148-7-2159

García-Delgado C, Yunta F, Eymar E (2015) Bioremediation of multi-polluted soil by spent mushroom (*Agaricus bisporus*) substrate: polycyclic aromatic hydrocarbons degradation and Pb availability. J Hazard Mater 300:281–288. https://doi.org/10.1016/j.jhazmat.2015.07.008

Giubilei MA, Leonardi V, Federici E, Covino S, Šašek V, Novotny C, Federici F, D'Annibale A, Petruccioli M (2009) Effect of mobilizing agents on mycoremediation and impact on the indigenous microbiota. J Chem Technol Biotechnol 84(6):836–844. https://doi.org/10.1002/jctb.2126

Guo M, Gong Z, Allinson G, Tai P, Miao R, Li X, Jia C, Zhuang J (2016) Variations in the bioavailability of polycyclic aromatic hydrocarbons in industrial and agricultural soils after bioremediation. Chemosphere 144:1513–1520. https://doi.org/10.1016/j.chemosphere.2015.10.027

Haritash AK, Kaushik CP (2009) Biodegradation aspects of polycyclic aromatic hydrocarbons (PAHs): a review. J Hazard Mater 169(1–3):1–15. https://doi.org/10.1016/j.jhazmat.2009.03.137

Hatakka A (2001) Biodegradation of lignin. In: Steinbüchel A (ed) Biopolymers. Biology, chemistry, biotechnology, applications, vol 1. Wiley-VCH Verlag, Weinheim, pp 129–180. https://doi.org/10.1002/3527600035.bpol1005

Heinfling A, Martínez MJ, Martínez AT, Bergbauer M, Szewzyk U (1998) Purification and characterization of peroxidases from the dye-decolorizing fungus *Bjerkandera adusta*. FEMS Microbiol Lett 165(1):43–50. https://doi.org/10.1111/j.1574-6968.1998.tb13125.x

Helmfrid I, Berglund M, Löfman O, Wingren G (2012) Health effects and exposure to polychlorinated biphenyls (PCBs) and metals in a contaminated community. Environ Int 44:53–58. https://doi.org/10.1016/j.envint.2012.01.009

Hestbjerg H, Brinch UC, Nielsen T (2001) Potential of *Pleurotus ostreatus* for bioremediation: temperature requirements and competitiveness. Abstracts, SETAC Organic Soil Contaminants, Copenhagen, Denmark, September 2–5, p 24

Hestbjerg H, Willumsen PA, Christensen M, Andersen O, Jacobsen CS (2003) Bioaugmentation of tar-contaminated soils under field conditions using *Pleurotus ostreatus* refuse from commercial mushroom production. Environ Toxicol Chem 22(4):692–698. https://doi.org/10.1002/etc.5620220402

Hofrichter M, Ullrich R (2014) Oxidations catalyzed by fungal peroxygenases. Curr Opin Chem Biol 19:116–125. https://doi.org/10.1016/j.cbpa.2014.01.015

Huff JE, Moore JA, Saracci R, Tomatis L (1980) Long-term hazards of polychlorinated dibenzodioxins and polychlorinated dibenzofurans. Environ Health Perspect 36:221–240. https://doi.org/10.1289/ehp.8036221

in der Wiesche C, Martens R, Zadrazil F (1996) Two-step degradation of pyrene by white-rot fungi and soil microorganisms. Appl Microbiol Biotechnol 46(5–6):653–659. https://doi.org/10.1007/s002530050876

Ishii K, Furuichi T (2007) Development of bioreactor system for treatment of dioxin-contaminated soil using *Pseudallescheria boydii*. J Hazard Mater 148(3):693–700. https://doi.org/10.1016/j.jhazmat.2007.03.032

Ishii K, Furuichi T, Tanikawa N, Kuboshima M (2009) Estimation of the biodegradation rate of 2,3,7,8-tetrachlorodibenzo-*p*-dioxin by using dioxin-degrading fungus, *Pseudallescheria boydii*. J Hazard Mater 162(1):328–332. https://doi.org/10.1016/j.jhazmat.2008.05.043

Jeffries TW, Choi S, Kirk TK (1981) Nutritional regulation of lignin degradation by *Phanerochaete chrysosporium*. Appl Environ Microbiol 42(2):290–296

Kamei I, Kondo R (2005) Biotransformation of dichloro-, trichloro-, and tetrachlorodibenzo-*p*-dioxin by the white-rot fungus *Phlebia lindtneri*. Appl Microbiol Biotechnol 68(4):560–566. https://doi.org/10.1007/s00253-005-1947-9

Kamei I, Suhara H, Kondo R (2005) Phylogenetical approach to isolation of white-rot fungi capable of degrading polychlorinated dibenzo-*p*-dioxin. Appl Microbiol Biotechnol 69:358–366. https://doi.org/10.1007/s00253-005-0052-4

Kamei I, Kogura R, Kondo R (2006a) Metabolism of 4,4′-dichlorobiphenyl by white-rot fungi *Phanerochaete chrysosporium* and *Phanerochaete* sp. MZ142. Appl Microbiol Biotechnol 72(3):566–575. https://doi.org/10.1007/s00253-005-0303-4

Kamei I, Sonoki S, Haraguchi K, Kondo R (2006b) Fungal bioconversion of toxic polychlorinated biphenyls by white-rot fungus, *Phlebia brevispora*. Appl Microbiol Biotechnol 73(4):932–940. https://doi.org/10.1007/s00253-006-0529-9

Kohlmeier S, Smits TMH, Ford RM, Keel C, Harms H, Lukas YW (2005) Taking the fungal highway: mobilization of pollutant-degrading bacteria by fungi. Environ Sci Technol 39(12):4640–4646. https://doi.org/10.1021/es047979z

Kramer S, Moller Hikel S, Adams K, Hinds D, Moon K (2012) Current status of the epidemiologic evidence linking polychlorinated biphenyls and non-Hodgkin lymphoma and the role of immune dysregulation. Environ Health Perspect 120(8):1067–1075. https://doi.org/10.1289/ehp.1104652

Krčmář P, Kubátová A, Votruba J, Erbanová P, Novotný Č, Šašek V (1999) Degradation of polychlorinated biphenyls by extracellular enzymes of *Phanerochaete chrysosporium* produced in a perforated plate bioreactor. World J Microbiol Biotechnol 15(2):269–276. https://doi.org/10.1023/A:1008994912875

Kubátová A, Erbanová P, Eichlerová I, Homolka L, Nerud F, Šašek V (2001) PCB congener selective biodegradation by the white rot fungus *Pleurotus ostreatus* in contaminated soil. Chemosphere 43(2):207–215. https://doi.org/10.1016/S0045-6535(00)00154-5

Lamar RT, White RB, Ashley KC (2002) Evaluation of white-rot fungi for the remediation of creosote-contaminated soil. Remediation 12(4):97–106. https://doi.org/10.1002/rem.10048

Lang E, Eller G, Zadrazil F (1997) Lignocellulose decomposition and production of ligninolytic enzymes during interaction of white rot fungi with soil microorganisms. Microb Ecol 34(1):1–10. https://doi.org/10.1007/s002489900029

Lang E, Nerud F, Zadrazil F (1998) Production of ligninolytic enzymes by *Pleurotus* sp. and *Dichomitus squalens* in soil and lignocellulose substrate as influenced by soil microorganisms. FEMS Microbiol Lett 167(2):239–244. https://doi.org/10.1016/S0378-1097(98)00395-4

Lang E, Gonser A, Zadrazil F (2000) Influence of incubation temperature on activity of ligninolytic enzymes in sterile soil by *Pleurotus* sp. and *Dichomitus squalens*. J Basic Microbiol 40 (1):33–39. https://doi.org/10.1002/(SICI)1521-4028(200002)40:1<33::AID-JOBM33>3.0.CO;2-Q

Latawiec AE, Swindell AL, Simmons P, Reid BJ (2010) Bringing bioavailability into contaminated land decision making: the way forward? Crit Rev Environ Sci Technol 41(1):52–77. https://doi.org/10.1080/00102200802641780

Leonowicz A, Cho NS, Luterek J, Wilkolazka A, Wojtas-Wasilewska M, Matuszewska A, Hofrichter M, Wesenberg D, Rogalski J (2001) Fungal laccase: properties and activity on lignin. J Basic Microbiol 41(3–4):185–227. https://doi.org/10.1002/1521-4028(200107)41:3/4<185::AID-JOBM185>3.0.CO;2-T

Li P, Sun T, Stagnitti F, Zhang C, Zhang H, Xiong X, Allinson G, Ma X, Allinson M (2004) Field-scale bioremediation of soil contaminated with crude oil. Environ Eng Sci 19(5):277–289. https://doi.org/10.1089/10928750260418926

Lladó S, Covino S, Solanas AM, Viñas M, Petruccioli M, D'Annibale A (2013) Comparative assessment of bioremediation approaches to highly recalcitrant PAH degradation in a real industrial polluted soil. J Hazard Mater 248–249:407–414. https://doi.org/10.1016/j.jhazmat.2013.01.020

Lundell TK, Mäkelä MR, Hildén K (2010) Lignin-modifying enzymes in filamentous basidiomycetes – ecological, functional and phylogenetic review. J Basic Microbiol 50(1):5–20. https://doi.org/10.1002/jobm.200900338

Manji S, Ishihara A (2004) Screening of tetrachlorodibenzo-*p*-dioxin-degrading fungi capable of producing extracellular peroxidases under various conditions. Appl Microbiol Biotechnol 63 (4):438–444. https://doi.org/10.1007/s00253-003-1356-x

Martínez AT (2002) Molecular biology and structure-function of lignin-degrading heme peroxidases. Enzym Microb Technol 30(4):425–444. https://doi.org/10.1016/S0141-0229(01)00521-X

Michels J (ed) (1998) Processes for the bioremediation of soil. Compilation of current projects of the joint research group. Editor: Federal Environmental Agency, Section III 3.6, BMBF Project Management Agency for Waste Management and Remediation of Hazardous Abandoned Sites (PT AWAS), Berlin, Germany

Mori T, Kondo R (2002) Degradation of 2,7-dichlorodibenzo-*p*-dioxin by wood-rotting fungi, screened by dioxin degrading ability. FEMS Microbiol Lett 213(1):127–131. https://doi.org/10.1111/j.1574-6968.2002.tb11296.x

Mori T, Watanabe M, Taura H, Kuno T, Kamei I, Kondo R (2015) Degradation of chlorinated dioxins and polycyclic aromatic hydrocarbons (PAHs) and remediation of PAH-contaminated soil by the entomopathogenic fungus, *Cordyceps militaris*. J Environ Chem Eng 3 (4):2317–2322. https://doi.org/10.1016/j.jece.2015.08.020

Mougin C, Kollmann A, Jolivalt C (2002) Enhanced production of laccase in the fungus *Trametes versicolor* by the addition of xenobiotics. Biotechnol Lett 24(2):139–142. https://doi.org/10.1023/A:1013802713266

Mouhamadou B, Faure M, Sage L, Marçais J, Souard F, Geremia RA (2013) Potential of autochthonous fungal strains isolated from contaminated soils for degradation of polychlorinated biphenyls. Fungal Biol 117(4):268–274. https://doi.org/10.1016/j.funbio.2013.02.004

Naidu R, Channey R, McConnell S, Johnston N, Semple KT, McGrath S, Dries V, Nathanail P, Harmsen J, Pruszinski A, MacMillan J, Palanisami T (2015) Towards bioavailability-based soil criteria: past, present and future perspectives. Environ Sci Pollut Res 22(12):8779–8785. https://doi.org/10.1007/s11356-013-1617-x

Nakamiya K, Furuichi T, Ishii K (2002) Isolation of a fungus from denitrifying activated sludge that degrades highly chlorinated dioxins. J Mater Cycles Waste Manag 4(2):127–134. https://doi.org/10.1007/s10163-002-0062-6

Nakamiya K, Hashimoto S, Ito H, Edmonds JS, Yasuhara A, Morita M (2005) Degradation of dioxins by cyclic ether degrading fungus, *Cordyceps sinensis*. FEMS Microbiol Lett 248 (1):17–22. https://doi.org/10.1016/j.femsle.2005.05.013

Nam I-H, Kim Y-M, Murugesan K, Jeon J-R, Chang Y-Y, Chang Y-S (2008) Bioremediation of PCDD/Fs-contaminated municipal solid waste incinerator fly ash by a potent microbial biocatalyst. J Hazard Mater 157(1):114–121. https://doi.org/10.1016/j.jhazmat.2007.12.086

Novotný Č, Vyas BRM, Erbanová P, Kubátová A, Šašek V (1997) Removal of PCBs by various white rot fungi in liquid cultures. Folia Microbiol 42(2):136–140. https://doi.org/10.1007/BF02898723

Novotný Č, Cajthaml T, Svobodová K, Šušla M, Šašek V (2009) *Irpex lacteus*, a white-rot fungus with biotechnological potential – *review*. Folia Microbiol 54(5):375–390. https://doi.org/10.1007/s12223-009-0053-2

Phan C-W, Sabaratnam V (2012) Potential uses of spent mushroom substrate and its associated lignocellulosic enzymes. Appl Microbiol Biotechnol 96(4):863–873. https://doi.org/10.1007/s00253-012-4446-9

Pinedo-Rivilla C, Aleu J, Collado IG (2009) Pollutants biodegradation by fungi. Curr Org Chem 13 (12):1194–1214. https://doi.org/10.2174/138527209788921774

Podgornik H, Podgornik A, Milavec P, Perdih A (2001) The effect of agitation and nitrogen concentration on lignin peroxidase (LiP) isoform composition during fermentation of *Phanerochaete chrysosporium*. J Biotechnol 88(2):173–176. https://doi.org/10.1016/S0168-1656(01)00270-X

Pointing SB (2001) Feasibility of bioremediation by white-rot fungi. Appl Microbiol Biotechnol 57 (1–2):20–33. https://doi.org/10.1007/s002530100745

Rabinovich ML, Bolobova AV, Vasil'chenko LG (2004) Fungal decomposition of natural aromatic structures and xenobiotics: a review. Appl Biochem Microbiol 40(1):5–23. https://doi.org/10.1023/B:ABIM.0000010343.73266.08

Ruiz-Aguilar GML, Fernández-Sánchez JM, Rodríguez-Vázquez R, Poggi-Varaldo H (2002) Degradation by white-rot fungi of high concentrations of PCB extracted from a contaminated soil. Adv Environ Res 6(4):559–568. https://doi.org/10.1016/S1093-0191(01)00102-2

Ruiz-Dueñas FJ, Camarero S, Pérez-Boada M, Martínez MJ, Martínez AT (2001) A new versatile peroxidise from *Pleurotus*. Biochem Soc Trans 29(2):116–122. https://doi.org/10.1042/bst0290116

Sack U, Heinze TM, Deck J, Cerniglia CE, Martens R, Zadrazil F, Fritsche W (1997) Comparison of phenanthrene and pyrene degradation by different wood-decaying fungi. Appl Environ Microbiol 63(10):3919–3925

Sage L, Périgon S, Faure M, Gaignaire C, Abdelghafour M, Mehu J, Geremia RA, Mouhamadou B (2014) Autochthonous ascomycetes in depollution of polychlorinated biphenyls contaminated soil and sediment. Chemosphere 110:62–69. https://doi.org/10.1016/j.chemosphere.2014.03.013

Šašek V (2003) Why mycoremediations have not yet come into practice. In: The utilization of bioremediation to reduce soil contamination: problems and solutions, NATO science series: IV, vol 19. Springer, Dordrecht, pp 247–266. https://doi.org/10.1007/978-94-010-0131-1_22

Scheel T, Höfer M, Ludwig S, Hölker U (2000) Differential expression of manganese peroxidase and laccase in white-rot fungi in the presence of manganese or aromatic compounds. Appl Microbiol Biotechnol 54(5):686–691. https://doi.org/10.1007/s002530000427

Seto M, Nishibori K, Masai E, Fukuda M, Ohdaira Y (1999) Degradation of polychlorinated biphenyls by a 'Maitake' mushroom, *Grifola frondosa*. Biotechnol Lett 21(1):27–31. https://doi.org/10.1023/A:1005457201291

Singh H (2006) Mycoremediation: fungal bioremediation. Wiley-Interscience, Hoboken

Siracusa G, Becarelli S, Lorenzi R, Gentini A, Di Gregorio S (2017) PCB in the environment: bio-based processes for soil decontamination and management of waste from the industrial production of *Pleurotus ostreatus*. New Biotechnol 39:232–239. https://doi.org/10.1016/j.nbt.2017.08.011

Steffen KT, Schubert S, Tuomela M, Hatakka A, Hofrichter M (2007) Enhancement of bioconversion of high-molecular mass polycyclic aromatic hydrocarbons in contaminated non-sterile soil by litter-decomposing fungi. Biodegradation 18(3):359–369. https://doi.org/10.1007/s10532-006-9070-x

Stella T, Covino S, Křesinová Z, D'Annibale A, Petruccioli M, Čvančarová M, Cajthaml T (2013) Chlorobenzoic acid degradation by *Lentinus* (*Panus*) *tigrinus*: *In vivo* and *in vitro* mechanistic study-evidence for P-450 involvement in the transformation. J Hazard Mater 260:975–983. https://doi.org/10.1016/j.jhazmat.2013.07.004

Stella T, Covino S, Čvančarová M, Filipová A, Petruccioli M, D'Annibale A, Cajthaml T (2017) Bioremediation of long-term PCB-contaminated soil by white-rot fungi. J Hazard Mater 324:701–710. https://doi.org/10.1016/j.jhazmat.2016.11.044

Takada S, Nakamura M, Matsueda T, Kondo R, Sakai K (1996) Degradation of polychlorinated dibenzo-*p*-dioxins and polychlorinated dibenzofurans by the white rot fungus *Phanerochaete sordida* YK-624. Appl Environ Microbiol 62(12):4323–4328

Takagi S, Shirota C, Sakaguchi K, Suzuki J, Sue T, Nagasaka H, Hisamatsu S, Sonoki S (2007) Exoenzymes of *Trametes versicolor* can metabolize coplanar PCB congeners and hydroxy PCB. Chemosphere 67(9):S54–S57. https://doi.org/10.1016/j.chemosphere.2006.05.090

Thomas DR, Carswell KS, Georgiou G (1992) Mineralization of biphenyl and PCBs by the white rot fungus *Phanerochaete chrysosporium*. Biotechnol Bioeng 40(11):1395–1402. https://doi.org/10.1002/bit.260401114

Thurston CF (1994) The structure and function of fungal laccases. Microbiology 140:19–26

Tigini V, Prigione V, Di Toro S, Fava F, Varese GC (2009) Isolation and characterisation of polychlorinated biphenyl (PCB) degrading fungi from a historically contaminated soil. Microb Cell Factories 8:5. https://doi.org/10.1186/1475-2859-8-5

Tortella GR, Diez MC, Durán N (2005) Fungal diversity and use in decomposition of environmental pollutants. Crit Rev Microbiol 31(4):197–212. https://doi.org/10.1080/10408410500304066

Tuomela M, Steffen KT, Kerko E, Hartikainen H, Hofrichter M, Hatakka A (2005) Influence of Pb contamination in boreal forest soil on the growth and ligninolytic activity of litter-decomposing fungi. FEMS Microbiol Ecol 53(1):179–186. https://doi.org/10.1016/j.femsec.2004.11.008

Tuomela M, Jørgensen K, Winquist E, Björklöf K, Schultz E, Anasonye F, Häkkinen L, Räsänen M, Sorvari J, Hartikainen ES, Steffen K, Valentin L (2012) Mycoremediation of contaminated soil in field scale. Conference abstract. Environ Eng Manag J 11(3):S38

Tuor U, Winterhalter K, Fiechter A (1995) Enzymes of white-rot fungi involved in lignin degradation and ecological determinants for wood decay. J Biotechnol 41(1):1–17. https://doi.org/10.1016/0168-1656(95)00042-O

Ullrich R, Nüske J, Scheibner K, Spantzel J, Hofrichter M (2004) Novel haloperoxidase from the agaric basidiomycete *Agrocybe aegerita* oxidizes aryl alcohols and aldehydes. Appl Environ Microbiol 70(8):4575–4581. https://doi.org/10.1128/AEM.70.8.4575-4581.2004

US EPA (1993) Handbook: approaches for the remediation of federal facility sites contaminated with explosive and radioactive wastes. Office of Research and Development, Cincinnati, Ohio. EPA/625/R-93/013

US EPA (2004) Exposure and human health reassessment of 2,3,7,8-tetrachlorodibenzo-*p*-dioxin (TCDD) and related compounds. Part I: Estimating exposure to dioxin-like compounds. Vol. 3. EPA/600/P-00/001

Valentín L, Feijoo G, Moreira MT, Lema JM (2006) Biodegradation of polycyclic aromatic hydrocarbons in forest and salt marsh soils by white-rot fungi. Int Biodeterior Biodegrad 58 (1):15–21. https://doi.org/10.1016/j.ibiod.2006.04.002

Valli K, Wariishi H, Gold MH (1992) Degradation of 2,7-dichlorodibenzo-*p*-dioxin by the lignin-degrading basidiomycete *Phanerochaete chrysosporium*. J Bacteriol 174(7):2131–2137. https://doi.org/10.1128/jb.174.7.2131-2137.1992

Van den Berg M, Birnbaum LS, Denison M, De Vito M, Farland W, Feeley M, Fiedler H, Hakansson H, Hanberg A, Haws L, Rose M, Safe S, Schrenk D, Tohyama C, Tritscher A, Tuomisto J, Tysklind M, Walker N, Peterson RE (2006) The 2005 World Health Organization reevaluation of human and mammalian toxic equivalency factors for dioxins and dioxin-like compounds. Toxicol Sci 93(2):223–241. https://doi.org/10.1093/toxsci/kfl055

van den Brink HM, van Gorcom RFM, van den Hondel CAMJJ, Punt PJ (1998) Cytochrome P450 enzyme systems in fungi. Fungal Genet Biol 23(1):1–17. https://doi.org/10.1006/fgbi.1997.1021

Vyas BRM, Šašek V, Matucha M, Bubner M (1994) Degradation of 3,3′,4,4′-tetrachlorobiphenyl by selected white rot fungi. Chemosphere 28(6):1127–1134. https://doi.org/10.1016/0045-6535(94)90331-X

Walker CH, Hopkin SP, Sibly RM, Peakall DB (2006) Principles of ecotoxicology, 3rd edn. CRC Press/Taylor & Francis, Boca Raton

Walter M, Boyd-Wilson K, Boul L, Ford C, McFadden D, Chong B, Pinfold J (2005) Field-scale bioremediation of pentachlorophenol by *Trametes versicolor*. Int Biodeterior Biodegrad 56(1):51–57. https://doi.org/10.1016/j.ibiod.2005.05.003

Winquist E, Björklöf K, Schultz E, Räsänen M, Salonen K, Anasonye F, Cajthaml T, Steffen KT, Jørgensen KS, Tuomela M (2014) Bioremediation of PAH-contaminated soil with fungi – from laboratory to field scale. Int Biodeterior Biodegrad 86:238–247. https://doi.org/10.1016/j.ibiod.2013.09.012

Yadav JS, Quensen JFIII, Tiedje JM, Reddy CA (1995) Degradation of polychlorinated biphenyl mixtures (Aroclors 1242, 1254, and 1260) by the white rot fungus *Phanerochaete chrysosporium* as evidenced by congener-specific analysis. Appl Environ Microbiol 61(7):2560–2565

Yamaguchi M, Kamei I, Nakamura M, Takano M, Sekiya A (2007) Selection of *Pleurotus pulmonarius* from domestic basidiomycetous fungi for biodegradation of chlorinated dioxins as environmentally persistent organopollutants. Shinrin Sogo Kenkyusho Kenkyu Hokoku 6(4):231–237

Zafra G, Absalón ÁE, Anducho-Reyes MÁ, Fernandez FJ, Cortés-Espinosa DV (2017) Construction of PAH-degrading mixed microbial consortia by induced selection in soil. Chemosphere 172:120–126. https://doi.org/10.1016/j.chemosphere.2016.12.038

Zeddel A, Majcherczyk A, Hüttermann A (1993) Degradation of polychlorinated biphenyls by white-rot fungi *Pleurotus ostreatus* and *Trametes versicolor* in a solid state system. Toxicol Environ Chem 40(1–4):255–266. https://doi.org/10.1080/02772249309357947

Chapter 23
Composting Practices for the Remediation of Matrices Contaminated by Recalcitrant Organic Pollutants

Ondřej Lhotský, Stefano Covino, and Tomáš Cajthaml

Abstract Composting has been known since ancient times. Nowadays this process is widely used for stabilization of biodegradable wastes and preparation of organic fertilizers. Due to low substrate selectivity and high biodiversity, the compost-inhabiting microbial consortia are capable of breaking down organic matter of different nature including those artificial chemical compounds that are persistent in the natural environment. Therefore, composting practices can be utilized for ex situ remediation of different matrices contaminated with recalcitrant organic pollutants. Composting and compost addition are a sustainable and effective bioremediation option, especially for the treatment of soil contaminated with mixtures of hydrocarbons (fuels, lubricating oils, creosote, etc.), explosives, phenols, some pesticides, and emerging pollutants. In order to understand the bioremediation technology based on composting processes, it is necessary to introduce the general principles of composting organic wastes. Therefore, this chapter consists of two parts. The first part concerns the general principles of composting organic wastes including information on the composting process, stages of composting, factors affecting the composting process, and composting systems. The second part focuses on the actual

O. Lhotský (✉)
DEKONTA a.s., Prague, Czech Republic

Institute for Environmental Studies, Faculty of Science, Charles University, Prague, Czech Republic
e-mail: lhotsky@dekonta.cz

S. Covino
Institute for Environmental Studies, Faculty of Science, Charles University, Prague, Czech Republic

Department of Chemistry, Biology and Biotechnology, University of Perugia, Perugia, Italy

T. Cajthaml
Institute of Microbiology of the Czech Academy of Sciences, Prague, Czech Republic

Institute for Environmental Studies, Faculty of Science, Charles University, Prague, Czech Republic

J. Filip et al. (eds.), *Advanced Nano-Bio Technologies for Water and Soil Treatment*, Applied Environmental Science and Engineering for a Sustainable Future,
https://doi.org/10.1007/978-3-030-29840-1_23

bioremediation of contaminated solid wastes. Factors and mechanisms affecting the co-composting process are discussed, and a review of the co-composting applications for different contaminant groups is also provided.

Keywords Composting practices · POPs · Bioremediation

23.1 Introduction

Composting is a bioprocess by means of which organic residues of different typology and origin are transformed into stable, humus-rich soil fertilizers. This technology has become a robust and reliable management option for biowastes deriving from either municipal, agricultural, or industrial areas. For the purpose of this chapter, however, the emphasis will be put on the use of composting for the bioremediation of soil (and other matrices) contaminated by recalcitrant organic pollutants.

Prior to dealing with these topics, it is deemed necessary to introduce the issue of wastes generation and management. This is important because both bioresidues, which are feedstock for producing soil fertilizers, and polluted matrices, which generally arise from the contaminated sites and are therefore considered hazardous refuses, are wastes that modern societies must deal with.

The types of wastes that are considered central to this chapter are:

(i) *organic residues* (often termed biowaste) from either municipal, agricultural, or industrial sources including food and kitchen waste from households, restaurants, caterers, retailers, garden and park waste, and comparable industrial organic residues from food processing plants.
(ii) *hazardous wastes*, chiefly soil and other solid matrices from (industrial) sites contaminated with persistent organic pollutants, but also sludge and sludge biosolids from wastewater treatment.

23.1.1 Background on Waste Streams and Management: Biowaste and Hazardous Waste

Waste production has always accompanied human activities. In the antique period, the density of population and the exploitation of natural resources was low enough to pose either minimal or no risk at all to both humans and the environment. The steady and progressive demographic growth experienced in the last centuries along with the formation of large conurbations and the increasing rate of industrialization have compelled communities to significantly improve their wastes management strategies.

At present, developed countries are making the effort to transform their economic and environmental policies into more sustainable ones. To fulfill this aim, the old-fashioned concept of "take, make, dispose", i.e., the underlying basis of the linear economy model is being progressively replaced by the idea of circular economy. Within this new frame, wastes are considered in terms of their reuse, recycling, recovery, or whether or not their volume can be reduced since they represent real danger to humankind and the environment (4R approach).

With regard to this, the EU waste policy has evolved over the last 30 years through a series of environmental action plans and a framework of legislation that aims to reduce the negative environmental and health impacts and create an energy and resource-efficient economy. The Council Directive 1999/31/EC of 26 April 1999 specifically addresses the issue of landfilling. It aims at preventing or considerably reducing any negative impact of landfilling on human health and natural ecosystems and, besides, it has introduced stringent technical requirements for biowaste landfilling. Directive 2008/98/EC on waste (Waste Framework Directive) provides a general framework of waste management requirements and sets the basic waste management definitions for the EU. For instance, it outlines different categories of waste and a hierarchic order of action for waste prevention, management, and reuse.

According to the classification of wastes by the EU legislation (Directive 2008/98/EC; Eurostat yearbook 2010) soils and sediments (river-, lake-, or seabed sediments) that are contaminated with hazardous chemicals are classified as hazardous wastes. Many of these are present on contaminated sites and could pose significant environmental hazards to terrestrial and aquatic ecosystems as they are major sources of pollution, which may result in ecotoxicological effects.

23.2 General Principles of Composting of Organic Wastes

Composting is a process fully compliant with the 4R approach as it implies the volumetric and quantitative reduction of wastes; it promotes the recycling of elements within the composting substrates, and it eventually generates a product that can be reused on agricultural fields and in other applications (Manderson 2009).

Over the last decades composting has become increasingly important as a management option for the biowastes arising from the most diverse human activities. Reports on waste management strategies from both world-leading economies and developing countries have confirmed this trend.

The present section will provide an overview of the technological development this technology has achieved so far and a description of the characteristic phases of composting with an analysis of the key parameters affecting the process. The considerations made here will be useful to the reader in view of the next section (s) where the composting process or the application of its final product (compost) are presented as bioremediation approaches.

23.2.1 Outline of the Composting Process

Composting is a biowaste management option used worldwide. Solid organic materials of agricultural, industrial, or municipal origin are transformed through composting into a stable, highly humified material that improves soil fertility and structure. The conversion of untreated (raw) wastes (*starting material* or *substrate*) into a stable material is carried out for the most part by aerobic microorganisms (*active biomass*), namely bacteria and fungi. Overall, composting is a result of the co-occurrence of mineralization and humification processes. The former is highest in the early phases (although it continues until full compost maturity, see below) and it gives rise to by-products such as water and carbon dioxide, while the latter happens chiefly in the mid–late stages and results in the formation of organic complexes rich in humic and fulvic acids. Moreover, as bio-oxidative reactions at expenses of the substrate are exothermic, the activity of microorganisms generates heat, thus influencing the temperature levels being attained within the compost mixture. The simplified scheme reported below can be used to sum up the inputs and outputs of the composting process:

$$starting\ material + microbes + O_2 \Rightarrow heat + CO_2 + H_2O + humified\ material$$

Microbial communities, in turn, are influenced by the variation of matrix core temperature; consequently, the overall process is a dynamic succession of microbial biocenoses in response to the changing environmental factors. The stages clearly distinguishable during composting are discussed in Sect. 23.2.2.

23.2.2 Temperature-Dependent Stages of Composting

Regardless of the system that is employed, the composting process starts as soon as the starting material has been mixed up and set on the ground or deposited in the reactor-like vessels. This happens as the substrate mixture is naturally endowed with a fair amount of microbial biomass to initiate the decay process. Nonetheless, microbial inoculants have been tested in order to speed-up or improve the effectiveness of composting processes (Ahmad et al. 2007; Karnchanawong and Nissaikla 2014).

Early colonizers of the matrix are mesophilic bacteria, actinomycetes, and fungi (Fig. 23.1) that can grow in temperatures ranging between 10 and 40–45 °C. The initial *mesophilic phase* is generally quite short, lasting a few hours or days because the temperature rises steadily owing to the exergonic reactions anticipated above. This phase is characterized by the high availability of soluble compounds such as simple carbohydrates (mono- or disaccharides), peptides, and amino acids as well as other easily utilizable compounds (e.g., starch), which support the rapid growth of mesophilic microbes. As bio-oxidation processes progress, the partial pressure of oxygen tends to decrease within the compost pile. Contrarily, CO_2 and volatile fatty acids are released because of mineralization and organic matter decomposition,

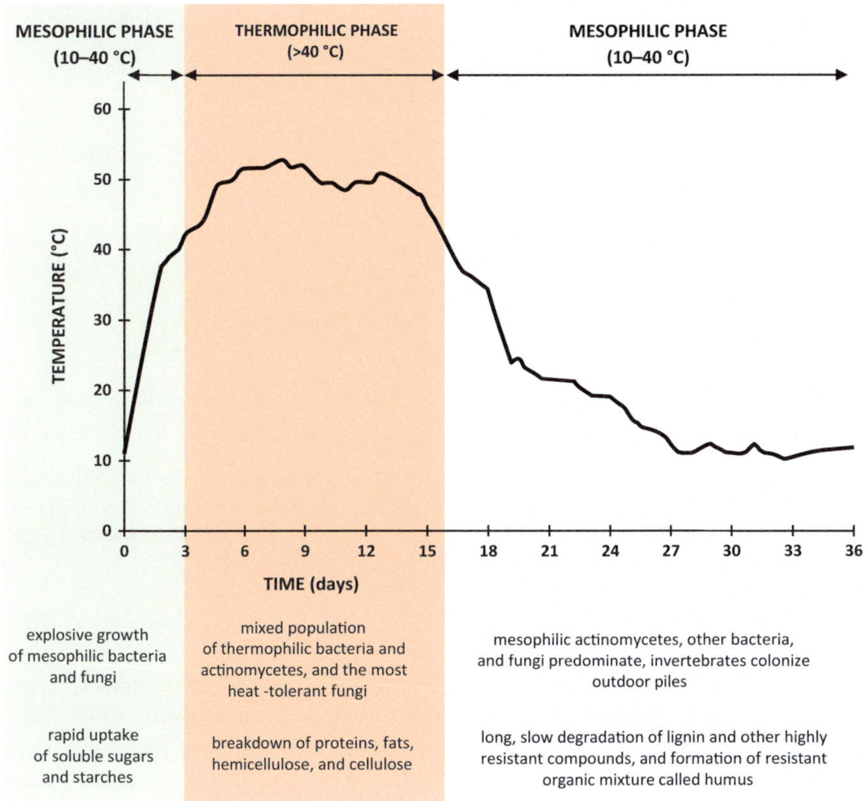

Fig. 23.1 Schematic representation of temperature-dependent stages of composting. (Adapted from Trautmann and Krasny 1998)

respectively, with an ensuing drop in the pH values. A slight acidification of the bulk material is often observed at the onset of the decomposition (Gajalakshmi and Abbasi 2008; Manderson 2009); however, this is a transient phenomenon as the pH rises again to sub-neutral values when organic acids are further utilized by microbes, and CO_2 is removed by aeration (Fig. 23.2).

When the temperatures rise above 40–45 °C, thermotolerant and thermophilic microbes begin to thrive in the substrate mixture, gradually outnumbering the mesophilic ones (Fig. 23.1). During this second stage, which is called *thermophilic phase* and is considered the distinctive feature of composting, the temperature can reach values as high as 65–70 °C for a period of days to weeks. The length of this phase and the temperature attained are largely dependent on the composting system having been adopted, the composition of the starting material, and the aeration regimes. Most authors and composting practitioners stress the importance of attaining elevated temperatures ($T > 55$ °C for some hours) to achieve an effective inactivation of weed seeds, insect larvae, and, more importantly, plant and human pathogens (Ahmad et al. 2007). Despite this, the biological hazard posed by compost

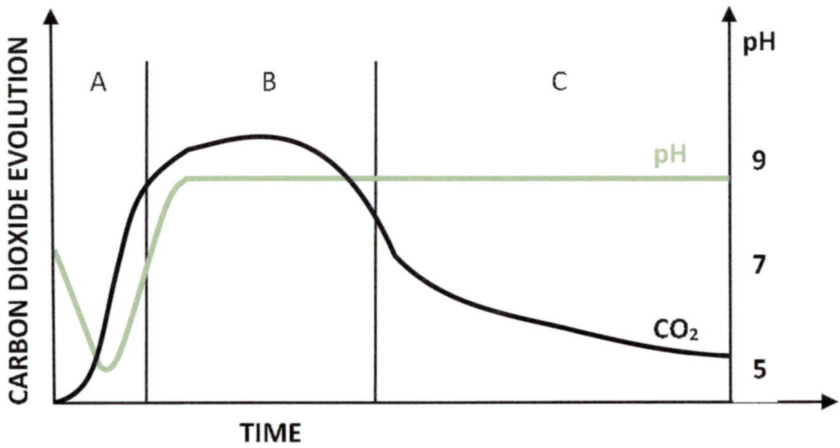

Fig. 23.2 Typical temperature courses of CO_2 production and a pH curve in a batch composting process. The pH falls initially and the start of the high-rate phase coincides with a rise in the pH. *A*—initial phase, *B*—high-rate phase, *C*—curing phase. (Adapted from Trautmann and Krasny 1998)

and compost-derived bioaerosols is still subject to discussion (Manderson 2009; Wéry 2014; De Gannes et al. 2013a, b; Briones 2014).

The thermophilic phase corresponds to the peak of fresh organic matter decomposition, i.e., the highest utilization of water-soluble substrates readily utilizable by bacteria. At the same time, however, the thermophilic conditions foster the degradation of water-insoluble polymeric structures (cellulose, hemicellulose, and lignin), proteinaceous substrates, and fats (Fig. 23.3) that were not a form of accessible nourishment in the previous stage. This process is aided by extracellular hydrolytic enzymes produced by specific microorganisms that can attack large molecules and polymeric structures yielding breakdown substrates beneficial to other microorganisms (Gajalakshmi and Abbasi 2008; Manderson 2009). According to Tuomela et al. (2000), lignin degradation also starts under thermophilic conditions. Indeed, the optimum temperature for thermophilic microfungi and actinomycetes that mainly degrade lignin is 40–50 °C (Hellmann et al. 1997).

Biomolecules less amenable to microbial activity become relatively enriched in the compost mix at this stage and their degradation or biotransformation is a much slower process compared to the assimilation of water-soluble nourishments (Fig. 23.3).

In addition, the active decomposers able to withstand thermophilic conditions represent a minor fraction of the overall compost cenosis; consequently, the microbial activity decreases, and the temperature drops to the mesophilic conditions of the *cooling phase* (Fig. 23.1). The latter phase is a transient period occurring between active composting phases (i.e., early mesophilic and thermophilic stages) and the *maturation* (or *curing*) *phase*, which starts when the temperature of the compost substrate falls to that of ambient air (Gajalakshmi and Abbasi 2008; Manderson 2009). During the cooling and maturation phase, mesophilic microorganisms recolonize the compost heap, and the humification process reaches its maximum.

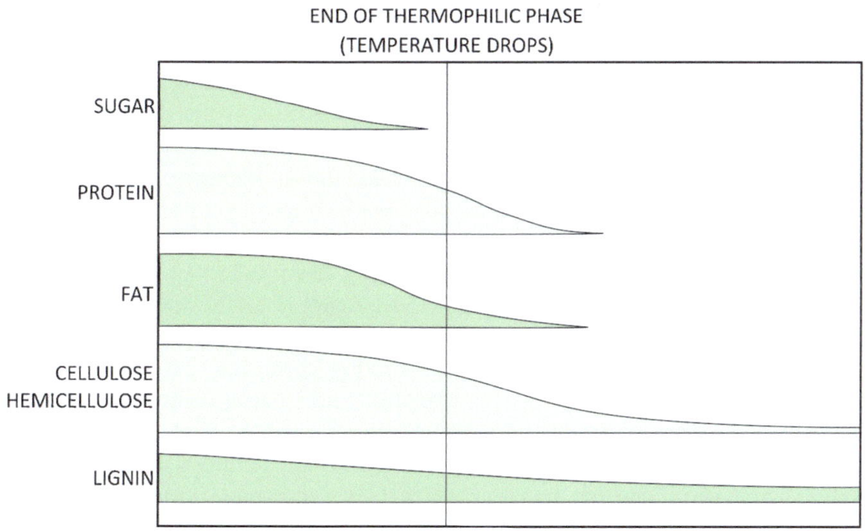

Fig. 23.3 Time course of biomolecules decomposition during composting. (Modified according to Trautmann and Krasny 1998)

According to the consolidated literature, key microbial players in this process are actinomycetes and fungi rather than bacteria (Anastasi et al. 2005; Tuomela et al. 2000; Wang et al. 2015); there are several reasons behind this. On the one hand, the composition of the substrate at this point—when main components are high molecular weight compounds with polymeric and polydispersed structures—favors the growth of filamentous fungi and actinomycetes. These microorganisms produce extracellular enzymes able to hydrolyze wood biopolymers, while bacteria mostly rely on the transport of their substrates intracellularly. On the other hand, the fact that maturation is generally achieved statically promotes the colonization of the substrate by hyphae (and pseudohyphae), which are otherwise disrupted by periodical turning of the compost matrix. It is to be pointed out, however, that bacterial degradation of structural plant macromolecules has been overlooked too often. A growing body of evidence indicates that several bacterial taxa harbor genes encoding plant cell wall-degrading enzymes, including ligninolytic enzymes (Brown and Chang 2014), and that these prokaryotes are abundant in natural ecosystems where humification takes place. Therefore, it is reasonable to think that bacteria participate in the overall compost humification process.

23.2.3 *Factors Affecting the Composting Process*

The *composition of the starting material* is of crucial importance for the composting process, as the organic components represent the natural growth substrate for microorganisms. One of the crucial parameters concerning the input material is the carbon to nitrogen ratio (*C/N ratio*). The optimum ratio fulfilling the metabolic requirement of microorganisms is about 25–35 parts C to one part N (Gajalakshmi and Abbasi 2008; Manderson 2009). Higher C/N proportion would result in a shortage of N causing general slowdown of the process, while the excess of N gives rise to the production of odorous compounds, mainly ammonia and amines, with an ensuing negative impact on operators and the environment. Therefore, care should be taken when the compost biodegradable "ingredients" are being mixed, considering their characteristic elemental composition. For instance, lignocellulosic residues generally have high C/N ratio, ranging from values of 400–600:1—in sawdust, mulch, and wood chips—to 50–150:1 found in cereal straw, hays, silage, and leaves. Significant exceptions are represented by grass clippings, leguminous plant residues, and alfalfa hay, having C/N ratios lower than 25:1. Compostable wastes particularly rich in nitrogen are manures (cattle, horse, sheep), poultry litter, and activated sludge biosolids arising from sewage treatment plants.

Besides the C/N ratio, the input materials must be carefully screened for their texture, dimension, and the speed at which they decompose during the process. These are in fact paramount features of the entire process as they drive the bulk density, porosity, and the *aeration*. Compost jargon defines a *bulking agent* as a (lignocellulose) material that improves the porosity and prevents the formation of anaerobic pockets within the composting material. Bulking agents are fundamental to composting of sludge biosolids since their use entails improved aeration and adjustment of the C/N ratio. In case of poor degradation of the bulking material, coarse particles present can be sieved off at the end of the process and recycled.

Moisture is one of the composting variables that affect the microbial activity, as most of the decomposition occurs in the thin liquid films on the surfaces of the substrate's particles. At start, the moisture content is adjusted to about 60–70%, while lower values (50–60%) are ideal in the later stages of the decomposition. A low level of moisture results in a significant slowdown of the microbial activity (especially bacteria, as fungi and actinomycetes are less dependent on water activity—a_w), while the excess of water causes clogging the pores, reduced air exchange, and subsequent anaerobiosis. Moisture is negatively affected by elevated temperature and an excess of forced aeration; therefore, it is important to moisten the substrate, especially during the thermophilic phase when the water content can quickly fall down. The lower extent of moisture in the curing material is one of the factors prompting the fungal and actinobacterial growth during the maturation phase.

23.2.4 Composting Systems

Composting is nowadays feasible in various systems, but care should be taken in defining the various approaches as the terminology used by compost scientists can be rather confusing (Gajalakshmi and Abbasi 2008; Manderson 2009). Biowaste processing systems can be categorized on the basis of the system configuration or, less effectively, depending on the means by which aeration is provided to the bulk substrate. Hereby, the former approach is used in agreement with Gajalakshmi and Abbasi (2008).

Windrow composting is probably the most traditional and widespread approach. In this system, the input material is reduced in average size by means of bio-shredders and is moistened to approximately 50–60%. The resulting composting substrate is heaped up to form rows of defined width (3–4 m) and height (2–2.5 m), while the length varies greatly according to the available space in the composting facility. The specified cross-sectional dimensions are dictated primarily by the nature of the substrate; the taller windrows would in fact result in "packing down" the matrix, and, therefore, in anaerobiosis, while the smaller ones would dissipate heat and dry out too quickly. Aeration is generally provided by mechanized turning equipment (Fitzpatrick et al. 2005; Gajalakshmi and Abbasi 2008; Manderson 2009) that allows oxygen to access the matrix and promotes the homogenization of the bulk material. Turnings are generally more frequent (twice a week) during the initial stages when the oxygen demand is high and the temperature still exceeds 50 °C, while turning on a weekly basis is typical of the cooling phase. Complete composting is accomplished within 3–4 weeks. Thereafter the compost is allowed to cure for another 3–4 weeks without turning (Gajalakshmi and Abbasi 2008). One variant of the windrow system, where turning is not envisaged, consists of embedding pipes in the basement underlying the windrow in order to increase the air exchange within the matrix. This, however, resembles aerated static piles approaches, which are discussed next.

(Aerated) static pile composting is akin to windrow systems, except that turning is not contemplated and the heaps are far larger than those specified above. This allows to accommodate a consistent amount of material and keep the operational costs low. On the other hand, static piles are "passively aerated" and the extent of the gas exchange highly depends on the matrix porosity, thus resulting in prolonged incubation periods. One means to accelerate the composting process is the aerated static pile system that was developed by the Beltsville Agricultural Research Center (MD, USA) (Manderson 2009). A network of perforated pipes is allocated longitudinally, along the base of the pile; blowers can be connected at the pipes' end for an effective delivery of oxygen to the substrate. When the blowers are turned off, the aeration is "passive". However, the "forced" aeration is achieved by running blowers. Blowers can be operated both in negative and positive mode. In the former case, air is forced to enter the pile and the collected off-gases are further treated through ripe compost-based biofilters; in the latter case, however, air is delivered to the compost heap and a layer of mature compost is usually overlaid to the heap to prevent emission of odorous off-gases.

An additional variant of the static pile is the "contained" pile, otherwise called "enclosed static" pile (Gajalakshmi and Abbasi 2008) or *"composting chamber"* (Šašek et al. 2003). In such a system the compost substrate is contained between walls over a perforated floor through which air is blown. Advantages offered by this approach include: efficient aeration of the composting substrate, effective collection of leachate, and capturing of odorous emission.

As for more sophisticated approaches, *rotary drum* composting devices are probably the oldest mechanical processing systems developed to treat organic wastes (Fitzpatrick et al. 2005), although they have come into fashion again recently (Gajalakshmi and Abbasi 2008). Here, the starting material is shredded and moistened prior to being fed into the chamber, i.e., a downward-slanted cylindrical drum (3–4 m in diameter and up to 30 m long), which is slowly rotating (1 rpm). Moisture, air, and out-gases are controlled throughout the process and a fair grade of decomposition is achieved within a few days (Manderson 2009). Therefore, further stabilization of the outgoing material is accomplished by piling or windrowing the material outdoor. Rotating drums and tunnel composters with moving beds are the best examples of continuous, plug-flow processes (Gajalakshmi and Abbasi 2008).

The maximum control over the process parameters is secured using the *in-vessel* systems. Due to the level of engineering reached in these composting approaches, in-vessel composters bear striking resemblance with solid-state "reactors" used in other industrial application. These systems differ in their size, their shape and the devices used to force aeration and mechanical mixing; however, they share a common feature, i.e., they are batch composters where decomposition processes occur in a confined space under optimal conditions. This makes in-vessel systems the most efficient and advanced composting methods, but at the same time the least cost-effective treatment option.

23.3 Composting Practices for the Bioremediation of Solid Wastes Contaminated with Recalcitrant Organic Pollutants

The first scientific reports describing composting approaches for soil bioremediation purposes date back to the 80s (Antizar-Ladislao et al. 2004; Loick et al. 2009). Since then a consistent (Antizar-Ladislao et al. 2005) number of studies have been published on this subject, and several literature reviews have critically discussed the feasibility of composting for bioremediation practices (Ro et al. 1998; Semple et al. 2001; Antizar-Ladislao et al. 2004; Chen et al. 2015; Kästner and Miltner 2016). Many of such studies have reported on laboratory-scale approaches mainly aimed at optimizing the process parameters and the operating conditions to maximize the contaminant removal but composting methodologies have proven effective towards a wide range of contaminants even at higher operational scales (i.e., pilot- and field-scale) (Antizar-Ladislao et al. 2004; US EPA 1997).

The principle of ***co-composting*** consists of mixing the contaminated matrix (e.g., soil) with non-hazardous organic amendments, which are typically biowastes used as the input material for conventional composting (e.g., garden and agricultural wastes, manure, sludge biosolids). The ratio in which the two wastes are mixed can vary, but generally the proportion of the non-contaminated substrate is higher than 20–30% of the total (Loick et al. 2009). Different composition of the organic substrate has been used to optimize the C/N balance and the matrix porosity, thus sustaining the development of microbial communities. The growth of microorganisms is boosted by the presence of the organic substrates, and the co-metabolic, besides metabolic, degradation of pollutants is achieved through the various temperature-dependent stages of the composting process. Co-composting has been performed in windrows, static piles, and as in-vessel (Antizar-Ladislao et al. 2004).

The process of co-composting is altogether very similar to the process of conventional biowaste composting, with bio-oxidation reactions causing a prompt increase in the temperature and attainment of the thermophilic conditions, followed by cooling and maturation phases. Thermophilic conditions represent the fingerprint feature of co-composting. Spontaneous thermogenesis marks in fact the difference between co-composting and other bioremediation approaches (e.g., land farming or biopiles), where the addition of amendments/bulking agents does not cause the temperature to rise above 40–45 °C. Due to this fact, the discussion concerning the usefulness of attaining such temperatures for the sake of the contaminant removal has been open for several years. On the one hand, the increase in the temperature during the composting process could potentially support the biodegradation of the contaminants, particularly with regard to lipophilic compounds (e.g., high molecular weight polycyclic aromatic hydrocarbons—PAHs) that tend to be adsorbed on soil organic matter; therefore, a consistent rise in the temperature might facilitate desorption kinetics from the organic matter to water and, in turn, biodegradation by microbial consortia as a result of the increased contaminant bioavailability (Margesin and Schinner 2001; Wong et al. 2004; Viamajala et al. 2007). On the other hand, the thermophilic stage represents an ecological bottleneck because oxygen solubility tends to decrease when the temperature increases, and most microbes, including contaminant-degrading microbes, cannot grow when the temperature rises above 55–60 °C (Antizar-Ladislao et al. 2005, 2006).

23.3.1 Regulatory Conditions

Regulatory limits and targets connected with remediation of contaminated soil significantly differ in each country. Contaminant concentrations in soil acceptable in certain countries may far exceed the acceptability limits in other ones. The situation concerning the approach to the actual process of composting of biologically degradable waste is similar. Requirements for composting plants, composting process parameters as well as for the treated biologically degradable waste differ in individual countries. A trend towards unification of the regulations may be seen within the EU currently, however, the situation is still relatively unclear.

Figure 23.4 summarizes the rules that have to be taken into account in the various phases of utilization of contaminated waste composting. Part A concerns regulatory limits for contaminant concentrations in soil and other materials to be decontaminated. Part B relates to regulations concerning organic substrates. Utilization of co-composting is possible only in the case that the conditions given in the decision tree are met (Antizar-Ladislao 2010).

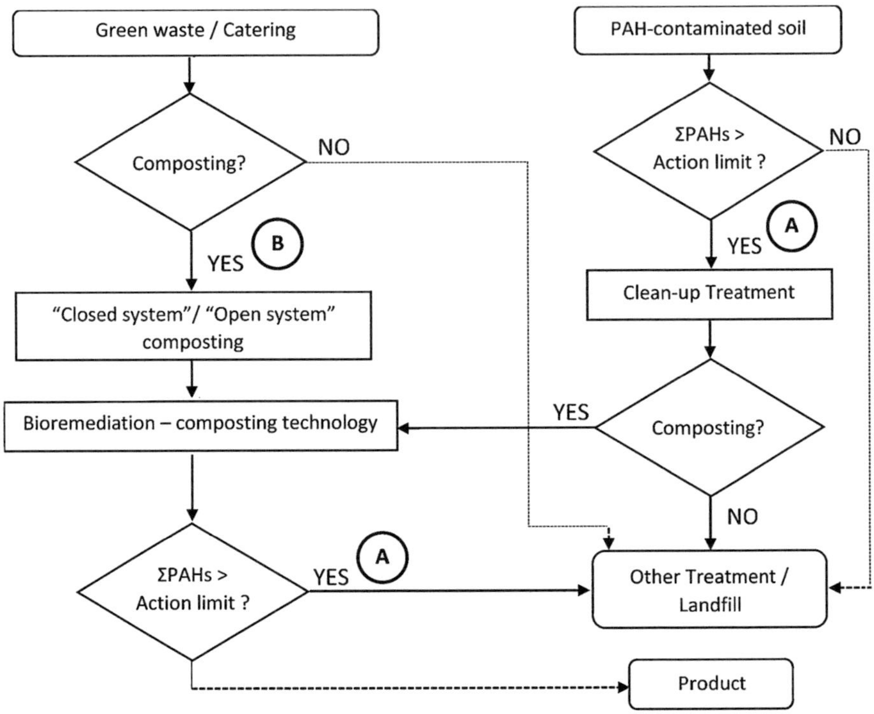

Fig. 23.4 Example of a flow diagram for use of composting in case of PAH-contaminated matrix. (Modified according to Antizar-Ladislao 2010)

23.3.2 Contaminant Removal in the Course of the Co-composting Process—Factors and Mechanisms

23.3.2.1 Sorption Mechanisms

The main physicochemical mechanism of contaminant removal in the course of the composting process is sorption of contaminants especially to the organic fraction of the compost, resulting usually in reducing their biological availability, and also their toxic action on plants and microorganisms (Antizar-Ladislao et al. 2004). This process, on the one hand, enables relatively quick removal of the contaminants, but, on the other hand, it complicates the biological degradation of organic contaminants, which may represent a problem from the point of view of compliance with the regulatory limits. In majority of cases it is necessary that the contaminant is released from the sorbed phase into the aqueous phase in order to be biodegraded (Harms and Zehnder 1994; Shelton and Doherty 1997). This desorption might be assisted by the biosurfactants produced by specific microorganisms, which actually means bioavailability of contaminants is species specific. However, it has been observed that there are bacteria capable of contaminants degradation without prior desorption. For example, Singh and Sharma (2003) observed that *Brevibacterium* sp. degraded the pesticide fenamiphos that was intercalated into the cationic-surfactant-modified montmorillonite clay (CTMA–Mt–fenamiphos complex). It seems that this process is less significant within bioremediation practice.

The sorption processes depend on a number of properties of the contaminants to be removed, the used organic substrates, and the contaminated soil, and they may be caused by several mechanisms, such as hydrogen and ionic bonds, van der Waals forces, hydrophobic interactions, and trapping in cavities of organic or inorganic matter (Pignatello and Xing 1995). A number of publications deal with the topic of sorption of the individual substances to particles of soil and humic substances (Chefetz and Xing 2009; Xing et al. 1996). However, in practice, complex mixtures of various contaminants that mutually influence their sorption properties are found in most of the contaminated localities. For example, Luthy et al. (1993) showed that the transfer of PAHs contained in tar into water, and thus also to microorganisms, may be considerably reduced by consolidated interfaces of non-aqueous-phase liquids (NAPLs), such as tar oils. According to the Raoult's law, the chemical activity, which is the force causing the transfer from the non-aqueous to the aqueous phase, is significantly higher in the case when only one substance is present than in the case with a diverse mixture (Eberhardt and Grathwohl 2002; Ghoshal et al. 1996; Kästner 2000; Lee et al. 1992; Luthy et al. 1994; Wehrer et al. 2013). Further, NAPLs, in view of their hydrophobic properties, absorb hydrophobic substances present in their surroundings, such as, fatty acids, waxes, and resins. NAPL enrichment by these molecules usually results in reduction of NAPL viscosity on its interface with the water phase, which further reduces the transport of mass into the water phase, and its bioavailability (Kästner and Miltner 2016). In many localities influenced by coal gas production, PAHs dissolved in tar oils occur in forms bound to coal or coke as a consequence of their joint disposal (Luthy et al. 1994, Rein et al. 2016; Wiesmann 1994). These bonds reduce the PAH bioavailability considerably, and may cause considerable problems during the decontamination of materials contaminated in this way.

Similar interactions occur between the individual components of contaminant mixtures, for example, also in the case of petroleum substances and explosives.

On the other hand, the presence of water-soluble compounds, such as humic and fulvic acids as well as small biomolecules, such as peptides and fatty acids may cause desorption, solubilization, and/or complexation of organic contaminants, which usually results in increasing their mobility (Mayer et al. 2005, 2007; Pignatello et al. 2006; Smith et al. 2009, 2011; Vacca et al. 2005). Further, these processes cause the increase in bioavailability, and thus accelerate the biodegradation of the mobilized contaminants (Haderlein et al. 2001; Ortega-Calvo and Saiz-Jimenez 1998; Puglisi et al. 2007). Within the framework of the biodegradation techniques, this process has to be controlled because if excessive contaminant mobilization occurs, unwanted toxic effects on the degradation organisms might appear (Puglisi et al. 2007).

Certain bacteria, for example Actinobacteria of the orders *Nocardia*, *Gordona*, *Rhodococci*, and *Mycobacteria* have markedly hydrophobic cell wall, causing increased bioavailability and also biodegradations of PAHs (Stringfellow and Alvarez-Cohen 1999).

Sorption occurs also in the case of fungal mycelium, for example, in the case of pesticides and chlorophenols (Benoit et al. 1998).

23.3.2.2 Biodegradation

Basically, aerobic degradation of organic contaminants may proceed thanks to three different principles (Kästner 2000):

(i) Metabolic degradation of organic contaminants as a source of carbon and energy, and their complete mineralization. Usually, the contaminant is processed inside a cell, and microbial biomass growth occurs thanks to its processing. CO_2 and H_2O are released as by-products. In an ideal case, accumulation of other metabolites does not occur.

(ii) Co-metabolic degradation of organic contaminants, when an organic contaminant is degraded non-selectively by an action of enzymes (often monooxygenases) participating in degradation of other organic sources of carbon. It usually proceeds inside a cell too, but it can proceed also out of it, and it often results in production of partially oxidized metabolites that are accumulated subsequently. It does not result in biomass growth, and the contaminant energy potential is not used.

(iii) Non-specific radical oxidation is a process that proceeds out of a cell, and is characterized by formation of radicals as oxidation intermediates due to an electron abstraction. Gradual secondary oxidation reactions of the primary metabolites may follow, up to CO_2. These processes are typical, in particular, for ligninolytic fungi (Hofrichter 2002).

In addition to the presence of microorganisms having the proper enzymatic properties for decomposition of organic contaminants, and ensuring availability of the contaminants, there are several further preconditions that have to be ensured in order that the co-composting process is efficient. One of the often neglected factors during the direct metabolic degradation of contaminants is the fact that the contaminant must be present in a sufficient concentration in the treated material, enabling to meet the energy needs of the degradation microorganisms and ensuring their mobility into the untreated areas of the material (Adam et al. 2014; Rein et al. 2016; Semple et al. 2001). The moment the contaminant concentration decreases under certain minimum limit its biodegradation stops. In real systems, biodegradation proceeds in a dynamic state when contaminants are slowly released from sorption places or NAPL in the direction to the relevant microorganisms degrading them. For this reason, the microbial activity depends on the substrate flow not on its concentration. This is true, in particular, in the case of hydrophobic compounds, such as petroleum substances, tars, and similar organic compounds when biodegradation is usually limited by the transfer of mass (Adam et al. 2014; Bosma et al. 1997; Marchal et al. 2013a, b; Rein et al. 2016; Volkering et al. 1992; Wick et al. 2001).

From this point of view, utilization of compost is ideal for biodegradation of these substances as it comprises bigger particles of slowly biologically degradable substrates, humic acids, and other complicated biomolecules. Moreover, mixing contaminated soils with organic substrates results in increasing the mass flow of the contaminants partially thanks to the actual mechanical processing of the material, but also thanks to the temperature growth and the increase of the concentration gradient, which often results in destruction of the interfaces between the phases preventing mass transfers. Destruction of the interfaces by mixing and adding organic substances results in the increase in biological availability, and thus in much quicker degradation of the organic contaminants (Kästner and Miltner 2016).

The addition of organic substrates to the contaminated materials in the course of co-composting usually also provides sufficient amounts of primary substrates and nutrients for the co-metabolic processes (Alburquerque et al. 2009; Chen et al. 2015; Gandolfi et al. 2010; Megharaj et al. 2011) that depend, in addition to the mass flow of the contaminant, also on the mass flow of these substances. The addition of organic substrates to the soil increases nitrogen and phosphorus concentrations and at the same time considerably improves the environmental properties of the decontaminated material thanks to the change of its structure, the increase in the pore volume, and the increase in the ability to absorb water. This is essential for ensuring sufficient transport of gases through the decontaminated material in order that the degradation microorganisms have sufficient oxygen supply and, simultaneously, the CO_2 accumulation does not occur. At the same time, sufficient water concentration is ensured too, which is generally necessary for the growth of microorganisms (Potts 1994). One often neglected fact during composting is the production of water as a final product of the microbial metabolism. In certain cases (for example, during composting of a grass substrate without addition of a bulking agent) this may result in oversaturating the material with water within the course of the process, which further prevents efficient exchange of gases, and causes CO_2 accumulation and O_2 deficiency. On the other hand, in the case of too dry or intensively ventilated materials, the water production is positive. The addition of an organic substrate may, because of its buffering capacity, also act as acidity regulation (Gandolfi et al. 2010; Zhang et al. 2011). Hernández et al. (2015) proved that the positive effects of the addition of an organic substrate on the activity of the microbial communities are long-term. Specifically, they persisted for more than 5 years in his studies.

Positive changes caused by organic substrate additions to the contaminated materials are often designated as the so-called "matrix effect" that is caused by the above-described mechanisms, such as addition of primary organic substrates, solubilization of the contaminants, and also changes of the material structure (Haderlein et al. 2001; Marchal et al. 2013a, b; Mayer et al. 2005; Ortega-Calvo and Saiz-Jimenez 1998; Puglisi et al. 2007; Smith et al. 2011). In addition to that, the matrix effect may manifest itself through the fact that it results in the desorption of the organic contaminants from the soil material, with subsequent sorption to biologically degradable solid organic substrates. Subsequently, in the course of the degradation of these organic substrates, the contaminants adsorbed on them are degraded too. In order to preserve these positive effects of adding the organic substrate to the contaminated materials, it is necessary to ensure the presence of a minimum amount of the organic substrate in the final mixture. This minimum amount is usually 10–50% of fresh weight, wherein preserving moisture and oxygen contents at suitable levels is no less important (Antizar-Ladislao et al. 2004; Li et al. 2010; Megharaj et al. 2011).

Naturally, the organic substrate having been added to the contaminated materials should also contain a high amount of microbial biomass, and, in particular, specific microorganisms that are able to degrade the target contaminants. Thus, from this point of view, this is bioaugmentation of the contaminated materials with proper degradation microorganisms. For example, Gandolfi et al. (2010) proved in his study that the organic substrate addition to the contaminated soil caused transfer from *Alphaproteobacteria* and *Gammaproteobacteria* to *Bacteroidetes* and *Firmicutes*. The size of the microbial communities is usually estimated on the basis of cultivation tests and/or by means of analysis of phospholipid fatty acids (PLFAs) (Adam et al. 2015; Covino et al. 2016; Gandolfi et al. 2010; Puglisi et al. 2007). On the other hand, organic substrate additions, and the improvement of conditions for microbial biomass growth resulting from them, also naturally support the increase in activity of microorganisms that were already present in the contaminated material before the addition. Moreover, these microorganisms usually have suitable degradation mechanisms at their disposal due to the long-term contaminant presence. Thus, from this point of view, the support of the original microorganisms may be more important than the bioaugmentation with new species, in some cases.

23.3.2.3 Microbial Communities

As described above, a wide spectrum of microorganisms is present in the compost in the course of the composting process, and thus the organic substrate addition usually results in significant increase in organic pollutant biodegradation potential in the contaminated materials (Dees and Ghiorse 2001; Loick et al. 2009; Thummes et al. 2007). Moreover, the composition of the microbial communities changes greatly in the course of the composting process, depending on the temperature and available organic substrates in particular. A number of lists of specific bacteria and fungi that are able to degrade specific organic contaminants, particularly PAHs, have already been published (Antizar-Ladislao et al. 2004; Cerniglia 1984, 1992; Heitkamp and Cerniglia 1989; Cerniglia and Sutherland 2010; Juhasz and Naidu 2000; Loick et al. 2009; Lu et al. 2011; Peng et al. 2008), showing the biodiversity of these biodegradation potentials. On the other hand, the lists are usually very general and do not deal with degradation of organic contaminants in specific conditions in the course of co-composting. There are only few studies describing in detail the development of microbial communities in the course of the co-composting process. An in-vessel composting bioremediation of pyrene-4,5,9,10-$^{13}C_4$ and unlabeled pyrene spiked soil amended with fresh wastes was investigated by Peng et al. (2013), DNA-based stable isotope probing (SIP) of active bacteria was involved as well. The highest dissipation of labeled pyrene was detected at 55 °C after 42 days of composting. The active bacterial communities in the composting changed over time, showing a distinct difference among different stages. α, β, γ Proteobacteria, and Actinobacteria were detected mainly while involved in pyrene degradation at 38 °C over 14 days of composting. *Streptomyces* appeared to dominate the pyrene degradation at 55 °C. β and γ Proteobacteria and Actinobacteria were the dominant pyrene degraders at 70 °C after 42 days of composting and at 38 °C after 60 days of composting. However, these results also indicate that instead of a specific key degrader more complex communities may be involved during the degradation of pyrene.

Covino et al. (2016) assessed the microbial shifts during the decontamination of creosote-treated wood by co-composting with agricultural wastes using two bulking agents, grass cuttings (GC), and broiler litter (BL). GC composting was characterized by high microbial biomass growth in the early phases as suggested by phospholipid fatty acid analyses. Based on the 454-pyrosequencing results, fungi (mostly Saccharomycetales) constituted an important portion of the microbial community, and bacteria were characterized by rapid shifts (from *Firmicutes* (*Bacilli*) and *Actinobacteria* to *Proteobacteria*). However, during BL composting, larger amounts of prokaryotic and eukaryotic PLFA markers were observed during the cooling and the maturation phases, which were dominated by *Proteobacteria* and the fungi belonging to the *Ascomycota* and those putatively related to the *Glomeromycota*.

The role of fungi in the co-composting process has not been satisfactorily studied yet. In addition to *Saccharomycetales*, the increase in the amounts of ligninolytic fungi may occur in the composts in the later phases, too. These fungi are able, thanks to their exoenzymes (peroxidases and laccases), to cause non-specific radical oxidation of organic contaminants (e.g., heavy PAHs) the biodegradation of which would be very difficult otherwise (Harms et al. 2011; Kües 2015; Steffen et al. 2002). In spite of that, some authors proved that the degradation of PAHs in contaminated soils did not correlate directly with the detected activities of ligninolytic enzymes (Eggen 1999). It is likely that the synergic metabolic cooperation between the present microorganisms and ligninolytic fungi occurs in practice when the primary products of oxidation by ligninolytic enzymes are further degraded by the present microorganisms (Ahlawat et al. 2010; García-Delgado et al. 2015). However, some authors draw attention to the danger of forming trace amounts of highly toxic metabolites (such as chlorophenols) as a consequence of the degradation of chlorinated aromatic structures by ligninolytic enzymes (Laine et al. 1997; Morimoto and Tatsumi 1997; Öberg et al. 1990).

An inexhaustible spectrum of organic substances with specific properties may be used as organic substrates for co-composting. Analyses of their composition usually concern only the basic parameters that are important for their composting (carbon, nitrogen, and moisture contents) because a detailed analysis of their composition is basically impossible in higher volumes. Naturally, this diversity is reflected in the microbial communities that develop in the course of the degradation of these substrates. Because of this fact, the efforts to identify relevant organisms participating in degradation of contaminants in the course of co-composting are a very demanding task as the microorganisms may differ significantly on the basis of the used substrate.

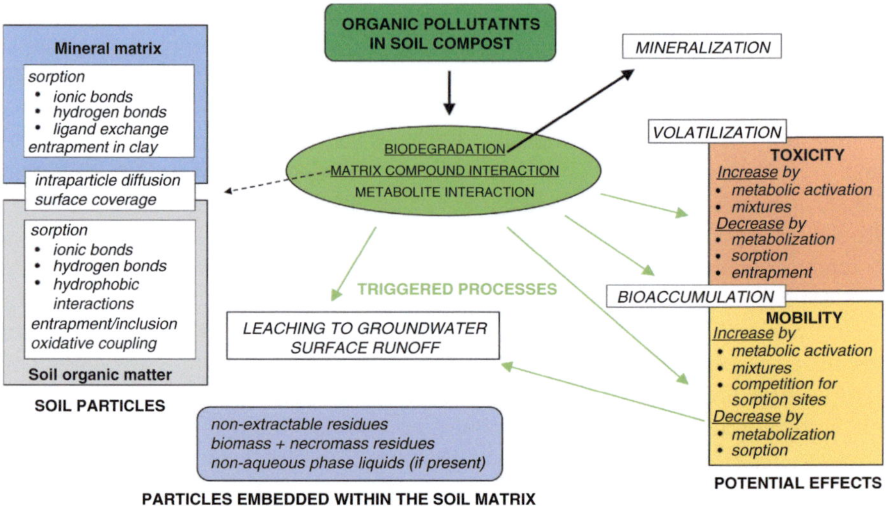

Fig. 23.5 The relevant processes and factors influencing the degradation of organic contaminants in the course of the co-composting process. (Modified according to Kästner and Miltner 2016)

There are further uncertainties concerning the fact whether the degradation of organic contaminants occurs more thanks to the presence of specific degradation microorganisms or the presence of complex microbial communities (Peng et al. 2013). In order to optimize the utilization of the co-composting process in practice, it will be necessary to verify, in the future, what the relations of the individual members of the microbial communities participating in degradation of the contaminants are, and what conditions are ideal for these communities/members (Fig. 23.5).

23.3.3 Co-composting Applications

23.3.3.1 Co-composting of Organic Contaminants

As described in the previous chapters, the co-composting method is suitable, in particular, in the case of organic contaminants that may be degraded in the course of the composting process. Typical organic contaminants removable by the co-composting method include substances of petroleum origin, PAHs, explosives, and pesticides. However, composting has recently begun to find use also within the framework of removing newly occurring contaminants, such as personal care chemicals, pharmaceuticals including antibiotics, phthalates, etc. (Youngquist et al. 2016; Buyuksonmez and Sekeroglu 2005). From the point of view of the contaminated materials, the co-composting process is most frequently used for

Table 23.1 Examples of composting studies dealing with petroleum-contaminated materials

Contaminant	Organic substrate/ treatment method	Results	References
Oil refinery sludge	The sludge mixed with shredded green wastes in ratios of 1:1 v/v (RI) and 1:3 v/v (RII), total volume of approximately 1 m^3, 120 days of co-composting	At the end of the experiment in RI 52.1% reduction of mineral oil concentration was observed where in RII it was 62.1% reduction	Fountoulakis (2009)
Oil sludge from petrol stations and petroleum residues from a refinery	Horse manure mixed with 1.8 to 7.1% of oil wastes in dry matter (280 L vessels)	Oil wastes decomposed to 78–93% during 4.5 months of composting	Kirchmann and Ewnetu (1998)
Diesel contaminated soil	Simultaneous degradation of diesel oil in soil and the organic matter in food waste in laboratory scale aerated vessels	Removal efficiencies of petroleum hydrocarbons and food waste after 15 days were 79% and 77%, respectively	Joo (2007)
Diesel contaminated soil	Soil spiked with diesel mixed with biowaste (vegetable, fruit, and garden waste) at a 1:10 ratio (fresh weight) and composted in a monitored composting bin system for 12 weeks	The first-order rate constant of diesel degradation in the biowaste mixture was four times higher than in the soil at room temperature, and 1.2 times higher than in the soil at composting temperature	Van Gestel (2003)

decontamination of soil and sludge; however, it may be used also in the case of other materials that may be colonized by microorganisms, such as railway sleepers (Covino et al. 2016). Nonetheless, the co-composting process is not suitable for inert materials, such as construction waste. The following paragraphs deal with the typical contaminants degradable during the co-composting process, together with references to the selected documents dealing with co-composting of these substances.

23.3.3.2 Substances of Petroleum Origin

Soil contamination by petroleum substances is often removed by different bioremediation methods than co-composting because petroleum substances may be removed by simpler and cheaper methods frequently. Thus, the co-composting process particularly finds its use in the cases when utilization of standard biodegradation methods is insufficient, for example, because of excess input concentrations of the contaminants, unsuitably contaminated matrix, or the presence of heavy petroleum fractions. It is often used, for example, in the case of petroleum sludge with high concentrations of petroleum substances. The table below mentions selected studies dealing with composting of petroleum substances (Table 23.1).

23.3.3.3 Polycyclic Aromatic Hydrocarbons

The co-composting method is very suitable in the case of PAHs, in view of the fact that these contaminants may be hardly removed using other biodegradation methods. The volume of literature dealing with this topic, and also the number of cases of application in practice corresponds to this. Especially a comprehensive review by Antizar-Ladislao (2010) may be recommended. A summary table describing a number of practical applications of the co-composting method may also be found in review by Loick (2009). Generally, it can be stated that co-composting of PAH-contaminated materials is a very efficient bioremediation approach.

23.3.3.4 Explosives

One of the first applications of the co-composting method in practice was its utilization for removing explosives by the United States Army. The first tests of this method were carried out already in 1986 (Doyle 1986). At present, it is one of the methods recommended by the United States Army for removing explosives from contaminated soil (Jenkins and Vogel 2014). A US EPA study contributed to that significantly. US EPA (1997) has evaluated composting as an ex situ solid phase biological treatment technology to degrade nitroaromatic and nitramine compounds in soils. Treatability study at the Umatilla Army Depot Activity site in Hermiston, Oregon demonstrated that composting is an alternative treatment to incineration for remediating these compounds. Composting has been selected as the treatment for 15,000 tons of TNT (2,4,6-trinitrotoluene), RDX (hexahydro-1,3,5-trinitro-1,3,5-triazine, Royal Demolition eXplosives) contaminated soils at Umatilla. Using composting instead of incineration, Umatilla saved approximately $2.6 million. Cleanup goals for Umatilla were established at concentrations of less than 30 mg/kg for TNT and RDX. The project exceeded these expectations by achieving undetectable levels of explosives. Contaminant by-products were either destroyed or permanently bound to soil or humus (US EPA 1994).

23.3.3.5 Pesticides

Pesticides are often present in agricultural waste to be composted. In the course of the composting process, the pesticides should be degraded because of their possible toxic effects on people as well as on other environmental components, especially on plants that may be endangered by the presence of herbicides in the compost.

During the composting process, a number of pesticides are degraded indeed, which also results in utilization of the co-composting method for the target decontamination of materials with pesticide content. However, certain recalcitrant substances, such as the majority of chlorinated aromatic pesticides (like DDT, lindane, aldrin, etc.), are not degraded much in the compost because the chlorine removal from the aromatic ring occurs in anaerobic conditions. Büyüksönmez et al. (1999)

stated in his review that the pesticide analyses of finished composts from various farms in the USA revealed the presence of only a few pesticides, usually chlorinated pesticides, including substances that were already banned in the USA years ago. Other pesticides of the organophosphate and carbamate types were detected only rarely. The comparison of pesticide concentrations before and after the composting showed that chlorinated substances were mostly resistant to biological degradation during the composting. Pesticides of other kinds were degraded relatively well, with certain exceptions. The studies examining the degradation mechanisms showed that mineralization was responsible only for partial removal of the pesticides. Other important mechanisms were adsorption and volatilization.

23.3.3.6 Micropollutants

Less attention was paid to the removal of newly emerging pollutants, for instance, pharmaceuticals and personal care products. Although the subject is rather wide owing to numerous substances that are nowadays present in the environment, many of typical micropollutants are trapped in wastewater treatment sludge that is composted for hygienization purposes. Nevertheless, the subject has been reviewed, e.g., in the paper by Semblante et al. (2015). The composting process represents an ideal method for elimination of micropollutants transfer into the environment. Many of the newly emerging micropollutants are not as recalcitrant as typically persistent organic pollutants and the process is probably able to substantially reduce the concentrations of the micropollutants only with moderate condition adjustment. An overview of studies dealing with this subject is shown in Table 23.2.

23.4 Conclusions

Composting and compost addition are sustainable and effective bioremediation option, especially for the treatment of soil contaminated with mixtures of hydrocarbons (fuels, lubricating oils, creosote, etc.), explosives, phenols, some pesticides, and emerging pollutants. In order to design full-scale bioremediation technology based on composting processes, preliminary investigations on the process variables, i.e., co-substrate composition, bulking agents, aeration or turning, ideal ratio of non-contaminated to contaminated waste as well as physical, chemical, and microbiological characteristics of the matrix to be remediated are necessary.

Beside the degradation of contaminants initially present in the polluted waste that is chiefly used to check whether remedial goals were met, a set of toxicological analyses will help assessing the effectiveness of the process and evaluate the residual eco- and genotoxicity associated to the final product. Unfortunately, it is not always clear what the potential reuse of the soil that has undergone such treatment is, especially considering the relatively high content of organic matter still present that prevents its use.

Table 23.2 Examples of successful composting studies dealing with PPCPs and other micropollutants documenting feasibility of the method toward these micropollutants

Compounds	Composting substrate	% of removal	References
Sulfadimethoxine, sulfamethoxazole (sulfonamide antimicrobial antibiotics)	10:1:0.2 v/v/v sewage sludge, spruce bark and coarse garden waste, and mature sludge compost	26–48	Vasskog et al. (2009)
Salinomycin (antibacterial and coccidiostat ionophore)	Poultry manure, hay C:N—25:1	99	Ramaswamy et al. (2010)
Chlortetracycline (synthetic antibiotic)	Broiler manure or hen manure	90	Bao et al. (2009)
Ciprofloxacin, norfloxacin, ofloxacin (fluoroquinolone antibiotic)	3:4 v/v peat and anaerobically digested sludge	46–95	Vasskog et al. (2009)
Fluoxetine, paroxetine, sertraline, and citalopram (antidepressants)	Anaerobically digested sludge	0–83	Martín et al. (2015)
Di(2-ethylhexyl)phthalate (plastifier, plastic additive)	2:1 w/w raw sludge and secondary sludge with 10% sheep manure	60–97	Pakou et al. (2009)
Nonylphenols (plastifier, components of surfactants)	10:1 w/w anaerobically digested sludge and dry wheat	88	Gibson et al. (2007)
Nonylphenols (plastifier, components of surfactants)	43:57, 65:35, and 84:16 v/v anaerobically digested sludge and wood shavings	50–80	Xia et al. (2005)
Nonylphenols (plastifier, components of surfactants)	1:1:1 v/v/v sewage sludge, straw, and garden park waste	75–95	Moeller and Reeh (2003)

Adapted from Semblante et al. (2015) and extended
All the composting processes included a thermophilic phase with temperatures higher than 50 °C

Both methodologies have proven effective in promoting the biodegradation and detoxification of several types of hazardous waste, as witnessed by the consistent body of literature on this subject. Therefore, composting and compost addition are mature stage bioremediation technologies provided that a meticulous preliminary characterization of the polluted matrix and a tight control over environmental variables ensure that the process runs correctly.

References

Adam IKU, Rein A, Miltner A, Fulgêncio ACD, Trapp S, Kästner M (2014) Experimental results and integrated modeling of bacterial growth on an insoluble hydrophobic substrate (Phenanthrene). Environ Sci Technol 48(15):8717–8726. https://doi.org/10.1021/es500004z

Adam IKU, Miltner A, Kästner M (2015) Degradation of ^{13}C-labeled pyrene in soil-compost mixtures and fertilized soil. Appl Microbiol Biotechnol 99(22):9813–9824. https://doi.org/10.1007/s00253-015-6822-8

Ahlawat OP, Gupta P, Kumar S, Sharma DK, Ahlawat K (2010) Bioremediation of fungicides by spent mushroom substrate and its associated microflora. Indian J Microbiol 50(4):390–395. https://doi.org/10.1007/s12088-011-0067-8

Ahmad R, Jilani G, Arshad M, Zahir ZA, Khalid A (2007) Bio-conversion of organic wastes for their recycling in agriculture: an overview of perspectives and prospects. Ann Microbiol 57 (4):471–479. https://doi.org/10.1007/BF03175343

Alburquerque JA, Gonzálvez J, Tortosa G, Baddi GA, Cegarra J (2009) Evaluation of "alperujo" composting based on organic matter degradation, humification and compost quality. Biodegradation 20(2):257–270. https://doi.org/10.1007/s10532-008-9218-y

Anastasi A, Varese GC, Filipello Marchisio V (2005) Isolation and identification of fungal communities in compost and vermicompost. Mycologia 97(1):33–44. https://doi.org/10.1080/15572536.2006.11832836

Antizar-Ladislao B (2010) Bioremediation: working with Bacteria. Elements 6(6):389–394. https://doi.org/10.2113/gselements.6.6.389

Antizar-Ladislao B, Lopez-Real J, Beck A (2004) Bioremediation of polycyclic aromatic hydrocarbon (PAH)-contaminated waste using composting approaches. Crit Rev Environ Sci Technol 34(3):249–289. https://doi.org/10.1080/10643380490434119

Antizar-Ladislao B, Lopez-Real J, Beck AJ (2005) Laboratory studies of the remediation of polycyclic aromatic hydrocarbon contaminated soil by in-vessel composting. Waste Manag 25(3):281–289. https://doi.org/10.1016/j.wasman.2005.01.009

Antizar-Ladislao B, Lopez-Real J, Beck AJ (2006) Degradation of polycyclic aromatic hydrocarbons (PAHs) in an aged coal tar contaminated soil under in-vessel composting conditions. Environ Pollut 141(3):459–468. https://doi.org/10.1016/j.envpol.2005.08.066

Bao Y, Zhou Q, Guan L, Wang Y (2009) Depletion of chlortetracycline during composting of aged and spiked manures. Waste Manag 29(4):1416–1423. https://doi.org/10.1016/j.wasman.2008.08.022

Benoit P, Barriuso E, Calvet R (1998) Biosorption characterization of herbicides, 2,4-D and atrazine, and two chlorophenols on fungal mycelium. Chemosphere 37(7):1271–1282. https://doi.org/10.1016/S0045-6535(98)00125-8

Bosma TNP, Middeldorp PJM, Schraa G, Zehnder AJB (1997) Mass transfer limitation of biotransformation: quantifying bioavailability. Environ Sci Technol 31(1):248–252. https://doi.org/10.1021/es960383u

Briones A (2014) Commentary on De Gannes et al. (2013): "Insights into fungal communities in composts revealed by 454-pyrosequencing: implications for human health and safety". Front Microbiol 5:372. https://doi.org/10.3389/fmicb.2014.00372

Brown ME, Chang MCY (2014) Exploring bacterial lignin degradation. Curr Opin Chem Biol 19:1–7. https://doi.org/10.1016/j.cbpa.2013.11.015

Buyuksonmez F, Sekeroglu S (2005) Presence of pharmaceuticals and personal care products (PPCPs) in biosolids and their degradation during composting. J Residuals Sci Technol 2:31–40

Büyüksönmez F, Rynk R, Hess TF, Bechinski E (1999) Occurrence, degradation and fate of pesticides during composting. Part I: Composting, pesticides, and pesticide degradation. Compost Sci Util 7(4):66–82. https://doi.org/10.1080/1065657X.1999.10701986

Cerniglia CE (1984) Microbial metabolism of polycyclic aromatic hydrocarbons. Adv Appl Microbiol 30:31–71. https://doi.org/10.1016/S0065-2164(08)70052-2

Cerniglia CE (1992) Biodegradation of polycyclic aromatic hydrocarbons. In: Rosenberg E (ed) Microorganisms to combat pollution. Springer, Dordrecht, pp 227–244. https://doi.org/10.1007/978-94-011-1672-5_16

Cerniglia CE, Sutherland JB (2010) Degradation of polycyclic aromatic hydrocarbons by Fungi. In: Timmis KN (ed) Handbook of hydrocarbon and lipid microbiology. Springer, Berlin, Heidelberg, pp 2079–2110. https://doi.org/10.1007/978-3-540-77587-4_151

Chefetz B, Xing B (2009) Relative role of aliphatic and aromatic moieties as sorption domains for organic compounds: a review. Environ Sci Technol 43(6):1680–1688. https://doi.org/10.1021/es803149u

Chen M, Xu P, Zeng G, Yang C, Huang D, Zhang J (2015) Bioremediation of soils contaminated with polycyclic aromatic hydrocarbons, petroleum, pesticides, chlorophenols and heavy metals by composting: applications, microbes and future research needs. Biotechnol Adv 33 (6):745–755. https://doi.org/10.1016/j.biotechadv.2015.05.003

Covino S, Fabianová T, Křesinová Z, Čvančarová M, Burianová E, Filipová A, Voříšková J, Baldrian P, Cajthaml T (2016) Polycyclic aromatic hydrocarbons degradation and microbial community shifts during co-composting of creosote-treated wood. J Hazard Mater 301:17–26. https://doi.org/10.1016/j.jhazmat.2015.08.023

de Gannes V, Eudoxie G, Hickey WJ (2013a) Insights into fungal communities in composts revealed by 454-pyrosequencing: implications for human health and safety. Front Microbiol 4:164. https://doi.org/10.3389/fmicb.2013.00164

de Gannes V, Eudoxie G, Hickey WJ (2013b) Prokaryotic successions and diversity in composts as revealed by 454-pyrosequencing. Bioresour Technol 133:573–580. https://doi.org/10.1016/j.biortech.2013.01.138

Dees PM, Ghiorse WC (2001) Microbial diversity in hot synthetic compost as revealed by PCR-amplified rRNA sequences from cultivated isolates and extracted DNA. FEMS Microbiol Ecol 35(2):207–216. https://doi.org/10.1016/S0168-6496(01)00092-7

Doyle RC, Isbister JD, Anspach GL, Kitchens JF (1986) Composting explosives/organics contaminated soils

Eberhardt C, Grathwohl P (2002) Time scales of organic contaminant dissolution from complex source zones: coal tar pools vs. blobs. J Contam Hydrol 59(1–2):45–66. https://doi.org/10.1016/S0169-7722(02)00075-X

Eggen T (1999) Application of fungal substrate from commercial mushroom production — *Pleuorotus ostreatus* — for bioremediation of creosote contaminated soil. Int Biodeterior Biodegrad 44(2–3):117–126. https://doi.org/10.1016/S0964-8305(99)00073-6

Fitzpatrick GE, Worden EC, Vendrame WA (2005) Historical development of composting technology during the 20th century. HortTechnology 15(1):48–51

Fountoulakis MS, Terzakis S, Georgaki E, Drakopoulou S, Sabathianakis I, Kouzoulakis M, Manios T (2009) Oil refinery sludge and green waste simulated windrow composting. Biodegradation 20(2):177–189. https://doi.org/10.1007/s10532-008-9211-5

Gajalakshmi S, Abbasi SA (2008) Solid waste management by composting: state of the art. Crit Rev Environ Sci Technol 38(5):311–400. https://doi.org/10.1080/10643380701413633

Gandolfi I, Sicolo M, Franzetti A, Fontanarosa E, Santagostino A, Bestetti G (2010) Influence of compost amendment on microbial community and ecotoxicity of hydrocarbon-contaminated soils. Bioresour Technol 101(2):568–575. https://doi.org/10.1016/j.biortech.2009.08.095

García-Delgado C, D'Annibale A, Pesciaroli L, Yunta F, Crognale S, Petruccioli M, Eymar E (2015) Implications of polluted soil biostimulation and bioaugmentation with spent mushroom substrate (*Agaricus bisporus*) on the microbial community and polycyclic aromatic hydrocarbons biodegradation. Sci Total Environ 508:20–28. https://doi.org/10.1016/j.scitotenv.2014.11.046

Ghoshal S, Ramaswami A, Luthy RG (1996) Biodegradation of naphthalene from coal tar and heptamethylnonane in mixed batch systems. Environ Sci Technol 30(4):1282–1291. https://doi.org/10.1021/es950494d

Gibson RW, Wang M-J, Padgett E, Lopez-Real JM, Beck AJ (2007) Impact of drying and composting procedures on the concentrations of 4-nonylphenols, di-(2-ethylhexyl)phthalate and polychlorinated biphenyls in anaerobically digested sewage sludge. Chemosphere 68 (7):1352–1358. https://doi.org/10.1016/j.chemosphere.2007.01.020

Haderlein A, Legros R, Ramsay B (2001) Enhancing pyrene mineralization in contaminated soil by the addition of humic acids or composted contaminated soil. Appl Microbiol Biotechnol 56 (3–4):555–559. https://doi.org/10.1007/s002530000520

Harms H, Zehnder AJB (1994) Influence of substrate diffusion on degradation of dibenzofuran and 3-chlorodibenzofuran by attached and suspended bacteria. Appl Environ Microbiol 60 (8):2736–2745

Harms H, Schlosser D, Wick LY (2011) Untapped potential: exploiting fungi in bioremediation of hazardous chemicals. Nat Rev Microbiol 9:177–192. https://doi.org/10.1038/nrmicro2519
Heitkamp MA, Cerniglia CE (1989) Polycyclic aromatic hydrocarbon degradation by a *Mycobacterium* sp. in microcosms containing sediment and water from a pristine ecosystem. Appl Environ Microbiol 55(8):1968–1973
Hellmann B, Zelles L, Palojärvi A, Bai Q (1997) Emission of climate-relevant trace gases and succession of microbial communities during open-windrow composting. Appl Environ Microbiol 63(3):1011–1018
Hernández T, Garcia E, García C (2015) A strategy for marginal semiarid degraded soil restoration: A sole addition of compost at a high rate. A five-year field experiment. Soil Biol Biochem 89:61–71. https://doi.org/10.1016/j.soilbio.2015.06.023
Hofrichter M (2002) Review: lignin conversion by manganese peroxidase (MnP). Enzym Microb Technol 30(4):454–466. https://doi.org/10.1016/S0141-0229(01)00528-2
Jenkins T, Vogel C (2014) Department of defense best management practices for munitions constituents on operational ranges
Joo H-S, Shoda M, Phae C-G (2007) Degradation of diesel oil in soil using a food waste composting process. Biodegradation 18:597–605. https://doi.org/10.1007/s10532-006-9092-4
Juhasz AL, Naidu R (2000) Bioremediation of high molecular weight polycyclic aromatic hydrocarbons: a review of the microbial degradation of benzo[*a*]pyrene. Int Biodeterior Biodegrad 45 (1–2):57–88. https://doi.org/10.1016/S0964-8305(00)00052-4
Karnchanawong S, Nissaikla S (2014) Effects of microbial inoculation on composting of household organic waste using passive aeration bin. Int J Recycl Org Waste Agric 3(4):113–119. https://doi.org/10.1007/s40093-014-0072-0
Kästner M (2000) Degradation of aromatic and polyaromatic compounds. In: Rehm H-J, Reed G (eds) Biotechnology, vol 11b Environmental processes II. Wiley-VCH Verlag GmbH, Weinheim, pp 211–239. https://doi.org/10.1002/9783527620951.ch9
Kästner M, Miltner A (2016) Application of compost for effective bioremediation of organic contaminants and pollutants in soil. Appl Microbiol Biotechnol 100(8):3433–3449. https://doi.org/10.1007/s00253-016-7378-y
Kirchmann H, Ewnetu W (1998) Biodegradation of petroleum-based oil wastes through composting. Biodegradation 9(2):151–156. https://doi.org/10.1023/A:1008355825404
Kües U (2015) Fungal enzymes for environmental management. Curr Opin Biotechnol 33:268–278. https://doi.org/10.1016/j.copbio.2015.03.006
Laine MM, Ahtiainen J, Wågman N, Öberg LG, Jørgensen KS (1997) Fate and toxicity of chlorophenols, polychlorinated dibenzo-*p*-dioxins, and dibenzofurans during composting of contaminated sawmill soil. Environ Sci Technol 31(11):3244–3250. https://doi.org/10.1021/es970233z
Lee LS, Rao PSC, Okuda I (1992) Equilibrium partitioning of polycyclic aromatic hydrocarbons form coal tar into water. Environ Sci Technol 26(11):2110–2115. https://doi.org/10.1021/es00035a006
Li X, Lin X, Zhang J, Wu Y, Yin R, Feng Y, Wang Y (2010) Degradation of polycyclic aromatic hydrocarbons by crude extracts from spent mushroom substrate and its possible mechanisms. Curr Microbiol 60(5):336–342. https://doi.org/10.1007/s00284-009-9546-0
Loick N, Hobbs PJ, Hale MDC, Jones DL (2009) Bioremediation of poly-aromatic hydrocarbon (PAH)-contaminated soil by composting. Crit Rev Environ Sci Technol 39(4):271–332. https://doi.org/10.1080/10643380701413682
Lu X-Y, Zhang T, Fang HH-P (2011) Bacteria-mediated PAH degradation in soil and sediment. Appl Microbiol Biotechnol 89(5):1357–1371. https://doi.org/10.1007/s00253-010-3072-7
Luthy RG, Ramaswami A, Ghoshal S, Merkel W (1993) Interfacial films in coal tar nonaqueous-phase liquid–water systems. Environ Sci Technol 27(13):2914–2918. https://doi.org/10.1021/es00049a035

Luthy RG, Dzombak DA, Peters CA, Roy SB, Ramaswami A, Nakles DV, Nott BR (1994) Remediating tar-contaminated soils at manufactured gas plant sites. Environ Sci Technol 28 (6):266A–276A. https://doi.org/10.1021/es00055a002

Manderson GJ (2009) Composting agricultural and industrial waste. Biotechnology VIII:41–44

Marchal G, Smith KEC, Rein A, Winding A, Trapp S, Karlson UG (2013a) Comparing the desorption and biodegradation of low concentrations of phenanthrene sorbed to activated carbon, biochar and compost. Chemosphere 90(6):1767–1778. https://doi.org/10.1016/j.chemosphere.2012.07.048

Marchal G, Smith KEC, Rein A, Winding A, Wollensen de Jonge L, Trapp S, Karlson UG (2013b) Impact of activated carbon, biochar and compost on the desorption and mineralization of phenanthrene in soil. Environ Pollut 181:200–210. https://doi.org/10.1016/j.envpol.2013.06.026

Margesin R, Schinner F (2001) Biodegradation and bioremediation of hydrocarbons in extreme environments. Appl Microbiol Biotechnol 56(5–6):650–663. https://doi.org/10.1007/s002530100701

Martín J, Santos JL, Aparicio I, Alonso E (2015) Pharmaceutically active compounds in sludge stabilization treatments: anaerobic and aerobic digestion, wastewater stabilization ponds and composting. Sci Total Environ 503–504:97–104. https://doi.org/10.1016/j.scitotenv.2014.05.089

Mayer P, Karlson U, Christensen PS, Johnsen AR, Trapp S (2005) Quantifying the effect of medium composition on the diffusive mass transfer of hydrophobic organic chemicals through unstirred boundary layers. Environ Sci Technol 39(16):6123–6129. https://doi.org/10.1021/es050556s

Mayer P, Fernqvist MM, Christensen PS, Karlson U, Trapp S (2007) Enhanced diffusion of polycyclic aromatic hydrocarbons in artificial and natural aqueous solutions. Environ Sci Technol 41(17):6148–6155. https://doi.org/10.1021/es070495t

Megharaj M, Ramakrishnan B, Venkateswarlu K, Sethunathan N, Naidu R (2011) Bioremediation approaches for organic pollutants: a critical perspective. Environ Int 37(8):1362–1375. https://doi.org/10.1016/j.envint.2011.06.003

Moeller J, Reeh U (2003) Degradation of Nonylphenol Ethoxylates (NPE) in sewage sludge and source separated municipal solid waste under bench-scale composting conditions. Bull Environ Contam Toxicol 70(2):248–254. https://doi.org/10.1007/s00128-001-0184-x

Morimoto K, Tatsumi K (1997) Effect of humic substances on the enzymatic formation of OCDD from PCP. Chemosphere 34(5–7):1277–1283. https://doi.org/10.1016/S0045-6535(97)00425-6

Öberg LG, Glas B, Swanson SE, Rappe C, Paul KG (1990) Peroxidase-catalyzed oxidation of chlorophenols to polychlorinated dibenzo-*p*-dioxins and dibenzofurans. Arch Environ Contam Toxicol 19(6):930–938. https://doi.org/10.1007/BF01055064

Ortega-Calvo J-J, Saiz-Jimenez C (1998) Effect of humic fractions and clay on biodegradation of Phenanthrene by a *Pseudomonas fluorescens* strain isolated from soil. Appl Environ Microbiol 64(8):3123–3126

Pakou C, Kornaros M, Stamatelatou K, Lyberatos G (2009) On the fate of LAS, NPEOs and DEHP in municipal sewage sludge during composting. Bioresour Technol 100(4):1634–1642. https://doi.org/10.1016/j.biortech.2008.09.025

Peng R-H, Xiong A-S, Xue Y, Fu X-Y, Gao F, Zhao W, Tian Y-S, Yao Q-H (2008) Microbial biodegradation of polyaromatic hydrocarbons. FEMS Microbiol Rev 32(6):927–955. https://doi.org/10.1111/j.1574-6976.2008.00127.x

Peng J, Zhang Y, Su J, Qiu Q, Jia Z, Zhu Y-G (2013) Bacterial communities predominant in the degradation of $^{13}C_4$-4,5,9,10-pyrene during composting. Bioresour Technol 143:608–614. https://doi.org/10.1016/j.biortech.2013.06.039

Pignatello JJ, Xing B (1995) Mechanisms of slow sorption of organic chemicals to natural particles. Environ Sci Technol 30(1):1–11. https://doi.org/10.1021/es940683g

Pignatello JJ, Kwon S, Lu Y (2006) Effect of natural organic substances on the surface and adsorptive properties of environmental black carbon (char): attenuation of surface activity by humic and fulvic acids. Environ Sci Technol 40(24):7757–7763. https://doi.org/10.1021/es061307m

Potts M (1994) Desiccation tolerance of prokaryotes. Microbiol Rev 58(4):755–805
Puglisi E, Cappa F, Fragoulis G, Trevisan M, Del Re AAM (2007) Bioavailability and degradation of phenanthrene in compost amended soils. Chemosphere 67(3):548–556. https://doi.org/10.1016/j.chemosphere.2006.09.058
Ramaswamy J, Prasher SO, Patel RM, Hussain SA, Barrington SF (2010) The effect of composting on the degradation of a veterinary pharmaceutical. Bioresour Technol 101(7):2294–2299. https://doi.org/10.1016/j.biortech.2009.10.089
Rein A, Adam IKU, Miltner A, Brumme K, Kästner M, Trapp S (2016) Impact of bacterial activity on turnover of insoluble hydrophobic substrates (phenanthrene and pyrene)—model simulations for prediction of bioremediation success. J Hazard Mater 306:105–114. https://doi.org/10.1016/j.jhazmat.2015.12.005
Ro KS, Preston KT, Seiden S, Bergs MA (1998) Remediation composting process principles: focus on soils contaminated with explosive compounds. Crit Rev Environ Sci Technol 28 (3):253–282. https://doi.org/10.1080/10643389891254223
Šašek V, Bhatt M, Cajthaml T, Malachová K, Lednická D (2003) Compost-mediated removal of polycyclic aromatic hydrocarbons from contaminated soil. Arch Environ Contam Toxicol 44 (3):336–342. https://doi.org/10.1007/s00244-002-2037-y
Semblante GU, Hai FI, Huang X, Ball AS, Price WE, Nghiem LD (2015) Trace organic contaminants in biosolids: impact of conventional wastewater and sludge processing technologies and emerging alternatives. J Hazard Mater 300:1–17. https://doi.org/10.1016/j.jhazmat.2015.06.037
Semple KT, Reid BJ, Fermor TR (2001) Impact of composting strategies on the treatment of soils contaminated with organic pollutants. Environ Pollut 112(2):269–283. https://doi.org/10.1016/S0269-7491(00)00099-3
Shelton DR, Doherty MA (1997) A model describing pesticide bioavailability and biodegradation in soil. Soil Sci Soc Am J 61(4):1078–1084. https://doi.org/10.2136/sssaj1997.03615995006100040013x
Singh A, Sharma S (2003) Effect of microbial Inocula on mixed solid waste composting, vermicomposting and plant response. Compost Sci Util 11(3):190–199. https://doi.org/10.1080/1065657X.2003.10702127
Smith KEC, Thullner M, Wick LY, Harms H (2009) Sorption to humic acids enhances polycyclic aromatic hydrocarbon biodegradation. Environ Sci Technol 43(19):7205–7211. https://doi.org/10.1021/es803661s
Smith KEC, Thullner M, Wick LY, Harms H (2011) Dissolved organic carbon enhances the mass transfer of hydrophobic organic compounds from nonaqueous phase liquids (NAPLs) into the aqueous phase. Environ Sci Technol 45(20):8741–8747. https://doi.org/10.1021/es202983k
Steffen K, Hatakka A, Hofrichter M (2002) Removal and mineralization of polycyclic aromatic hydrocarbons by litter-decomposing basidiomycetous fungi. Appl Microbiol Biotechnol 60 (1–2):212–217. https://doi.org/10.1007/s00253-002-1105-6
Stringfellow WT, Alvarez-Cohen L (1999) Evaluating the relationship between the sorption of PAHs to bacterial biomass and biodegradation. Water Res 33(11):2535–2544. https://doi.org/10.1016/S0043-1354(98)00497-7
Thummes K, Kämpfer P, Jäckel U (2007) Temporal change of composition and potential activity of the thermophilic archaeal community during the composting of organic material. Syst Appl Microbiol 30(5):418–429. https://doi.org/10.1016/j.syapm.2007.01.006
Trautmann NM, Krasny ME (1998) Composting in the classroom: scientific inquiry for high school students. Kendall/Hunt Publishing Company, Dubuque
Tuomela M, Vikman M, Hatakka A, Itävaara M (2000) Biodegradation of lignin in a compost environment: a review. Bioresour Technol 72(2):169–183. https://doi.org/10.1016/S0960-8524(99)00104-2
US EPA (1994) Tech trends: the applied technologies Journal for Superfund Removals and Remedial Actions and RCRA Corrective Actions, November 1994
US EPA (1997) Innovative uses of compost: composting of soils contaminated by explosives. US EPA Publ. EPA530-F-97-045, 1–4

Vacca DJ, Bleam WF, Hickey WJ (2005) Isolation of soil Bacteria adapted to degrade humic acid-sorbed phenanthrene. Appl Environ Microbiol 71(7):3797–3805. https://doi.org/10.1128/AEM.71.7.3797-3805.2005

Van Gestel K, Mergaert J, Swings J, Coosemans J, Ryckeboer J (2003) Bioremediation of diesel oil-contaminated soil by composting with biowaste. Environ Pollut 125(3):361–368. https://doi.org/10.1016/S0269-7491(03)00109-X

Vasskog T, Bergersen O, Anderssen T, Jensen E, Eggen T (2009) Depletion of selective serotonin reuptake inhibitors during sewage sludge composting. Waste Manag 29(11):2808–2815. https://doi.org/10.1016/j.wasman.2009.06.010

Viamajala S, Peyton BM, Richards LA, Petersen JN (2007) Solubilization, solution equilibria, and biodegradation of PAH's under thermophilic conditions. Chemosphere 66(6):1094–1106. https://doi.org/10.1016/j.chemosphere.2006.06.059

Volkering F, Breure AM, Sterkenburg A, van Andel JG (1992) Microbial degradation of polycyclic aromatic hydrocarbons: effect of substrate availability on bacterial growth kinetics. Appl Microbiol Biotechnol 36(4):548–552. https://doi.org/10.1007/BF00170201

Wang X, Cui H, Shi J, Zhao X, Zhao Y, Wei Z (2015) Relationship between bacterial diversity and environmental parameters during composting of different raw materials. Bioresour Technol 198:395–402. https://doi.org/10.1016/j.biortech.2015.09.041

Wehrer M, Rennert T, Totsche KU (2013) Kinetic control of contaminant release from NAPLs – experimental evidence. Environ Pollut 179:315–325. https://doi.org/10.1016/j.envpol.2013.03.041

Wéry N (2014) Bioaerosols from composting facilities – a review. Front Cell Infect Microbiol 4:42. https://doi.org/10.3389/fcimb.2014.00042

Wick LY, Colangelo T, Harms H (2001) Kinetics of mass transfer-limited bacterial growth on solid PAHs. Environ Sci Technol 35(2):354–361. https://doi.org/10.1021/es001384w

Wiesmann U (1994) Biological nitrogen removal from wastewater. In: Biotechnics/wastewater. Advances in biochemical engineering/biotechnology, vol 51. Springer, Berlin/Heidelberg, pp 113–154. https://doi.org/10.1007/BFb0008736

Wong JWC, Fang M, Zhao Z, Xing B (2004) Effect of surfactants on solubilization and degradation of phenanthrene under thermophilic conditions. J Environ Qual 33(6):2015–2025. https://doi.org/10.2134/jeq2004.2015

Xia K, Bhandari A, Das K, Pillar G (2005) Occurrence and fate of pharmaceuticals and personal care products (PPCPs) in biosolids. J Environ Qual 34(1):91–104. https://doi.org/10.2134/jeq2005.0091

Xing B, Pignatello JJ, Gigliotti B (1996) Competitive sorption between atrazine and other organic compounds in soils and model sorbents. Environ Sci Technol 30(8):2432–2440. https://doi.org/10.1021/es950350z

Youngquist CP, Mitchell SM, Cogger CG (2016) Fate of antibiotics and antibiotic resistance during digestion and composting: a review. J Environ Qual 45(2):537–545. https://doi.org/10.2134/jeq2015.05.0256

Zhang Y, Zhu Y-G, Houot S, Qiao M, Nunan N, Garnier P (2011) Remediation of polycyclic aromatic hydrocarbon (PAH) contaminated soil through composting with fresh organic wastes. Environ Sci Pollut Res 18(9):1574–1584. https://doi.org/10.1007/s11356-011-0521-5

Chapter 24
Modern Bioremediation Approaches: Use of Biosurfactants, Emulsifiers, Enzymes, Biopesticides, GMOs

Martin Halecký and Evguenii Kozliak

Abstract New trends in bioremediation are reviewed with the major focus on applications of both synthetic and biological surfactants including high molecular weight bioemulsifiers. The use of eco-friendly biosurfactants is discussed including their applications for removal of heavy metals from soil in addition to traditional organic contaminants. Both success stories and limitations of biosurfactant applications are described on the basis of current literature. Another newly developed technology, the use of enzymes (free or immobilized) instead of/in combination with microorganisms is discussed with respect to removal of both organic contaminants and metals from soil along with addressing the key disadvantage of the enzyme application, which is its high cost. Application of genetically modified organisms, i.e., microorganisms and plants, for soil bioremediation is reviewed focusing on introduction into the environment as well as contained use in closed reactors. Finally, a brief review is provided on the current research and application of biopesticides as promising agents for prevention of soil contamination.

Keywords Biosurfactants · Synthetic surfactants · Emulsifiers · Enzymes · Biopesticides · GMOs

24.1 Introduction

As mentioned in the introduction to Part V, several decontamination methods have been successfully used for soil and sediment cleaning for decades. Some of them, however, require application of drastic conditions and/or use of toxic agents causing a negative impact on the local wildlife and the environment. In addition, these traditional decontamination methods in many cases ended up incurring rather high

---

M. Halecký (✉)
Department of Biotechnology, University of Chemistry and Technology, Prague, Czech Republic
e-mail: martin.halecky@vscht.cz

E. Kozliak
Department of Chemistry, University of North Dakota, Grand Forks, ND, USA

J. Filip et al. (eds.), *Advanced Nano-Bio Technologies for Water and Soil Treatment*, Applied Environmental Science and Engineering for a Sustainable Future,
https://doi.org/10.1007/978-3-030-29840-1_24

costs. To make the traditional methods cost-effective and eco-friendly, their optimization is required, namely, the application of new eco-friendly yet efficient agents as well as an appropriate eco-friendly enhanced environment attenuation.

As another new development, the high potential of GMOs, with either their direct introduction into the environment or the contained use for a cost-effective production of decontamination agents in closed bioreactors, is being considered but not sufficiently used at this point. Currently the emphasis is also put on preventing contamination while using Integrated Pest Management and Integrated Crop Management and favoring biopesticides over traditional pesticides in agriculture as legislatively established in the framework for Community action to achieve sustainable use of pesticides by Directive 2009/128/EC of the European Parliament. Although novel agents enhancing current technologies, e.g., biosurfactants and biopesticides have been experimentally tested for decades and are currently highly supported by the governments, their wide commercial use is limited because of their high price, fragmented (regional) market, a lack of awareness of the benefits of biosurfactants and biopesticides and, in some cases, quality fluctuations and/or poor customer support.

24.2 Perspectives of the Use of Synthetic Surfactants and Biosurfactants in Remediation Protocols

24.2.1 Surfactant Properties and Types

Generally, surfactants (either synthetic or biosurfactants) are amphiphilic compounds that lower the surface tension between two immiscible liquids or between the liquid and solid phases. Surfactants accumulate at the liquid–air, liquid–liquid, and liquid–solid interface and reduce the repulsive forces between those phases, allowing them to mix and interact more readily (Soberón-Chávez and Maier 2011). Thus, they may be used as wetting agents, detergents, foaming agents, emulsifiers, and dispersants. To have these specific abilities, surfactants always contain two domains/moieties—hydrophobic and hydrophilic.

The most important characteristics of any surfactant are critical micelle concentration (CMC), hydrophilic–lipophilic balance (HLB), chemical structure, and charge. CMC is the concentration of surfactants above which micelles start to form and the newly added surfactant molecules incorporate into the micelles instead of being dispersed throughout the solution. The surface tension strongly depends on the surfactant concentration below the CMC, then leveling off once the CMC is reached/exceeded (Christofi and Ivshina 2002). CMC characterizes the degree of influence on the surface tension; the lower the CMC value, the more efficient the surfactant. The HLB defines the extent of surfactant hydrophilicity or lipophilicity. All surfactants belong to one of two possible emulsion types. High HLB value indicates the propensity toward the formation of an oil-in-water emulsion, while low HLB indicates the likely stabilization of a water-in-oil emulsion (Christofi and Ivshina 2002; Van Hamme et al. 2006).

Synthetic surfactants are frequently used worldwide in bioremediation technologies as solvent washing enhancers and emulsifying agents for removal of organic pollutants. Those technologies have been verified in both lab-scale and field-scale studies; they have been applied in situ and ex situ (Sekhon et al. 2011). The hydrophobic moiety of synthetic surfactants is represented by alkanes (paraffins), alkenes (olefins), alkylphenols, alkylbenzenes, and alcohols. Poly(oxyethylenes), sucrose, or polypeptides are common polar domains in nonionic surfactants, while anionic surfactants contain sulfate, sulfonate, or carboxylate groups and cationic surfactants contain a quaternary ammonium group (Sekhon et al. 2011). Sodium dodecyl sulfate (SDS, anionic), Triton X-100, Tween 80, Brij 35, and Tergitol NP10 (neutral/non-ionic) are the most common synthetic surfactants successfully used in remediation technologies (Bustamante et al. 2012).

Biosurfactants are natural agents produced predominantly by microorganisms, but also by plants, e.g., fruit pericarp from *Sapindus mukurossi* (Roy et al. 1997), soapberry-derived saponin (Maity et al. 2013), and soya lecithin (Fava et al. 2004). They could also be in the form of specific natural molecules such as humic acids (Conte et al. 2005) or waste materials such as humic substrates from compost (Kulikowska et al. 2015a). As opposed to low molecular weight compounds commonly defined as biosurfactants, the amphiphilic molecules with high molecular weight are called bioemulsifiers because of their specific behavior in oil/water systems; owing to their high HLB values, biosurfactants usually make stable oil-in-water emulsions (Calvo et al. 2009).

Similarly to synthetic surfactants, biosurfactants can be classified by their charge; however, only few are cationic as anionic or neutral are the most common types. The hydrophobic domains of biosurfactants are those of long-chain fatty acids or hydroxylated fatty acids, while the hydrophilic moieties are represented by amino acids, cyclic peptides, carboxylic acids, alcohols, phosphates, or carbohydrates (Mulligan 2005). Bioemulsifiers, high-molecular-mass biosurfactants, consist of amphipathic proteins, polysaccharides, lipoproteins, lipopolysaccharides, or complexes/combinations of those biopolymers (Calvo et al. 2009).

Microbial biosurfactants can be divided into several basic groups according to their molecular structure including glycolipids (rhamnolipids, trehalolipids, and sophorolipids); fatty acids, phospholipids; and neutral lipids (corynomycolic acid, spiculisporic acid, and phosphatidylethanolamine); lipopeptides (surfactin and lichenysin); and polymeric biosurfactants (emulsan, alasan, biodispersan, liposan, and mannoprotein) (Pacwa-Płociniczak et al. 2011). Biosurfactants have a higher level of specificity than chemical surfactants, which is caused by the wide diversity of their basic chemical structures, e.g., lipids, glycolipids, phospholipids, fatty acids, or peptide structures as well as by the broad variety of their functional groups and detailed chemical structure, e.g., branching, number of functional groups or chains, and carbon chain length.

Specifically, microbial biosurfactants may also be classified by their producers. Bacteria, e.g., *Pseudomonas*, *Bacillus*, *Acinetobacter*, *Rhodococcus*; and fungi, e.g., *Candida* and *Saccharomyces* are the most frequently mentioned biosurfactant-producing microbial strains as reviewed in detail previously (Pacwa-Płociniczak et al. 2011; de Cássia et al. 2014).

24.2.2 Synthetic Surfactants vs. Biosurfactants

The advantages of synthetic surfactants are their availability via synthesis, low costs, and precisely defined chemical composition, all leading to their predictable effects. On the other hand, their major disadvantage compared to biosurfactants is higher toxicity. Another fundamental problem of biosurfactants is the inherently low yield of bioprocesses required for their production, making their market prices relatively high (Lau et al. 2014). In addition, some of their other properties rightly mentioned above as benefits should at the same time be considered as potentially unfavorable. Namely, their biodegradability usually leads to the necessity of periodical biosurfactant additions in order to maintain their effective concentration within the non-sterile processing medium (Maslin and Maier 2000). Also, significant variability caused by rather small changes in the cultivation environment can be observed among the batch cultivations, thus making the composition of the final product and its properties less certain. However, there are many important inherent advantages of biosurfactants, e.g., low toxicity, low CMC values, ecological acceptability, biodegradability, high selectivity, high yet specific activity at extreme temperatures, pH, and salinity as reviewed elsewhere (Bustamante et al. 2012; Bezza and Chirwa 2015).

Several comparative studies stated either a minimal negative effect, if any, or even a boost in cell growth and degradation activity by biosurfactants as opposed to chemical surfactants. Solubilization and/or emulsification of biosurfactants were also similar to those of their synthetic analogs, in many cases even higher. For example, *Sphingomonas* sp. GF2B was able to mineralize up to 83.6% of phenantrene (PHE) within 10 days without the addition of surfactants. Biosurfactants facilitated PHE biodegradation to make it as high as 99.5%, whereas the addition of Tween 80, conversely, inhibited the PHE biodegradation to merely 33.5% (Pei et al. 2010). Oil dispersion by several chemical surfactants and biosurfactants with simultaneous bioremediation of a marine offshore oil spill was studied by Pi et al. (2017). The results showed that the enhancement of petroleum hydrocarbon bioremediation by rhamnolipids was twice as efficient compared to commercial chemical dispersants, GM-2, without any toxic effect on microorganisms. Moreover, when assessing the effect of surfactant combination, a consistent negative effect of GM-2 on rhamnolipids' efficiency was observed.

Because of the above-mentioned disadvantages of synthetic surfactants, especially higher toxicity, low selectivity, and higher CMC values, their effective eco-friendly replacement with biosurfactants is sought. However, synthetic surfactants remain dominant because of the disadvantages of biosurfactants, particularly their high price (Lau et al. 2014). Thus, the most feasible cost-effective alternative to synthetic surfactants may be the use of in situ biosurfactant-producing (and pollutant-biodegrading) microorganisms in bioremediation technologies (Ángeles and Refugio 2013; Hosseininoosheri et al. 2016). Another promising approach to making biosurfactants cost-competitive is the optimization of cultivation conditions (Soares dos Santos et al. 2016) or the use of improved GMO producers (Sekhon et al. 2011; Martins Das Neves et al. 2007).

24.2.3 Specific Properties of Biosurfactants

Generally, biosurfactants show very low or no toxicity, thus they can be widely used in remediation protocols. On the other hand, biosurfactants induce an increase in bioavailability and mobility of toxic compounds that can cause secondary toxic effects against some biota. For example, tests evaluating the seed germination and growth inhibition showed an increase in the phytotoxicity of diesel oil for four terrestrial plant species (alfalfa, sorghum, mustard, and cuckooflower) after the addition of rhamnolipids (Marecik et al. 2012).

Rhamnolipids, as surfactants, can negatively affect microbial cell adhesion and/or biofilm development, thus they can be used to control the microbial biofilm. On the other hand, biosurfactants may exhibit toxicity against some pathogenic microorganisms; therefore, they can counteract human (Das et al. 2008) or plant (Borah et al. 2016) pathogens.

The action of biosurfactants or bioemulsifiers on the environment intensifies the cell–pollutant interactions, thus increasing the pollutant bioavailability. Microorganisms gain direct access to water-solubilized hydrocarbons, heavy metals, or pesticides; their cells have direct contact with released oil as a separate phase in the form of large drops, pseudosolubilized, or emulsified oil (Zhong et al. 2014; Alvarez Yela et al. 2016; Yang et al. 2016). Microorganisms usually uptake exterior water-solubilized and pseudosolubilized hydrocarbon molecules; however, an alternate mechanism similar to active pinocytosis in appearance, by which small rhamnolipid-coated hexadecane droplets were transported through the membrane to cell interior, was proved to exist for *Pseudomonas* sp. (Cameotra and Singh 2009).

Incensement of pollutants' mobility and bioavailability can fundamentally increase the efficiency of remediation techniques like washing or microorganism-based remediation, e.g., bioremediation or bioaugmentation. Depending on the concentration and molecular weight, three basic modes of biosurfactant action can be distinguished, i.e., mobilization, solubilization, and emulsification (Pacwa-Płociniczak et al. 2011). Mobilization occurs under a biosurfactant concentration below CMC when the adsorbed organic compounds are desorbed from a matrix by means of biosurfactants breaking free as separate molecules to become (bio)available in the aqueous phase.

Schematic mechanisms of the action of both low- and high-molecular-mass biosurfactants in soil are shown in Figs. 24.1 and 24.2, respectively. By contrast, solubilization occurs under biosurfactant concentrations above CMC and leads to

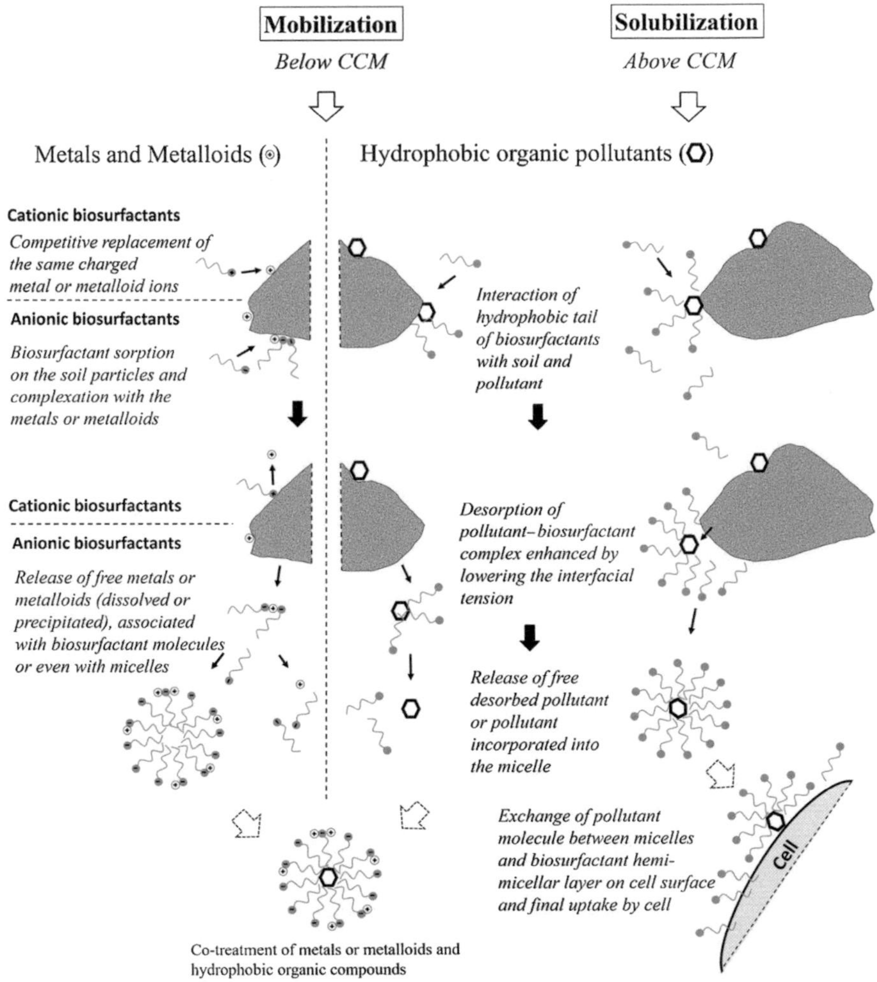

Fig. 24.1 Action of low-molecular-mass biosurfactants. Cationic biosurfactants provide competitive replacement of metal ions while anionic biosurfactants form ionic or coordination bonds with heavy metals enhanced by lowering the interfacial tension. Biosurfactants increase the solubility, bioavailability, and mobility of hydrophobic organic pollutants via their mediated desorption (mobilization) and formation of micelles (solubilization) (Mulligan 2005; Li and Chen 2009; Pacwa-Płociniczak et al. 2011)

incorporating the hydrophobic pollutants into the micelles as a microphase. In this case, the hydrophobic pollutants become protected against the water-based chemicals occurring in the aqueous phase and show an increased formal pollutant concentration in water above its normal equilibrium value (so-called pseudosolubility). Mobilization and solubilization are biosurfactant actions that increase the mobility and bioavailability of not only organic compounds present as single molecules, e.g., pesticides and PAHs or other hydrocarbons but also liquid

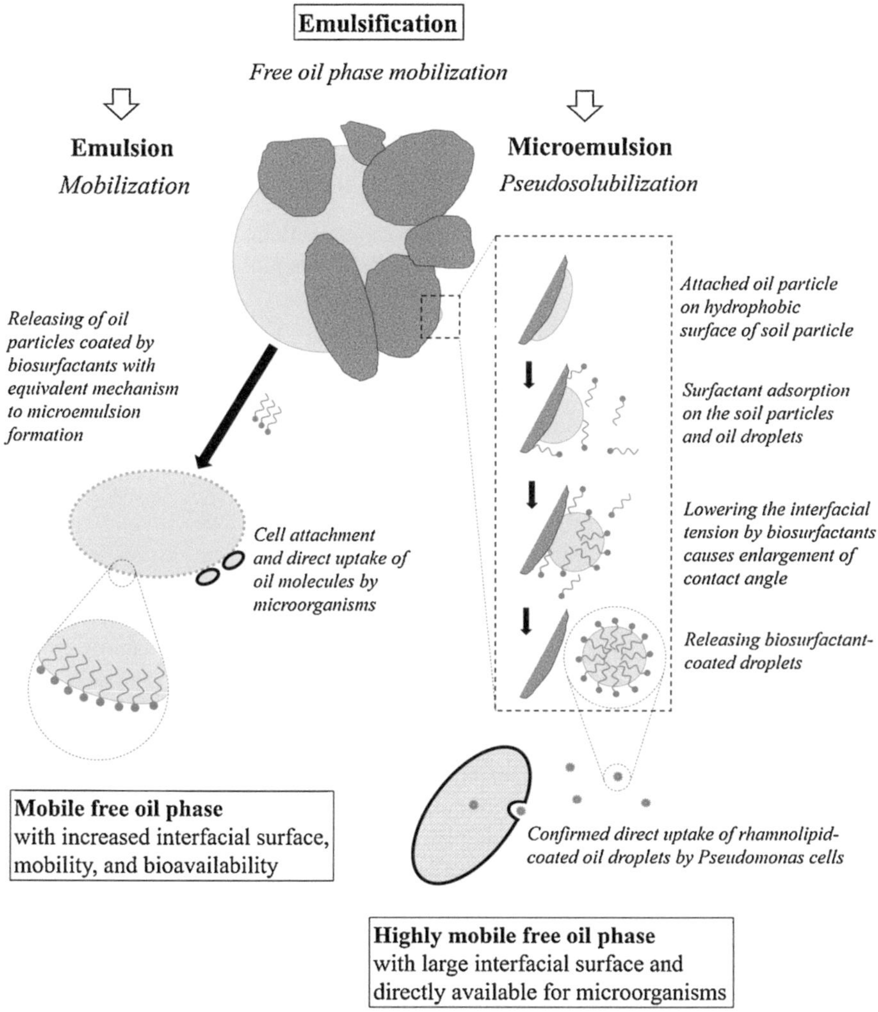

Fig. 24.2 Action of high-molecular-mass biosurfactants. Biosurfactant-coated oil particles (emulsion) or droplets (microemulsions) are released with either an increased interfacial surface, mobility, and bioavailablability or even become directly available for microorganisms (da Rosa et al. 2015; Cameotra and Singh 2009)

hydrocarbons occurring as separate phases forming emulsions or microemulsions. High molecular weight biosurfactants, i.e., bioemulsifiers physically interact with both the immiscible liquid, typically oil occurring as small droplets, and water, thus stabilizing the interface between these two phases in the suspension.

Many biosurfactants are anionic; therefore, they may form ionic or coordination bonds with heavy metals that are stronger than weak metal–soil interactions. This process is enhanced by lowering the interfacial tension. On the other hand, cationic biosurfactants can replace the same charged metal ions by competition and any types of biosurfactants can simply release metals bonded to micelles (Pacwa-Płociniczak et al. 2011).

While in the bulk aqueous phase, the above-mentioned processes are carried out as described; the sorption on particles may reduce the biosurfactant effective concentration in more complex systems with water–solid interfaces (represented by wet soil or sediment). Singh and Cameotra (2013) stated a 50% adsorption of a lipopeptide biosurfactant on soil. This phenomenon combined with the biosurfactant biodegradation should be taken into account when determining the required amount of a biosurfactant for the given task (Maslin and Maier 2000; Ochoa-Loza et al. 2007). To address this problem, a pulse strategy (with a periodic addition of biosurfactant into the system) is often used (Maslin and Maier 2000). However, whenever possible, the best strategy is an in situ (natural or enhanced) production of biosurfactants by suitable species (Hosseininoosheri et al. 2016). The other common setback is that biosurfactants accumulated at the water–gas interface significantly decrease the oxygen mass transfer from the gas to aqueous phase (Sheppard and Cooper 1990), thus potentially causing an oxygen depletion, which can negatively affect remediation bioprocesses.

The above-mentioned influence of biosurfactants on pollutant availability could be accompanied by their influence on microorganisms. The protective function of biosurfactants against toxic compounds described in several publications can be utilized in remediation techniques (Maslin and Maier 2000; Chrzanowski et al. 2009). In addition, biosurfactants can increase the cell wall hydrophobicity, thus enhancing the degradation of hydrophobic compounds by *Pseudomonas aeruginosa* strains with a low cell wall hydrophobicity. However, the action of other strains with low cell wall hydrophobicity was not enhanced (Zhang and Miller 1994). The mechanism of this process was studied in detail by Al-Tahhan et al. (2000). They found that rhamnolipids increase the cell surface hydrophobicity by releasing the genus-specific lipopolysaccharides from the outer membrane of *Pseudomonas aeruginosa*. Another mechanism for *Pseudomonas aeruginosa* was described by Sotirova et al. (2009) when the cell surface hydrophobicity was increased by changing the composition of outer membrane proteins in the presence of a rhamnolipid.

24.2.4 Application Potential of Biosurfactants

24.2.4.1 Treatment of the Sites Contaminated by Organic Pollutants

Hydrocarbon contamination is one of the major environmental problems that result from the activities related to the petrochemical industry. Small-scale field experiments conducted by Tahseen et al. (2016) showed a significant improvement in oil-contaminated soil remediation as a result of rhamnolipid addition. The addition of rhamnolipids caused a greater effect than the addition of mineral nutrients in the form of an NPK (nitrogen, phosphorus, and potassium) fertilizer (32% vs. 26% of degraded oil in 90 days, respectively). Similarly to those results, a significant improvement was observed for the rhamnolipids + bacterial consortium treatment compared to the rhamnolipids + nutrients addition (53% vs. 36% of degraded oil in 90 days, respectively). As expected, the combined rhamnolipids + nutrients + bacterial consortium treatment led to an even greater efficiency, degradation of oil by 77% in 90 days.

Besides petroleum contamination, PAHs are among the most common toxic soil contaminants. Because the environmental PAHs usually occur as a mixture, the biosurfactants' specificity must be taken into account. Portet-Koltalo et al. (2013) conducted a comparative study on the removal of seven PAHs from contaminated soil by either SDS (a synthetic surfactant) or two cyclolipopeptide-based biosurfactants (amphisin and viscosin-like mixture), both produced by two *Pseudomonas fluorescens* strains. SDS was able to release all the studied PAHs while both biosurfactants were only effective toward the desorption of the lowest molecular weight PAHs (naphthalene to fluorene). This selectivity was confirmed by Bezza and Chirwa (2015) when they reported a higher efficiency and faster kinetics of the three-ring phenanthrene desorption from soil in comparison to four-ring pyrene by a lipopeptidal biosurfactant produced by *Paenibacillus dendritiformis*.

Pesticides are widespread, dangerous, and persistent pollutants that tend to be strongly adsorbed in contaminated soils. A glucolipid-based biosurfactant produced by *Burkholderia cenocepacia* BSP3 noticeably enhanced the pesticide solubilization making it available for either washing or bioremediation treatments (Wattanaphon et al. 2008). Also an in situ release of a rhamnolipid produced by *Pseudomonas aeruginosa* CH7, a β-cypermethrin-degrading bacterium, has the potential for enhancing the degradation of pesticides by increasing their bioavailibility (Zhang et al. 2011).

24.2.4.2 Treatment of Sites Contaminated with Heavy Metals

Heavy metals can be efficiently removed from contaminated water, sludge, and soils by a biosurfactant action. Das et al. (2008) reported an efficient chelation of Pb and Cd followed by precipitation by an anionic lipopeptide biosurfactant of a marine origin. Tang et al. (2017) applied saponin, which is an anionic plant biosurfactant having carboxyl groups at the hydrophilic end of its molecule, and showed its ability to bind heavy metals (Zn, Ni, Cu, Mn, Cr, and Pb) for their ultimate removal from water sludge. A specific feature of this surfactant is that its metal binding facilitates the micelle formation by saponin molecules. It was shown that Cr exhibits the highest (65%), while Pb and Mn the lowest (25–35%) extraction efficiency.

Luna et al. (2016) described an efficient treatment of soil and aqueous solutions contaminated from the automotive battery industry using an anionic biosurfactant from *Candida sphaerica*. The results showed the removal rates of 95%, 90%, and 79% for Fe, Zn, and Pb, respectively. The addition of HCl increased the metal removal rate. Additionally, the re-use of a recycled biosurfactant after metal precipitation yielded a reduced but still suitable removal efficiencies of 70%, 62%, and 45% for Fe, Zn, and Pb, respectively.

A foam-enhanced washing technique has been developed to reduce the channeling effect characteristic for soil treatment, thus achieving a homogenous flow, even in a heterogeneous porous medium. Foam can be successfully created and stabilized by biosurfactants. Maity et al. (2013) demonstrated a foam-enhanced removal of Cu, Pb, and Zn from contaminated soil using a plant-based biosurfactant saponin from soapberry *Sapindus mukorossi* and surfactin from *Bacillus subtilis* (BBK006). The optimized foam-enhanced technology resulted in removing 98%, 95%, and 56% of Pb, Cu, and Zn, respectively, while the efficiency of simple washing was twice less efficient.

24.2.4.3 Treatment of Sites Co-contaminated with Heavy Metals and Hydrocarbons

Industrial areas are often co-contaminated with heavy metals and hydrocarbons. Remediation of heavy metal- and hydrocarbon-contaminated soil could be difficult because of the different chemical nature of these two contaminant types. Nevertheless, several recent studies showed the potential of biosurfactants for addressing this problem. One study on washing soil contaminated with high concentrations of heavy metals (Fe, Pb, Ni, Cd, Cu, Co, and Zn) and petroleum hydrocarbons using alipopeptide biosurfactant consisting of surfactin and fengycin originated from *Bacillus subtilis* A21 was published by Singh and Cameotra (2013). Bisurfactant-based washing removed significant amounts of both petroleum hydrocarbons (64.5%) and metals, namely Cd (44.2%), Co (35.4%), Pb (40.3%), Ni (32.2%), Co (26.2%), and Zn (32.07%). The treated soil exhibited a non-inhibited (100%) germination of brown Indian mustard (*Brassica juncea*), whereas the simply washed soil yielded 0% germination. Maslin and Maier (2000) concluded that the biodegradation of phenanthrene can be rhamnolipid-enhanced in organic-metal co-contaminated soils by masking (complexing) toxic cadmium. This protective effect was further improved during the lab-scale simulated degradation of phenanthrene in the aqueous medium and two co-contaminated soils.

24.2.4.4 Long-Term Contaminated Sites—Aged Pollutants

The sorption of pollutants often becomes nearly irreversible over time. Such "aging" of both types of pollutants, either organic compounds (Ncibi et al. 2007) or heavy metals (Liang et al. 2016) causes their low bioavailability, thus prolonging the bioremediation time and decreasing its efficiency. However, it was found that biosurfactants help to effectively release such "irreversibly" adsorbed PAHs (Bezza and Chirwa 2016, 2017; Sánchez-Trujillo et al. 2013), pesticides (Wattanaphon et al. 2008), PCBs (Fava et al. 2003), and heavy metals (Kulikowska et al. 2015b), making the remediation process shorter and more efficient.

Humic substances extracted from compost were used as a washing agent for a simultaneous removal of Cu, Cd, Zn, Pb, and Ni from artificially contaminated soils aged for 1 month, 12 months, and 24 months using both single and multiple washing. The washing efficiency was as high as 80%. Regardless of the applied washing mode, the removal of Cd and Pb was not affected by the contamination age, whereas the removal of Cu, Ni, and Zn was higher in soils that had been aged for a shorter time (Kulikowska et al. 2015b). Bezza and Chirwa (2016) investigated the biosurfactant-enhanced bioremediation of aged PAHs when an additional biosurfactant from *Pseudomonas aeruginosa* and stimulation of the biosurfactant in situ production were applied. The biosurfactant-enhanced degradation of PAHs was 86.5% and 57% in the controls with no second biosurfactant. The additional biosurfactant increased the PAHs bioavailability, thus stimulating the growth of degraders. Then, the population growth led to the depletion of bioavailable PAHs, which triggered an in situ biosurfactant production.

24.2.4.5 Biosurfactant in Prevention of Soil Contamination

As stated above, the assumed low biosurfactant toxicity is not necessarily displayed under all conditions. However, the specific toxicity of some biosurfactants can be appropriately used in many fields. Biosurfactants can actually be used as biopesticides, thus replacing chemical pesticides in order to reduce the pollution of soil, water, and sediment. Increasing amount of studies have already proved the possibility of using biosurfactants against pathogens in agriculture. For example, potato late blight disease caused by pathogen *Phytophthora infestans* was significantly reduced by biosurfactant-producing strain *Pseudomonas koreensis* and its biosurfactant (Hultberg et al. 2010). On a similar token, rhamnolipids can be effectively used against *Fusarium verticillioides* to control the stalk and ear rot disease of maize (Borah et al. 2016).

24.3 Perspectives of Enzymes in Remediation Techniques

Enzymes are naturally present in soil ecosystems. They are produced by soil microorganisms as well as by plants and soil invertebrates. Typical enzymes with potential for bioremediation are listed in Table 24.1. Many enzymes, e.g., peroxidases, laccases, tyrosinases, organophosphorous hydrolases and dehalogenases, lipases, proteases, phosphotriesterases, nitrile- and cyanide-degrading enzymes, and mono- and dioxygenases have been studied for decades to treat the targeted organic compounds, e.g., hydrocarbons; phenols; polyaromatic, nitroaromatic, and chlorinated compounds; dyes; organophosphorous pesticides; nerve/paralyzing agents; or inorganic compounds, e.g., those of metals or metalloids (Piotrowska-Długosz 2017). Fungi and bacteria are the most studied producers of these enzymes (Piotrowska-Długosz 2017); however, marine microorganisms producing enzymes with specific and promising properties are also widely tested (Sivaperumal et al. 2017).

Table 24.1 Enzymes and their actions for soil bioremediation purposes

Enzyme	Producer	Degradation/detoxication action	References
Laccase	*Coriolopsis gallica*	Conversion of bisphenol A into carboxylic acid derivatives such as tartaric acid, β-hydroxybutyric acid and pyroglutamic acid in presence of 1-hydroxybenzotriazole as a laccase mediator	Daâssi et al. (2016)
Manganese peroxidase	*Anthracophyllum discolor*	Conversion of pyrene, anthracene, fluoranthene and phenanthrene in presence of $MnSO_4$	Acevedo et al. (2010)
Cyanide hydratase	*Aspergillus niger* K10	Cyanide decomposition into formate and ammonia	Rinágelová et al. (2014)
Rhodanese (engineered)	*Pseudomonas aeruginosa* (*Escherichia coli* as producer)	Cyanide decomposition in presence of thiosulfate into thiocyanate	Cipollone et al. (2006)
Nitrilase	*Rhodococcus rhodochrous* J1	Conversion of benzonitrile into benzoic acid and acrylonitrile into acrylic acid	Nagasawa et al. (2000)
Nitrile hydratase	*Rhodococcus* sp. RHA1	Conversion of acetonitrile and propionitrile, acrylonitrile and butyronitrile into their amides	Okamoto and Eltis (2007)
Chromate reductase	*Bacillus amyloliquefaciens*	Reduction of Cr(VI) into Cr(III)	Rath et al. (2014)

24.3.1 Specifics of the Use of Enzymes for Soil Bioremediation

The efficiency of enzymes' application, similarly to biosurfactants, strongly depends on the soil composition and the physical properties. Thus the porosity and composition of both mineral and biotic fractions as well as grain fraction distribution, inhomogeneities, pH, and pollutant availability must be evaluated on the contaminated site to choose the best strategy for application of enzymes, which are still considered expensive bioagents (Tuomela and Hatakka 2011; Quiquampoix et al. 2002).

Soil is a multiphase time-variable system consisting of mineral matter, organic matter, water, soil air/gas, and organisms—both invertebrates and microorganisms. Many physical, chemical, and biological interactions occur in soil on several levels of soil structure. In addition, inhomogeneities in both structure and composition as well as local specific processes may further hinder precise soil characterization and application of enzyme-based bioremediation procedures (Zimmerman and Ahn 2010).

Enzymes as protein-based molecules are naturally susceptible to interactions with mineral and organic soil matter (Quiquampoix et al. 2002). Several mechanisms, e.g., covalent attachment, physical entrapment, nonpolar and electrostatic interactions, hydrogen and ionic bonding, result in the formation of enzyme–humus or enzyme–clay complexes (Quiquampoix et al. 2002; Zimmerman and Ahn 2010). Sorption on soil components protects enzymes against inhibitory factors in the environment, thus maintaining a suitable enzyme activity at non-ideal pH, ionic strength, and temperature as well as in the presence of light (photodegradation), proteolytic enzymes, and toxic heavy metals (Tietjen and Wetzel 2003; Quiquampoix et al. 2002; Wang et al. 2017; Wang et al. 2014).

On the other hand, many adverse effects resulting from enzyme–soil interactions decreasing enzyme activity have been described as well, namely: conformational changes, interfacial pH affection, diffusion-limiting kinetics, orientation effects (steric factors), restriction of co-enzyme availability, enzyme encapsulation or denaturation by organic matter or toxic metals, and enzymes' degradation by oxidative minerals (Quiquampoix et al. 2002; Zimmerman and Ahn 2010).

24.3.2 Evaluation of Using Enzymes for Bioremediation

Advantages and disadvantages of enzymes over microorganisms for their use in bioremediation are summarized in Table 24.2. While free enzymes tend to lower mass transfer limitations resulting in faster reactions, immobilized enzymes, either the artificially attached to suitable particles or the naturally immobilized on soil particles, show significant mass transfer limitations (Datta et al. 2017; Quiquampoix et al. 2002). Enzymes can be applied under the lack of substances or conditions essential for live cell functioning, thus eliminating the limitations of nutrient availability, cell acclimation, formation of metabolic by-products, inhibition by other present cell-toxic chemicals. This creates the possibility of their application to toxic or concentrated cell-inhibiting compounds as well as to harsh operational conditions, e.g., extreme pH, temperature, and ionic strength/salinity (Huang et al. 2009; Gianfreda et al. 2016). While the introduction of non-native microbial species is common and mostly non-controversial, yet a subject to approval by state authorities, the introduction of genetically modified microorganisms is strictly regulated and rarely publically acceptable. However, both non-native and genetically modified species can be replaced by enzymes, either native or engineered, presumably with no ecological impact.

The key disadvantage is still the high cost of enzyme application, despite the advances in enzyme production, isolation and purification, potential of the use of genetically engineered enzymes and their hyperproduction by genetically modified species (Eibes et al. 2015). Total mineralization of the majority of complex polluting molecules is undermined by the existence of complex degradation pathways involving many different enzymes. Single enzymes usually only "detoxify" a given pollutant by a single-step chemical modification and thus reduce its toxicity by cleaving a specific group (Acevedo et al. 2010; Daâssi et al. 2016). This limitation can be addressed by applying either mixtures of free enzymes or enzymatic nanoparticles (constructed or natural) that have all the necessary enzymes attached to them. However, the cost of such an advanced bioremediation application is enormous. In addition, it must be ensured that the action of each essential enzyme occurs with the same substrate molecule but at a different time.

Table 24.2 Comparison of advantages and disadvantages of enzymes application in comparison to live microorganisms (Datta et al. 2017; Zimmerman and Ahn 2010; Eibes et al. 2015; Gianfreda et al. 2016; Wang 2006; Acevedo et al. 2010; Wang et al. 2014)

Advantages	Disadvantages
Lower mass transfer limitations (for free enzymes)	Still expensive despite the advances in production
Higher availability of pollutants adsorbed in soil compared to microbial cells	Some need specific cofactors, inductors or mediators
Minimum ecological impact	Necessity of using enzyme mixtures for complex pollutants
Easily controlled process	Risk of mutual decomposition, e.g., proteolysis in enzyme mixtures
Effective in small quantity	Potentially quick decomposition in soil by microbial cells or free enzymes, e.g., peptidases
Long-term storage	Short lifetime
	Competition effects in soil (organic matter)
No DNA	Sorption on soil particles
No introduction of foreign (microbial) species at site	
Publicly acceptable even if GMO-produced	
Overcoming of microorganisms' limits:	Lack of advantageous features of live cells:
No nutrient requirements	Self-replication
No need for acclimation characteristic for whole cells	Active movement (chemotaxis)
No formation of metabolic by-products	Adaptation (changes in cell transport and metabolism)
Enabled application to toxic or concentrated (cell-inhibiting) compounds	Evolution (mutation, horizontal gene transfer)
Potential application under harsh operational conditions (pH, temperature, and ionic strength/ salinity)	Production of supporting substances (biosurfactants for pollutants' release)
No inhibition by other cell-toxic chemicals present on site	

Additional compounds essential for enzymatic reactions may also be required. Mediators and/or co-factors, whether naturally present or added, are compounds with redox potential serving as redox shuttles between the enzyme and substrate. For example, laccase mediators, e.g., *p*-coumaric acid, syringaldehyde, or acetosyringone are essential for catalysis of redox reactions catalyzed by this enzyme (Ji et al. 2016). Similarly, hydrogen peroxide for peroxidases, organic hydroperoxides for peroxygenases, and Mn^{2+} for manganese peroxidase are necessary for enzyme action (Acevedo et al. 2010). Ensuring that all necessary substances are lined up at the same spot for the effective course of a given enzyme-catalyzed reaction is problematic; this problem, in particular, can be a major obstacle in static, heterogeneous systems with high and varied sorption properties such as soil (Datta et al. 2017).

A different yet similarly hampering limitation of enzymatic remediation is characteristic for metal and metalloid contaminated sites. Enzymes as well as microorganisms can convert toxic metals and metalloids into their less toxic forms, e.g., reduced or insoluble, but they cannot physically remove them from a contaminated site; furthermore, such conversion is reversible (Wall and Krumholz 2006).

One more limitation is that the enzymes used for pollution bioremediation, e.g., chitinase, carboxymethyl cellulase, β-glucosidase, protease, acid phosphatase, polyphenol oxidase, laccase, and guaiacol oxidase also exhibit an undesired activity against organic soil colloids (Wang et al. 2014), thus decreasing the efficiency of the attack on the target pollutant because of competition.

Natural biodegradability of enzymes can shorten the lifetime of the applied enzymes because of their decomposition in soil by naturally occurring peptidases and microorganisms. Also the risk of mutual decomposition, e.g., proteolysis, should be taken into account when mixtures of enzymes are applied; therefore, repetitive enzyme application can be essential (Gianfreda et al. 2016). In contrast, specific features of live cells lacking in enzyme application, namely self-replication, active movement (chemotaxis), adaptation (changes in cell transport and metabolisms), evolution (mutation, horizontal gene transfer), and production of supporting substances (biosurfactants for pollutants' release) may favor their use.

24.3.3 *Application Potential of Enzymes in Bioremediation*

Although the list of potential disadvantages is relatively long, a number of promising lab-scale applications of enzymes, either free or immobilized, have been described, as summarized below. The current focus addressing most of the above-listed limitations is on using nanoscale carriers for bioremediation of toxic and recalcitrant pollutants.

24.3.3.1 Treatment of Organic Pollutants

As stated above, enzymatic attacks cause detoxification of complex pollutants rather than their mineralization. For example, bisphenol A was completely removed within 3 h by the action of *Coriolopsis gallica* laccase; however, GC-MS analyses revealed its conversion into carboxylic acid derivatives such as tartaric, β-hydroxybutyric, and pyroglutamic acids (Daâssi et al. 2016). Similarly, several main metabolites, e.g., 10,11-dihydro-10,11-dihydroxy-CBZ, 10,11-dihydro-10,11-epoxy-CBZ, and acridone were identified resulting from the laccase action on carbamazepine. The concomitant decrease of biotoxicity was confirmed (Ji et al. 2016). In another study, both free and immobilized laccases decreased the concentration of sulfathiazole and sulfamethoxazole as well as their biotoxicity in the presence of 1-hydroxybenzotriazole as a laccase mediator. Intermediates were not determined; however, detoxification rather than mineralization was assumed (Rahmani et al. 2015).

In these cases, the ultimate products turned out to be non-toxic, i.e., the enzyme application was justified; however, in general this may not be the case. Another way of enzymatic detoxication is the formation of polymers. For example, triclosan was converted by an integrated nano-bio redox process into non-toxic dimer and trimer products (Bokare et al. 2010).

Enzymes can also be effectively applied to detoxification of cyanide and nitriles. Although rhodanese can be used to detoxicate cyanide in the presence of thiosulfate forming less toxic thiocyanate (Cipollone et al. 2006), a more promising way is to use cyanide hydratases that detoxify cyanide in the sole presence of water. For example, cyanide hydratases convert cyanide into formamide and then cyanide dihydratase converts this intermediate into ammonia and formate. Some bacteria, e.g., *Pseudomonas stutzeri*, and many fungi, e.g., *Fusarium lateritium*, *Aspergillus niger*, *Botryotinia fuckeliana*, produce a variety of cyanide (di)hydratases that were successfully tested for cyanide detoxification (Martínková et al. 2015).

Two main pathways were described for nitrile hydrolysis, either direct hydrolysis to the corresponding carboxylic acids and ammonia by nitrilase (Nagasawa et al. 2000) or a two-step conversion into their amides by nitrile hydratase and further into their carboxylic acids and ammonia by amidase (Okamoto and Eltis 2007). The nitrilase isolated from *Rhodococcus rhodochrous* J1 was able to catalyze a quick conversion of both aromatic and aliphatic nitriles, namely benzonitrile into benzoic acid and acrylonitrile into acrylic acid (Nagasawa et al. 2000). The isolated nitrile hydratase of *Rhodococcus* sp. RHA1 effectively converted acetonitrile, propionitrile, acrylonitrile, and butyronitrile into the corresponding amides. Bacteria containing both essential enzymes to complete the nitrile group detoxification, namely both nitrile hydratase and amidase, could be isolated, too (Okamoto and Eltis 2007). For real-world field scale remediation, however, nitrilases converting nitriles directly to carboxylic acids are much more practical than nitrile hydratases, which require a second step and another enzyme to complete the conversion.

24.3.3.2 Detoxification of Heavy Metals

Detoxification of metals is based mainly on their reduction that usually lowers both their toxicity and solubility. The redox reactions catalyzed by enzymes require an electron donor, usually NAD(P)H originated from aerobic respiration. The second common metabolic pathway occurs when metals/metalloids serve as terminal electron acceptors in anaerobic respiration. In either case, the electrons for metal reduction are produced by complex metabolism. Thus, extracellular reduction by free enzymes (as opposed to extracellular reduction conducted by externally bounded membrane reductases) suffers from a deficiency of essential co-factors (Thatoi et al. 2014). Therefore, the addition of NAD(P)H, a mixture of enzymes or the use of the whole cell metabolism is essential for most cases of enzymatic metal reduction. Payne et al. (2002) explored a promising approach when a mixture of hydrogenase and cytochrome c3 from *Desulfovibrio vulgaris* was used in combination with lactate, pyruvate, and hydrogen as electron donors for the reduction of uranium(VI) to uranium(IV). Cytochrome c3 regenerates the hydrogenase using those electron donors (Payne et al. 2002). However, being merely detoxified rather than removed, metals can be reoxidized chemically by oxygen or can serve as electron (energy) sources for endogenous lithotrophic microorganisms, thus such detoxification can be viewed as only temporary (Wall and Krumholz 2006).

24.3.3.3 Enzyme-Enhanced (Bio)Remediation

Enzymes can be used for enhancing other bioremediation methods. In many cases, enzymatic attacks on toxic pollutants lead to both reduction of their toxicity and increase in their biodegradability. Thus, microbial bioremediation can be fundamentally improved, especially in the case of recalcitrant pollutants with heteroatoms, e.g., organophosphates, chlorinated and nitro compounds. For example, undefined mixtures of extracellular enzymes produced by a microbial consortium under induction by diesel as a substrate were applied during microbial degradation of hydrocarbons on three different oil-contaminated sites. A diesel hydrocarbon degradation of approximately 90% was achieved in 60 h for enzyme-enhanced tests, whereas as many as 175 h were needed to match this efficiency for the degradation tests involving only microbial cells (Jiménez-T et al. 2011).

An application of nano-bio decontamination based on sequential reduction–oxidation was reported by Bokare et al. (2010) for triclosan detoxification under anaerobic conditions (Bokare et al. 2010). A rapid reductive dechlorination by palladized zero-valent iron nanoparticles generated 2-phenoxyphenol, followed by its transformation into a non-toxic polymer using laccase derived from *Trametes versicolor* in the presence of a natural redox mediator, syringaldehyde (Bokare et al. 2010).

24.3.3.4 Enhanced Applications of Enzymes

A number of ways to immobilize/stabilize enzymes on (nano)carriers, e.g., entrapment, encapsulation, adsorption, covalent binding, and self-immobilization as well as many types of nanocarriers, both natural or artificial, e.g., mesostructured silica materials, magnetite nanoparticles, caolinite, porous carbon tubes, and alginate have been described so far (Fernández-Fernández et al. 2013). For example, the laccase of *Trametes versicolor* was immobilized with only a slight loss of activity on functionalized nanoparticles SBA-15 with the average diameter smaller than 10 nm, which are suitable for applications to soil bioremediation (Fernando Bautista et al. 2010). To decrease the anthracene toxicity by conversion to anthroquinone, the laccase from *Trametes versicolor* was immobilized on functionalized (silicated) kaolinite using glutaraldehyde. In both cases, significant enzyme stabilization was shown.

Among other applications, Hong et al. (2017) developed an enzyme-silicate conjugate material consisting of a self-assembled molecular-size thin silicate network cage encasing each individual enzyme molecule (α-chymotrypsin and lipase). Due to the near molecular size of these conjugates, both minimized substrate diffusion limitations and high enzyme stability were achieved. While the k_{cat}/K_m

ratio of α-chymotrypsin slightly decreased from 7.6 ± 0.2 to $6.0 \pm 0.1 \times 10^5$ 1/(M s) comparing the native and the coated enzymes, the stability of the coated enzymes increased drastically retaining ca 90% of its initial activity after 3 days of the use, whereas the activity of native enzymes at that point was only about 10%. A similar strong stabilization effect was observed for lipase (Hong et al. 2017).

In addition to artificial nanoparticles, Ng et al. (2015) investigated the production of natural nanoparticles with multiple enzymatic activities by algae *Shewanella* (Ng et al. 2015). *Shewanella xiamenensis* BC01 produced nanoparticles with oxidoreductases including catalase, manganese peroxidase, laccase, NADH dehydrogenase, flavin reductase, azoreductase, and Fe reductase that were generally spherical in shape with a particle size of 7–8 nm (Ng et al. 2015).

Engineered enzymes as well as their producers are a rapidly expanding field of research allowing for production of large quantities of more stable, active, and/or selective enzymes, yielding cheaper products with a higher utility value (Zhang et al. 2016b).

24.3.4 Enzymes in Prevention and Detection of Contamination

Enzymes, e.g., chitinases, proteases, and lipases, can be used as eco-friendly biopesticides because of their anti-insect and anti-fungal potential. Chitinases–bacterial, fungal, or plant–are promising agents for controlling harmful fungi and plants owing to the presence of chitin in the exoskeletal and intestinal linings of insects and in the fungal cell walls (Yan et al. 2008; Liu et al. 2002). Binod et al. (2007) reported that chitinase of *Trichoderma harzianum* negatively affects the growth and the metamorphosis of *Helicoverpa armigera* larvae, with up to 70% mortality rate. Proteases cleaving the peptide bonds of proteins are naturally produced by insect and nematode pathogens in order to penetrate cuticles made up of 70% of protein (Goettel et al. 1989). Serine proteases isolated from a nematophagous fungus, *Lecanicillium psalliotae*, were able to eradicate nematode *Panagrellus redivivus* with 81–100% efficiency (Yang et al. 2005).

On a similar token, lipases may be used for insect control by hydrolyzing ester bonds of lipoproteins, fats, and waxes in the interior parts of the insect body. A purified extracellular lipase from a fungus applied for biocontrol, *Nomuraea rileyi* MJ, was used, as a single agent, to both promote the fungus spore germination and enhance the mortality of an agricultural pest, *Spodoptera litura* (Supakdamrongkul et al. 2010).

Currently, synergistic effects of enzymes and toxin-derived biopesticides, e.g., δ-endotoxins of *Bacillus thuringiensis*, are investigated when these agents are applied in combination. Protease and chitinase of *Paecilomyces lilacinus* were studied by

Khan et al. (2004) to evaluate their activity against the plant-pathogenic nematode *Meloidogyne javanica*. The individual enzymes showed high activities against eggs and juveniles of this nematode, which were further enhanced in their mixture. Similarly, a synergistic effect was shown for chitinase-producing *B. thuringiensis* enhancing the insecticidal activity of *B. thuringiensis* strain DL5789 against agricultural pest insects *Spodoptera exigua* larvae by more than twofold (Liu et al. 2002).

Enzymes can be used as active agents for selective biosensors detecting a wide range of pollutants, both organic and inorganic. For example, a change in conductivity of the enzyme membrane occurs in a biosensor when tyrosinase either interacts with 4-chlorophenol substrate or is inhibited by toxic compounds including diuron, atrazine, copper, lead, and zinc ions (Anh et al. 2006). Cyanide dihydratase converting cyanide into ammonia, which is then detected by an ammonia-selective electrode, was used for constructing a fast-responding and accurate biosensor (Keusgen et al. 2004).

24.4 Perspectives of Using Genetically Modified Organisms in Bioremediation Techniques

In Europe, three possible ways are distinguished for using GMOs, namely, a contained use, introduction of GMO into the environment, and placing GMO or genetically enhanced products on the market. The contained use, i.e., the GMM cultivation in closed reactor systems, is the only use of GMOs unanimously accepted by the general public. A number of studies have proved a high bioremediation potential of genetically modified microorganisms and plants. However, due to possible risks and low public acceptance, the applications of GMOs for remediation technologies are still scarce (Singh et al. 2011; Kolseth et al. 2015).

There are several ways to use GMOs in bioremediation techniques. The first and most logical approach is the introduction into the environment of modified microorganisms or plants having the ability to degrade a broad range of organic soil pollutants as well as to accumulate in biomass and/or to transform toxic heavy metals (usually via reduction). Also, a cost-effective production of eco-friendly bioremediation agents, e.g., biosurfactants and enzymes is a promising way of how genetically modified microorganisms could be used. In addition, enzymes with enhanced original or new unique properties can be created by genetic engineering and used in bioremediation. Prevention of soil contamination using cost-effectively GMO-produced biopesticides or insect-resistant and disease-resistant transgenic crops is another promising way to maintain soil and aquifer clean.

24.4.1 *Use of GMO in Bioremediation (Introduction into the Environment)*

24.4.1.1 Genetically Modified Microorganisms

A number of studies have been conducted on the bacterial metabolism of engineered microorganisms, e.g., *Escherichia coli*. However, for real-world bioremediation application, the strains that were isolated from the exact contaminated location are considered being more suitable, with a higher likelihood of long-term survival (Lan et al. 2006).

Natural bioremediation processes promoted by microorganisms, e.g., metal reduction, complexation, precipitation, and promotion of the plant growth on the contaminated site (phytoremediation) can be further improved by genetic engineering. For example, *Pichia pastoris* plants were genetically engineered to overexpress a metal-resistant variant of cytochrome b5 reductase to provide a high-throughput bioaccumulation and biotransformation of silver and selenium. Their ions were enzymatically reduced to form stable 70–180 nm elemental nanoparticles. These nanoparticles exhibited at least a tenfold lower cytotoxicity toward HDF, EPG85–257, and T47D cells than silver nitrate and selenium dioxide (Elahian et al. 2017).

Multipurpose approaches can be enabled by enhanced engineering. For example, a cadmium-resistant bacterium *Pseudomonas aeruginosa* was isolated from a Cd-contaminated oil field and further engineered to overexpress targeting metallothioneins on the cell surface to immobilize Cd^{2+}. Introduction of engineered bacteria with improved cadmium tolerance and accumulation enabled the growth promotion of green peas, *Pisum sativum* L., significantly elevating the shoot and root biomass production and leaf chlorophyll content (Huang et al. 2016). *Pseudomonas putida* X3 was engineered by introducing methyl parathion (MP)-degrading gene to obtain the ability to degrade methyl parathion. The application of *Pseudomonas putida* X3 into Cd-contaminated soil reduced the amount of bioavailable Cd by its conversion into a less soluble/exchangeable form and organic-bound Cd (Zhang et al. 2016a).

Enhanced bioremediation of pesticides and chlorinated organic compounds is a common target of genetic engineering. Yang et al. (2010) genetically engineered a native soil bacterium *Stenotrophomonas* sp. strain YC-1 producing methyl parathion hydrolase by a surface anchor system derived from the truncated ice nucleation protein from *Pseudomonas syringae* to possess a broader substrate specificity combined with an enhanced degradation rate of organophosphates. As a result, a mixture of six organophosphate pesticides was completely degraded within 5 h.

24.4.1.2 Genetically Modified Plants

Genetic modification can be used for either improving existing (accumulation of heavy metals) or introducing new abilities (degradation of chlorinated hydrocarbons, explosives, and pesticides). The natural ability of some plants to bioaccumulate heavy metals is used in phytoremediation of contaminated sites.

Overproduction of metal transporters (metal bioaccumulation), overproduction of the enzymes involved in glutathione synthesis and sulfur utilization (cell protection against heavy metals), overproduction of phytochelatin synthase (metal phytochelation), and enhanced phytovolatization of Hg and Se are the most common approaches to using genetically modified organisms for enhanced phytoremediation of heavy metals (Kotrba et al. 2009). For example, the cpSL transgenic Indian mustard *Brassica juncea* accumulated twice as much Se in shoots, it had a 1.8 times higher leaf Se concentration, and was more metal-tolerant (showing a faster growth) than the wild-type plants growing on selenium- and boron-contaminated saline sediments (Bañuelos et al. 2007). Transgenic *Arabidopsis thaliana* plants overexpressing yeast protein YCF1 (detoxifying cadmium by transporting it into vacuoles in yeast) showed a 2.2 times higher biomass yield and 1.5 times higher Cd and Pb accumulation in shoots as compared to wild-type plants (Song et al. 2003).

Introduction of mammalian cytochrome P450 2E1 can drastically increase the degradation of trichloroethylene, ethylene dibromide, carbon tetrachloride, chloroform, and vinyl chloride in the engineered plants (Gohel et al. 2006). The transgenic plants showed a dramatic, up to 640-fold, enhancement of the metabolism of trichloroethylene compared to null vector control plants. They also showed an increased uptake and debromination of ethylene dibromide (Doty et al. 2000). A field trial of trichloroethylene phytoremediation by transgenic poplars expressing cytochrome P450 2E1 was conducted by Legault et al. (2017). The trichloroethylene biodegradation was improved (but not as much as in the corresponding laboratory-scale study), via the evapotranspiration, i.e., evaporation facilitated by plant translocation, so the trichloroethylene content of transgenic leaves was reduced by 80% and its diffusion from transgenic stems (measured as a loss of volatiles from the stem) was reduced by 90% compared to the wild-type poplars.

Explosives and pesticides can also be effectively treated by phytoremediation using genetically engineered plants. Expression of a bacterial nitroreductase gene in *Arabidopsis thaliana* was used to enhance its tolerance to the uptake and degradation of 2,4,6-trinitrotoluene (Kurumata et al. 2005). Genetically modified *Arabidopsis thaliana* plants expressing mammalian cytochrome P450 enzyme CYP1A2 were able to efficiently degrade the herbicide chlortoluron and showed an improved herbicide resistance (Kebeish et al. 2014). Enhanced tolerance to two herbicides, atrazine and metolachlor, as well as their uptake and degradation, were described by Kawahigashi et al. (2006) for transgenic rice expressing human CYP1A1, CYP2B6, and CYP2C19.

A well-functioning rhizosphere is essential for normal plant growth and health in natural environments as well as for the efficient phytoremediation. Symbiosis between plants and microbes can be enhanced using genetic modification tools to make phytoremediation more efficient. For instance, the addition of a plant growth promoting transgenic rhizobacterium *Pseudomonas putida* to the rhizosphere of sunflower (*Helianthus annuus*), cowpea (*Vigna unguiculata*), wheat (*Triticum sativum*), and corn (*Zea* spp.) was tested. The expression of a metal-binding peptide (EC20) in the rhizosphere led to a decrease in cadmium phytotoxicity and up to 40% increase in cadmium accumulation in the root (Wu et al. 2006).

A double genetically modified symbiotic system (both rhizobacterium and plant) was used to improve the Cu phytostabilization in legume roots (Pérez-Palacios et al. 2017). *Medicago truncatula* plants expressing the metallothionein gene *mt4a* from *Arabidopsis thaliana* in roots were used in a symbiotic system to improve the plant Cu tolerance while the genetically modified rhizobacterium *Ensifer medicae*, expressing copper resistance genes *copAB* from *Pseudomonas fluorescens,* was used to improve the plant root Cu accumulation. Results suggested a reduced oxidative stress and further improved root Cu accumulation without altering the metal loading to shoots, thus leading to diminished values of the metal translocation from roots to shoots (Pérez-Palacios et al. 2017).

24.4.2 Production of Useful Chemicals for Remediation Technologies by GMOs (Contained Use)

Surfactant-producing microorganisms can be genetically modified to become hyperproducing; an unpretentious yet well-cultivable strain can be advantageously modified to become a producer. Another application may be a producer modification to enable the use of cheap waste materials as substrates.

Currently the most promising approach is probably the in situ production in a closed reactor, as it is acceptable by the general public. This production poses no risk of adverse environmental effects and may enable production of such biosurfactants that would be cost-competitive to synthetic surfactants (Sekhon et al. 2011; Rashid et al. 2015). For example, there was a twofold increase in the biosurfactant and esterase activities after a successful cloning of the biosurfactant genes from *Bacillus subtilis* SK320 into *E. coli* using olive oil as a substrate. Moreover, the obtained biosurfactant–esterase complex turned out to be a powerful emulsifier (exhibiting the reduction of the surface tension of water from 72 dyn/cm to as low as 30.7 dyn/cm), thus showing promise for bioremediation, hydrocarbon biodegradation, and pharmaceutical applications (Sekhon et al. 2011).

Not only can useful natural enzymes be overexpressed in hyperproduction strains but also modified enzymes may be obtained this way, with enhanced selectivity and/or reaction kinetics. For example, using protein engineering through combinatorial active site saturation testing (CASTing), a 5000-fold increase in k_{cat}/K_M

(measuring the catalytic efficiency) of a specific phosphotriesterase was achieved, enabling its application for the detoxification of an organophosphate insecticide, malathion (Naqvi et al. 2014). Brissos et al. (2015) reported the major improvement of metallo-oxidase McoA from a hyperthermophilic bacterium *Aquifex aeolicus* for aromatic compounds through direct evolution. The k_{cat}/K_M of the obtained enzyme was found to be two orders of magnitude higher than that of the wild-type enzyme for a typical laccase substrate, ABTS (2,2′-azino-bis(3-ethylbenzothiazoline-6-sulfonic acid)), along with a higher activity for phenolics and synthetic aromatic dyes.

24.4.3 GMOs in Prevention of Contamination

To prevent or reduce the use of pesticides, genetically modified insect-resistant and disease-resistant transgenic crops have been engineered. Insect-resistant crops, e.g., maize, potato, and cotton are mostly based on a manipulation resulting in expressing the genes encoding the entomocidal δ-endotoxin from *Bacillus thuringiensis* (Bt-expressing crops). However, besides these currently successfully commercialized crops, other modification approaches, e.g., a RNAi-mediated crop protection, crops with multiple resistance genes, and crops expressing protease inhibitors were applied (Scott et al. 2013; Christou et al. 2006). In addition, many variants of disease-resistant transgenic crops (virus-, fungi-, and bacteria-resistant) have been engineered and some of them have been commercially applied (Galvez et al. 2014; Collinge et al. 2008).

Another way of applying genetic modification is enhanced high-yield production of biopesticides to obtain cost-competitive products thus replacing pesticides in agriculture (Gohel et al. 2006).

24.5 Biopesticides

Advanced production technologies, advanced waste water and waste air cleaning technologies to prevent the deposition of either soluble or volatile contaminants in soil or sediments, advanced risk management, prevention and mitigation of the consequences of environmental disasters are the most common measures taken to minimize soil contamination. However, there is one group of important soil contaminants directly applied to the agricultural land, namely pesticides. To minimize the land contamination by pesticides, the Integrated Pest Management and Integrated Crop Management strategies are preferred in the modern agriculture introducing specific crop cultivation practices, eco-friendly fertilization techniques, and favoring biopesticides over pesticides. Therefore, biopesticides protecting plants against insect pests, weeds, and pathogens as well as biostimulants enhancing nutrition efficiency, abiotic stress tolerance, and/or crop quality traits are promising agents for prevention of soil contamination (du Jardin 2015; Mishra et al. 2015).

The International Biocontrol Manufacturers' Association (IBMA) classifies biocontrol agents into four groups: (1) macrobial, (2) microbial, (3) natural products, and (4) semiochemicals. The application of genetically manipulated plants (GMPs) with enhanced resistance against pests is another promising way. van Lenteren et al. (2018) reviewed in detail the current state of the art in the area of commercially available macrobial and microbial biopesticides. Summarizing these data, microorganisms account for 63% and invertebrates add the remaining 37% of the commercially used biopesticides. Individual microorganism types are represented within the microorganism group as follows: bacteria 45%, fungi 40%, viruses 10%, yeasts 4%, and bacteriophages 1%. The target organisms of microbial biopesticides are a variety of insect pests, weeds, and pathogens. The most important microbial agent is *Bacillus thuringiensis*, which produces dozens of different biopesticides and, in fact, the first commercial biopesticide.

Macrobial biological control agents (natural enemies) and their percentage terms are as follows: Hymenoptera 46%, Acari 18%, Coleoptera 12%, Neuroptera 10%, Hemiptera 6%, Diptera 4%, Nematode 2%, Mantodea 2%, Thysanoptera 1%. They are effective against a wide spectrum of insect pests, e.g., Aphids, Mites, Dipterans, Thrips, Pseudococcids, Lepidopterans.

Natural products/biochemicals are represented by a variety of secondary metabolites produced by plants and some microorganisms, e.g., actinomicetes, which either deter or kill microorganisms, insects, and/or plants. Semiochemicals serve as insect behavior-modifying agents, mostly insect sex pheromones, for crop protection. They are used for either monitoring or pest control by mass trapping, lure-and-kill systems, and mating disruption (Chandler et al. 2011).

Biological control is especially successful and currently plays the central role in the production of many greenhouse crops (Chandler et al. 2011). For example, the predatory mite *Amblyseius swirskii* was successfully used as a biological control agent for whitefly and thrips in sweet pepper greenhouses in Spain and was further integrated into Integrated Pest Management in commercial greenhouses (Calvo et al. 2012). Although biopesticides are known, experimentally tested, commercially used for decades, and highly supported by governments, the global market of biological control agents, e.g., invertebrates, microorganisms, and biochemicals is less than 2% of the pesticide market (van Lenteren et al. 2018).

Although biopesticides are supported by the framework for Community action to achieve the sustainable use of pesticides by Directive 2009/128/EC of the European Parliament, in Europe, unlike the USA, the registration of new biopesticides is still slow and the procedure is expensive, particularly for small manufacturers. The changes in registration procedures will result in faster registration of more microbial biological control agents and, consequently, in lower product costs (van Lenteren et al. 2018). Generally, the wide use of biopesticides is limited because of their high price, expensive and long research, expensive and slow registration, fragmented (regional) market, and lack of awareness about their benefits, poor customer support, and, in some cases, poor and/or fluctuating quality and shelf life, and inconsistent field performance (Mishra et al. 2015). Recently, several company mergers and acquisitions that may facilitate a healthy and sustainable development of the biopesticide industry have occurred. In addition, the largest agrochemical firms, e.g., Bayer CropScience and Syngenta have turned to the biopesticide market investing into the corresponding research, production capacities, and distribution as well as the acquisitions of smaller producers of biopestides.

This biopesticide industry consolidation could lead to a new round of commercial and academic research on novel biopesticides and biostimulants directly connected to cost-effective production, established worldwide distribution and market as well as full customer support, which are the essential prerequisites for a significant increase of the share of these eco-friendly biological agents at the expense of chemical agents.

References

Acevedo F, Pizzul L, Castillo MP, González ME, Cea M, Gianfreda L, Diez MC (2010) Degradation of polycyclic aromatic hydrocarbons by free and nanoclay-immobilized manganese peroxidase from *Anthracophyllum discolor*. Chemosphere 80(3):271–278. https://doi.org/10.1016/j.chemosphere.2010.04.022

Al-Tahhan RA, Sandrin TR, Bodour AA, Maier RM (2000) Rhamnolipid-induced removal of lipopolysaccharide from *Pseudomonas aeruginosa*: effect on cell surface properties and interaction with hydrophobic substrates. Appl Environ Microbiol 66(8):3262–3268. https://doi.org/10.1128/AEM.66.8.3262-3268.2000

Alvarez Yela AC, Tibaquirá Martínez MA, Rangel Piñeros GA, López VC, Villamizar SH, Núñez Vélez VL, Abraham W-R, Vives Flórez MJ, González Barrios AF (2016) A comparison between conventional *Pseudomonas aeruginosa* rhamnolipids and *Escherichia coli* transmembrane proteins for oil recovery enhancing. Int Biodeterior Biodegrad 112:59–65. https://doi.org/10.1016/j.ibiod.2016.04.033

Ángeles M-T, Refugio R-V (2013) *In situ* biosurfactant production and hydrocarbon removal by *Pseudomonas putida* CB-100 in bioaugmented and biostimulated oil-contaminated soil. Braz J Microbiol 44(2):595–605. https://doi.org/10.1590/S1517-83822013000200040

Anh TM, Dzyadevych SV, Prieur N, Duc CN, Pham TD, Jaffrezic Renault N, Chovelon J-M (2006) Detection of toxic compounds in real water samples using a conductometric tyrosinase biosensor. Mater Sci Eng C 26(2–3):453–456. https://doi.org/10.1016/j.msec.2005.10.025

Bañuelos G, LeDuc DL, Pilon-Smits EAH, Terry N (2007) Transgenic Indian mustard overexpressing selenocysteine lyase or selenocysteine methyltransferase exhibit enhanced potential for selenium phytoremediation under field conditions. Environ Sci Technol 41 (2):599–605. https://doi.org/10.1021/es061152i

Bezza FA, Chirwa EMN (2015) Production and applications of lipopeptide biosurfactant for bioremediation and oil recovery by *Bacillus subtilis* CN2. Biochem Eng J 101:168–178. https://doi.org/10.1016/j.bej.2015.05.007

Bezza FA, Chirwa EMN (2016) Biosurfactant-enhanced bioremediation of aged polycyclic aromatic hydrocarbons (PAHs) in creosote contaminated soil. Chemosphere 144:635–644. https://doi.org/10.1016/j.chemosphere.2015.08.027

Bezza FA, Chirwa EMN (2017) The role of lipopeptide biosurfactant on microbial remediation of aged polycyclic aromatic hydrocarbons (PAHs)-contaminated soil. Chem Eng J 309:563–576. https://doi.org/10.1016/j.cej.2016.10.055

Binod P, Sukumaran RK, Shirke SV, Rajput JC, Pandey A (2007) Evaluation of fungal culture filtrate containing chitinase as a biocontrol agent against *Helicoverpa armigera*. J Appl Microbiol 103(5):1845–1852. https://doi.org/10.1111/j.1365-2672.2007.03428.x

Bokare V, Murugesan K, Kim Y-M, Jeon J-R, Kim E-J, Chang YS (2010) Degradation of triclosan by an integrated nano-bio redox process. Bioresour Technol 101(16):6354–6360. https://doi.org/10.1016/j.biortech.2010.03.062

Borah SN, Goswami D, Sarma HK, Cameotra SS, Deka S (2016) Rhamnolipid biosurfactant against *Fusarium verticillioides* to control stalk and ear rot disease of maize. Front Microbiol 7:1505. https://doi.org/10.3389/fmicb.2016.01505

Brissos V, Ferreira M, Grass G, Martins LO (2015) Turning a hyperthermostable metallo-oxidase into a laccase by directed evolution. ACS Catal 5(8):4932–4941. https://doi.org/10.1021/acscatal.5b00771

Bustamante M, Durán N, Diez MC (2012) Biosurfactants are useful tools for the bioremediation of contaminated soil: a review. J Soil Sci Plant Nutr 12(4):667–687. https://doi.org/10.4067/S0718-95162012005000024

Calvo C, Manzanera M, Silva-Castro GA, Uad I, González-López J (2009) Application of bioemulsifiers in soil oil bioremediation processes. Future prospects. Sci Total Environ 407(12):3634–3640. https://doi.org/10.1016/j.scitotenv.2008.07.008

Calvo FJ, Bolckmans K, Belda JE (2012) Biological control-based IPM in sweet pepper greenhouses using *Amblyseius swirskii* (Acari: Phytoseiidae). Biocontrol Sci Tech 22(12):1398–1416. https://doi.org/10.1080/09583157.2012.731494

Cameotra SS, Singh P (2009) Synthesis of rhamnolipid biosurfactant and mode of hexadecane uptake by *Pseudomonas* species. Microb Cell Factories 8:16. https://doi.org/10.1186/1475-2859-8-16

Chandler D, Bailey AS, Tatchell GM, Davidson G, Greaves J, Grant WP (2011) The development, regulation and use of biopesticides for integrated pest management. Philos Trans R Soc B 366(1573):1987–1998. https://doi.org/10.1098/rstb.2010.0390

Christofi N, Ivshina IB (2002) Microbial surfactants and their use in field studies of soil remediation. J Appl Microbiol 93(6):915–929. https://doi.org/10.1046/j.1365-2672.2002.01774.x

Christou P, Capell T, Kohli A, Gatehouse JA, Gatehouse AMR (2006) Recent developments and future prospects in insect pest control in transgenic crops. Trends Plant Sci 11(6):302–308. https://doi.org/10.1016/j.tplants.2006.04.001

Chrzanowski Ł, Wick LY, Meulenkamp R, Kaestner M, Heipieper HJ (2009) Rhamnolipid biosurfactants decrease the toxicity of chlorinated phenols to *Pseudomonas putida* DOT-T1E. Lett Appl Microbiol 48(6):756–762. https://doi.org/10.1111/j.1472-765X.2009.02611.x

Cipollone R, Ascenzi P, Frangipani E, Visca P (2006) Cyanide detoxification by recombinant bacterial rhodanese. Chemosphere 63(6):942–949. https://doi.org/10.1016/j.chemosphere.2005.09.048

Collinge DB, Søgaard Lund O, Thordal-Christensen H (2008) What are the prospects for genetically engineered, disease resistant plants? Eur J Plant Pathol 121(3):217–231. https://doi.org/10.1007/s10658-007-9229-2

Conte P, Agretto A, Spaccini R, Piccolo A (2005) Soil remediation: humic acids as natural surfactants in the washings of highly contaminated soils. Environ Pollut 135(3):515–522. https://doi.org/10.1016/j.envpol.2004.10.006

da Rosa CFC, Freire DMG, Ferraz HC (2015) Biosurfactant microfoam: application in the removal of pollutants from soil. J Environ Chem Eng 3(1):89–94. https://doi.org/10.1016/j.jece.2014.12.008

Daâssi D, Prieto A, Zouari-Mechichi H, Martínez MJ, Nasri M, Mechichi T (2016) Degradation of bisphenol A by different fungal laccases and identification of its degradation products. Int Biodeterior Biodegrad 110:181–188. https://doi.org/10.1016/j.ibiod.2016.03.017

Das P, Mukherjee S, Sen R (2008) Antimicrobial potential of a lipopeptide biosurfactant derived from a marine *Bacillus circulans*. J Appl Microbiol 104(6):1675–1684. https://doi.org/10.1111/j.1365-2672.2007.03701.x

Datta R, Anand S, Moulick A, Baraniya D, Pathan SI, Rejsek K, Vranova V, Sharma M, Sharma D, Kelkar A, Formanek P (2017) How enzymes are adsorbed on soil solid phase and factors limiting its activity: a review. Int Agrophys 31(2):287–302. https://doi.org/10.1515/intag-2016-0049

de Cássia FS Silva R, Almeida DG, Rufino RD, Luna JM, Santos VA, Sarubbo LA (2014) Applications of biosurfactants in the petroleum industry and the remediation of oil spills. Int J Mol Sci 15(7):12523–12542. https://doi.org/10.3390/ijms150712523

Doty SL, Shang TQ, Wilson AM, Tangen J, Westergreen AD, Newman LA, Strand SE, Gordon MP (2000) Enhanced metabolism of halogenated hydrocarbons in transgenic plants containing mammalian cytochrome P450 2E1. Proc Natl Acad Sci U S A 97(12):6287–6291. https://doi.org/10.1073/pnas.97.12.6287

du Jardin P (2015) Plant biostimulants: definition, concept, main categories and regulation. Sci Hortic 196:3–14. https://doi.org/10.1016/j.scienta.2015.09.021

Eibes G, Arca-Ramos A, Feijoo G, Lema JM, Moreira MT (2015) Enzymatic technologies for remediation of hydrophobic organic pollutants in soil. Appl Microbiol Biotechnol 99 (21):8815–8829. https://doi.org/10.1007/s00253-015-6872-y

Elahian F, Reiisi S, Shahidi A, Mirzaei SA (2017) High-throughput bioaccumulation, biotransformation, and production of silver and selenium nanoparticles using genetically engineered *Pichia pastoris*. Nanomedicine 13(3):853–861. https://doi.org/10.1016/j.nano.2016.10.009

Fava F, Bertin L, Fedi S, Zannoni D (2003) Methyl-β-cyclodextrin-enhanced solubilization and aerobic biodegradation of polychlorinated biphenyls in two aged-contaminated soils. Biotechnol Bioeng 81(4):381–390. https://doi.org/10.1002/bit.10579

Fava F, Berselli S, Conte P, Piccolo A, Marchetti L (2004) Effects of humic substances and soya lecithin on the aerobic bioremediation of a soil historically contaminated by polycyclic aromatic hydrocarbons (PAHs). Biotechnol Bioeng 88(2):214–223. https://doi.org/10.1002/bit.20225

Fernández-Fernández M, Sanromán MÁ, Moldes D (2013) Recent developments and applications of immobilized laccase. Biotechnol Adv 31(8):1808–1825. https://doi.org/10.1016/j.biotechadv.2012.02.013

Fernando Bautista L, Morales G, Sanz R (2010) Immobilization strategies for laccase from *Trametes versicolor* on mesostructured silica materials and the application to the degradation of naphthalene. Bioresour Technol 101(22):8541–8548. https://doi.org/10.1016/j.biortech.2010.06.042

Galvez LC, Banerjee J, Pinar H, Mitra A (2014) Engineered plant virus resistance. Plant Sci 228:11–25. https://doi.org/10.1016/j.plantsci.2014.07.006

Gianfreda L, Rao MA, Scelza R, de la Luz Mora M (2016) Chapter 6: Role of enzymes in environment cleanup/remediation. In: Dhillon GS, Kaur S (eds) Agro-industrial wastes as feedstock for enzyme production. Academic, San Diego, p 133–155. doi:https://doi.org/10.1016/B978-0-12-802392-1.00006-X

Goettel MS, St Leger RJ, Rizzo NW, Staples RC, Roberts DW (1989) Ultrastructural localization of a cuticle-degrading protease produced by the entomopathogenic fungus *Metarhizium anisopliae* during penetration of host (*Manduca sexta*) cuticle. J Gen Microbiol 135:2233–2239

Gohel V, Singh A, Vimal M, Ashwini P, Chhatpar HS (2006) Bioprospecting and antifungal potential of chitinolytic microorganisms. Afr J Biotechnol 5(2):54–72

Hong S-G, Kim BC, Na HB, Lee J, Youn J, Chung S-W, Lee C-W, Lee B, Kim HS, Hsiao E, Kim SH, Kim B-G, Park HG, Chang HN, Hyeon T, Dordick JS, Grate JW, Kim J (2017) Single enzyme nanoparticles armored by a thin silicate network: single enzyme caged nanoparticles. Chem Eng J 322:510–515. https://doi.org/10.1016/j.cej.2017.04.022

Hosseininoosheri P, Lashgari HR, Sepehrnoori K (2016) A novel method to model and characterize in-situ bio-surfactant production in microbial enhanced oil recovery. Fuel 183:501–511. https://doi.org/10.1016/j.fuel.2016.06.035

Huang Q, Zhu J, Qiao X, Cai P, Rong X, Liang W, Chen W (2009) Conformation, activity and proteolytic stability of acid phosphatase on clay minerals and soil colloids from an Alfisol. Colloids Surf B Biointerfaces 74(1):279–283. https://doi.org/10.1016/j.colsurfb.2009.07.031

Huang JL, Liu ZB, Li SY, Xu B, Gong YH, Yang Y, Sun HX (2016) Isolation and engineering of plant growth promoting rhizobacteria *Pseudomonas aeruginosa* for enhanced cadmium bioremediation. J Gen Appl Microbiol 62(5):258–265. https://doi.org/10.2323/jgam.2016.04.007

Hultberg M, Bengtsson T, Liljeroth E (2010) Late blight on potato is suppressed by the biosurfactant-producing strain *Pseudomonas koreensis* 2.74 and its biosurfactant. BioControl 55(4):543–550. https://doi.org/10.1007/s10526-010-9289-7

Ji C, Hou JW, Wang K, Zhang YT, Chen V (2016) Biocatalytic degradation of carbamazepine with immobilized laccase-mediator membrane hybrid reactor. J Membr Sci 502:11–20. https://doi.org/10.1016/j.memsci.2015.12.043

Jiménez-T RG, Moliterni E, Rodríguez L, Fernández FJ, Villaseñor J (2011) Feasibility of mixed enzymatic complexes to enhanced soil bioremediation processes. Procedia Environ Sci 9:54–59. https://doi.org/10.1016/j.proenv.2011.11.010

Kawahigashi H, Hirose S, Ohkawa H, Ohkawa Y (2006) Phytoremediation of the herbicides atrazine and metolachlor by transgenic rice plants expressing human *CYP1A1*, *CYP2B6*, and *CYP2C19*. J Agric Food Chem 54(8):2985–2991. https://doi.org/10.1021/jf052610u

Kebeish R, Azab E, Peterhaensel C, El-Basheer R (2014) Engineering the metabolism of the phenylurea herbicide chlortoluron in genetically modified *Arabidopsis thaliana* plants expressing the mammalian cytochrome P450 enzyme CYP1A2. Environ Sci Pollut Res 21 (13):8224–8232. https://doi.org/10.1007/s11356-014-2710-5

Keusgen M, Kloock JP, Knobbe D-T, Jünger M, Krest I, Goldbach M, Klein W, Schöning MJ (2004) Direct determination of cyanides by potentiometric biosensors. Sensors Actuators B Chem 103(1–2):380–385. https://doi.org/10.1016/j.snb.2004.04.067

Khan A, Williams KL, Nevalainen HKM (2004) Effects of *Paecilomyces lilacinus* protease and chitinase on the eggshell structures and hatching of *Meloidogyne javanica* juveniles. Biol Control 31(3):346–352. https://doi.org/10.1016/j.biocontrol.2004.07.011

Kolseth A-K, D'Hertefeldt T, Emmerich M, Forabosco F, Marklund S, Cheeke TE, Hallin S, Weih M (2015) Influence of genetically modified organisms on agro-ecosystem processes. Agric Ecosyst Environ 214:96–106. https://doi.org/10.1016/j.agee.2015.08.021

Kotrba P, Najmanova J, Macek T, Ruml T, Mackova M (2009) Genetically modified plants in phytoremediation of heavy metal and metalloid soil and sediment pollution. Biotechnol Adv 27 (6):799–810. https://doi.org/10.1016/j.biotechadv.2009.06.003

Kulikowska D, Gusiatin ZM, Bułkowska K, Kierklo K (2015a) Humic substances from sewage sludge compost as washing agent effectively remove Cu and Cd from soil. Chemosphere 136:42–49. https://doi.org/10.1016/j.chemosphere.2015.03.083

Kulikowska D, Gusiatin ZM, Bułkowska K, Klik B (2015b) Feasibility of using humic substances from compost to remove heavy metals (Cd, Cu, Ni, Pb, Zn) from contaminated soil aged for different periods of time. J Hazard Mater 300:882–891. https://doi.org/10.1016/j.jhazmat.2015.08.022

Kurumata M, Takahashi M, Sakamoto A, Ramos JL, Nepovim A, Vanek T, Hirata T, Morikawa H (2005) Tolerance to, and uptake and degradation of 2,4,6-trinitrotoluene (TNT) are enhanced by the expression of a bacterial nitroreductase gene in *Arabidopsis thaliana*. Z Naturforsch C 60 (3–4):272–278. https://doi.org/10.1515/znc-2005-3-412

Lan WS, Gu JD, Zhang JL, Shen BC, Jiang H, Mulchandani A, Chen W, Qiao CL (2006) Coexpression of two detoxifying pesticide-degrading enzymes in a genetically engineered bacterium. Int Biodeterior Biodegrad 58(2):70–76. https://doi.org/10.1016/j.ibiod.2006.07.008

Lau EV, Gan S, Ng HK, Poh PE (2014) Extraction agents for the removal of polycyclic aromatic hydrocarbons (PAHs) from soil in soil washing technologies. Environ Pollut 184:640–649. https://doi.org/10.1016/j.envpol.2013.09.010

Legault EK, James CA, Stewart K, Muiznieks I, Doty SL, Strand SE (2017) A field trial of TCE phytoremediation by genetically modified poplars expressing cytochrome P450 2E1. Environ Sci Technol 51(11):6090–6099. https://doi.org/10.1021/acs.est.5b04758

Li J-L, Chen B-H (2009) Surfactant-mediated biodegradation of polycyclic aromatic hydrocarbons. Materials 2(1):76–94. https://doi.org/10.3390/ma2010076

Liang S, Guan D-X, Li J, Zhou C-Y, Luo J, Ma LQ (2016) Effect of aging on bioaccessibility of arsenic and lead in soils. Chemosphere 151:94–100. https://doi.org/10.1016/j.chemosphere.2016.02.070

Liu M, Cai QX, Liu HZ, Zhang BH, Yan JP, Yuan ZM (2002) Chitinolytic activities in *Bacillus thuringiensis* and their synergistic effects on larvicidal activity. J Appl Microbiol 93 (3):374–379. https://doi.org/10.1046/j.1365-2672.2002.01693.x

Luna JM, Rufino RD, Sarubbo LA (2016) Biosurfactant from *Candida sphaerica* UCP0995 exhibiting heavy metal remediation properties. Process Saf Environ Prot 102:558–566. https://doi.org/10.1016/j.psep.2016.05.010

Maity JP, Huang YM, Hsu C-M, Wu C-I, Chen C-C, Li C-Y, Jean J-S, Chang Y-F, Chen C-Y (2013) Removal of Cu, Pb and Zn by foam fractionation and a soil washing process from contaminated industrial soils using soapberry-derived saponin: a comparative effectiveness assessment. Chemosphere 92(10):1286–1293. https://doi.org/10.1016/j.chemosphere.2013.04.060

Marecik R, Wojtera-Kwiczor J, Ławniczak Ł, Cyplik P, Szulc A, Piotrowska-Cyplik A, Chrzanowski Ł (2012) Rhamnolipids increase the phytotoxicity of diesel oil towards four common plant species in a terrestrial environment. Water Air Soil Pollut 223(7):4275–4282. https://doi.org/10.1007/s11270-012-1190-9

Martínková L, Veselá AB, Rinágelová A, Chmátal M (2015) Cyanide hydratases and cyanide dihydratases: emerging tools in the biodegradation and biodetection of cyanide. Appl Microbiol Biotechnol 99(21):8875–8882. https://doi.org/10.1007/s00253-015-6899-0

Martins Das Neves LC, Miazaki Ohara Miyamura TT, Junji Kobayashi M, Vessoni Penna TC, Converti A (2007) Production of biosurfactant by a genetically-modified strain of *Bacillus subtilis* expressing green fluorescent protein. Ann Microbiol 57(3):377–381. https://doi.org/10.1007/BF03175076

Maslin P, Maier RM (2000) Rhamnolipid-enhanced mineralization of phenanthrene in organic-metal co-contaminated soils. Biorem J 4(4):295–308. https://doi.org/10.1080/10889860091114266

Mishra J, Tewari S, Singh S, Arora NK (2015) Biopesticides: where we stand? In: Arora NK (ed) Plant microbes Symbiosis: applied facets. Springer, New Delhi, pp 37–75. https://doi.org/10.1007/978-81-322-2068-8_2

Mulligan CN (2005) Environmental applications for biosurfactants. Environ Pollut 133 (2):183–198. https://doi.org/10.1016/j.envpol.2004.06.009

Nagasawa T, Wieser M, Nakamura T, Iwahara H, Yoshida T, Gekko K (2000) Nitrilase of *Rhodococcus rhodochrous* J1. Eur J Biochem 267(1):138–144. https://doi.org/10.1046/j.1432-1327.2000.00983.x

Naqvi T, Warden AC, French N, Sugrue E, Carr PD, Jackson CJ, Scott C (2014) A 5000-fold increase in the specificity of a bacterial phosphotriesterase for malathion through combinatorial active site mutagenesis. PLoS One 9(4):e94177. https://doi.org/10.1371/journal.pone.0094177

Ncibi MC, Mahjoub B, Gourdon R (2007) Effects of aging on the extractability of naphthalene and phenanthrene from Mediterranean soils. J Hazard Mater 146(1–2):378–384. https://doi.org/10.1016/j.jhazmat.2006.12.032

Ng I-S, Xu F, Zhang X, Ye C (2015) Enzymatic exploration of catalase from a nanoparticle producing and biodecolorizing algae *Shewanella xiamenensis* BC01. Bioresour Technol 184:429–435. https://doi.org/10.1016/j.biortech.2014.09.079

Ochoa-Loza FJ, Noordman WH, Jannsen DB, Brusseau ML, Maier RM (2007) Effect of clays, metal oxides, and organic matter on rhamnolipid biosurfactant sorption by soil. Chemosphere 66 (9):1634–1642. https://doi.org/10.1016/j.chemosphere.2006.07.068

Okamoto S, Eltis LD (2007) Purification and characterization of a novel nitrile hydratase from *Rhodococcus* sp. RHA1. Mol Microbiol 65(3):828–838. https://doi.org/10.1111/j.1365-2958.2007.05834.x

Pacwa-Płociniczak M, Płaza GA, Piotrowska-Seget Z, Cameotra SS (2011) Environmental applications of biosurfactants: recent advances. Int J Mol Sci 12(1):633–654. https://doi.org/10.3390/ijms12010633

Payne RB, Gentry DM, Rapp-Giles BJ, Casalot L, Wall JD (2002) Uranium reduction by *Desulfovibrio desulfuricans* strain G20 and a cytochrome c_3 mutant. Appl Environ Microbiol 68(6):3129–3132. https://doi.org/10.1128/aem.68.6.3129-3132.2002

Pei X-H, Zhan X-H, Wang S-M, Lin Y-S, Zhou L-X (2010) Effects of a biosurfactant and a synthetic surfactant on phenanthrene degradation by a sphingomonas strain. Pedosphere 20 (6):771–779. https://doi.org/10.1016/s1002-0160(10)60067-7

Pérez-Palacios P, Romero-Aguilar A, Delgadillo J, Doukkali B, Caviedes MA, Rodríguez-Llorente ID, Pajuelo E (2017) Double genetically modified symbiotic system for improved Cu phytostabilization in legume roots. Environ Sci Pollut Res 24(17):14910–14923. https://doi.org/10.1007/s11356-017-9092-4

Pi Y, Chen B, Bao M, Fan F, Cai Q, Ze L, Zhang B (2017) Microbial degradation of four crude oil by biosurfactant producing strain *Rhodococcus* sp. Bioresour Technol 232:263–269. https://doi.org/10.1016/j.biortech.2017.02.007

Piotrowska-Długosz A (2017) The use of enzymes in bioremediation of soil xenobiotics. In: Hashmi MZ, Kumar V, Varma A (eds) Xenobiotics in the soil environment: monitoring, toxicity and management. Springer, Cham, pp 243–265. https://doi.org/10.1007/978-3-319-47744-2_17

Portet-Koltalo F, Ammami MT, Benamar A, Wang H, Le Derf F, Duclairoir-Poc C (2013) Investigation of the release of PAHs from artificially contaminated sediments using cyclolipopeptidic biosurfactants. J Hazard Mater 261:593–601. https://doi.org/10.1016/j.jhazmat.2013.07.062

Quiquampoix H, Servagent-Noinville S, Baron M-H (2002) Enzyme adsorption on soil mineral surfaces and consequences for the catalytic activity. In: Burns RG, Dick RP (eds) Enzymes in the environment: activity, ecology, and applications. CRC Press, Boca Raton, pp 285–306

Rahmani K, Faramarzi MA, Mahvi AH, Gholami M, Esrafili A, Forootanfar H, Farzadkia M (2015) Elimination and detoxification of sulfathiazole and sulfamethoxazole assisted by laccase immobilized on porous silica beads. Int Biodeter Biodegrad 97:107–114. https://doi.org/10.1016/j.ibiod.2014.10.018

Rashid N-FM, Azemi M-AFM, Amirul A-AA, Wahid MEA, Bhubalan K (2015) Simultaneous production of biopolymer and biosurfactant by genetically modified *Pseudomonas Aeruginosa* UMTKB-5. Int Proc Chem, Biol Environ Eng 90(3):16–21. https://doi.org/10.7763/ipcbee

Rath BP, Das S, Mohapatra PKD, Thatoi H (2014) Optimization of extracellular chromate reductase production by *Bacillus amyloliquefaciens* (CSB 9) isolated from chromite mine environment. Biocatal Agric Biotechnol 3(3):35–41. https://doi.org/10.1016/j.bcab.2014.01.004

Rinágelová A, Kaplan O, Veselá AB, Chmátal M, Křenková A, Plíhal O, Pasquarelli F, Cantarella M, Martínková L (2014) Cyanide hydratase from *Aspergillus niger* K10: overproduction in *Escherichia coli*, purification, characterization and use in continuous cyanide degradation. Process Biochem 49(3):445–450. https://doi.org/10.1016/j.procbio.2013.12.008

Roy D, Kommalapati RR, Mandava SS, Valsaraj KT, Constant WD (1997) Soil washing potential of a natural surfactant. Environ Sci Technol 31(3):670–675. https://doi.org/10.1021/es960181y

Sánchez-Trujillo MA, Morillo E, Villaverde J, Lacorte S (2013) Comparative effects of several cyclodextrins on the extraction of PAHs from an aged contaminated soil. Environ Pollut 178:52–58. https://doi.org/10.1016/j.envpol.2013.02.029

Scott JG, Michel K, Bartholomay LC, Siegfried BD, Hunter WB, Smagghe G, Zhu KY, Douglas AE (2013) Towards the elements of successful insect RNAi. J Insect Physiol 59 (12):1212–1221. https://doi.org/10.1016/j.jinsphys.2013.08.014

Sekhon KK, Khanna S, Cameotra SS (2011) Enhanced biosurfactant production through cloning of three genes and role of esterase in biosurfactant release. Microb Cell Factories 10(1):49. https://doi.org/10.1186/1475-2859-10-49

Sheppard JD, Cooper DG (1990) The effects of a biosurfactant on oxygen transfer in a cyclone column reactor. J Chem Technol Biotechnol 48(3):325–336. https://doi.org/10.1002/jctb.280480308

Singh AK, Cameotra SS (2013) Efficiency of lipopeptide biosurfactants in removal of petroleum hydrocarbons and heavy metals from contaminated soil. Environ Sci Pollut Res 20 (10):7367–7376. https://doi.org/10.1007/s11356-013-1752-4

Singh JS, Abhilash PC, Singh HB, Singh RP, Singh DP (2011) Genetically engineered bacteria: an emerging tool for environmental remediation and future research perspectives. Gene 480 (1–2):1–9. https://doi.org/10.1016/j.gene.2011.03.001

Sivaperumal P, Kamala K, Rajaram R (2017) Chapter 8: Bioremediation of industrial waste through enzyme producing marine microorganisms. Adv Food Nutr Res 80:165–179. https://doi.org/10.1016/bs.afnr.2016.10.006

Soares dos Santos A, Pereira N Jr, Freire DMG (2016) Strategies for improved rhamnolipid production by *Pseudomonas aeruginosa* PA1. PeerJ 4:e2078. https://doi.org/10.7717/peerj.2078

Soberón-Chávez G, Maier RM (2011) Biosurfactants: a general overview. In: Soberón-Chávez G (ed) Biosurfactants: from genes to applications. Springer, Berlin/Heidelberg, pp 1–11. https://doi.org/10.1007/978-3-642-14490-5_1

Song W-Y, Ju Sohn E, Martinoia E, Jik Lee Y, Yang Y-Y, Jasinski M, Forestier C, Hwang I, Lee Y (2003) Engineering tolerance and accumulation of lead and cadmium in transgenic plants. Nat Biotechnol 21(8):914–919. https://doi.org/10.1038/nbt850

Sotirova A, Spasova D, Vasileva-Tonkova E, Galabova D (2009) Effects of rhamnolipid-biosurfactant on cell surface of *Pseudomonas aeruginosa*. Microbiol Res 164(3):297–303. https://doi.org/10.1016/j.micres.2007.01.005

Supakdamrongkul P, Bhumiratana A, Wiwat C (2010) Characterization of an extracellular lipase from the biocontrol fungus, *Nomuraea rileyi* MJ, and its toxicity toward *Spodoptera litura*. J Invertebr Pathol 105(3):228–235. https://doi.org/10.1016/j.jip.2010.06.011

Tahseen R, Afzal M, Iqbal S, Shabir G, Khan QM, Khalid ZM, Banat IM (2016) Rhamnolipids and nutrients boost remediation of crude oil-contaminated soil by enhancing bacterial colonization and metabolic activities. Int Biodeterior Biodegrad 115:192–198. https://doi.org/10.1016/j.ibiod.2016.08.010

Tang J, He J, Liu T, Xin X (2017) Removal of heavy metals with sequential sludge washing techniques using saponin: optimization conditions, kinetics, removal effectiveness, binding intensity, mobility and mechanism. RSC Adv 7(53):33385–33401. https://doi.org/10.1039/C7RA04284A

Thatoi H, Das S, Mishra J, Rath BP, Das N (2014) Bacterial chromate reductase, a potential enzyme for bioremediation of hexavalent chromium: a review. J Environ Manag 146:383–399. https://doi.org/10.1016/j.jenvman.2014.07.014

Tietjen T, Wetzel RG (2003) Extracellular enzyme-clay mineral complexes: enzyme adsorption, alteration of enzyme activity, and protection from photodegradation. Aquat Ecol 37 (4):331–339. https://doi.org/10.1023/B:AECO.0000007044.52801.6b

Tuomela M, Hatakka A (2011) 6.16 – oxidative fungal enzymes for bioremediation. In: Moo-Young M (ed) Comprehensive biotechnology, 2nd edn. Academic, Burlington, pp 183–196. https://doi.org/10.1016/B978-0-08-088504-9.00370-6

Van Hamme JD, Singh A, Ward OP (2006) Physiological aspects. Part 1 in a series of papers devoted to surfactants in microbiology and biotechnology. Biotechnol Adv 24(6):604–620. https://doi.org/10.1016/j.biotechadv.2006.08.001

van Lenteren JC, Bolckmans K, Köhl J, Ravensberg WJ, Urbaneja A (2018) Biological control using invertebrates and microorganisms: plenty of new opportunities. BioControl 63(1):39–59. https://doi.org/10.1007/s10526-017-9801-4

Wall JD, Krumholz LR (2006) Uranium reduction. Annu Rev Microbiol 60:149–166. https://doi.org/10.1146/annurev.micro.59.030804.121357

Wang P (2006) Nanoscale biocatalyst systems. Curr Opin Biotechnol 17(6):574–579. https://doi.org/10.1016/j.copbio.2006.10.009

Wang WJ, Li YH, Wang HM, Zu YG (2014) Differences in the activities of eight enzymes from ten soil fungi and their possible influences on the surface structure, functional groups, and element composition of soil colloids. PLoS One 9(11):e111740. https://doi.org/10.1371/journal.pone.0111740

Wang ZQ, Li YB, Tan XP, He WX, Xie W, Megharaj M, Wei GH (2017) Effect of arsenate contamination on free, immobilized and soil alkaline phosphatases: activity, kinetics and thermodynamics. Eur J Soil Sci 68(1):126–135. https://doi.org/10.1111/ejss.12397

Wattanaphon HT, Kerdsin A, Thammacharoen C, Sangvanich P, Vangnai AS (2008) A biosurfactant from *Burkholderia cenocepacia* BSP3 and its enhancement of pesticide solubilization. J Appl Microbiol 105(2):416–423. https://doi.org/10.1111/j.1365-2672.2008.03755.x

Wu CH, Wood TK, Mulchandani A, Chen W (2006) Engineering plant-microbe symbiosis for rhizoremediation of heavy metals. Appl Environ Microbiol 72(2):1129–1134. https://doi.org/10.1128/AEM.72.2.1129-1134.2006

Yan RX, Hou JH, Ding DF, Guan WQ, Wang CY, Wu ZQ, Li MG (2008) *In vitro* antifungal activity and mechanism of action of chitinase against four plant pathogenic fungi. J Basic Microbiol 48(4):293–301. https://doi.org/10.1002/jobm.200700392

Yang J, Huang X, Tian B, Wang M, Niu Q, Zhang K (2005) Isolation and characterization of a serine protease from the nematophagous fungus, *Lecanicillium psalliotae*, displaying nematicidal activity. Biotechnol Lett 27(15):1123–1128. https://doi.org/10.1007/s10529-005-8461-0

Yang C, Song C, Mulchandani A, Qiao C (2010) Genetic engineering of *Stenotrophomonas* strain YC-1 to possess a broader substrate range for organophosphates. J Agric Food Chem 58 (11):6762–6766. https://doi.org/10.1021/jf101105s

Yang Z, Zhang Z, Chai L, Wang Y, Liu Y, Xiao R (2016) Bioleaching remediation of heavy metal-contaminated soils using *Burkholderia* sp. Z-90. J Hazard Mater 301:145–152. https://doi.org/10.1016/j.jhazmat.2015.08.047

Zhang YM, Miller RM (1994) Effect of a *Pseudomonas* rhamnolipid biosurfactant on cell hydrophobicity and biodegradation of octadecane. Appl Environ Microbiol 60(6):2101–2106

Zhang C, Wang S, Yan Y (2011) Isomerization and biodegradation of beta-cypermethrin by *Pseudomonas aeruginosa* CH7 with biosurfactant production. Bioresour Technol 102 (14):7139–7146. https://doi.org/10.1016/j.biortech.2011.03.086

Zhang R, Xu XJ, Chen WL, Huang QY (2016a) Genetically engineered *Pseudomonas putida* X3 strain and its potential ability to bioremediate soil microcosms contaminated with methyl parathion and cadmium. Appl Microbiol Biotechnol 100(4):1987–1997. https://doi.org/10.1007/s00253-015-7099-7

Zhang X, Ng I-S, Chang J-S (2016b) Cloning and characterization of a robust recombinant azoreductase from *Shewanella xiamenensis* BC01. J Taiwan Inst Chem Eng 61:97–105. https://doi.org/10.1016/j.jtice.2016.01.002

Zhong H, Liu Y, Liu Z, Jiang Y, Tan F, Zeng G, Yuan X, Yan M, Niu Q, Liang Y (2014) Degradation of pseudo-solubilized and mass hexadecane by a *Pseudomonas aeruginosa* with treatment of rhamnolipid biosurfactant. Int Biodeterior Biodegrad 94:152–159. https://doi.org/10.1016/j.ibiod.2014.07.012

Zimmerman AR, Ahn M-Y (2010) Organo-mineral enzyme interaction and soil enzyme activity. In: Shukla G, Varma A (eds) Soil enzymology. Springer, Berlin/Heidelberg, pp 271–292. https://doi.org/10.1007/978-3-642-14225-3_15

Chapter 25
Field Study IX: Pilot-Scale Composting of PAH-Contaminated Materials: Two Different Approaches

Petra Innemanová and Tomáš Cajthaml

Abstract Results of this field study describe composting of two different PAH-contaminated materials (soils originated from four industrial sites and creosote-treated railway crossties). Sequential mild supercritical fluid extraction (SFE) was employed to estimate the fast-release fractions (F) of PAHs in order to predict bioaccessible portion of PAHs in the materials. These values were consequently compared with the real biodegradation data obtained from the two pilot-scale experiments in order to investigate whether the bioaccessible data can be used as a prediction parameter of the degradation efficiency. The final results of the PAH elimination ranged between 62.4% and 92.9% in the case of the soil samples and between 81.4% and 97.0% for two different treatments of contaminated railway crossties. The results showed that the elimination efficiency decreased with the increasing number of aromatic rings in the PAH molecules. Estimation of the bioaccessible fraction of PAHs using the SFE method has been proven to be a useful tool for prediction of the bioremediation process efficiency. This fact is particularly important because if the composting method is employed in the full scale, any addition of the necessary bulking agent leads to the final increase in the total amount of the treated hazardous wastes if the remediation process fails.

Keywords Composting · Bioremediation · PAHs · Supercritical fluid extraction · Bioavailability

---

P. Innemanová (✉)
DEKONTA a.s, Stehelčeves, Czech Republic

Institute for Environmental Studies, Faculty of Science, Charles University, Prague, Czech Republic
e-mail: innemanova@dekonta.cz

T. Cajthaml
Institute of Microbiology of the Czech Academy of Sciences, Prague, Czech Republic

Institute for Environmental Studies, Faculty of Science, Charles University, Prague, Czech Republic

J. Filip et al. (eds.), *Advanced Nano-Bio Technologies for Water and Soil Treatment*, Applied Environmental Science and Engineering for a Sustainable Future,
https://doi.org/10.1007/978-3-030-29840-1_25

25.1 Introduction

Composting is a natural biological process carried out under controlled aerobic conditions. In this process, various microorganisms including bacteria and fungi break down organic matter into simpler substances. The effectiveness of the composting process depends on the environmental conditions present within the composting system, i.e., oxygen, temperature, moisture, material disturbance, organic matter content, and the size and activity of microbial populations.

Composting has been proved to be efficient for remediation of soils contaminated with PAHs. For remediation purposes, the composting substrate (based on an agricultural-waste mixture of optimum nutrient ratio) is mixed with contaminated soil. Due to a wide range of acting microorganisms, the process is quite robust and the outcome appears to depend mainly on the bioavailability of the target pollutant, which seems to be a crucial factor in biodegradability and ecotoxicity of PAHs in environmental matrices (Cajthaml and Šašek 2005; Čvančarová et al. 2013).

This chapter describes two different approaches of PAH elimination using a composting process tested under pilot-scale conditions:

1. PAH-contaminated soil matrices collected from four different long-term contaminated sites were co-composted with raw materials, such as straw, gypsum, and chicken manure, used for cultivation of white button mushroom—Treatment 1.
2. Shaved creosote-treated railway crossties (CTW), which as PAH-contaminated wood represent a bulking agent, were co-composted with two different additional substrates, namely fresh grass cuttings (GC) and broiler litter straw-based substrate (BL)—Treatment 2.

Supercritical fluid extraction (SFE) was used as a rapid test to predict the bioremediation results of composting for both treatments.

25.2 Materials and Methods

Soil Samples The PAH-contaminated soils for Treatment 1 were collected from the following sites: former gasworks, Prague (sample A); former tar-processing plant, Ostrava (sample B); former wood-treatment-plant, Soběslav (sample C); and former gas-holder site, Prague (sample D). The soils were air-dried, sieved through a 5 mm mesh screen and homogenized. The waste creosote-treated railway crossties (CTW) for Treatment 2 were obtained from EKO-BIO VYSOČINA, spol. s r.o., Czech Republic and shaved to fine debris (3 cm).

Preparation of Composting Experiments Treatment 1 was performed in a thermally insulated composting chamber using stabilized broiler litter (BL) that is commercially used for cultivation of *Agaricus bisporus* and consists of chopped wheat straw, chicken manure, and gypsum (Šašek et al. 2003). The raw material was obtained from a local company producing spent mushroom substrate. The compost (400 kg wet weight) was mixed with 120 kg of dry contaminated soil. The ratio of contaminated soil to compost was approximately 1:1 (dry weight basis). The samples were analysed after a cooling phase (6–8 weeks) and after another 14 and 42 weeks of maturation.

The second experiment (Treatment 2) was conducted outdoors in garden composters with a waterproof bottom tray and a total working volume of 400 L. Two different organic co-substrates were used as bulking agents in two independent variants of Treatment 2, namely fresh GC originated from the garden of the Institute of Microbiology and BL (described above). BL was pre-wetted with tap water in a ratio of 1:2.2 (w/w), while GC was used as provided. The two composters were filled in a multi-stratum fashion, overlapping layers of either BL or GC (14 kg wet weight) with layers of CTW (5 kg, 3 cm in size) until an overall weight of 128 kg (13 layers) was achieved within each of the reaction vessel. In both cases, the overall initial moisture content was approximately 60% w/w. The samples for chemical analyses were collected during the mild-thermophilic (day 5), the mesophilic/cooling phase (days 12 and 40), and the maturation phase (days 140 and 240).

Analytical Procedure The evaluation of the PAHs extraction and quantification procedures used in Treatment 1 has been published by Cajthaml and Šašek (2005). The obtained data were recalculated according to the content of the soil in the samples (soil/substrate mixture) using the amount of ash after combustion at 650 °C as an internal standard.

All of the extractions and quantifications of PAHs in Treatment 2 were performed according to Covino et al. (2016). To calculate the amount of CTW in the collected samples of non-homogenous mixture of CTW and co-substrate, heavy metals such as Co and Pb were used as internal standards because of their presence only in CTW. Contrary to Treatment 1, the ash as an internal standard was not usable in the case of CTW. The metals were determined after total acid digestion and atomic absorption spectrometry conducted by Analytical Laboratories, Plzeň (Czech Republic) according to ISO 8288:1986 (ISO 1986) and EN 13657:2002 (CEN 2002).

The bioaccessible fraction of PAHs, which represents the portion of the target chemical that is bioavailable for biotransformation processes in soils (Hawthorne and Grabanski 2000), was estimated using a combination of sequential mild SFE and by calculating the slow- and fast-release (F) fractions. The SFE extractor (Labio, a.s., Czech Republic) operated at 50 °C and 200 bar (Cajthaml and Šašek 2005; Covino et al. 2016). Three parallel samples (1 g) underwent sequential supercritical CO_2 extraction over the time range 0–200 min, and the fractions (collected after 5, 10, 20, 40, 60, 80, 120, 160, and 200 min) were analysed separately to complete the desorption-kinetic profiles. The so-called "F-fraction" is usually assumed to be representative of the equilibrium release conditions, and the remaining slowly released portion is considered to be kinetically rate-limited. Therefore, the F-fraction represents the bioaccessible portion of the target chemical that is available for biotransformation processes in the soil (Covino et al. 2016).

25.3 Results

25.3.1 Treatment 1

The biodegradability of 13 PAHs out of the 16 compounds according to the U.S. EPA Method 610 was investigated in Treatment 1. Naphthalene, acenaphthylene, and acenaphthene were below the detection limit in all soils A–D. Figure 25.1 illustrates the evolution of the PAH concentration over time. The evaluated compounds are divided into groups covering 2–3-ring, 4-ring, and 5–6-ring PAHs for greater clarity of the graphical display.

The results indicate that there are great differences in the effectiveness of the process among the treated soils. The remediation outcomes in the cases of soils C and D demonstrate that even between days 100 and 300 of maturation there is still a good potential for PAH degradation. However, after day 100, further degradation in soils A and B was not significant.

Comparison of real degradation efficiency in composted soils and prediction calculated on the basis of bioavailable F-fraction measurement (using SFE method) is demonstrated in Table 25.1. The results show good correlation between the estimated and the real removal of PAHs in the case of 2–3-ring and 4-ring PAH groups. Nevertheless, the estimation of the whole process efficiency based on the assessment of bioavailable PAH fraction was underestimated in all the treated soils. The most notable difference was observed for the 5–6-ring PAHs.

25.3.2 Treatment 2

All the 16 compounds according to the U.S. EPA Method 610 were investigated in the case of Treatment 2, which represents quite a different PAH composting approach in comparison with Treatment 1. Wooden chips, generally used as a bulking agent in the composting process, were made of creosote-treated railway crossties and as such represent PAH-contaminated material. The overall concentration of the detected PAHs in pure CTW was as high as 26,498 ± 1548. Using GC as the composting co-substrate resulted in a significantly higher PAH degradation compared with BL after 240 days of incubation (Fig. 25.2 and Table 25.2). This

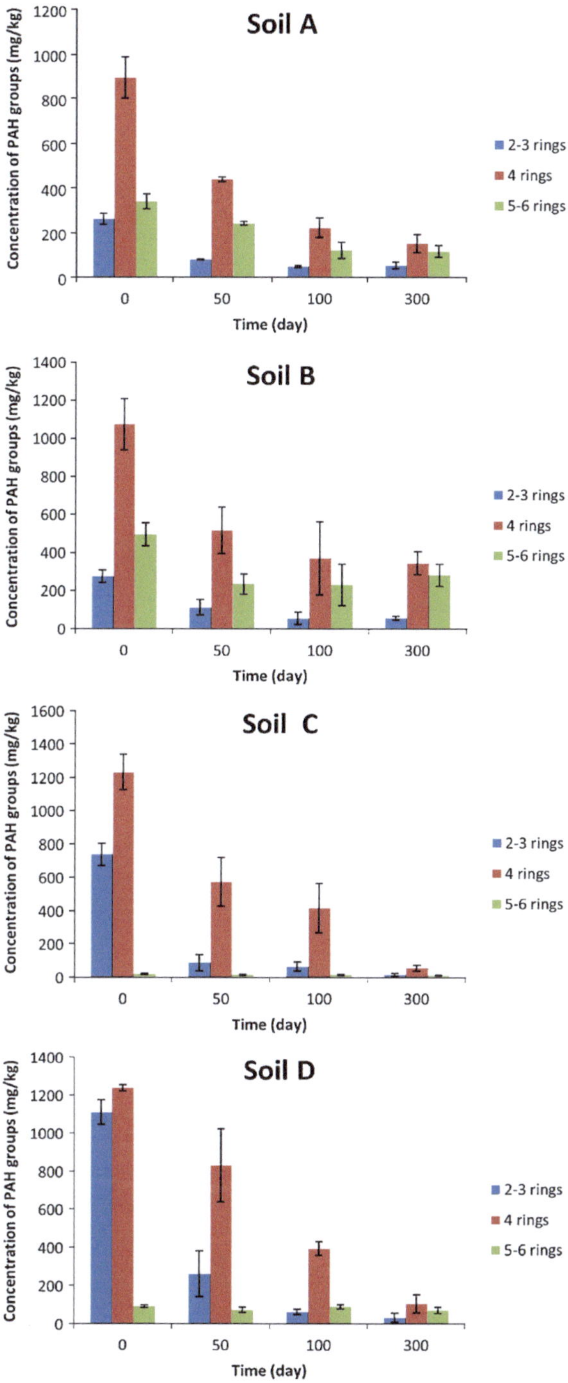

Fig. 25.1 Concentration of PAH groups in composted PAH-contaminated soils collected from the following sites: former gasworks, Prague (sample A); former tar-processing plant, Ostrava (sample B); former wood-treatment-plant, Soběslav (sample C); and former gas-holder site, Prague (sample D). The data are the mean ± standard deviation of five independent replicates. (Cajthaml and Šašek 2005)

Table 25.1 Comparison of real PAH biodegradation efficiency and predicted efficiency calculated according to F-fraction—Treatment 1: PAH-contaminated soils from sites: former gasworks, Prague (sample A); former tar-processing plant, Ostrava (sample B); former wood-treatment-plant, Soběslav (sample C); and former gas-holder site, Prague (sample D)

	2–3-ring PAHs	4-ring PAHs	5–6-ring PAHs	Σ PAHs
Soil A				
Original conc. (mg/kg)	262.0	893.5	339.5	1495.0
Degradation efficiency (%)	79.7	83.1	66.5	78.5
F-fraction (%)	72.9	66.5	35.9	60.7
Soil B				
Original conc. (mg/kg)	275.7	1074.2	496.4	1846.2
Degradation efficiency (%)	78.7	67.8	43.3	62.4
F-fraction (%)	60.2	50.6	9.1	40.9
Soil C				
Original conc. (mg/kg)	736.5	1231.0	19.2	1986.7
Degradation efficiency (%)	98.0	95.5	42.0	95.9
F-fraction (%)	93.6	93.6	21.1	92.9
Soil D				
Original conc. (mg/kg)	1110.8	1237.3	91.0	2439.0
Degradation efficiency (%)	97.1	91.6	22.0	91.5
F-fraction (%)	89.0	83.4	11.1	83.2

difference could be explained by a probably higher content of easily degradable organic compounds in BL beside GC. Preferential degradation of these substances may have led only to partial PAH removal in the variation with BL as a composting co-substrate.

A prediction of the biodegradation efficiency based on F-fraction determination well correlated with the real results at BL compost (Table 25.2). Nevertheless, the values predicted for the GC variation were underestimated in the cases of the 4- and 5–6-ring PAHs.

25.4 Conclusion

The presented results suggest that composting is a very robust bioremediation method by means of which bioavailable amounts of PAHs can be degraded regardless of the character of the contaminated material. A typical feature of composting is manipulation with a bulk volume of the materials. Since it is based on mixing non-contaminated materials with contaminated soil (or wood), it may result in a far greater quantity of the contaminated material if the bioremediation proves not to be successful. The all above-described remediation processes demonstrate how important this consideration is. Though the final biodegradation rates were significant, ranging between 62.4% (soil B) and 97% (CTW—GC), the resulting material

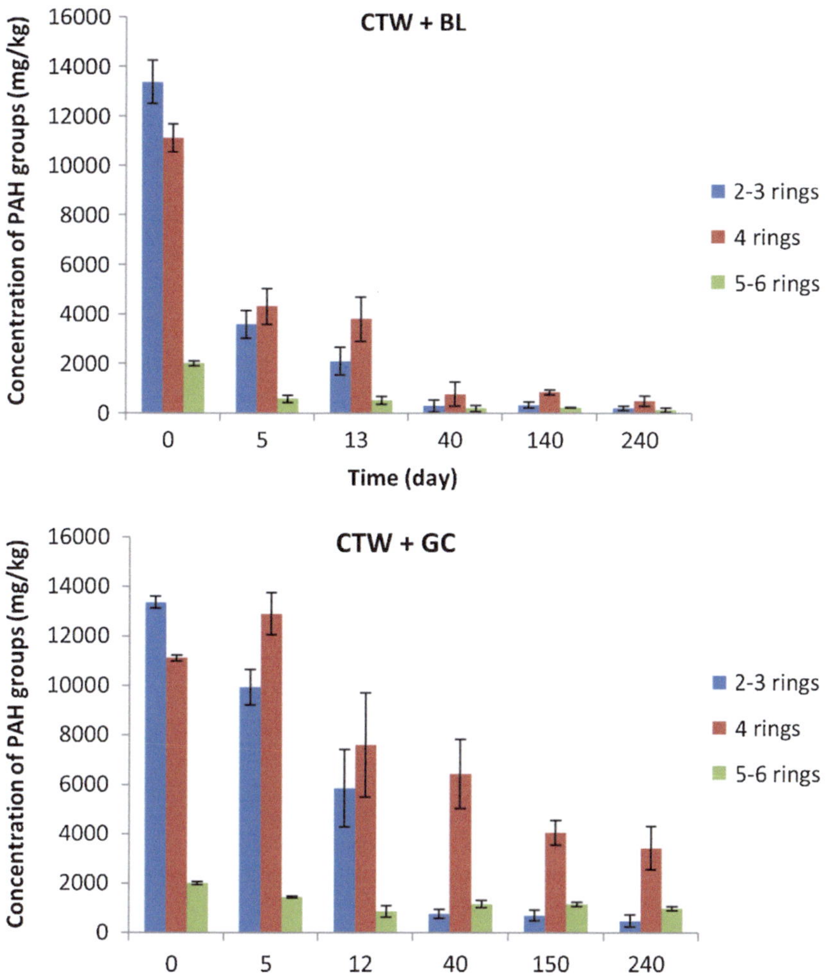

Fig. 25.2 Concentration of PAH groups in composted creosote-treated wood (CTW) with either broiler litter (BL) or grass-cuttings (GC) co-substrate. The data are the mean ± standard deviation of five independent replicates (Covino et al. 2016)

Table 25.2 Comparison of real PAH biodegradation efficiency and predicted efficiency calculated according to F-fraction—Treatment 2: Creosote-treated wood co-composted with broiler litter (CTW + BL) or grass cuttings (CTW + GC)

	2–3-ring PAHs	4-ring PAHs	5–6-ring PAHs	Σ PAHs
Original conc. (mg/kg)	13,371	11,123	2004	26,498
F-fraction (%)	96.2	64.4	50.6	79.4
CTW – BL				
Degradation efficiency (%)	96.3	69.1	50.8	81.4
CTW – GC				
Degradation efficiency (%)	98.2	95.7	93.9	97.0

still ranks among hazardous waste according to the Czech legislation system and can be dumped only in landfills designated for this purpose.

The use of SFE for estimation of the contaminant bioavailability in soil prior to a large-scale composting treatment seems to be a very promising tool to avoid this risk despite the fact that the predicted PAHs removal rates can be underestimated.

References

Cajthaml T, Šašek V (2005) Application of supercritical fluid extraction (SFE) to predict bioremediation efficacy of long-term composting of PAH-contaminated soil. Environ Sci Technol 39 (21):8448–8452. https://doi.org/10.1021/es050023j

CEN (2002) Characterization of waste – digestion for subsequent determination of aqua regia soluble portion of elements. EN 13657:2002

Covino S, Fabianová T, Křesinová Z, Čvančarová M, Burianová E, Filipová A, Voříšková J, Baldrian P, Cajthaml T (2016) Polycyclic aromatic hydrocarbons degradation and microbial community shifts during co-composting of creosote-treated wood. J Hazard Mater 301:17–26. https://doi.org/10.1016/j.jhazmat.2015.08.023

Čvančarová M, Křesinová Z, Cajthaml T (2013) Influence of the bioaccessible fraction of polycyclic aromatic hydrocarbons on the ecotoxicity of historically contaminated soils. J Hazard Mater 254–255:116–124. https://doi.org/10.1016/j.jhazmat.2013.03.060

Hawthorne SB, Grabanski CB (2000) Correlating selective supercritical fluid extraction with bioremediation behavior of PAHs in a field treatment plot. Environ Sci Technol 34 (19):4103–4110. https://doi.org/10.1021/es001178o

ISO (1986) Water quality—determination of cobalt, nickel, copper, zinc, cadmium and lead—flame atomic absorption spectrometric methods. ISO 8288:1986

Šašek V, Bhatt M, Cajthaml T, Malachová K, Lednická D (2003) Compost-mediated removal of polycyclic aromatic hydrocarbons from contaminated soil. Arch Environ Contam Toxicol 44 (3):0336–0342. https://doi.org/10.1007/s00244-002-2037-y

Chapter 26
Field Study X: Oil Waste Processing Using Combination of Physical Pretreatment and Bioremediation

Petra Najmanová and Robert Raschman

Abstract In this chapter, a field study from an oil field in Kazakhstan is described. This study ranges from lab- and pilot-scale tests to the application of a new on-site technology. Although it is related to biodegradation only marginally, this study shows a new approach to the remediation of sites with enormous oil contamination. The dense medium separation technology is based on the physical pretreatment (gravity separation in heavy suspensions), followed by bioremediation. Contaminated soil heavily polluted with petroleum substances, including macroscopic petroleum conglomerates with a fraction ≥ 1 mm, is initially treated by grading. The fine-grained fraction is decontaminated by biodegradation. The coarse-grained fraction is divided by the heavy-suspension separation into a heavy product and a light product. The light product consists of large and light petroleum conglomerates, which are treated by combustion or other processing. The heavy product made up of soil with other petroleum substances is treated by crushing and biodegradation together with the fine-grain fraction. Part of the process comprises recycling of process water and heavy suspension, where the process water is also used to wet the material during biodegradation.

Keywords Bioremediation · Oil waste · Heavy suspension · Soil separation

26.1 Introduction

Waste emerging from upstream oil industry usually contains a mixture of contaminated soil, drilling muds, and oil in various states of weathering. A massive contamination level of 10% petroleum hydrocarbons excludes direct processing by bioremediation technology. A new technology for the removal of petroleum

P. Najmanová (✉)
DEKONTA, a.s., Stehelčeves, Czech Republic

Department of Biotechnology, University of Chemistry and Technology, Prague, Czech Republic
e-mail: najmanova@dekonta.cz

R. Raschman
DEKONTA, a.s., Stehelčeves, Czech Republic

J. Filip et al. (eds.), *Advanced Nano-Bio Technologies for Water and Soil Treatment*, Applied Environmental Science and Engineering for a Sustainable Future,
https://doi.org/10.1007/978-3-030-29840-1_26

hydrocarbons from oil waste was used in an oil field in Kazakhstan. This technology is based on the combination of physical pretreatment (gravity separation in heavy suspensions) and bioremediation.

Contaminated soil heavily polluted with petroleum substances, including macroscopic petroleum conglomerates with a fraction ≥ 1 mm, is initially treated by screening. The fine-grained fraction is then treated by bioremediation. The coarse-grained fraction is divided by the heavy-suspension separation into a product of high specific gravity (sink) and a product of low specific gravity (float). The float product consists of weathered oil particles (tar balls) and conglomerates of oil and soil with high content of petroleum hydrocarbons (typically over 30%). The sink product, made up of a less contaminated material, is crushed and mixed with the fine-grain fraction and sludge generated by the heavy-suspension separation system. The produced mixture is shipped to an on-site treatment facility and treated by bioremediation. The costly dewatering of sludge generated by the heavy-suspension separation system is not necessary as the watery sludge is utilized as a source of water in the bioremediation process. This technology will mean considerable cost savings, making it the main advantage over others.

26.2 Feasibility Lab Test of Bioremediation

The aim of the test performed on a material contaminated with petroleum hydrocarbons was to determine and propose the most suitable and effective bioremediation process as well as to specify the optimal technological parameters for the full-scale treatment process. For the laboratory bioremediation test, a 5 kg sample of soil was collected on-site from the oil field in Kazakhstan. The homogenized sample was placed into a plastic container and the mineral nutrients, in the form of NP fertilizer (YaraMila NP 26–14 + 2S), were added in a ratio of 3 g/kg. Every second week a BIOTECH bacterial suspension containing the following bacterial strains—*Pseudomonas fluorescens, Pseudomonas stutzeri, Pseudomonas veronii, Pseudomonas putida, and Shewanella haliotis*—was applied into the soil (10 mL of mixed solution per kg, with an initial concentration of 1.0×10^8 CFU/mL). The bioremediation test was performed under aerobic conditions (mixed every week) at 20 °C (Innemanová et al. 2018). The control variant was only mixed once a week and set at 20 °C with no addition of fertilizers or BIOTECH. The results of the biodegradation lab test are shown in Fig. 26.1.

A significant decrease in a total petroleum hydrocarbons (TPH) concentration in the soil sample treated by bioremediation was achieved during the test period, which lasted about three months. From the initial value over 80 g/kg, the contamination was reduced to approx. 15 g/kg TPH, which represents approx. 81% efficiency.

On the other hand, it is obvious from Fig. 26.1 that the decrease in TPH starts to stagnate and only a limited additional decrease in the TPH contamination could be expected. Probably the biodegradation efficiency reachable in this case will approach 90%, which still represents a high level of contamination (approx. 10 g/kg TPH). It should be taken into consideration that a full-scale bioremediation process is usually less effective than the process carried out under laboratory conditions.

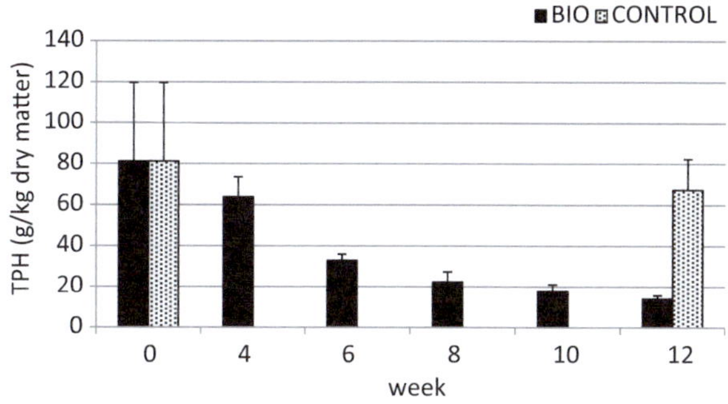

Fig. 26.1 Progress of TPH degradation within the laboratory test

Fig. 26.2 Construction of the bioremediation platform

Nevertheless, on the basis of the results obtained during the feasibility lab test, it was concluded that bioremediation was a potentially suitable technology for the treatment of contaminated soil collected at the dumping site of Kazakh oil field. Following this, it was suggested a pilot-scale test should be carried out.

26.3 Pilot-Scale Test of Bioremediation

With regard to the results of the lab test, a pilot-scale test was suggested and performed on-site with approx. 150 m^3 of contaminated soil from an oil waste deposit in Kazakhstan within five months. A temporary biodegradation platform, using a protective geotextile and high-density polyethylene (HDPE) geomembrane, was constructed on-site, with an area of 20 m × 20 m (see Fig. 26.2). The barriers around the platform were installed to prevent the release of the treated soil and the contaminated water from the bioremediation platform. The barrier was designed as a soil dike approx. 1 m high. The dike was covered with the same sealing system as the platform (HDPE geomembrane + protective geotextile). The removal of storm water from the platform was ensured by the inclination of the surface (approx. 1% slope). The removed water was drained to a separately constructed water collection tank and used for moistening within the bioremediation process.

The contaminated soil was divided into two piles. The first one was treated using a BIOTECH bacterial solution called "bioaugmentation". The biodegradation of the

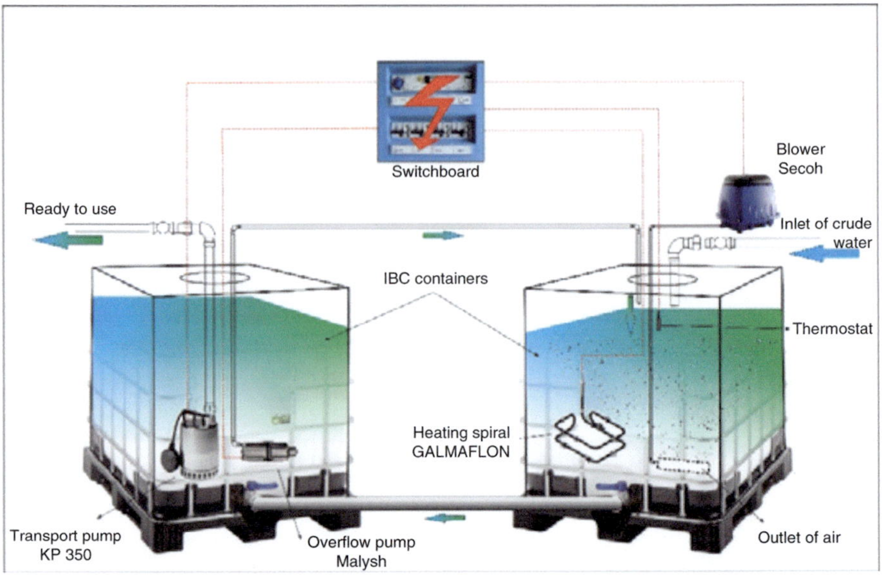

Fig. 26.3 On-site bioreactors for bacteria cultivation

second pile, called "biostimulation", was enhanced by using an organic substrate (50% vol. of camel dung) to improve the material characterization and addition of the nutrients.

The intensive soil aeration, to ensure aerobic conditions in the whole volume of the contaminated material, was performed by front loaders that moved the soil from one place to another once a month.

The freeze-dried BIOTECH mixture was used as an inoculum for cultivation in the bioreactors on-site (see Fig. 26.3). The nutrients (YARA fertilizer; 1% w/w) and a source of carbon (molasses; 1% vol.) were also added into the bioreactors. After 24 h of cultivation under aeration and a temperature of 30 °C, the desired bacteria concentration was achieved (10^8–10^9 CFU/mL) in a volume of 2 m^3.

The BIOTECH bacterial solution was applied to the contaminated soil (bioaugmentation biopile) by spraying once a month, so that the concentration of bacteria in the surface soil layer (approx. 0.5 m thick) was 10^6–10^7 per 1 g.

26.3.1 Results

The initial sampling and the following chemical analyses showed a variety of different types of contaminated material (see Table 26.1) with a higher content of petroleum hydrocarbons, having more than 500 g/kg in dry matter on average. The contamination was spread out very heterogeneously, the TPH content ranged from 10% (sandy material) to 90% (tar).

Table 26.1 Description of mixture of contaminated materials

Contaminated material	Description	Content (% w/w)
Sand	Yellow-brown color; fine-grained with pebbles; weak hydrocarbon smell	5
Soil	Grey color; fine-grained with pebbles; weak to intense hydrocarbon smell	30
Drilling cuttings	Dark grey to black color; hard grain size and agglomerates; intense hydrocarbon smell	30
Tar	Black color; hard grain size, melting at higher temperature; intense hydrocarbon smell	30
Construction debris	White-grey color; hard pieces; weak hydrocarbon smell	5

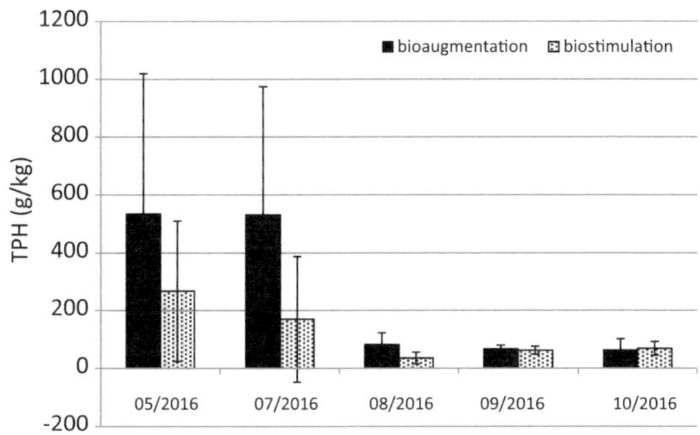

Fig. 26.4 Progress of TPH degradation within the pilot-scale test

The initial TPH concentrations of the contaminated soil were very high and very heterogenic. On average, five samples per each pile, it was more than 500 g/kg in the case of bioaugmentation, and nearly 300 g/kg for biostimulation. Nevertheless, a rapid decrease of petroleum hydrocarbons was observed (see Fig. 26.4). The efficiency of bioremediation in the bioaugmentation pile, resp. biostimulation pile was 88%, resp. 74%, after five months of on-site treatment the final concentrations of TPH were 63 g/kg, resp. 69 g/kg on average.

The optimal conditions of the bioremediation process were monitored by additional analyses, such as the concentration of heterotrophic bacteria and respiration intensity (ASTM 2018), see Fig. 26.5. The concentration of heterotrophic bacteria increased by more than 100× (bioaugmentation), resp. 1000× (biostimulation) during the biological treatment and kept around 10^7 CFU/g, which was sufficient for effective biodegradation.

The respiration intensity, measured as the production of carbon dioxide as an end product of bioremediation (see Fig. 26.6), increased during the first three months and started to slowly decline afterwards (in the case of biostimulation pile).

The ecotoxicity of treated soil was determined using *Vibrio fischeri*, bioluminescent bacteria (ISO 2007). The relative inhibition was measured at the beginning and

Fig. 26.5 The concentration of heterotrophic bacteria within the pilot-scale test

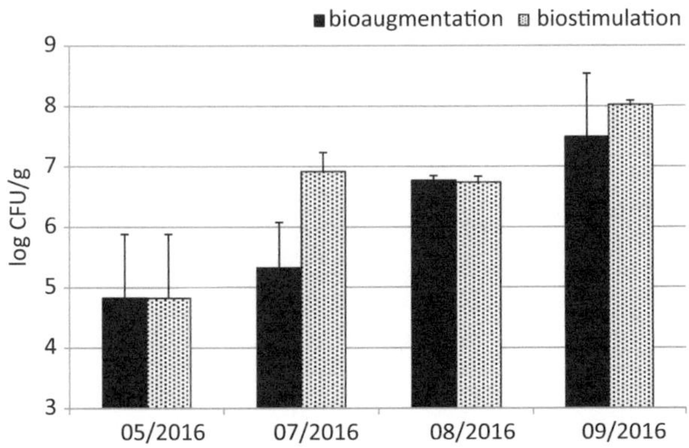

Fig. 26.6 The respiration intensity within the pilot-scale test

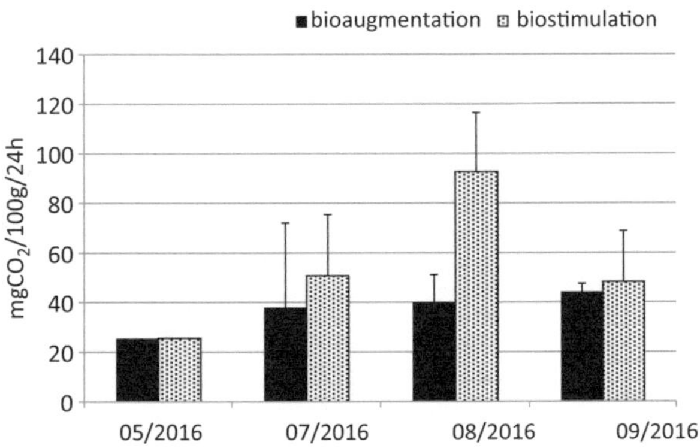

Fig. 26.7 Ecotoxicity using *Vibrio fischeri* within the pilot-scale test

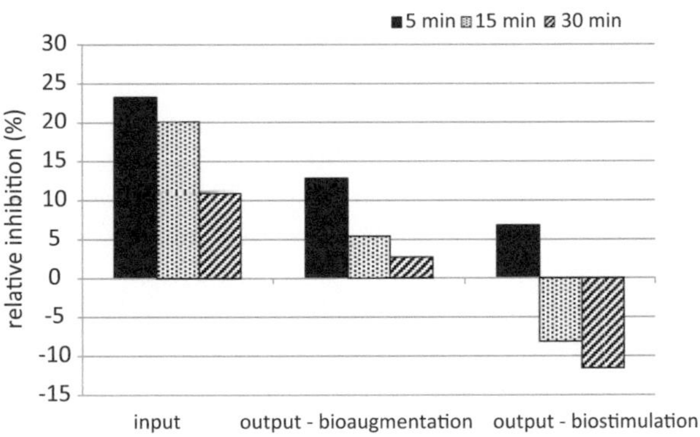

at the end of the bioremediation process. Ecotoxicity had rapidly dropped in both piles after five months of the treatment. The biodegradation technology led to a decrease in the material toxicity achieving the legislatively acceptable level (see Fig. 26.7).

Fig. 26.8 Pilot-scale test of (**a**) separation and (**b**) floating parts

Despite the promising efficiency of the bioremediation pilot-scale test (around 80%), the output concentrations (approx. 6–7 g/kg) were still very high and achieving the target limits (2% of TPH) would have been costly, even if possible. So the alternative method of physical pretreatment followed by bioremediation was suggested and performed on-site.

26.4 Pilot-Scale Test of Physical Pretreatment

As shown in Sect. 26.3.1, the biotechnological methods do not allow sufficient decomposition of the soil parts with a high content of petroleum products and pieces of pure bitumen. However, the contamination in oil sludge is usually spread out in a heterogenic way so there are still some parts of the contaminated soil suitable for a biodegradation process. The suggested technological process of physical pretreatment was based on separating highly contaminated parts of sludge and soil from parts with lower contamination. To remove these parts from the soil before biodegradation, an original separation method was developed.

The pilot test was performed on-site in Kazakhstan using a 4.4 m $\times$ 2.3 m $\times$ 0.9 m container (see Fig. 26.8). About 20 m^3 of contaminated material was processed. The separation technology consisted of two steps: (i) screening (vibrating screen) when the removal of fine fractions below 5 mm was ensured and (ii) heavy media separation, where the removal of tar balls was achieved. With respect to the specific gravity of tar (approx. 1.1 g/cm^3) and the specific gravity of sand (2.7 g/cm^3), the solution of sodium silicate (water glass), with a specific gravity 1.5 g/cm^3, was used as a dense medium in the separation process. The results of the separation process are shown in the following table (Table 26.2) and scheme (Fig. 26.9).

After the separation test the yield of the floating part (bitumen) was 25% (w/w) with approx. 80% content of TPH and was supposed to be used as an alternative fuel in a local cement plant. The soil was separated as a sinking part with higher density and with the TPH content less than 10%. This part is supposed to be treated using

Table 26.2 Yields and TPH content after the separation process

Product	Yield (% w/w)	TPH content (% w/w)
Floating part - tar balls	25	81.0
Sinking part - soil	75	8.9

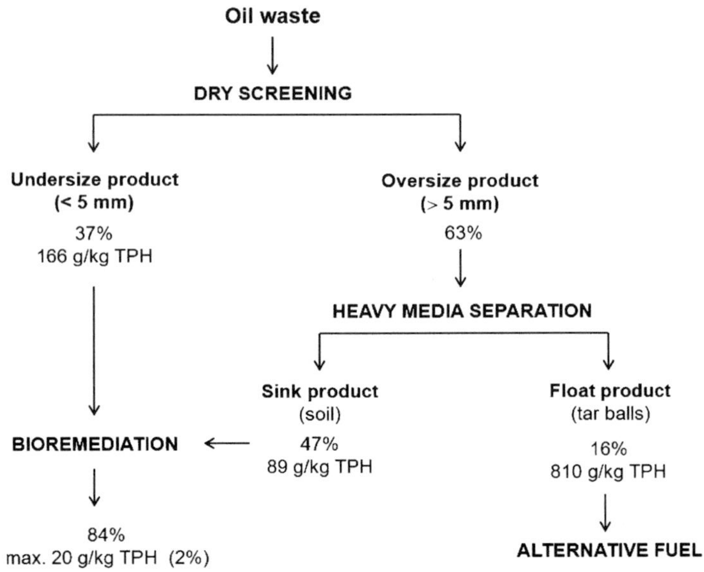

Fig. 26.9 Technological scheme of oil sludge treatment

bioremediation technology as shown above. In that case, we can ensure that the target limit under 2% w/w of TPH is achieved.

26.5 Dense Media Separation (DMS) Technology

On the basis of the results from the pilot-scale tests on the oil field site, a new technology for full-scale application of dense media separation was developed (Raschman and Najmanová 2018).

Decontamination of large volumes of soils polluted by petroleum hydrocarbons is most often carried out by excavation and subsequent ex situ bioremediation. In case that the soil contains macroscopic particles of weathered oil or conglomerates formed by mixture of weathered oil and sand/soil admixtures, efficient bioremediation of such material is not possible because petroleum hydrocarbons concentrated in the weathered oil particles/conglomerates are inaccessible and toxic for bacteria.

Therefore, it is necessary to separate the weathered oil particles/conglomerates from the contaminated soil prior to bioremediation.

To solve this problem, DEKONTA has developed an innovative separation method utilizing the difference in specific gravity of weathered oil particles/conglomerates (typically 1.1–1.5 g/cm^3) and sand/stone (typically 2.2–2.7 g/cm^3).

Dense media separation process is commonly used for processing of coal and some minerals. Nevertheless, it has not been applied to remediation purposes so far.

The principal of the process is that if a mixture of particles differing in their specific gravity is submersed in heavy liquid of suitable density, lighter density particles naturally float, while those denser than the liquid sink. Mixture of fine media material, such as magnetite or ferrosilicon, suspended in water is used to produce media (slurry) of appropriate specific gravity.

Prescreened input material is fed to the heavy media vessel that has the required media specific gravity. The light density material floats and is discharged from the float section of the vessel, while the higher specific gravity material sinks and is discharged from the sink section of the vessel. The material is discharged on drain and rinse screens in order to separate the material that drains through from the remainder on the screen. The washed medium needs to be re-concentrated and is sent back to the magnetic separator circuit for recovery. From the magnetic separator, the medium reports to the densifier or the media thickener, where it is pumped back to the heavy media separator to maintain the proper specific gravity of the vessel and replace the medium continuously lost with the discharge of the float and sink material.

The sink product from the DMS process (contaminated soil/sand) is subsequently treated by bioremediation. The float product (weathered oil particles/conglomerates) with the average concentration of petroleum hydrocarbons 30% as minimum is used as an alternative fuel.

Technological scheme of the DMS process is provided below (Fig. 26.10).

26.6 Conclusion

Decontamination of large volumes of soils polluted by petroleum hydrocarbons is most often carried out by excavation and subsequent ex situ bioremediation. In case that the soil contains macroscopic particles of weathered oil or conglomerates formed by mixture of weathered oil and soil admixtures, efficiency of bioremediation of such material has its limitations. In that case the new technology of dense media separation can be used as the first step to separate a highly contaminated material and make the material suitable for biodegradation.

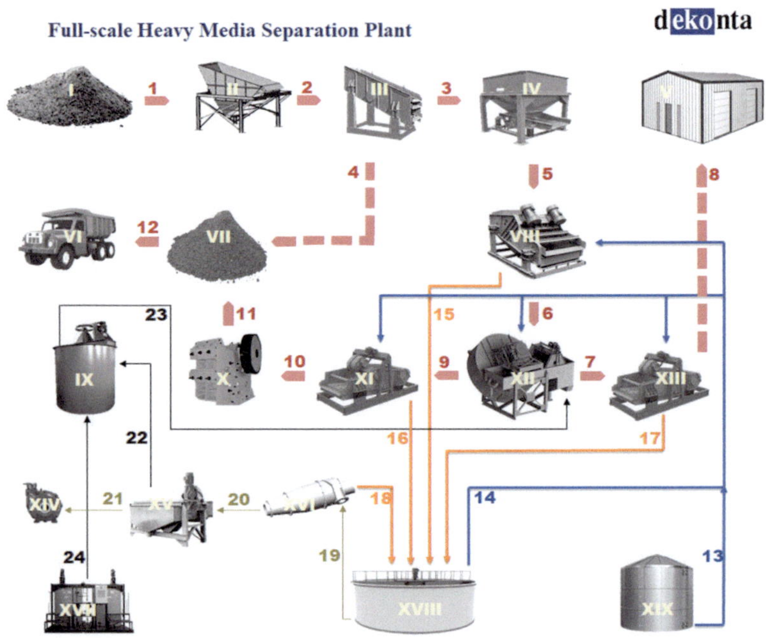

Equipment:

- I. Contaminated material storage
- II. Storage hopper with sloping grizzly and vibrating feeder
- III. Dry vibrating screen
- IV. Storage hopper with belt feeder
- V. Separated tar storage
- VI. Shipment of pretreated soil to a bioremediation facility
- VII. Storage of pretreated soil
- VIII. Vibrating screen for washing material before heavy media separation
- IX. Tank for preparation of magnetite suspension
- X. Crusher
- XI. Vibrating screen for sink product washing
- XII. Heavy media separator
- XIII. Vibrating screen for float product washing
- XIV. Pump for delivery of slurry to a bioremediation facility
- XV. Magnetic separator
- XVI. Hydrocyclone
- XVII. Magnetite storage tank
- XVIII. Sedimentation tank
- XIX. Process water tank

Material streams:

1. Input material (mixture of weathered oil sludge and soil/sand/clay/stones)
2. Input material after removal of oversized particles (+ 300 mm) on sloping grizzly
3. Coarse fraction of input material
4. Fine fraction of input material
5. Coarse fraction of input material fed by belt feeder
6. Coarse fraction of input material after washing
7. Float product of dense media separation (weathered oil particles and conglomerates)
8. Dewatered float product
9. Sink product of dense media separation (sand/soil/stones)
10. Dewatered sink product
11. Crushed sink product (sand/soil/stones)
12. Sink product shipped to a bioremediation facility
13. Process water
14. Clarified water
15. Sludge from input material washing
16. Sludge from sink material washing
17. Sludge from float material washing
18. Hydrocyclone overflow
19. Sediment from a sedimentation tank (input to a hydrocyclone)
20. Hydrocyclone underflow
21. Sludge after magnetite separation
22. Magnetite concentrate
23. Heavy suspension ready for use
24. Fresh magnetite

Fig. 26.10 Full-scale heavy media separation plant

The DMS process offers highly efficient and zero waste technology with many advantages. The process water is reused for soil moistening during bioremediation, the most contaminated particles after heavy media separation can serve as an alternative fuel for cement plants and the treated soil is used as a recultivation material. It leads to a rapid reduction of costs compared with currently used technologies (e.g., thermal desorption).

References

ASTM (2018) Standard test method for determining aerobic biodegradation of plastic materials in soil. ASTM D5988–18. ASTM International, West Conshohocken. https://doi.org/10.1520/D5988-18

Innemanová P, Filipová A, Michalíková K, Wimmerová L, Cajthaml T (2018) Bioaugmentation of PAH-contaminated soils: a novel procedure for introduction of bacterial degraders into contaminated soil. Ecol Eng 118:93–96. https://doi.org/10.1016/j.ecoleng.2018.04.014

ISO (2007) Water quality—determination of the inhibitory effect of water samples on the light emission of Vibrio fischeri (Luminescent bacteria test)—part 3: method using freeze-dried bacteria. ISO 11348-3:2007

Raschman R, Najmanová P (2018) A method of decontamination of soils contaminated with petroleum substances and a line for implementing this method. Patent no. 307139

Part V
Ecotoxicology of Both Environmental Pollutants and Nanomaterials Used for Remediation

Chapter 27
Ecotoxicology of Environmental Pollutants

Luděk Bláha and Jakub Hofman

Abstract The most discussed hazardous properties of chemicals with respect to environmental impacts include persistency, tendency to bioaccumulate in biota, human toxicity, and ecotoxicity to living biota. In this chapter, aimed primarily at students and non-professionals in ecotoxicology such as environmental engineers or remediation experts, brief introduction to ecotoxicology is provided. The chapter brings introductory insights into the concepts of ecotoxicology and effects of chemicals in biological systems including the discussion of concentration–response relationships. Further, we introduce methods and tools in the testing of ecotoxicity, i.e., bioassays, and discuss their applications in both prospective and retrospective hazard assessment. Finally, the chapter provides a simple overview of major contaminants and their documented ecotoxic impacts on living organisms in both aquatic and terrestrial ecosystems.

Keywords Ecotoxicology · Environmental toxicology · Ecotoxicity testing · Bioassays · Chemical effects

27.1 Introduction

Environmental pollutants pose diverse hazards (i.e., inherited properties), which may under specific conditions represent threats to the environment or human health. The most broadly discussed hazardous properties of chemicals with respect to environmental impacts include persistency, tendency to bioaccumulate in biota, and (eco)toxicity. The assessment criteria and methods for measuring persistency and bioaccumulative properties of chemicals are relatively well defined within regulatory frameworks (Mackay et al. 2014), and global treaties such as Stockholm Convention on Persistent Organic Pollutants (POPs) (Stockholm Convention 2008). On the other hand, understanding adverse effects of chemicals on humans and

L. Bláha (✉) · J. Hofman
RECETOX, Faculty of Science, Masaryk University in Brno, Brno, Czech Republic
e-mail: blaha@recetox.muni.cz

J. Filip et al. (eds.), *Advanced Nano-Bio Technologies for Water and Soil Treatment*,
Applied Environmental Science and Engineering for a Sustainable Future,
https://doi.org/10.1007/978-3-030-29840-1_27

ecosystems is more complex because of the diversity of biota, i.e., diversity of biological "targets" of chemicals (Blaha and Holoubek 2013).

Toxicity and ecotoxicity are two types of chemical hazards that are currently dealt with and evaluated separately. Despite many similarities and parallels, toxicology and ecotoxicology pursue different goals and maintain different focuses.

- The **effects of environmental pollutants on human health** include relevant exposure scenarios like use of contaminated drinking water, inhalation of polluted air, or ingestion of food with toxicants accumulated through food chains (US EPA 2011). Documented **adverse outcomes** in exposed individuals may be acute health effects and even human death but—more commonly—chronic impacts such as developmental or metabolic diseases, reproductive failure, immunosuppression or cancer can occur. Hazards to **individual humans** are characterized by traditional toxicological models like rodents, dogs, pigs, or monkeys and the results of toxicological testing are expressed as LD50 (lethal dose to 50% of the tested organisms) or NOAEL (no observed adverse effect level).
- Environmental contaminants also cause adverse effects to **aquatic and soil biota** including microorganisms (Cravo-Laureau et al. 2017; Ghiglione et al. 2016; Roose-Amsaleg and Laverman 2016; van Beelen and Doelman 1997; Yadav et al. 2016), algae and higher plants (Sandermann 2004), invertebrates (Cattaneo et al. 2009; Cortet et al. 1999; Frampton et al. 2006; Kammenga et al. 2000; Korsloot et al. 2004; Lagadic and Caquet 1998) as well as vertebrates like fish, amphibians, birds, and even mammals (Carvalho et al. 2014, Maier et al. 2015, Oprsal et al. 2015, Isidori et al. 2016). By affecting natural biodiversity and functionality of ecosystems, the effects may propagate further by impairing "ecosystem services". Thus, the assessment of ecotoxicity, i.e., hazards of chemicals to integrity and functionality of natural populations and communities of non-human biota, is of essential importance (Newman 2014; Walker et al. 2012).

In this chapter, **ecotoxicology** of environmental pollutants is covered with the aims to (i) provide introductory insights into the effects of chemicals in biological systems, (ii) outline the concepts and methods in the assessment of ecotoxicity, and (iii) provide an overview of reported impacts of major water and soil contaminants.

27.2 Biological Effects of Environmental Pollutants

Before discussing practical issues of ecotoxicity testing, it is essential to have good and common understanding of relevant terms such as toxic effect, adverse outcome, etc.

First, the central so-called **cause–effect paradigm** is the red line for both toxicology and ecotoxicology. "All things are poison and nothing without poison. Solely, the dose determines that a thing is not a poison" (Paracelsus 1493–1541). Whenever "toxicity" or "effect" is mentioned in this chapter (e.g., death of an earthworm owing to acute poisoning by arsenic, cytotoxicity caused by benzo[*a*]

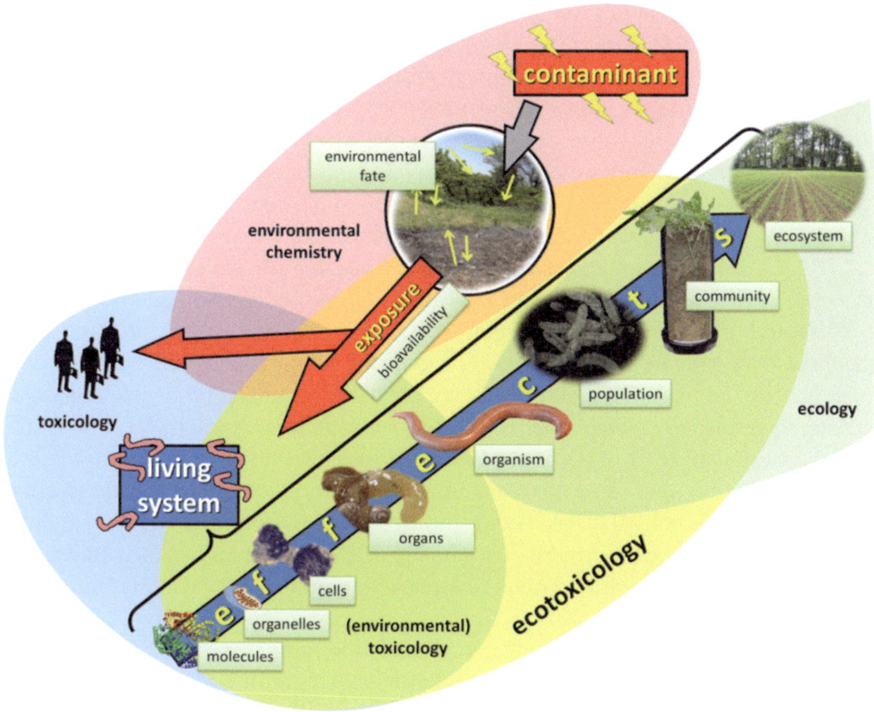

Fig. 27.1 Scheme showing overlaps of toxicology and ecotoxicology, environmental chemistry and ecology, and propagation of (eco)toxic effects through biological systems

pyrene, or disruption of offspring production in springtails induced by cadmium) the reader should understand that such an effect occurs only under certain favorable conditions, most importantly at certain specific and sufficiently high **concentration or dose**. This concentration of toxic compounds is relatively lower in comparison to other, i.e., less toxic compounds. However, even the other "less toxic" compounds may induce the same effects when the conditions have been changed (e.g., increased doses, prolonged exposure durations).

According to current understanding, the **adverse outcome**, i.e., in vivo manifestation of the effect, results from a chain of events formalized by the term "adverse outcome pathway" (AOP). This chain starts at the molecular level, i.e., the interaction of a pollutant with its partnering biological molecule, next it propagates through higher levels (cellular, tissues, organs) and becomes apparent in vivo as a systemic or organ-specific toxicity. This disruption may then affect viability of the population, structure and processes of the community and ecosystem (Fig. 27.1). The adverse outcomes manifest under favorable conditions, e.g., sensitivity of early developmental life stage, when the chain of events is complete and not repaired by natural processes through toxicokinetics (ADME—absorption, distribution, metabolism, and excretion).

The observed in vivo effects thus always start at molecular level (molecular initiating event, MIE, within the frame of an AOP concept), which may be either **specific (key-lock principle)** through binding the toxicant to a specific place on DNA or a protein (e.g., activation of estrogenic receptor by endocrine disruptive compounds) or—more commonly—**nonspecific**, when the chemical does not have a specific target and broadly interacts with diverse molecules in a biological system. Key examples include accumulation of lipophilic molecules in the cell membranes (narcotic or basal toxicity), which is inherited to all organic compounds including, e.g., PAHs, BTEX, aliphatic hydrocarbons, or damage to biological molecules—proteins, nucleic acids, phospholipids—caused by (pro)oxidative conditions.

It should also be understood that—depending on the actual situation—one chemical may activate several different modes of actions, which could be of both "specific" or "non-specific" nature. These events can occur and propagate in parallel causing synergistic or antagonistic effects. For example, chromium(VI) may cause direct damage to DNA (leading potentially to mutation and eventually tumor development) but the same toxicant also activates intracellular repair systems, which remove the potential risk of developing cancer by controlled death—apoptosis—of the affected cell. Similarly, simultaneous presence of zinc in the soil leads to a decreased uptake of toxic cadmium to biota lowering thus its potential toxic effects and hazards (Murtaza et al. 2017).

The outlined complexity increases when addressing the effects of complex samples (chemical mixtures) such as contaminated water or soil, wastes, effluents, etc. (Carvalho et al. 2014). In the exposed biota, many diverse and hardly predictable events may occur simultaneously and are likely to be modulated by various components present within the sample. For example, the effects of pesticide chloroacetanilide on earthworms highly differ depending on levels of H^+ (pH), concentrations of ions, salts, or non-toxic metals; fragments of naturally present dead organic matter (e.g., humic acids); and many small organic molecules like peptides, aminoacids, nucleotides, phospholipids, DNA or RNA, etc. All the components—mentioned as an example—affect both bioavailability and effects of the chemical on the studied organism (Hofman et al. 2014).

27.3 Principles of Ecotoxicity Testing

Despite the years of research, it is still difficult to reach the ultimate goal of ecotoxicology, i.e., thorough understanding as well as predicting and protecting populations in communities from the effects of environmental pollutants. Researchers continue studying the complex systems and developing advanced approaches such as mechanistic high-throughput chemical screenings in vivo, utilization of "omic" responses and biomarkers, computational in silico ecotoxicology, or higher tier modelling of populations, communities, and ecosystems (Cao et al. 2017; Fernandez et al. 2011; Groth et al. 2016). However, such approaches are still not ready to be used in practice. Therefore regulators, risk assessors, and environmental engineers have adopted a pragmatic and scientific-based approach in assessing and predicting the ecotoxicological effects.

Trophic Levels & Batteries of Assays The approach considers the key structural pillars of biological communities, i.e., trophic levels. In principle, representatives of biological species from the groups of **producers** (autotrophic organisms like algae and plants), **consumers** (invertebrate and vertebrate animals), and **decomposers** (microorganisms)[1] are selected as model organisms and used in **batteries of bioassays** for testing of ecotoxicity.

The bioassays recommended for ecotoxicity testing within regulatory frameworks (including chemical safety as well as risk management and remediation) need to be **validated** to assure repeatability and quality of the results using **standardized guidelines** that are currently available through international organizations such as OECD or ISO.

The broadly used bioassays recommended by regulations include usually **short-term exposure (acute) tests**. These have both advantages (e.g., possibility to maintain the model organisms in laboratory, good standardization, time and cost-effectiveness) and disadvantages (lower sensitivity in comparison to wild species, lower representativeness for environmentally relevant long-term chronic exposures, etc.).

- **Aquatic Environment** The most commonly used assays are 72–96 h algal growth inhibition test with, e.g., *Scenedesmus subspicatus* (OECD 2011), 24–48 h immobilization test with crustacean *Daphnia magna* (OECD 2004), 96 h acute test for lethality with adult fish (OECD 1992), currently promoted alternative 96 h toxicity test with zebrafish *Danio rerio* embryos (OECD 2013), and 16 h bacterial growth inhibition test with *Pseudomonas putida* (ISO 1995)
- **Terrestrial and Soil Environment** Commonly used bioassays include 7–14-day acute test for lethality with earthworms *Eisenia fetida* (OECD 1984; ISO 2012a), 96 h root elongation test with plants, e.g., *Lactuca sativa* (OECD 2006a; ISO 2012c), 24 h acute test for lethality of nematode *Caenorhabditis elegans* (ASTM 2014), or 6 h test for effects on microbial nitrification in soil (ISO 2012e).

Standardized longer exposure assays testing for population-relevant endpoints such as reproduction are also available. For example, 21-day reproduction test with *D. magna* (OECD 2012) or reproduction test with enchytraeids (ISO 2014b; OECD 2016a), springtails (ISO 2014a; OECD 2009), and earthworms (ISO 2012b; OECD 2016b), life-long tests with terrestrial plants (OECD 2006b) or assays with long term cultivation of soil microbial communities for effects on carbon and nitrogen transformations (ISO 2012d; OECD 2000a, 2000b). However, because of large time and

[1]Because of their fast life cycle, high plasticity and adaptability, assays with bacteria and other microorganisms are not usually among the basic regulatory batteries for ecotoxicity testing. Nevertheless, many bacterial models exist and play inevitable roles. For example, in testing for genotoxicity (Ames test with *Salmonella typhimurium*) or in rapid screenings (30-min assay with bioluminescent bacteria *Aliivibrio* (formerly *Vibrio*) *fisheri*, also known as Microtox test).

cost demands, they are required and performed only under specific and priority situations. Additional requirements may also apply to specific areas such as pesticides, where ecotoxicity characterization is required for other important species such as honey bees or birds.

27.4 Prospective Assessment: Ecotoxicity of Chemicals

The assessment of ecotoxicity is required and performed within two different frameworks, i.e., prospective assessment (testing of individual chemicals or materials) and retrospective assessment (testing of contaminated natural samples).

Prospective approach relates mainly to the assessment of individual chemicals that are brought to the market by industry. Various regulatory frameworks exist for groups of chemicals with different use like industrial compounds (European Parliament and the Council of the European Union 2006, 2008), plant protection products—pesticides (US EPA 1996; European Parliament and the Council of the European Union 2009), biocides (European Parliament and the Council of the European Union 2012), or pharmaceuticals. Regulations impose variable requirements regarding ecotoxicity ranging from few simple laboratory assays for industrial chemicals to high-tier field experiments for pesticides. It should be highlighted that all the toxicity and ecotoxicity bioassays have primarily been developed and designed for this purpose, i.e., standardized assessment of individual chemicals.

Chemical ecotoxicity should be tested in both aquatic and soil arrangements. Aquatic set-up allows very good standardization of the testing media as well as good homogenization and distribution of the chemical in the test system. The latter factors are however more complicated for soil. Chemicals can either be externally added (spiked) into "standard" natural soils with different properties, which can be commercially purchased (such as (LUFA 2018)), or tested in the artificial soil prepared according to standardized protocol (OECD 1984, 2009, 2016a, b). However, as recently shown on different artificial soils collected within the international laboratory ringtest, even the standard artificial soils greatly differ in their final composition, which affected the results of the ecotoxicity tests of externally added toxicants by orders of magnitude (Bielská et al. 2017; Hofman et al. 2014; Vašíčková et al. 2015).

Ecotoxicity testing of individual chemicals includes exposure of test organisms in parallel to a negative control (blank) and to a series of chemical concentrations. After a given exposure time, the endpoint of interest is evaluated (e.g., growth, lethality, number of offspring, etc.), the effects are related to control (e.g., percentage of change), and the ecotoxicological parameters are derived from a concentration–response curve (Fig. 27.2) including typically concentration of a chemical causing 50% effect (lethal, effective, inhibitory—LC50, EC50, IC50, respectively), no-observed-effect concentrations (NOEC), and lowest-observed-effect concentration (LOEC).

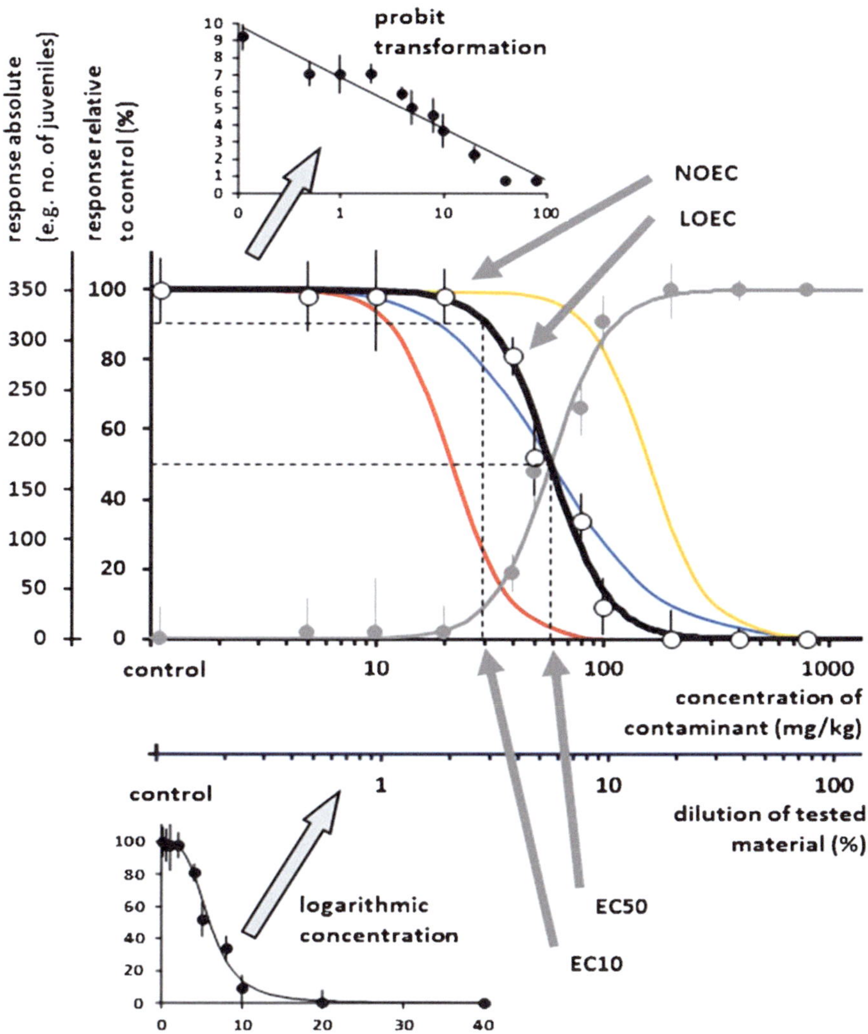

Fig. 27.2 Evaluation and visualization of ecotoxicity results. Tested concentrations are usually expressed log-transformed, the response of biological system to increasing concentration can be either binary (e.g., mortality) or continuous (e.g., number of juveniles). Ecotoxicity parameters (NOEC, LOEC, EC50, or LC50, etc.) are derived from the modelled concentration–response curve

Under current legislation, ecotoxicity results obtained by the industry researchers usually remain proprietary to the owner (who paid for testing), and may not be fully disclosed or available to public. In parallel, ecotoxicity testing is performed also within the frame of research or public funded studies, and the results are released in research reports or scientific papers. Publicly available ecotoxicity data are organized in databases, where US EPA ECOTOX (US EPA 2018) provides the most comprehensive set of information. Information derived from the ecotoxicity tests (i.e., hazard of a chemical described by ECx/LCx, NOEC, LOEC, or similar values) may be used for the assessment of the potential risks, such as setting the environmental quality standards (EQS) for different matrices like water or soil.

In addition to individual chemicals, also "materials" containing mixtures of known or unknown chemicals are prospectively tested for their ecotoxic properties. An example can be wastes that should be tested before their further use (Moser and Römbke 2009). Other examples include dredged sediments, which should be evaluated for ecotoxicity before their application on agricultural soils as fertilizers (Vašíčková et al. 2013). Unfortunately, relevant legislations are country-specific and not harmonized internationally.

27.5 Retrospective Assessment: Ecotoxicology of Contaminated Samples

Retrospective ecotoxicity testing includes the assessment of contaminated environmental samples like water, sediments, or soil. Most commonly, risk assessment of contaminated samples is based on **chemical monitoring**, i.e., targeted instrumental or bioanalytical analyses and quantification of known toxic components such as toxic metals or various organic compounds. With respect to remediation and removal of chemicals from contaminated environments, a number of methods have been developed ranging from AAS or ICP-MS/MS for metals to diverse gas and liquid chromatographic approaches (GC, LC) coupled with diverse detectors for organic pollutants. However, the traditionally used chemical assessment has a number of limitations and some efforts implement ecotoxicity and toxicity testing into risk assessment schemes. One of the major advantages of ecotoxicity testing is the response of the organism to all (potentially) toxic components in the sample (including eventually also by-products formed during the sample treatment, etc.). The bioassay or sets of bioassays (batteries) thus provide an integrative biological response, which reflects the "overall toxic potential" of the sample.

The assessment of complex contaminated samples by bioassays has a number of specificities, which further depend on the purpose. Namely, whether the bioassays are intended (i) for the use in **distributed monitoring of environmental quality**, i.e., assessment and comparison of multiple samples from different sites (effect-based environmental monitoring), or (ii) for the use during the **site-specific remediation**.

As mentioned above, the bioassays have primarily been developed and used for testing of individual chemicals where standard concentrations and conditions can be assured. In complex samples, even the composition of matrix may vary considerably. This is important for both water (variable pH, salinity, hardness, OC content, etc.), but even more for soils with enormous variability in parameters like texture, granularity, content, and composition of the clay and the organic material, etc. All these parameters affect **bioavailability** of present toxicants and consequently their biological effects. For example, in the whole effluent toxicity (WET) testing, nutrients such as nitrogen or phosphorus (usually enriched in municipal waters) may stimulate the growth of algae in the bioassay, which often outweighs the effect of other toxicants leading to "false negative" responses (Sandermann 2004). Similarly, variability of the composition of even standardized artificial soils affected the ecotoxicity of PAHs and other toxicants by orders of the magnitude (Bielská et al. 2017; Hofman et al. 2014; Vašíčková et al. 2015).

Despite the promotion by international authorities including, e.g., EU, routine application of bioassays in regulatory monitoring or field surveys is still at the stage of research or ad hoc risk assessment studies. Nevertheless, some environmental monitoring programs implemented **toxicity testing with bioassay(s)** such as whole effluent toxicity testing in the USA or Germany (Escher and Leusch 2011) or EU-supported effect-based monitoring programs (Tousova et al. 2017). Additional flagship initiatives include, e.g., effect-based assessment of estrogen endocrine disrupters in the effluents (Centre Ecotox 2018).

A number of open issues remain to be thoroughly addressed before broader acceptance of bioassays in distributed monitoring, including:

- Validation of the bioassay responses for samples with different composition and variable situations. Responses of the bioassay should be characterized across gradients of environmentally realistic conditions (e.g., in water for concentrations of ions, pH, OC levels, oxygen, ammonia). This should be done for both model laboratory samples (factorial design) and testing diverse field samples of variable composition.
- Quantitative characterization of the bioassay sensitivity and performance for both model laboratory and field samples should be provided, i.e., characterization of limits of detection, risks/frequency of false positives/negatives.
- Variability should be assessed and characterized by dedicated inter-laboratory exercise(s) that should demonstrate the repeatability and the robustness of the bioassay responses to standardized, yet chemically complex samples in different laboratories and conditions.
- Definition of thresholds or trigger values for each bioassay is needed. For example, it should be clearly defined what response would be considered significant or acceptable or what biological response would indicate non-acceptable risk. Examples of such trigger values could be, e.g., quantitative inhibition—percentage of immobilization of *Daphnia magna* in undiluted sample (or 5× or 10× diluted) or IC20 (IC50, etc.) expressed as dilution factor of the original sample.

Unlike the effect-based monitoring, **ecotoxicity testing of the contaminated samples** before, during, and after remediation provides invaluable information on changes of the overall toxicity during the remediation process. The problem of the matrix interference discussed above is less relevant because of the specificity to certain site. Parameters of the soil or water matrix usually do not greatly change during the treatment, which should allow for clear observation of changes in the eventual ecotoxicity during the treatment process.

27.6 Ecotoxicity of Major Pollutant Classes

The following sections provide a brief overview and examples of major classes of contaminants and their ecotoxicological hazards.

27.6.1 *Crude Oil, Oil, Petrol, Gasoline*

Petroleum is a naturally occurring substance of global importance. Petroleum itself as well as its products are complex mixtures (Fig. 27.3) composing of various saturated aliphatic (paraffins) compounds, saturated cyclic hydrocarbons (naphthenes) as well as mono- and polycyclic aromatic chemicals ranging from benzene through naphthalene to larger polycyclic aromatic hydrocarbons (PAHs). It also contains smaller amounts of S-, N-, and O-containing compounds and traces of metallic compounds. The sum of concentrations of benzene, toluene, ethylbenzene, and the xylenes (*o*, *m*, and *p*)—referred to as BTEX—is commonly used as a surrogate for the total amount of volatile monocyclic aromatic hydrocarbons (MAHs) in petroleum. It is very difficult to estimate the impact of these complex mixtures on different organisms or ecosystems and there is little information on the eventual interactive effects like synergism or antagonism among various

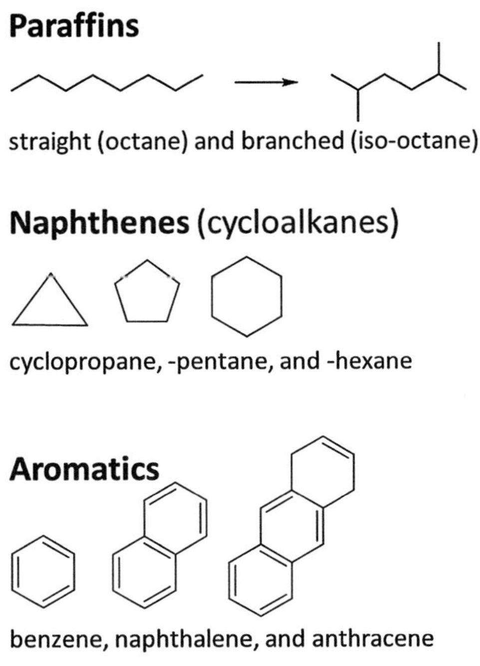

Fig. 27.3 Illustrative structures of crude oil components

components. Major environmentally related risks include spills in marine systems, soil, surface water and ground water contamination. Such accidents are known to have a deteriorating impact on ecosystem functioning. Although ecotoxicological studies with individual components of oil or its products are available (e.g., acute 48 h LC50 values for different fish species range 2–175 mg/L for different types of oil), it is considered that PAHs (see Sect. 27.6.3) provide the largest contribution to the overall toxicity. They are easily bioaccumulated through food chains and they are prone to photoactivation in the environment or metabolic bioactivation in the organisms, both of which create reactive species targeting important macromolecules like proteins, phospholipids as well as nucleic acids DNA or RNA. The effects in organisms include acute lethality, developmental malformations as well as chronic effects like immunotoxicity or behavioral toxicity, which have mostly been documented in vertebrates like fish, amphibians, or other wildlife.

27.6.2 Benzene

Benzene and its derivatives are broadly distributed in the environment due to their use as solvents or intermediates in chemical and pharmaceutical industry. The often reported toxic effects in mammals—including humans—are acute neurotoxicity or cardiotoxicity. Chronic exposures, when metabolic activation in liver and production of ROS have been described, result in hepatotoxicity or bone marrow toxicity (hematotoxicity, anaemia). Acute aquatic ecotoxicity is within low mg/L concentrations (see Table 27.1). Although benzene-related toxicity in man is a matter of concern, its direct threat to ecosystems is relatively lower also because of its fast (photo)transformation and lower stability.

27.6.3 Polycyclic Aromatic Hydrocarbons (PAHs)

PAHs are a structurally diverse group, which contains two or more fused benzene rings and may often be substituted by alkyls and other substituents. They are ubiquitous, often occurring in high concentrations around sources of combustion,

Table 27.1 Examples of comparable 96 h ecotoxicity values of benzene to invertebrate and vertebrate aquatic organisms (US EPA 2018)

Species	96 h LC50 (mg/L)
Grass shrimp (*Palaemonetes pugio*)	27
Crab larvae stage 1 (*Cancer magister*)	1108
Pacific herring (*Clupea harengus pallasi*)	40–45
Fathead minnow (*Pimephales promelas*)	34–35
Leopard frog (*Rana pipiens*)	3.7

petroleum operations, municipal and industrial wastewater, automobile transport, etc. Hundreds of different PAH compounds have been identified so far; therefore, routine monitoring primarily focuses on frequently occurring compounds—16 so-called US EPA PAHs or on an extended list of 40 compounds that includes less common but carcinogenic compounds. The majority of dominant and frequently found PAHs have molecular weight ranging 120–300 and relatively high hydrophobicity, which allows their enrichment in organic-rich matrices like sediments as well as bioaccumulation in the organisms and trophic transfer through food webs. Most of the information on ecotoxicity of individual PAHs is based on studies of prototypical and broadly occurring compounds like benzo[*a*]pyrene, anthracene, pyrene. Direct accumulation in membranes (narcotic toxicity) is one of the common toxicity mechanisms. Many PAHs easily undergo oxidation upon exposure to UV irradiation, which creates reactive products causing so-called photoactivation (photoenhanced) toxicity. Reactive oxidized PAHs are also created in the organisms (bioactivation) during the metabolism, and they cause either direct toxicity to major biomolecules like proteins or DNA or they indirectly stimulate cellular oxidative stress. Other important modes of toxicity, shared also with highly hazardous PCDD/Fs, include activation of aryl hydrocarbon receptor AhR or modulations of metabolic or hormonal pathways (endocrine disruption). Ecotoxic effects of PAHs have been studied extensively and include growth and metabolic inhibitions in algae and higher plants, acute lethality of many aquatic and soil invertebrates (LC50s span across several orders of magnitude depending on actual molecule and studied species but may be as low as few μg/g). Highly sensitive are especially fish embryos where toxicity may manifest at concentrations below μg/L. Other known effects include mutagenicity and development of neoplasia, immunotoxicity, etc.

27.6.4 Phenols

Phenols are mono- (phenol, cresol, naphthol, xylenol) or poly- (e.g., pyrocatechol, resorcin, pyrogallol, etc.) hydroxy derivatives of benzene and its alkyl substituents. They are broadly used in agriculture, pharmacy, or plastic industry; therefore, they are often released through industrial wastewater. Chlorinated and nitro-phenols are also used as fungicides or herbicides. Main modes of toxic action of phenols include basal narcotic toxicity via accumulation to membranes and effects on energy metabolism in mitochondria. In vertebrates they are often toxic to nervous system. For example, the acute ecotoxicity of phenol to freshwater organisms ranges from few mg/L (trout, *Daphnia magna*) up to hundreds mg/L (algae). There are also many natural polyphenolic compounds, where biological effects like estrogenicity have been discussed.

Table 27.2 Illustrative examples of ecotoxicity of halogenated compounds to *D. magna* (US EPA 2018)

Species	48 h LC50 (μg/L)
1,2-dichloromethane	363
1,1-dichloropropane	23
1,1,2-trichloroethane	70
1,1,1-trichloroethane	530
1,1-dichloroethylene	135
Chlorobenzene	17
1,2-dichlorobenzene	0.042
Pentachlorobenzene	5.3
2-chlorophenol	5.24
2-chlorobiphenyl	0.43
DDT	0.003
Lindane	0.899
2,2′,4,4'-tetrachloro-1,1-biphenyl	30

27.6.5 Halogenated Hydrocarbons (HHCs)

HHCs, i.e., mostly chlorinated but also brominated derivatives of aliphatic or aromatic hydrocarbons, have many inevitable applications in industry as well as in our daily lives. Many hazardous HHCs can also be formed during combustion processes. In this chapter, we briefly discuss ecotoxicity of the main classes including **aliphatic HHCs**, **monoaromatic halogenated compounds** (chlorinated benzenes and similar), **polychlorinated dibezo-*p*-dioxins and -furans** (PCDD/Fs), **polychlorinated biphenyls** (PCBs), and **organochlorine pesticides**. Selected examples of ecotoxicity to a model organism *Daphnia magna* are presented in Table 27.2.

Selected chemicals from this group (originally 12 "dirty dozen" compounds and others newly added) have been identified as a global threat to the environment and health due to their persistency, bioaccumulation, long-range transport, and toxicity; thus, they are globally regulated by UNEP Stockholm Convention on POPs (Stockholm Convention 2008).

27.6.5.1 Halogenated Aliphatic Hydrocarbons (HACs)

Huge amounts of HACs are produced globally as solvents or intermediates in organic synthesis, agrochemicals, pharmaceuticals, etc. Some of the main representatives include chloro- or bromoalkanes, −alkenes, and -alkynes, vinyl chloride (precursor for the most broadly used polymer PVC), phosgene, chlorinated paraffins (large C10–C30 chains). Also some pesticides like toxaphene, mirex, and many others belong among HACs (see below). HACs are typically volatile and reactive (which decreases hazards of bioaccumulation and environmental persistence). The greatest (eco)toxicological concerns relate to occupational exposures and toxicity to humans, but HACs are also among major contaminants of soils and underground water (related to industrial activities, spills, or landfills), and may cause diverse acute and chronic effects to biota and wildlife. They act mostly through non-specific modes of action like membrane disruption or reactive toxicity with biological molecules.

27.6.5.2 Halogenated (Monocyclic) Aromatic Hydrocarbons

There are many different structural variants and applications of halo-derivatives of **benzene**, **toluene**, **phenol**, and **similar monocyclic aromatic compounds**. For example, *m*- and *p*-dichlorobenzenes or nitrated trichlorobenzenes are used as chemical intermediates and solvents, *p*-dichlorobenzene is an insecticide fumigant and disinfectant, mixtures of trichlorobenzenes are used for controlling termites, hexachlorobenzene (HCB) is a fungicide and intermediate in chemical industry, etc. Toxicity within this group of molecules is highly variable ranging from acute and chronic toxicity through nonspecific modes of action (membrane and reactive toxicity) to specific effects on the particular biota for insecticides and fungicides.

27.6.5.3 Polychlorinated Dibenzo-P-Dioxins and -Furans (PCDD/Fs)

PCDD/Fs are planar molecules with varying substituents. They are extremely stable and are among the most toxic compounds ever identified, both through acute and chronic exposures. They are also the major problematic compounds regulated by POPs Stockholm Convention. These are not commercial products but exist as by-products formed during chlorophenol synthesis and during incineration. Their environmental concentrations are in general (also in comparison to other HHCs) very low, which—fortunately—minimizes their acute toxic effects. Nevertheless, PCDD/Fs due to high lipophilicity bioaccumulate and biomagnify in food chain affecting thus mostly terminal predators (birds, mammals) after long-term chronic exposures. The key mechanism of toxic action of dioxins and furans and also other structurally related molecules (so-called "dioxin-like" toxicity—including, e.g., PCBs discussed below as well as PAHs and others) involves activation of the intracellular aryl hydrocarbon receptor (AhR), which leads to synthesis of detoxification CYP450 enzymes and deregulation of various biological functions. The main types of toxicities ascribed to dioxins and dioxin-like compounds include immune suppression (e.g., documented mass mortalities of seals and dolphins due to sensitivity to ubiquitous viruses); porphyria (disruption of synthesis of components of blood pigment hemoglobin associated with sensory disorders, paralysis, or neurotoxicity); cancer promotion; endocrine disruption (effects on synthesis, action, and degradation of sex hormones estrogens and androgens); or disruption of vitamin A metabolism (teratogenicity, growth disruption, sterility). The above-mentioned effects were widely documented in fish, amphibians, reptiles, birds as well as mammals, and also humans. Nevertheless, PCDD/Fs are highly toxic also, e.g., to invertebrates through chronic mechanisms like endocrine disruption. Because of their importance, broad diversity of mixtures, and shared toxicity mechanism through AhR, their toxicity and risks are assessed through internationally accepted mechanism using toxic equivalency factors (TEFs). Individual TEFs reflect toxic potency relative to 2,3,7,8-TCDD, the most potent dioxin. The concentrations detected in the environmental samples are then multiplied by respective TEFs, and their sum (toxic equivalents—TEQs) provides comparative information on dioxin-like risks associated with the sample.

27.6.5.4 Polychlorinated Biphenyls (PCBs)

PCBs are a group of industrial chemicals that have been produced since 1929 and broadly utilized in electrical equipment (transformers and capacitors) as well as in open applications like liquids, sealants, lubricants, paints, adhesives, plasticizers. Because of their persistent, bioaccumulative, and toxic (PBT) properties, their production and use is now banned globally, but they still undergo environmental cycling and pose environmental and health hazards. Similarly to PCDD/Fs, PCBs are highly bioaccumulative affecting primarily organisms at the top of the food chain. Many isomers (coplanar compounds, i.e., non-ortho-substituted like PCB126) are known to share toxic modes of action with PCDD/Fs with assigned TEFs. However, chronic toxic effects (like tumor promotion, endocrine disruptive properties, neurotoxicity, developmental toxicity, etc.) have also been documented at "non-coplanar" isomers, which have orders of magnitude higher levels in both products and the environment (e.g., PCB153 and others). Many of these effects could be attributed to hydroxy derivatives of PCBs formed by the metabolism.

27.6.5.5 Organochlorine Pesticides (OCPs)

As already mentioned, some of the HHCs can be used as pesticides, i.e., OCPs (Fig. 27.4). They are primarily toxic to their target pests—mainly insects. However, they also cause high acute toxicity to non-target aquatic and soil invertebrates as well

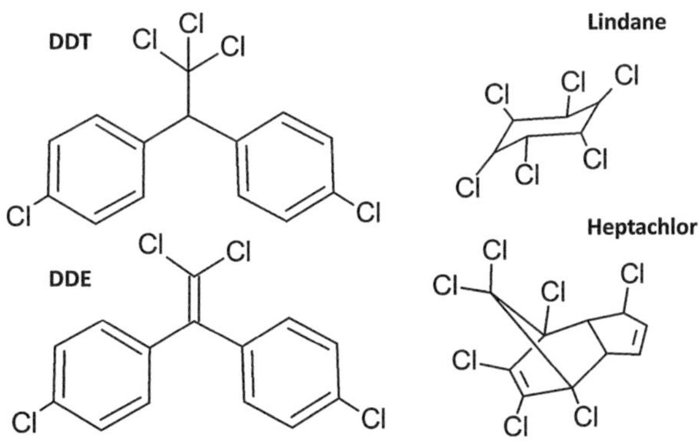

Fig. 27.4 Examples of chemical structures of organochlorine pesticides

as to fish. Because of the PBT properties, important is also their bioaccumulation and chronic toxic effects in non-target organisms at the top of the food chain like birds or mammals. The group of "old" legacy OCPs (regulated by UNEP Stockholm Convention) includes well-known DDT, which is transformed in the environment to its metabolites DDE and DDD. Other examples are so-called "drin" pesticides (aldrin, dieldrin, chlordane, endrin, mirex, toxaphene) or gamma isomer of hexachlorocyclohexane known as lindane. In addition to the acute toxicity, systemic toxic effects upon chronic exposures include (like other POP compounds) endocrine disruptive effects, developmental toxicity and teratogenicity, tumor promotion, immunosuppressions, etc.

27.6.6 *Toxic Metals*

Anthropogenic use of many metals and metalloids, such as arsenic, As, results in major environmental loads. In the following paragraphs, we briefly introduce relevant information on ecotoxicity of the selected representatives—Hg and also Cd, Pb, Cu.

27.6.6.1 Mercury

Hg is found in small concentrations in many rocks, and, therefore, certain natural background levels can be detected in soils, air, and water around the world. Its mining and industrial applications have increased significantly since the industrial revolution. Other releases typically include waste incineration, coal combustion, base metal smelting, chlor-alkali industry, small-scale gold mining, use in dental amalgam as well as use in artisanal jewelry and trinkets for tourists in certain regions. Mercury has been recognized as a serious threat within the UN Environment Minamata Convention. It is known for its characteristic environmental fate that includes formation of organo-mercury species (e.g., methyl Hg) by microorganisms in abiotic conditions like sediments, which may easily enter the food chain and become bioaccumulated. Moreover, they have also been shown more toxic than the inorganic mercury. Some well-known effects in humans include neuropathies, developmental disruptions (Minamata disease), or kidney problems. Invertebrates display variable sensitivity, and concentrations as low as 0.04 μg/L were shown to inhibit reproduction in *Daphnia magna* (Biesinger et al. 1982). For freshwater fish, inorganic mercury is not as toxic as some other metals like Cu, Pg, Zn, or Cd with 96 h LC50 ranging in hundreds of μg/L (Table 27.3).

Table 27.3 Examples of ecotoxicity values (96 h LC50) for inorganic and organic mercury to fish (US EPA 2018)

Organism	96 h LC50 (μg/L)
Inorganic mercury (mercuric chloride)	
Rainbow trout	33
Brook trout	0.3–0.9
Carp	180
Striped bass	90
Organic mercury	
Rainbow trout	24–42
Brook trout	65–75
Lamprey	48

27.6.6.2 Cadmium, Lead, and Copper

Cadmium (Cd) is a rare element, which enters the environment via several routes: refining and use of Cd, Cu and Ni smelting, and fuel combustion. It can accumulate in biota with high bioconcentration factors of 2000–4000 especially for mollusks. Ecotoxicity of Cd to aquatic biota depends on the pH and the water hardness. Acute LC50s may range from 2 to 3200 μg/L (US EPA 2018). The chronic effects in both aquatic invertebrates and vertebrates include, e.g., reduced growth or reproduction impairment occurring at low μg/L concentrations. Embryos and early life stages are commonly more sensitive under the same conditions than adults.

Lead (Pb) is found in all media and biota. Its elevated concentrations result from diverse human activities like base metal mining and smelting, combustion of leaded gasoline, use of Pb in hunting or in paints, and also from the uncontrolled disposal of Pb-containing products like batteries and electronic devices. Background environmental concentrations are low, usually below μg/L, but they may increase depending on the sources of the pollution and the environmental conditions, where the low pH and the ion content greatly affect Pb^{2+} bioavailability. Very high concentrations can be found in soil near the sources of the contamination (up to gram per kg of soil). Among all the studied species, diatoms (algae) and crustaceans were found highly sensitive (US EPA 2018), but the pronounced toxic effects like neurotoxicity or anaemia have been demonstrated in humans and other vertebrates such as birds, which may accumulate high Pb levels through the food chain.

Copper (Cu) is nowadays mined on every continent except Antarctica. Its application as an electrical conductor in wires and cables accounts for about 75% of its usage. In the form of alloys, copper can also be used in building construction (tubes, sheets, rods) or fine electroindustry. Copper is essential to various enzymes in biota, but the elevated environmental concentrations have been found highly toxic to many aquatic and marine species, especially to many plants and algae. Similarly to Cd, bioavailability and toxicity of Cu are strongly influenced by some environmental conditions like pH or water hardness. It is known for its high ecotoxicity to aquatic species with EC50 within low μg/L concentrations, and it is often used as herbicide or fungicide.

27.6.7 Synthetic Polymers and Additives

From all the possible combinations of monomers that could be used for synthesis of **polymers**, several classes are dominantly used in diverse applications—namely polycarboxylates, polyesters, styrine-acrylics, charged polymers (anionic, cationic, amphoteric), or polyamines. The major environmental concerns are related to ubiquitous occurrence and heavy pollution even in remote ocean areas, poor (bio) degradability, and currently diverse (mostly mechanic) biotic effects of small fragments of solid materials (microplastics). Direct acute ecotoxicity of the final polymers is usually low due to low bioavailability, (EC50s cannot be reached, i.e., >1000 mg/L in standard assays), but during long-term aging, leaching of both small mono- to oligomers as well as additives may cause hazards.

A special group are perfluorinated high molecular weight polymers with various advantageous properties (sold under trademarks like Teflon or Goretex). Their precursors (also referred as telomers) are **perfluorooctane sulfonic acid (PFOS)** and **perfluorooctanoic acid (PFOA)**, which are ubiquitously distributed in the environment and biota. Acute ecotoxicity of these compounds is low, but due to the stability of the perfluorochains, they may manifest chronic impacts on biota. In addition they have also been associated with chronic diseases in humans, and their global regulation is being considered under the UNEP Stockholm Convention.

Properties of polymeric materials are advanced by additions of numerous plasticizers, stabilizers, or dyes. Additives may form up to dozens of percents of the total polymeric mass. Especially **phthalates** (alkyl derivatives of phthalic acid) are ubiquitously used and distributed in the air (bound to particles), water, or soil, hence the attention of environmental scientists. Their acute toxicity and ecotoxicity is relatively low. However, many phthalates are a matter of concern due to their endocrine disruptive properties and possible chronic effects in vertebrates, mammals, and humans (Gennings et al. 2014; Howdeshell et al. 2017).

Another class of commonly used chemicals are **bisphenols**, namely bisphenol A, whose toxicity has been broadly documented in literature (e.g., reviewed in WHO/UNEP 2013). Many studies indicate toxicity in humans (miscarriages, decreased birth weight, reproductive and sexual dysfunctions, breast and prostate cancer). Similarly, chronic ecotoxicity has been documented also in other biota (Canesi and Fabbri 2015). Other bisphenols have been much less studied (Rochester and Bolden 2015) despite their relevant exposure levels.

27.6.8 Micropollutants, Emerging Contaminants, Currently Used Pesticides

Over the last decades, the occurrence of micropollutants in the environment has become a worldwide issue. Micropollutants, also termed contaminants of emerging concern, include both anthropogenic as well as natural substances such as pharmaceuticals, personal care products, steroid hormones, industrial chemicals, pesticides, and many others. Micropollutants are commonly present in the environment at trace concentrations that may range from few ng/L to several μg/L (Luo et al. 2014; Loos et al. 2010, 2013). Complex description of the ecotoxicology of these diverse compounds is beyond the scope of the present chapter; therefore, we discuss only the most prominent and well-documented examples.

Brominated flame retardants (BFRs) are polybrominated compounds broadly used as additives—flame retardants—to materials and products like plastics, textile, furniture, electronics, etc. The most important representatives are polybrominated diphenyl ethers (PBDEs), hexabromocyclododecane (HBCD), tetrabromobisphenol A (TBBPA), or polybrominated biphenyls (PBBs). They are ubiquitous contaminants with high hydrophobicity, stability, bioaccumulative potential, and toxicity. The main documented effects are—similarly to many POPs—related to chronic exposures of larger animals and include reproduction toxicity, disruption of development and growth, immunotoxicity, or other endocrine-disruptive effects. Because of their hazardous properties, some of the mentioned classes like PBDEs or HBCD have recently been added to the list of UNEP Stockholm Convention.

Pharmaceuticals and personal care products (PPCPs) have become of great research concern, especially with respect to their potential chronic toxic effects that are not usually captured in standardized acute toxicity effects (Ortiz de García et al. 2014). The most extensively studied examples include estradiol and 17-alpha-ethinylestradiol known to cause reproductive disruptions in fish at very low ng/L concentrations. Other highly potent fish endocrine and reproductive disrupters are trenbolone (androgen) or levonogestrel. Some broadly used non-steroidal anti-inflammatory drugs (pain killers) like diclofenac or ibuprofen are also known to cause kidney and liver toxicity or affect spawning in fish at low concentrations (LOEC <1 μg/L). Other examples include, e.g., transgenerational effects of antibiotics in *D. magna* (Dalla Bona et al. 2015) or ecotoxicity and genotoxicity of anticancer drugs (Isidori et al. 2016; Zounková et al. 2011).

Although legacy pesticides like DDT and other OCPs have been mostly banned, they are being replaced with **currently used pesticides (CUPs)** to assure necessary agriculture production. CUPs are used annually in large amounts and a majority enters the soil forming thus short- or long-term residues. Acute ecotoxicity of CUPs is considered low as strict procedures and risk assessment is required before their authorization. However, chronic toxicity to non-target species is of concern as recent surveys indicate non-negligible levels of residues, e.g., in arable soils (Hvězdová et al. 2018) or ground and surface waters (Loos et al. 2010, 2013). The most commonly found compounds are, e.g., triazine herbicides, conazole fungicides, chloroacetanilides, and many others.

References

ASTM (2014) E2172–01(2014) standard guide for conducting laboratory soil toxicity tests with the nematode Caenorhabditis elegans. ASTM International, West Conshohocken. https://doi.org/10.1520/E2172-01R14

Bielská L, Hovorková I, Kuta J, Machát J, Hofman J (2017) The variability of standard artificial soils: cadmium and phenanthrene sorption measured by a batch equilibrium method. Ecotoxicol Environ Saf 135:17–23. https://doi.org/10.1016/j.ecoenv.2016.09.015

Biesinger KE, Anderson LE, Eaton JG (1982) Chronic effects of inorganic and organic mercury on *Daphnia magna*: toxicity, accumulation, and loss. Arch Environ Contam Toxicol 11 (6):769–774. https://doi.org/10.1007/BF01059166

Blaha L, Holoubek I (2013) Emerging issues in ecotoxicology: persistent organic pollutants (POPs). In: Férard J-F, Blaise C (eds) Encyclopedia of aquatic ecotoxicology. Springer, Dordrecht, pp 429–436. https://doi.org/10.1007/978-94-007-5704-2_40

Canesi L, Fabbri E (2015) Environmental effects of BPA: focus on aquatic species. Dose-Response 13(3):1559325815598304. https://doi.org/10.1177/1559325815598304

Cao JL, Feng YZ, He SY, Lin XG (2017) Silver nanoparticles deteriorate the mutual interaction between maize (*Zea mays* L.) and arbuscular mycorrhizal fungi: a soil microcosm study. Appl Soil Ecol 119:307–316. https://doi.org/10.1016/j.apsoil.2017.04.011

Carvalho RN, Arukwe A, Ait-Aissa S, Bado-Nilles A, Balzamo S, Baun A, Belkin S, Blaha L, Brion F, Conti D, Creusot N, Essig Y, Ferrero VEV, Flander-Putrle V, Fürhacker M, Grillari-Voglauer R, Hogstrand C, Jonáš A, Kharlyngdoh JB, Loos R, Lundebye A-K, Modig C, Olsson P-E, Pillai S, Polak N, Potalivo M, Sanchez W, Schifferli A, Schirmer K, Sforzini S, Stürzenbaum SR, Søfteland L, Turk V, Viarengo A, Werner I, Yagur-Kroll S, Zounková R, Lettieri T (2014) Mixtures of chemical pollutants at European legislation safety concentrations: how safe are they? Toxicol Sci 141(1):218–233. https://doi.org/10.1093/toxsci/kfu118

Cattaneo AG, Gornati R, Chiriva-Internati M, Bernardini G (2009) Ecotoxicology of nanomaterials: the role of invertebrate testing. ISJ - Invertebr Surviv J 6(1):78–97

Centre Ecotox (2018) Effect-based and chemical analytical monitoring for the steroidal estrogens. http://www.ecotoxcentre.ch/projects/aquatic-ecotoxicology/monitoring-of-steroidal-estrogens. Accessed Mar 2018

Cortet J, Gomot-De Vauflery A, Poinsot-Balaguer N, Gomot L, Texier C, Cluzeau D (1999) The use of invertebrate soil fauna in monitoring pollutant effects. Eur J Soil Biol 35(3):115–134. https://doi.org/10.1016/S1164-5563(00)00116-3

Cravo-Laureau C, Cagnon C, Lauga B, Duran R (eds) (2017) Microbial Ecotoxicology. Springer, Cham. https://doi.org/10.1007/978-3-319-61795-4

Dalla Bona M, Zounková R, Merlanti R, Blaha L, De Liguoro M (2015) Effects of enrofloxacin, ciprofloxacin, and trimethoprim on two generations of *Daphnia magna*. Ecotoxicol Environ Saf 113:152–158. https://doi.org/10.1016/j.ecoenv.2014.11.018

Escher B, Leusch F (2011) Water quality assessment and whole effluent toxicity testing. In: Bioanalytical tools in water quality assessment, vol 10. IWA Publishing, pp 33–46. https://doi.org/10.2166/9781780400778

European Parliament and the Council of the European Union (2006) Regulation (EC) No 1907/2006 of the European Parliament and of the Council of 18 December 2006 concerning the Registration, Evaluation, Authorisation and Restriction of Chemicals (REACH), establishing a European Chemicals Agency, amending Directive 1999/45/EC and repealing Council Regulation (EEC) No 793/93 and Commission Regulation (EC) No 1488/94 as well as Council Directive 76/769/EEC and Commission Directives 91/155/EEC, 93/67/EEC, 93/105/EC and 2000/21/EC (Text with EEA relevance)

European Parliament and the Council of the European Union (2008) Regulation (EC) No 1272/2008 of the European Parliament and of the Council of 16 December 2008 on classification, labelling and packaging of substances and mixtures, amending and repealing Directives 67/548/EEC and 1999/45/EC, and amending Regulation (EC) No 1907/2006 (Text with EEA relevance)

European Parliament and the Council of the European Union (2009) Regulation (EC) No 1107/2009 of the European Parliament and of the Council of 21 October 2009 concerning the placing of plant protection products on the market and repealing Council Directives 79/117/EEC and 91/414/EEC

European Parliament and the Council of the European Union (2012) Regulation (EU) No 528/2012 of the European Parliament and of the Council of 22 May 2012 concerning the making available on the market and use of biocidal products Text with EEA relevance

Fernández MD, Pro J, Alonso C, Aragonese P, Tarazona JV (2011) Terrestrial microcosms in a feasibility study on the remediation of diesel-contaminated soils. Ecotoxicol Environ Saf 74 (8):2133–2140. https://doi.org/10.1016/j.ecoenv.2011.08.009

Frampton GK, Jänsch S, Scott-Fordsmand JJ, Römbke J, van den Brink PJ (2006) Effects of pesticides on soil invertebrates in laboratory studies: a review and analysis using species sensitivity distributions. Environ Toxicol Chem 25(9):2480–2489. https://doi.org/10.1897/05-438R.1

Gennings C, Hauser R, Koch HM, Kortenkamp A, Lioy PJ, Mirkes PE, Schwetz BA (2014) Report to the U.S. Consumer Product Safety Commission by the chronic Hazard advisory panel on phthalates and phthalate alternatives. U.S. Consumer Product Safety Commission, Bethesda

Ghiglione J-F, Martin-Laurent F, Pesce S (2016) Microbial ecotoxicology: an emerging discipline facing contemporary environmental threats. Environ Sci Pollut Res 23(5):3981–3983. https://doi.org/10.1007/s11356-015-5763-1

Groth VA, Carvalho-Pereira T, da Silva EM, Niemeyer JC (2016) Ecotoxicological assessment of biosolids by microcosms. Chemosphere 161:342–348. https://doi.org/10.1016/j.chemosphere.2016.07.029

Hofman J, Hovorková I, Semple KT (2014) The variability of standard artificial soils: behaviour, extractability and bioavailability of organic pollutants. J Hazard Mater 264:514–520. https://doi.org/10.1016/j.jhazmat.2013.10.039

Howdeshell KL, Hotchkiss AK, Gray LE Jr (2017) Cumulative effects of antiandrogenic chemical mixtures and their relevance to human health risk assessment. Int J Hyg Environ Health 220 (2, Part A):179–188. https://doi.org/10.1016/j.ijheh.2016.11.007

Hvězdová M, Kosubová P, Košíková M, Scherr KE, Šimek Z, Brodský L, Šudoma M, Škulcová L, Sáňka M, Svobodová M, Krkošková L, Vašíčková J, Neuwirthová N, Bielská L, Hofman J (2018) Currently and recently used pesticides in central European arable soils. Sci Total Environ 613–614:361–370. https://doi.org/10.1016/j.scitotenv.2017.09.049

Isidori M, Piscitelli C, Russo C, Smutná M, Bláha L (2016) Teratogenic effects of five anticancer drugs on *Xenopus laevis* embryos. Ecotoxicol Environ Saf 133:90–96. https://doi.org/10.1016/j.ecoenv.2016.06.044

ISO (1995) ISO 10712:1995 Water quality — Pseudomonas putida growth inhibition test (Pseudomonas cell multiplication inhibition test)

ISO (2012a) ISO 11268-1:2012 Soil quality — Effects of pollutants on earthworms — Part 1: Determination of acute toxicity to Eisenia fetida/Eisenia andrei

ISO (2012b) ISO 11268-2:2012 Soil quality — Effects of pollutants on earthworms — Part 2: Determination of effects on reproduction of Eisenia fetida/Eisenia andrei
ISO (2012c) ISO 11269-1:2012 Soil quality — Determination of the effects of pollutants on soil flora — Part 1: Method for the measurement of inhibition of root growth
ISO (2012d) ISO 14238:2012 Soil quality — Biological methods — Determination of nitrogen mineralization and nitrification in soils and the influence of chemicals on these processes
ISO (2012e) ISO 15685:2012 Soil quality — Determination of potential nitrification and inhibition of nitrification — Rapid test by ammonium oxidation
ISO (2014a) ISO 11267:2014 soil quality — inhibition of reproduction of Collembola (Folsomia candida) by soil contaminants
ISO (2014b) ISO 16387:2014 Soil quality — Effects of contaminants on Enchytraeidae (Enchytraeus sp.) — Determination of effects on reproduction
Kammenga JE, Dallinger R, Donker MH, Köhler HR, Simonsen V, Triebskorn R, Weeks JM (2000) Biomarkers in terrestrial invertebrates for Ecotoxicological soil risk assessment. Rev Environ Contam Toxicol 164:93–147
Korsloot A, van Gestel CAM, van Straalen NM (2004) Environmental stress and cellular response in arthropods. CRC Press, Hoboken
Lagadic L, Caquet T (1998) Invertebrates in testing of environmental chemicals: are they alternatives? Environ Health Perspect 106(Suppl 2):593–611
Loos R, Locoro G, Comero S, Contini S, Schwesig D, Werres F, Balsaa P, Gans O, Weiss S, Blaha L, Bolchi M, Gawlik BM (2010) Pan-European survey on the occurrence of selected polar organic persistent pollutants in ground water. Water Res 44(14):4115–4126. https://doi.org/10.1016/j.watres.2010.05.032
Loos R, Carvalho R, António DC, Comero S, Locoro G, Tavazzi S, Paracchini B, Ghiani M, Lettieri T, Blaha L, Jarosova B, Voorspoels S, Servaes K, Haglund P, Fick J, Lindberg RH, Schwesig D, Gawlik BM (2013) EU-wide monitoring survey on emerging polar organic contaminants in wastewater treatment plant effluents. Water Res 47(17):6475–6487. https://doi.org/10.1016/j.watres.2013.08.024
LUFA (2018) Speyer Standardböden https://www.lufa-speyer.de/index.php/dienstleistungen/standardboeden/8-dienstleistungen/artikel/57-standard-soils. Accessed July 2018
Luo Y, Guo W, Ngo HH, Nghiem LD, Hai FI, Zhang J, Liang S, Wang XC (2014) A review on the occurrence of micropollutants in the aquatic environment and their fate and removal during wastewater treatment. Sci Total Environ 473–474:619–641. https://doi.org/10.1016/j.scitotenv.2013.12.065
Mackay D, Hughes DM, Romano ML, Bonnell M (2014) The role of persistence in chemical evaluations. Integr Environ Assess Manage 10(4):588–594. https://doi.org/10.1002/ieam.1545
Maier D, Blaha L, Giesy JP, Henneberg A, Köhler H-R, Kuch B, Osterauer R, Peschke K, Richter D, Scheurer M, Triebskorn R (2015) Biological plausibility as a tool to associate analytical data for micropollutants and effect potentials in wastewater, surface water, and sediments with effects in fishes. Water Res 72:127–144. https://doi.org/10.1016/j.watres.2014.08.050
Moser H, Römbke J (eds) (2009) Ecotoxicological characterization of waste: results and experiences of an international ring test. Springer, New York. https://doi.org/10.1007/978-0-387-88959-7
Murtaza G, Javed W, Hussain A, Qadir M, Aslam M (2017) Soil-applied zinc and copper suppress cadmium uptake and improve the performance of cereals and legumes. Int J Phytoremediation 19(2):199–206. https://doi.org/10.1080/15226514.2016.1207605
Newman MC (2014) Fundamentals of ecotoxicology: the science of pollution, 4th edn. CRC Press, Boca Raton
OECD (1984) Test no. 207: earthworm, acute toxicity tests. In: OECD guidelines for the testing of chemicals, section 2. OECD Publishing, Paris. https://doi.org/10.1787/9789264070042-en
OECD (1992) Test no. 203: fish, acute toxicity test. In: OECD guidelines for the testing of chemicals, section 2. OECD Publishing, Paris. https://doi.org/10.1787/9789264069961-en

OECD (2000a) Test no. 216: soil microorganisms: nitrogen transformation test. In: OECD guidelines for the testing of chemicals, section 2. OECD Publishing, Paris. https://doi.org/10.1787/9789264070226-en

OECD (2000b) Test no. 217: soil microorganisms: carbon transformation test. In: OECD guidelines for the testing of chemicals, section 2. OECD Publishing, Paris. https://doi.org/10.1787/9789264070240-en

OECD (2004) Test no. 202: Daphnia sp. acute immobilisation test. In: OECD guidelines for the testing of chemicals, section 2. OECD Publishing, Paris. https://doi.org/10.1787/9789264069947-en

OECD (2006a) Test no. 208: terrestrial plant test: seedling emergence and seedling growth test. In: OECD guidelines for the testing of chemicals, section 2. OECD Publishing, Paris. https://doi.org/10.1787/9789264070066-en

OECD (2006b) Test no. 227: terrestrial plant test: vegetative vigour test. In: OECD guidelines for the testing of chemicals, section 2. OECD Publishing, Paris. https://doi.org/10.1787/9789264067295-en

OECD (2009) Test no. 232: collembolan reproduction test in soil. OECD Publishing, Paris. https://doi.org/10.1787/9789264076273-en

OECD (2011) Test no. 201: freshwater alga and cyanobacteria, growth inhibition test. In: OECD guidelines for the testing of chemicals, section 2. OECD Publishing, Paris. https://doi.org/10.1787/9789264069923-en

OECD (2012) Test no. 211: Daphnia magna reproduction test. In: OECD guidelines for the testing of chemicals, section 2. OECD Publishing, Paris. https://doi.org/10.1787/9789264185203-en

OECD (2013) Test no. 236: fish embryo acute toxicity (FET) test. In: OECD guidelines for the testing of chemicals, section 2. OECD Publishing, Paris. https://doi.org/10.1787/9789264203709-en

OECD (2016a) Test no. 220: Enchytraeid reproduction test. In: OECD guidelines for the testing of chemicals, section 2. OECD Publishing, Paris. https://doi.org/10.1787/9789264264472-en

OECD (2016b) Test no. 222: earthworm reproduction test (Eisenia fetida/Eisenia andrei). In: OECD guidelines for the testing of chemicals, section 2. OECD Publishing, Paris. https://doi.org/10.1787/9789264264496-en

Oprsal J, Blaha L, Pouzar M, Knotek P, Vlcek M, Hrda K (2015) Assessment of silver nanoparticle toxicity for common carp (*Cyprinus carpio*) fish embryos using a novel method controlling the agglomeration in the aquatic media. Environ Sci Pollut Res 22(23):19124–19132. https://doi.org/10.1007/s11356-015-5120-4

Ortiz de García SA, Pinto Pinto G, García-Encina PA, Irusta-Mata R (2014) Ecotoxicity and environmental risk assessment of pharmaceuticals and personal care products in aquatic environments and wastewater treatment plants. Ecotoxicology 23(8):1517–1533. https://doi.org/10.1007/s10646-014-1293-8

Rochester JR, Bolden AL (2015) Bisphenol S and F: a systematic review and comparison of the hormonal activity of Bisphenol a substitutes. Environ Health Perspect 123(7):643–650. https://doi.org/10.1289/ehp.1408989

Roose-Amsaleg C, Laverman AM (2016) Do antibiotics have environmental side-effects? Impact of synthetic antibiotics on biogeochemical processes. Environ Sci Pollut Res 23(5):4000–4012. https://doi.org/10.1007/s11356-015-4943-3

Sandermann H Jr (2004) Molecular ecotoxicology of plants. Trends Plant Sci 9(8):406–413. https://doi.org/10.1016/j.tplants.2004.06.001

Stockholm Convention (2008) Stockholm Convention on Persistent Organic Pollutants http://chm.pops.int/. Accessed Mar 2018

Tousova Z, Oswald P, Slobodnik J, Blaha L, Muz M, Hu M, Brack W, Krauss M, Di Paolo C, Tarcai Z, Seiler T-B, Hollert H, Koprivica S, Ahel M, Schollée JE, Hollender J, Suter MJF, Hidasi AO, Schirmer K, Sonavane M, Ait-Aissa S, Creusot N, Brion F, Froment J, Almeida AC, Thomas K, Tollefsen KE, Tufi S, Ouyang X, Leonards P, Lamoree M, Torrens VO, Kolkman A, Schriks M, Spirhanzlova P, Tindall A, Schulze T (2017) European demonstration program on the effect-based and chemical identification and monitoring of organic pollutants in European surface waters. Sci Total Environ 601–602:1849–1868. https://doi.org/10.1016/j.scitotenv.2017.06.032

US EPA (1996) Federal Insecticide, Fungicide, and Rodenticide Act. 7 U.S.C. §136 et seq. https://www.epa.gov/laws-regulations/summary-federal-insecticide-fungicide-and-rodenticide-act. Accessed Aug 2018

US EPA (2011) Exposure Factors Handbook 2011 Edition (Final Report). EPA/600/R-09/052F. National Center for Environmental Assessment, Washington, DC

US EPA (2018) ECOTOX user guide: ECOTOXicology knowledgebase system. Version 5.0. Available: https://cfpub.epa.gov/ecotox/. Accessed Mar 2018

van Beelen P, Doelman P (1997) Significance and application of microbial toxicity tests in assessing ecotoxicological risks of contaminants in soil and sediment. Chemosphere 34 (3):455–499. https://doi.org/10.1016/S0045-6535(96)00388-8

Vašíčková J, Kalábová T, Komprdová K, Priessnitz J, Dymák M, Lána J, Škulcová L, Šindelářová L, Sáňka M, Čupr P, Vácha R, Hofman J (2013) Comparison of approaches towards ecotoxicity evaluation for the application of dredged sediment on soil. J Soils Sediments 13 (5):906–915. https://doi.org/10.1007/s11368-013-0670-x

Vašíčková J, Váňa M, Komprdová K, Hofman J (2015) The variability of standard artificial soils: effects on the survival and reproduction of springtail (*Folsomia candida*) and potworm (*Enchytraeus crypticus*). Ecotoxicol Environ Saf 114:38–43. https://doi.org/10.1016/j.ecoenv.2015.01.007

Walker CH, Sibly RM, Hopkin SP, Peakall DB (2012) Principles of ecotoxicology, 4th edn. CRC Press

WHO/UNEP (2013) State of the Science of Endocrine Disrupting Chemicals – 2012

Yadav T, Mungray AA, Mungray AK (2016) Effect of multiwalled carbon nanotubes on UASB microbial consortium. Environ Sci Pollut Res 23(5):4063–4072. https://doi.org/10.1007/s11356-015-4385-y

Zounková R, Klimešová Z, Nepejchalová L, Hilscherová K, Bláha L (2011) Complex evaluation of ecotoxicity and genotoxicity of antimicrobials oxytetracycline and flumequine used in aquaculture. Environ Toxicol Chem 30(5):1184–1189. https://doi.org/10.1002/etc.486

Chapter 28
Ecotoxicity of Nanomaterials Used for Remediation

Claire Coutris, Alena Ševců, and Erik J. Joner

Abstract Remediation using nanoparticles depends on proper documentation of safety aspects, one of which is their ecotoxicology. Ecotoxicology of nanoparticles has some special features; while traditional ecotoxicology aims at measuring possible negative effects of more or less soluble chemicals or dissolved elements, nanoecotoxicology aims at measuring the toxicity of particles, and its main focus is on effects that are unique to nano-sized particles, as compared to larger particles or solutes. One of the main challenges when testing the ecotoxicity of nanoparticles lies in maintaining stable and reproducible exposure conditions, and adapting these to selected test organisms and endpoints. Another challenge is to use test media that are relevant to the matrices to be treated. Testing of nanoparticles used for remediation, particularly redox-active Fe-based nanoparticles, should also make sure to exclude confounding effects of altered redox potential that are not nanoparticle-specific. Yet another unique aspect of nanoparticles used for remediation is considerations of ageing of nanoparticles in soil or water, leading to reduced toxicity over field-relevant time scales. This review discusses these and other aspects of how to design and interpret appropriate tests and use these in hazard descriptions for subsequent risk assessments.

Keywords Environment · Nanoparticles · Organic pollutants · Polluted soil · Toxicity

---

C. Coutris · E. J. Joner (✉)
Division of Environment and Natural Resources, Norwegian Institute of Bioeconomy Research, Ås, Norway
e-mail: erik.joner@nibio.no

A. Ševců
Institute for Nanomaterials, Advanced Technologies and Innovation, Technical University of Liberec, Liberec, Czech Republic

J. Filip et al. (eds.), *Advanced Nano-Bio Technologies for Water and Soil Treatment*, Applied Environmental Science and Engineering for a Sustainable Future, https://doi.org/10.1007/978-3-030-29840-1_28

28.1 Introduction

Ecotoxicology for evaluating possible negative effects of nanomaterials (nanoecotoxicology) has some special features; while traditional ecotoxicology aims at measuring possible negative effects of more or less soluble chemicals or dissolved elements, nanoecotoxicology aims at measuring the toxicity of particles. In addition, nanoecotoxicology has its main focus on those effects that are unique to nano-sized particles, as compared to larger particles or solutes. For this reason, experiments in nanoecotoxicology usually compare the results with effects caused by larger particles with similar composition. Another common comparison is that to the effects of dissolved ions of the same elements constituting the nanoparticles tested, since many metal-based nanoparticles may partly dissolve, and the toxic effects can be due to their soluble ionic component. These so-called control treatments are not always easy to establish, or they may result in imperfect comparisons, as larger-scale particles (often referred to as "bulk material") may behave quite differently due to their larger size, and soluble salts of elements found in many nanomaterials may not exist, or may precipitate during the tests (Kahru and Dubourguier 2010; Handy et al. 2012a; Sørensen et al. 2015).

Exposing organisms to nanomaterials requires stable suspensions of these nanomaterials. This is typically obtained through the use of surface-active agents reducing the attractive forces between particles (Labille and Brant 2010), causing them to remain suspended in water or other media for a period of time that would permit absorption or other interactions causing harm to the test organism. These surfactants may themselves affect the test organisms, and thereby the test outcome. Control treatments used for comparisons should therefore take this into account.

28.2 Toxicity of Particles

Particle toxicity can be rather different from toxicity of soluble substances or ions. This is due to the strong barriers against the uptake of particles that most organisms possess. Ions enter organisms through channels (transporters) in the cell membranes, which can discriminate the uptake based on characteristics like charge and size. Organic molecules may either pass through uptake channels for organic nutrients or cross bilayer membranes due to their hydrophobicity. Nanoparticles fit none of these routes of transport, and are therefore relegated to entering cells by random "back doors", like compromised cell membranes or accidental passive uptake. The size of the particles in question is then of course of major importance for such uptake. Most nanoparticles used for remediation are found among the larger nanoparticles, typically >50–200 nm. To put this in perspective, a 20 nm silver nanoparticle would contain 750,000 atoms (Oughton et al. 2008), each atom approximately twice as big as, e.g., a zinc ion that enters cells through ion channels.

Needless to say, most of the regular paths for entering into cells are not permitting the entry of nanoparticles. One exception here may be endocytosis.

Yet, some nanoparticles find their way through cell walls and membranes, and end up inside cells. Here they represent quite a different type of toxicity than dissolved ions of the same elements, partly because they constitute discrete particles rather than diffusive ions that may spread among cells. A particle is likely to stay inside the cell it has entered and end up in lysosomes (in the case of eukaryotic cells), but may, e.g., dissolve and represent a steady source of dissolved ions that may be harmful. The concentration of such ions is likely to be substantially higher in a cell containing a nanoparticle than in a cell being exposed (along with neighboring cells) to a similar level of dissolved elements that may move more or less freely within the exposed tissue.

Nanoparticle toxicity mechanisms include effects of dissolved ions from metallic nanoparticles (common for Ag nanoparticles that release Ag^{+}), induction of reactive oxygen species (ROS and other types of oxidative stress; such effects have been shown for iron-based nanoparticles (Lewinski et al. 2008)) and damages directly related to the surface and shape of nanoparticles (e.g., nanotubes and their asbestos-like induction of damages on cells). For Fe-based nanoparticles, only mechanisms related to ROS-formation and oxidative stress have been described as direct effects. These have been reviewed recently (Lei et al. 2018), and will not be detailed further here. Indirect effects on O_2 availability is another mechanism, but as we argue below, this is not a nanoparticle-specific toxic effect.

28.2.1 Ageing and Other Time-Dependent Modifications of Toxicity

A particular aspect of nanoparticle toxicity testing that is relevant for nano-sized zero-valent iron (nZVI) and other nanoparticles to be used for remediation is reduction in toxicity over time. While all nanoparticles are subject to changes in surface properties as a result of interactions with environmental matrices, nanoparticles for remediation are frequently designed to lose their reactivity by interacting with environmental pollutants. Including temporal changes in toxicity during testing is therefore a particularly relevant aspect that should be assessed for such materials. Reduced toxicity as a result of ageing in soil has indeed been demonstrated for nZVI aged for 30 days in soil, using growth (body weight) of earthworms as an endpoint (El-Temsah and Joner 2012a) and partly for rice after 2–4 weeks ageing (Wang et al. 2016). In the natural environment, living organisms will mostly be exposed to aged nZVI, and not to pristine particles. This is important to keep in mind when designing toxicity studies. Not only may such tests show that adverse effects of the nanoparticles are short-lived, but it may also be helpful in designing nanoparticles for remediation as toxicity and reactivity against pollutants are likely to be strongly linked (Hjorth et al. 2017).

Different types of nanomaterials may affect organisms differently. This is yet another aspect to consider when choosing how a given nanomaterial is tested with respect to ecotoxicity. For approval of new nanomaterials or for conducting risk assessments, a set of minimum three tests with contrasting organisms must be carried out (Baun et al. 2009).

28.3 Choice of Test Organisms

The choice of test organisms is important for several reasons, and may ultimately determine the outcome of a testing scheme. First, the choice of organisms must be relevant for the matrix to be treated. If nanomaterials for treating polluted soil are to be tested, soil organisms should be chosen. Similarly, freshwater and marine organisms are relevant to their native habitats. Within these three major organism habitats, there may be some overlap, or it may be relevant to include organisms from two groups as a remediation situation can affect more than one matrix: treated soil may lead to nanomaterials ending up in nearby ponds and streams, or streams and rivers may reach brackish or saltwater habitats.

When choosing test organisms within these major groups, there are at least three key aspects to consider:

- How contact with the tested material may occur
- Which endpoints are available to assess effects
- Which trophic level the organism belongs to (and how this will affect exposure).

Ecotoxicity can be strongly affected by the mode of exposure. Dermal contact is commonly affecting an organism less than ingestion or interference with respiratory organs. This distinction is less relevant for, e.g., microorganisms and plants, but even for microorganisms and plants that have no intestines, internalization may occur and cause different toxicity than surface contact. In many cases, the nature of the organism's natural habitat and the test design will determine the mode of exposure. Plants may, e.g., be exposed in an aqueous suspension (seed germination tests and hydroponic plant cultures), or in solid matrices with more or less resemblance to a real soil at a site to be treated (El-Temsah and Joner 2012b). While exposure in aqueous suspensions may say something about the inherent toxicity of the nanoparticles tested, it will give a far higher exposure than equivalent tests using soil, and should thus include appropriate exposure estimates when questions of risks are addressed. When testing toxicity of nanoparticles to plants or soil organisms using soil as an exposure medium, the choice of a test soil is also decisive, as the relative amount of different soil constituents may vary considerably and affect both bioavailability of particles and whether plants or soil organisms thrive in them. Using a soil with a minimum of soil organic matter will go a long way to ensure that plants germinate and grow in them, or that earthworms are active in ingesting soil during a test. But organic matter in soil may also result in a different availability of nanoparticles compared to a sub-soil void of humus, which is far more representative of soils being remediated using such particles. This is a trade-off situation where test organisms and test media should be selected as to be appropriate for the purpose of the test.

Another example of exposure control concerns earthworms. Dermal exposure of earthworms is usually measured by dissolving or suspending the material to be tested in water that imbibes a filter paper lining a glass vial where worms are placed (OECD 1984). Exposure through ingestion, on the other hand, uses a soil matrix where the material to be tested is mixed in. As in the example of exposing plant roots to nanoparticles in water or soil, exposure conditions for earthworms also differ greatly between water and soil. However, for worms the exposure matrix also determines mode of contact: nanoparticles suspended in water mainly result in dermal exposure, while nanoparticles mixed into soil or feed result in intestinal exposure plus dermal contact (Lapied et al. 2010). To relate data from the rapid and inexpensive dermal tests to test made with soil where bioavailability of nanoparticles is reduced by interactions with the soil components, one may perform dermal contact tests in soil by preventing worms from ingesting soil by gluing shut their mouths using super glue. The contribution from intestinal exposure may then be found by comparing worms with and without glued mouths.

For many test organisms, exposure through ingested material may differ widely according to how nanoparticles are introduced into the test system. Here, nanoparticles mixed into feed may result in far higher exposure than if directly mixed into soil. While certain earthworms (epigeic and anecic worms) seek out organic debris when they forage, other worms (endogeic worms) ingest soil and feed on the evenly distributed organic matter therein. Thus, exposing earthworms, e.g., to nanoparticles contained in organic feed or mixed homogeneously into soil that represents a volume that would frequently be at least 50 times higher may result in very different rates of uptake. Similarly, nematodes may be exposed to nanoparticles adsorbed onto bacteria upon which they feed, or through a suspension of nanoparticles where no prior association between bacteria and nanoparticles has occurred (Kleiven et al. 2018). To maximize ingestion by nematodes or other particle feeders (e.g., *Daphnia*), the test may omit the feed (e.g., bacteria and algae, respectively), but this may cause constipation and blockage of the digestive tract of the test organisms, and adverse outcomes that are caused by excessively high availability of nanoparticles (Roberts et al. 2007). In a more realistic exposure situation, the organisms would ingest mainly digestible particles that would ensure normal gut passage.

Aggregation (including agglomeration) of nanoparticles during exposure in aqueous media is a major determinant of exposure when particle uptake is size-dependent. Both medium constituents, particularly divalent ions like Ca^{2+}, excretions from test organisms and pH changes may cause this (Keller et al. 2012; Baker et al. 2014). Benthic organisms typically feeding on larger particles may potentially experience higher exposure due to aggregation, as discrete nanoparticles may be too small to be perceived as food. As mentioned above, surface-active compounds may reduce aggregation and counteract aggregation effects, and even organisms may cause dispersion by producing organics that stabilize nanoparticles in suspension (Unrine et al. 2012).

28.3.1 Endpoint Selection

An ecotoxicological endpoint is the parameter measured as a response to a potentially toxic substance. Numerous endpoints may be used when assessing the effects on a test organism. For many organisms, mortality or growth rate are rather coarse endpoints used for testing acute toxicity, while enzymatic activities (e.g., of anti-oxidative enzymes), genetic mutations, or expression of genes related to damage repair or stress are gradually more sensitive test endpoints, permitting detection of more subtle and chronic adverse effects at lower concentrations (Walker et al. 2001). The interpretations of the responses to the endpoints with less obvious toxicity functions are however a minefield. Are they, e.g., altered behavior, avoidance, expression of stress-related genes, or enhanced frequencies of apoptosis indicators of toxicity? For example, if nZVI is introduced into soil or a test system, the redox conditions may change rapidly as to cause oxidative stress (or irreversible organismal damage) due to reactions of free O_2 with nZVI. But frequently such conditions are short-lived (El-Temsah et al. 2013), and oxygen will diffuse in from the border zones and re-establish oxygenated conditions and alleviate the stress caused by the reduced O_2 availability. For those organisms that have survived the period of reduced O_2 availability, the effects may be fully reversible, with no negative impacts on populations or communities (Nguyen et al. 2018a). Thus, if the exposure ceases and the organism has avoided it or only passed through a period with sub-optimal living conditions (stress) due to the nanoparticle exposure, it is not appropriate to interpret this as toxicity.

A particular consideration to make when it comes to testing of nZVI and other nanoparticles for remediation that may affect oxygen availability to organisms is the fact that these nanoparticles may cause a lack of oxygen needed by aerobic organisms during respiration. nZVI may, e.g., react with the available O_2 in the test medium as to render the conditions anoxic, thus asphyxiating the test organisms. This is particularly relevant for exposure in water and wet soil where O_2 diffusion and replenishment is slow. Such induction of anaerobic conditions and its detrimental effects on aerobic organisms is not a nano-specific effect (though the dynamics of O_2 consumption may differ between nanoparticles and similarly reductive chemicals/bulk-size particles due to the specific surface area and chemical reactivity). The effects nZVI may have on alternative electron acceptors (NO_3, SO_4, oxidized forms of Mn, etc.) can similarly preclude the use of anaerobic test organisms to circumvent the need for O_2 during testing.

Testing for nanoparticle-specific effects when the nanoparticles to be tested cause changes in the availability of O_2 or other electron acceptors require the use of control treatments that have comparable redox conditions, or the use of test systems or exposure matrices that buffer against such changes, coupled with appropriate monitoring of redox potentials during the tests.

Nanoparticles that form colored suspensions may lead to a particular set of confounding effects related to shading of light. This is relevant for algae and other photosynthetic organisms that may experience lower light availability when used in tests where nanoparticle suspensions are dense enough to reduce transmittance (Handy et al. 2012b; Hjorth et al. 2016; Nguyen et al. 2018b). Algal growth rates or measurements of photosynthesis, chlorophyll content and related endpoints should thus account for confounding effects of shading.

28.3.2 *Trophic Interactions*

The trophic level of an organism can determine the way it is exposed to nanomaterials. This is partly due to the feeding habits of organisms at different trophic levels. Free-living microalgae may absorb nanoparticles directly from the water suspending the particles, but particles may also affect the algae by affecting the amount of available nutrient ions, or by shading the algae from light as to reduce photosynthesis. A filter feeder grazing on these algae may experience a similar concentration of nanomaterials through contact with water, but will in addition ingest, e.g., the aforementioned microalgae that may contain nanomaterials. Depending on whether bioaccumulation (increased concentrations in organisms with increasing lifetime) or biomagnification (predators accumulating higher concentrations than found in their prey) occurs, the next level predator may experience different exposure through the ingested food. So far, bioaccumulation has been observed for some nanomaterials (Petersen et al. 2008; Wang et al. 2013), and in some cases even biomagnification (Judy et al. 2011; Majumdar et al. 2016; Gupta et al. 2017). No such studies have been made with nanomaterials used for remediation. In some cases, the bioaccumulation may be due to the fact that an element contained in the nanoparticles in question is a micronutrient that the organism needs and scavenges for, as observed for cobalt nanoparticles (Coutris et al. 2012). Iron, found in many nanoparticles used for remediation, is likely to behave similarly if test organisms are experiencing sub-optimal iron supply.

28.4 Standardized or Non-standardized Tests?

A major part of the research on nanoparticle toxicity has been made using non-standardized tests, in the sense that they do not follow test protocols approved by standardization organizations like the Organisation for Economic Co-operation and Development (OECD) and International Organization for Standardization (ISO). Non-standardized tests have the advantage of choosing freely among organisms, endpoints, and exposure media. This allows for exposure optimization and exploitation of the vast knowledge on biota, ranging from their behavior, physiology, metabolism, reproduction, and genetic peculiarities to community dynamics and ecosystem functions when interacting with their habitat. Non-standardized tests may thus be best suited to elucidate toxicity mechanisms or describe pertinent environmental consequences of spreading potentially toxic nanoparticles. In comparison, standardized tests are limited to easily culturable organisms exposed under well-defined conditions, using a limited number of rather crude endpoints. The advantages of using standardized tests are that the results can easily be compared with those obtained for other chemicals, which in turn permits hazard classification, and that standardized test results can easily be used for product documentation when chemicals are used in commercial products requiring approval regarding possible negative environmental effects.

28.4.1 Standardized Testing Methods

OECD and ISO have published a number of test guidelines (TGs) that describe in all details how chemicals testing for approval of new chemicals should be conducted. These tests have been developed for soluble chemicals and have not taken into account considerations that may be important for toxicity testing of nanoparticles. Yet, the OECD has concluded that the approaches for testing and assessment of traditional chemicals are in general appropriate for assessing nanomaterials, but that the tests may have to be adapted to the specificities of nanomaterials (see ref. OECD). This concerns, e.g., methods of sample preparation, particularly regarding homogenization and distribution of the nanoparticles in the test media. Similarly, adaptations may be needed for certain test guidelines.

The first step is the preparation of a stable suspension of nanomaterials (see references in Hund-Rinke et al. 2016). This can be obtained through the use of surface-active agents reducing the attractive forces between particles and causing them to remain suspended in water or other media for a period of time that would permit absorption or other interactions causing harm to the test organism. These surfactants may themselves affect the test organisms, and thereby the test outcome, and should therefore be included in control treatments.

It is a common requirement of standardized test guidelines that exposure concentrations should remain stable (often stated as no more than 20% deviation between exposure concentrations and nominal concentrations) over the duration of the test. This is a challenge with nanomaterials, which easily agglomerate and sediment out of a water column (to mention the case of aquatic tests), exposing pelagic organisms to lower concentrations and benthic organisms to higher concentrations than originally intended. Several factors influence the agglomerating and settling behavior of particles, such as agitation of the test system, ionic strength, pH, presence of specific ligands/chelating agents, and organic matter content of the exposure medium. The following modifications have been proposed by Hund-Rinke et al. (2016) for maintaining (more) constant exposure conditions in test systems:

- Conduct the OECD TG 202—acute immobilization of *Daphnia magna* test (OECD 2004) at pH values enabling more stable dispersions of nanomaterials and use a growth medium with very low ionic strength, e.g., very soft EPA medium. This medium has been shown to allow normal growth and reproduction of *D. magna*.
- In the OECD TG 210—fish, early-life stage toxicity test (OECD 2013) with zebrafish, improve nanomaterial dispersion by using exposure chambers coupled with water changes every 24 h.
- In tests using spiked soil or sediment, i.e., OECD TG 216—nitrogen transformation test (OECD 2000a); OECD TG 217—carbon transformation test (OECD 2000b); OECD TG 220—enchytraeid reproduction test (OECD 2016a); OECD TG 222—earthworm reproduction test *Eisenia fetida/Eisenia andrei* (OECD 2016b); OECD TG 225—sediment-water lumbriculus toxicity test (OECD 2007), add nanomaterials to each replicate, to ensure homogeneity of spiking. Exceptions can be made for low concentrations, where this modification can be difficult to implement.

There has been a concern that some OECD test guidelines were not suited for the detection of toxic effects, in the sense that they would underestimate the potential toxicity of some nanomaterials. One way this underestimation could occur is by reduction of the bioavailability of nanomaterials and their transformation products due to sorption to organic matter or the elevated pH. Underestimation of the toxicity of nanomaterials can also occur when the duration of the test is too short compared to the slow transformation of nanomaterials in soil, which can be the source of toxic chemical species. The following modifications proposed by (Hund-Rinke et al. 2016) may minimize the interference of the nanomaterials with the components of the test media or the toxicity endpoints:

- OECD TG 201—freshwater alga and cyanobacteria, growth inhibition test (OECD 2011): the chelating agent EDTA can interfere with metal nanomaterials and a modified EDTA-free version of the OECD algal medium (OECD-M) for *Raphidocelis subcapitata* is proposed.
- For the OECD TG 201 (OECD 2011), it is recommended to measure biomass by determination of *in vitro* chlorophyll a, instead of optical density and *in vivo* fluorescence measurements or cell counting by hemocytometry.
- OECD TG 216—nitrogen transformation test (OECD 2000a): for the testing of ion-releasing metal nanomaterials, the pH of the soils should be at the lower end of the range accepted according to the test guideline (pH 5.5). It is also proposed to extend the duration of the test to 56 days, since some nanomaterials only show effects after ageing, and to include multiple short-term measurements of the potential ammonium activity, instead of single measurements at the start and the end of the test.

- OECD TG 217—carbon transformation test (OECD 2000b): similar modifications are proposed for this test, except the multiple short-term measurements.

28.4.2 Fe-Based Nanoparticles Exempt from Nano-Fear

nZVI and Fe-oxide-based nanoparticles have, to some extent, dodged the skepticism that clings to other types of nanoparticles. This is partly because nanoparticles for remediation are used to treat and remove the harmful effects of toxic pollutants, thus reducing the exposure of humans and the environment to highly toxic and mobile chemicals like TCE (trichloroethylene) for which there are no doubts of adverse effects or environmental exposure. Further, Fe-based nanoparticles have repeatedly been shown to have limited or even very limited mobility in the environment, restricting movement away from the treated areas (Johnson et al. 2013), which are often fenced in and unavailable to the public. Thus, risks appear confined to the treated areas. A third point in favor of Fe-based nanoparticles comes from the fact that most natural environments contain ample amounts of Fe, even in forms similar to those coming out of nanoremediation treatments. The products of nZVI aged in aerated water are mainly Fe_3O_4 (magnetite) and γ-Fe_2O_3 (maghemite), accompanied by γ-FeOOH (goethite). If corrosion continues, the products are predominantly γ-FeOOH, with small amounts of Fe_3O_4 and γ-Fe_2O_3 (Liu et al. 2015). The final aqueous corrosion product of nZVI is FeOOH (Pullin et al. 2017; Lei et al. 2018). Finally, Fe-based nanoparticles have been scrutinized in several research projects in parallel to their development, and the outcome of the ecotoxicity measurements as well as practitioners feedbacks indicate that Fe-based nanoparticles are causing low concern, if any (Bardos et al. 2011; Hjorth et al. 2017).

28.4.3 Ecotoxicity Does Not Equal Risk

Risk is the product of hazard (ecotoxicity) multiplied by the probability of encountering hazard (exposure). As mobility of nZVI and other Fe-based nanoparticles is limited, and as their prescribed use targets subsoils at several meters depth in industrial brownfields, the risk to humans (apart from workers exposed during production, transport, and deployment) and wildlife is extremely low or inexistent. The low hazard level due to the low inherent toxicity of the Fe-based nanomaterials of course also contributes to the low risks.

References

Baker TJ, Tyler CR, Galloway TS (2014) Impacts of metal and metal oxide nanoparticles on marine organisms. Environ Pollut 186:257–271. https://doi.org/10.1016/j.envpol.2013.11.014

Bardos P, Bone B, Elliott D, Hartog N, Henstock J, Nathanail P (2011) A risk/benefit approach to the application of Iron nanoparticles for the remediation of contaminated sites in the environment. Defra, London

Baun A, Hartmann NB, Grieger KD, Hansen SF (2009) Setting the limits for engineered nanoparticles in European surface waters – are current approaches appropriate? J Environ Monit 11(10):1774–1781. https://doi.org/10.1039/B909730A

Coutris C, Hertel-Aas T, Lapied E, Joner EJ, Oughton DH (2012) Bioavailability of cobalt and silver nanoparticles to the earthworm *Eisenia fetida*. Nanotoxicology 6(2):186–195. https://doi.org/10.3109/17435390.2011.569094

El-Temsah YS, Joner EJ (2012a) Ecotoxicological effects on earthworms of fresh and aged nano-sized zero-valent iron (nZVI) in soil. Chemosphere 89(1):76–82. https://doi.org/10.1016/j.chemosphere.2012.04.020

El-Temsah YS, Joner EJ (2012b) Impact of Fe and Ag nanoparticles on seed germination and differences in bioavailability during exposure in aqueous suspension and soil. Environ Toxicol 27(1):42–49. https://doi.org/10.1002/tox.20610

El-Temsah YS, Oughton DH, Joner EJ (2013) Effects of nano-sized zero-valent iron on DDT degradation and residual toxicity in soil: a column experiment. Plant Soil 368(1–2):189–200. https://doi.org/10.1007/s11104-012-1509-8

Gupta GS, Kumar A, Senapati VA, Pandey AK, Shanker R, Dhawan A (2017) Laboratory scale microbial food chain to study bioaccumulation, biomagnification, and Ecotoxicity of cadmium telluride quantum dots. Environ Sci Technol 51(3):1695–1706. https://doi.org/10.1021/acs.est.6b03950

Handy RD, van den Brink N, Chappell M, Mühling M, Behra R, Dušinská M, Simpson P, Ahtiainen J, Jha AN, Seiter J, Bednar A, Kennedy A, Fernandes TF, Riediker M (2012a) Practical considerations for conducting ecotoxicity test methods with manufactured nanomaterials: what have we learnt so far? Ecotoxicology 21(4):933–972. https://doi.org/10.1007/s10646-012-0862-y

Handy RD, Cornelis G, Fernandes T, Tsyusko O, Decho A, Sabo-Attwood T, Metcalfe C, Steevens JA, Klaine SJ, Koelmans AA, Horne N (2012b) Ecotoxicity test methods for engineered nanomaterials: practical experiences and recommendations from the bench. Environ Toxicol Chem 31(1):15–31. https://doi.org/10.1002/etc.706

Hjorth R, Sørensen SN, Olsson ME, Baun A, Hartmann NB (2016) A certain shade of green: can algal pigments reveal shading effects of nanoparticles? Integr Environ Assess Manage 12 (1):200–202. https://doi.org/10.1002/ieam.1728

Hjorth R, Coutris C, Nguyen NHA, Sevcu A, Gallego-Urrea JA, Baun A, Joner EJ (2017) Ecotoxicity testing and environmental risk assessment of iron nanomaterials for sub-surface remediation – recommendations from the FP7 project NanoRem. Chemosphere 182:525–531. https://doi.org/10.1016/j.chemosphere.2017.05.060

Hund-Rinke K, Baun A, Cupi D, Fernandes TF, Handy R, Kinross JH, Navas JM, Peijnenburg W, Schlich K, Shaw BJ, Scott-Fordsmand JJ (2016) Regulatory ecotoxicity testing of nanomaterials – proposed modifications of OECD test guidelines based on laboratory experience with silver and titanium dioxide nanoparticles. Nanotoxicology 10(10):1442–1447. https://doi.org/10.1080/17435390.2016.1229517

Johnson RL, Nurmi JT, O'Brien Johnson GS, Fan D, O'Brien Johnson RL, Shi Z, Salter-Blanc AJ, Tratnyek PG, Lowry GV (2013) Field-scale transport and transformation of Carboxymethylcellulose-stabilized Nano zero-Valent Iron. Environ Sci Technol 47 (3):1573–1580. https://doi.org/10.1021/es304564q

Judy JD, Unrine JM, Bertsch PM (2011) Evidence for biomagnification of gold nanoparticles within a terrestrial food chain. Environ Sci Technol 45(2):776–781. https://doi.org/10.1021/es103031a

Kahru A, Dubourguier H-C (2010) From ecotoxicology to nanoecotoxicology. Toxicology 269 (2–3):105–119. https://doi.org/10.1016/j.tox.2009.08.016

Keller AA, Garner K, Miller RJ, Lenihan HS (2012) Toxicity of Nano-zero Valent Iron to freshwater and marine organisms. PLoS One 7(8):e43983. https://doi.org/10.1371/journal.pone.0043983

Kleiven M, Rossbach LM, Gallego-Urrea JA, Brede DA, Oughton DH, Coutris C (2018) Characterizing the behavior, uptake, and toxicity of NM300K silver nanoparticles in *Caenorhabditis elegans*. Environ Toxicol Chem 37(7):1799–1810. https://doi.org/10.1002/etc.4144

Labille J, Brant J (2010) Stability of nanoparticles in water. Nanomedicine 5(6):985–998. https://doi.org/10.2217/nnm.10.62

Lapied E, Moudilou E, Exbrayat J-M, Oughton DH, Joner EJ (2010) Silver nanoparticle exposure causes apoptotic response in the earthworm *Lumbricus terrestris* (Oligochaeta). Nanomedicine 5(6):975–984. https://doi.org/10.2217/nnm.10.58

Lei C, Sun Y, Tsang DCW, Lin D (2018) Environmental transformations and ecological effects of iron-based nanoparticles. Environ Pollut 232:10–30. https://doi.org/10.1016/j.envpol.2017.09.052

Lewinski N, Colvin V, Drezek R (2008) Cytotoxicity of nanoparticles. Small 4(1):26–49. https://doi.org/10.1002/smll.200700595

Liu A, Liu J, W-x Z (2015) Transformation and composition evolution of nanoscale zero valent iron (nZVI) synthesized by borohydride reduction in static water. Chemosphere 119:1068–1074. https://doi.org/10.1016/j.chemosphere.2014.09.026

Majumdar S, Trujillo-Reyes J, Hernandez-Viezcas JA, White JC, Peralta-Videa JR, Gardea-Torresdey JL (2016) Cerium biomagnification in a terrestrial food chain: influence of particle size and growth stage. Environ Sci Technol 50(13):6782–6792. https://doi.org/10.1021/acs.est.5b04784

Nguyen NHA, Špánek R, Kasalický V, Ribas D, Vlková D, Řeháková H, Kejzlar P, Ševců A (2018a) Different effects of nano-scale and micro-scale zero-valent iron particles on planktonic microorganisms from natural reservoir water. Environ Sci Nano 5(5):1117–1129. https://doi.org/10.1039/C7EN01120B

Nguyen NHA, Von Moos NR, Slaveykova VI, Mackenzie K, Meckenstock RU, Thűmmler S, Bosch J, Ševců A (2018b) Biological effects of four iron-containing nanoremediation materials on the green alga *Chlamydomonas* sp. Ecotoxicol Environ Saf 154:36–44. https://doi.org/10.1016/j.ecoenv.2018.02.027

OECD (1984) Test No. 207: Earthworm, acute toxicity tests. In: OECD guidelines for the testing of chemicals, section 2. OECD Publishing, Paris. https://doi.org/10.1787/9789264070042-en

OECD (2000a) Test No. 216: Soil microorganisms: nitrogen transformation test. In: OECD Guidelines for the testing of chemicals, section 2. OECD Publishing, Paris. https://doi.org/10.1787/9789264070226-en

OECD (2000b) Test No. 217: Soil microorganisms: carbon transformation test. In: OECD Guidelines for the testing of chemicals, section 2. OECD Publishing, Paris. https://doi.org/10.1787/9789264070240-en

OECD (2004) Test No. 202: Daphnia sp. Acute immobilisation test. In: OECD Guidelines for the testing of chemicals, section 2. OECD Publishing, Paris. https://doi.org/10.1787/9789264069947-en

OECD (2007) Test No. 225: Sediment-water Lumbriculus toxicity test using spiked sediment. In: OECD Guidelines for the testing of chemicals, section 2. OECD Publishing, Paris. https://doi.org/10.1787/9789264067356-en

OECD (2011) Test No. 201: Freshwater alga and cyanobacteria, growth inhibition test. In: OECD Guidelines for the testing of chemicals, section 2. OECD Publishing, Paris. https://doi.org/10.1787/9789264069923-en

OECD (2013) Test No. 210: Fish, early-life stage toxicity test. In: OECD Guidelines for the testing of chemicals, section 2. OECD Publishing, Paris. https://doi.org/10.1787/9789264203785-en

OECD (2016a) Test No. 220: Enchytraeid reproduction test. In: OECD Guidelines for the testing of chemicals, section 2. OECD Publishing, Paris. https://doi.org/10.1787/9789264264472-en

OECD (2016b) Test No. 222: Earthworm reproduction test (Eisenia fetida/Eisenia andrei). In: OECD Guidelines for the testing of chemicals, section 2. OECD Publishing, Paris. https://doi.org/10.1787/9789264264496-en

OECD Assessment of chemicals. http://www.oecd.org/chemicalsafety/risk-assessment/. Accessed 09.07.2019

Oughton DH, Hertel-Aas T, Pellicer E, Mendoza E, Joner EJ (2008) Neutron activation of engineered nanoparticles as a tool for tracing their environmental fate and uptake in organisms. Environ Toxicol Chem 27(9):1883–1887. https://doi.org/10.1897/07-578.1

Petersen EJ, Huang Q, Weber WJ Jr (2008) Bioaccumulation of radio-labeled carbon nanotubes by *Eisenia foetida*. Environ Sci Technol 42(8):3090–3095. https://doi.org/10.1021/es071366f

Pullin H, Crane RA, Morgan DJ, Scott TB (2017) The effect of common groundwater anions on the aqueous corrosion of zero-valent iron nanoparticles and associated removal of aqueous copper and zinc. J Environ Chem Eng 5(1):1166–1173. https://doi.org/10.1016/j.jece.2017.01.038

Roberts AP, Mount AS, Seda B, Souther J, Qiao R, Lin S, Ke PC, Rao AM, Klaine SJ (2007) In vivo biomodification of lipid-coated carbon nanotubes by *Daphnia magna*. Environ Sci Technol 41(8):3025–3029. https://doi.org/10.1021/es062572a

Sørensen SN, Hjorth R, Delgado CG, Hartmann NB, Baun A (2015) Nanoparticle ecotoxicity—physical and/or chemical effects? Integr Environ Assess Manage 11(4):722–724. https://doi.org/10.1002/ieam.1683

Unrine JM, Colman BP, Bone AJ, Gondikas AP, Matson CW (2012) Biotic and abiotic interactions in aquatic microcosms determine fate and toxicity of Ag nanoparticles. Part 1. Aggregation and dissolution. Environ Sci Technol 46(13):6915–6924. https://doi.org/10.1021/es204682q

Walker CH, Hopkin SP, Sibly RM, Peakall DB (2001) Principles of ecotoxicology. Taylor & Francis, London

Wang Y, Miao A-J, Luo J, Wei Z-B, Zhu J-J, Yang L-Y (2013) Bioaccumulation of CdTe quantum dots in a freshwater alga *Ochromonas danica*: a kinetics study. Environ Sci Technol 47 (18):10601–10610. https://doi.org/10.1021/es4017188

Wang J, Fang Z, Cheng W, Tsang PE, Zhao D (2016) Ageing decreases the phytotoxicity of zero-valent iron nanoparticles in soil cultivated with *Oryza sativa*. Ecotoxicology 25(6):1202–1210. https://doi.org/10.1007/s10646-016-1674-2

Part VI
Future Prospects

Chapter 29
Future Prospects for Treating Contaminants of Emerging Concern in Water and Soils/Sediments

Carmen Mihaela Neculita, Lucie Coudert, Eric Rosa, and Catherine N. Mulligan

Abstract The definition and classification of contaminants of emerging concerns (CECs) in water and soils/sediments have evolved over time. Stricter environmental regulations in many countries and consciousness for sustainable development have led to adapted solutions to reduce the footprint and potential impacts of anthropogenic activities. Characterization and treatment technologies for CECs have evolved consequently. Major actual challenges in CECs treatment include the lack of analytical methodologies for their adequate quantification in complex matrices, uncertainties on operational costs, limited creation of residual salinity, low volumes of stable sludge, and easy and safe management with limited potential impact of treated effluents on receiving natural waters. Other needs include the use of low toxicity, highly biodegradable, and green additives, reduction of emissions and energy requirements, and adequate management of the solids produced during the application of sorption and/or biotechnologies that contain CECs. Therefore, present gaps and research needs for the CECs include mainly analytical determination, toxicity evaluation of CECs and their degradation products, as well as the distribution, fate and degradation of CECs in the environment.

C. M. Neculita (✉) · L. Coudert
Research Institute on Mines and Environment (RIME), Université du Québec en Abitibi-Témiscamingue (UQAT), Rouyn-Noranda, QC, Canada
e-mail: Carmen-Mihaela.Neculita@uqat.ca

E. Rosa
Research Institute on Mines and Environment (RIME), Université du Québec en Abitibi-Témiscamingue (UQAT), Rouyn-Noranda, QC, Canada

Groupe de Recherche sur l'Eau Souterraine (GRES—Groundwater Research Group), UQAT, Amos, QC, Canada

C. N. Mulligan
Building, Civil, and Environmental Engineering, Concordia University, Montréal, QC, Canada

J. Filip et al. (eds.), *Advanced Nano-Bio Technologies for Water and Soil Treatment*, Applied Environmental Science and Engineering for a Sustainable Future,
https://doi.org/10.1007/978-3-030-29840-1_29

Keywords Organic/inorganic CECs · Advanced oxidation · Biological treatment · Residual toxicity

29.1 Sources and Characteristics of Contaminants of Emerging Concern in Water and Soils/Sediments

Historically, the classification of contaminants of emerging concern (CECs) evolved with industrial development, contaminant release, wastewater treatment issues and challenges, scientific knowledge and regulations, etc. At the beginning of the 1990s, priority pollutants referred to nonpolar hazardous compounds and heavy metals, and intensive monitoring programs and treatment processes were developed to reduce their release, and their consequent negative impacts on the environment (Petrović et al. 2003). Today, these contaminants are better managed in industrialized countries. However, new unregulated or emerging contaminants have become a focus of interest and awareness because of their potentially negative impact on the environment. There is a consensus within the scientific community that these contaminants require legislative intervention; moreover, improvement of actual wastewater treatment facilities is needed as most of them have not been designed to treat these new CECs and their degradation products (Fig. 29.1).

In the area of municipal and industrial wastewater, CECs are defined as contaminants deriving from natural or anthropogenic sources, both persistent in the environment and potentially toxic to humans, wildlife, and flora (Petrie et al. 2015; Sauvé and Desrosiers 2014). The CECs could be divided into two/three groups: (1) newly identified and therefore not yet fully apprehended, and (2) and more common and documented (Sauvé and Desrosiers 2014). They originate from several sources and typical concentrations found in wastewater range from few ng/L to several mg/L depending on the nature of the contaminant (Bolong et al. 2009;

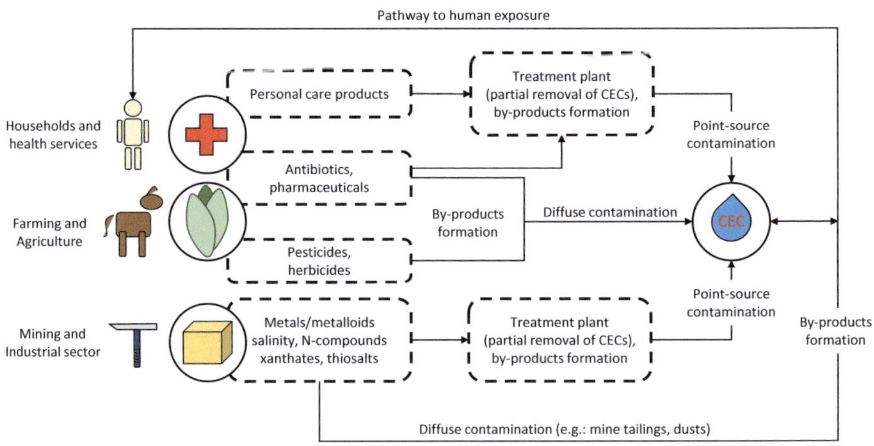

Fig. 29.1 Pathways of human exposure to CECs

Gogoi et al. 2018; Wilkinson et al. 2017). The main CECs identified to date in municipal and industrial wastewater include organic compounds such as personal care products (fragrance, detergents, lotions, deodorants, cleaning products, sunscreens and ultraviolet filters, etc.), pharmaceuticals (analgesics, anesthetics, antimicrobials, anti-inflammatory drugs, sedatives, etc.), illicit drugs, pesticides, nanomaterials, flame retardants, cyanotoxins, and hormones. Some of these compounds are known or suspected to have adverse effects on wildlife endocrine systems and human health (Shaver 2011; Wilkinson et al. 2017). This list is non-exhaustive as the variety and the number of the compounds released to the environment is in constant evolution and expected to significantly grow in the future because of advancements in chemical, medical, and pharmaceutical industries and improvements of analytical methods to detect them and their degradation products/metabolites in complex matrix (Rodriguez-Narvaez et al. 2017).

Similarly to municipal wastewater, in mine water, CECs can be classified into at least four categories (Neculita et al. 2018): (1) new contaminants (e.g., rare earth elements—REE, Se, Mn); (2) contaminants that are difficult to treat (e.g., sulfates, salinity); (3) known contaminants, but with low target/regulatory criteria in sensitive environments, particularly in the northern climate (e.g., As, Cu), and (4) nitrogenous contaminants (NH_3, NO_2^-, NO_3^-), for which either regulatory criteria are clearly defined or they are controlled via the monitoring of aquatic toxicity. Potential sources of toxicity in mine effluents include free acidity, depressed or high pH, dissolved metals, nitrogen compounds, thiosalts, xanthates, etc. (ESG International Inc. and Lakefield Research 2002). In gold mine effluents, toxicants also include cyanides and derivatives, As, and total suspended solids (TSS).

Finally, soil and sediment contamination by CECs results from various inputs. These may be divided into two different areas: (1) dumping materials, bankrupt and abandoned manufacturing plants, insufficient methods for waste storage, treatment of soils/sediments and urban and industrial disposal facilities, and (2) use of chemicals for nutrient supplementation, weed and pest control, and other agricultural practices resulting in pollution of surface water and excessive nitrogen and phosphate loadings (Yong et al. 2014).

The toxicity of CECs is poorly known (Sauvé and Desrosiers 2014). For example, toxicity associated with ion imbalance can occur if ion concentrations exceed or do not meet the physiological tolerance of the organisms present in the receiving ecosystem (Goodfellow et al. 2000). The internal ionic concentration of the cells must be maintained, so that the cellular functions are also maintained (Cormier et al. 2013), these being often based on temporary changes of the ionic equilibria implying osmotic and cellular processes. If the external ion concentration increases, a disruption of ionic equilibrium occurs and other effects can also be entailed, including an osmotic shock or oxidative stress that can lead to the death of organisms (Cormier et al. 2013; Goodfellow et al. 2000). Among other factors, salinity/total dissolved solids (TDS) is a major contributor to the toxicity (Banks et al. 1997; Goodfellow et al. 2000; Kennedy et al. 2005; van Dam et al. 2014). The sensitivity to salinity is organism-dependent and the effects can be seen on algal growth, growth and reproduction of pelagic (e.g., *Daphnia*) and benthic (e.g., amphipod) invertebrates, on fish embryos, etc. Growth and reproduction appear to be more sensitive to the energy requirements of osmoregulation than parameters such as survival (Goodfellow et al. 2000). These processes may be more sensitive to osmotic stress. Several studies have already demonstrated that the natural variability of biological diversity, environmental conditions, stream chemistry and flow could entail toxicity after the addition of a saline effluent (van Dam et al. 2014).

29.2 Treatment of Inorganic/Organic CECs in Waters and Soils/Sediments

Responsible management of contaminated water is a continuous challenge of wastewater treatment plants (WWTPs) as well as mining activities due to their evolutive composition. Petrović et al. (2003) reviewed the performances of conventional municipal WWTPs. They highlighted that elimination of CECs through activated sludge treatment (AST) was relatively low in most cases. According to their review, the degradation of pharmaceuticals and personal care products such as ibuprofen, diclofenac, naproxen, and triclosan ranged from 35 to 90%, while carbamazepine showed a low removal, with only 7%. Low removal yields can result from the fact that at neutral pH, pharmaceuticals such as ibuprofen or diclofenac are present in wastewater as ions that are not adsorbed on the activated sludge.

The degradation of pesticides such as chlorinated phenoxyacids seems to be ineffective at real WWTPs even if laboratory tests showed that nearly 100% of these compounds could be degraded. This difference of performances can be explained by the long adaptation period of activated sludge (~ 4 months) and the short period of extensive use of pesticides (Petrović et al. 2003). The degradation of alkylphenolic surfactants (mainly used in personal care products) is efficient using AST, with removal yields reaching 90–95%. However, the formation of treatment-resistant metabolites is problematic as studies showed that these compounds have the ability to mimic natural hormones by interacting with estrogen receptors. Consequently, CECs found in municipal and industrial wastewater can enter receiving waters because of their incomplete degradation/removal through AST and present a risk for the environment and human health (Bolong et al. 2009; Sophia and Lima 2018).

Some characteristics and drawbacks of the main physicochemical treatment processes for CECs in mine water are presented in Fig. 29.2. In mine water, the treatment of CECs is challenging because of their complex chemistry and persistency, the lack of a proven best available technology economically achievable (BATEA), and insufficient knowledge as to the optimal efficiency conditions with mixtures of contaminants and variable effluent quality. Serious issues related to mine water treatment can lead to delayed approvals of projects, regulatory uncertainty, increasing complexity and operating costs, constraints on production, property damage, and reputational impacts (Ringwood et al. 2016). As a result, the long-term satisfactory performance of water treatment processes and sustainable management of sludge produced represent a priority for governments and the mining

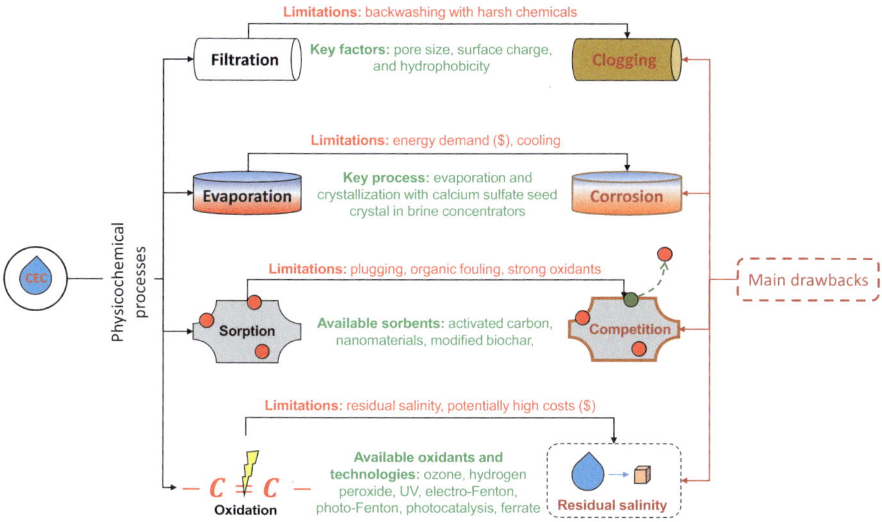

Fig. 29.2 Characteristics and drawbacks of main physicochemical treatment processes for CECs in mine water

industry. In the Canadian context, water treatment challenges include high flow rates and contamination, remoteness and extreme cold climate conditions, sensitive environments as well as evolving water quality standards and stricter regulations. Rapid degradation kinetics and/or complete reactions avoid by-products formation and allow a balance between efficiency, costs, and regulations. However, most of the treatment processes create residual salinity.

Therefore, the priorities in water management include (1) improving/developing of sustainable treatment technologies for CECs, (2) limiting residual salinity (from chemicals used during the treatment of mine water), and (3) producing stable sludge.

A variety of *in situ* and *ex situ* remediation techniques exists to manage the contaminated sites (Yong et al. 2014). *Ex situ* techniques include excavation or dredging, contaminant solidification/stabilization, incineration, vitrification, washing, and biological treatment processes. *In situ* processes include: (1) bioremediation, air or steam stripping for volatile compounds, (2) extraction methods for soluble components, (3) chemical treatments for oxidation, and (4) stabilization/solidification for metals or organic contaminants. Phytoremediation, although less documented, has also been used. Other technologies related to nanotechnologies are also being developed. Site characterization and conceptual model development, and laboratory and pilot testing are essential for considering alternate remediation approaches. To enhance sustainability, new efforts are required and being developed to reduce resource requirements and emissions and waste generation to protect the general public and the environment.

29.3 Potential Solutions

To address some of the abovementioned needs, research must focus on physicochemical processes that create low residual salinity. Promising options fitting into this requirement involve physicochemical-based processes (membranes, evaporation, sorption), biotechnologies, and advanced oxidation processes (AOPs) using ferrates and ozone microbubbles. Some of the most promising approaches meeting the identified requirements have been summarized before, with the emphasis on their principle, potential applications, performances, and challenges.

29.3.1 Physicochemical Processes

1. **Filtration** uses membranes, which are classified into microfiltration (MF), ultrafiltration (UF), nanofiltration (NF), and reverse osmosis (RO). Specific filtering features of these membranes (pore size, surface charge, and hydrophobicity) determine the contaminant type that can be retained. Several mechanisms including adsorption, electrostatic repulsion, and sieving are involved in the removal of CECs from municipal and industrial wastewater. MF is widely used to treat municipal and industrial wastewater as it can be used at atmospheric pressure. However, MF is not useful for the removal of CECs from municipal and industrial wastewater as this system is not able to remove contaminants of size <1 μm (Rodriguez-Narvaez et al. 2017). With a pore size ranging from 1 to 100 nm, UF is able to remove a wide variety of CECs including plasticizers (bisphenol A, bis(2-ethylhexyl)phthalate, butyl benzyl phthalate), herbicides (atrazine), pharmaceuticals (17α-ethynylestradiol, naproxen, diclofenac) (Rodriguez-Narvaez et al. 2017; Yoon et al. 2007). Operating at low pressure, NF can also be used for CECs removal from municipal and industrial wastewater because of its small pore size (between 1 and 10 nm), with removal yields higher than 50% for most of CECs and 90% for pharmaceuticals (Bolong et al. 2009; Gogoi et al. 2018). Relative to previous membrane systems, the use of hydraulic pressure as the driving force to separate CECs from wastewater allowed better efficiency as such systems are able to remove particles of size <1 nm. For example, ibuprofen, naproxen, fenoprofen, gemfibrozil, and ketoprofen have been removed for up to 95–99% (Rodriguez-Narvaez et al. 2017). To prevent the fouling of membranes, commonly encountered during wastewater treatment, specific conditions must be maintained and backwashing using harsh chemicals is used to clean the membrane when required. Another limitation of RO is the necessity to remineralize the water after the treatment because of the nonselective removal of the elements and to adjust the pH, which increases the treatment costs.

   Given the molecular weight and size of Se species, only NF and RO can be effective in Se removal (Sandy and DiSante 2010). Indeed, RO was found effective in Se removal below 5 ppb, at pilot and full scale, while NF was less effective than RO but still performant at high contamination (Santos et al. 2015). Scaling is a main limitation of membranes (RO/NF) with mine water as salt formation could damage and impair their proper function. As low temperature favors the crystallization of minerals, this effect is enhanced. TDS, TSS, alkalinity, and bacterial-algal fouling potential are also factors to influence the efficiency of the membranes, along with energy consumption, which increases the treatment costs. The pre-treatment of contaminated water is therefore required. The biosurfactant—rhamnolipid—addition was studied in micellar-enhanced ultrafiltration (MEUF) of heavy metals from contaminated waters (El Zeftawy and Mulligan 2011). The MEUF system performance was successfully investigated for copper, zinc, nickel, lead, and cadmium retention by two different membranes with molecular weight cutoffs (MWCO) of 10,000 Da and 30,000 Da.

2. **Evaporation** allows the separation of contaminants from a concentrate (brine), by thermal water removal. Mechanical evaporation/crystallization shows the greatest prospects. Brine concentrators use a calcium sulfate seed crystal that acts as a nucleation promoter for gypsum removal. Although efficient, evaporation is mechanically complex and expensive, requires energy and maintenance, as well as steam and cooling water (Sandy and DiSante 2010). Scale and corrosion are additional constraints, especially when dealing with high ionic -strength effluents.
3. **Sorption** binds undesirable ions in water to solid particles, which can be natural and residual materials. Over the last decades, various studies have evaluated the performances of activated carbon, modified biochar, and/or nanomaterials to remove both organic and inorganic contaminants from municipal and/or industrial wastewater (Dubey et al. 2017; Rodriguez-Narvaez et al. 2017; Sophia and Lima 2018; Thines et al. 2017). The main advantage of using natural or residual sorbents to remove organic CECs from municipal and industrial wastewater is the nonproduction of toxic or treatment of recalcitrant degradation products. According to the literature review of Rivera-Utrilla et al. (2013), activated carbons demonstrate high adsorption capacities to remove pharmaceuticals from waters, with values ranging from 25 to 260 mg/g depending on the sorbent used and the CEC of interest. The performances of carbon-based nanomaterials depending on the treatment used for their production are discussed by Adeleye et al. (2016), Sophia and Lima (2018), and Zhao et al. (2018). For example, Peng et al. (2012) were able to achieve 100% removal of norfloxacin, an antibiotic, using single-walled nanotubes (SWNT) and only 35% removal when using multi-walled nanotubes (MWNT). Similarly, the removal of tetracycline achieved 92% with SWNT, whereas only 16% of this antibiotic was removed with MWNT (Ji et al. 2010). Halogenated organics such as trichloroethylene, trichlorobenzene (TCB), dichlorobenzene (DCB), and polychlorinated biphenyls (PCBs) can efficiently be removed using carbon-based nanomaterials (Adeleye et al. 2016). In their review, Zhao et al. (2018) compared the performances of several nanomaterials activated or not to remove CECs from water as well as their

maximum adsorption capacities. Compared to low-cost or residual materials, nanomaterials present several advantages including faster and more efficient removal of metals and organic CECs under a large range of pH and temperature, a higher surface area-to-volume ratio, and a lower dose requirement. Recent advances reported the potential use of biochar modified by KOH or H_2SO_4 to remove tetracycline from water with adsorption capacities of 23 and 59 mg/g for acidic and basic treatments of biochar (Liu et al. 2012). Jing et al. (2014) showed that the removal of tetracycline was improved by 46% in the presence of MeOH-modified biochar compared to the unmodified adsorbent.

Sulfate, as a major constituent of mine effluents acts as a competitor for Se oxyanions leading to the rapid exhaustion of the ion exchange capacity of materials, but Fe(III)-loaded sorbent, such as biochar, improves the sorption capacity (Holmes and Gu 2016). Sulfate pre-treatment (e.g., precipitation as gypsum) is necessary to minimize this limitation. Other factors to be considered are TSS-induced plugging, organic fouling, and strong oxidants (Microbial Technologies, Inc. 2006). Recent findings [synthetic effluents; pH 6, 1 mg/L As (V) and 1 mg/L Sb(III)] showed that Fe-loaded biochar (by evaporation) tripled and quintupled the raw biochar sorption capacity for As(V) and Sb(III), respectively (Calugaru et al. 2019). Other studies confirmed that Fe(III) impregnation of biochar enhanced its As(V) removal capacity by more than 250 times (Calugaru et al. 2018). Other approaches include the use of granular activated carbon supported nanoscale zero-valent iron (Chowdhury and Mulligan 2013) and other nanomaterials (such as Fe-Cu) (Babaee et al. 2018).

4. **AOPs** operate at near-ambient temperature and pressure, and involve the generation of powerful oxidants, which are able to break –C=C– double bonds and aromatic rings for effective treatment of organic contaminants in water (Bokare and Choi 2014; Gogate and Pandit 2004). There is a growing interest in developing AOPs to remove CECs from wastewater because of their capabilities to degrade organic contaminants and/or oxidize metals/metalloids. Common AOPs include the use of ozone (O_3), hydrogen peroxide (H_2O_2), and UV, or their combination. AOPs are applied whenever conventional oxidation is insufficient, kinetics are slow, and the contaminants are refractory or partially oxidized, yielding stable by-products with greater toxicity than the original species. AOPs have been used to enhance the degradation of pharmaceuticals and pesticides (Ikehata et al. 2006). For example, ***electro-Fenton and photo-Fenton processes***, which are based on the generation of hydroxyl radicals through electrochemical reactions or UV irradiation, have been successfully applied for the degradation of chlorophenols and pesticides (Adeleye et al. 2016; Brillas et al. 1995; Gözmen et al. 2003; Oturan et al. 2001). The use of ***ozonation*** alone or in combination with H_2O_2 or UV is of high interest because of the potential to degrade CECs from municipal and industrial wastewater. As discussed by Rodriguez-Narvaez et al. (2017), these processes have been applied to the degradation of several CECs with satisfying efficiencies (> 80%). The degradation of CECs occurred through the direct oxidation of ozone or its decomposition

onto hydroxyl radicals that oxidize CECs. Alteration of dimethyl amino groups and aniline of antimicrobials by ozonation leads to the deactivation of the bactericidal functions of these compounds (Gogoi et al. 2018). Efficiency higher than 90% was obtained when using ozonation onto non-aromatic alkenes, aromatic compounds that are abundant in electrons (sulfamethoxazoles—SMXs), and aromatic compounds with deprotonated amine groups (trimethoprim—TMP) (Gogoi et al. 2018). The combination of ozone and UV irradiation has been studied to remove pesticides (Adeleye et al. 2016; Ikehata et al. 2006; Lafi and Al-Qodah 2006), pharmaceuticals (Gebhardt and Schröder 2007), and antibiotics (Akmehmet Balcioğlu and Ötker 2004) from wastewater. ***Photocatalysis*** is another promising AOP that has been found to be highly effective in removing CECs and their derivatives through the formation of free hydroxyl radicals in the presence of a metal oxide semiconductor such as TiO_2 and a light source (Fast et al. 2017). Removal efficiencies greater that 90% have been reported in some studies presented in Fast et al. (2017) and Rodriguez-Narvaez et al. (2017). The dosage of photocatalyst to be used is of huge importance as under- or overdosing loads may result in slower degradation rates of CECs. The fouling of TiO_2 catalyst or UV lamps can occur, resulting in loss of CECs degradation efficiencies (Fast et al. 2017). In recent years, hexavalent state of tetraoxy iron [$Fe(VI)O_4{}^{2-}$], also known as ***ferrate***, has been used as a strong oxidant to remove pollutants from municipal and hospital wastewater including pharmaceuticals, personal care products, endocrine disrupting chemicals, and cyanotoxins (Jiang 2013; Jiang et al. 2014; Li 2017; Sharma et al. 2005; Sharma et al. 2015). According to several studies, ferrates showed satisfactory disinfectant properties, inactivating a wide variety of microorganisms even at low Fe(VI) concentrations (Sharma et al. 2005). Ferric ions produced from Fe(VI) represent a powerful coagulant that is suitable for the removal of humic acids, radionuclides, and metals from wastewater (Sharma et al. 2005). Sharma et al. (2015) reported that complete degradation of two antibiotics, sulfamethoxazole and trimethoprim, was observed when using Fe(VI) with molar ratio of SMX:Fe(VI) or TMP:Fe(VI) between 4:1 and 5:1. Moreover, the authors observed that by-products with no residual antibiotic activity were produced during the oxidation of TMP using Fe(VI). The degradation of fluoroquinolone antibiotics (FQAs) from synthetic and natural waters was studied at pH 7 using Fe(VI). The results highlighted that, except for flumequine, all FQAs were degraded within 2 min at molar ratio Fe(VI):FQA lower than 20:1. The authors mentioned that both multivalent cations (e.g., Ca^{2+}, Mg^{2+}, Cu^{2+}, and Fe^{3+}) and humic acid decreased the degradation of fluoroquinolone (Feng et al. 2016). Yang et al. (2012) investigated ferrate performance to degrade 68 endocrine disrupting chemicals and personal care products from spiked wastewater. Li (2017) showed that 44% of mefenamic acid ($C_0 = 1$ mg/L), a pharmaceutical, can be removed using 2 mg/L of ferrate.

Previous studies on mine water with AOPs aimed at cyanides and derivatives (including NH_3-N), sulfur, sulfides and thiosalts, and metal removal using ferrates [Fe(VI)] (Gonzalez-Merchan et al. 2016; Sharma 2011) and ozone microbubbles

(Khuntia et al. 2013, 2014; Ryskie et al. 2019). Previous findings showed good efficiency of wet Fe(VI) in NH_3 removal from synthetic and actual gold mine effluents (Gonzalez-Merchan et al. 2016), but increased residual salinity in treated water from the oxidant (Gonzalez-Merchan et al. 2018). Noteworthy, wet Fe(VI) preparation involves the oxidation of an Fe(III) salt (sulfate, chloride, or sulfate) under alkaline conditions (NaOH) using an oxidant (NaOCl). Previous findings also showed good efficiency of ozone microbubbles in NH_3 and metal removal, from synthetic and actual effluents (Ryskie et al. 2019). A major drawback of the AOPs is often the creation of residual salinity. Further tests are required for evaluating the influence of bubbles size, the simultaneous removal of other oxidizable contaminants present in the water, as well as aquatic toxicity of final effluent.

29.3.2 Biological Processes

Biological processes are commonly used to remove organic compounds from municipal and industrial wastewater and inorganic contaminants from mine water (precipitation as sulfides). Performances of biological treatments are dependent on microorganism activities. Various factors including water composition, loading rate, type of media, temperature, and aeration rate can alter the microbial activity and therefore the efficiency of treatment. Consequently, biological treatment is susceptible to the presence of toxin or excess of nutrients in water, seasonal and loading rate changes, and shock. The main drawbacks of biological treatments are related to fouling and filter clogging phenomenon as well as careful control of operating conditions to favor microbial activities. As discussed earlier, conventional biological processes such as activated sludge are not very effective in removing recalcitrant organic CECs from municipal and industrial wastewater. Anaerobic processes can be effective but the degradation of organic compounds is quite slow. Moreover, several studies reported that anaerobic processes are not efficient for the removal of pharmaceuticals, personal care products, and endocrine disrupting compounds. This can be explained because these CECs are present in much lower concentrations than other organic contaminants, reducing the portion of these contaminants that can be effectively degraded (Adeleye et al. 2016). To improve the performances of biological treatment to remove organic CECs from water, it can be combined with physical treatments such as MF or UF. For example, studies reported that membrane bioreactors (MBRs), which combined an activated sludge process with a membrane process (MF or UF) are extensively used in wastewater treatment plants to remove organic compounds from municipal and industrial wastewater (Rodriguez-Narvaez et al. 2017). However, there are only few papers reporting on the performances of MBRs for the degradation of CECs (Sui et al. 2011; Yang et al. 2011). According to these studies, only easily biodegradable CECs (such as caffeine, diclofenac, trimethoprim, acetaminophen, ibuprofen, tetracycline) can be efficiently removed from wastewater (60–100%), whereas low biodegradable ones (such as metoprolol, bezafibrate, primidone, carbamazepine) may not be eliminated through biological processes (<40%) (Yang et al. 2011). Actual research focuses on the development of alternative solutions of biological treatments to remove recalcitrant CECs from municipal and industrial wastewater and some interesting processes include bioelectrochemical systems (BESs). In such systems, organic CECs can be oxidized by the bacteria forming a biofilm while electrons are transferred from the microorganisms to the electrode surface. BESs have demonstrated satisfying performances to remove some CECs (Rodriguez-Narvaez et al. 2017; Werner et al. 2015). Another promising technology is based on the use of electrochemical membrane bioreactors with the integration of bioelectrogenesis. Compared to BESs, it reduces the fouling phenomenon and improves the CECs removals (Ma et al. 2015; Rodriguez-Narvaez et al. 2017). Biosurfactant applications for remediation of contaminated soil and water are promising because of their biodegradability, low toxicity, and effectiveness in enhancing biodegradation and affinity for metals/metalloids including As. Various studies have been performed with regard to the enhanced bioremediation, the flushing and washing of soils, sediments, mining residues, and water contaminated with metals alone or with hydrocarbons (Mulligan 2017). Removal of metals by biosurfactants occurs through solubilization, complexation, and ion exchange.

29.4 Research Needs

Current challenges in the treatment of CECs include some of the following: (1) lack of analytical methodologies to adequately quantify CECs in complex matrices such as industrial, municipal, and mine water; (2) uncertainties on operational costs (necessary infrastructure, chemicals availability); (3) limited creation of residual salinity (source of toxicity); (4) low volumes of stable sludge, easy and safe to manage; (5) potential impact of treated effluents on receiving natural water; (6) use of low toxicity and highly biodegradable, green additives for treatment; 7) reduction of emissions and energy requirements; and 8) management of the solids produced during the application of sorption and/or biotechnologies that contain CECs.

Gaps and research needs for the CECs found in municipal and industrial wastewater include: (1) ***Analytical determination***: Improve the knowledge related to the determination of CECs and their degradation products or metabolites produced after physicochemical or biological treatments in complex matrices even at very low concentrations; (2) ***Toxicity of CECs and their degradation products***: Investigate the toxic effects of many of CECs and their degradation by-products on living organisms; (3) ***Distribution and fate of CECs in the environment***: Determine the behavior and fate of CECs and their degradation pathways on the environment to implement regulations in order to protect surface and groundwater quality; (4) ***Degradation of CECs***: Test the performances of the new physicochemical and/or biological technologies on real wastewater at pilot plant scale and in the presence of dissolved organic matter and other coexisting substances to confirm/rebut the results obtained at laboratory scale.

Operating mines apply active-treatment strategies, using BATEA, which can be engineered to accommodate variable flow rates and effluent quality, but with a cost-dependent flexibility. With evolving water quality standards and stricter regulations, new treatment approaches are required to ensure satisfactory performance and stable sludge for refractory contaminants at high flow rates and high concentrations, especially under remote and extreme cold-climate conditions. Mine effluents, including gold mining of low grade refractory ores, are complex and notoriously challenging to treat, mainly because of: (1) incomplete characterization (e.g., metastable contaminants, such as thiosalts, or contaminants in low concentrations, such as Se) especially with mixed effluents (drainage water from mining works, supernatant and exfiltrations from tailings ponds, etc.); (2) low temperature (and higher solubility of carbonates under such condition with accelerated exhaustion) and freeze–thaw cycles (with enhanced kinetics of some contaminants generation); and (3) poorly understood transient aquatic toxicity. Chemical treatment systems are usually efficient but tend to generate increasing residual salinity. New and/or adapted treatment approaches must be developed for these specific conditions.

Gaps and research needs for the four groups of CECs in mine effluents are the following: (1) **Salinity/TDS** originates from several sources: ores, chemicals (ore processing, water treatment), water recycling, saline water pumping, de-icing agents, permafrost alteration, etc. Salinity, a highly soluble and persistent contamination, is defined as an integrated measure of major, biologically essential ions (Na^+, Ca^{2+}, Mg^{2+}, K^+, Cl^-, SO_4^{2-} and HCO_3^-) and generally viewed to possess low individual toxicity (van Dam et al. 2014). (2) **Thiosalts**, as metastable sulfur oxyanions (up to 60% produced during sulfide flotation), are correlated to aquatic toxicity due to the pH depression following their oxidation to sulfuric acid (Miranda-Trevino et al. 2013) and to chemical degradation of xanthates (dithiocarbonates used in the flotation processes for most sulfide minerals) under acidic conditions (Alto et al. 1978; Novak and Holtze 2010; Xu et al. 1988). (3) **Se**, which is characterized as having a narrow window between essentiality and toxicity (Sharma et al. 2017). A predicted no-effect concentration (PNEC) for Se is set by Environment Canada, for freshwater, at 1 ppb (and by British Colombia at 2 ppb for freshwater and marine environment), while US EPA, Australia, and New Zealand have set values at 5 ppb or higher (Sharma et al. 2017). In addition, the US EPA criterion maximum concentrations (CMCs) for freshwater organisms are 257 ppb for Se(IV) and 417 ppb for Se(VI) (Sharma et al. 2017). A 5 ppb Se concentration was also recently issued as a trigger concentration for Canadian mine effluents (Government of Canada 2017). Precisely, if an effluent has >5 ppb of Se, the mine must conduct fish tissue sampling in the receiving environment to evaluate tissue Se concentrations. At the same time, the toxicity differences for Se species, i.e., oxyanions (selenite and selanite) vs. Se-based nanoparticles (released to the environment and produced by bioremediation of soluble Se) suggest the needs for additional research (Sharma et al. 2017) for the proper analysis of Se speciation (Pettine et al. 2015), for the environmental alteration of Se nanoparticles and for the correlation with associated toxicity. (4) **Nitrogen compounds** (NH_3, NO_2^-, NO_3^-) are also toxic (Baker et al. 2017; Jermakka et al. 2015).

Reduction of contaminant levels should not be the only consideration for soil or water remediation. Innovative approaches are needed to perform the treatment in a sustainable manner to reduce energy requirements and greenhouse gas emissions. Waste generation reduction and resource protection are essential. Consultation with stakeholders, which is essential for sustainability incorporation, should be maintained across the entire project life cycle which includes the evaluation of the contamination; development of alternatives; process design; and implementation, operation, and decommission of the process.

References

Adeleye AS, Conway JR, Garner K, Huang Y, Su Y, Keller AA (2016) Engineered nanomaterials for water treatment and remediation: costs, benefits and applicability. Chem Eng J 286:640–662. https://doi.org/10.1016/j.cej.2015.10.105

Akmehmet Balcioğlu I, Ötker M (2004) Pre-treatment of antibiotic formulation wastewater by O_3, O_3/H_2O_2, and O_3/UV processes. Turk J Eng Environ Sci 28(5):325–331

Alto K, Broderius S, Smith LL (1978) Toxicity of xanthates to freshwater fish and invertebrates. Minnesota Environmental Quality Council, Minneapolis

Babaee Y, Mulligan CN, Rahaman MS (2018) Removal of arsenic (III) and arsenic (V) from aqueous solutions through adsorption by Fe/cu nanoparticles. J Chem Technol Biotechnol 93 (1):63–71. https://doi.org/10.1002/jctb.5320

Baker JA, Gilron G, Chalmers BA, Elphick JR (2017) Evaluation of the effect of water type on the toxicity of nitrate to aquatic organisms. Chemosphere 168:435–440. https://doi.org/10.1016/j.chemosphere.2016.10.059

Banks D, Younger PL, Arnesen R-T, Iversen ER, Banks SB (1997) Mine-water chemistry: the good, the bad and the ugly. Environ Geol 32(3):157–174. https://doi.org/10.1007/s002540050204

Bokare AD, Choi W (2014) Review of iron-free Fenton-like systems for activating H_2O_2 in advanced oxidation processes. J Hazard Mater 275:121–135. https://doi.org/10.1016/j.jhazmat.2014.04.054

Bolong N, Ismail AF, Salim MR, Matsuura T (2009) A review of the effects of emerging contaminants in wastewater and options for their removal. Desalination 239(1–3):229–246. https://doi.org/10.1016/j.desal.2008.03.020

Brillas E, Bastida RM, Llosa E, Casado J (1995) Electrochemical destruction of aniline and 4-chloroaniline for wastewater treatment using a carbon-PTFE O_2-fed cathode. J Electrochem Soc 142(6):1733–1741. https://doi.org/10.1149/1.2044186

Calugaru IL, Neculita CM, Genty T, Zagury GJ (2018) Metals and metalloids treatment in contaminated neutral effluents using modified materials. J Environ Manag 212:142–159. https://doi.org/10.1016/j.jenvman.2018.02.002

Calugaru IL, Neculita CM, Genty T, Zagury GJ (2019) Removal efficiency of as(V) and Sb(III) in contaminated neutral drainage by Fe-loaded biochar. Environ Sci Pollut Res 26(9):9322–9332. https://doi.org/10.1007/s11356-019-04381-1

Chowdhury R, Mulligan CN (2013) Removal of arsenate from contaminated water by granular carbon embedded with nano scale zero-valent iron. Paper presented at the CGS Annual Conference, Montreal, Canada, 29 Sept – 2 Oct 2013

Cormier SM, Sutter GW II, Zheng L, Pond GJ (2013) Assessing causation of the extirpation of stream macroinvertebrates by a mixture of ions. Environ Toxicol Chem 32(2):277–287. https://doi.org/10.1002/etc.2059

Dubey S, Banerjee S, Upadhyay SN, Sharma YC (2017) Application of common nano-materials for removal of selected metallic species from water and wastewaters: a critical review. J Mol Liq 240:656–677. https://doi.org/10.1016/j.molliq.2017.05.107

El Zeftawy MAM, Mulligan CN (2011) Use of rhamnolipid to remove heavy metals from wastewater by micellar-enhanced ultrafiltration (MEUF). Sep Purif Technol 77(1):120–127. https://doi.org/10.1016/j.seppur.2010.11.030

ESG International Inc., Lakefield Research (2002) Guidance document for conducting toxicity reduction evaluation (TRE) investigations of Canadian metal mining effluents. Prepared for: Environment Canada and Mining Association of Canada. Guelph and Lakefield

Fast SA, Gnaneswar Gude V, Truax DD, Martin J, Magbanua BS (2017) A critical evaluation of advanced oxidation processes for removing contaminants removal. Environ Processes 4 (1):283–302. https://doi.org/10.1007/s40710-017-0207-1

Feng M, Wang X, Chen J, Qu R, Sui Y, Cizmas L, Wang Z, Sharma VK (2016) Degradation of fluoroquinolone antibiotics by ferrate(VI): effects of water constituents and oxidized products. Water Res 103:48–57. https://doi.org/10.1016/j.watres.2016.07.014

Gebhardt W, Schröder HF (2007) Liquid chromatography–(tandem) mass spectrometry for the follow-up of the elimination of persistent pharmaceuticals during wastewater treatment applying biological wastewater treatment and advanced oxidation. J Chromatogr A 1160(1–2):34–43. https://doi.org/10.1016/j.chroma.2007.05.075

Gogate PR, Pandit AB (2004) A review of imperative technologies for wastewater treatment II: hybrid methods. Adv Environ Res 8(3–4):553–597. https://doi.org/10.1016/S1093-0191(03)00031-5

Gogoi A, Mazumder P, Kumar Tyagi V, Tushara Chaminda GG, Kyoungjin An A, Kumar M (2018) Occurrence and fate of emerging contaminants in water environment: a review. Groundwater Sustainable Dev 6:169–180. https://doi.org/10.1016/j.gsd.2017.12.009

Gonzalez-Merchan C, Genty T, Bussière B, Potvin R, Paquin M, Benhammadi M, Neculita CM (2016) Ferrates performance in thiocyanates and ammonia degradation in gold mine effluents. Miner Eng 95:124–130. https://doi.org/10.1016/j.mineng.2016.06.022

Gonzalez-Merchan C, Genty T, Paquin M, Gervais M, Bussière B, Potvin R, Neculita CM (2018) Influence of ferric iron source on ferrate's performance and residual contamination during the treatment of gold mine effluents. Miner Eng 127:61–66. https://doi.org/10.1016/j.mineng.2018.07.014

Goodfellow WL, Ausley LW, Burton DT, Denton DL, Dorn PB, Grothe DR, Heber MA, Norberg-King TJ, Rodgers JH Jr (2000) Major ion toxicity in effluents: a review with permitting recommendations. Environ Toxicol Chem 19(1):175–182. https://doi.org/10.1002/etc.5620190121

Government of Canada (2017) Risk management approach for selenium and its compounds under the selenium-containing substance grouping. Environment and Climate Change Canada, Health Canada

Gözmen B, Oturan MA, Oturan N, Erbatur O (2003) Indirect electrochemical treatment of biphenol a in water via electrochemically generated Fenton's reagent. Environ Sci Technol 37 (16):3716–3723. https://doi.org/10.1021/es034011e

Holmes AB, Gu FX (2016) Emerging nanomaterials for the application of selenium removal for wastewater treatment. Environ Sci Nano 3(5):982–996. https://doi.org/10.1039/C6EN00144K

Ikehata K, Naghashkar NJ, El-Din MG (2006) Degradation of aqueous pharmaceuticals by ozonation and advanced oxidation processes: a review. Ozone Sci Eng 28(6):353–414. https://doi.org/10.1080/01919510600985937

Jermakka J, Wendling L, Sohlberg E, Heinonen H, Vikman M (2015) Potential technologies for the removal and recovery of nitrogen compounds from mine and quarry waters in subarctic conditions. Crit Rev Environ Sci Technol 45(7):703–748. https://doi.org/10.1080/10643389.2014.900238

Ji L, Chen W, Bi J, Zheng S, Xu Z, Zhu D, Alvarez PJ (2010) Adsorption of tetracycline on single-walled and multi-walled carbon nanotubes as affected by aqueous solution chemistry. Environ Toxicol Chem 29(12):2713–2719. https://doi.org/10.1002/etc.350

Jiang J-Q (2013) The role of ferrate(VI) in the remediation of emerging micro pollutants. Procedia Environ Sci 18:418–426. https://doi.org/10.1016/j.proenv.2013.04.056

Jiang W, Chen L, Rani Batchu S, Gardinali PR, Jasa L, Marsalek B, Zboril R, Dionysiou DD, O'Shea KE, Sharma VK (2014) Oxidation of microcystin-LR by ferrate(VI): kinetics, degradation pathways, and toxicity assessments. Environ Sci Technol 48(20):12164–12172. https://doi.org/10.1021/es5030355

Jing X-R, Wang Y-Y, Liu W-J, Wang Y-K, Jiang H (2014) Enhanced adsorption performance of tetracycline in aqueous solutions by methanol-modified biochar. Chem Eng J 248:168–174. https://doi.org/10.1016/j.cej.2014.03.006

Kennedy AJ, Cherry DS, Zipper CE (2005) Evaluation of ionic contribution to the toxicity of a coal-mine effluent using *Ceriodaphnia dubia*. Arch Environ Contam Toxicol 49(2):155–162. https://doi.org/10.1007/s00244-004-0034-z

Khuntia S, Majumder SK, Ghosh P (2013) Removal of ammonia from water by ozone microbubbles. Ind Eng Chem Res 52(1):318–326. https://doi.org/10.1021/ie302212p

Khuntia S, Majumder SK, Ghosh P (2014) Oxidation of as(III) to as(V) using ozone microbubbles. Chemosphere 97:120–124. https://doi.org/10.1016/j.chemosphere.2013.10.046

Lafi WK, Al-Qodah Z (2006) Combined advanced oxidation and biological treatment processes for the removal of pesticides from aqueous solutions. J Hazard Mater 137(1):489–497. https://doi.org/10.1016/j.jhazmat.2006.02.027

Li N (2017) Ferrate as a new treatment chemical for removal of effluent organic matter (EfOM) and emerging micro-pollutants in treated municipal wastewater for water reuse. Dissertation, Montclair State University

Liu P, Liu W-J, Jiang H, Chen J-J, Li W-W, Yu H-Q (2012) Modification of bio-char derived from fast pyrolysis of biomass and its application in removal of tetracycline from aqueous solution. Bioresour Technol 121:235–240. https://doi.org/10.1016/j.biortech.2012.06.085

Ma J, Wang Z, Mao B, Zhan J, Wu Z (2015) Electrochemical membrane bioreactors for sustainable wastewater treatment: principles and challenges. Curr Environ Eng 2(1):38–49. https://doi.org/10.2174/2212717802011508311145842

Microbial Technologies, Inc. (2006) Evaluation of treatment options to reduce water-borne selenium at coal mines in West-Central Alberta. Edmonton

Miranda-Trevino JC, Pappoe M, Hawboldt K, Bottaro C (2013) The importance of thiosalts speciation: review of analytical methods, kinetics, and treatment. Crit Rev Environ Sci Technol 43(19):2013–2070. https://doi.org/10.1080/10643389.2012.672047

Mulligan CN (2017) Biosurfactants for remediation of metal contamination. In: Das S, Dash HR (eds) Handbook of metal-microbe interactions and bioremediation. CRC Press, pp 299–314

Neculita CM, Coudert L, Genty T, Drapeau M, Ryskie S, Delay-Fortier S (2018) Emerging contaminants in mine effluents: operational challenges of their treatment and research needs. Paper presented at the 6th Mines and the Environment Symposium, Rouyn-Noranda, QC, Canada, 17–20 June 2018

Novak L, Holtze K (2010) Toxicity investigations. Case studies with metal mining effluent. http://bc-mlard.ca/files/presentations/2010-9-toxicity-case-studies-metal-mining-effluents.pdf. Accessed 20 Feb 2019

Oturan MA, Oturan N, Lahitte C, Trevin S (2001) Production of hydroxyl radicals by electrochemically assisted Fenton's reagent: application to the mineralization of an organic micropollutant, pentachlorophenol. J Electroanal Chem 507(1–2):96–102. https://doi.org/10.1016/S0022-0728(01)00369-2

Peng H, Pan B, Wu M, Liu Y, Zhang D, Xing B (2012) Adsorption of ofloxacin and norfloxacin on carbon nanotubes: hydrophobicity- and structure-controlled process. J Hazard Mater 233–234:89–96. https://doi.org/10.1016/j.jhazmat.2012.06.058

Petrie B, Barden R, Kasprzyk-Hordern B (2015) A review on emerging contaminants in wastewaters and the environment: current knowledge, understudied areas and recommendations for future monitoring. Water Res 72:3–27. https://doi.org/10.1016/j.watres.2014.08.053
Petrović M, Gonzalez S, Barceló D (2003) Analysis and removal of emerging contaminants in wastewater and drinking water. TrAC Trends Anal Chem 22:685–696. https://doi.org/10.1016/S0165-9936(03)01105-1
Pettine M, McDonald TJ, Sohn M, Anquandah GAK, Zboril R, Sharma VK (2015) A critical review of selenium analysis in natural water samples. Trend Environ Anal Chem 5:1–7. https://doi.org/10.1016/j.teac.2015.01.001
Ringwood K, Douglas B, Cote C, Harold M, McIntyre N, Butcher M (2016) Water stewardship – a practical handbook for operations. In: Proceedings of the 5th International Congress on Water Management in Mining, Santiago, Chile, 18–20 May 2016
Rivera-Utrilla J, Sánchez-Polo M, Ferro-García MÁ, Prados-Joya G, Ocampo-Pérez R (2013) Pharmaceuticals as emerging contaminants and their removal from water. A review. Chemosphere 93(7):1268–1287. https://doi.org/10.1016/j.chemosphere.2013.07.059
Rodriguez-Narvaez OM, Peralta-Hernandez JM, Goonetilleke A, Bandala ER (2017) Treatment technologies for emerging contaminants in water: a review. Chem Eng J 323:361–380. https://doi.org/10.1016/j.cej.2017.04.106
Ryskie S (2017) Treatment of ammonia nitrogen in mine effluents using advanced oxidation processes. Master's thesis, RIME-UQAT-Polytechnique
Ryskie S, Gonzalez-Merchan C, Neculita CM, Genty T (2019) Efficiency of ozone microbubbles for ammonia removal from mine effluents. Miner Eng (accepted)
Sandy T, DiSante C (2010) Review of available technologies for removal of selenium from water. Prepared for North American Metals Council. http://www.namc.org/docs/00062756.pdf. Accessed 20 Feb 2019
Santos S, Ungureanu G, Boaventura R, Botelho C (2015) Selenium contaminated waters: An overview of analytical methods, treatment options and recent advances in sorption methods. Sci Total Environ 521–522:246–260. https://doi.org/10.1016/j.scitotenv.2015.03.107
Sauvé S, Desrosiers M (2014) A review of what is an emerging contaminant. Chem Cent J 8:15. https://doi.org/10.1186/1752-153X-8-15
Sharma VK (2011) Oxidation of inorganic contaminants by ferrates (VI, V, and IV)–kinetics and mechanisms: a review. J Environ Manag 92(4):1051–1073. https://doi.org/10.1016/j.jenvman.2010.11.026
Sharma VK, Kazama F, Jiangyong H, Ray AK (2005) Ferrates (iron(VI) and iron(V)): environmentally friendly oxidants and disinfectants. J Water Health 3(1):45–58. https://doi.org/10.2166/wh.2005.0005
Sharma VK, Zboril R, Varma RS (2015) Ferrates: greener oxidants with multimodal action in water treatment technologies. Acc Chem Res 48(2):182–191. https://doi.org/10.1021/ar5004219
Sharma VK, McDonald TJ, Sohn M, Anquandah GAK, Pettine M, Zboril R (2017) Assessment of toxicity of selenium and cadmium selenium quantum dots: a review. Chemosphere 188:403–413. https://doi.org/10.1016/j.chemosphere.2017.08.130
Shaver D (2011) Sources and fate of emerging contaminants in municipal wastewater treatment. MASc thesis, University of Guelph
Sophia AC, Lima EC (2018) Removal of emerging contaminants from the environment by adsorption. Ecotoxicol Environ Saf 150:1–17. https://doi.org/10.1016/j.ecoenv.2017.12.026
Sui Q, Huang J, Deng S, Chen W, Yu G (2011) Seasonal variation in the occurrence and removal of pharmaceuticals and personal care products in different biological wastewater treatment processes. Environ Sci Technol 45(8):3341–3348. https://doi.org/10.1021/es200248d
Thines RK, Mubarak NM, Nizamuddin S, Sahu JN, Abdullah EC, Ganesan P (2017) Application potential of carbon nanomaterials in water and wastewater treatment: a review. J Taiwan Inst Chem Eng 72:116–133. https://doi.org/10.1016/j.jtice.2017.01.018
van Dam RA, Harford AJ, Lunn SA, Gagnon MM (2014) Identifying the cause of toxicity of a saline mine water. PLoS One 9(9):e106857. https://doi.org/10.1371/journal.pone.0106857

Werner CM, Hoppe-Jones C, Saikaly PE, Logan BE, Amy GL (2015) Attenuation of trace organic compounds (TOrCs) in bioelectrochemical systems. Water Res 73:56–67. https://doi.org/10.1016/j.watres.2015.01.013

Wilkinson J, Hooda PS, Barker J, Barton S, Swinden J (2017) Occurrence, fate and transformation of emerging contaminants in water: an overarching review of the field. Environ Pollut 231:954–970. https://doi.org/10.1016/j.envpol.2017.08.032

Xu Y, Lay JP, Korte F (1988) Fate and effects of xanthates in laboratory freshwater systems. Bull Environ Contam Toxicol 41:683–689

Yang X, Flowers RC, Weinberg HS, Singer PC (2011) Occurrence and removal of pharmaceuticals and personal care products (PPCPs) in an advanced wastewater reclamation plant. Water Res 45 (16):5218–5228. https://doi.org/10.1016/j.watres.2011.07.026

Yang B, Ying G-G, Zhao J-L, Liu S, Zhou L-J, Chen F (2012) Removal of selected endocrine disrupting chemicals (EDCs) and pharmaceuticals and personal care products (PPCPs) during ferrate(VI) treatment of secondary wastewater effluents. Water Res 46(7):2194–2204. https://doi.org/10.1016/j.watres.2012.01.047

Yong RN, Mulligan CN, Fukue M (2014) Sustainable practices in Geoenvironmental engineering, 2nd edn. CRC Press, Boca Raton

Yoon Y, Westerhoff P, Snyder SA, Wert EC, Yoon J (2007) Removal of endocrine disrupting compounds and pharmaceuticals by nanofiltration and ultrafiltration membranes. Desalination 202(1–3):16–23. https://doi.org/10.1016/j.desal.2005.12.033

Zhao L, Deng J, Sun P, Liu J, Ji Y, Nakada N, Qiao Z, Tanaka H, Yang Y (2018) Nanomaterials for treating emerging contaminants in water by adsorption and photocatalysis: systematic review and bibliometric analysis. Sci Total Environ 627:1253–1263. https://doi.org/10.1016/j.scitotenv.2018.02.006

Part VII
Technical Chapters

Chapter 30
Tool I: Characterization of nZVI Mobility in 1D and Cascade Columns by Ferromagnetic Susceptibility Sensor

Petr Parma, Alena Ševců, and Miroslav Černík

Abstract Mobility of nZVI is a vital parameter for designing effective in situ remediation strategy. This chapter describes the utilization of a laboratory device developed for the characterization of migration of differently modified nZVI materials based on their ferromagnetic properties. The one-dimensional column is suitable for comparing the migration properties of different types of nZVI and is less demanding for the material, while the cascade column simulates the process after nZVI injection into the treated well more realistically.

Keywords Nanoparticle · Iron · Magnetic susceptibility · Migration · Column

30.1 Introduction

Assessment of migration ability of different nZVI types or ferromagnetic nanoparticles under laboratory conditions is, in the vast majority of cases, limited to a one-dimensional arrangement (He et al. 2009; Kocur et al. 2013; Saleh et al. 2008). Experiments in a column setup represent an efficient and reliable option for mobility studies, especially when comparing nZVI materials of different particle size and surface modifications. It is possible to use a variety of columns filled with a suitable material, which could be monodisperse and chemically homogeneous material such as sorted or specially purified quartz sand (Saleh et al. 2008; Raychoudhury et al. 2012) or a polydisperse soil material taken directly from the studied site. The simplest column tests can be performed by pumping defined dispersion of nZVI through sand or soil, usually in short columns (about 10–25 cm) (Zhang et al. 2017).

P. Parma (✉) · A. Ševců
Institute for Nanomaterials, Advanced Technologies and Innovation, Technical University of Liberec, Liberec, Czech Republic
e-mail: petr.parma@tul.cz

M. Černík
Institute for Nanomaterials, Advanced Technologies and Innovation, Technical University of Liberec, Liberec, Czech Republic

AQUATEST a.s., Prague, Czech Republic

J. Filip et al. (eds.), *Advanced Nano-Bio Technologies for Water and Soil Treatment*, Applied Environmental Science and Engineering for a Sustainable Future,
https://doi.org/10.1007/978-3-030-29840-1_30

A more convenient option is to use devices based on ferromagnetic susceptibility (Vecchia et al. 2009; Strutz et al. 2016) that possess higher sensitivity and enable realistic simulation of field conditions. The one-dimensional column device and the cascade column with ferromagnetic susceptibility detector described in this chapter were both developed at the Technical University of Liberec.

30.2 One-Dimensional Column Device

This device enables mobility tests in long columns where, in a single experiment, it is possible to combine a standard assay for migration of nZVI particles in suspension with a washing test by simply replacing the original suspension with nanoparticle-free water.

For one-dimensional experimental arrangement, the basis of the device is a short length coil, freely encircling a column with the possibility of defined shifting in the direction of the axis of the column. The coil inductance varies reflecting the changes in the ferromagnetic susceptibility of the environment in which the coil is located. The coil is connected to the resonant circuit, which uses the coil inductance (***L***) dependence on the susceptibility (χ) of the environment where the coil is located. This dependency is described in Eq. 30.1:

$$L(\chi) = L_0 \cdot (1 + \chi) \tag{30.1}$$

The detection coil is part of a resonant circuit with a constant capacitance C_0 and the frequency of this circuit is described in Eq. 30.2:

$$f(\chi) = \frac{1}{2\pi\sqrt{L(\chi) \cdot C_0}} = \frac{1}{2\pi\sqrt{L_0 \cdot C_0} \cdot \sqrt{1+\chi}} \tag{30.2}$$

For very low χ values, it is possible to calculate frequency using Taylor series (Eq. 30.3):

$$f(\chi) = \frac{1}{2\pi\sqrt{L(\chi) \cdot C_0}} \approx \frac{1}{2\pi\sqrt{L_0 \cdot C_0}} \cdot \left(1 - \frac{\chi}{2} + 3\frac{\chi^2}{8} - 5\frac{\chi^3}{16} + O\chi^4\right) \tag{30.3}$$

When neglecting items from the second order above, then the relationship in Eq. (30.2) changes to the relationship in Eq. 30.4:

$$f(\chi) \approx \frac{1}{2\pi\sqrt{L_0 \cdot C_0}} \cdot \left(1 - \frac{\chi}{2}\right) \tag{30.4}$$

By comparing the resonant frequency in the air and in the presence of nZVI (ferromagnetic material), we get a relationship in Eq. 30.5:

$$\frac{f(\chi)}{f(0)} = \left(1 - \frac{\chi}{2}\right) \tag{30.5}$$

which can be adjusted to the final Eq. 30.6:

$$\chi = -2\frac{f(\chi) - f(0)}{f(0)} \tag{30.6}$$

The device developed for the evaluation of nZVI mobility is able to measure values of the numerator and denominator of Eq. 30.6.

The entire LC circuit is integrated directly in the measuring head. The device measures the frequency change of the LC circuit, which operates at frequency of approximately 4 kHz and is able to measure changes at the level of units of mHz. The device is equipped with a software, which allows to (i) control the drive unit of the measuring head on the basis of the selected program, (ii) repeat measurements at predetermined intervals, (iii) change the speed of the measuring head, and (iv) collect and store data from the analyzer.

The column is positioned vertically with the suspension inlet from the underside (Fig. 30.1). The column is filled with a suitable material and subsequently with water before the migration experiment. A suspension of the tested nZVI particles is pushed into the column using a peristaltic pump for accurate dosing of the suspension, which is constantly stirred with a rod stirrer in the storage vessel to prevent sedimentation of

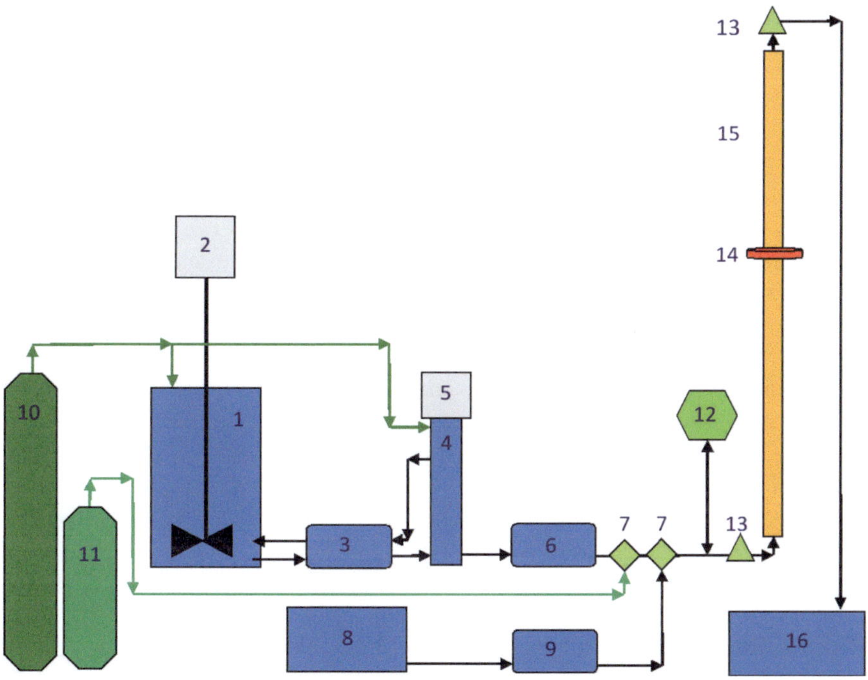

Fig. 30.1 Scheme of the device for one-dimensional measurement of nZVI migration. **1**—vessel for stock nZVI suspension, **2**—laboratory rod mixer, **3, 6, 9**—peristaltic pump, **4**—dispersing unit, **5**—laboratory disperser, **7**—three-way valve, **8**—vessel for water, **10**—gas bottle N_2, **11**—gas bottle CO_2, **12**—manometer, **13**—conductivity meter, **14**—measurement head (detector), **15**—experimental column, **16**—vessel for waste

Fig. 30.2 Migration curves of NANOFER 25 repeatedly measured every 10 min

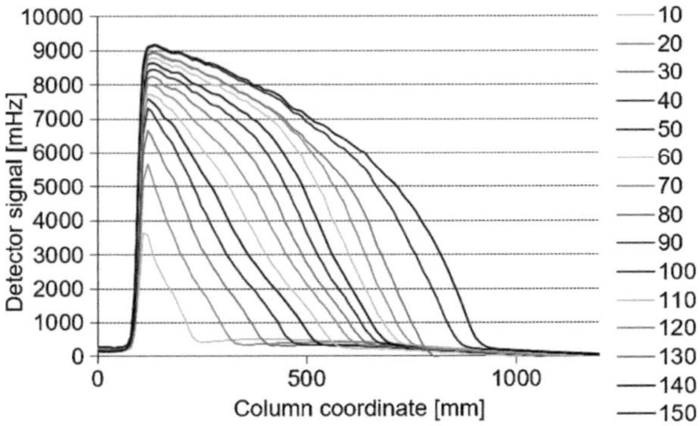

Fig. 30.3 Values measured by detector at position 130 mm during whole experiment with NANOFER 25

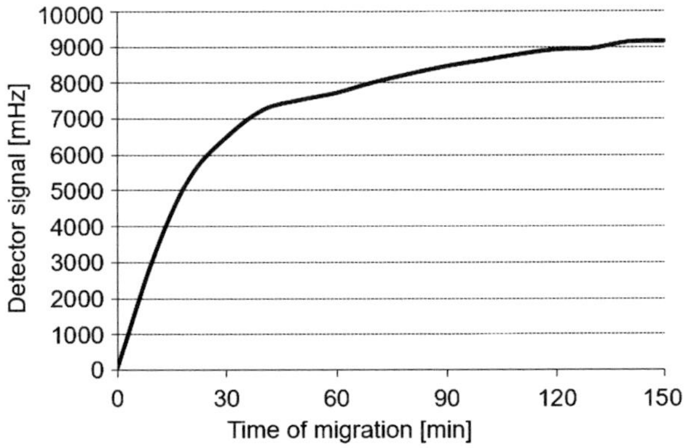

the particles. A container with a laboratory disperser is included to prevent particle aggregation. Moreover, inert gas is supplied throughout the experiment above the level of the suspension. The device is further equipped with a pressure detector at the inlet to the column, which provides additional information about the ongoing processes.

It is possible to measure the concentration at the coordinates outside the base of the column to obtain accurate information about the boundary conditions.

Example migration curves of commercial nZVI particles NANOFER 25 (NANO IRON s.r.o., Czech Republic) measured over a 10-minute interval are shown in Fig. 30.2. The nZVI concentration was 10 g/L, the column load was quartz sand with a grain of 1.2 mm. The dosing rate corresponds to a leakage rate of 100 m/24 h. Selected values can also be shown in the defined column positions (Fig. 30.3).

30.3 Cascade Column as an Advanced Alternative for Mobility Experiments

When nZVI particles are injected into a homogeneous isotropic environment, the treated area gradually forms an expanding cylindrical zone around the application well. The area at the bottom of the well is spherical. Therefore, the nZVI mobility

Fig. 30.4 Decrease in migration speed with increasing distance from the well (radius of the well 5 cm, pore speed on the shell of the well 100 m/24 h)

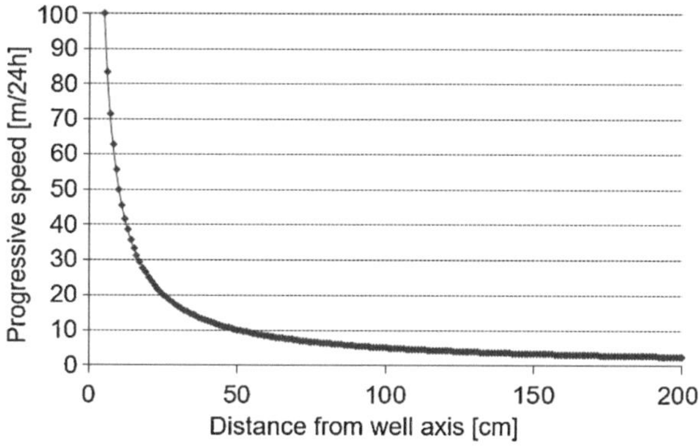

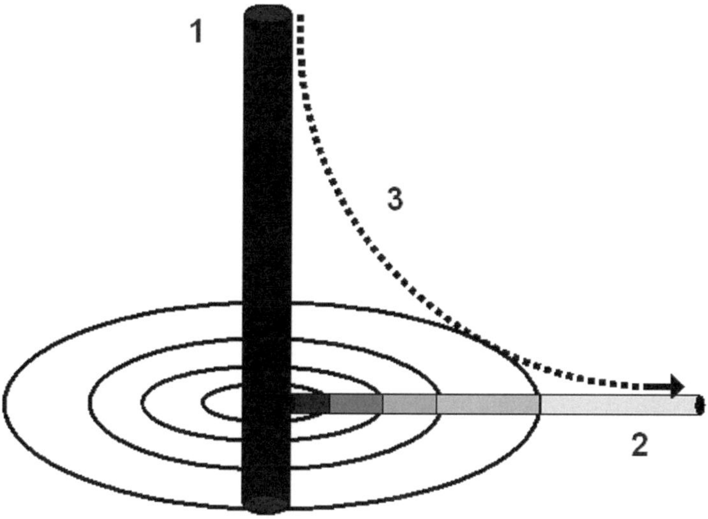

Fig. 30.5 Schematic interpretation of the cascade column related to the situation in the treated area. 1—application well, 2—profile of the collector simulated by the cascade column, 3—decreasing migration speed with increasing distance from the well

speed decreases with increasing distance from the borehole (Strutz et al. 2016). The relationship is hyperbolic in the radial direction (Fig. 30.4). Thus, a column with a constant transport rate of the suspension does not simulate the conditions of a real application. Therefore, a column with gradual flow rate reduction (cascade) was designed to at least partially simulate the decrease in the speed of the injected nZVI suspension. The connection with a real situation is demonstrated in Fig. 30.5.

The radial zone is divided into defined layers, with each layer is approximated by one section of column with a constant flow. A cascade column allows performing the entire measurement continuously by achieving a gradual reduction in the flow rate in defined column sections.

It is necessary to select the length of individual sections and the corresponding flow rate of the suspension before constructing the cascade column. If the transport velocity in the homogeneous isotropic environment decreases with the inverted value of the diameter, then we can use a relationship for the velocity v, at a distance r from the injection well axis, where v_p is the velocity on the well surface and r_p is the well diameter (Eq. 30.7):

$$v = \frac{v_p \cdot r_p}{r} \tag{30.7}$$

Since the flow rate is constant throughout each section of the column, it is necessary to choose the method of calculating this velocity for the most appropriate approximation of the real curve. To accurately define the flow rate of the suspension through the calculated section of the column that corresponds to a real section of the collector, the velocity in a given section of the column ($v_{1,2}$) is defined by the velocity at the beginning (v_1) and at the end (v_2) of each section (Eq. 30.8):

$$v_{1,2} = \frac{2}{\frac{1}{v_1} + \frac{1}{v_2}} \tag{30.8}$$

Equation 30.8 corresponds to the equation for the harmonic average.

The fundamental part of the cascade column consists of a selected poly(methyl methacrylate) column having an outer diameter of 70 mm and a wall thickness of 4 mm (Fig. 30.6). The actual measurement of the nZVI content is performed by the device described in the previous section with an internal diameter of the detector corresponding to the diameter of the cascade column. An inner tube ($d = 20$ mm) serves as a technical channel for the inlet and outlet of the suspension that is pumped into the column through a series of thin tubes (Fig. 30.6a). Importantly, the nZVI in the suspension flowing through the thin tubes may contribute to the overall signal. However, this can be ignored because (i) the diameter of the tubes is considerably smaller than the diameter of the column, (ii) the concentration of nZVI in the suspension at a specific site is always lower than its concentration in the porous water, and (iii) the tubes are near the detector axis and the sensitivity of the detector around the axis is considerably smaller than near the coil.

Mobility of nZVI particles (type NANOFER 25) can serve as an example experiment (Fig. 30.6b). The column was filled with quartz sand with a grain size of 1 mm. The suspension of nZVI particles NANOFER 25 (20 g/L) with carboxymethylcellulose (10 g/L) was dosed for more than 8 h using a pressure pump. As a starting point, the migration rate in the theoretical well was presumed to be 100 m/24 h and the diameter of the well was 10 cm. The boundaries of the column sections were set to 21.3 cm, 31.3 cm, 67.2 cm, and 185 cm from the beginning of the column, these values correspond to a flow rate of 32 m/24 h, 16 m/24 h, 9.22 m/24 h, and 3.89 m/24 h.

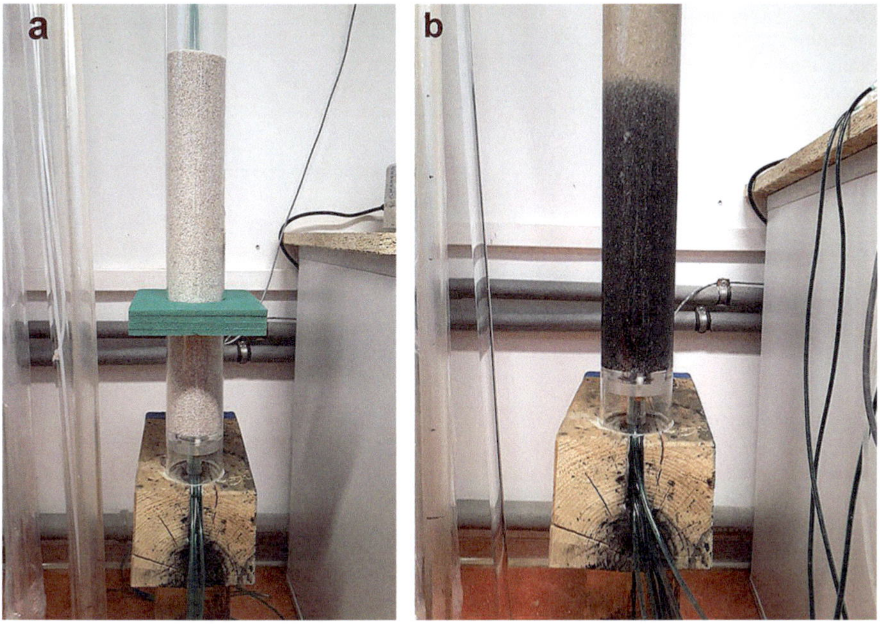

Fig. 30.6 (**a**) Cascade column with visible inner tube serving as a technical channel for the inlet and outlet of the suspension that is pumped into the column through series of thin tubes. (**b**) Cascade column during an experiment where the black zone indicates mobility of nZVI

The measurement was done using the device described in the previous section. The migration curves obtained at selected times of the experiment were compared with an experiment performed under the same conditions in a standard one-dimensional column with an internal diameter of 20 mm. A comparison of the migration curves is displayed in Figs. 30.7 and 30.8. The results are plotted on separate axes of both graphs because a detector with different sensitivity was used for each column.

The cascade column simulates the process during nZVI injection into the well more precisely than the standard one-dimensional column. However, experiments in standard columns with constant flow are much less demanding for the material (sand or soil filling, nZVI suspension, eventual surface modifiers, etc.). For this reason, the one-dimensional column is more suitable for comparing the migration properties of different types of nZVI while the results obtained using the cascade column are suitable for more realistic testing of the migration of nZVI.

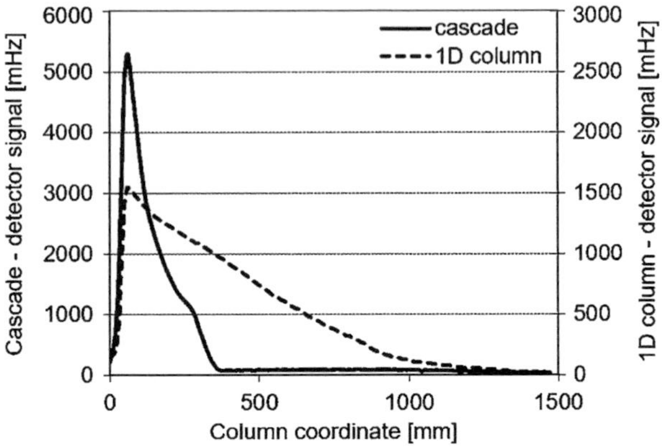

Fig. 30.7 Comparison of migration curves measured using standard one-dimensional (1D) column and in a cascade column after 30 min

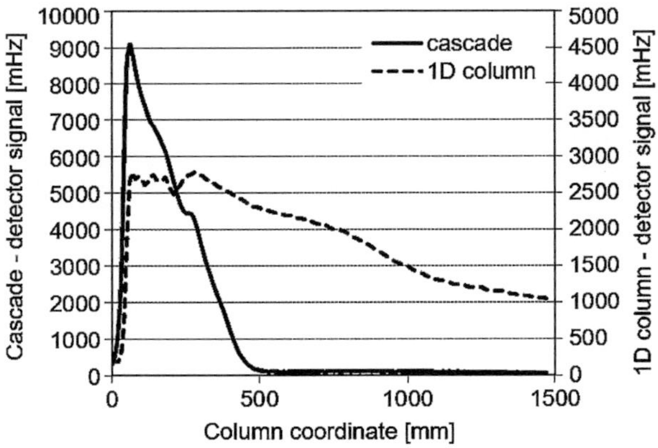

Fig. 30.8 Comparison of migration curves measured using standard one-dimensional (1D) column and in a cascade column after 120 min

References

He F, Zhang M, Qian T, Zhao D (2009) Transport of carboxymethyl cellulose stabilized iron nanoparticles in porous media: column experiments and modeling. J Colloid Interface Sci 334 (1):96–102. https://doi.org/10.1016/j.jcis.2009.02.058

Kocur CM, O'Carroll DM, Sleep BE (2013) Impact of nZVI stability on mobility in porous media. J Contam Hydrol 145:17–25. https://doi.org/10.1016/j.jconhyd.2012.11.001

Raychoudhury T, Tufenkji N, Ghoshal S (2012) Aggregation and deposition kinetics of carboxymethyl cellulose-modified zero-valent iron nanoparticles in porous media. Water Res 46(6):1735–1744. https://doi.org/10.1016/j.watres.2011.12.045

Saleh N, Kim H-J, Phenrat T, Matyjaszewski K, Tilton RD, Lowry GV (2008) Ionic strength and composition affect the mobility of surface-modified Fe^0 nanoparticles in water-saturated sand columns. Environ Sci Technol 42(9):3349–3355. https://doi.org/10.1021/es071936b

Strutz TJ, Hornbruch G, Dahmke A, Köber R (2016) Effect of injection velocity and particle concentration on transport of nanoscale zero-valent iron and hydraulic conductivity in saturated porous media. J Contam Hydrol 191:54–65. https://doi.org/10.1016/j.jconhyd.2016.04.008
Vecchia ED, Luna M, Sethi R (2009) Transport in porous media of highly concentrated iron micro- and nanoparticles in the presence of xanthan gum. Environ Sci Technol 43(23):8942–8947. https://doi.org/10.1021/es901897d
Zhang M, He F, Zhao D, Hao X (2017) Transport of stabilized iron nanoparticles in porous media: effects of surface and solution chemistry and role of adsorption. J Hazard Mater 322:284–291. https://doi.org/10.1016/j.jhazmat.2015.12.071

Chapter 31
Tool II: Membrane Interface Probe

Vladislav Knytl

Abstract Detailed investigation of soil and sediment pollution is a crucial aspect of remediation. Direct-push methods (direct sensing tools) represent a powerful tool for characterization of the contaminated site. One of these methods is Membrane Interface Probe suitable for detection of volatile organic compounds in soils.

Keywords Membrane interface probe · Direct sensing tools · Contamination · Investigation

31.1 Introduction

Detailed investigation of soil and sediment pollution is a crucial aspect of remediation. Direct-push methods (direct sensing tools) represent a powerful tool for characterization of the contaminated site. These methods are an essential part of the so-called high-resolution site characterization (HRSC). This strategy developed by the US EPA (Environmental Protection Agency) enables to obtain the most detailed information regarding the actual distribution of pollution within the site (US EPA 2016).

The data obtained from the direct investigation helps to create a more accurate model of the contamination and eliminate blind paths and unexpected barriers that arise from a lack of knowledge of the distribution of contaminants. It is important to deliver the remediation agent directly into the contaminated layers in order to enhance the remediation efficiency. This is especially useful in the case of remediation agents based on direct chemical reaction with contaminants. One of the major limitations of these materials is their limited migration potential through the aquifer. Therefore, a lot of research effort goes into enhancing its migration capabilities (Soukupova et al. 2015).

V. Knytl (✉)
DEKONTA a.s., Stehelčeves, Czech Republic
e-mail: knytl@dekonta.cz

J. Filip et al. (eds.), *Advanced Nano-Bio Technologies for Water and Soil Treatment*, Applied Environmental Science and Engineering for a Sustainable Future,
https://doi.org/10.1007/978-3-030-29840-1_31

31.2 Membrane Interface Probe (MIP)

One of the core direct investigation technologies is MIP—membrane interface probe (Geoprobe, USA). MIP is a probing system designed to delineate the contamination plume and be capable of depicting distribution and relative magnitude of both halogenated and nonhalogenated volatile organic compound (VOC) contaminants in real time (McAndrews et al. 2003; Bronders et al. 2009). It can be used in both saturated and vadose soils.

The principle of the operation is summarized in the following text. The probe is equipped with a heating block with special semipermeable membrane, which comprises of a thin film polymer impregnated into a stainless steel screen. The probe is pushed into the subsurface by a suitable drilling rig and the block is heated up to 120 °C. Mobilized contaminants diffuse across the membrane on the basis of a concentration gradient between the contaminated soil and the clean carrier gas, which sweeps behind the membrane (see Fig. 31.1). The analytical solution for contaminated gas is a gas chromatograph (GC) equipped with three detectors. Each detector provides sensitivity to a particular group of contaminants: PID (aromatic hydrocarbons, unsaturated hydrocarbons); FID (ethenes, light hydrocarbons C1–C9); and XSD (chlorinated hydrocarbons).

MIP technology can be enhanced by addition of various components. The most common is electric conductivity (EC) dipole for measuring electric conductivity of the environment in the vicinity of the probe. This is useful for determination of the soil composition (mainly coarser-or finer-sized particles) and localization of

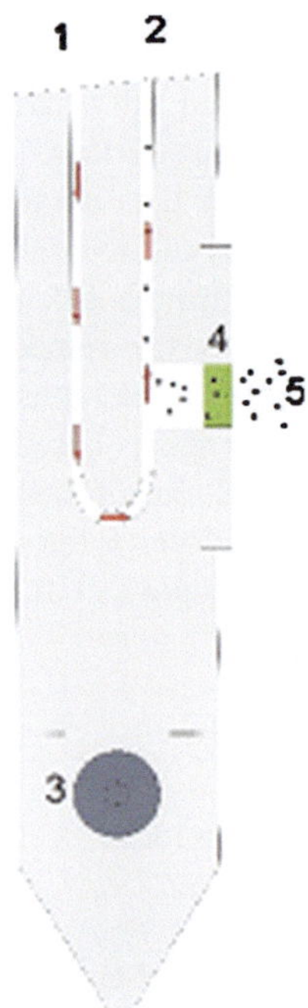

Fig. 31.1 A schematic illustration of MIP probe (1) teflon tube with flowing carrier gas (N_2), (2) return PEEK tube with contamination, (3) dipole for electric conductivity measurements, (4) semipermeable membrane, (5) VOCs in sediments

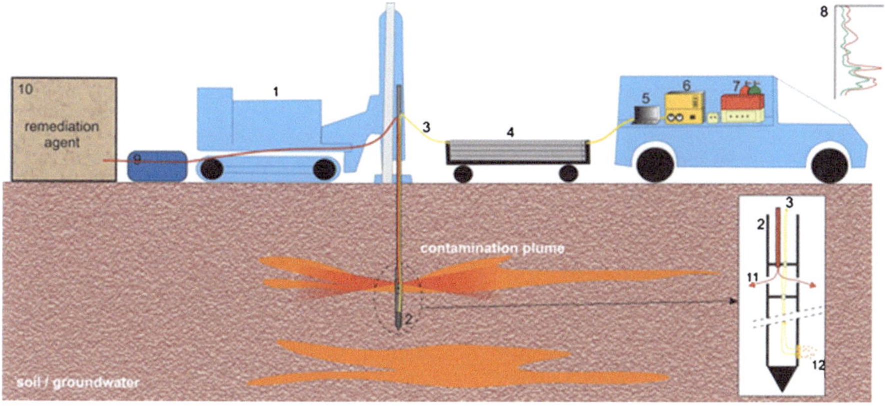

Fig. 31.2 Schematic representation of the MIP-IN detection-injection device, (1) drilling machine, 2) MIP-IN probe, (3) tubes with the carrier gas and MIP cables (so-called trunkline), (4) drilling rods holder, (5) PC for MIP data processing, (6) MIP control unit, (7) gas chromatograph with three detectors and gas cylinders (N_2, H_2), (8) measured signals of MIP probe, (9) high pressure injecting pump, (10) mixer tank with the liquid remediation reagent, (11) reagent injected via jets, (12) contamination entering through the semipermeable membrane. (Adapted from Kukačka et al. 2015)

lithological boundaries, e.g., sand/gravel, sand/clay. A proper understanding of lithology is crucial for tracking possible routes of contaminant migration and possible area of its accumulation. The other useful additions are cone penetration testing (CPT), hydraulic profiling tool (HPT), or possible upgrade of GC to consider results more quantitative (McCall et al. 2014; Adamson et al. 2014).

There are other technologies useful for direct investigation based on different principles. One of them is optical image profiler for petroleum hydrocarbons (McCall et al. 2016).

Research conducted in recent years has pushed direct logging tools even further, linking investigation directly to the remediation process. An example of such an interconnection is technology MIP-IN, which represents the combination of an MIP technology and direct-push injection of the remediation agent (Fig. 31.2). This approach allows treating the contaminated zone within the investigation process immediately (Kukačka et al. 2015; Bastiaens and Stubdrup 2012).

31.3 Results from the Field

The MIP usability in the field can be demonstrated by the results obtained from the chemical factory located in the east of the Czech Republic. The site is located in the north of a city, close to drinking water resources used as a reservoir for the city. The entire factory grounds are significantly contaminated, mainly by BTEX, monochlorobenzene (MCB), chlorinated ethenes (ClE), and drug residuals that are present in the soil and groundwater. The aquifer is shallow and well permeable, with water table at 3–4 m below ground level. This site served well for detailed

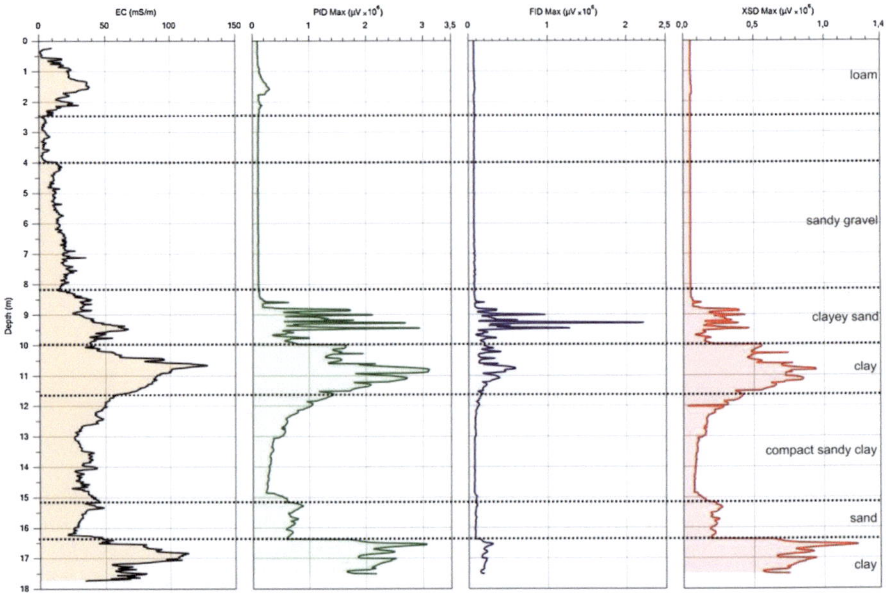

Fig. 31.3 Typical example of MIP log from the Czech Republic site shows contamination with chlorinated solvents in upper and lower aquifer. The first column represents electric conductivity (EC), followed by signals (mV) of PID, FID, and XSD detector. Simplified geological formations match well with EC values, and the transitions zones (e.g., sand/clay, gravel/sand) are well visible

investigation. The MIP probing has helped to identify two migration zones through which the contamination of ClE is spreading. A revelation of a lower contaminated zone (below 15 m bgs) was an important finding because the depth of all historical monitoring wells within the area reached only 10 m bgs. Therefore, the extent of this contamination was unclear before the MIP tool was employed. The typical MIP log from this site is depicted in Fig. 31.3.

As mentioned earlier, the data obtained from several MIP probes can be interpolated and serve as a valuable model of contamination. For example, cuts through the contamination plume can identify the source zones of the contamination and help to direct the injection of the oxidizing agent. One of the profiles from the same site is shown in Fig. 31.4. Probing was performed in the area contaminated with both BTEX and chlorinated solvents.

The data obtained from large MIP investigation can represent a valuable source for creation of 3D models of contaminated sites. Models of contamination plumes can significantly help to estimate the total amount of contaminants and reveal potential sources zones or migration paths. The large MIP investigation performed on the site of an old dump near Opava, a town in the Czech Republic, is supposed to be a good example of using a 3D model (Fig. 31.5). The survey has significantly helped to understand the extent of the contaminated area and also distinguish between different kinds of pollution. This model has served well in designing a spatial distribution of wells and sampling points. After such characterization the site

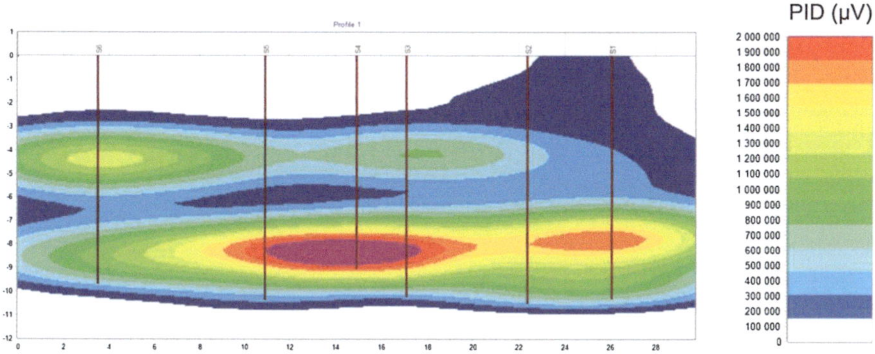

Fig. 31.4 An example of a profile obtained by interpolation of MIP data; in this case signals from a PID detector, which is sensitive for BTEX

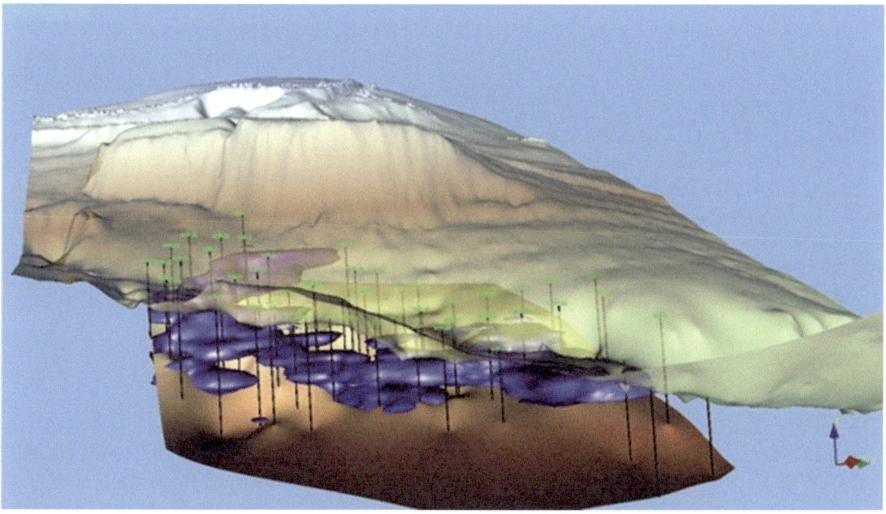

Fig. 31.5 An example of a 3D model of chlorinated solvents plume leaking from an old dump site. The plume is vertically bordered by a surface model of the terrain and sandstone eluvium (maximum depth accessible to direct-push drilling rig)

is prepared for the remediation process; in this particular case, it is excavation and direct-push application of remediation agents.

Despite a number of advantages, the MIP technology has several limitations. MIP is a semiquantitative tool and therefore not supposed to replace traditional soil bores and monitoring wells. It is a suitable solution for delineation of contamination plume, defining source zones, and a better understanding of the relation between contamination and lithology. The other limitation is the actual depth level which the probe is capable of reaching. It depends on the power of the drilling rig and geological conditions. Detection limits of each detector are also an important issue. Tracking dissolved (first hundreds of μg/L) contamination in very permeable aquifers could be problematic. In order to enhance the performance of the detection limits, the manufacturer offers an updated technology called low-level MIP. Correct evaluation of MIP logs is crucial. If the probe enters a highly contaminated zone, it will cause significant "drag off" or "tailing", so the clean intervals could appear contaminated (Bumberger et al. 2012). However, an experienced operator should recognize this and interpret the data correctly.

References

Adamson DT, Chapman S, Mahler N, Newell C, Parker B, Pitkin S, Rossi M, Singletary M (2014) Membrane interface probe protocol for contaminants in low-permeability zones. Groundwater 52(4):550–565. https://doi.org/10.1111/gwat.12085

Bastiaens L, Stubdrup OP (2012) Soil treatment device and use thereof for treating contaminated soil and/or groundwater contained therein. Patent WO 2012/123518 A3

Bronders J, Van Keer I, Touchant K, Vanermen G, Wilczek D (2009) Application of the membrane interphase probe (MIP): an evaluation. J Soils Sediments 9(1):74–82. https://doi.org/10.1007/s11368-008-0054-9

Bumberger J, Radny D, Berndsen A, Goblirsch T, Flachowsky J, Dietrich P (2012) Carry-over effects of the membrane interface probe. Groundwater 50(4):578–584. https://doi.org/10.1111/j.1745-6584.2011.00879.x

Kukačka J, Bastiaens L, De Vos J, Andersen B (2015) MIP-IN: new device for combined detection of pollutants and injection of reagents. Waste Forum 4:232–241

McAndrews B, Heinze K, DiGuiseppi W (2003) Defining TCE plume source areas using the membrane interface probe (MIP). Soil Sediment Contam 12(6):799–813. https://doi.org/10.1080/10588330390254838

McCall W, Christy TM, Pipp D, Terkelsen M, Christensen A, Weber K, Engelsen P (2014) Field application of the combined membrane-interface probe and hydraulic profiling tool (MiHpt). Groundwater Monit Rem 34(2):85–95. https://doi.org/10.1111/gwmr.12051

McCall W, Christy T, Goodrich J (2016) Application of the optical image profiler (OIP) to characterize the Grimm oil site, Kalona, Iowa. https://doi.org/10.13140/RG.2.2.10431.30888

Soukupova J, Zboril R, Medrik I, Filip J, Safarova K, Ledl R, Mashlan M, Nosek J, Cernik M (2015) Highly concentrated, reactive and stable dispersion of zero-valent iron nanoparticles: direct surface modification and site application. Chem Eng J 262:813–822. https://doi.org/10.1016/j.cej.2014.10.024

US EPA (2016) Contaminated site clean-up information. https://clu-in.org/characterization/technologies/hrsc/hrscintro.cfm. Accessed 10 July 2019

Chapter 32
Tool III: Fracturing for Enhanced Delivery of In Situ Remediation Substances in Contaminated Sediments

Jan Kukačka and Petr Kvapil

Abstract This chapter summarises the basic information on methods for fracturing the geological environment for the purposes of permeability increase and thus aquifer remediation enhancement. The technology allows to increase the infiltration of remediation substances into contaminated geological environment showing low permeability that can be hardly remediated without fracturing.

Keywords Hydraulic fracturing · Pneumatic fracturing · Aquifer remediation · Fracture propagation

32.1 Introduction

Treatment of contaminated geological environments with increasing permeability employing fracturing techniques may be regarded as a promising method for enhancing the efficiency of in situ remediation of aquifers. These techniques are based on creating a network of artificially made cracks in the profile of sedimentary rocks, which subsequently facilitate the pumping of the contaminants out of the aquifer and injecting reactive substances into the soil with the purpose of supporting in situ decomposition of the contaminants. Hydraulic and pneumatic fracturing methods are verified in the long run. These methods have already been used for more than 50 years for various purposes, especially in oil mining and water industries.

Currently, various fracturing methods are available for groundwater and soil remediation, used both in saturated and unsaturated zones. Pneumatic fracturing enables creation of cracks in the soil profile by means of controlled impulses of high

J. Kukačka (✉)
DEKONTA a.s., Stehelčeves, Czech Republic
e-mail: kukacka@dekonta.cz

P. Kvapil
Photon Water Technology s.r.o., Liberec, Czech Republic

J. Filip et al. (eds.), *Advanced Nano-Bio Technologies for Water and Soil Treatment*, Applied Environmental Science and Engineering for a Sustainable Future,
https://doi.org/10.1007/978-3-030-29840-1_32

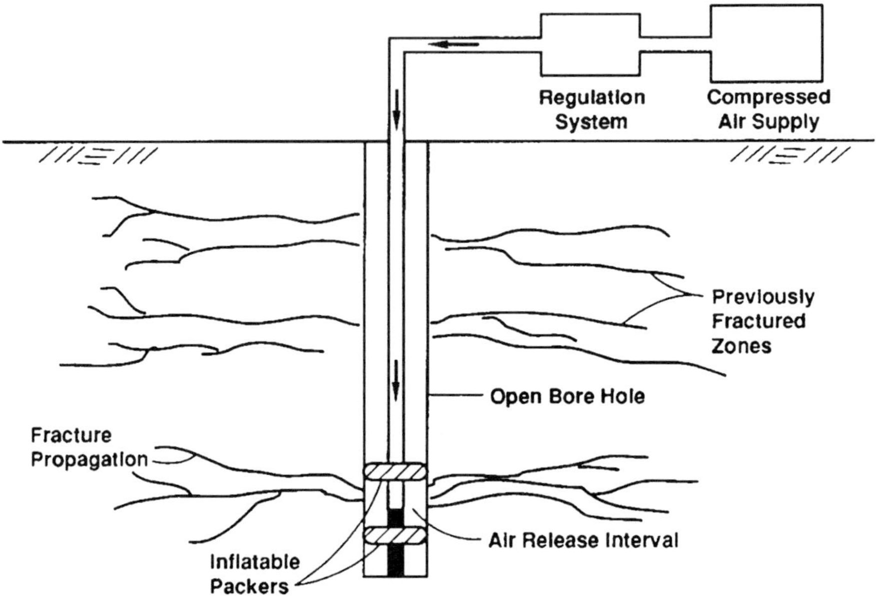

Fig. 32.1 Pneumatic fracturing principle. (Reproduced from Suthersan 1997)

pressure of air or another gas. In the case of hydraulic fracturing, the cracks are produced by pressure injection of a liquid into the geological environment. Pneumatic and hydraulic fracturing methods are further described below.

The speed of crack formation differs by orders of magnitude in the individual methods. The crack created by means of hydraulic fracturing is formed by the lowest speed (ca 0.1 m/s); in the case of pneumatic fracturing the speed is ca 2 m/s.

32.2 Pneumatic Fracturing

Pneumatic fracturing is carried out through high-pressure injection of gas into the borehole so that it causes formation of new cracks and more permeable areas in the geological environment. Fracturing may be carried out by means of either boreholes with special casing (without packing) or penetration probes without casing (direct push probes). It can be performed either in one level or in more levels, with the use of packers. The pressurized air source is usually a compressor or high-pressure containers (Schuring 2002).

In the case of multilevel fracturing, cracks are formed gradually and the fractured horizon is usually delimited by two packers (Fig. 32.1). During the process the packers are gradually moved in order to carry out geological environment fracturing with the required number of formed cracks. The vertical interval between the cracks is usually in the range from 0.2 to 1.0 m. The speed of crack formation is usually

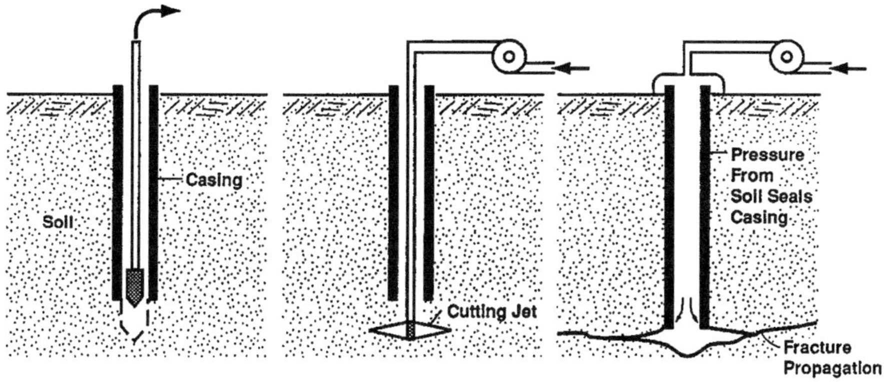

Fig. 32.2 Hydraulic fracturing principle. (Reproduced from Suthersan 1997)

relatively high (ca 2 m/s). The equipment for pneumatic fracturing may be placed into a car trailer. In addition to what is stated above, complementary equipment may be used for concurrent injection of liquid or solid complementary substances necessary for the remediation purposes.

In the case of pneumatic fracturing applications, the ability of a geological environment to spontaneously keep the newly formed crack open is relied on. It has been proved that relatively high amounts of liquid may flow through cracks of even very small sizes. In the case of fragile geological materials, such as rocks of the bedrock material, the formed cracks usually do not close back mainly owing to irregular and coarse surface of the fractures and small movements of the individual rock blocks. In the case of plastic clays, it has been proved that their swelling causes the cracks to close. Within the framework of the pneumatic fracturing, the formed cracks may be filled with the so-called permeable proppant (sand or a ceramic material). The use of a proppant is particularly suitable in fine-grained incohesive materials (silty sand). The proppant is transported into the geological environment during the pneumatic fracturing directly by the injected gas flowing with high speed.

32.3 Hydraulic Fracturing

Hydraulic fracturing is carried out through a high-pressure injection of water or polymer-based liquids into the borehole so that it causes artificial formation of a new, more permeable area (crack) in the given depth. Similarly to pneumatic fracturing, hydraulic fracturing may be carried out in one of two ways. For hydraulic fracturing, either a system of boreholes with special casing or penetration probes without casing (Fig. 32.2), formed by the direct push method, may be used (Schuring 2002).

Before the beginning of the actual hydraulic fracturing, the cracks in question may start in the required depth by means of a high-pressure water beam, optionally with the addition of an abrasive. This is followed by an injection of the fracturing liquid, formed, in the majority of cases, by a mixture of a cross-linked or linear polymer with water as the base and sand or another suitable proppant. Further fracturing boreholes are gradually added to the site, till the target region is fully fractured. Theoretically, hydraulic fracturing may be carried out in boreholes without casing, using a system of packers. In this way, even several fractured areas located one above the other may be formed, wherein the required vertical interval between the individual areas is usually in the range from 1.0 to 2.0 m. The expected speed of crack formation is usually at the level of ca 0.1 m/s. The necessary technological equipment is commonly available in oil drilling and mining companies. However, the equipment is usually intended for much higher requested outputs than the ones necessary in the remediation practice. Thus, a smaller and simpler equipment has been developed for remediation purposes.

Depending on the application depth, either a vertically or horizontally oriented crack is formed. At a depth up to 30 m, which is common for remediation practice, formation of a vertically oriented crack is practically excluded. The formed crack may be oriented in a different than horizontal way only if the natural geological environment layers are thus oriented.

In the case of unconsolidated geological materials, a proppant is always added into the fracturing mixture. For a pressure injection, the proppant must be always kept in a suspension. This is achieved either through intensive mixing or through addition of other additives and the fracturing liquid (Zeidler and Pašiak 1986). In this way, a cross-linked polymer (gel) is formed, which is able to maintain its structure for the necessary time period (ca 12–24 hours). After this time period, it disintegrates to a liquid showing viscosity similar to water, which enables its subsequent washing out of the geological environment. A various spectrum of liquid as well as solid materials may be used for hydraulic fracturing thanks to which a very quick single injection of suitable auxiliary substances for implementation of in situ remediation methods may be achieved (Siegrist et al. 2001).

32.4 Feasibility Limits of Fracturing Methods

Suitability of utilization of the fracturing for supporting remediation measures is influenced, in particular, by the following geotechnical conditions of the geological environment:

- Kind of the environment: The final effect of the fracturing may be estimated in advance. The highest efficiencies are achieved in a compact environment showing low permeability, such as fine-grained sedimentation materials.
- Permeability of the environment: The most frequent application of fracturing is a targeted increase in aquifer permeability, and thus the impact of remediation. The general rule is that the smaller the permeability, the higher efficiency of procedures for fracturing is achieved, and vice versa.
- Depth of the area of interest: Theoretically, there is no depth limit for creating new fractures in the geological environment that would ensure sufficient flow and pressure applied to the area of interest. In the remediation practice, the depth can

reach up to 30 m below ground level. Further, various orientation of the created fractures may be expected depending on the depth. In the case of the usual depth (ca 30 m below ground level), subhorizontal crack orientation may be expected (as described above).

- Plasticity: In highly plastic and cohesive materials with a high proportion of clayey minerals, plastic deformations of the geological material may take place during the fracturing. The likelihood that a fracture will form due to cracking of the geological material is reduced. Plasticity of the material is connected with the amount and type of the present clayey minerals (an increased plasticity is shown by unconsolidated geological materials with clayey mineral content >30%). Plastic materials change volume with a change of water content. Generally, plastic geological materials with a low water content behave better during the fracturing than geological materials with a high water content. In plastic geological materials, there is greater likelihood that the created crack would be closed back so it is recommended that the created cracks are filled with a proppant.
- Secondary structures: They are further important factors influencing the crack direction and speed of its formation. They particularly include natural layering, cracks in the rock, foliation areas, and, optionally, the presence of root channels and/or anthropogenic disturbances of the rock block (old excavations, boreholes, underground lines, etc.). Secondary structures in the geological bedrock may behave as preferential ways. They may play a dominant role in spreading pressure changes in the material, and they usually reduce the fracturing efficiency. While planning the locations and depth of the fractured areas, these structures should be taken into account.
- Groundwater level: The geological environment may be fractured regardless of the fact whether it is in a saturated or unsaturated zone. Groundwater presence slightly influences the course of the work and the behavior of the environment after the fracturing. For example, in the case of pneumatic fracturing, the presence of water in the geological material increases tightness of the environment preventing penetration of gases, and, thus, higher reach of the fracturing may be expected.

32.5 Fracturing Feasibility Risks

During planning and implementing the procedures based on geological environment fracturing, it is necessary to take into account several important factors that could influence the further remediation process.

- Risk of contaminant mobilization due to fracturing: Geological environment fracturing causes an increase in its permeability, and thus also increase in the environment ability to transport contamination that showed low mobility, or was out of reach of flowing groundwater, before the fracturing. Increase in permeability of the geological environment as well as simplification of access to the contaminant are the aims of the described methods. Therefore, it is recommended that possible impacts on the contaminant's increasing mobility are evaluated, and implementation of the measures is incorporated into the time schedule of the project properly.

- Risk of impacts on stability of buildings and above-ground structures: During fracturing, deformation of the geological material in the surroundings of the newly formed crack takes place. Vertical movements on the surface and lifting of the surface may take place, as well as propagation of vibrations and seismic waves. Hence, it is recommended that the possible impacts on underground, as well as above-ground structures, are thoroughly assessed before the fracturing implementation. The extent of vertical movements on the surface of the ground depends on the fracturing method, depth of the horizon of interest, number of cracks, lithology of the geological environment, cohesion and strength of the geological material, plasticity, density, consistency, weathering level, layering of the environment, locations of aquifers and insulators, fracturing pressure and gradient, speed of seismic wave propagation, etc. Thus, fracturing preparation and implementation in gradual steps is necessary: investigation drilling, laboratory and field tests, pilot fracturing tests, project and implementation of full-scale fracturing.

32.6 Fracturing in Remediation Applications

For the remediation purposes, hydraulic fracturing may be used for enhancing remediation of unsaturated and saturated zones, as well as for remediation of perched aquifers. In a majority of cases, fracturing is combined with a selected decontamination method, for the purpose of increasing its efficiency (pump-and-treat approach, venting, in situ oxidation/reduction, methods based on the principles of in situ biodegradation, etc.). Geological environment fracturing may increase its effective hydraulic and pneumatic conductivity. Especially in an environment showing low permeability, such as clayey and silty geological materials, transport of contaminants is controlled by diffusion, and, thus, the process is very slow. Creation of a network of artificial cracks in the geological environment causes reduction of length of the diffusion paths, and, simultaneously, advection share of the transport increases. As a result, the remediation intervention is accelerated. Concurrently, this method may ensure access to places that would be left untouched in the case of utilization of standard remediation boreholes. The radius of influence of a borehole increases, and, thus, the number of boreholes necessary for the remediation is reduced. Simultaneously, fracturing may cause reduction of the level of soil heterogeneity.

The results stated in Table 32.1 may be achieved by fracturing.

The possibility of direct injection of various kinds of liquids and crystalline materials into the fractured soils and rocks during the actual fracturing process together or instead of regular proppant, for the purpose of increasing the remediation efficiency, represents an advantage of utilization of hydraulic and pneumatic fracturing (Siegrist et al. 2001). Generally, this may concern, for example, substances of the following kinds: (1) nutrients, buffers, and bacterial strains for supporting

Table 32.1 Results achieved by application of hydraulic and pneumatic fracturing

	Hydraulic fracturing		Pneumatic fracturing	
Parameter	Range of values	Average value	Range of values	Average value
Environment permeability increase	5–153×	34×	1.5–175×	28×
Efficiency increase	5–10×	8×	3–25×	10×
Fractured area radius	1–7.6 m	4.9 m	1.4–10.7 m	4.9 m
Increase of borehole reach radius	1.4–9×	5×	1.4–30×	8×

Reproduced from Schuring (2002)

Table 32.2 Possibility of combining the individual fracturing methods with the remediation methods

Remediation method	Contamination kind	Pneumatic fracturing	Hydraulic fracturing
Venting/bioventing	Volatile organic compounds	A	A
Two-phase extraction	Volatile organic compounds	A	A
Pump-and-treat	Dissolved contaminants	A	A
Skimming	Non-aqueous phase liquid	A	A
Biodegradation	Organic contaminants	A	A
Reactive walls	Without limitation	A	A
Chemical oxidation/ reduction	Organic and inorganic	A	A
Air sparging	Volatile organic compounds	A	A
Electrokinetics	Organic and inorganic	P	A
In situ vitrification	Without limitation	A	P
Thermal methods	Organic compounds generally	A	A
Soil washing with surfactants	Organic and inorganic	P	P

Reproduced from Schuring (2002)
A – implemented at least as a field pilot test, P – not implemented, but feasible

biodegradation, (2) reactive liquid and crystalline substances (e.g., iron filings, iron nanoparticles, $KMnO_4$, etc.) for supporting in situ chemical oxidation/reduction of the contaminants, (3) electrically conductive substances for supporting in situ electrokinetics, and in situ vitrification, (4) solid proppants for ensuring long-term effect of increasing hydraulic conductivity of the geological environment.

Hydraulic and pneumatic fracturing may be also combined in order to ensure efficient transport of the remediation substance into the contaminated place (Table 32.2). The pneumatic phase is used for initiation and opening of cracks in

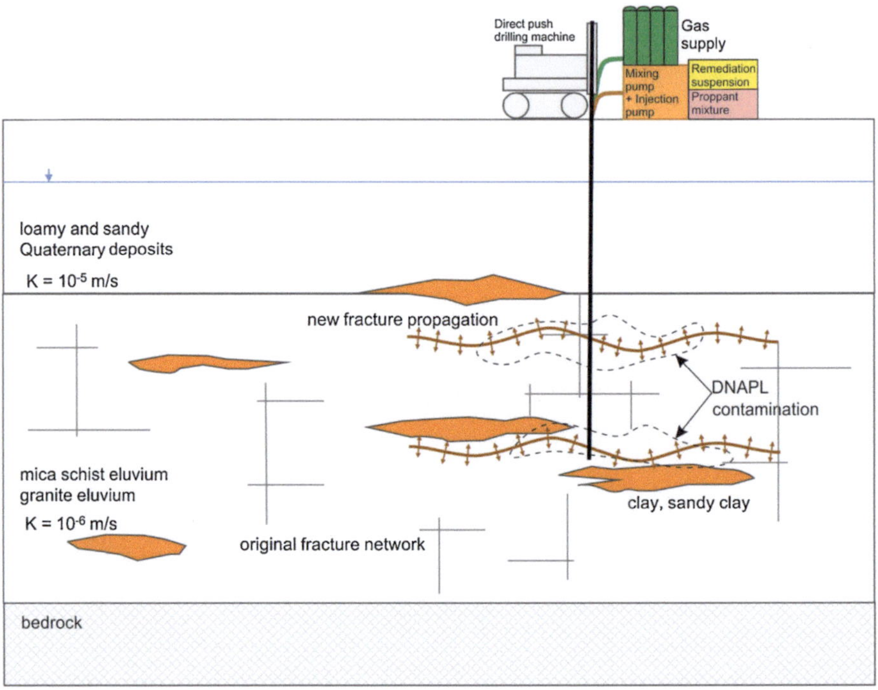

Fig. 32.3 Fracturing for enhanced delivery of in situ remediation substances in contaminated soil using combination of pneumatic and hydraulic fracturing

the environment, and the subsequent hydraulic phase ensures transport of a suspension of a remediation mixture and a solid proppant in the fractured environment. This ensures that the created cracks will remain open in the long term. Such a combination of pneumatic and hydraulic fracturing may be suitably technically implemented by means of penetration probes without casing, formed by the direct push method. Figure 32.3 schematically shows the whole setup for enhanced delivery of in situ remediation substances in contaminated soil.

References

Schuring JR (2002) Fracturing technologies to enhance remediation. Technology evaluation report. GWRTAC

Siegrist RL, Urynowicz MA, West OR, Crimi ML, Lowe KS (2001) Principles and practices of in situ chemical oxidation using permanganate. Battelle Press, Columbus

Suthersan SS (1997) Hydraulic and pneumatic fracturing. In: Remediation engineering: design concepts. CRC Press LLC, Boca Raton, pp 237–254

Zeidler I, Pašiak J (1986) Zhodnotenie výsledkov hydraulického štěpenia nízkopriepustných kolektorov. Zemní plyn a nafta 31(2):251–260

Chapter 33
Tool IV: Monitoring of nZVI Migration and Fate in the Groundwater Conditions

Petra Skácelová and Jan Filip

Abstract An analytical toolbox suitable for monitoring of fate and transformation of nZVI after its application into the groundwater is suggested in the following chapter. The methods are assorted into two main groups classified as "indirect" and "direct". Each technique is described in detail and discussed with respect to its employment for tracing iron nanoparticles.

Keywords nZVI · In-situ remediation · Monitoring · Fate · Transformation · Analytical toolbox

33.1 Introduction

When industrial nanoparticles are injected into the environment, including nanoscale zero-valent iron (nZVI) particles used for groundwater remediation, their fast and precise tracking in the groundwater flowing through porous media is vital for two different reasons. Firstly, despite the unquestionable benefits of nanoremediation, complete risk assessment is required, as it is in the case of other remediation techniques (Shi et al. 2015). While considering the pros and cons of nZVI usage, the following aspects should be taken into account: the influence of nanoparticles (NPs) on the environment and indigenous biota; the scope of the impact; the potential risk of NPs spreading into drinking water sources. Secondly, the

P. Skácelová
Regional Centre of Advanced Technologies and Materials, Palacký University Olomouc, Olomouc, Czech Republic

NANO IRON, s.r.o., Židlochovice, Czech Republic

J. Filip (✉)
Regional Centre of Advanced Technologies and Materials, Palacký University Olomouc, Olomouc, Czech Republic
e-mail: jan.filip@upol.cz

J. Filip et al. (eds.), *Advanced Nano-Bio Technologies for Water and Soil Treatment*, Applied Environmental Science and Engineering for a Sustainable Future,
https://doi.org/10.1007/978-3-030-29840-1_33

information about nZVI behavior, migration, and fate below the ground serves as a tool for the evaluation of their remediation efficiency.

Nanoscale zero-valent iron is a progressive material for in situ remediation of contaminated sites. Besides testing the reactivity of iron nanoparticles with a broad range of diverse contaminants, e.g., Cr(VI), chlorinated ethenes, lindane, heavy metals, it is equally important to find out how far the nanoparticles are able to migrate to ensure contact with the contaminant in the subsurface. Migration of zero-valent iron in groundwater is a task that is investigated from different points of view. The experimental equipment is usually arranged as a one-dimensional assignment (laboratory column) in various setups with different ways of detecting migrating substances (see Chap. 30)

The two- and three-dimensional experimental equipment (VEGAS research facility, University of Stuttgart), preferably with large-scale dimension, enables to conduct a detailed study of the migration of nZVI particles in a homogeneous artificial aquifer under conditions simulating real situations. The experimental setup at VEGAS has a scale from $1 \times 0.12 \times 0.7$ m to $8 \times 1 \times 3$ m (L $\times$ W $\times$ H) (Pešková et al. 2016) enabling the visualization of nZVI migration by injecting the nanoparticles into the source zone of contamination (direct-push), as well as comparing the migration properties of different nZVI particles. Moreover, such a large-scale 3D experiment allows a calculation of the spread efficiency of the particles in the subsurface and, in particular, the quantification of the whole remedial action.

The basic steps in all studies focused on the evaluation of fate and migration of nZVI during groundwater remediation include detailed field research and proper design of the monitoring system. In order to explore hydrological and geological profile of the remediated site, it is necessary to create a sufficient number of monitoring wells of the appropriate depth and position. Another crucial step of nZVI tracing is groundwater/soil/sediment sampling itself. A detailed sampling plan has to be proposed in advance, considering the main targets and outputs of nZVI tracing analysis (Laborda et al. 2016).

Distinguishing between injected nZVI particles (including their transformation products) and naturally occurring iron-bearing colloids represents perhaps the most complex problem in the whole process described (Černík 2010).

In the next paragraphs, we focus on technical solutions for reliable and practically usable tracing of nZVI particles in the environment during/after the nanoremediation action. Different methods and approaches are discussed and experience of their on-site application is presented.

33.2 Sampling Procedure

In order to precisely track the injected nZVI particles at the remediated site, a complex system of monitoring wells is needed. The specific number and position of the wells depends on a particular (hydro)geological profile, groundwater flow direction and velocity as well as general site situation. Nevertheless, special attention should always be paid to designing and constructing the monitoring system.

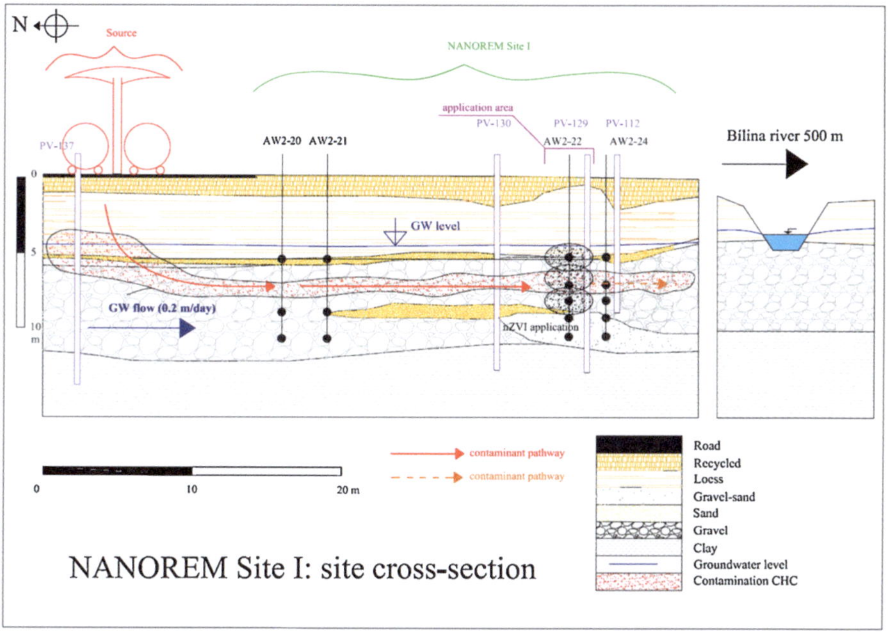

Fig. 33.1 Example of treated site (Spolchemie Site I, Czech Republic) with an array of application and monitoring wells (both open-screen wells and wells equipped with micropumping and magnetic susceptibility measurement)

There are two different types of monitoring wells that can be prepared for the purpose of groundwater/sediment sampling (see Fig. 33.1). The open-screen wells enable the sampling of large volumes of groundwater and sediment samples but the setback is the impossibility of differentiating (sampling) various groundwater levels. In a simplified way, a mixture of groundwater from all the horizons situated above the well bottom is sampled from the open-screen well. Conversely, the second type of monitoring wells, equipped with micropumps, enables the sampling of groundwater from various horizons below the ground level. Unfortunately, this type of wells is not suitable for sediment sampling and does not make pumping of larger volume of groundwater possible.

Although the particulate monitoring plan must be established in accordance with the application arrangement and temporal order of remediation, there are a few general rules that should be followed, e.g., (a) more than one sampling event before nZVI injection is necessary to create the baseline of the studied characteristics, (b) sampling period should be constant over long-term monitoring but the period of sampling should be shortened right after the nZVI injection.

Sampling in a dynamic mode is usually performed using peristaltic or submersible automatic pumps (Fig. 33.2). After a specific volume of groundwater is pumped from the well, different physicochemical parameters such as pH, oxidation-reduction potential (ORP), conductivity, dissolved oxygen, and temperature can be measured on site. The samplers designed for sampling in static mode are typically mechanical and simple in terms of construction and use (Fig. 33.3). Static sampling of groundwater/sediment is sufficient for identification of nZVI particles at laboratories (i.e., using combination of microscopic, diffraction, and/or spectroscopic techniques) and observation of their structural transformation (see Sect. 33.3).

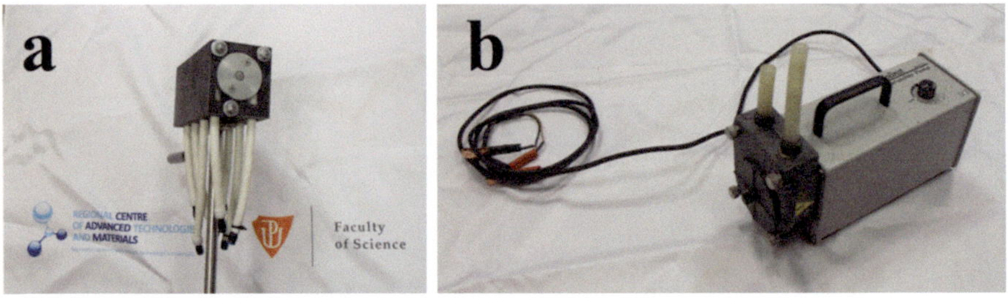

Fig. 33.2 Peristaltic pumps for dynamic sampling: (**a**) manual 4-head peristaltic pump connectable to a micropump system and (**b**) automatic battery-powered peristaltic pump

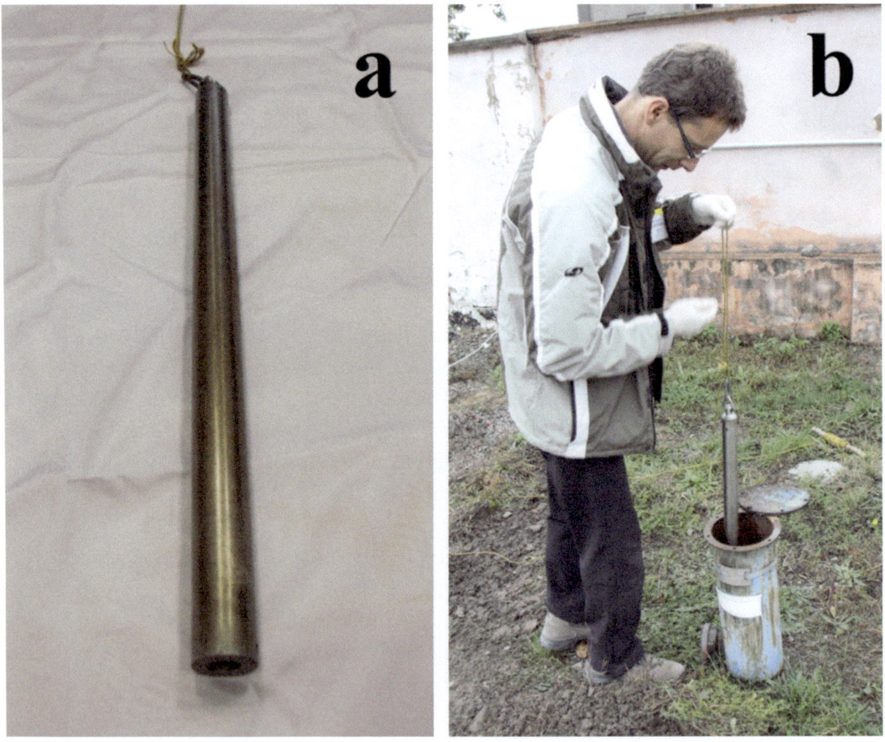

Fig. 33.3 Stainless steel bailer for sampling of groundwater and sediment in (**a**) a static mode and (**b**) its usage on site

33.3 Methods for Nzvi Tracing

Methods suitable for nZVI fate and migration monitoring can be divided into two main groups. *Indirect methods*, such as pH, ORP, H_2, and dissolved Fe or tracer concentration measurement, provide a fast and reliable insight into the current situation at the site and enable to monitor rapid changes in groundwater chemistry. However, they are lacking in the ability to reliably distinguish between the changes caused by injected nZVI and the natural fluctuations of the measured parameters or divergences caused by other factors. By contrast, *direct methods*, such as spectroscopic, diffraction, and microscopic techniques, serve for a confident identification of nZVI within the subsurface. These methods are usually more demanding and costly.

33.3.1 Indirect Methods

33.3.1.1 pH

Concentration of OH^- ions in groundwater is influenced by nZVI due to its reaction with water. Therefore, a rapid increase in pH can be observed in the area of direct contact of nZVI with groundwater. Depending on the buffering capacity of a particular aquifer, the pH values can raise by 1–3 units in the vicinity of nZVI injection. Moreover, pH tends to drop down slowly to the original values when all the active nZVI particles are fully oxidized or their surface is passivated. pH should be measured in a dynamic sampling mode, requiring pumping of elevated amount of groundwater prior to the measurement.

33.3.1.2 ORP

Similarly to pH, decrease in redox potential, as a response to the reducing conditions caused by nZVI introduction, is typically fast. Immediately after the nZVI application, ORP can decrease by 100–400 mV, even in the distant monitoring wells. Nevertheless, ORP tends to increase back to the original value after partial nZVI consumption. Similarly to the case of pH, ORP measurements require dynamic sampling and the values should not be taken into account before their complete stabilization.

33.3.1.3 Magnetic Susceptibility

Magnetic susceptibility probes proved to be suitable for indirect nZVI tracing at a larger scale (Shi et al. 2015). Measurements can be performed by gradually measuring a single borehole (Fig. 33.4) or by placing the field probes into the ground (Fig. 33.5). The subsurface sensors allow monitoring the changes in magnetic properties in the vicinity of the susceptibility probe in a continuous mode. This

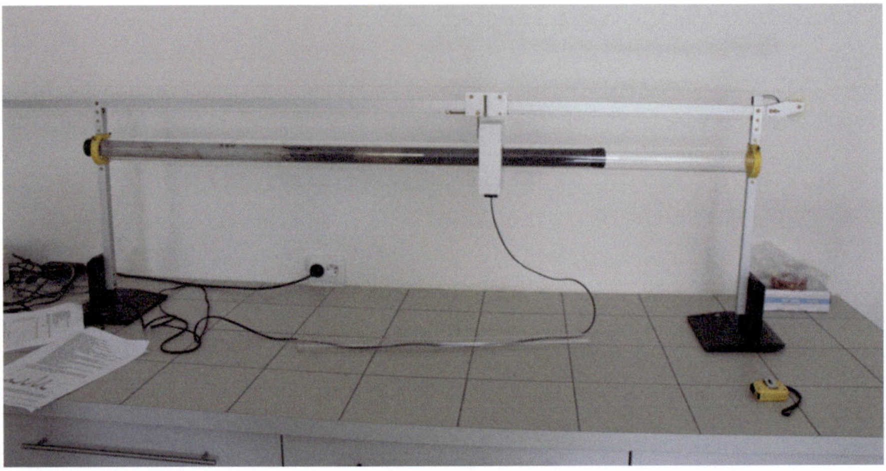

Fig. 33.4 Magnetic susceptibility measurement on the boreholes drilled on Spolchemie Site I

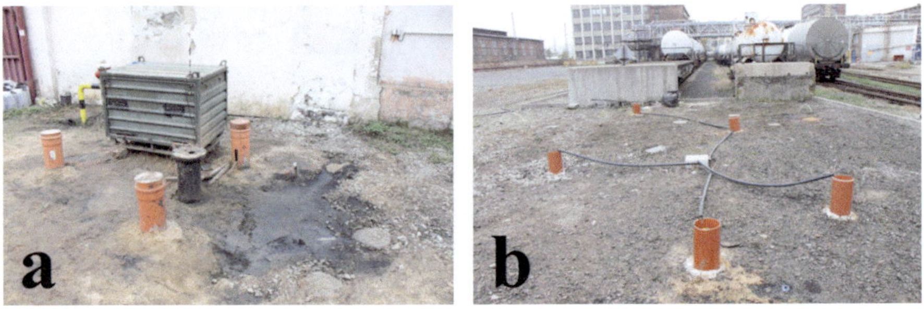

Fig. 33.5 Susceptibility measurement employing underground probes at (**a**) the Spolchemie Site I and (**b**) Spolchemie Site II

monitoring system can also bring unique information about distribution of the injected liquids on the basis of measuring the temperature. A disadvantage of the susceptibility method is its inclination to the falsely positive signals caused by pieces of magnetically susceptible rocks, e.g., basalts, and magnetite-containing rocks. A complex in situ system of susceptibility probes was developed at University Stuttgart (Wiener et al. 2016) and tested at Spolchemie sites (see Fig. 33.5 and Chap. 5).

33.3.1.4 X-Ray Fluorescence

X-ray fluorescence spectroscopy (XRF) belongs to the elemental analysis methods. It is based on the interaction of X-rays with the solid material and detection of secondary X-rays, which is characteristic for each particular element. XRF can serve for qualitative as well as quantitative analysis of majority of chemical elements and can be designed not only for laboratory usage (e.g., desktop instruments) but also for on-site analysis (e.g., portable spectrometers, see Fig. 33.6).

Fig. 33.6 Portable XRF analyzer used for investigation of the boreholes drilled on Spolchemie Site I

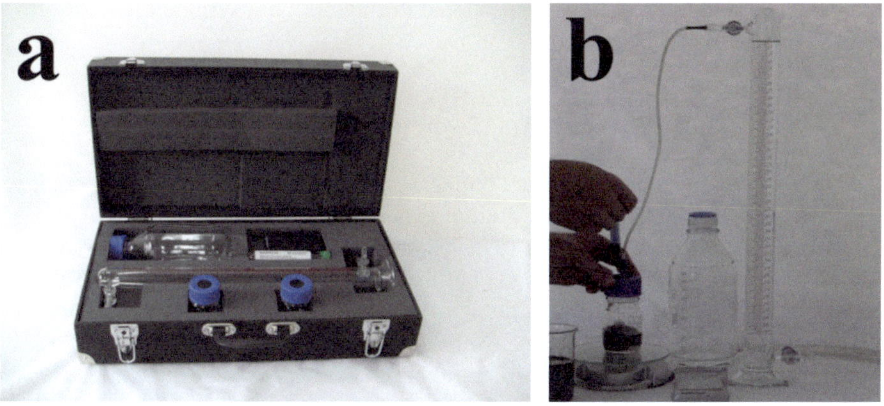

Fig. 33.7 nZVI tester can be transported either (**a**) disassembled in a case or (**b**) assembled and used for indirect measurement of Fe^0 content in the sample

33.3.1.5 Hydrogen Evolution

Hydrogen gas measurement can be used for in situ detection of nZVI reaction directly within the monitoring wells as well as for characterization of nZVI slurry or nZVI transformation products after their acid digestion (Filip et al. 2018). Sophisticated portable gas detectors can serve for in situ H_2 concentration measurements. Evaluation of H_2 content in the wells is an easy and fast method but the drawback is the low sensitivity and the fact that it can be influenced by microbial activity resulting in hydrogen production. On the other hand, simple gas volumetric apparatus (nZVI tester developed by NANO IRON s.r.o., Czech Republic) enables volumetric measurements of the gas evolved from the reaction of nZVI with acids (Fig. 33.7).

33.3.1.6 Fe Concentration

Determination of concentration of total and dissolved iron in groundwater (or in sediment collected from the bottom of monitoring/injection wells) is probably the most straightforward method for tracing the nZVI distribution in the aquifer. It can immediately show the elevated amount of nanoparticles under the assumption that the background Fe concentration is low. On-site Fe determination can be performed by portable spectrophotometer for measurement absorbance of Fe^0 and Fe^{2+} in groundwater after complexation reaction with ferrozine and phenanthroline, respectively. If higher precision and more reliable results are needed, iron content can be measured using atomic absorption spectrometry (AAS) or inductively coupled plasma (ICP) spectroscopic techniques in laboratories. In this case, additional steps as sample conservation, transport, and pretreatment are required. The samples are usually processed according to a standard EPA method 3005. Prior to the determination of dissolved Fe, nZVI and its transformation products can be removed magnetically and the sample is then filtered through 0.2 μm syringe filter.

33.3.1.7 Tracer Concentration

Another way to elucidate the impact of nZVI application is the use of a conservative tracer. An inert substance (usually lithium iodide, bromide ions, or contrast dye, e.g., fluorescein) is added to the nZVI suspension prior to its injection and the concentration of this element in groundwater is subsequently determined. An ideal tracer is highly soluble in the water and does not undergo substantial transformation or sorption to the soil and rocks. Due to the fact that this substance is dissolved in a liquid part of nZVI suspension, its migration does not exactly follow the migration of nanoparticles, but it rather reflects movement of the injected fluid and displays changes in the flow of groundwater within the aquifer.

33.3.1.8 Concentration of Contaminants and Their Degradation Products

The position of reactive nanoparticles within the aquifer can also be deduced indirectly from measuring the levels of target contaminants and/or their degradation products. It is possible to monitor the total concentration of the target contaminant as well as its speciation (e.g., oxidation states of metals). In the case of chlorinated hydrocarbons (CHCs), particular pollutants as well as their sum can be measured. Concentration of volatile organic compounds (VOCs, common pollutants treated by nZVI) can be determined directly at the site using membrane interface probe—MIP (find more in Chap. 31). Nevertheless, much more common protocol involves groundwater sampling and its subsequent analysis in a laboratory employing gas chromatography–mass spectrometry (GC-MS, Fig. 33.8). Therefore, special

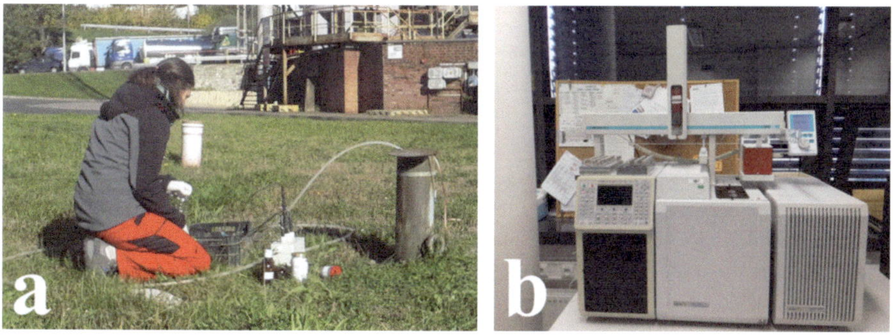

Fig. 33.8 (**a**) Sampling of groundwater containing CHCs at Spolchemie Site I, Czech Republic, and (**b**) subsequent analyzing by HS-GC-MS

attention should be paid to the sampling procedure and subsequent handling of the sample. Groundwater samples containing volatile compounds, including CHCs and their degradation products, must be treated with particular care. The best practice of VOCs-containing samples processing encompasses, besides other things, transport in glass vials and direct introduction of the headspace (HS) into the gas chromatograph.

33.3.2 *Direct Methods*

Unlike all the previously discussed (*indirect*) methods, the analytical characterization of the collected samples presented below cannot be performed directly at the site because it requires usage of sophisticated analytical devices (microscopes, diffractometers, and/or spectroscopes).

33.3.2.1 ICP-MS-Based Techniques

Inductively coupled plasma mass spectrometry (ICP-MS) seems to be a promising direct method for nZVI tracing, as it is able to distinguish between industrial nanoparticles and naturally occurring colloids and/or minerals. There are three suggested approaches to differentiating the injected particles from those of natural origin. Field-flow fractionation belongs to the standard separation techniques and if coupled with ICP-MS (FFF-ICP-MS), this technique sorts nanoparticles into size fractions and analyzes their chemical composition subsequently. Single-particle ICP-MS (sp-ICP-MS) enables to perform both of the above-mentioned steps (size separation and chemical analysis) at once. The instrument collects MS-signal for each single particle and the signal intensity correlates with the size and mass fraction of analyzed nanoparticles. Another option to recognize the injected nZVI particles is to utilize their "trace-element fingerprint", i.e., identification of a characteristic chemical profile based on the content of the selected elements and their mutual ratios.

33.3.2.2 Electron Microscopy

The advanced microscopic techniques, such as field emission gun scanning electron microscopy (FE-SEM) and transmission electron microscopy (TEM), enable the visualization of nZVI particles and their transformation products directly after their separation from sediment (see Figs. 33.9 and 33.10). For SEM/TEM as well as for

Fig. 33.9 FE-SEM micrograph of the sediment sample showing nZVI transformation products

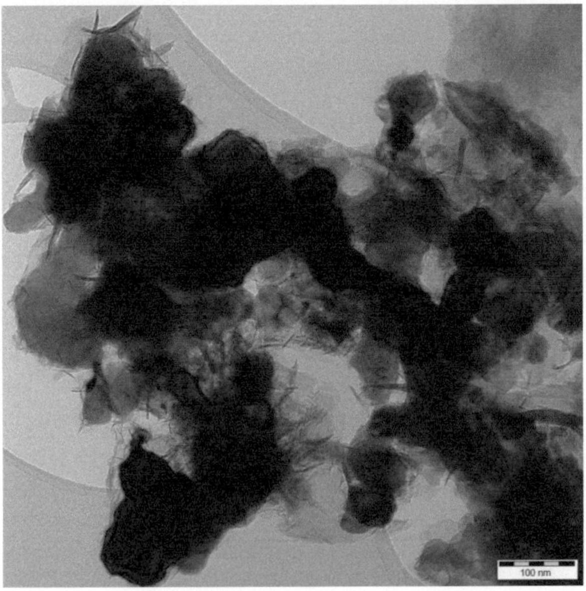

Fig. 33.10 TEM micrograph of the sediment sample displaying nZVI particles surrounded by their transformation products

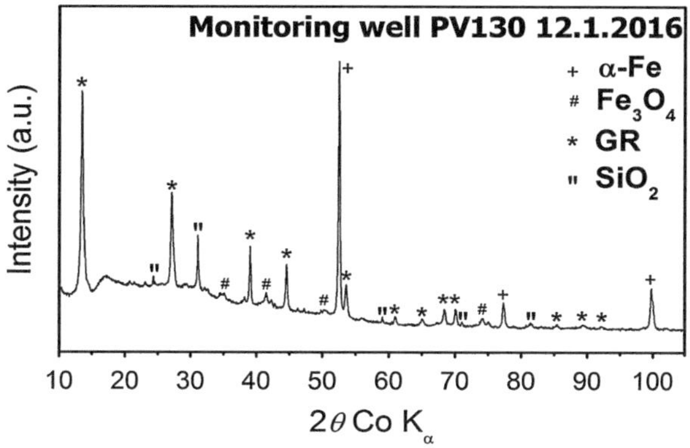

Fig. 33.11 Typical XRD pattern of a sediment sample containing nZVI particles (and their reaction products) showing all the identified crystalline phases (GR—green rust)

the other methods mentioned hereafter, nanoparticles must be isolated from the sample in order to avoid analysis of ballast material. Magnetic separation proved to be simple and suitable for samples of groundwater-sediment mixture containing nZVI and its transformation products. This additional step in the sample preparation procedure allows easy identification of nanoparticles and their particle-size and morphology characterization. Additional instrumental tools coupled with microscope (e.g., energy-dispersive spectroscopy, EDS) can serve for chemical analysis of the solid sample. Further investigation can be performed employing other sophisticated extensions such as high-resolution techniques (HR-TEM, HR-SEM).

33.3.2.3 X-Ray Powder Diffraction

X-ray powder diffraction (XRD) measurements are crucial in terms of qualitative and quantitative analysis of (nano)crystalline solid samples. XRD is suitable for fast identification and structural characterization of nZVI particles and their transformation products (Fig. 33.11) that can be easily distinguished from an amorphous matrix. They also enable to estimate a mean size of particles according to the width of diffraction peaks. Moreover, for reliable data collection, it is necessary to avoid sample oxidation/transformation during the XRD measurement by, e.g., fast collection of successive scans, protecting the sample with transparent foil, or measurements in special chambers with protective inert atmosphere. For the subsequent phase analysis utilizing full-profile fitting (i.e., Rietveld refinement), either the first fast scan or a sum of all recorded scans (i.e., if no phase change took place in the course of successive XRD measurements) is processed.

33.3.2.4 Mössbauer Spectroscopy

Mössbauer spectroscopy on isotope ^{57}Fe is based on a Mössbauer effect and uses gamma radiation to identify valence state and speciation of Fe atoms and quantify each form of Fe in particular sample. Due to its high selectivity, this analytical

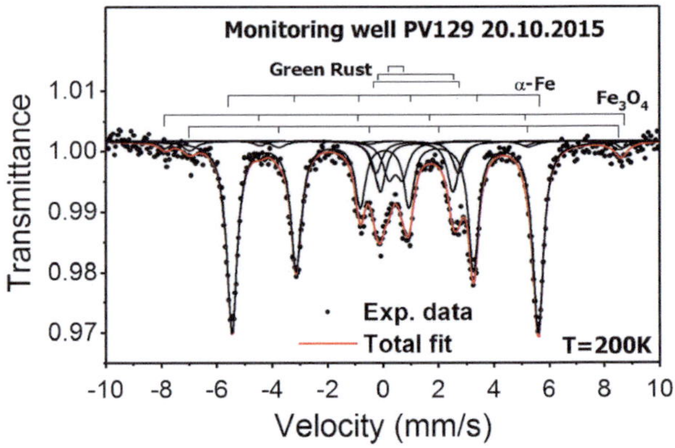

Fig. 33.12 Typical Mössbauer spectrum of the sample containing Fe^0 (sextet assigned as α-Fe) and its transformation products magnetite (two sextets assigned to Fe_3O_4) and green rust (three doublets)

method has proved to be a useful tool for the characterisation of nZVI nanoparticles and it represents a unique technique for determination of the Fe^0/Fe_{tot} ratio in the solid samples, including the identification of nZVI in complex environmental matrices (Fig. 33.12). The main disadvantage of Mössbauer spectroscopy lies in relatively long data-collection time—it usually takes 1 day to analyze fresh nZVI particles but it can take up to 1 week for environmental samples with a low Fe content. Therefore, gradual oxidation of Fe^0 during the data collection becomes an imminent issue and the samples need to be analyzed in a protective atmosphere or deep-frozen. For these reasons, the Mössbauer spectrometer is not used directly on site, despite being compact and portable.

References

Černík M (2010) Chemicky podporované *in situ* sanační technologie, 1st edn. VŠCHT Praha, Prague

Filip J, Soukupova J, Kaslik J, Slunsky J, Zboril R (2018) Nanoscale zerovalent iron particles for groundwater and soil treatment: monitoring and control of their solid-state synthesis, stability, and activity. Iron nanomaterials for water and soil treatment. Pan Stanford Publishing Pte. Ltd, Singapore

Laborda F, Bolea E, Cepriá G, Gómez MT, Jiménez MS, Pérez-Arantegui J, Castillo JR (2016) Detection, characterization and quantification of inorganic engineered nanomaterials: a review of techniques and methodological approaches for the analysis of complex samples. Anal Chim Acta 904:10–32. https://doi.org/10.1016/j.aca.2015.11.008

Pešková K, Miyajima K, Braun J (2016) Study of the migration of nanoiron particles in the 2- and 3-D homogeneous artifical aquifer. In: NANOCON 2016 – 8th international conference on Nanomaterials – Research and Application, Brno, 2016. TANGER, spol. s.r.o., Ostrava, pp 240–245

Shi Z, Fan D, Johnson RL, Tratnyek PG, Nurmi JT, Wu Y, Williams KH (2015) Methods for characterizing the fate and effects of nano zerovalent iron during groundwater remediation. J Contam Hydrol 181:17–35. https://doi.org/10.1016/j.jconhyd.2015.03.004

Wiener A, Müller F, Klaas N, Braun J (2016) Cascading column system: improved susceptibility sensors. Paper presented at the NanoRem final conference, Frankfurt am Main

Chapter 34
Tool V: Microbiological Methods for Monitoring nZVI Performance in Groundwater Conditions

Alena Ševců, Iva Dolinová, Tomáš Cajthaml, Jana Steinová, and Roman Špánek

Abstract Biodegradation is a cost-effective tool in the remediation of persistent organic pollutants but can take place only when the concentrations of toxic substances are decreased to levels suitable for microbial growth, i.e., after nZVI application. Assessment of the impact of nZVI on indigenous microbial communities is, therefore, an essential step in designing a successful remediation strategy. Here, we summarize the most common microbiological methods for monitoring the effect of nZVI on microorganisms such as the cultivation of bacteria, fluorescence in situ hybridization (FISH), phospholipid fatty acid (PLFA) analysis, real-time PCR, and sequencing.

Keywords Microbial monitoring · Cultivation · FISH · PLFA · Real-time PCR · Sequencing

34.1 Introduction

The high efficiency of nZVI in reducing persistent organic pollutants in aquifers is well known. The application of this highly reactive material inevitably influences indigenous microbial communities. These communities can adapt to harsh conditions in polluted environments and often utilize present pollutants for their metabolic

A. Ševců (✉) · I. Dolinová · J. Steinová · R. Špánek
Institute for Nanomaterials, Advanced Technologies and Innovation, Technical University of Liberec, Liberec, Czech Republic
e-mail: alena.sevcu@tul.cz

T. Cajthaml
Institute of Microbiology of the Czech Academy of Sciences, Prague, Czech Republic

Institute for Environmental Studies, Faculty of Science, Charles University, Prague, Czech Republic

J. Filip et al. (eds.), *Advanced Nano-Bio Technologies for Water and Soil Treatment*, Applied Environmental Science and Engineering for a Sustainable Future,
https://doi.org/10.1007/978-3-030-29840-1_34

activity, thus significantly improving the nZVI performance. Results from the monitoring process reflecting the effect of nZVI on microbial communities are therefore an important criterion for successfully evaluating in situ remediation.

The presence, quantity, and/or diversity of microorganisms can be monitored by either simple microbiological methods such as cultivation techniques or using more advanced methods such as epifluorescence microscopy or molecular genetics. Researchers and remediation engineers should thoroughly consider various specific aspects of these methods to choose the most appropriate one. The key criteria are time, cost effectiveness, and the diagnostic value. Here, we summarize the most commonly utilized microbiological methods for monitoring the effect of nZVI on microorganisms.

34.2 Cultivation

Cultivation methods are used to estimate the number of microorganisms in the given sample volume which are able to grow on the given medium. It is important to know that only a small portion of microorganisms can be cultivated (0.1–2%). The most commonly used methods are colony-forming unit (CFU) counting (spread-plate technique and the membrane filter cultivation on agar plates) and the end-point dilution method (most probable number—MPN). The end-point dilution method presumes that the distribution of cells in the liquid medium is uniform, and the cells do not cluster together, which is not typical for environmental samples from polluted sites. Therefore only the CFU method is described here.

The CFU method is based on the assumption that a single colony develops from each cell. However, cells can sometimes clump and a single colony may grow from several cells clustered together. The viable count could be thus invariably lower than the real total cell number present in a sample. A suspension must be prepared from soil or water samples and then diluted. Usually a ten-fold dilution series is applied. There is a wide range of cultivation media available supporting the growth of different functional groups of microorganisms. Since the application of nZVI might favor iron-reducing or iron-oxidizing bacteria depending on environmental conditions (Emerson et al. 2010; Weber et al. 2006) selective media might be used in addition to more general media in order to focus on these specific groups of microorganisms (APHA 2005). Particularly, Citrate Agar (HiMedia) was proved to be suitable and highly selective for iron-oxidizing bacteria in our laboratory (Fig. 34.1).

In brief, a known volume (e.g., 0.1 mL) of the dilutions is plated onto the surface of a suitable growth medium pre-prepared in Petri dishes. The incubation is held under defined conditions (temperature), and after 3–7 days the average CFU on the plates is determined. Plates containing between 20–200 colonies are optimal for counting. The number of CFU per mL (or g) of the initial sample can be calculated from the average CFU on parallel plates and the known dilution factor (CFU/mL or g).

Fig. 34.1 Colony of iron-oxidizing bacteria from polluted groundwater grown on citrate agar (*Sphaerotilus*-like group), obtained in bright-field microscope (Zeiss Axio Imager, magnification 25×) by K. Bobčíková, TUL

The membrane filter method is often used to estimate the CFU counts from samples with low bacterial numbers. A known volume of sample is filtered through a membrane filter, then the filter is placed onto a Petri dish with a growth medium. The nutrients diffuse through the filter from the medium and enable the development of colonies. After an adequate incubation period, the CFU count on the surface of the filter can be determined.

34.3 Fluorescence Microscopy

Fluorescence microscopy techniques can be used in order to examine the effect of nZVI on organisms living in soil or groundwater environments. Such techniques include fluorescence staining for total counts, viability counts, and fluorescence in situ hybridization (FISH). For instance, the total cell number, e.g., expressed as the number of bacteria per mL or L in liquid samples or per g of dry weight in soil samples, can be determined by direct counting using DAPI (4′,6-diamidino-2-phenylindole) staining as a DNA fluorescence agent.

Cell viability, besides other methods, is assessed using a two-dye fluorescent bacterial viability kit that distinguishes viable (commonly green fluorescence) and dead (red fluorescence) cells under a fluorescence microscope (Fig. 34.2). Afterwards, it is possible to calculate the ratio of live and dead cells in the sample.

The composition of the soil or groundwater microbial communities is often analyzed by FISH. A key advantage of this technique is the ability to observe individual cells using fluorescence microscopy or flow cytometry. Thus, it can give detailed information about morphology, number, and spatial distribution of the targeted microbial groups. In brief, FISH detects nucleic acid sequences using fluorescently labeled probe that hybridizes specifically to its complementary target sequence (in rRNA, mRNA, and functional genes). The choice of the probes has to be based on specificity, sensitivity, and ease of cell penetration. The most common and commercially available are single-molecule-labeled oligonucleotide probes. A polynucleotide probe has several fluorophores, resulting in higher sensitivity than

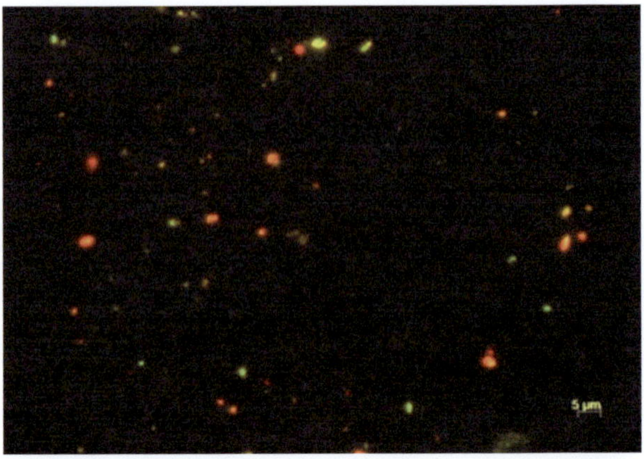

Fig. 34.2 Fluorescently labeled viable (green) and dead (red) bacterial cells in a natural water sample, obtained using fluorescece microscope (Zeiss Axio Imager) by N.H.A. Nguyen, TUL. Scale bar is 5 μm

oligonucleotide probes; however, the specificity is lower. The sensitivity of FISH in natural water or soil samples considerably increased when combined with catalyzed reporter deposition (CARD)-FISH (Kubota 2013). Recently direct-geneFISH, which allows a direct cell identification based on rRNA-targeted oligonucleotide probes and gene detection with dsDNA polynucleotide probes, has been developed for environmental samples (Barrero-Canosa et al. 2017). The advantage is in a single hybridization, taking far less time than CARD-FISH.

Here, a study on the nZVI effect on natural soil bacteria using FISH is described as an example. First, soil mesocosms were prepared with 5% of nZVI materials (NANOFER 25S and biochar-embedded (BC)-nZVI). After incubation, the bacteria were separated from soil using density gradient centrifugation, filtrated, and hybridized with specific probes for Eubacteria (EUB338 probe), Alfaproteobacteria (ALF1B), Betaproteobacteria (BET42a), Gammaproteobacteria (GAM42a), and Actinobacteria (HGC69A). After the hybridization, each filter was mounted using antifadent Citifluor solution on a glass slide and typically ten images for each probe were taken using fluorescence Axio Imager microscope and AxioVision software (Zeiss). The evaluation of the FISH images was then performed in Matlab (The Mathworks) with Image Processing Toolbox. The results were finally expressed as relative abundance of the studied bacterial groups (Fig. 34.3).

34.4 Phospholipid Fatty Acid (PLFA) Analysis

PLFA analysis represents another suitable method for studying microorganisms in environmental samples. The principle lies in a quantitative analysis of fatty acids originating from phospholipids, which are an essential structural component of all microbial cellular membranes. However, this application is shadowed by a rapid development and availability of more effective molecular biology methods, which are also described in this chapter. On the other hand, due to certain specificity of

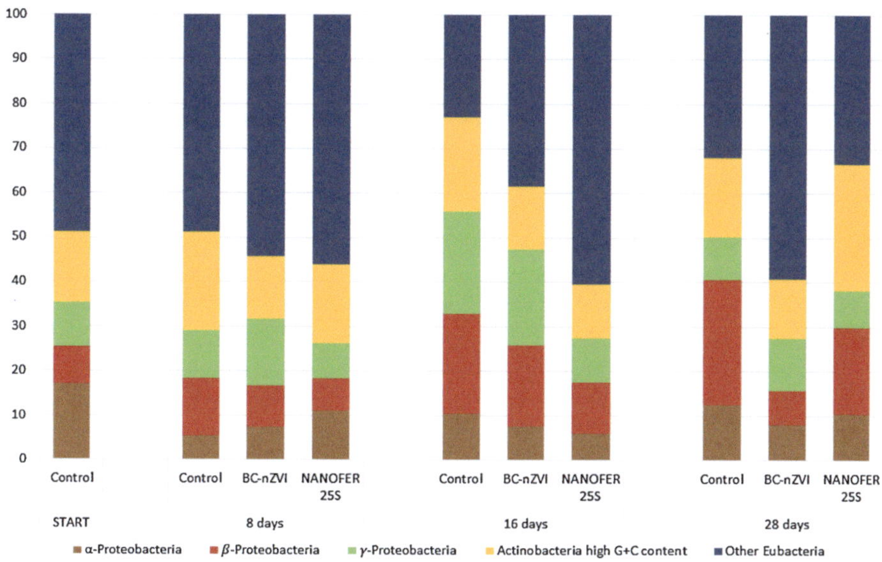

Fig. 34.3 Effect of nZVI on microbial community structure in soil obtained by FISH. Data provided by F. Hrnčiřík, TUL

PLFAs with respect to bacterial groups, this method is nowadays widely used for rapid quantification of microbial biomass in samples (e.g., in soil and sludge) with an insight into the composition of basic bacterial groups. In this way, PLFA analysis also enables to observe extensive changes in the community composition of the living microbes in soil and water environments. Despite certain concerns about the assignment of individual PLFA to bacterial taxonomic groups, some typical PLFAs represent dominant groups of bacteria and in this way the biomass of these groups can be quantified. The main advantage of the PLFA method is its simple use and precise analytical protocols for determination of the monitored analytes without the need for cultivation steps. Another important positive aspect is the fact that PLFAs decompose relatively quickly via phosphate removal from the original structure, and, therefore, PLFAs represent only the active and living biomass.

Typical protocol for the determination of PLFAs encompasses extraction of the total lipids, separation of the PLFA fraction, derivatization (mostly methylation), and analysis by gas chromatography. The total lipids from the samples are usually extracted via the Bligh and Dyer method (Bligh and Dyer 1959) using a mixture of chloroform, methanol, and phosphate buffer (1:2:0.8; v/v/v). Once the total lipids are extracted, the samples are separated using a simple column chromatography with available solid-phase extraction silica gel columns. Due to their polarity, PLFAs are easily separated from other lipids. The samples are typically subjected to mild alkaline methanolysis resulting in methyl esters of fatty acids originating from the phospholipid fraction (Sampedro et al. 2009). Methyl esters of fatty acids are routinely analyzed using high-resolution capillary gas chromatography with either flame ionization detector or any hyphenated mass detector.

The obtained data are typically processed by summarization of groups of typical fatty acids describing individual groups of microbes. The fungal biomass is usually quantified on the basis of the 18:2ω6,9 content, which represents a eukaryotic fatty acid; however, in case of the presence of other eukaryotic organisms, other fungal chemical markers can be used, e.g., ergosterol (Baldrian et al. 2013). The total bacterial biomass can be quantified as the sum of i14:0; i15:0; a15:0; 16:1ω9; 16:1ω7; 10Me-16:0; i17:0; a17:0; cy17:0; 17:0; 10Me-17:0; 10Me-18:0; and cy19:0 fatty acids, Actinobacteria are determined on the basis of 10Me-17:0; 10Me-16:0; and 10Me-18:0, Gram-positive (G+) bacteria are quantified as the sum of i14:0; i15:0; a15:0; i16:0; i17:0; and a17:0, and Gram-negative (G-) bacteria as the sum of 16:1w7; 16:1w5; 18:1w7; cy19:0; and cy17:0 (Stella et al. 2015).

There are certain limitations regarding the interpretation of the results. The data do not provide easily comprehensible picture about the biomass in comparison with, e.g., the CFU. Moreover, the individual groups cannot be compared with one another because various numbers of different fatty acids are used for the quantification. However, the data clearly show the changes in the individual populations, e.g., over the time or during the process of different types of decontamination actions. Additionally, ratios of the individual groups are often used (e.g., fungi/bacteria). Typical examples of respective references are shown below (Figs. 34.4 and 34.5).

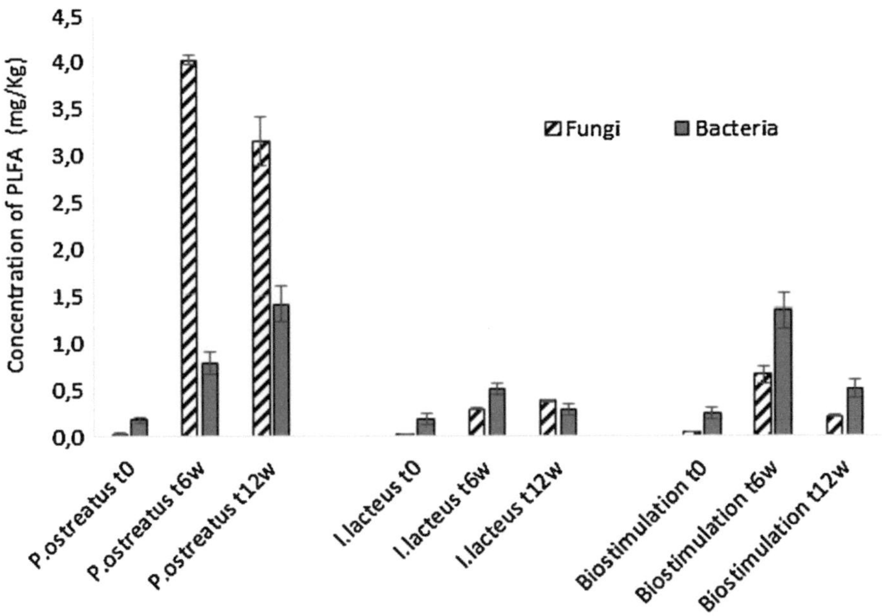

Fig. 34.4 Concentrations of fungal PLFA and sum of bacterial PLFA markers in soil after bioaugmentation of polychlorinated biphenyl (PCB)-contaminated soil. The data show survival of the white-rot fungi used in the contaminated environment. (Adapted from Stella et al. 2017)

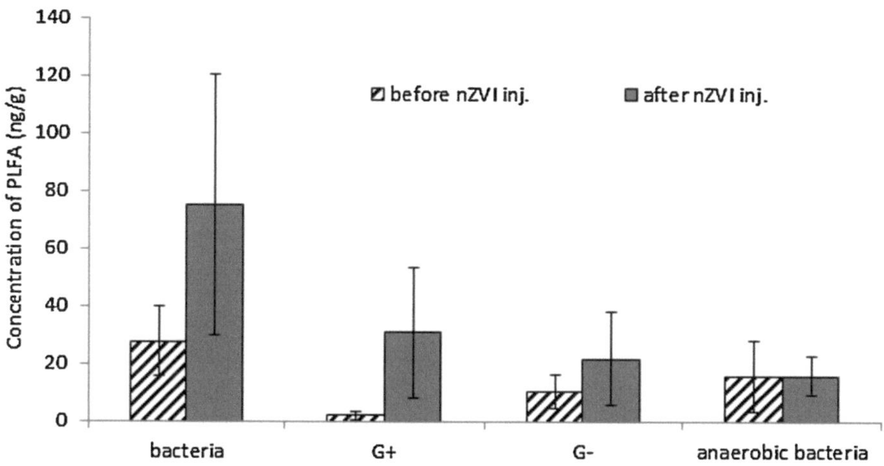

Fig. 34.5 Concentration of specific and total bacterial PLFA in soil from the saturated zone before and after the application of nZVI into the aquifer contaminated by Cr(VI). The data show improvement of ecotoxicological properties toward the present bacteria. (Adapted from Něměček et al. 2014)

34.5 Molecular Biology Approaches

Real-time PCR (quantitative PCR, qPCR) method is routinely used for quantification of microorganisms present in contaminated soil and water samples (Dolinová et al. 2016, 2017; Něměček et al. 2017). Different DNA primers (short oligonucleotides used to amplify specific regions of DNA) are used to detect species-specific and functional groups of microorganisms in order to evaluate the nZVI impact over certain periods of time (Table 34.1).

In principle, qPCR is based on the conventional PCR technology but it differs in how the amplification of a targeted DNA molecule during the PCR in real time is monitored (not at its end, as in conventional PCR). The qPCR method combines PCR amplification and detection into a single step. This approach eliminates the need for detection of products using gel electrophoresis, and more importantly, it enables to quantify the template gene. There are two commonly used methods for how the PCR products can be detected. First, sequence-specific DNA probes that are labeled with a fluorescent reporter can be used. This method allows detection of only specific products after hybridization of the probe with its complementary sequence. The second commonly used approach is based on nonspecific fluorescent dyes that intercalate with any double-stranded DNA. Real-time PCR instruments measure the accumulation of fluorescent signal during the exponential phase of the reaction enabling the fast and precise quantification of the PCR products.

Positive and negative DNA controls should be added to every experiment to ensure that a specific product was amplified. Positive control serves as the amplification control when all the tested DNA samples are negative. On the other hand, including the negative control (which contains only reaction mixture without any

Table 34.1 Example of specific primers for target genes. Adapted from Dolinová et al. (2017)

Target gene	Target microorganism	Primer	Primer sequence (5′-3′)
pceA	*Dehalospirillum multivorans*	Primer F Primer R	GCTGCTGGATGGACCTTAGA CATAGCGATACCTGCAACGA
	Dehalobacter spp.; *Dehalococcoides mccartyi*; *Desulfitobacterium* spp.; *Desulfomonile* spp.; *Desulfuromonas* spp.	Primer F Primer R	ACCGAAACCAGTTACGAACG GACTATTGTTGCCGGCACTT
tceA	*Dehalobacter* spp.; *Dehalococcoides mccartyi; Desulfitobacterium* spp.; *Desulfuromonas* spp.; *Geobacter* spp.; *Sulfurospirillum* spp.;	Primer F Primer R	ACGCCAAAGTGCGAAAAGC TAATCTATTCCATCCTTTCTC
vcrA	*Dehalobacter* spp.; *Dehalococcoides mccartyi*; *Desulfitobacterium* spp.; *Desulfuromonas* spp.; *Geobacter* spp.; *Sulfurospirillum* spp.	Primer F Primer R	TGCTGGTGGCGTTGGTGCTCT TGCCCGTCAAAAGTGGTAAAG
bvcA	*Dehalobacter* spp.; *Dehalococcoides mccartyi*; *Desulfitobacterium* spp.; *Desulfuromonas* spp.; *Geobacter* spp.; *Sulfurospirillum* spp.;	Primer F Primer R	TGCCTCAAGTACAGGTGGT ATTGTGGAGGACCTACCT
etnC	VC-assimilating bacteria	Primer F Primer R	CAGGAGTCSCTKGACCGTCA CARACCGCCGTAKGACTTTGT
etnE	VC-assimilating bacteria	Primer F Primer R	AACTACCCSAAYCCSCGCTGGTACGAC GTCGGCAGTTTCGGTGATCGTGCTCTTGAC

VC vinyl chloride

added template DNA) enables to check whether or not the reaction mixture was contaminated. All the raw fluorescent results are later analyzed using software connected to the instrument.

There are two basic approaches to the interpretation of the output data. First, the absolute quantification gives the exact number of the targeted DNA molecules by comparison with DNA standards using a calibration curve. On the other hand,

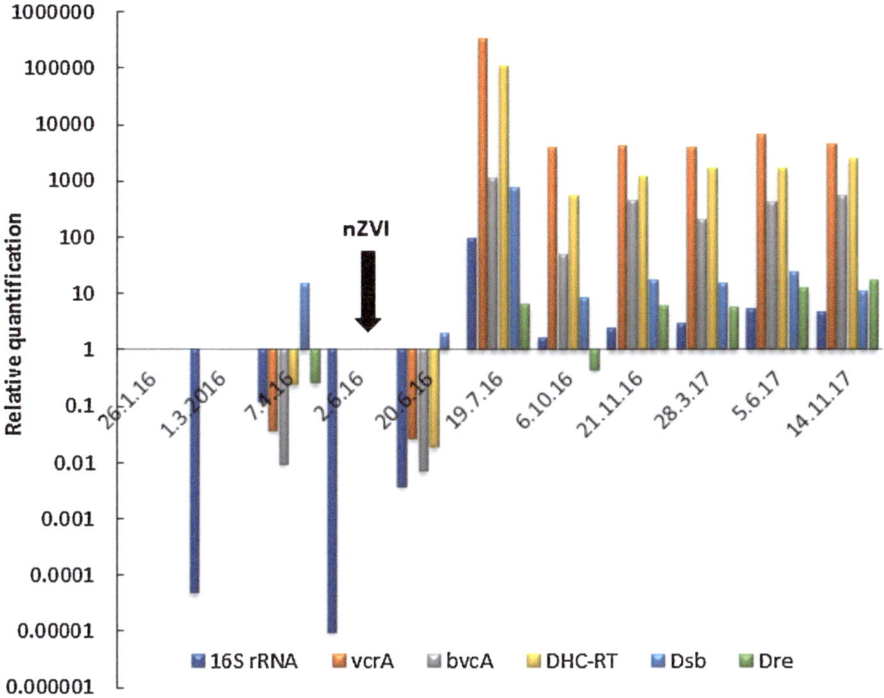

Fig. 34.6 Example of detecting relative abundance of specific genes and bacteria before and after nZVI injection. Legend: vcrA and bvcA—vinyl chloride reductases, DHC-RT–*Dehalococcoides*, Dsb—*Desulfitobacterium*, Dre–*Dehalobacter*. A relative quantification of specific genes is calculated by comparing the Ct values of a sample with the Ct value of a reference sample, with amplification efficiency (EF) as a base and the difference between the Ct values as an exponent

relative quantification determines fold differences in the abundance of the target gene. Trends in relative abundance of functional genes and organohalide-respiring bacteria are shown in Fig. 34.6, when the nZVI treatment had clearly positive effect on increasing the abundance of the monitored markers.

The qPCR results can be also expressed semiquantitatively, e.g., as a heat map showing comparison of relative quantity of each marker. This can be a very useful, easily understandable, and graphically attractive outcome of the qPCR analysis for the routine monitoring process of relevant bacterial groups during remediation activities. The heat maps can be created by different procedures. Here, an example of a heat map based on the in-house developed protocol is shown in Table 34.2. The relative abundance of the total bacterial biomass (monitored by 16S rDNA gene), vinyl chloride reductases, and organohalide-respiring bacteria were monitored before and after the nZVI application at the site contaminated with chlorinated ethenes. It is evident that during the first days of the nZVI treatment the abundance of all markers decreased compared with the state before the nZVI application. Approximately 3 weeks after the nZVI application the numbers of all genes considerably increased and this situation lasted for a few months.

Table 34.2 Relative quantity of bacterial markers before and after nZVI application, color represents quantity of individual markers. The average Ct values (normalized to volume of the sample) of one marker are divided into two sets; values lower than 36 and values equal to or higher than 36 that are classified as 'low quantity'. The second set of values is divided into five intervals of the same size. The interval of lowest values is classified as 'high quantity'. Ct values equal to 40 are classified as 'not detected'. Calculation and data provided by M. NechanickÃ, TUL

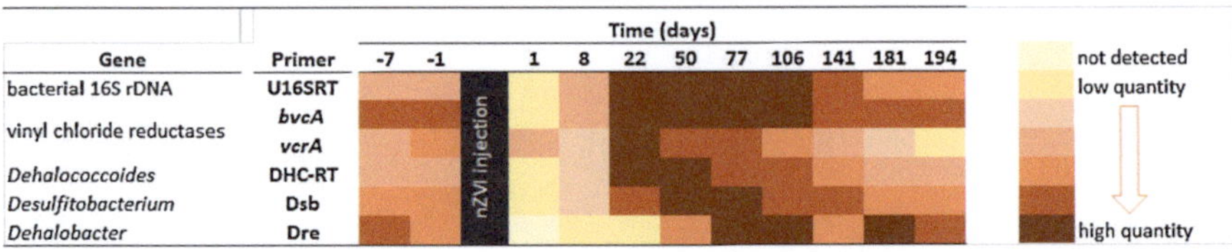

With respect to the adopted approach, it is necessary to consider each molecular biological marker separately and see the trends in relative abundance rather than comparing the absolute numbers between each other.

Droplet Digital PCR (ddPCR) is a relatively new and dynamically developing method based on a water-oil emulsion droplet technology. A sample is fractionated into a high number of tiny droplets, and PCR amplification of the template molecules occurs in each individual droplet. It enables an absolute count of the targeted DNA copies per input sample without the need for running standard curves. To our best knowledge, this promising method has not been used yet for monitoring the nZVI effect on natural microbial communities although it has already been successfully used for a high number of environmental as well as clinical applications. Its main advantages include higher efficiency, specificity, and sensitivity compared with qPCR.

Next-generation sequencing (NGS), also known as high-throughput sequencing, includes several different modern sequencing technologies such as Illumina sequencing (iSeq, MiniSeq, MiSeq, NextSeq, HiSeq, HiSeq X, and NovaSeq), Thermo Fisher Scientific platforms (Ion Torrent, Ion S5, Ion Proton, and SOLiD), or Pacific Biosciences platform (PacBio RSII and the Sequel system). These technologies can be used for various applications, including genome sequencing, amplicon sequencing, or metagenome sequencing. From the perspective of the remediation practices, the biggest advantage of NGS is that microorganisms can be efficiently identified within a mixed community without the need for culturing the samples or preparing clone libraries. Therefore, the effect of nZVI application can be monitored on the whole microbial community present in the treated sites (Nguyen et al. 2018; Vogel et al. 2018). Moreover, it is possible to analyze the profiles of active genes for specific enzymes that catalyze degradation of certain toxic compounds based on mRNA sequencing. This information can help in diagnostics of bioremediation capabilities of present microbes.

The only drawback is the quality and quantity of the DNA isolated from the environmental samples, which should be higher than 2 ng/μL. Lower DNA concentrations might lead to inaccurate results and misleading interpretation.

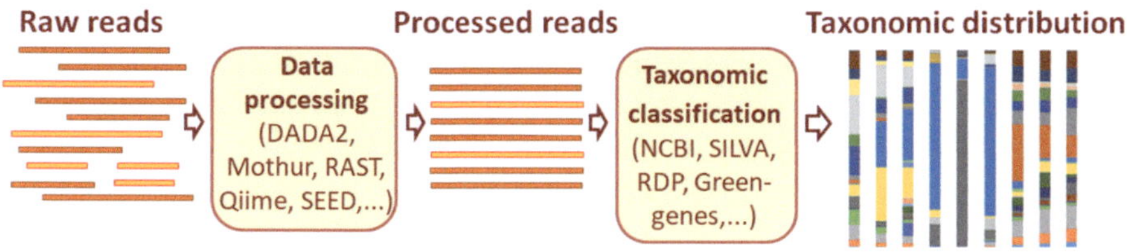

Fig. 34.7 Simplified workflow of taxonomic assignment

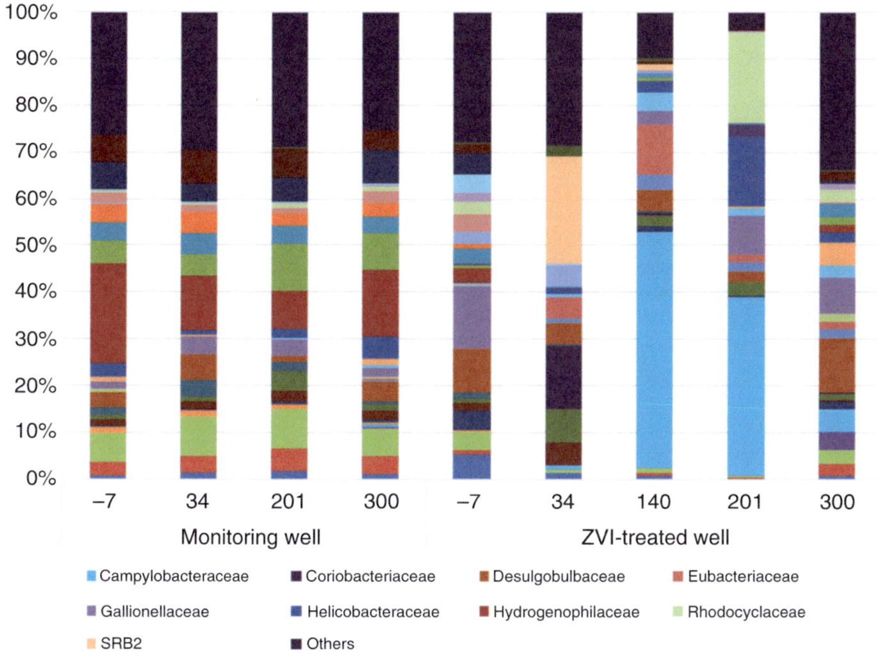

Fig. 34.8 Phylum and class composition of bacterial populations present in the reference well and ZVI-treated well over a 300-day monitoring. Phyla with higher than 3% abundance are shown

Careful processing of raw data excluding all low-quality reads, chimeric sequences, or too short sequences is vital to obtaining only reliable values (Fig. 34.7). This data processing can be provided commercially by sequencing companies, but we highly recommend collaboration with experienced bioinformaticians, who are familiar with appropriate software and understand the principles of molecular biology. Sequences are then classified against various available databases (e.g., SILVA, NCBI, Greengenes, RDP) (Balvočiūtė and Huson 2017).

NGS data can be visualized by many various diagrams and here we only describe one of the commonly used methods. Phylum, class, or species abundance can be displayed in column plots simply to compare the effect of nZVI between the treated and the reference wells over certain periods of time (Fig. 34.8).

It is obvious that the bacterial composition in the reference well was relatively stable over the whole sampling period while the treatment strongly influenced the microbial community structure. When the physicochemical data are available, an experienced environmental microbiologist can interpret the NGS results in a broader context and can help to efficiently plan the next remediation strategy.

References

APHA (2005) Standard methods for the examination of water and wasterwater, 21st edn. American Public Health Association, Washington, DC

Baldrian P, Větrovský T, Cajthaml T, Dobiášová P, Petránková M, Šnajdr J, Eichlerová I (2013) Estimation of fungal biomass in forest litter and soil. Fungal Ecol 6(1):1–11. https://doi.org/10.1016/j.funeco.2012.10.002

Balvočiūtė M, Huson DH (2017) SILVA, RDP, Greengenes, NCBI and OTT—how do these taxonomies compare? BMC Genomics 18(2):114. https://doi.org/10.1186/s12864-017-3501-4

Barrero-Canosa J, Moraru C, Zeugner L, Fuchs BM, Amann R (2017) Direct-geneFISH: a simplified protocol for the simultaneous detection and quantification of genes and rRNA in microorganisms. Environ Microbiol 19(1):70–82. https://doi.org/10.1111/1462-2920.13432

Bligh EG, Dyer WJ (1959) A rapid method of total lipid extraction and purification. Can J Biochem Physiol 37(8):911–917. https://doi.org/10.1139/y59-099

Dolinová I, Czinnerová M, Dvořák L, Stejskal V, Ševců A, Černík M (2016) Dynamics of organohalide-respiring bacteria and their genes following in-situ chemical oxidation of chlorinated ethenes and biostimulation. Chemosphere 157:276–285. https://doi.org/10.1016/j.chemosphere.2016.05.030

Dolinová I, Štrojsová M, Černík M, Němeček J, Macháčková J, Ševců A (2017) Microbial degradation of chloroethenes: a review. Environ Sci Pollut Res 24(15):13262–13283. https://doi.org/10.1007/s11356-017-8867-y

Emerson D, Fleming EJ, McBeth JM (2010) Iron-oxidizing bacteria: an environmental and genomic perspective. Annu Rev Microbiol 64:561–583. https://doi.org/10.1146/annurev.micro.112408.134208

Kubota K (2013) CARD-FISH for environmental microorganisms: technical advancement and future applications. Microbes Environ 28(1):3–12. https://doi.org/10.1264/jsme2.ME12107

Němeček J, Lhotský O, Cajthaml T (2014) Nanoscale zero-valent iron application for in situ reduction of hexavalent chromium and its effects on indigenous microorganism populations. Sci Total Environ 485–486:739–747. https://doi.org/10.1016/j.scitotenv.2013.11.105

Němeček J, Dolinová I, Macháčková J, Špánek R, Ševců A, Lederer T, Černík M (2017) Stratification of chlorinated ethenes natural attenuation in an alluvial aquifer assessed by hydrochemical and biomolecular tools. Chemosphere 184:1157–1167. https://doi.org/10.1016/j.chemosphere.2017.06.100

Nguyen NHA, Špánek R, Kasalický V, Ribas D, Vlková D, Řeháková H, Kejzlar P, Ševců A (2018) Different effects of nano-scale and micro-scale zero-valent iron particles on planktonic microorganisms from natural reservoir water. Environ Sci Nano 5(5):1117–1129. https://doi.org/10.1039/C7EN01120B

Sampedro I, Giubilei M, Cajthaml T, Federici E, Federici F, Petruccioli M, D'annibale A (2009) Short-term impact of dry olive mill residue addition to soil on the resident microbiota. Bioresour Technol 100(23):6098–6106. https://doi.org/10.1016/j.biortech.2009.06.026

Stella T, Covino S, Burianová E, Filipová A, Křesinová Z, Voříšková J, Větrovský T, Baldrian P, Cajthaml T (2015) Chemical and microbiological characterization of an aged PCB-contaminated soil. Sci Total Environ 533:177–186. https://doi.org/10.1016/j.scitotenv.2015.06.019

Stella T, Covino S, Čvančarová M, Filipová A, Petruccioli M, D'Annibale A, Cajthaml T (2017) Bioremediation of long-term PCB-contaminated soil by white-rot fungi. J Hazard Mater 324:701–710. https://doi.org/10.1016/j.jhazmat.2016.11.044

Vogel M, Nijenhuis I, Lloyd J, Boothman C, Pöritz M, Mackenzie K (2018) Combined chemical and microbiological degradation of tetrachloroethene during the application of Carbo-Iron at a contaminated field site. Sci Total Environ 628–629:1027–1036. https://doi.org/10.1016/j.scitotenv.2018.01.310

Weber KA, Achenbach LA, Coates JD (2006) Microorganisms pumping iron: anaerobic microbial iron oxidation and reduction. Nat Rev Microbiol 4:752–764. https://doi.org/10.1038/nrmicro1490

MIX
Papier aus verantwortungsvollen Quellen
Paper from responsible sources
FSC® C105338

If you have any concerns about our products,
you can contact us on
ProductSafety@springernature.com

In case Publisher is established outside the EU,
the EU authorized representative is:
Springer Nature Customer Service Center GmbH
Europaplatz 3, 69115 Heidelberg, Germany

Printed by Libri Plureos GmbH
in Hamburg, Germany